The Methods of Distances in the Theory of Probability and Statistics

Svetlozar T. Rachev • Lev B. Klebanov
Stoyan V. Stoyanov • Frank J. Fabozzi

The Methods of Distances in the Theory of Probability and Statistics

Springer

Svetlozar T. Rachev
Department of Applied
Mathematics and Statistics
College of Business
Stony Brook University
Stony Brook, NY, USA

FinAnalytica
New York, NY, USA

Stoyan V. Stoyanov
EDHEC Business School
EDHEC-Risk Institute
Singapore

Lev B. Klebanov
Department of Probability and Statistics
Charles University
Prague, Czech Republic

Frank J. Fabozzi
EDHEC Business School
EDHEC-Risk Institute
Nice, France

ISBN 978-1-4614-4868-6 ISBN 978-1-4614-4869-3 (eBook)
DOI 10.1007/978-1-4614-4869-3
Springer New York Heidelberg Dordrecht London

Library of Congress Control Number: 2012953265

Mathematics Subject Classification (2000): 60F05, 60F17, 60K25, 62G35, 62E20

Printed on acid-free paper

Springer is part of Springer Science+Business Media (www.springer.com)

STR
To my grandchildren Iliana, Zoya,
and Zari

LBK
To my wife Marina

SVS
To my wife Petya

FJF
To my wife Donna
and my children Francesco, Patricia,
and Karly

Preface

The development of the theory of probability metrics – a branch of probability theory – began with the study of problems related to limit theorems in probability theory. In general, the applicability of limit theorems stems from the fact that they can be viewed as an approximation to a given stochastic model and, consequently, can be accepted as an approximate substitute. The key question that arises in adopting the approximate model is the magnitude of the error that must be accepted. Because the theory of probability metrics studies the problem of measuring distances between random quantities or stochastic processes, it can be used to address the key question of how good the approximate substitute is for the stochastic model under consideration. Moreover, it provides the fundamental principles for building probability metrics – the means of measuring such distances.

The theory of probability metrics has been applied and has become an important tool for studying a wide range of fields outside of probability theory such as statistics, queueing theory, engineering, physics, chemistry, information theory, economics, and finance. The principal reason is that because distances are not influenced by the particular stochastic model under consideration, the theory of probability metrics provides some universal principles that can be used to deal with certain kinds of large-scale stochastic models found in these fields.

The first driving force behind the development of the theory of probability metrics was Andrei N. Kolmogorov and his group. It was Kolmogorov who stated that every approximation problem has its own distance measure in which the problem can be solved in a most natural way. Kolmogorov also contended that without estimates of the rate of convergence in the central limit theorem (CLT) (and similar limit theorems such as the functional limit theorem and the max-stable limit theorem), limit theorems provide very limited information. An example worked out by Y.V. Prokhorov and his students is as follows. Regardless of how slowly a function $f(n) > 0$, $n = 1, \ldots$, decays to zero, there exists a corresponding distribution function $F(x)$ with finite variance and mean zero, for which the CLT is valid at a rate slower than $f(n)$. In other words, without estimates for convergence in the CLT, such a theorem is meaningless because the convergence to the normal law of the normalized sum of independent, identically distributed random variables with distribution function $F(x)$ can be slower than any given rate $f(n) \to 0$. The problems associated with finding the appropriate rate of convergence invoked a variety of probability distances in which the speed of convergence (i.e., convergence rate) was estimated. This included the works of Yurii V. Prokhorov, Vladimir V. Sazonov, Vladimir M. Zolotarev, Vygantas Paulauskas, Vladimir V. Senatov, and others.

The second driving force in the development of the theory of probability metrics was mass-transportation problems and duality theorems. This started with the work of Gaspard Monge in the eighteenth century and Leonid V. Kantorovich in the 1940s – for which he was awarded the Nobel Prize in Economics in 1975 – on optimal mass transportation, leading to the birth of linear programming. In mathematical terms, Kantorovich's result on mass transportation can be formulated in the following metric way. Given the marginal distributions of two probability measures P and Q on a general (separable) metric space (U, d), what is the minimal expected value – referred to as $\kappa(P, Q)$ or the Kantorovich metric – of a distance $d(X, Y)$ over the set of all probability measures on the product space $U \times U$ with marginal distributions $P_X = P$ and $P_Y = Q$? If the measures P and Q are discrete, then this is the classic transportation problem in linear programming. If U is the real line, then $\kappa(P, Q)$ is known as the Gini statistical index of dissimilarity formulated by Corrado Gini. The Kantorovich problem has been used in many fields of science – most notably statistical physics, information theory, statistics, and probability theory. The fundamental work in this field was done by Leonid V. Kantorovich, Johannes H. B. Kemperman, Hans G. Kellerer, Richard M. Dudley, Ludger Rüschendorf, Volker Strassen, Vladimir L. Levin, and others. Kantorovich-type duality theorems established the main relationships between metrics in the space of random variables and metrics in the space of probability laws/distributions. The unifying work on those two directions was done by V. M. Zolotarev and his students.

In this book, we concentrate on four specialized research directions in the theory of probability metrics, as well as applications to different problems of probability theory. These include:

- Description of the basic structure of probability metrics,
- Analysis of the topologies in the space of probability measures generated by different types of probability metrics,
- Characterization of the ideal metrics for a given problem, and
- Investigation of the main relationships between different types of probability metrics.

Our presentation in this book is provided in a general form, although specific cases are considered as they arise in the process of finding supplementary bounds or in applications to important special cases.

The target audience for this book is graduate students in the areas of functional analysis, geometry, mathematical programming, probability, statistics, stochastic analysis, and measure theory. It may be partially used as a source of material for lectures for students in probability and statistics. As noted earlier in this preface, the theory of probability metrics has been applied to fields outside of probability theory such as engineering, physics, chemistry, information theory, economics, and finance. Specialists in these areas will find the book to be a useful reference to gain a greater understanding of this specialized area and its potential application.

New York, USA — Svetlozar T. Rachev
Prague, Czech Republic — Lev B. Klebanov
Singapore — Stoyan V. Stoyanov
Nice, France — Frank J. Fabozzi

Contents

Chapter 1
Main Directions in the Theory of Probability Metrics

1.1 Introduction

Increasingly, the demands of various real-world applications in the sciences, engineering, and business have resulted in the creation of new, more complicated probability models. In the construction and evaluation of these models, model builders have drawn on well-developed limit theorems in probability theory and the theory of random processes. The study of limit theorems in general spaces and a number of other questions in probability theory make it necessary to introduce functionals – defined on either classes of probability distributions or classes of random elements – and to evaluate their nearness in one or another probabilistic sense. Thus various metrics have appeared including the well-known Kolmogorov (uniform) metric, L_p metrics, the Prokhorov metric, and the metric of convergence in probability (Ky Fan metric). We discuss these measures and others in the chapters that follow.

1.2 Method of Metric Distances and Theory of Probability Metrics

The use of metrics in many problems in probability theory is connected with the following fundamental question: is the proposed stochastic model a satisfactory approximation to the real model, and if so, within what limits? To answer this question, an investigation of the qualitative and quantitative stability of a proposed stochastic model is required. Analysis of quantitative stability assumes the use of metrics as measures of distances or deviations. The main idea of the *method of metric distances* (MMD) – developed by Vladmir M. Zolotarev and his students to solve stability problems – is reduced to the following two problems.

S.T. Rachev et al., *The Methods of Distances in the Theory of Probability and Statistics*,
DOI 10.1007/978-1-4614-4869-3_1,

Problem 1.2.1 (Choice of ideal metrics). Find the most appropriate (i.e., ideal) metrics for the stability problem under consideration and then solve the problem in terms of these ideal metrics.

Problem 1.2.2 (Comparisons of metrics). If the solution of the stability problem must be written in terms of other metrics, then solve the problem of comparing these metrics with the chosen (i.e., ideal) metrics.

Unlike Problem 1.2.1, Problem 1.2.2 does not depend on the specific stochastic model under consideration. Thus, the independent solution of Problem 1.2.2 allows its application in any particular situation. Moreover, by addressing the two foregoing problems, a clear understanding of the specific regularities that form the stability effect emerges.

Questions concerning the bounds within which stochastic models can be applied (as in all probabilistic limit theorems) can only be answered by investigation of qualitative and quantitative stability. It is often convenient to express such stability in terms of a metric. The *theory of probability metrics* (TPM) was developed to address this. That is, TPM was developed to address Problems 1.2.1 and 1.2.2, thus providing a framework for the MMD. Figure 1.1 summarizes the problems concerning MMD and TPM.

1.3 Probability Metrics Defined

The term *probability metric*, or *p. metric*, means simply a semimetric in a space of random variables (taking values in some separable metric space). In probability theory, sample spaces are usually not fixed, and one is interested in those metrics whose values depend on the joint distributions of the pairs of random variables. Each such metric can be viewed as a function defined on the set of probability measures on the product of two copies of a probability space. Complications connected with the question of the existence of pairs of random variables on a given space with given probability laws can be easily avoided. Fixing the marginal distributions of the probability measure on the product space, one can find the infimum of the values of our function on the class of all measures with the given marginals. Under some regularity conditions, such an infimum is a metric on the class of probability distributions, and in some concrete cases (e.g., for the L_1 distance in the space of random variables – Kantorovich's theorem; for the Ky Fan metric – Strassen–Dudley's theorem; for the indicator metric – Dobrushin's theorem) were found earlier [giving, respectively, the Kantorovich (or Wasserstein) metric, the Prokhorov metric, and the total variation distance].

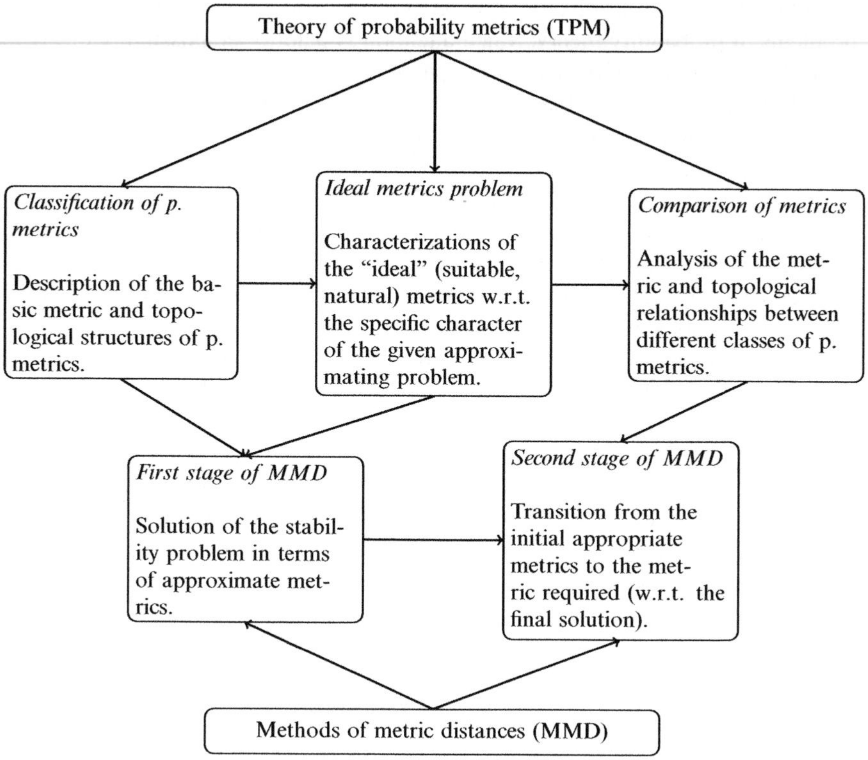

Fig. 1.1 Theory of probability metrics as a necessary tool to investigate the method of metric distances

1.4 Main Directions in the Theory of Probability Metrics

The necessary classification of the set of p. metrics is naturally carried out from the point of view of a metric structure and generating topologies. That is why the following two research directions arise:

Direction 1. Description of basic structures of p. metrics.

Direction 2. Analysis of topologies in space of probability measures generated by different types of p. metrics; such an analysis can be carried out with the help of convergence criteria for different metrics.

At the same time, more specialized research directions arise. Namely:

Direction 3. Characterization of ideal metrics for a given problem.

Direction 4. Investigations of main relationships between different types of p. metrics.

In this book, all four directions are covered as well as applications to different problems in probability theory. Much attention is paid to the possibility of giving equivalent definitions of p. metrics (e.g., in direct and dual terms and in terms of the Hausdorff metric for sets). Indeed, in concrete applications of p. metrics, the use of different equivalent variants of the definitions in different steps of the proof is often a decisive factor.

One of the main classes of metrics considered in this book is the class of minimal metrics, an idea that goes back to the work of Kantorovich in the 1940s dealing with transportation problems in linear programming. Such metrics have been found independently by many authors in several fields of probability theory (e.g., Markov processes, statistical physics).

Another useful class of metrics studied in this book is the class of *ideal* metrics that satisfy the following properties:

1. $\mu(P_c, Q_c) \leq |c|^r \mu(P, Q)$ for all $c \in [-C, C], c \neq 0$,
2. $\mu(P_1 * Q, P_2 * Q) \leq \mu(P_1, P_2)$,

where $P_c(A) := P((1/c)A)$ for any Borel set A on a Banach space U and where $*$ denotes convolution. This class is convenient for the study of functionals of sums of independent random variables, giving nearest bounds of the distance to limit distributions.

The presentation we provide in this book is given in a general form, although specific cases are considered as they arise in the process of finding supplementary bounds or in applications to important special cases.

1.5 Overview of the Book

The book is divided into five parts. In Part I, we set forth general topics in the TPM. Following the definition of a probability metric in Chap. 2, different examples of probability metrics are provided and the application of the Kolmogorov metric in mathematical statistics is discussed. Then the axiomatic construction of probability metrics is defined. There is also a discussion of an interesting property about the Kolmogorov metric, a property that is used to prove biasedness in the classic Kolmogorov test. More definitions and examples are provided in Chap. 3, where primary, simple, and compound distances and minimal and maximal distances and norms are provided and motivated. The introduction and motivation of three classifications of probability metrics according to their metric structure, as well as examples of probability metrics belonging to a particular structural group, are explained in Chap. 4. The generic properties of the structural groups and the links between them are also covered in the chapter.

In Part II, we concern ourselves with the study of the dual and explicit representations of minimal distances and norms, as well as the topologies that these metric structures induce in the space of probability measures. We do so by examining further the concepts of primary, simple, and compound distances, in particular focusing on their relationship to each other. The Kantorovich and the Kantorovich–

Rubinstein problems are introduced and illustrated in a one-dimensional and multidimensional setting in Chap. 5. These problems – more commonly referred to as the mass transportation and mass transshipment problems, respectively – are abstract formulations of optimization problems. Although the applications are important in areas such as job assignments, classification problems, and best allocation policy, our purpose for covering them in this book is due to their link to the TPM. In particular, an application leading to an explicit representation for a class of minimal norms is provided. Continuing with our coverage of Kantorovich and the Kantorovich–Rubinstein functionals in Chap. 6, we look at the conditions under which there is equality and inequalities between these two functionals. Because these two functionals generate minimal distances (Kantorovich functional) and minimal norms (Kantorovich–Rubinstein functional), the relationship between the two can be quantified, allowing us to provide criteria for convergence, compactness, and completeness of probability measures in probability spaces, as well as to analyze the problem of uniformity between these two functionals. The discussions in Chaps. 5 and 6 demonstrate that the notion of minimal distance represents the main relationship between compound and simple distances. Our focus in Chap. 7 is on the notion of K-minimal metrics, and we describe their general properties and provide representations with respect to several particular metrics such as the Lévy metric and the Kolmogorov metric. The relationship between the multidimensional Kantorovich theorem and the work by Strassen on minimal probabilistic functionals is also covered. In Chap. 8, we discuss the relationship between minimal and maximal distances, comparing them to the corresponding dual representations of the minimal metric and minimal norm, providing closed-form solutions for some special cases and studying the topographical structures of minimal distances and minimal norms. The general relations between compound and primary probability distances, which are similar to the relations between compound and simple probability distances, are the subject of Chap. 9.

The application of minimal probability distances is the subject of the five chapters in Part III. Chapter 10 contains definitions, properties, and some applications of moment distances. These distances are connected to the property of definiteness of the classic problem of moments, and one of them satisfies an inequality that is stronger than the triangle inequality. In Chap. 11, we begin with a discussion of the convergence criteria in terms of a simple metric between characteristic functions, assuming they are analytic. We then turn to providing estimates of a simple metric between characteristic functions of two distributions in terms of moment-based primary metrics and discussing the inverse problem of estimating moment-based primary metrics in terms of a simple metric between characteristic functions. In Chaps. 11 through 14, we then use our understanding of minimal distances explained in Chap. 7 to demonstrate how the minimal structure is especially useful in problems of approximations and stability of stochastic models. We explain how to apply the topological structure of the space of laws generated by minimal distance and minimal norm functionals in limit-type theorems, which provide weak convergence together with convergence of moments. We study vague convergence in Chap. 11, the Glivenko–Cantelli theorem in Chap. 12, queueing systems in Chap. 13, and optimal quality in Chap. 14.

Any concrete stochastic approximation problem requires an *appropriate* or *natural* metric (e.g., topology, convergence, uniformities) having properties that are helpful in solving the problem. If one needs to develop the solution to the approximation problem in terms of other metrics (e.g., topology), then the transition is carried out using general relationships between metrics (e.g., topologies). This two-stage approach, described in Sect. 1.2 (selection of the appropriate metric, which we labeled Problem 1.2.1, and comparison of metrics, labeled Problem 1.2.2) is the basis of the TPM. In Part IV – Chaps. 15 through 20 – we determine the structure of *appropriate* or, as we label it in this book, *ideal* probability distances for various probabilistic problems. The fact that a certain metric is (or is not) appropriate depends on the concrete approximation (or stability) problem we are dealing with; that is, any particular approximation problem has its own "ideal" probability distance (or distances) on which terms we can solve the problem in the most "natural" way. In the opening chapter to this part of the book, Chap. 15, we describe the notion of ideal probability metrics for summation of independent and identically distributed random variables and provide examples of ideal probability metrics. We then study the structure of such "ideal" metrics in various stochastic approximation problems such as the convergence of random motions in Chap. 16, the stability of characterizations of probability distributions in Chaps. 17 and 20, stability in risk theory in Chap. 18, and the rate of convergence for the sums and maxima of random variables in Chap. 19.

Part V is devoted to a class of distances – Euclidean-type distances. In this part of the book, we provide definitions, properties, and applications of such distances. The space of measures for these distances is isometric to a subset of a Hilbert space. We give a description of all such metrics. Some of the distances appear to be ideal with respect to additive operations on random vectors. Subclasses of the distances are very useful to obtain a characterization of distributions and especially to recover a distribution from its potential. All Euclidean-type distances are very useful for constructing nonparametric, two-sample multidimensional tests. As background material for the discussion in this part of the book, in Chap. 21 we introduce the mathematical concepts of positive and negative definite kernels, describe their properties, and provide theoretical results that characterize coarse embeddings in a Hilbert space. Because kernel functions are central to the notion of potential of probability measures, in Chap. 22 we introduce special classes of probability metrics through negative definite kernel functions and show how, for strongly negative definite kernels, a probability measure can be uniquely determined by its potential. Moreover, the distance between probability measures can be bounded by the distance between their potentials; that is, under some technical conditions, a sequence of probability measures converges to a limit if and only if the sequence of their potentials converges to the potential of the limiting probability measure. Also as explained in Chap. 22, the problem of characterizing classes of probability distributions can be reduced to the problem of recovering a measure from potential. The problem of parameter estimation by the method of minimal distances and the study of the properties of these estimators are the subject of Chap. 23. In Chap. 24, we construct multidimensional statistical tests based on the theory of distances generated by negative definite kernels in the set of probability measures described in Chap. 23. The connection between distances generated by negative definite kernels and zonoids is the subject of Chap. 25. In Chap. 26, we discuss multidimensional statistical tests of uniformity based on the theory of distances generated by negative definite kernels and calculate the asymptotic distribution of these test statistics.

Part I
General Topics in the Theory of Probability Metrics

Chapter 2
Probability Distances and Probability Metrics: Definitions

The goals of this chapter are to:

- Provide examples of metrics in probability theory;
- Introduce formally the notions of a probability metric and a probability distance;
- Consider the general setting of random variables (RVs) defined on a given probability space $(\Omega, \mathcal{A}, \Pr)$ that can take values in a separable metric space U in order to allow for a unified treatment of problems involving random elements of a general nature;
- Consider the alternative setting of probability distances on the space of probability measures $\mathcal{P}_2$ defined on the σ-algebras of Borel subsets of $U^2 = U \times U$, where U is a separable metric space;
- Examine the equivalence of the notion of a probability distance on the space of probability measures $\mathcal{P}_2$ and on the space of joint distributions $\mathcal{L}\mathfrak{X}_2$ generated by pairs of RVs (X, Y) taking values in a separable metric space U.

S.T. Rachev et al., *The Methods of Distances in the Theory of Probability and Statistics*, DOI 10.1007/978-1-4614-4869-3_2, © Springer Science+Business Media, LLC 2013

Notation introduced in this chapter:

Notation	Description
EN	Engineer's metric
$\mathfrak{X}^p$	Space of real-valued random variables with $E\|X\|^p < \infty$
$\boldsymbol{\rho}$	Uniform (Kolmogorov) metric
$\mathfrak{X} = \mathfrak{X}(\mathbb{R})$	Space of real-valued random variables
L	Lévy metric
$\boldsymbol{\kappa}$	Kantorovich metric
$\boldsymbol{\theta}_p$	L_p-metric between distribution functions
K, **K***	Ky Fan metrics
$\mathcal{L}_p$	L_p-metric between random variables
$\mathbf{MOM}_p$	Metric between pth moments
(S, ρ)	Metric space with metric ρ
$\mathbb{R}^n$	n-dimensional vector space
$r(C_1, C_2)$	Hausdorff metric (semimetric between sets)
$s(F, G)$	Skorokhod metric
$\mathbb{K} = \mathbb{K}_\rho$	Parameter of a distance space
$\mathcal{H}$	Class of Orlicz's functions
ρ_H	Birnbaum–Orlicz distance
Kr	Kruglov distance
(U, d)	Separable metric space with metric d
s.m.s.	Separable metric space
U^k	k-fold Cartesian product of U
$\mathcal{B}_k = \mathcal{B}_k(U)$	Borel σ-algebra on U^k
$\mathcal{P}_k = \mathcal{P}_k(U)$	Space of probability laws on $\mathcal{B}_k$
$T_{\alpha,\beta,\ldots,\gamma}P$	Marginal of $P \in \mathcal{P}_k$ on coordinates $\alpha, \beta, \ldots, \gamma$
$\Pr_X$	Distribution of X
μ	Probability semidistance
$\mathfrak{X} := \mathfrak{X}(U)$	Set of U-valued RVs
$\mathcal{L}\mathfrak{X}_2 := \mathcal{L}\mathfrak{X}_2(U)$	Space of $\Pr_{X,Y}$, $X, Y \in \mathfrak{X}(U)$
u.m.	Universally measurable
u.m.s.m.s.	Universally measurable separable metric space

2.1 Introduction

Generally speaking, a functional that measures the distance between random quantities is called a *probability metric*.[1] In this chapter, we provide different examples of probability metrics and discuss an application of the Kolmogorov

[1] Mostafaei and Kordnourie (2011) is a more recent general publication on probability metrics and their applications.

metric in mathematical statistics. Then we proceed with the axiomatic construction of probability metrics on the space of probability measures defined on the twofold Cartesian product of a separable metric space U. This definition induces by restriction a probability metric on the space of joint distributions of random elements defined on a probability space $(\Omega, \mathcal{A}, \Pr)$ and taking values in the space U. Finally, we demonstrate that under some fairly general conditions, the two constructions are essentially the same.

2.2 Examples of Metrics in Probability Theory

Below is a list of various metrics commonly found in probability and statistics.

1. *Engineer's metric*:

$$\mathbf{EN}(X,Y) := |\mathbb{E}(X) - \mathbb{E}(Y)|, \quad X, Y \in \mathfrak{X}^1, \tag{2.2.1}$$

   where $\mathfrak{X}^p$ is the space of all real-valued RVs) with $\mathbb{E}|X|^p < \infty$.
2. *Uniform (or Kolmogorov) metric*:

$$\boldsymbol{\rho}(X,Y) := \sup\{|F_X(x) - F_Y(x)| : x \in \mathbb{R}\}, \quad X, Y \in \mathfrak{X} = \mathfrak{X}(\mathbb{R}), \tag{2.2.2}$$

   where F_X is the distribution function (DF) of X, $\mathbb{R} = (-\infty, +\infty)$, and $\mathfrak{X}$ is the space of all real-valued RVs.
3. *Lévy metric*:

$$\mathbf{L}(X,Y) := \inf\{\varepsilon > 0 : F_X(x-\varepsilon) - \varepsilon \leq F_Y(x) \leq F_X(x+\varepsilon) + \varepsilon, \quad \forall x \in \mathbb{R}\}. \tag{2.2.3}$$

Remark 2.2.1. We see that $\boldsymbol{\rho}$ and $\mathbf{L}$ may actually be considered metrics on the space of all distribution functions. However, this cannot be done for $\mathbf{EN}$ simply because $\mathbf{EN}(X,Y) = 0$ does not imply the coincidence of F_X and F_Y, while $\boldsymbol{\rho}(X,Y) = 0 \iff \mathbf{L}(X,Y) = 0 \iff F_X = F_Y$. The Lévy metric metrizes weak convergence (convergence in distribution) in the space $\mathcal{F}$, whereas $\boldsymbol{\rho}$ is often applied in the central limit theorem (CLT).[2]

4. *Kantorovich metric*:

$$\kappa(X,Y) = \int_{\mathbb{R}} |F_X(x) - F_Y(x)| \mathrm{d}x, \qquad X, Y \in \mathfrak{X}^1.$$

[2]See Hennequin and Tortrat (1965).

5. *L_p-metrics between distribution functions*:

$$\boldsymbol{\theta}_p(X,Y) := \left(\int_{-\infty}^{\infty} |F_X(t) - F_Y(t)|^p \mathrm{d}t\right)^{1/p}, \quad p \geq 1, \quad X,Y \in \mathfrak{X}^1. \quad (2.2.4)$$

Remark 2.2.2. Clearly, $\boldsymbol{\kappa} = \boldsymbol{\theta}_1$. Moreover, we can extend the definition of $\boldsymbol{\theta}_p$ when $p = \infty$ by setting $\boldsymbol{\theta}_\infty = \boldsymbol{\rho}$. One reason for this extension is the following dual representation for $1 \leq p \leq \infty$:

$$\theta_p(X,Y) = \sup_{f \in \mathcal{F}_p} |Ef(X) - Ef(Y)|, \quad X,Y \in \mathfrak{X}^1,$$

where $\mathcal{F}_p$ is the class of all measurable functions f with $\|f\|_q < 1$. Here, $\|f\|_q (1/p + 1/q = 1)$ is defined, as usual, by[3]

$$\|f\|_q := \begin{cases} \left(\int |f|^q\right)^{1/q}, 1 \leq q < \infty, \\ \operatorname*{ess\,sup}_{\mathbb{R}} |f|, \quad q = \infty. \end{cases}$$

6. *Ky Fan metrics*:

$$\mathbf{K}(X,Y) := \inf\{\varepsilon > 0 : \Pr(|X - Y| > \varepsilon) < \varepsilon\}, \qquad X,Y \in \mathfrak{X}, \qquad (2.2.5)$$

and

$$\mathbf{K}^*(X,Y) := E\frac{|X-Y|}{1+|X-Y|}. \qquad (2.2.6)$$

Both metrics metrize convergence in probability on $\mathfrak{X} = \mathfrak{X}(\mathbb{R})$, the space of real RVs.[4]

7. *L_p-metric*:

$$\mathcal{L}_p(X,Y) := \{E|X-Y|^p\}^{1/p}, \quad p \geq 1, \quad X,Y \in \mathfrak{X}^p. \qquad (2.2.7)$$

Remark 2.2.3. Define

$$m^p(X) := \{E|X|^p\}^{1/p}, \quad p > 1, \quad X \in \mathfrak{X}^p. \qquad (2.2.8)$$

and

$$\mathbf{MOM}_p(X,Y) := |m^p(X) - m^p(Y)|, \quad p \geq 1, \quad X,Y \in \mathfrak{X}^p. \qquad (2.2.9)$$

---

[3]The proof of this representation is given by (Dudley, 2002, p. 333) for the case $p = 1$.

[4]See Lukacs (1968, Chap. 3) and Dudley (1976, Theorem 3.5).

Then we have, for $X_0, X_1, \ldots \in \mathfrak{X}^p$,

$$\mathcal{L}_p(X_n, X_0) \to 0 \iff \begin{cases} \mathbf{K}(X_n, X_0) \to 0, \\ \mathbf{MOM}_p(X_n, X_0) \to 0 \end{cases} \tag{2.2.10}$$

[see, e.g., Lukacs (1968, Chap. 3)].

Other probability metrics in common use include the discrepancy metric, the Hellinger distance, the relative entropy metric, the separation distance metric, the χ^2-distance, and the f-divergence metric. These probability metrics are summarized in Gibbs and Su (2002).

All of the aforementioned (semi-)metrics on subsets of $\mathfrak{X}$ may be divided into three main groups: primary, simple, and compound (semi-)metrics. A metric μ is *primary* if $\mu(X, Y) = 0$ implies that certain moment characteristics of X and Y agree. As examples, we have **EN** (2.2.1) and $\mathbf{MOM}_p$ (2.2.9). For these metrics

$$\begin{aligned} \mathbf{EN}(X, Y) = 0 &\iff EX = EY, \\ \mathbf{MOM}_p(X, Y) = 0 &\iff m^p(X) = m^p(Y). \end{aligned} \tag{2.2.11}$$

A metric μ is *simple* if

$$\mu(X, Y) = 0 \iff F_X = F_Y. \tag{2.2.12}$$

Examples are $\boldsymbol{\rho}$ (2.2.2), $\mathbf{L}$ (2.2.3), and $\boldsymbol{\theta}_p$ (2.2.4). The third group, the *compound* (semi-)metrics, has the property

$$\mu(X, Y) = 0 \iff \Pr(X = Y) = 1. \tag{2.2.13}$$

Some examples are $\mathbf{K}$ (2.2.5), $\mathbf{K}^*$ (2.2.6), and $\mathcal{L}_p$ (2.2.7).

Later on, precise definitions of these classes will be given as well as a study of the relationships between them. Now we will begin with a common definition of probability metric that will include the types mentioned previously.

2.3 Kolmogorov Metric: A Property and an Application

In this section, we consider a paradoxical property of the Kolmogorov metric and an application in the area of mathematical statistics.

Consider the metric space $\mathfrak{F}$ of all one-dimensional distributions metrized by the Kolmogorov distance

$$\boldsymbol{\rho}(F, G) = \sup_{x \in \mathbb{R}} |F(x) - G(x)|, \tag{2.3.1}$$

which we define now in terms of the elements of $\mathfrak{F}$ rather than in terms of RVs as in the definition in (2.2.2). Denote by $B(F, r)$ an open ball of radius $r > 0$ centered

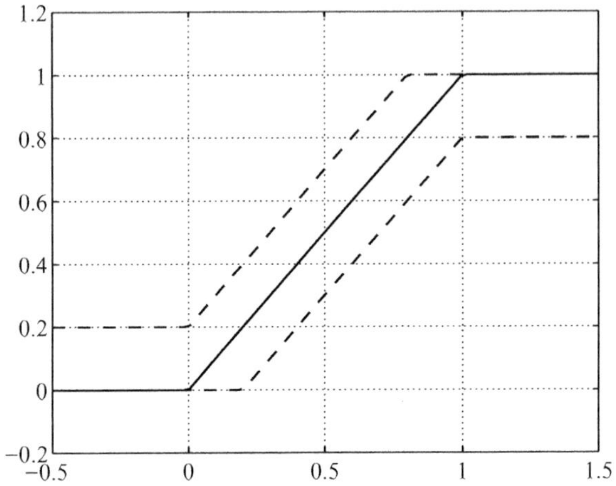

Fig. 2.1 The ball $B(F_o, \delta_\alpha)$. The *solid line* is the center of the ball and the *dashed line* represents the boundary of the ball

at F in the metric space $\mathfrak{F}$ with $\boldsymbol{\rho}$-distance and let F_o be a continuous distribution function (DF). The following result holds.

Theorem 2.3.1. *For any $r > 0$ there exists a continuous DF F_r such that*

$$B(F_r, r) \subset B(F_o, r) \tag{2.3.2}$$

and

$$B(F_r, r) \neq B(F_o, r).$$

Proof. Let us show that there are F_o and F_a such that (2.3.2) holds. Without loss of generality we may choose

$$F_o(x) = \begin{cases} 0, & x < 0, \\ x, & 0 \leq x < 1, \\ 1, & x \geq 1. \end{cases}$$

For a given (but fixed) n define δ_α such that (2.3.1) is true.

Figure 2.1 provides an illustration of the ball $B(F_o, \delta_\alpha)$. The boundary of the ball is shown by means of a dashed line, the center of the ball is the solid line, and the radius δ_α equals 0.2.

Consider now F_a defined in the following way:

$$F_a(x) = \begin{cases} 0, & x < \delta_\alpha/2, \\ 2x - \delta_\alpha, & \delta_\alpha/2 \leq x < \delta_\alpha, \\ x, & \delta_\alpha \leq x < 1 - \delta_\alpha, \\ 2x - (1 - \delta_\alpha), & 1 - \delta_\alpha \leq x < 1 - \delta_\alpha/2, \\ 1, & x \geq 1 - \delta_\alpha/2. \end{cases}$$

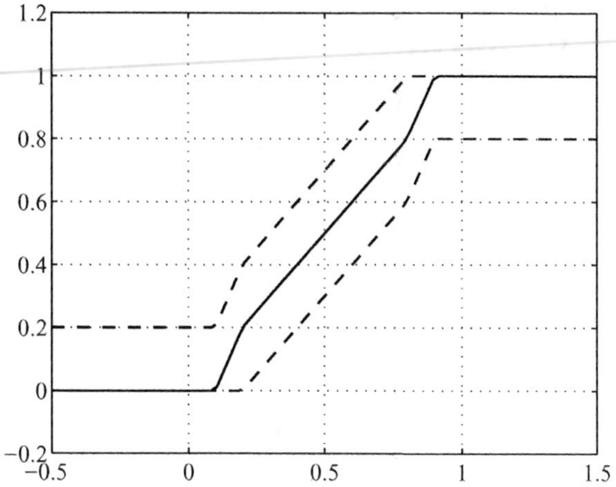

Fig. 2.2 The ball $B(F_a, \delta_\alpha)$. The *solid line* is the center of the ball and the *dashed line* represents the boundary of the ball

An illustration is given in Fig. 2.2. Comparing Figs. 2.1 and 2.2, we can see that

$$B(F_a, \delta_\alpha) \subset B(F_o, \delta_\alpha)$$

and

$$B(F_a, \delta_\alpha) \neq B(F_o, \delta_\alpha). \qquad \square$$

We demonstrate that this property leads to biasedness of the Kolmogorov goodness-of-fit tests. Suppose that $X_1, \ldots, X_n$ are independent and identically distributed (i.i.d.) RVs (observations) with (unknown) DF F. Based on the observations, one needs to test the hypothesis

$$H_o : F = F_o,$$

where F_o is a fixed DF.

Definition 2.3.1. For a specific alternative hypothesis, a test is said to be unbiased if the probability of rejecting the null hypothesis

(a) Is greater than or equal to the significance level α when the alternative is true and
(b) Is less than or equal to the significance level when the null hypothesis is true.

A test is said to be biased for an alternative hypothesis if it is not unbiased for this alternative.

Let d be a distance in the space of all probability distributions on the real line. Below we consider a test with the following properties:

1. We reject the null hypothesis H_o if

$$d(G_n, F_o) > \delta_\alpha,$$

where G_n is an empirical DF constructed on the basis of the observations $X_1, \ldots, X_n$ and δ_α satisfies

$$\Pr\{d(G_n, F_o) > \delta_\alpha\} \leq \alpha. \tag{2.3.3}$$

2. The test is distribution free, i.e.,

$$\Pr_F\{d(G_n, F_o) > \delta_\alpha\}$$

does not depend on continuous DF F.

We refer to such tests as *distance-based tests*.

Theorem 2.3.2. *Suppose that for some $\alpha > 0$ there exists a continuous DF F_a such that*

$$B(F_a, \delta_\alpha) \subset B(F_o, \delta_\alpha) \tag{2.3.4}$$

and

$$\Pr_{F_o}\{G_n \in B(F_o, \delta_\alpha) \setminus B(F_a, \delta_\alpha)\} > 0. \tag{2.3.5}$$

Then the distance-based test is biased for the alternative F_a.

Proof. Let $X_1, \ldots, X_n$ be a sample from F_a and G_n be the corresponding empirical DF. Then

$$\Pr_{F_o}\{G_n \in B(F_o, \delta_\alpha)\} \geq 1 - \alpha.$$

In view of (2.3.4) and (2.3.5), we have

$$\Pr_{F_o}\{G_n \in B(F_o, \delta_\alpha)\} > 1 - \alpha,$$

that is,

$$\Pr_{F_o}\{d(G_n, F_o) > \delta_\alpha\} < \alpha.$$ □

Now let us consider the Kolmogorov goodness-of-fit test. Clearly, it is a distance-based test for the distance

$$d(F, G) = \boldsymbol{\rho}(F, G).$$

From Theorem 2.3.1 it follows that (2.3.4) holds. The relation (2.3.5) is almost obvious. From Theorem 2.3.2 it follows that the Kolmogorov goodness-of-fit test is biased.

Remark 2.3.1. The biasedness of the Kolmogorov goodness-of-fit test is a known fact.[5] The same property holds for the Cramér–von Mises goodness-of-fit test.[6]

---

[5]See Massey (1950) and Thompson (1979).

[6]See Thompson (1966).

2.4 Metric and Semimetric Spaces, Distance, and Semidistance Spaces

Let us begin by recalling the notions of metric and semimetric spaces. Generalizations of these notions will be needed in the theory of probability metrics (TPM).

Definition 2.4.1. A set $S := (S, \rho)$ is said to be a *metric* space with the metric ρ if ρ is a mapping from the product $S \times S$ to $[0, \infty)$ having the following properties for each $x, y, z \in S$:

(1) *Identity property*: $\rho(x, y) = 0 \iff x = y$;
(2) *Symmetry*: $\rho(x, y) = \rho(y, x)$;
(3) *Triangle inequality*: $\rho(x, y) \le \rho(x, z) + \rho(z, y)$.

Here are some well-known examples of metric spaces:

(a) *The n-dimensional vector space* $\mathbb{R}^n$ endowed with the metric $\rho(x, y) := \|x - y\|_p$, where

$$\|x\|_p := \left(\sum_{i=1}^{n} |x_i|^p\right)^{\min(1,1/p)}, \quad x = (x_1, \ldots, x_n) \in \mathbb{R}^n, \quad 0 < p < \infty,$$

$$\|x\|_\infty := \sup_{1 \le i \le n} |x_i|.$$

(b) *The Hausdorff metric between closed sets*

$$r(C_1, C_2) = \max\left\{\sup_{x_1 \in C_1} \inf_{x_2 \in C_2} \rho(x_1, x_2),\ \sup_{x_2 \in C_2} \inf_{x_1 \in C_1} \rho(x_1, x_2)\right\},$$

where the C_i are closed sets in a bounded metric space (S, ρ).[7]

(c) *The H-metric*. Let $D(\mathbb{R})$ be the space of all bounded functions $f : \mathbb{R} \to \mathbb{R}$, continuous from the right and having limits from the left, $f(x-) = \lim_{t \uparrow x} f(t)$. For any $f \in D(\mathbb{R})$ define the graph Γ_f as the union of the sets $\{(x, y) : x \in \mathbb{R}, y = f(x)\}$ and $\{(x, y) : x \in \mathbb{R}, y = f(x-)\}$. The H-metric $H(f, g)$ in $D(\mathbb{R})$ is defined by the Hausdorff distance between the corresponding graphs, $H(f, g) := r(\Gamma_f, \Gamma_g)$. Note that in the space $\mathcal{F}(\mathbb{R})$ of distribution functions, H metrizes the same convergence as the *Skorokhod metric*:

[7] See Hausdorff (1949).

$$s(F,G) = \inf\Big\{\varepsilon > 0 : \text{there exists a strictly increasing continuous function } \lambda : \mathbb{R} \to \mathbb{R} \text{ such that } \lambda(\mathbb{R}) = \mathbb{R},\ \sup_{t\in\mathbb{R}} |\lambda(t) - t| < \varepsilon, \text{ and } \sup_{t\in\mathbb{R}} |F(\lambda(t)) - G(t)| < \varepsilon\Big\}.$$

Moreover, H-convergence in $\mathcal{F}$ implies convergence in distributions (the weak convergence). Clearly, $\boldsymbol{\rho}$-convergence [see (2.2.2)] implies H-convergence.[8]

If the identity property in Definition 2.4.1 is weakened by changing property (1) to

$$x = y \Rightarrow \rho(x,y) = 0, \tag{1*}$$

then S is said to be a *semimetric space* (or *pseudometric space*) and ρ a *semimetric* (or *pseudometric*) in S. For example, the Hausdorff metric r is only semimetric in the space of all Borel subsets of a bounded metric space (S,ρ).

Obviously, in the space of real numbers, **EN** [see (2.2.1)] is the usual uniform metric on the real line $\mathbb{R}$ [i.e., $\mathbf{EN}(a,b) := |a-b|$, $a,b \in \mathbb{R}$]. For $p \geq 0$, define $\mathcal{F}^p$ as the space of all distribution functions F with $\int_{-\infty}^0 F(x)^p \mathrm{d}x + \int_0^\infty (1 - F(x))^p \mathrm{d}x < \infty$. The distribution function space $\mathcal{F} = \mathcal{F}^0$ can be considered a metric space with metrics $\boldsymbol{\rho}$ and $\mathbf{L}$, while $\boldsymbol{\theta}_p (1 \leq p < \infty)$ is a metric in $\mathcal{F}^p$. The Ky Fan metrics [see (2.2.5), (2.2.6)], resp. $\mathcal{L}_p$-metric [see (2.2.7)], may be viewed as semimetrics in $\mathfrak{X}$ (resp. $\mathfrak{X}^1$) as well as metrics in the space of all Pr-equivalence classes

$$\widetilde{X} := \{Y \in \mathfrak{X} : \Pr(Y = X) = 1\}, \quad \forall X \in \mathfrak{X} \text{ [resp. } \mathfrak{X}^p]. \tag{2.4.1}$$

EN, $\mathbf{MOM}_p$, $\boldsymbol{\theta}_p$, and $\mathcal{L}_p$ can take infinite values in $\mathfrak{X}$, so we will assume, in the next generalization of the notion of metric, that ρ may take infinite values; at the same time, we will also extend the notion of triangle inequality.

Definition 2.4.2. The set $\mathbf{S}$ is called a *distance space* with distance ρ and parameter $\mathbb{K} = \mathbb{K}_\rho$ if ρ is a function from $S \times S$ to $[0,\infty]$, $\mathbb{K} \geq 1$, and for each $x,y,z \in S$ the identity property (1) and the symmetry property (2) hold, as does the following version of the triangle inequality: (3*) (*Triangle inequality with parameter* $\mathbb{K}$)

$$\rho(x,y) \leq \mathbb{K}[\rho(x,z) + \rho(z,y)]. \tag{2.4.2}$$

If, in addition, the identity property (1) is changed to (1*), then S is called a semidistance space and ρ is called a semidistance (with parameter $\mathbb{K}_\rho$).

Here and in what follows we will distinguish the notions *metric* and *distance*, using *metric* only in the case of *distance with parameter* $\mathbb{K} = 1$, *taking finite or infinite values.*

[8] A more detailed analysis of the metric H will be given in Sect. 4.2.

Remark 2.4.1. It is not difficult to check that each distance ρ generates a topology in S with a basis of open sets $B(a,r) := \{x \in S; \rho(x,a) < r\}, \in S, r > 0$. We know, of course, that every metric space is normal and that every separable metric space has a countable basis. In much the same way, it is easily shown that the same is true for distance space. Hence, by Urysohn's metrization theorem,[9] every separable distance space is metrizable.

Actually, distance spaces have been used in functional analysis for a long time, as shown by the following examples.

Example 2.4.1. Let $\mathcal{H}$ be the class of all nondecreasing continuous functions H from $[0,\infty)$ onto $[0,\infty)$, which vanish at the origin and satisfy Orlicz's condition

$$K_H := \sup_{t>0} \frac{H(2t)}{H(t)} < \infty. \tag{2.4.3}$$

Then $\widetilde{\rho} := H(\rho)$ is a distance in S for each metric ρ in S and $\mathbb{K}_{\widetilde{\rho}} = K_H$.

Example 2.4.2. The *Birnbaum–Orlicz space* L^H ($H \in \mathcal{H}$) consists of all integrable functions on $[0,1]$ endowed with the Birnbaum–Orlicz distance[10]

$$\rho_H(f_1, f_2) := \int_0^1 H(|f_1(x) - f_2(x)|)\mathrm{d}x. \tag{2.4.4}$$

Obviously, $\mathbb{K}_{\rho_H} = K_H$.

Example 2.4.3. Similarly to (2.4.4), Kruglov (1973) introduced the following distance in the space of distribution functions:

$$\mathbf{Kr}(F,G) = \int \phi(F(x) - G(x))\mathrm{d}x, \tag{2.4.5}$$

where the function ϕ satisfies the following conditions:

(a) ϕ is even and strictly increasing on $[0,\infty)$, $\phi(0) = 0$;
(b) For any x and y and some fixed $A \geq 1$

$$\phi(x+y) \leq A(\phi(x) + \phi(y)). \tag{2.4.6}$$

Obviously, $\mathbb{K}_{\mathbf{Kr}} = A$.

---

[9]See Dunford and Schwartz (1988, Theorem 1.6.19).

[10]Birnbaum and Orliz (1931) and Dunford and Schwartz (1988, p. 400)

2.5 Definitions of Probability Distance and Probability Metric

Let U be a separable metric space (s.m.s.) with metric d, $U^k = \underbrace{U \times \cdots \times U}_{k \text{ times}}$ the k-fold Cartesian product of U, and $\mathcal{P}_k = \mathcal{P}_k(U)$ the space of all probability measures defined on the σ-algebra $\mathcal{B}_k = \mathcal{B}_k(U)$ of Borel subsets of U^k. We will use the terms *probability measure* and *law* interchangeably. For any set $\{\alpha, \beta, \ldots, \gamma\} \subseteq \{1, 2, \ldots, k\}$ and for any $P \in \mathcal{P}_k$ let us define the marginal of P on the coordinates $\alpha, \beta, \ldots, \gamma$ by $T_{\alpha,\beta,\ldots,\gamma}P$. For example, for any Borel subsets A and B of U, $T_1 P(A) = P(A \times U \times \cdots \times U)$, $T_{1,3}P(A \times B) = P(A \times U \times B \times \cdots \times U)$. Let $\mathbb{B}$ be the operator in U^2 defined by $\mathbb{B}(x, y) := (y, x)$ $(x, y \in U)$. All metrics $\mu(X, Y)$ cited in Sect. 2.2 [see (2.2.1)–(2.2.9)] are completely determined by the joint distributions $\Pr_{X,Y}$ $(\Pr_{X,Y} \in \mathcal{P}_2(\mathbb{R}))$ of the RVs $X, Y \in \mathfrak{X}(\mathbb{R})$.

In the next definition we will introduce the notion of probability distance, and thus we will describe the primary, simple, and compound metrics in a uniform way. Moreover, the space where the RVs X and Y take values will be extended to U, an arbitrary s.m.s.

Definition 2.5.1. A mapping μ defined on $\mathcal{P}_2$ and taking values in the extended interval $[0, \infty]$ is said to be a *probability semidistance with parameter* $\mathbb{K} := \mathbb{K}_\mu \geq 1$ (or *p. semidistance* for short) in $\mathcal{P}_2$ if it possesses the following three properties:

(1) (*Identity property* (**ID**)). If $P \in \mathcal{P}_2$ and $P(\cup_{x \in U}\{(x, x)\}) = 1$, then $\mu(P) = 0$;
(2) (*Symmetry* (**SYM**)). If $P \in \mathcal{P}_2$, then $\mu(P \circ \mathbb{B}^{-1}) = \mu(P)$;
(3) (*Triangle inequality* (**TI**)). If $P_{13}, P_{12}, P_{23} \in \mathcal{P}_2$ and there exists a law $Q \in \mathcal{P}_3$ such that the following "consistency" condition holds:

$$T_{13}Q = P_{13}, \qquad T_{12}Q = P_{12}, \qquad T_{23}Q = P_{23}, \tag{2.5.1}$$

then

$$\mu(P_{13}) \leq \mathbb{K}[\mu(P_{12}) + \mu(P_{23})].$$

If $\mathbb{K} = 1$, then μ is said to be a *p. semimetric*. If we strengthen the condition **ID** to

$\widetilde{\textbf{ID}}$: if $P \in \mathcal{P}_2$, then

$$P(\cup\{(x, x) : x \in U\}) = 1 \iff \mu(P) = 0,$$

then we say that μ is a *probability distance with parameter* $\mathbb{K} = \mathbb{K}_\mu \geq 1$ (or *p. distance* for short).

Definition 2.5.1 acquires a visual form in terms of RVs, namely: let $\mathfrak{X} := \mathfrak{X}(U)$ be the set of all RVs on a given probability space $(\Omega, \mathcal{A}, \Pr)$ taking values in $(U, \mathcal{B}_1)$. By $\mathcal{L}\mathfrak{X}_2 := \mathcal{L}\mathfrak{X}_2(U) := \mathcal{L}\mathfrak{X}_2(U; \Omega, \mathcal{A}, \Pr)$ we denote the space of all joint distributions $\Pr_{X,Y}$ generated by the pairs $X, Y \in \mathfrak{X}$. Since $\mathcal{L}\mathfrak{X}_2 \subseteq \mathcal{P}_2$, the notion of a p. (semi-)distance is naturally defined on $\mathcal{L}\mathfrak{X}_2$. Considering μ on the subset $\mathcal{L}\mathfrak{X}_2$, we will put

$$\mu(X,Y) := \mu(\Pr_{X,Y})$$

and call μ a *p. semidistance on* $\mathfrak{X}$. If μ is a p. distance, then we use the phrase *p. distance* on $\mathfrak{X}$. Each p. semidistance μ on $\mathfrak{X}$ is a semidistance on $\mathfrak{X}$ in the sense of Definition 2.4.2.[11] Then the relationships **ID**, $\widetilde{\mathbf{ID}}$, **SYM**, and **TI** have simple "metrical" interpretations:

$$\mathbf{ID}^{(*)} \qquad \Pr(X=Y)=1 \Rightarrow \mu(X,Y)=0,$$

$$\widetilde{\mathbf{ID}}^{(*)} \qquad \Pr(X=Y)=1 \iff \mu(X,Y)=0,$$

$$\mathbf{SYM}^{(*)} \quad \mu(X,Y)=\mu(Y,X),$$

$$\mathbf{TI}^{(*)} \qquad \mu(X,Z) < \mathbb{K}[\mu(X,Z)+\mu(Z,Y)].$$

Definition 2.5.2. A mapping $\mu : \mathcal{L}\mathfrak{X}_2 \to [0,\infty]$ is said to be a *p. semidistance* on $\mathfrak{X}$ (resp. *distance) with parameter* $\mathbb{K} := \mathbb{K}_\mu \geq 1$ if $\mu(X,Y) = \mu(\Pr_{X,Y})$ satisfies the properties $\mathbf{ID}^{(*)}$ [resp. $\widetilde{\mathbf{ID}}^{(*)}$], $\mathbf{SYM}^{(*)}$, and $\mathbf{TI}^{(*)}$ for all RVs $X, Y, Z \in \mathfrak{X}(U)$.

Example 2.5.1. Let $H \in \mathcal{H}$ (Example 2.4.1) and (U,d) be an s.m.s. Then $\mathcal{L}_H(X,Y) = EH(d(Z,V))$ is a p. distance in $\mathfrak{X}(U)$. Clearly, $\mathcal{L}_H$ is finite in the subspace of all X with finite moment $EH(d(X,a))$ for some $a \in U$. Kruglov's distance $\mathbf{Kr}(X,Y) := \mathbf{Kr}(F_X,F_Y)$ is a p. semidistance in $\mathfrak{X}(\mathbb{R})$.

Examples of p. metrics in $\mathfrak{X}(U)$ are the Ky Fan metric

$$\mathbf{K}(X,Y) := \inf\{\varepsilon>0 : \Pr(d(X,Y)>\varepsilon)<\varepsilon\}, \quad X,Y \in \mathfrak{X}(U), \tag{2.5.2}$$

and the $\mathcal{L}_p$*-metrics* $(0 \leq p \leq \infty)$

$$\mathcal{L}_p(X,Y) := \{Ed^p(X,Y)\}^{\min(1,1/p)}, \quad 0<p<\infty, \tag{2.5.3}$$

$$\mathcal{L}_\infty(X,Y) := \operatorname{ess\,sup} d(X,Y) := \inf\{\varepsilon>0 : \Pr(d(X,Y)>\varepsilon)=0\}, \tag{2.5.4}$$

$$\mathcal{L}_0(X,Y) := EI\{X,Y\} := \Pr(X,Y). \tag{2.5.5}$$

The engineer's metric **EN**, Kolmogorov metric $\boldsymbol{\rho}$, Kantorovitch metric $\boldsymbol{\kappa}$, and the Lévy metric **L** (Sect. 2.2) are p. semimetrics in $\mathfrak{X}(\mathbb{R})$.

Remark 2.5.1. Unlike Definition 2.5.2, Definition 2.5.1 is free of the choice of the initial probability space and depends solely on the structure of the metric space U. The main reason for considering not arbitrary but separable metric spaces (U,d) is that we need the measurability of the metric d to connect the metric structure of U with that of $\mathfrak{X}(U)$. In particular, the measurability of d enables us to handle, in a well-defined way, probability metrics such as the Ky Fan metric **K** and $\mathcal{L}_p$-metrics.

[11] If we replace "semidistance" with "distance," then the statement continues to hold.

Note that $\mathcal{L}_0$ does not depend on the metric d, so one can define $\mathcal{L}_0$ on $\mathfrak{X}(U)$, where U is an arbitrary measurable space, while in (2.5.2)–(2.5.4) we need $d(X,Y)$ to be an RV. Thus the natural class of spaces appropriate for our investigation is the class of s.m.s.

2.6 Universally Measurable Separable Metric Spaces

What follows is an exposition of some basic results regarding universally measurable separable metric spaces (u.m.s.m.s.). As we will see, the notion of u.m.s.m.s. plays an important role in TPM.

Definition 2.6.1. Let P be a Borel probability measure on a metric space (U,d). We say that P is *tight* if for each $\varepsilon > 0$ there is a compact $K \subseteq U$ with $P(K) \geq 1 - \varepsilon$.[12]

Definition 2.6.2. An s.m.s. (U,d) is *universally measurable* (u.m.) if every Borel probability measure on U is tight.

Definition 2.6.3. An s.m.s. (U,d) is *Polish* if it is topologically complete [i.e., there is a topologically equivalent metric e such that (U,e) is complete]. Here the topological equivalence of d and e simply means that for any $x, x_1, x_2, \ldots$ in U

$$d(x_n, x) \to 0 \iff e(x_n, x) \to 0.$$

Theorem 2.6.1. *Every Borel subset of a Polish space is u.m.*

Proof. See Billingsley (1968, Theorem 1.4), Cohn (1980, Proposition 8.1.10), and Dudley (2002, p. 391). □

Remark 2.6.1. Theorem 2.6.1 provides us with many examples of u.m. spaces but does not exhaust this class. The topological characterization of u.m.s.m.s. is a well-known open problem.[13]

In his famous paper on measure theory, Lebesgue (1905) claimed that the projection of any Borel subset of $\mathbb{R}^2$ onto $\mathbb{R}$ is a Borel set. As noted by Souslin and his teacher Lusin (1930), this is in fact not true. As a result of the investigations surrounding this discovery, a theory of such projections (the so-called analytic or Souslin sets) was developed. Although not a Borel set, such a projection was shown to be Lebesgue-measurable; in fact it is u.m. This train of thought leads to the following definition.

---

[12] See (Dudley, 2002, Sect. 11.5).

[13] See Billingsley (1968, Appendix III, p. 234)

Definition 2.6.4. Let S be a Polish space, and suppose that f is a measurable function mapping S onto an s.m.s. U. In this case, we say that U is *analytic*.

Theorem 2.6.2. *Every analytic s.m.s. is u.m.*

Proof. See Cohn (1980, Theorem 8.6.13, p. 294) and Dudley (2002, Theorem 13.2.6). □

Example 2.6.1. Let $\mathbb{Q}$ be the set of rational numbers with the usual topology. Since $\mathbb{Q}$ is a Borel subset of the Polish space R, then $\mathbb{Q}$ is u.m.; however, $\mathbb{Q}$ is not itself a Polish space.

Example 2.6.2. In any uncountable Polish space, there are analytic (hence u.m.) non-Borel sets.[14]

Example 2.6.3. Let $C[0,1]$ be the space of continuous functions $f : [0,1] \to \mathbb{R}$ under the uniform norm. Let $E \subseteq C[0,1]$ be the set of f that fail to be differentiable at some $t \in [0,1]$. Then a theorem of Mazukiewicz (1936) says that E is an analytic, non-Borel subset of $C[0,1]$. In particular, E is u.m.

Recall again the notion of *Hausdorff metric* $r := r_\rho$ in the space of all subsets of a given metric space (S,ρ)

$$r(A,B) = \max\left\{\sup_{x\in A}\inf_{y\in B}\rho(x,y), \sup_{y\in B}\inf_{x\in A}\rho(x,y)\right\}$$

$$= \inf\{\varepsilon > 0 : A^\varepsilon \supseteq B, B^\varepsilon \supseteq A\}, \tag{2.6.1}$$

where A^ε is the open ε-neighborhood of A, $A^\varepsilon = \{x : d(x.A) < \varepsilon\}$.

As we noticed in the space 2^S of all subsets $A \neq \emptyset$ of S, the Hausdorff distance r is actually only a semidistance. However, in the space $\mathcal{C} = \mathcal{C}(S)$ of all closed nonempty subsets, r is a metric (Definition 2.4.1) and takes on both finite and infinite values, and if S is a bounded set, then r is a finite metric on $\mathcal{C}$.

Theorem 2.6.3. *Let (S,ρ) be a metric space, and let $(\mathcal{C}(S), r)$ be the space described previously. If (S,ρ) is separable (or complete, or totally bounded), then $(\mathcal{C}(S), r)$ is separable (or complete, or totally bounded).*

Proof. See Hausdorff (1949, Sect. 29) and Kuratowski (1969, Sects. 21 and 23). □

Example 2.6.4. Let $S = [0,1]$, and let ρ be the usual metric on S. Let $\mathcal{R}$ be the set of all finite complex-valued Borel measures m on S such that the Fourier transform

$$\widehat{m}(t) = \int_0^1 \exp(iut)m(du)$$

[14] See Cohn (1980, Corollary 8.2.17) and Dudley (2002, Proposition 13.2.5).

vanishes at $t = \pm\infty$. Let $\mathcal{M}$ be the class of sets $E \in \mathcal{C}(S)$ such that there is some $m \in \mathcal{R}$ concentrated on E. Then $\mathcal{M}$ is an analytic, non-Borel subset of $(\mathcal{C}(S), r_\rho)$.[15]

We seek a characterization of u.m.s.m.s. in terms of their Borel structure.

Definition 2.6.5. A measurable space M with σ-algebra $\mathcal{M}$ is *standard* if there is a topology τ on M such that (M, τ) is a compact metric space and the Borel σ-algebra generated by τ coincides with $\mathcal{M}$.

An s.m.s. is standard if it is a Borel subset of its completion.[16] Obviously, every Borel subset of a Polish space is standard.

Definition 2.6.6. Say that two s.m.s. U and V are called Borel-isomorphic if there is a one-to-one correspondence f of U onto V such that $B \in \mathcal{B}(U)$ if and only if $f(B) \in \mathcal{B}(V)$.

Theorem 2.6.4. *Two standard s.m.s. are Borel-isomorphic if and only if they have the same cardinality.*

Proof. See Cohn (1980, Theorem 8.3.6) and Dudley (2002, Theorem 13.1.1). □

Theorem 2.6.5. *Let U be an s.m.s. The following are equivalent:*

(1) U is u.m.
(2) For each Borel probability m on U there is a standard set $S \in \mathcal{B}(U)$ such that $m(S) = 1$.

Proof. $1 \Rightarrow 2$: Let m be a law on U. Choose compact $K_n \subseteq U$ with $m(K_n) \geq 1 - 1/n$. Put $S = \cup_{n\geq 1} K_n$. Then S is σ-compact and, hence, standard. Thus, $m(S) = 1$, as desired.

$2 \Leftarrow 1$: Let m be a law on U. Choose a standard set $S \in \mathcal{B}(U)$ with $m(S) = 1$. Let $\overline{U}$ be the completion of U. Then S is Borel in its completion $\overline{S}$, which is closed in $\overline{U}$. Thus, S is Borel in $\overline{U}$. It follows from Theorem 2.6.1 that

$$1 = m(S) = \sup\{m(K) : K \text{ compact}\}.$$

Thus, every law m on U is tight, so that U is u.m. □

Corollary 2.6.1. *Let (U, d) and (V, e) be Borel-isomorphic separable metric spaces. If (U, d) is u.m., then so is (V, e).*

Proof. Suppose that m is a law on V. Define a law n on U by $n(A) = m(f(A))$, where $f : U \to V$ is a Borel isomorphism. Since U is u.m., there is a standard set $\subseteq U$ with $n(S) = 1$. Then $f(S)$ is a standard subset of V with $m(f(S)) = 1$. Thus, by Theorem 2.6.5, V is u.m. □

---

[15] See Kaufman (1984).

[16] See Dudley (2002, p. 347).

The following result, which is in essence due to Blackwell (1956), will be used in an important way later on.[17]

Theorem 2.6.6. *Let U be a u.m. separable metric space, and suppose that* Pr *is a probability measure on U. If $\mathcal{A}$ is a countably generated sub-σ-algebra of $\mathcal{B}(U)$, then there is a real-valued function $P(B|x)$, $B \in \mathcal{B}(U)$, $x \in U$, such that*

(1) For each fixed $B \in \mathcal{B}(U)$ the mapping $x \to P(B|x)$ is an $\mathcal{A}$-measurable function on U;
(2) For each fixed $x \in U$ the set function $B \to P(B|x)$ is a law on U;
(3) For each $A \in \mathcal{A}$ and $B \in \mathcal{B}(U)$ we have $\int_A P(B|x)\,\mathrm{Pr}(\mathrm{d}x) = \mathrm{Pr}(A \cap B)$;
(4) There is a set $N \in \mathcal{A}$ with $\mathrm{Pr}(N) = 0$ such that $P(B|x) = 1$ whenever $x \in U - N$.

Proof. Choose a sequence $F_1, F_2, \ldots$ of sets in $\mathcal{B}(U)$ that generates $\mathcal{B}(U)$ and is such that a subsequence generates $\mathcal{A}$. We will prove that there exists a metric e on U such that (U, d) and (U, e) are Borel-isomorphic and for which the sets $F_1, F_2, \ldots$ are *clopen*, i.e., open and closed. □

Claim. If (U, d) is an s.m.s. and $A_1, A_2, \ldots$ is a sequence of Borel subsets of U, then there is some metric e on U such that

(i) (U, e) is an s.m.s. isometric with a closed subset of $\mathbb{R}$;
(ii) $A_1, A_2, \ldots$ are clopen subsets of (U, e);
(iii) (U, d) and (U, e) are Borel-isomorphic (Definition 2.6.6).

Proof of claim. Let $B_1, B_2, \ldots$ be a countable base for the topology of (U, d). Define sets $C_1, C_2, \ldots$ by $C_{2n-1} = A_n$ and $C_{2n} = B_n$ $(n = 1, 2, \ldots)$ and $f : U \to \mathbb{R}$ by $f(x) = \sum_{n=1}^{\infty} 2I_{C_n}(x)/3^n$. Then f is a Borel isomorphism of (U, d) onto $f(U) \subseteq K$, where K is the Cantor set

$$K := \left\{ \sum_{n=1}^{\infty} \alpha_n/3^n : \alpha_n \text{ take value 0 or 2} \right\}.$$

Define the metric e by $e(x, y) = |f(x) - f(y)|$, so that (U, e) is isometric with $f(U) \subseteq K$. Then $A_n = f^{-1}\{x \in K; x(n) = 2\}$, where $x(n)$ is the nth digit in the ternary expansion of $x \in K$. Thus, A_n is clopen in (U, e), as required.

Now (U, e) is (Corollary 2.6.1) u.m., so there are compact sets $K_1 \subseteq K_2 \subseteq \cdots$ with $\mathrm{Pr}(K_n) \to 1$. Let $\mathcal{G}_1$ and $\mathcal{G}_2$ be the (countable) algebras generated by the sequences $F_1, F_2, \ldots$ and $F_1, F_2, \ldots, K_1, K_2, \ldots$, respectively. Then define $P_1(B|x)$ so that (1) and (3) are satisfied for $B \in \mathcal{G}_2$. Since $\mathcal{G}_2$ is countable, there is some set $N \in \mathcal{A}$ with $\mathrm{Pr}(N) = 0$ and such that for $x \in N$,

(a) $P_1(\cdot|x)$ is a finitely additive probability on $\mathcal{G}_2$,
(b) $P_1(A|x) = 1$ for $A \in \mathcal{A} \cap \mathcal{G}_2$ and $x \in A$,
(c) $P_1(K_n|x) \to 1$ as $n \to \infty$.

[17] See Theorem 3.3.1 in Sect. 3.3.

Claim. For $x \in N$ the set function $B \to P_1(B|x)$ is countably additive on $\mathcal{G}_1$.

Proof of claim. Suppose that $H_1, H_2, \ldots$ are disjoint sets in $\mathcal{G}_1$ whose union is U. Since the H_n are clopen and the K_n are compact in (U, e), there is, for each n, some $M = M(n)$ such that $K_n \subseteq H_1 \cup H_2 \cup \cdots \cup H_M$. Finite additivity of $P_1(x, \cdot)$ on $\mathcal{G}_2$ yields, for $x \notin N$, $P_1(K_n|x) \leq \sum_{i=1}^{M} P_1(H_i|x) \leq \sum_{i=1}^{\infty} P_1(H_i|x)$. Let $n \to \infty$, and apply (c) to obtain $\sum_{i=1}^{\infty}(P_1(H_i|x) = 1$, as required.

In view of the claim, for each $x \in N$ we define $B \to P(B|x)$ as the unique countably additive extension of P_1 from $\mathcal{G}_1$ to $\mathcal{B}(U)$. For $x \in N$ put $P(B|x) = \Pr(B)$. Clearly, (2) holds. Now the class of sets in $\mathcal{B}(U)$ for which (1) and (3) hold is a monotone class containing $\mathcal{G}_1$, and so coincides with $\mathcal{B}(U)$.

Claim. Condition (4) holds.

Proof of claim. Suppose that $A \in \mathcal{A}$ and $x \in A - N$. Let A_0 be the $\mathcal{A}$-atom containing x. Then $A_0 \subseteq A$, and there is a sequence $A_1, A_2, \ldots$ in $\mathcal{G}_1$ such that $A_0 = A_1 \cap A_2 \cap \cdots$. From (b), $P(A_n|x) = 1$ for $n \geq 1$, so that $P(A_0|x) = 1$, as desired. □

Corollary 2.6.2. *Let U and V be u.m.s.m.s., and let* Pr *be a law on $U \times V$. Then there is a function $P : \mathcal{B}(V) \times U \to \mathbb{R}$ such that*

(1) For each fixed $B \in \mathcal{B}(V)$ the mapping $x \to P(B|x)$ is measurable on U;
(2) For each fixed $x \in U$ the set function $B \to P(B|x)$ is a law on V;
(3) For each $A \in \mathcal{B}(U)$ and $B \in \mathcal{B}(V)$ we have

$$\int_{\mathcal{A}} P(B|x) P_1(\mathrm{d}x) = \Pr(A \cap B),$$

where P_1 is the marginal of Pr *on U.*

Proof. Apply the preceding theorem with $\mathcal{A}$ the σ-algebra of rectangles $A \times U$ for $A \in \mathcal{B}(U)$. □

2.7 Equivalence of the Notions of Probability (Semi-) distance on $\mathcal{P}_2$ and on $\mathfrak{X}$

As we saw in Sect. 2.5, every p. (semi-)distance on $\mathcal{P}_2$ induces (by restriction) a p. (semi-)distance on $\mathfrak{X}$. It remains to be seen whether every p. (semi-)distance on $\mathfrak{X}$ arises in this way. This will certainly be the case whenever

$$\mathcal{L}\mathfrak{X}_2(U, (\Omega, \mathcal{A}, \Pr)) = \mathcal{P}_2(U). \tag{2.7.1}$$

Note that the left member depends not only on the structure of (U, d) but also on the underlying probability space.

In this section we will prove the following facts.

1. There is some probability space $(\Omega, \mathcal{A}, \Pr)$ such that (2.7.1) holds for every separable metric space U.
2. If U is a separable metric space, then (2.7.1) holds for every nonatomic probability space $(\Omega, \mathcal{A}, \Pr)$ if and only if U is u.m.

We need a few preliminaries.

Definition 2.7.1. [18] If $(\Omega, \mathcal{A}, \Pr)$ is a probability space, then we say that $A \in \mathcal{A}$ is an *atom* if $\Pr(A) > 0$ and $\Pr(B) = 0$ or $\Pr(A)$ for each measurable $B \subseteq A$. A probability space is *nonatomic* if it has no atoms.

Lemma 2.7.1. [19] *Let ν be a law on a complete s.m.s. (U, d) and suppose that $(\Omega, \mathcal{A}, \Pr)$ is a nonatomic probability space. Then there is a U-valued RV X with distribution $\mathcal{L}(X) = \nu$.*

Proof. Denote by d^* the following metric on U^2: $d^*(x, y) := d(x_1, x_2) + d(y_1, y_2)$ for $x = (x_1, y_1)$ and $y = (x_2, y_2)$. For each k there is a partition of U^2 comprising nonempty Borel sets $\{A_{ik} : i = 1, 2, \ldots\}$ with $\operatorname{diam}(A_{ik}) < 1/k$ and such that A_{ik} is a subset of some $A_{j,k-1}$.

Since $(\Omega, \mathcal{A}, \Pr)$ is nonatomic, we see that for each $C \in \mathcal{A}$ and for each sequence p_i of nonnegative numbers such that $p_1 + p_2 + \cdots = \Pr(C)$ there exists a partitioning $C_1, C_2, \ldots$ of C such that $\Pr(C_i) = p_i$, $i = 1, 2, \ldots$ [20]

Therefore, there exist partitions $\{B_{ik} : i = 1, 2, \ldots\} \subseteq \mathcal{A}$, $k = 1, 2, \ldots$, such that $B_{ik} \subseteq B_{jk-1}$ for some $j = j(i)$ and $\Pr(B_{ik}) = \nu(A_{ik})$ for all i, k. For each pair (i, j) let us pick a point $x_{ik} \in A_{ik}$ and define U^2-valued $X_k(\omega) = x_{ik}$ for $\omega \in B_{ik}$. Then $d^*(X_{k+m}(\omega), X_k(\omega)) < 1/k$, $m = 1, 2, \ldots$, and since (U^2, d^*) is a complete space, there exists the limit $X(\omega) = \lim_{k\to\infty} X_k(\omega)$. Thus

$$d^*(X(\omega), X_k(\omega)) \le \lim_{m\to\infty} [d^*(X_{k+m}(\omega), X(\omega)) + d^*(X_{k+m}(\omega), X_k(\omega))] \le \frac{1}{k}.$$

Let $P_k := \Pr_{X_k}$ and $P^* := \Pr_X$. Further, our aim is to show that $P^* = \nu$. For each closed subset $A \subseteq U$

$$P_k(A) = \Pr(X_k \in A) \le \Pr(X \in A^{1/k}) = P^*(A^{1/k}) \le P_k(A^{2/k}), \tag{2.7.2}$$

where $A^{1/k}$ is the open $1/k$-neighborhood of A. On the other hand,

$$P_k(A) = \sum\{P_k(x_{ik}) : x_{ik} \in A\} = \sum\{\Pr(B_{ik}) : x_{ik} \in A\}$$

---

[18] See Loeve (1963, p. 99) and Dudley (2002, p. 82).

[19] See Berkes and Phillip (1979).

[20] See, for example, Loeve (1963, p. 99).

$$= \sum\{\nu(A_{ik}) : x_{ik} \in A\} \le \sum\{\nu(A_{ik} \cap A^{1/k}) : x_{ik} \in A\}$$
$$\le \nu(A^{1/k}) \le \sum\{\nu(A_{ik}) : x_{ik} \in A^{2/k}\} \le P_k(A^{2/k}). \tag{2.7.3}$$

Further, we can estimate the value $P_k(A^{2/k})$ in the same way as in (2.7.2) and (2.7.3), and thus we get the inequalities

$$P^*(A^{1/k}) \le P_k(A^{2/k}) \le P^*(A^{2/k}), \tag{2.7.4}$$
$$\nu(A^{1/k}) \le P_k(A^{2/k}) \le \nu(A^{3/k}). \tag{2.7.5}$$

Since $\nu(A^{1/k})$ tends to $\nu(A)$ with $k \to \infty$ for each closed set A and, analogously, $P^*(A^{1/k}) \to P^*(A)$ as $k \to \infty$, then by (2.7.4) and (2.7.5) we obtain the equalities

$$P^*(A) = \lim_{k\to\infty} P_k(A^{2/k}) = \nu(A)$$

for each closed A, and hence $P^* = \nu$. □

Theorem 2.7.1. *There is a probability space* $(\Omega, \mathcal{A}, \Pr)$ *such that for every s.m.s.* U *and every Borel probability* μ *on* U *there is an RV* $X : \Omega \to U$ *with* $\mathcal{L}(X) = \mu$.

Proof. Define $(\Omega, \mathcal{A}, \Pr)$ as the measure-theoretic (von Neumann) product[21] of the probability spaces $(C, \mathcal{B}(C), \nu)$, where C is some nonempty subset of $\mathbb{R}$ with Borel σ-algebra $\mathcal{B}(C)$ and ν is some Borel probability on $(C, \mathcal{B}(C))$.

Now, given an s.m.s. U, there is some set $C \subseteq \mathbb{R}$ Borel-isomorphic with U (Claim 2.6 in Theorem 2.6.6). Let $f : C \to U$ supply the isomorphism. If μ is a Borel probability on U, then let ν be a probability on C such that $f(\nu) := \nu f^{-1} = \mu$. Define $X : \Omega \to U$ as $X = f \circ \pi$, where $\pi : \Omega \to C$ is a projection onto the factor $(C, \mathcal{B}(C), \nu)$. Then $\mathcal{L}(X) = \mu$, as desired. □

Remark 2.7.1. The preceding result establishes claim (i) made at the beginning of the section. It provides one way of ensuring (2.7.1): simply insist that all RVs be defined on a "superprobability space" as in Theorem 2.7.1. We make this assumption throughout the sequel.

The next theorem extends the Berkes and Phillips's lemma 2.7.1 to the case of u.m.s.m.s. U.

Theorem 2.7.2. *Let* U *be an s.m.s. The following statements are equivalent.*

(1) U *is u.m.*
(2) If $(\Omega, \mathcal{A}, \Pr)$ *is a nonatomic probability space, then for every Borel probability* P *on* U *there is an RV* $X : \Omega \to U$ *with law* $\mathcal{L}(X) = P$.

[21] See Hewitt and Stromberg (1965, Theorems 22.7 and 22.8, p. 432–133).

Proof. 1 ⇒ 2: Since U is u.m., there is some standard set $S \in \mathcal{B}(U)$ with $P(S) = 1$ (Theorem 2.6.5). Now there is a Borel isomorphism f mapping S onto a Borel subset B of $\mathbb{R}$ (Theorem 2.6.4). Then $f(P) := P \circ f^{-1}$ is a Borel probability on $\mathbb{R}$. Thus, there is an RV $g : \Omega \to \mathbb{R}$ with $\mathcal{L}(g) = f(P)$ and $g(\Omega) \subseteq B$ (Lemma 2.7.1 with $(U, d) = (\mathbb{R}, |\cdot|)$). We may assume that $g(\Omega) \subseteq B$ since $\Pr(g^{-1}(B)) = 1$. Define $x : \Omega \to U$ by $x(\omega) = f^{-1}(g(\omega))$. Then $\mathcal{L}(X) = \nu$, as claimed.

2 ⇒ 1: Now suppose that ν is a Borel probability on U. Consider an RV $X : \Omega \to U$ on the (nonatomic) probability space $((0,1), \mathcal{B}(0,1), \lambda)$ with $\mathcal{L}(X) = \nu$. Then range(X) is an analytic subset of U with $\nu^*(\text{range}(X)) = 1$. Since range$(X)$ is u.m. (Theorem 2.6.2), there is some standard set $S \subseteq \text{range}(X)$ with $P(S) = 1$. This follows from Theorem 2.6.5. The same theorem shows that U is u.m. □

Remark 2.7.2. If U is a u.m.s.m.s., we operate under the assumption that all U-valued RVs are defined on a nonatomic probability space. Then (2.7.1) will be valid.

References

Berkes I, Phillip W (1979) Approximation theorems for independent and weakly independent random vectors. Ann Prob 7:29–54

Billingsley P 1968 Convergence of probability measures. Wiley, New York

Birnbaum ZW, Orliz W (1931) Über die verallgemeinerung des begriffes der zueinander Konjugierten Potenzen. Stud Math 3:1–67

Blackwell D (1956) On a class of probability spaces. In: Proceedings of the 3rd Berkeley symposium on mathematical statistics and probability, vol 2, pp 1–6

Cohn DL (1980) Measure theory. Birkhauser, Boston

Dudley RM (1976) Probabilities and metrics: convergence of laws on metric spaces, with a view to statistical testing. In: Aarhus University Mathematics Institute lecture notes series no. 45, Aarhus

Dudley RM (2002) Real analysis and probability, 2nd edn. Cambridge University Press, New York

Dunford N, Schwartz J (1988) Linear operators, vol 1. Wiley, New York

Gibbs A, Su F (2002) On choosing and bounding probability metrics. Int Stat Rev 70:419–435

Hausdorff F (1949) Set theory. Dover, New York

Hennequin PL, Tortrat A (1965) Théorie des probabilités et quelques applications. Masson, Paris

Hewitt E, Stromberg K (1965) Real and abstract analysis. Springer, New York

Kaufman R (1984) Fourier transforms and descriptive set theory. Mathematika 31:336–339

Kruglov VM (1973) Convergence of numerical characteristics of independent random variables with values in a Hilbert space. Theor Prob Appl 18:694–712

Kuratowski K (1969) Topology, vol II. Academic, New York

Lebesgue H (1905) Sur les fonctions representables analytiquement. J Math Pures Appl V:139–216

Loeve M (1963) Probability theory, 3rd edn. Van Nostrand, Princeton

Lukacs E (1968) Stochastic convergence. D.C. Heath, Lexington, MA

Lusin N (1930) Lecons Sur les ensembles analytiaues. Gauthier-Villars, Paris

Massey FJ Jr (1950) A note on the power of a non-parametric test. Annal Math Statist 21:440–443

Mazukiewicz S (1936) Uberdie Menge der differenzierbaren Funktionen. Fund Math 27:247–248

Mostafaei H, S Kordnourie (2011) Probability metrics and their applications. Appl Math Sci 5(4):181–192

Thompson R (1966) Bias of the one-sample Cramér-Von Mises test. J Am Stat Assoc 61:246–247

Thompson R (1979) Bias and monotonicity of goodness-of-fit tests. J Am Stat Assoc 74:875–876

Chapter 3
Primary, Simple, and Compound Probability Distances and Minimal and Maximal Distances and Norms

The goals of this chapter are to:

- Formally introduce primary, simple, and compound probability distances;
- Provide examples of and study the relationship between primary, simple, and compound distances;
- Introduce the notions of minimal probability distance, minimal norms, cominimal functionals, and moment functions, which are needed in the study of primary, simple, and compound probability distances.

Notation introduced in this chapter:

Notation	Description
μ_h	Primary distance generated by a probability semidistance μ and mapping h
$\widetilde{\mu}_h$	Primary h-minimal distance
$m_i P = m_i^{(p)} P$	Marginal moment of order p
$\mathcal{M}_{H,p}(g)$	Primary distance generated by g, H, p
$\mathcal{M}(g)$	Primary metric generated by g
$\boldsymbol{\Omega}$	Discrete primary metric
$\mathbf{EN}(X, Y; H)$	Engineer's distance
$\mathbf{EN}(X, Y; p)$	L_p-engineer's metric
$\stackrel{w}{\to}$	Weak convergence of laws
$\widehat{\mu}$	Minimal distance w.r.t. μ
ℓ_H	Minimal distance w.r.t. $\mathcal{L}_H$ (Kantorovich distance)
ℓ_p	Minimal metric w.r.t. $\mathcal{L}_p$
$\boldsymbol{\sigma}$	Total variation metric
F^{-1}	Generalized inverse of distribution function F
$\boldsymbol{\pi}$	Prokhorov metric
$\boldsymbol{\pi}_\lambda$	Parametric version of Prokhorov metric
$\boldsymbol{\pi}_H$	Prokhorov distance
$\boldsymbol{\theta}_H$	Birnbaum–Orlicz distance
$\boldsymbol{\rho}_H$	Birnbaum–Orlicz uniform distance
$\overset{\circ}{\mu\nu}(P_1, P_2, \alpha)$	Cominimal metric functional w.r.t. the probability distances μ and ν

S.T. Rachev et al., *The Methods of Distances in the Theory of Probability and Statistics*, DOI 10.1007/978-1-4614-4869-3_3, © Springer Science+Business Media, LLC 2013

Notation	Description
$\overline{\mu\nu}(P_1, P_2, \alpha)$	Simple semidistance with $K_{\overline{\mu\nu}} = 2K_\mu K_\nu$
$\mu_c(m)$	Total cost of transportation of masses under plan m
$\overset{\circ}{\mu}_c$	Minimal norm w.r.t. μ_c
$\zeta_{\mathcal{F}}$	Zolotarev semimetric
$\mathcal{M}(X, Y)$	Moment metric
$\mathbb{L}_H$	H-average compound distance
$\mathbf{KF}_H$	Ky Fan distance
$\mathbf{K}_\lambda$	Parametric family of Ky Fan metrics
$\mathbf{\Theta}_H$	Birnbaum–Orlicz compound distance
$\mathbf{\Theta}_p$	Birnbaum–Orlicz compound metric
$\mathbf{R}_H$	Birnbaum–Orlicz compound average distance
$\breve{\mu}$	Maximal distance w.r.t. μ
$\overset{(s)}{\mu}$	μ-upper bound with marginal sum fixed
$\overset{(m,p)}{\mu}$	μ-upper bound with fixed pth marginal moments
$\underset{(m,p)}{\mu}$	μ-lower bound with fixed pth marginal moments
$\overline{\mu}$	μ-upper bound with fixed sum of pth marginal moments
$\underline{\mu}$	μ-lower bound with fixed sum of pth marginal moments

3.1 Introduction

The goal of Chap. 2 was to introduce the concept of measuring distances between random quantities and to provide examples of probability metrics. While we treated the general theory of probability metrics in detail, we did not provide much theoretical background on the distinction between different classes of probability metrics. We only noted that three classes of probability (semi-)metrics are distinguished – *primary*, *simple*, and *compound*. The goal of this chapter is to revisit these ideas but at a more advanced level.

When delving into the details of primary, simple, and compound probability metrics, we also consider a few related objects. They include cominimal functionals, minimal norms, minimal metrics, and moment functions. In the theory, these related functionals are used to establish upper and lower bounds to given families of probability metrics. They also help specify under what conditions a given probability metric is finite.

3.2 Primary Distances and Primary Metrics

Let $h : \mathcal{P}_1 \to \mathbb{R}^J$ be a mapping, where $\mathcal{P}_1 = \mathcal{P}_1(U)$[1] is the set of Borel probability measures (laws) for some s.m.s. (U, d) and J is some index set. This function h

---

[1]At times, when no confusion can arise, we suppress the subscript in $\mathcal{P}_1$ and use $\mathcal{P}$ instead.

induces a partition of $\mathcal{P}_2 = \mathcal{P}_2(U)$ (the set of laws on U^2) into equivalence classes for the relation

$$P \overset{h}{\sim} Q \iff h(P_1) = h(Q_1) \text{ and } h(P_2) = h(Q_2),$$
$$\text{where } P_i := T_i P,\ Q_i := T_i Q, \tag{3.2.1}$$

in which P_i and Q_i $(i = 1, 2)$ are the ith marginals of P and Q, respectively. Let μ be a probability semidistance (which we denote hereafter as p. semidistance) on $\mathcal{P}_2$ with parameter $\mathbb{K}_\mu$ (Definition 2.5.1 in Chap. 2), such that μ is constant on the equivalence classes of $\sim$; that is,

$$P \overset{h}{\sim} Q \iff \mu(P) = \mu(Q). \tag{3.2.2}$$

Definition 3.2.1. If the p. semidistance $\mu = \mu_h$ satisfies relation (3.2.2), then we call μ a *primary distance* (*with parameter* $\mathbb{K}_\mu$), which we abbreviate as p. distance. If $\mathbb{K}_\mu = 1$ and μ assumes only finite values, we say that μ is a primary metric.

Obviously, by relation (3.2.2), any primary distance is completely determined by the pair of marginal characteristics (hP_1, hP_2). In the case of primary distance μ, we will write $\mu(hP_1, hP_2) := \mu(P)$, and hence μ may be viewed as a distance in the image space $h(\mathcal{P}_1) \subseteq \mathbb{R}^J$, i.e., the following metric properties hold:

ID$^{(1)}$ $\quad hP_1 = hP2 \iff \mu(hP_1, hP_2) = 0;$

SYM$^{(1)}$ $\quad \mu(hP_1, hP_2) = \mu(hP_2, hP_1);$

TI$^{(1)}$ $\quad$ if the following marginal conditions are fulfilled :

$$a = h(T_1 P^{(1)}) = h(T_1 P^{(2)}),$$

$$b = h(T_2 P^{(2)}) = h(T_1 P^{(3)}), \text{ and}$$

$$c = h(T_2 P^{(1)}) = h(T_2 P^{(3)}) \text{ for some law } P^{(1)}, P^{(2)}, P^{(3)} \in \mathcal{P}_2,$$

then $\mu(a, c) \leq \mathbb{K}_\mu[\mu(a, b) + \mu(b, c)]$.

The notion of primary semidistance μ_h becomes easier to interpret assuming that a probability space $(\Omega, \mathcal{A}, \Pr)$ with property (2.7.1) is fixed (Remark 2.7.1). In this case μ_h is a usual distance (Definition 2.4.1) in the space

$$h(\mathfrak{X}) := \{hX := h\Pr_X, \text{ where } X \in \mathfrak{X}(U)\}, \tag{3.2.3}$$

and thus the metric properties of $\mu = \mu_h$ take the simplest form (Definition 2.4.2):

$$\mathbf{ID}^{(1*)} \qquad hX = hY \iff \mu(hX, hY) = 0,$$

$$\mathbf{SYM}^{(2*)} \qquad \mu(hX, hY) = \mu(hY, hX),$$

$$\mathbf{TI}^{(3*)} \qquad \mu(hX, hZ) \le \mathbb{K}_\mu[\mu(hX, hY) + \mu(hY, hZ)].$$

Further, we will consider several examples of primary distances and metrics.

Example 3.2.1 (Primary minimal distances). Each p. semidistance μ and each mapping $h : \mathcal{P}_1 \to \mathbb{R}^J$ determine a functional $\widetilde{\mu}_h : h(\mathcal{P}_1) \times h(\mathcal{P}_1) \to [0, \infty]$ defined by the following equality:

$$\widetilde{\mu}_h(\overline{a}_1, \overline{a}_2) := \inf\{\mu(P) : hP_i \equiv \overline{a}_i, i = 1, 2\} \tag{3.2.4}$$

(where P_i are the marginals of P) for any pair $(\overline{a}_1, \overline{a}_2) \in h(\mathcal{P}_1) \times h(\mathcal{P}_1)$.

Subsequently, we will prove (Chap. 5) that $\widetilde{\mu}_h$ is a primary distance for different special functions h and spaces U.

Definition 3.2.2. The functional $\widetilde{\mu}_h$ is called a *primary h-minimal distance* with respect to the p. semidistance μ.

Open Problem 3.2.1. In general it is not true that the metric properties of a p. distance μ imply that $\widetilde{\mu}_h$ is a distance. The following two examples illustrate this fact (see further Chap. 9).

(a) Let $U = \mathbb{R}$, $d(x, y) = |x - y|$. Consider the p. metric

$$\mu(X, Y) = \mathcal{K}_0(X, Y) = \Pr(X \neq Y) \quad X, Y \in \mathfrak{X}(\mathbb{R})$$

and the mapping $h : \mathfrak{X}(\mathbb{R}) \to [0, \infty]$ given by $hX = E|X|$. Then (see further Sect. 9.2 in Chap. 9)

$$\widetilde{\mu}_h(a, b) = \inf\{\Pr(X \neq Y) : E|X| = a, \ E|Y| = b\} = 0$$

for all $a \ge 0$ and $b \ge 0$. Hence in this case the metric properties of μ imply only semimetric properties for $\widetilde{\mu}_h$.

(b) Now let μ be defined as in (a) but $h : \mathfrak{X}(\mathbb{R}) \to [0, \infty] \times [0, \infty]$ be defined by $hX = (E|X|, EX^2)$. Then

$$\mu_h((a_1, a_2), (b_1, b_2)) = \inf\{\Pr(X \neq Y) : E|X| = a_1, \ EX^2 = a_2, E|Y| = b_1, EY^2 = b_2\}, \tag{3.2.5}$$

where $\widetilde{\mu}_h$ is not even a p. semidistance since the triangle inequality $TI^{(3*)}$ is not valid.

With respect to this, the following open problem arises: *under which condition on the space U, p. distance μ on $\mathfrak{X}(U)$, and transformation $h : \mathfrak{X}(U) \to R^J$ is the primary h-minimal distance $\widetilde{\mu}_h$ a primary p. distance in $h(\mathfrak{X})$?*

As we will see subsequently (Sect. 9.2), Examples 3.2.2–3.2.5 below of primary distances are special cases of primary h-minimal distances.

Example 3.2.2. Let $H \in \mathcal{H}$ (Example 2.4.1), and let $\overline{0}$ be a fixed point of an s.m.s. (U, d). For each $P \in \mathcal{P}_2$ with marginals $P_i = T_i P$, let $m_1 P$ and $m_2 P$ denote the *marginal moments of order* $p > 0$,

$$m_i P := m_i^{(p)} P := \left(\int_U d^p(x, \overline{0}) P_i(\mathrm{d}x) \right)^{p'} \quad p > 0 \quad p' := \min(1, 1/p).$$

Then

$$\mathcal{M}_{H,p}(P) := \mathcal{M}_{H,p}(m_1 P, m_2 P) := H(|m_1 P - m_2 P|) \tag{3.2.6}$$

is a primary distance. One can also consider $\mathcal{M}_{H,p}$ as a distance in the space

$$m^{(p)}(\mathcal{P}_1) := \left\{ m^{(p)} := \left(\int_U d^p(x, a) P(\mathrm{d}x) \right)^{p'} < \infty, P \in \mathcal{P}(U) \right\} \tag{3.2.7}$$

of moments $m^{(p)} P$ of order $p > 0$. If $H(t) = t$, then

$$\mathcal{M}(P) := \mathcal{M}_{H,1}(P) = \left| \int_U d(x, \overline{0})(P_1 - P_2)(\mathrm{d}x) \right|$$

is a primary metric in $m^{(p)}(\mathcal{P}_1)$.

Example 3.2.3. Let $g : [0, \infty] \to \mathbb{R}$ and $H \in \mathcal{H}$. Then

$$\mathcal{M}(g)_{H,p}(m_1 P, m_2 P) := H(|g(m_1 P) - g(m_2 P)|) \tag{3.2.8}$$

is a primary distance in $g \circ m(\mathcal{P}_1)$ and

$$\mathcal{M}(g)(m_1 P, m_2 P) := |g(m_1 P) - g(m_2 P)| \tag{3.2.9}$$

is a primary metric.

If U is a Banach space with norm $\| \cdot \|$, then we define the primary distance $\mathcal{M}_{H,p}(g)$ as follows:

$$\mathcal{M}_{H,p}(g)(m^{(p)} X, m^{(p)} Y) := H(|m^{(p)} P - m^{(p)} Y|), \tag{3.2.10}$$

where [see (2.2.8)] $m^{(p)} X$ is the *p-th moment (norm) of X*

$$m^{(p)}X := \{E\|X\|^p\}^{p'}.$$

By (3.2.9), $\mathcal{M}_{H,p}(g)$ may be viewed as a distance (Definition 2.4.2) in the space

$$g \circ m(\mathfrak{X}) := \{g \circ m(X) := g(\{E\|X\|^p\}^{p'}), X \in \mathfrak{X}\} \; p' = \min(1, p^{-1}), \; \mathfrak{X} = \mathfrak{X}(U) \tag{3.2.11}$$

of moments $g \circ m(X)$. If U is the real line $\mathbb{R}$ and $g(t) = H(t) = t$, where $t \geq 0$, then $\mathcal{M}_{H,p}(m^{(p)}X, m^{(p)}Y)$ is the usual deviation between moments $m^{(p)}X$ and $m^{(p)}Y$ [see (2.2.9)].

Example 3.2.4. Let J be an index set (with arbitrary cardinality) and g_i $(i \in J)$ real functions on $[0, \infty]$, and for each $P \in \mathcal{P}(U)$ define the set

$$hP := \{g_i(mP), i \in J\}. \tag{3.2.12}$$

Further, for each $P \in \mathcal{P}_2(U)$ let us consider hP_1 and hP_2, where the P_i are the marginals of P. Then

$$\Omega(hP_1, hP_2) = \begin{cases} 0 & \text{if } hP_1 \equiv hP_2 \\ 1 & \text{otherwise} \end{cases} \tag{3.2.13}$$

is a primary metric.

Example 3.2.5. Let U be the n-dimensional Euclidean space $\mathbb{R}^n$, $H \in \mathcal{H}$. Define the *engineer's distance*

$$\mathbf{EN}(X, Y; H) := H\left(\left|\sum_{i=1}^{n}(EX_i - EY_i)\right|\right), \tag{3.2.14}$$

where $X = (X_1, \dots, X_n)$ and $Y = (Y_1, \dots, Y_n)$ belong to the subset $\widetilde{\mathfrak{X}}(\mathbb{R}^n) \subseteq \mathfrak{X}(\mathbb{R}^n)$ of all n-dimensional random vectors that have integrable components. Then $\mathbf{EN}(\cdot, \cdot; H)$ is a p. semidistance in $\widetilde{\mathfrak{X}}(\mathbb{R}^n)$. Analogously, the L_p*-engineer metric*

$$\mathbf{EN}(X, Y, p) := \left[\sum_{i=1}^{n}|EX - EY|^p\right]^{\min(1,1/p)}, p > 0, \tag{3.2.15}$$

is a primary metric in $\widetilde{\mathfrak{X}}(\mathbb{R}^n)$. In the case $p = 1$ and $n = 1$, the metric $\mathbf{EN}(\cdot, \cdot; p)$ coincides with the engineer metric in $\mathfrak{X}(\mathbb{R})$ [see (2.2.1)].

3.3 Simple Distances and Metrics: Cominimal Functionals and Minimal Norms

Clearly, any primary distance $\mu(P)$ $(P \in \mathcal{P}_2)$ is completely determined by the pair of marginal distributions $P_i = T_i P$, where $i = 1, 2$, since the equality $P_1 = P_2$ implies $hP_1 = hP_2$ [see relations (3.2.1), (3.2.2), and Definition 3.2.1]. On the other hand, if the mapping h is "rich enough," then the opposite implication

$$hP_1 = hP_2 \Rightarrow P_1 = P_2$$

takes place. The simplest example of such "rich" $h : \mathcal{P}(U) \to \mathbb{R}^J$ is given by the equalities

$$h(P) := \{P(C),\ C \in \mathcal{C},\ P \in \mathcal{P}(U)\}, \tag{3.3.1}$$

where $J \equiv \mathcal{C}$ is the family of all closed nonempty subsets $C \subseteq U$. Another example is

$$h(P) = \left\{ Pf := \int_U f \mathrm{d}P,\ f \in C^b(U) \right\}, \quad P \in \mathcal{P}(U),$$

where $C^b(U)$ is the set of all bounded continuous functions on U. Keeping in mind these two examples we will define the notion of "simple" distance as a particular case of primary distance with h given by equality (3.3.1).

Definition 3.3.1. The p. semidistance μ is said to be a *simple semidistance* in $\mathcal{P} = \mathcal{P}(U)$ if for each $P \in \mathcal{P}_2$

$$\mu(P) = 0 \Leftarrow T_1 P = T_2 P.$$

If, in addition, μ is a p. semimetric, then μ will be called a *simple semimetric.* If the converse implication ($\Rightarrow$) also holds, then we say that μ is *simple distance.* If, in addition, μ is a p. semimetric, then μ will be called a *simple metric.*

Since the values of the simple distance $\mu(P)$ depend only on the pair marginals P_1, P_2, we will consider μ as a functional on $\mathcal{P}_1 \times \mathcal{P}_1$ and use the notation

$$\mu(P_1, P_2) := \mu(P_1 \times P_2) \quad (P_1, P_2 \in \mathcal{P}_1),$$

where $P_1 \times P_2$ means the measure product of laws P_1 and P_2. In this case, the metric properties of μ take the form (Definition 2.5.1) (for each P_1, P_2, $P_3 \in \mathcal{P}$)

$$\mathbf{ID}^{(2)}\ P_1 = P_2 \iff \mu(P_1, P_2) = 0,$$

$$\mathbf{SYM}^{(2)}\ \mu(P_1, P_2) = \mu(P_2, P_1),$$

$$\mathbf{TI}^{(2)}\ \mu(P_1, P_2) \leq \mathbb{K}_\mu(\mu(P_1, P_2) + \mu(P_2, P_3)).$$

Hence, the space $\mathcal{P}$ of laws P with a simple distance μ is a distance space (Definition 2.4.2). Clearly, each primary distance is a simple semidistance in $\mathcal{P}$. The Kolmogorov metric $\boldsymbol{\rho}$ (2.2.2), the Lévy metric $\mathbf{L}$ (2.2.3), and the $\boldsymbol{\theta}_p$-metrics (2.2.4) are simple metrics in $\mathcal{P}(\mathbb{R})$.

Let us consider a few more examples of simple metrics, which we will use later on.

Example 3.3.1. Minimal distances

Definition 3.3.2. For a given p. semidistance μ on $\mathcal{P}_2$ the functional $\widehat{\mu}$ on $\mathcal{P}_1 \times \mathcal{P}_1$ defined by the equality

$$\widehat{\mu}(P_1, P_2) := \inf\{\mu(P) : T_i P = P_i,\ i = 1,2\}, \quad P_1, P_2 \in \mathcal{P}_1 \tag{3.3.2}$$

is said to be a (simple) *minimal* (w.r.t. μ) *distance.*

As we showed in Sect. 2.7 that, for a "rich enough" probability space, the space $\mathcal{P}_2$ of all laws on U^2 coincides with the set of joint distributions $\Pr_{X,Y}$ of U-valued random variables. For this reason, $\mu(P) = \mu(\Pr_{X,Y})$ always holds for some $X, Y \in \mathfrak{X}(U)$, and therefore (3.3.2) can be rewritten as follows:

$$\widehat{\mu}(P_1, P_2) = \inf\{\mu(X, Y) : \Pr_X = P_1,\ \Pr_Y = P_2\}.$$

In this form, the equation is the definition of minimal metrics introduced by Zolotarev (1976).

In the next theorem, we will consider the conditions on U that guarantee $\widehat{\mu}$ to be a simple metric. We use the notation $\xrightarrow{\text{w}}$ to mean "weak convergence of laws."[2]

Theorem 3.3.1. *Let U be a u.m.s.m.s. (Definition 2.6.2) and let μ be a p. semidistance with parameter $\mathbb{K}_\mu$. Then $\widehat{\mu}$ is a simple semidistance with parameter $\mathbb{K}_{\widehat{\mu}} = \mathbb{K}_\mu$. Moreover, if μ is a p. distance satisfying the "continuity" condition*

$$\left.\begin{array}{l} P^{(n)} \in \mathcal{P}_2, \quad P^{(n)} \xrightarrow{\text{w}} P \in \mathcal{P}_2 \\ \mu(P^{(n)}) \to 0 \end{array}\right\} \Rightarrow \mu(P) = 0,$$

then $\widehat{\mu}$ is a simple distance with parameter $\mathbb{K}_{\widehat{\mu}} = \mathbb{K}_\mu$.

Remark 3.3.1. The continuity condition is not restrictive; in fact, all p. distances we will use satisfy this condition.

Remark 3.3.2. Clearly, if μ is a p. semimetric, then, by Theorem 3.3.1, $\widehat{\mu}$ is a simple semimetric.

Proof. $\mathbf{ID}^{(2)}$: If $P_1 \in \mathcal{P}_1$, then we let $X \in \mathfrak{X}(U)$ have the distribution P_1. Then, by $\mathbf{ID}^{(*)}$ (Definition 2.5.2),

[2]See Billingsley (1999).

$$\widehat{\mu}(P_1, P_1) \leq \mu(\Pr_{(X,X)}) = 0.$$

Suppose now that μ is a p. distance and the continuity condition holds. If $\widehat{\mu}(P_1, P_2) = 0$, then there exists a sequence of laws $P^{(n)} \in \mathcal{P}_2$ with fixed marginals $T_i P^{(n)} = P_i$ $(i = 1, 2)$ such that $\mu(P^{(n)}) \to 0$ as $n \to \infty$. Since P_i is a tight measure, then the sequence $\{P^{(n)}, n \geq 1\}$ is uniformly tight, i.e., for any $\varepsilon > 0$ there exists a compact $K_\varepsilon \subseteq U^2$ such that $P^{(n)}(K_\varepsilon) \geq 1 - \varepsilon$ for all $n \geq 1$,[3] Using Prokhorov compactness criteria[4] we choose a subsequence $P^{(n')}$ that weakly tends to a law $P \in \mathcal{P}_2$; hence, $T_i P = P_i$ and $\mu(P) = 0$. Since μ is a p. distance, P is concentrated on the diagonal $x = y$, and thus $P_1 = P_2$ as desired.

SYM$^{(2)}$: Obvious.

TI$^{(2)}$: Let $P_1, P_2, P_3 \in \mathcal{P}_1$. For any $\varepsilon > 0$ define a law $P_{12} \in \mathcal{P}_2$ with marginals $T_i P_{12} = P_i$ $(i = 1, 2)$ and a law $P_{23} \in \mathcal{P}_2$ with $T_i P_{23} = P_{i+1}$ $(i = 1, 2)$ such that $\widehat{\mu}(P_1, P_2) \geq \mu(P_{12}) - \varepsilon$ and $\widehat{\mu}(P_2, P_3) \geq \mu(P_{23}) - \varepsilon$. Since U is a u.m.s.m.s., there exist Markov kernels $P'(A|z)$ and $P''(A|z)$ defined by the equalities

$$P_{12}(A_1 \times A_2) := \int_{A_2} P'(A_1|z) P_2(\mathrm{d}z), \tag{3.3.3}$$

$$P_{23}(A_2 \times A_3) := \int_{A_2} P''(A_3|z) P_2(\mathrm{d}z) \tag{3.3.4}$$

for all $A_1, A_2, A_3 \in \mathcal{B}_1$ (Corollary 2.6.2). Then define a set function Q on the algebra $\mathcal{A}$ of finite unions of Borel rectangles $A_1 \times A_2 \times A_3$ by the equation

$$Q(A_1 \times A_2 \times A_3) := \int_{A_2} P'(A_1|z) P''(A_3|z) P_2(\mathrm{d}z). \tag{3.3.5}$$

It is easily checked that Q is countably additive on $\mathcal{A}$ and therefore extends to a law on U^3. We use "Q" to represent this extension as well. The law Q has the projections $T_{12}Q = P_{12}$, $T_{23}Q = P_{23}$. Since μ is a p. semidistance with parameter $\mathbb{K} = \mathbb{K}_\mu$, we have

$$\begin{aligned} \mu(P_1, P_3) &\leq \mu(T_{13}Q) \leq \mathbb{K}[\mu(P_{12}) + \mu(P_{13})] \\ &\leq \mathbb{K}[\widehat{\mu}(P_1, P_2) + \widehat{\mu}(P_2, P_3)] + 2\mathbb{K}\varepsilon. \end{aligned}$$

Letting $\varepsilon \to 0$ we complete the proof of **TI$^{(2)}$**. □

As will be shown in Part II, all simple distances in the next examples are actually simple minimal $\widehat{\mu}$ distances w.r.t. p. distances μ that will be introduced in Sect. 3.4 (see further Examples 3.4.1–3.4.3).

---

[3] See Dudley (2002, Sect. 11.5).

[4] See, for instance, Billingsley (1999, Sect. 5).

Example 3.3.2 (Kantorovich metric and Kantorovich distance). In Sect. 2.2, we introduced the Kantorovich metric κ and its "dual" representation

$$\kappa(P_1, P_2) = \int_{-\infty}^{+\infty} |F_1(x) - F_2(x)|\mathrm{d}x$$
$$= \sup\left\{\left|\int_{\mathbb{R}} f\mathrm{d}(P_1 - P_2)\right| : f : \mathbb{R} \to \mathbb{R},\, f' \text{ exists a.e. and } |f'| < 1 \text{ a.e.}\right\},$$

where the P_i are laws on $\mathbb{R}$ with distribution functions (DFs) F_i and a finite first absolute moment. From the preceding representation it also follows that

$$\kappa(P_1, P_2) = \sup\left\{\left|\int_{\mathbb{R}} f\mathrm{d}(P_1 - P_2)\right| : f : \mathbb{R} \to \mathbb{R},\, f \text{ is } (1,1)\text{-Lipschitz, i.e., } |f(x) - f(y)| \le |x - y| \forall x, y \in \mathbb{R}\right\}.$$

In this example, we will extend the definition of the foregoing simple p. metric of the set $\mathcal{P}(U)$ of all laws on an s.m.s. (U, d). For any $\alpha \in (0, \infty)$ and $\beta \in [0, 1]$ define the Lipschitz function class

$$\mathrm{Lip}_{\alpha\beta} := \{f : U \to \mathbb{R} : |f(x) - f(y)| \le \alpha \mathrm{d}^{\beta}(x, y)\ \forall x, y \in U\} \tag{3.3.6}$$

with the convention

$$\mathrm{d}^0(x, y) := \begin{cases} 1 \text{ if } x \ne y, \\ 0 \text{ if } x = y. \end{cases} \tag{3.3.7}$$

Denote the set of all bounded functions $f \in \mathrm{Lip}_{\alpha\beta}(U)$ by $\mathrm{Lip}^b_{\alpha\beta}(U)$. Let $\mathcal{G}_H(U)$ be the class of all pairs (f, g) of functions that belong to the set

$$\mathrm{Lip}^b(U) := \bigcup_{\alpha > 0} \mathrm{Lip}_{\alpha,1}(U) \tag{3.3.8}$$

and satisfy the inequality

$$f(x) + g(y) \le H(d(x, y)), \quad \forall x, y \in U, \tag{3.3.9}$$

where H is a convex function from $\mathcal{H}$. Recall that $H \in \mathcal{H}$ if H is a nondecreasing continuous function from $[0, \infty)$ onto $[0, \infty)$ and vanishes at the origin and $K_H := \sup_{t>0} H(2t)/H(t) < \infty$. For any two laws P_1 and P_2 on an s.m.s. (U, d) define

$$\ell_H(P_1, P_2) := \sup\left\{\int_U f\mathrm{d}P_1 + \int_U g\mathrm{d}P_2 : (f, g) \in \mathcal{G}_H(U)\right\}. \tag{3.3.10}$$

We will prove further that ℓ_H is a simple distance with $\mathbb{K}_{\ell_H} = \mathbb{K}_H$ in the space of all laws P with finite "H-moment," $\int H(\mathrm{d}(x,a))P(\mathrm{d}x) < \infty$. The proof is based on the representation of ℓ_H as a minimal distance $\ell_H = \widehat{\mathcal{L}}_H$ (Corollary 5.3.2) w.r.t. a p. distance (with $\mathbb{K}_{\mathcal{L}_H} = \mathbb{K}_H$) $\mathcal{L}_H(P) = \int_{U^2} H(\mathrm{d}(x,y))P(\mathrm{d}x,\mathrm{d}y)$ and then an appeal to Theorem 3.3.1 proves that ℓ_H is a simple p. distance if (U,d) is a universally measurable s.m.s. In the case $H(t) = t^p$ $(1 < p < \infty)$, define

$$\ell_p(P_1, P_2) := \ell_H(P_1, P_2)^{1/p}, \quad 1 < p < \infty. \tag{3.3.11}$$

In addition, for $p \in [0,1]$ and $p = \infty$ denote

$$\ell_p(P_1, P_2) := \sup\left\{ \left| \int_U f\mathrm{d}(P_1 - P_2) \right| : f \in \mathrm{Lip}^b_{1,p}(U) \right\},$$
$$p \in (0,1], \ P_1, P_2 \in \mathcal{P}(U), \tag{3.3.12}$$

$$\ell_0(P_1 - P_2) := \left\{ \left| \int_U f\mathrm{d}(P_1 - P_2) \right| : f \in \mathrm{Lip}_{1,0}(U) \right\}$$
$$= \sigma(P_1, P_2) := \sup_{A \in \mathcal{B}_1} |P_1(A) - P_2(A)|, \tag{3.3.13}$$

$$\ell_\infty(P_1, P_2) := \inf\{\varepsilon > 0 : P_1(A) \le P_2(A^\varepsilon)\ \forall A \in \mathcal{B}_1\}, \tag{3.3.14}$$

where, as above, $\mathcal{B}_1 = \mathcal{B}(U)$ is the Borel σ-algebra on an s.m.s. (U,d), and $A^\varepsilon := \{x : d(x,A) < \varepsilon\}$.

For any $0 \le p \le 1$, $p = \infty$, ℓ_p is a simple metric in $\mathcal{P}(U)$, which follows immediately from the definition. To prove that ℓ_p is a p. metric (taking possibly infinite values), one can use the equality

$$\sup_{A \in \mathcal{B}_1} [P_1(A) - P_2(A^\varepsilon)] = \sup_{A \in \mathcal{B}_1} [P_2(A) - P_1(A^\varepsilon)].$$

The equality $\ell_0 = \sigma$ in (3.3.13) follows from the fact that both metrics are minimal w.r.t. one and the same probability distance $\mathcal{L}_0(P) = P((x,y) : x \ne y)$ (see further Corollaries 6.2.1 and 7.5.2). We will prove also (Corollary 7.4.2) that $\ell_H = \widehat{\mathcal{L}}_H$, as a minimal distance w.r.t. $\mathcal{L}_H$ defined previously, admits the Birnbaum–Orlicz representation (Example 2.4.2)

$$\ell_H(P_1, P_2) = \ell_H(F_1, F_2) := \int_0^1 H(|F_1^{-1}(t) - F_2^{-1}(t)|)\mathrm{d}t \tag{3.3.15}$$

in the case of $U = \mathbb{R}$ and $d(x,y) = |x - y|$. In (3.3.15),

$$F_i^{-1}(t) := \sup\{x : F_i(x) \le t\} \tag{3.3.16}$$

is the (generalized) *inverse* of the DF F_i determined by P_i ($i = 1,2$). Letting $H(t) = t$ we claim that

$$\ell_1(P_1, P_2) = \int_0^1 |F_1^{-1}(t) - F_2^{-1}(t)|\mathrm{d}t$$
$$= \kappa(P_1, P_2) := \int_{-\infty}^{\infty} |F_1(x) - F_2(x)|\mathrm{d}x \ P_i \in \mathcal{P}(\mathbb{R}),\ i = 1,2. \tag{3.3.17}$$

Remark 3.3.3. Here and in the rest of the book, for any simple semidistance μ on $\mathcal{P}(\mathbb{R}^n)$ we will use the following notations interchangeably:

$$\mu = \mu(P_1, P_2),\ \forall P_1, P_2 \in \mathcal{P}(\mathbb{R}^n);$$
$$\mu = \mu(X_1, X_2) := \mu(\mathrm{Pr}_{X_1}, \mathrm{Pr}_{X_2}),\ \forall X_1, X_2 \in \mathfrak{X}(\mathbb{R}^n);$$
$$\mu = \mu(F_1, F_2) := \mu(P_1, P_2),\ \forall F_1, F_2 \in \mathcal{F}(\mathbb{R}^n),$$

where Pr_{X_i} is the distribution of X_i, F_i is the distribution function of P_i, and $\mathcal{F}(\mathbb{R}^n)$ stands for the class of distribution functions on $\mathbb{R}^n$.

The ℓ_1-metric (3.3.17) is known as the *average metric* in $\mathcal{F}(\mathbb{R})$ as well as the *first difference pseudomoment*, and it is also denoted by κ.[5] A great contribution in the investigation of ℓ_1-metric properties was made by Kantorovich (1942, 1948) and Kantorovich and Akilov (1984, Chap. VIII). That is why the metric ℓ_1 is called the Kantorovich metric. Considering ℓ_H as a generalization ℓ_1, we will call ℓ_H the *Kantorovich distance.*

Example 3.3.3 (Prokhorov metric and Prokhorov distance). Prokhorov (1956) introduced his famous metric

$$\pi(P_1, P_2) := \inf\{\varepsilon > 0 : P_1(C) \le P_2(C^\varepsilon) + \varepsilon,$$
$$P_2(C) \le P_1(C^\varepsilon) + \varepsilon, \quad \forall C \in \mathcal{C}\}, \tag{3.3.18}$$

where $\mathcal{C} := \mathcal{C}(U)$ is the set of all nonempty closed subsets of a Polish space U and

$$C^\varepsilon := \{x : d(x, C) < \varepsilon\}. \tag{3.3.19}$$

The metric π admits the following representations: for any laws P_1 and P_2 on an s.m.s. (U, d)

$$\pi(P_1, P_2) = \inf\{\varepsilon > 0 : P_1(C) \le P_2(C^\varepsilon) + \varepsilon, \text{ for any } C \in \mathcal{C}\}$$

---

[5]See Zolotarev (1976).

$$= \inf\{\varepsilon > 0 : P_1(C) \le P_2(C^{\varepsilon]}) + \varepsilon, \text{ for any } C \in \mathcal{C}\}$$

$$= \inf\{\varepsilon > 0 : P_1(A) \le P_2(A^{\varepsilon}) + \varepsilon, \text{ for any } A \in \mathcal{B}_1\}, \qquad (3.3.20)$$

where

$$C^{\varepsilon]} = \{x : d(x, C) < \varepsilon\} \qquad (3.3.21)$$

is the ε-closed neighborhood of C.[6]

Let us introduce a *parametric version of the Prokhorov metric*:

$$\pi_\lambda(P_1, P_2) := \inf\{\varepsilon > 0 : P_1(C) \le P_2(C^{\lambda\varepsilon}) + \varepsilon \text{ for any } C \in \mathcal{C}\}. \qquad (3.3.22)$$

The next lemma gives the main relationship between Prokhorov-type metrics and the metrics ℓ_0 and ℓ_∞ defined by equalities (3.3.13) and (3.3.14).

Lemma 3.3.1. *For any* $P_1, P_2 \in \mathcal{P}(U)$

$$\lim_{\lambda\to 0} \pi_\lambda(P_1, P_2) = \sigma(P_1, P_2) = \ell_0(P_1, P_2), \qquad (3.3.23)$$

$$\lim_{\lambda\to 0} \lambda\pi_\lambda(P_1, P_2) = \ell_\infty(P_1, P_2).$$

Proof. For any fixed $\varepsilon > 0$ the function $A_\varepsilon(\lambda) := \sup\{P_1(C) - P_2(C^{\lambda\varepsilon}) : C \in \mathcal{C}\}$, $\lambda \ge 0$, is nonincreasing on $\varepsilon > 0$, hence

$$\pi_\lambda(P_1, P_2) = \inf\{\varepsilon > 0 : A_\varepsilon(\lambda) \le \varepsilon\} = \max_{\varepsilon>0} \min(\varepsilon, A_\varepsilon(\lambda)).$$

For any fixed $\varepsilon > 0$, $A_\varepsilon(\cdot)$ is nonincreasing and

$$\lim_{\lambda\downarrow 0} A_\varepsilon(\lambda) = A_\varepsilon(0) = \sup_{C\in\mathcal{C}}(P_1(C) - P_2(C)) = \sup_{A\in\mathcal{B}(U)} (P_1(A) - P_2(A))$$

$$= \sup_{A\in\mathcal{B}(U)} |P_1(A) - P_2(A)| =: \sigma(P_1, P_2).$$

Thus

$$\lim_{\lambda\to 0} \pi_\lambda(P_1, P_2) = \max_{\varepsilon>0} \min\left(\varepsilon, \lim_{\lambda\to 0} A_\varepsilon(\lambda)\right)$$

$$= \max_{\varepsilon>0} \min(\varepsilon, \sigma(P_1, P_2)) = \sigma(P_1, P_2).$$

Analogously, as $\lambda \to \infty$,

$$\lambda\pi_\lambda(P_1, P_2) = \inf\{\lambda\varepsilon > 0 : A_\varepsilon(\lambda) \le \varepsilon\}$$

---

[6]See, for example, Dudley (1976, Theorem 8.1).

$$= \inf\{\varepsilon > 0 : A_\varepsilon(1) \leq \varepsilon/\lambda\} \to \inf\{\varepsilon > 0 : A_\varepsilon(1) \leq 0\}$$
$$= \ell_\infty(P_1, P_2).$$
□

As a generalization of $\boldsymbol{\pi}_\lambda$ we define the *Prokhorov distance*

$$\boldsymbol{\pi}_H(P_1, P_2) := \inf\{H(\varepsilon) > 0 : P_1(A^\varepsilon) \leq P_2(A) + H(\varepsilon),\ \forall A \in \mathcal{B}_1\} \quad (3.3.24)$$

for any strictly increasing function $H \in \mathcal{H}$. From (3.3.24),

$$\pi(P_1, P_2) = \inf\{\delta > 0 : P_1(A) \leq P_2(A^{H^{-1}(\delta)}) + \delta \text{ for any } A \in \mathcal{B}_1\}, \quad (3.3.25)$$

and it is easy to check that $\boldsymbol{\pi}_H$ is a simple distance with $\mathbb{K}_{\pi H} = \mathbb{K}_H$ [condition (2.4.3)]. The metric $\boldsymbol{\pi}_\lambda$ is a special case of $\boldsymbol{\pi}_H$ with $H(t) = t/\lambda$.

Example 3.3.4 (Birnbaum–Orlicz distance ($\boldsymbol{\theta}_H$) and $\boldsymbol{\theta}_p$-metric in $\mathcal{P}(\mathbb{R})$). Let $U = \mathbb{R}$, $\mathrm{d}(x, y) = |x - y|$. Following Example 2.4.2, we define the *Birnbaum–Orlicz average distance*

$$\boldsymbol{\theta}_H(F_1, F_2) := \int_{-\infty}^{+\infty} H(|F_1(t) - F_2(t)|)\mathrm{d}t \quad H \in \mathcal{H} \quad F_i \in \mathcal{F}(\mathbb{R}), \quad i = 1, 2, \quad (3.3.26)$$

and the *Birnbaum–Orlicz uniform distance*

$$\boldsymbol{\rho}_H(F_1, F_2) := H(\boldsymbol{\rho}(F_1, F_2)) = \sup_{x \in \mathbb{R}} H(|F_1(x) - F_2(x)|). \quad (3.3.27)$$

The $\boldsymbol{\theta}_p$*-metric* ($p > 0$)

$$\boldsymbol{\theta}_p(F_1, F_2) := \left\{\int_{-\infty}^{\infty} |F_1(t) - F_2(t)|^p \mathrm{d}t\right\}^{p'}, \quad p' := \min(1, 1/p), \quad (3.3.28)$$

is a special case of $\boldsymbol{\theta}_H$ with appropriate normalization that makes $\boldsymbol{\theta}_p$ a p. metric taking finite and infinite values in the DF space $\mathcal{F} := \mathcal{F}(\mathbb{R})$. In the case $p = \infty$, we denote $\boldsymbol{\theta}_\infty$ to be the Kolmogorov metric

$$\boldsymbol{\theta}_\infty(F_1, F_2) := \boldsymbol{\rho}(F_1, F_2) := \sup_{x \in \mathbb{R}} |F_1(x) - F_2(x)|. \quad (3.3.29)$$

In the case $p = 0$, we set

$$\boldsymbol{\theta}_0(F_1, F_2) := \int_{-\infty}^{\infty} I\{t : F_1(t) \neq F_2(t)\}\mathrm{d}t = \mathrm{Leb}(F_1 \neq F_2).$$

Here, as in what follows, $I(A)$ is the indicator of the set A.

Example 3.3.5 (Cominimal metrics). As we saw in Sect. 3.2, each primary distance $\mu(P) = \mu(h(T_1P), h(T_2P))$ $(P \in \mathcal{P}_2)$ determines a semidistance (Definition 2.4.2) in the space of equivalence classes

$$\{P \in \mathcal{P}_2 : h(T_1P) = a, h(T_2P) = b\}, \quad a, b \in \mathbb{R}^J. \tag{3.3.30}$$

Analogously, the minimal distance

$$\begin{aligned}\widehat{\mu}(P) &:= \widehat{\mu}(T_1P, T_2P)\\ &:= \inf\{\mu(\widetilde{P}) : \widetilde{P} \in \mathcal{P}_2(U), \widetilde{P} \text{ and } P \text{ have one and the same marginals,}\\ T_i\widetilde{P} &= T_iP, i = 1, 2\}, P \in \mathcal{P}_2(U),\end{aligned}$$

may be viewed as a semidistance in the space of classes of equivalence

$$\{P \in \mathcal{P}_2 : T_1P = P_1, T_2P = P_2\}, \quad P_1, P_2 \in \mathcal{P}_1. \tag{3.3.31}$$

Obviously, the partitioning (3.3.31) is more refined than (3.3.30), and hence each primary semidistance is a simple semidistance. Thus

$$\begin{aligned}&\{\text{the class of primary distances (Definition 3.2.1)}\}\\ \subset\ &\{\text{the class of simple semidistances (Definition 3.3.1)}\}\\ \subset\ &\{\text{the class of all p. semidistances (Definition 2.5.1)}\}.\end{aligned}$$

Open Problem 3.3.1. A basic open problem in the theory of probability metrics is to find a good classification of the set of all p. semidistances. Does there exist a "Mendeleyev periodic table" of p. semidistances?

One can get a classification of p. semidistances considering more and more refined partitions of $\mathcal{P}_2$. For instance, one can use a partition finer than (3.3.31), generated by

$$\{P \in \mathcal{P}_2 :\ T_1P = P_1,\ T_2P = P_2,\ P \in \mathcal{PC}_t\}, \quad t \in T, \tag{3.3.32}$$

where P_1 and P_2 are laws in $\mathcal{P}_1$ and $\mathcal{PC}_t$ $(t \in T)$ are subsets of $\mathcal{P}_2$ whose union covers $\mathcal{P}_2$. As an example of the set $\mathcal{PC}_t$ one could consider

$$\mathcal{PC}_t = \left\{P \in \mathcal{P}_2 : \int_{U^2} f_i \mathrm{d}P \le b_i,\ i \in J\right\}, \quad t = (J, \overline{b}, \overline{f}), \tag{3.3.33}$$

where J is an index set, $\overline{b} := (b_i, i \in J)$ is a set of reals, and $\overline{f} = \{f_i, i \in J\}$ is a family of bounded continuous functions on U^2.[7]

Another useful example of a set $\mathcal{PC}_t$ is constructed using a given probability metric $\nu(P)$ $(P \in \mathcal{P}_2)$ and has the form

[7] See Kemperman (1983) and Levin and Rachev (1990).

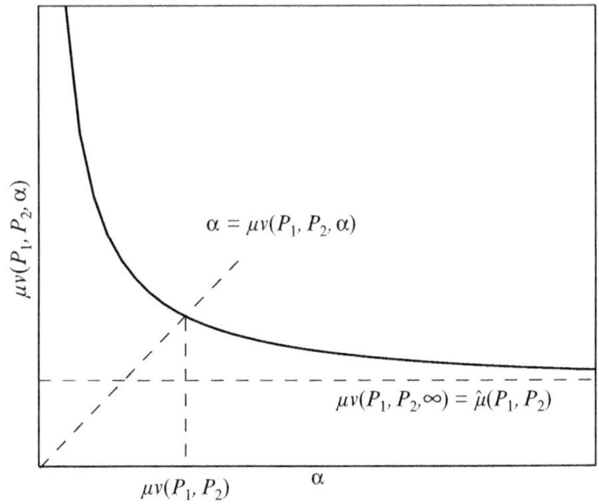

Fig. 3.1 Cominimal distance $\mu\nu(P_1, P_2)$

$$\mathcal{PC}_t = \{P \in \mathcal{P}_2 : \nu(P) \leq t\}, \tag{3.3.34}$$

where $t \in [0, \infty]$ is a fixed number.

Open Problem 3.3.2. Under what conditions is the functional

$$\mu(P_1, P_2; \mathcal{PC}_t) := \inf\{\mu(P) : P \in \mathcal{P}_2, T_i P = P_i (i = 1, 2), P \in \mathcal{PC}_t\}$$
$$(P_1, P_2 \in \mathcal{P}_1)$$

a simple semidistance (resp. semimetric) w.r.t. the given p. distance (resp. metric) μ?

Further, we will examine this problem in the special case of (3.3.34) (Theorem 3.3.2). Analogously, one can investigate the case of $\mathcal{PC}_t = l\{P \in \mathcal{P}_2 : \nu_i(P) \leq \alpha_i, i = 1, 2 \ldots\}$ $[t = (\alpha_1, \alpha_2, \ldots)]$ for fixed p. metrics ν_i, and $\alpha_i \in [0, \infty]$.

Following the main idea of obtaining primary and simple distances by means of minimization procedures of certain types (Definitions 3.2.2 and 3.3.2), we will present the notion of *cominimal distance*.

For given compound semidistances μ and ν with parameters $\mathbb{K}_\mu$ and $\mathbb{K}_\nu$, respectively, and for each $\alpha > 0$ denote

$$\mu\nu(P_1, P_2, \alpha) = \inf\{\mu(P) : P \in \mathcal{P}_2, T_1 P = P_1, T_2 P = P_2, \nu(P) \leq \alpha\},$$
$$P_1, P_2 \in \mathcal{P}_1 \tag{3.3.35}$$

[see (3.3.32) and (3.3.34)].

Definition 3.3.3. The functional $\mu\nu(P_1, P_2, \alpha)$ $(P_1, P_2 \in \mathcal{P}_1, \alpha > 0)$ will be called the *cominimal (metric) functional w.r.t. the p. distances μ and ν* (Fig. 3.1)

As we will see in the next theorem, the functional $\mu\nu(\cdot,\cdot,\alpha)$ has some metric properties, but nevertheless it is not a p. distance; however, $\mu\nu(\cdot,\cdot,\alpha)$ induces p. semidistances as follows.

Let $\mu\nu$ be the so-called *cominimal distance*

$$\mu\nu(P_1, P_2) = \inf\{\alpha > 0; \mu\nu(P_1, P_2, \alpha) < \alpha\} \tag{3.3.36}$$

(Fig. 3.1), and let

$$\overline{\mu\nu}(P_1, P_2) = \limsup_{\alpha\to 0} \alpha\mu\nu(P_1, P_2, \alpha).$$

Then the following theorem is true.

Theorem 3.3.2. *Let U be a u.m.s.m.s. and μ be a p. distance satisfying the "continuity" condition in Theorem 3.3.1. Then, for any p. distance ν,*

(a) $\mu\nu(\cdot,\cdot,\alpha)$ satisfies the following metric properties:

$$\begin{aligned}
&\mathbf{ID}^{(3)}: && \mu\nu(P_1, P_2, \alpha) = 0 \iff P_1 = P_2,\\
&\mathbf{SYM}^{(3)}: && \mu\nu(P_1, P_2, \alpha) = \mu\nu(P_2, P_1, \alpha),\\
&\mathbf{TI}^{(3)}: && \mu\nu(P_1, P_3, \mathbb{K}_\nu(\alpha+\beta)) \le \mathbb{K}_\mu(\mu\nu(P_1, P_2, \alpha) + \mu\nu(P_2, P_3, \beta))\\
& && \textit{for any } P_1, P_2, P_3 \in \mathcal{P}_1, \alpha \ge 0, \beta \ge 0;
\end{aligned}$$

(b) $\mu\nu$ is a simple distance with parameter $\mathbb{K}_{\mu\nu} = \max[\mathbb{K}_\mu, \mathbb{K}_\nu]$. In particular, if μ and ν are p. metrics, then $\mu\nu$ is a simple metric;

(c) $\overline{\mu\nu}$ is a simple semidistance with parameter $\mathbb{K}_{\overline{\mu\nu}} = 2\mathbb{K}_\mu\mathbb{K}_\nu$.

Proof. (a) By Theorem 3.3.1 and Fig. 3.1, $\mu\nu(P_1, P_2, \alpha) = 0 \Rightarrow \widehat{\mu}(P_1, P_2) = 0 \to P_1 = P_2$, and if $P_1 \in \mathcal{P}_1$ and X is an RV with distribution P_1, then $\mu\nu(P_1, P_2, \alpha) \le \mu(\Pr_{X,X}) = 0$. Thus, $\mathbf{ID}^{(3)}$ is valid. Let us prove $\mathbf{TI}^{(3)}$. For each P_1, P_2, $P_3 \in \mathcal{P}_1$ $\alpha \ge 0$, $\beta \ge 0$, and $\varepsilon \ge 0$ define laws $P_{12} \in \mathcal{P}_2$ and $P_{23} \in \mathcal{P}_2$ such that $T_i P_{12} = P_i$, $T_i P_{23} = P_{i+1}$ $(i = 1,2)$, $\nu(P_{12}) \le \alpha$, $\nu(P_{23}) \le \beta$, and $\mu\nu(P_1, P_2, \alpha) \ge \mu(P_{12}) - \varepsilon$, $\mu\nu(P_2, P_3, \alpha) \ge \mu(P_{23}) - \varepsilon$. Define a law $Q \in \mathcal{P}_3$ by (3.3.5). Then Q has bivariate marginals $T_{12}Q = P_{12}$ and $T_{23}Q = P_{23}$; hence, $\nu(T_{13}Q) \le \mathbb{K}_\nu[\nu(P_{12}) + \nu(P_{23})] \le \mathbb{K}_\nu(\alpha+\beta)$ and

$$\begin{aligned}
\mu\nu(P_1, P_3, \mathbb{K}_\nu(\alpha+\beta)) &\le \mu(T_{13}Q) \le \mathbb{K}[\mu(P_{12}) + \mu(P_{23})]\\
&\le \mathbb{K}_\mu[\mu\nu(P_1, P_2, \alpha) + \mu\nu(P_2, P_3, \beta) + 2\varepsilon].
\end{aligned}$$

Letting $\varepsilon \to 0$, we get $\mathbf{TI}^{(3)}$.

(b) If $\mu\nu(P_1, P_2) < \alpha$ and $\mu\nu(P_2, P_3) < \beta$, then there exists P_{12} (resp. P_{23}) with marginals P_1 and P_2 (resp. P_2 and P_3) such that $\mu(P_{12}) < \alpha$, $\nu(P_{12}) < \alpha$, $\mu(P_{23}) < \beta$. In a similar way, as in (a), we conclude that $\mu\nu(P_1, P_3, \mathbb{K}_\nu(\alpha+\beta)) < \mathbb{K}_\mu(\alpha+\beta)$; thus, $\mu\nu(P_1, P_2) < \max(\mathbb{K}_\mu, \mathbb{K}_\nu)(\alpha+\beta)$.

(c) Follows from (a) with $\alpha = \beta$.

□

Example 3.3.6 (Minimal norms). Each cominimal distance $\mu\nu$ is greater than the minimal distance $\widehat{\mu}$ (Fig. 3.3). We now consider examples of simple metrics $\overset{\circ}{\mu}$ corresponding to given p. distances $\widehat{\mu}$ that have (like $\mu\nu$) a "minimal" structure but $\overset{\circ}{\mu} \leq \widehat{\mu}$.

Let $\mathcal{M}_k$ be the set of all finite nonnegative measures on the Borel σ-algebra $\mathcal{B}_k = \mathcal{B}(U^k)$ (U is an s.m.s.). Let $\mathcal{M}_0$ denote the space of all finite signed measures ν on $\mathcal{B}_1$ with total mass $m(U) = 0$. Denote by $\mathcal{CS}(U^2)$ the set of all continuous, symmetric, and nonnegative functions on U^2. Define the functionals

$$\mu_c(m) := \int_{U^2} c(x,y)m(\mathrm{d}x,\mathrm{d}y), \quad m \in \mathcal{M}_2, \quad c \in \mathcal{CS}(U^2), \tag{3.3.37}$$

and

$$\overset{\circ}{\mu}_c(\nu) := \inf\{\mu_c(m) : T_1 m - T_2 m = \nu\}, \quad \nu \in \mathcal{M}_0, \tag{3.3.38}$$

where $T_i m$ denotes the ith marginal measure of m.

Lemma 3.3.2. *For any $c \in \mathcal{CS}(U^2)$ the functional $\overset{\circ}{\mu}_c$ is a seminorm in the space $\mathcal{M}_0$.*

Proof. Obviously, $\overset{\circ}{\mu}_c \geq 0$. For any positive constant a we have

$$\begin{aligned}\overset{\circ}{\mu}_c(a\nu) &= \inf\{\mu_c(m) : T_1(1/a)m - T_2(1/a)m = \nu\}\\ &= a\inf\{\mu_c((1/a)m) : T_1(1/a)m - T_2(1/a)m = \nu\}\\ &= a\overset{\circ}{\mu}_c(\nu).\end{aligned}$$

If $a \leq 0$ and $\widetilde{m}(A \times B) := m(B \times A)$, where $A, B \in \mathcal{B}_1$, then by the symmetry of c we get

$$\begin{aligned}\mu_c(a\nu) &= \inf\{\mu_c(m) : T_2(-1/a)m - T_1(-1/a)m = \nu\}\\ &= \inf\{\mu_c(\widetilde{m}) : T_1(-1/a)\widetilde{m} - T_2(-1/a)\widetilde{m} = \nu\}\\ &= |a|\overset{\circ}{\mu}_c(\nu).\end{aligned}$$

Let us prove now that $\overset{\circ}{\mu}_c$ is a subadditive function. Let ν_1, $\nu_2 \in \mathcal{M}_0$. For m_1, $m_2 \in \mathcal{M}_2$ with $T_1 m_i - T_2 m_i = \nu_i$ ($i = 1, 2$), let $m = m_1 + m_2$. Then we have $\mu_c(m) = \mu_c(m_1) + \mu_c(m_2)$ and $T_1 m - T_2 m = \nu_1 + \nu_2$; hence, $\overset{\circ}{\mu}_c(\nu_1 + \nu_2) \leq \overset{\circ}{\mu}_c(\nu_1) + \overset{\circ}{\mu}_c(\nu_2)$. □

In the next theorem, we give a sufficient condition for

$$\overset{\circ}{\mu}_c(P_1, P_2) := \overset{\circ}{\mu}_c(P_1 - P_2), \quad P_1, P_2 \in \mathcal{P}_1, \tag{3.3.39}$$

to be a simple metric in $\mathcal{P}_1$. In the proof we will make use of *Zolotarev's semimetric* $\zeta_{\mathcal{F}}$. That is, for a given class $\mathcal{F}$ of the bounded continuous function $f : U \to \mathbb{R}$ we define

$$\zeta\mathcal{F}(P_1, P_2) = \sup_{f \in \mathcal{F}} \left| \int_U f \mathrm{d}(P_1 - P_2) \right|, \quad P_i \in \mathcal{P}(U).$$

Clearly, $\zeta_{\mathcal{F}}$ is a simple semimetric. Moreover, if the class $\mathcal{F}$ is "rich enough" to preserve the implication $\zeta_{\mathcal{F}}(P_1, P_2) = 0 \iff P_1 = P_2$, then we have that $\zeta_{\mathcal{F}}$ is a simple metric.

Theorem 3.3.3. *The following statements hold:*

(i) For any $c \in \mathcal{CS}(U^2)$, $\overset{\circ}{\mu}_c(P_1, P_2)$, defined by equality (3.3.39)*, is a semimetric in $\mathcal{P}_1$.*

(ii) Further, if the class $\mathcal{F}_c := \{f : U \to \mathbb{R}, \ |f(x) - f(y)| \leq c(x, y), \ \forall x, y \in U\}$ contains the class $\mathcal{G}$ of all functions

$$f(x) := f_{k,C}(x) := \max\{0, 1/k - d(x, C)\}, \quad x \in U$$

(k is an integer greater than some fixed k_0, C is a closed nonempty set), then $\overset{\circ}{\mu}_c$ is a simple metric in $\mathcal{P}_1$.

Proof. (i) The proof follows immediately from Lemma 3.3.2 and the definition of semimetric (Definition 2.4.1).

(ii) For any $m \in \mathcal{M}_2$ such that $T_1 m - T_2 m = P_1 - P_2$ and for any $f \in \mathcal{F}_c$ we have

$$\left| \int_U f \mathrm{d}(P_1 - P_2) \right| = \left| \int_{U^2} f(x) - f(y) m(\mathrm{d}x, \mathrm{d}y) \right|$$
$$\leq \int_{U^2} |f(x) - f(y)| m(\mathrm{d}x, \mathrm{d}y) \leq \mu_c(m);$$

hence, the Zolatarev metric $\zeta_{\mathcal{F}_c}(P_1, P_2)$ is a lower bound for $\overset{\circ}{\mu}_c(P_1, P_2)$. On the other hand, by assumption, $\zeta_{\mathcal{F}_c} \geq \zeta_{\mathcal{G}}$. Thus, assuming $\overset{\circ}{\mu}_c(P_1, P_2) = 0$, we get $0 \leq \zeta_{\mathcal{G}}(P_1, P_2) \leq \zeta_{\mathcal{F}_c}(P_1, P_2) \leq \overset{\circ}{\mu}_c(P_1, P_2) = 0$. Next, for any closed nonempty set C we have

$$P_1(C) < k \int_U f_{k,C} \mathrm{d}P_1 \leq k\zeta_{\mathcal{G}}(P_1, P_2) + k \int_U f_{k,C} \mathrm{d}P_2 \leq P_2(C^{1/k}).$$

Letting $k \to \infty$ we get $P_1(C) \leq P_2(C)$, and hence, by symmetry, $P_1 = P_2$. □

Remark 3.3.4. Obviously, $\mathcal{F}_d \supseteq \mathcal{G}$, and hence $\overset{\circ}{\mu}_d$ is a simple metric in $\mathcal{P}_1$; however, if $p > 1$, then $\overset{\circ}{\mu}_{d^p}$ is not a metric in $\mathcal{P}_1$, as shown in the following example. Let $U = [0, 1]$, $d(x, y) = |x - y|$. Let P_1 be a law concentrated on the origin and P_2 a law concentrated on 1. For any $n = 1, 2, \ldots$ consider a measure $m^{(n)} \in \mathcal{M}_2$ with total mass $m^{(n)}(U^2) = 2n + 1$ and

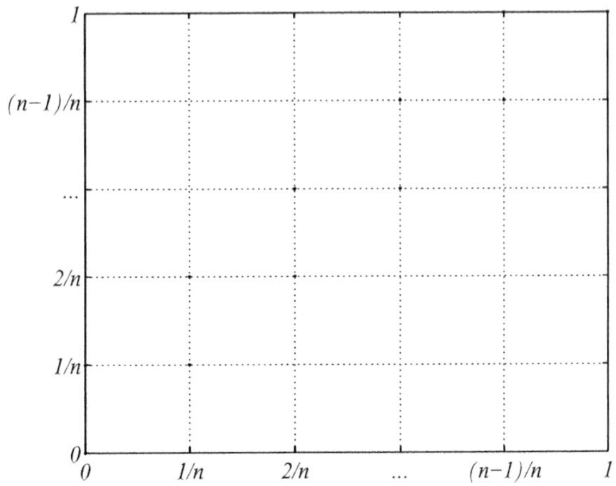

Fig. 3.2 Support of measure $m^{(n)}$ with marginals P_1 and P_2

$$m^{(n)}\left(\left\{\frac{i}{n}, \frac{i}{n}\right\}\right) = 1, \quad i = 0, \ldots, n,$$

$$m^{(n)}\left(\left\{\frac{i}{n}, \frac{(i+1)}{n}\right\}\right) = 1, \quad i = 0, \ldots, n-1$$

(Fig. 3.2). Then, obviously, $T_1 m^{(n)} - T_2 m^{(n)} = P_1 - P_2$ and

$$\int_{U\times U} |x-y|^p m^{(n)}(\mathrm{d}x, \mathrm{d}y) = \sum_{i=0}^{n-1} \left(\frac{1}{n}\right)^p = n^{1-p};$$

hence, if $p > 1$, then

$$\mathring{\mu}_d(P_1, P_2) \leq \inf_{n>0} \int_{U^2} |x-y|^p m^{(n)}(\mathrm{d}x, \mathrm{d}y) = 0,$$

and thus $\mathring{\mu}_{d^p}(P_1, P_2) = 0$.

Definition 3.3.4. The simple semimetric $\mathring{\mu}_c$ [see equality (3.3.39)] is said to be the *minimal norm w.r.t. the functional* μ_c.

Obviously, for any $c \in \mathcal{CS}$,

$$\mathring{\mu}_c(P_1, P_2) \leq \widehat{\mu}_c(P_1, P_2) := \inf\{\mu_c(P) : P \in \mathcal{P}_2, T_i P = P_i, i = 1, 2\},$$
$$P_1, P_2 \in \mathcal{P}_1; \tag{3.3.40}$$

hence, for each minimal metric of the type $\widehat{\mu}_c$ we can construct an estimate from below by means of $\overset{\circ}{\mu}_c$, but what is more important, $\overset{\circ}{\mu}_c$ is a *simple semimetric, even though μ_c is not a probability semidistance.* For instance, let $c_h(x, y) := d(x, y)h(\max(d(x, a), d(y, a)))$, where h is a nondecreasing nonnegative continuous function on $[\alpha, \infty)$ for some $\alpha > 0$. Then, as in Theorem 3.3.3, we conclude that $\zeta_{c_h} \le \overset{\circ}{\mu}_{c_h}$ and $\zeta_{c_h}(P_1, P_2) = 0 \Rightarrow P_1 = P_2$. Thus, $\overset{\circ}{\mu}_{c_h}$ is a simple metric, while if $h(t) = t^p$ $(p > 1)$, then μ_{c_h} is not a p. distance. Further (Sect. 5.4 in Chap. 5), we will prove that $\overset{\circ}{\mu}$ admits a dual formula: for any laws P_1 and P_2 on an s.m.s. (U, d), with finite $\int d(x, a)h(d(x, a))(P_1 + P_2)(\mathrm{d}x)$,

$$\overset{\circ}{\mu}_{c_h}(P_1, P_2) = \sup\left\{\left|\int_U f\mathrm{d}(P_1 - P_2)\right| : f : U \to \mathbb{R},\right.$$

$$\left. |f(x) - f(y)| \le c_h(x, y) \quad \forall x, y \in U\right\}. \qquad (3.3.41)$$

From equality (3.3.41) it follows that if $U = \mathbb{R}$ and $d(x, y) = |x - y|$, then $\overset{\circ}{\mu}_c$ may be represented explicitly as an average metric with weight $h(\cdot - a)$ between DFs

$$\overset{\circ}{\mu}_{c_h}(P_1, P_2) = \overset{\circ}{\mu}_{c_h}(F_1, F_2) := \int_{-\infty}^{\infty} |F_1(x) - F_2(x)| h(|x - a|)\mathrm{d}x, \qquad (3.3.42)$$

where F_i is the DF of P_i (Sect. 5.5).

3.4 Compound Distances and Moment Functions

We continue the classification of probability distances. Recall some basic examples of p. metrics on an s.m.s. (U, d):

(a) The *moment metric* (Example 3.2.2):

$$\mathcal{M}(X, Y) = |Ed(X, a) - Ed(Y, a)|, \quad X, Y \in \mathfrak{X}(U)$$

[$\mathcal{M}$ is a primary metric in the space $\mathfrak{X}(U)$ of U-valued RVs].

(b) The Kantorovich metric (Example 3.3.2):

$$\kappa(X, Y) = \sup\{|Ef(X) - Ef(Y)| : f : U \to \mathbb{R} \text{ bounded},$$

$$|f(x) - f(y)| \le \mathrm{d}(x, y) \quad \forall x \text{ and } y \in U\}$$

[κ is a simple metric in $\mathfrak{X}(U)$].

(c) The L_1-metric [see (2.5.3)]:

$$\mathcal{L}_1(X,Y) = Ed(X,Y), \qquad X,Y \in \mathfrak{X}(U).$$

The $\mathcal{L}_1$-metric is a p. metric in $\mathfrak{X}(U)$ (Definition 2.5.2). Since the value of $\mathcal{L}_1(X,Y)$ depends on the joint distribution of the pair (X,Y), we will call $\mathcal{L}_1$ a compound metric.

Definition 3.4.1. A *compound distance* (resp. metric) is any probability distance μ (resp. metric). See Definitions 2.5.1 and 2.5.2.

Remark 3.4.1. In many papers on probability metrics, *compound* metric stands for a metric that is not simple; however, all *nonsimple* metrics used in these papers are in fact *compound* in the sense of Definition 3.4.1. The problem of classification of p. metrics that are neither compound (in the sense of Definition 3.4.1) nor simple is open (see Open Problems 3.3.1 and 3.3.2).

Let us consider some examples of compound distances and metrics.

Example 3.4.1 (Average compound distances). Let (U,d) be an s.m.s. and $H \in \mathcal{H}$ (Example 2.4.1). Then

$$\mathcal{L}_H(P) := \int_{U^2} H(\mathrm{d}(x,y))P(\mathrm{d}x,\mathrm{d}y), \quad P \in \mathcal{P}_2, \tag{3.4.1}$$

is a compound distance with parameter K_H [see (2.4.3)] and will be called an *H-average compound distance*. If $H(t) = t^p$, $p > 0$, and $p' = \min(1, 1/p)$, then

$$\mathcal{L}_p(P) := [\mathcal{L}_H(P)]^{p'}, \quad P \in \mathcal{P}_2, \tag{3.4.2}$$

is a compound metric in

$$\mathcal{P}_2^{(p)} := \left\{ P \in \mathcal{P}_2 : \int_{U^2} d^p(x,a)[P(\mathrm{d}x,\mathrm{d}y) + P(\mathrm{d}y,\mathrm{d}x)] < \infty \right\}.$$

In the space

$$\mathfrak{X}^{(p)}(U) := \{X \in \mathfrak{X}(U) : Ed^p(X,a) < \infty\},$$

the metric $\mathcal{L}_p$ is the usual *p-average metric*

$$\mathcal{L}_p(X,Y) := [Ed^p(X,Y)]^{p'}, \quad 0 < p < \infty. \tag{3.4.3}$$

In the limit cases $p = 0$, $p = \infty$, we will define the compound metrics

$$\mathcal{L}_0(P) := P\left(\bigcup_{x \neq y}(x,y)\right), \quad P \in \mathcal{P}_2, \tag{3.4.4}$$

and

$$\mathcal{L}_\infty(P) := \inf\{\varepsilon > 0 : P(d(x,y) > \varepsilon) = 0\}, \quad P \in \mathcal{P}_2, \tag{3.4.5}$$

that on $\mathfrak{X}$ have the forms

$$\mathcal{L}_0(X,Y) := EI\{X \neq Y\} = \Pr(X \neq Y), \quad X, Y \in \mathfrak{X}, \tag{3.4.6}$$

and

$$\mathcal{L}_\infty(X,Y) := \operatorname{ess\,sup} d(X,Y) := \inf\{\varepsilon > 0 : \Pr(d(X,Y) > \varepsilon) = 0\}. \tag{3.4.7}$$

Example 3.4.2 (Ky Fan distance and Ky Fan metric). The Ky Fan metric $\mathbf{K}$ in $\mathfrak{X}(\mathbb{R})$ was defined by equality (2.2.5) in Chap. 2, and we will extend that definition considering the space $\mathcal{P}_2(U)$ for an s.m.s. (U,d). We define the Ky Fan metric in $\mathcal{P}_2(U)$ as follows:

$$\mathbf{K(P)} := \inf\{\varepsilon > 0 : P(d(x,y) > \varepsilon) < \varepsilon\}, \quad P \in \mathcal{P}_2$$

and on $\mathfrak{X}(U)$ by $\mathbf{K}(X,Y) = \mathbf{K}(\Pr_{X,Y})$. In this way, $\mathbf{K}$ takes the form of the *distance in probability* in $\mathfrak{X} = \mathfrak{X}(U)$:

$$\mathbf{K}(X,Y) := \inf\{\varepsilon > 0 : \Pr(d(X,Y) > \varepsilon) < \varepsilon\}, \quad X, Y \in \mathfrak{X}. \tag{3.4.8}$$

A possible extension of the metric structure of $\mathbf{K}$ is the *Ky Fan distance*:

$$\mathbf{KF}_H(P) := \inf\{\varepsilon > 0 : P(H(d(x,y)) > \varepsilon) < \varepsilon\} \tag{3.4.9}$$

for each $H \in \mathcal{H}$. It is easy to verify that $\mathbf{KF}_H$ is a compound distance with parameter $\mathbb{K}_{\mathbf{KF}} := K_H$ [see (2.4.3)]. A particular case of the Ky Fan distance is the *parametric family of Ky Fan metrics* given by

$$\mathbf{K}_\lambda(P) := \inf\{\varepsilon > 0 : P(d(x,y) > \lambda\varepsilon) < \varepsilon\}. \tag{3.4.10}$$

For each $\lambda > 0$

$$\mathbf{K}_\lambda(X,Y) := \inf\{\varepsilon > 0 : \Pr(d(X,Y) > \lambda\varepsilon) < \varepsilon\}, \quad X, Y \in \mathfrak{X},$$

metrizes the convergence "in probability" in $\mathfrak{X}(U)$, i.e.,

$$\mathbf{K}_\lambda(X_n, Y) \to 0 \iff \Pr(d(X_n, Y) > \varepsilon) \to 0 \text{ for any } \varepsilon > 0.$$

In the limit cases,

$$\lim_{\lambda \to 0} \mathbf{K}_\lambda = \mathcal{L}_0, \quad \lim_{\lambda \to \infty} \lambda \mathbf{K}_\lambda = \mathcal{L}_\infty, \tag{3.4.11}$$

we get, however, average compound metrics [see equalities (3.4.4)–(3.4.7)] that induce convergence, stronger than convergence in probability, i.e.,

$$\mathcal{L}_0(X_n, Y) \to 0 \underset{\not\Leftarrow}{\Rightarrow} X_n \to Y \text{ "in probability"}$$

and

$$\mathcal{L}_\infty(X_n, Y) \to 0 \underset{\not\Leftarrow}{\Rightarrow} X_n \to Y \text{ "in probability."}$$

Example 3.4.3 (Birnbaum–Orlicz compound distances). Let $U = \mathbb{R}$, $d(x, y) = |x - y|$. For each $p \in [0, \infty]$ consider the following compound metrics in $\mathfrak{X}(\mathbb{R})$:

$$\mathbf{\Theta}_p(X_1, X_2) := \left[\int_{-\infty}^{\infty} \tau^p(t; X_1, X_2)\mathrm{d}t\right]^{p'}, \quad 0 < p < \infty \; p' := \min(1, 1/p), \tag{3.4.12}$$

$$\mathbf{\Theta}_\infty(X_1, X_2) := \sup_{t\in\mathbb{R}} \tau(t; X_1, X_2), \tag{3.4.13}$$

$$\mathbf{\Theta}_0(X_12, X_2) := \int_{-\infty}^{\infty} I\{t : \tau(t; X_1, X_2) \neq 0\}\mathrm{d}t,$$

where

$$\tau(t; X_1, X_2) := \Pr(X_1 \leq t < X_2) + \Pr(X_2 \leq t < X_1). \tag{3.4.14}$$

If $H \in \mathcal{H}$, then

$$\mathbf{\Theta}_H(X_1, X_2) := \int_{-\infty}^{\infty} H(\tau(t; X_1, X_2))\mathrm{d}t \tag{3.4.15}$$

is a compound distance with $\mathbb{K}_{\mathbf{\Theta}_H} = K_H$. The functional $\mathbf{\Theta}_H$ will be called a *Birnbaum–Orlicz compound average distance*, and

$$\mathbf{R}_H(X_1, X_2) := H(\mathbf{\Theta}_\infty(X_1, X_2)) = \sup_{t\in\mathbb{R}} H(\tau(t; X_1, X_2)) \tag{3.4.16}$$

will be called a *Birnbaum–Orlicz compound uniform distance*.

Each example 3.3.i. is closely related to the corresponding example 3.2.i. In fact, we will prove (Corollary 5.3.2) that ℓ_H [see (3.3.10)] is a minimal distance (Definition 3.3.2) w.r.t. $\mathcal{L}_H$ for any convex $H \in \mathcal{H}$, i.e.,

$$\ell_H = \widehat{\mathcal{L}}_H. \tag{3.4.17}$$

Analogously, the simple metrics ℓ_p [see (3.3.11)–(3.3.14)], the Prokhorov metric $\boldsymbol{\pi}_\lambda$ [see (3.3.22)], and the Prokhorov distance $\boldsymbol{\pi}_H$ [see (3.3.24)] are minimal w.r.t. the $\mathcal{L}_p$-metric, Ky Fan metric $\mathbf{K}_\lambda$, and Ky Fan distance $\mathbf{KF}_H$, i.e.,

$$\ell_p = \widehat{\mathcal{L}}_p \; (p \in [0, \infty]), \quad \boldsymbol{\pi}_\lambda = \widehat{\mathbf{K}}_\lambda \; (\lambda > 0), \quad \boldsymbol{\pi}_H = \widehat{\mathbf{KF}}_H. \tag{3.4.18}$$

Finally, the Birnbaum–Orlicz metric and distance $\boldsymbol{\theta}_p$ and $\boldsymbol{\theta}_H$ [see (3.3.28) and (3.3.26)] and the Birnbaum–Orlicz uniform distance $\boldsymbol{\rho}_H$ [see (3.3.27)] are minimal w.r.t. their "compound versions" $\boldsymbol{\Theta}_p$, $\boldsymbol{\Theta}_H$, and $\mathbf{R}_H$, i.e.,

$$\boldsymbol{\theta}_p = \widehat{\boldsymbol{\Theta}}_p \ (p \in [0,\infty]), \quad \boldsymbol{\theta}_H = \widehat{\boldsymbol{\Theta}}_H, \quad \rho_H = \widehat{\mathbf{R}}_H. \tag{3.4.19}$$

Equalities (3.4.17)–(3.4.19) represent the main relationships between simple and compound distances (resp. metrics) and serve as a framework for the theory of probability metrics (Fig. 1.1, *comparison of metrics*).

Analogous relationships exist between primary and compound distances. For example, we will prove (Chap. 9) that the primary distance

$$\mathcal{M}_{H,1}(\alpha,\beta) := H(|\alpha-\beta|) \tag{3.4.20}$$

[see (3.2.6)] is a primary minimal distance (Definition 3.2.2) w.r.t. the p. distance $H(\mathcal{L}_1)$ $(H \in \mathcal{H})$, i.e.,

$$\mathcal{M}_{H,1}(\alpha,\beta) := \inf\left\{ H(\mathcal{L}_1(P)) : \int_{U^2} d(x,a)P(\mathrm{d}x,\mathrm{d}y) = \alpha, \right.$$
$$\left. \int_{U^2} d(a,y)P(\mathrm{d}x,\mathrm{d}y) = \beta \right\}. \tag{3.4.21}$$

Since a compound metric μ may take infinite values, we have to determine a concept of μ-boundedness. With that aim in view, we introduce the notion of a *moment function*, which differs from the notion of simple distance in the *identity* property only [Definition 3.3.1 and $\mathbf{ID}^{(2)}$, $\mathbf{TI}^{(2)}$].

Definition 3.4.2. A mapping $\mathbb{M} : \mathcal{P}_1 \times \mathcal{P}_1 \to [0,\infty]$ is said to be a *moment function* (with parameter $\mathbb{K}_{\mathbb{M}} \geq 1$) if it possesses the following properties for all P_1, P_2, $P_3 \in \mathcal{P}_1$:

$$\mathbf{SYM}^{(4)} : \mathbb{M}(P_1,P_2) = \mathbb{M}(P_2,P_1),$$
$$\mathbf{TI}^{(4)} : \mathbb{M}(P_1,P_3) \leq \mathbb{K}_{\mathbb{M}}[\mathbb{M}(P_1,P_2) + \mathbb{M}(P_2,P_3)].$$

We will use moment functions as upper bounds for p. distances μ. As an example, we will now consider μ to be the p. average distance [see equalities (3.4.2) and (3.4.3)]

$$\mathcal{L}_p(P) := \left[\int_{U\times U} d^p(x,y)P(\mathrm{d}x,\mathrm{d}y)\right]^{p'}, \quad p>0, \quad p' := \min(1,1/p),\ P \in \mathcal{P}_2. \tag{3.4.22}$$

For any $p > 0$ and $a \in U$ define the moment function

$$\Lambda_{p,a}(P_1,P_2) := \left[\int_U d^p(x,a)P_1(\mathrm{d}x)\right]^{p'} + \left[\int_U d^p(x,a)P_2(\mathrm{d}x)\right]^{p'}. \tag{3.4.23}$$

By the Minkowski inequality, we get our first (rough) upper bound for the value $\mathcal{L}_p(P)$ under the convention that the marginals $T_i P = P_i$ $(i = 1, 2)$ are known:

$$\mathcal{L}_p(P) \le \Lambda_{p,a}(P_1, P_2). \tag{3.4.24}$$

Obviously, by inequality (3.4.24), we can get a more refined estimate

$$\mathcal{L}_p(P) \le \Lambda_p(P_1, P_2), \tag{3.4.25}$$

where

$$\Lambda_p(P_1, P_2) := \inf_{a \in U} \Lambda_{p,a}(P_1, P_2). \tag{3.4.26}$$

Further, we will consider the following question.

Problem 3.4.1. What is the best possible inequality of the type

$$\mathcal{L}_p(P) \le \breve{\mathcal{L}}_p(P_1, P_2), \tag{3.4.27}$$

where $\breve{\mathcal{L}}_p$ is a functional that depends on the marginals $P_i = T_i P$ $(i = 1, 2)$ only?

Remark 3.4.2. Suppose (X, Y) is a pair of *dependent* random variables taking on values in the s.m.s. (U, d). Knowing only the marginal distributions $P_1 = \Pr_X$ and $P_2 = \Pr_Y$, what is the best possible improvement of the *triangle inequality* bound

$$\mathcal{L}_1(X, Y) := Ed(X, Y) \le Ed(X, a) + Ed(Y, a). \tag{3.4.28}$$

The answer is simple: the best possible upper bound for $Ed(X, Y)$ is given by

$$\breve{\mathcal{L}}_1(P_1, P_2) := \sup\{\mathcal{L}_1(X_1, X_2) : \Pr_{X_i} = P_i, i = 1, 2\}. \tag{3.4.29}$$

More difficult is to determine dual and explicit representations for $\breve{\mathcal{L}}_1$ similar to those of the minimal metric $\breve{\mathcal{L}}_1$ (Kantorovich metric). We will discuss this problem in Sect. 8.2 in Chap. 8.

More generally, for any compound semidistance $\mu(P)$ $(P \in \mathcal{P}_2)$ let us define the functional

$$\breve{\mu}(P_1, P_2) := \sup\{\mu(P) : T_i P = P_i, i = 1, 2\}, \quad P_1, P_2 \in \mathcal{P}_1. \tag{3.4.30}$$

Definition 3.4.3. The functional $\breve{\mu} : \mathcal{P}_1 \times \mathcal{P}_1 \to [0, \infty]$ will be called the *maximal distance* w.r.t. the given compound semidistance μ.

Note that, by definition, a maximal distance need not be a distance. We prove the following theorem.

Theorem 3.4.1. *If (U, d) is a u.m.s.m.s. and μ is a compound distance with parameter $\mathbb{K}_\mu$, then $\breve{\mu}$ is a moment function and $\mathbb{K}_{\breve{\mu}} = K_\mu$. Moreover, the following stronger version of the* **TI**$^{(4)}$ *is valid:*

$$\check{\mu}(P_1, P_3) \le \mathbb{K}_\mu[\widehat{\mu}(P_1, P_2) + \check{\mu}(P_2, P_3)], \quad P_1, P_2, P_3 \in \mathcal{P}_1, \tag{3.4.31}$$

where $\widehat{\mu}$ is the minimal metric w.r.t. μ.

Proof. We will prove inequality (3.4.31) only. For each $\varepsilon > 0$ define laws P_{12}, $P_{13} \in \mathcal{P}_2$ such that

$$T_1 P_{12} = P_1, \quad T_2 P_{12} = P_2, \quad T_1 P_{13} = P_1, \quad T_2 P_{13} = P_3$$

and

$$\widehat{\mu}(P_1, P_2) \ge \mu(P_{12}) - \varepsilon, \quad \check{\mu}(P_1, P_3) \le \mu(P_{13}) + \varepsilon.$$

As in Theorem 3.3.1, let us define a law $Q \in \mathcal{P}_3$ [see (3.3.5)] having marginals $T_{12}Q = P_{12}$, $T_{13}Q = P_{13}$. By Definitions 2.5.1, 3.3.2, and 3.4.3, we have

$$\begin{aligned}\check{\mu}(P_1, P_3) &\le \mu(T_{13}Q) + \varepsilon \le \mathbb{K}_\mu[\mu(P_{12}) + \mu(P_{23})] + \varepsilon \\ &\le \mathbb{K}_\mu[\widehat{\mu}(P_1, P_2) + \varepsilon + \check{\mu}(P_2, P_3)] + \varepsilon.\end{aligned}$$

Letting $\varepsilon \to 0$ we get (3.4.31). □

Definition 3.4.4. The moment functions $\check{\mu}$ will be called *a maximal distance with parameter* $\mathbb{K}_{\check{\mu}} = \mathbb{K}_\mu$, and if $\mathbb{K}_\mu = 1$, then $\check{\mu}$ will be called a *maximal metric*.

As before, we note that a maximal distance (resp. metric) may fail to be a distance (resp. metric). (The **ID** property may fail.)

Corollary 3.4.1. *If (U, d) is a u.m.s.m.s. and μ is a compound metric on $\mathcal{P}_2$, then*

$$|\check{\mu}(P_1, P_3) - \check{\mu}(P_2, P_3)| \le \check{\mu}(P_1, P_2) \tag{3.4.32}$$

for all P_1, P_2, $P_3 \in \mathcal{P}_1$.

Remark 3.4.3. By the triangle inequality $\mathbf{TI}^{(4)}$ we have

$$|\check{\mu}(P_1, P_3) - \check{\mu}(P_2, P_3)| \le \check{\mu}(P_1, P_2). \tag{3.4.33}$$

Inequality (3.4.32) thus gives us refinement of the triangle inequality for maximal metrics.

We will further investigate the following problem, which is related to a description of the minimal and maximal distances.

Problem 3.4.2. If c is a nonnegative continuous function on U^2 and

$$\mu_c(P) := \int_{U^2} c(x, y) P(\mathrm{d}x, \mathrm{d}x), \quad P \in \mathcal{P}_2, \tag{3.4.34}$$

then what are the best possible inequalities of the type

$$\phi(P_1, P_2) \leq \mu_c(P) \leq \psi(P_1, P_2) \tag{3.4.35}$$

when the marginals $T_i P = P_i, i = 1, 2$, are fixed?

If $c(x, y) = H(d(x, y))$, $H \in \mathcal{H}$, then $\mu_c = \mathcal{L}_H$ [see (3.4.1)] and the best possible lower and upper bounds for $\mathcal{L}_H(P)$ [with fixed $P_i = T_i P$ $(i = 1, 2)$] are given by the minimal distance $\phi(P_1, P_2) = \widehat{\mathcal{L}}(P_1, P_2)$ and the maximal distance $\psi(P_1, P_2) = \check{\mathcal{L}}_H(P_1, P_2)$. For more general functions c the dual and explicit representations of $\widehat{\mu}_c$ and $\widehat{\mu}_c$ will be discussed later (Chap. 8).

Remark 3.4.4. In particular, for any convex nonnegative function ψ on $\mathbb{R}$ and $c(x, y) = \psi(x - y)$ $(x, y \in \mathbb{R})$, the functionals of $\widehat{\mathcal{L}}_H$ and $\check{\mathcal{L}}_H$ have the following explicit forms:

$$\widehat{\mathcal{L}}_H(P_1, P_2) := \int_0^1 H(F_1^{-1}(t) - F_2^{-1}(t))\mathrm{d}t,$$

$$\check{\mathcal{L}}_H(P_1, P_2) := \int_0^1 H(F_1^{-1}(t) - F_2^{-1}(1 - t))\mathrm{d}t,$$

where F_i^{-1} is the generalized inverse function (3.3.16) w.r.t. the DF F_i (Sect. 8.2).

Another example of a moment function that is an upper bound for $\mathcal{L}_H$ $(H \in \mathcal{H})$ is given by

$$\Lambda_{H,\mathbf{0}}(P_1, P_2) := K_H \int_U H(d(x, \mathbf{0}))(P_1 + P_2)(\mathrm{d}x), \tag{3.4.36}$$

where $\mathbf{0}$ is a fixed point of U. In fact, since $H \in \mathcal{H}$, then $H(d(x, y)) \leq K_H[H(d(x, \mathbf{0})) + H(d(y, \mathbf{0}))]$ for all $x, y \in U$, and hence

$$\mathcal{L}_H(P) \leq \overline{\Lambda}_{H,\mathbf{0}}(P_1, P_2). \tag{3.4.37}$$

One can easily improve inequality (3.4.37) by the following inequality:

$$\mathcal{L}_H(P) \leq \overline{\Lambda}_H(P_1, P_2) := \inf_{a \in U} \overline{\lambda}_{H,a}(P_1, P_2). \tag{3.4.38}$$

The upper bounds $\overline{\Lambda}_{H,a}$, $\overline{\Lambda}_H$ of $\mathcal{L}_H$ depend on the sum $P_1 + P_2$ only; hence, if P is an unknown law in $\mathcal{P}_2$ and we know only the sum of marginals $P_1 + P_2 = T_1 P + T_2 P$, then the best improvement of inequality (3.4.38) is given by

$$\mathcal{L}_H(P) \leq \mathcal{L}_H^{(s)}(P_1 + P_2), \tag{3.4.39}$$

where

$$\mathcal{L}_H^{(s)}(P_1 + P_2) := \sup\{\mathcal{L}_H(P) : T_1 P + T_2 P = P_1 + P_2\}. \tag{3.4.40}$$

Remark 3.4.5. Following Remark 3.4.2, we have that if (X,Y) is a pair of dependent U-valued RVs, and we know only the sum of distributions $\Pr_X + \Pr_Y$, then $\mathcal{L}_1^{(s)}(\Pr_X + \Pr_Y)$ is the best possible improvement of the triangle inequality (3.4.28). Further (Sect. 8.2), we will prove that in the particular case $U = \mathbb{R}$, $d(x,y) = |x-y|$, and $p \geq 1$,

$$\mathcal{L}_p^{(s)}(P_1 + P_2) = \left(\int_0^1 |V^{-1}(t) - V^{-1}(1-t)|^p \mathrm{d}t\right)^{1/p},$$

where V^{-1} is the generalized inverse [see (3.3.16)] of $V(t) = \frac{1}{2}(F_1(t) + F_2(t))$, $t \in \mathbb{R}$, and F_i is the DF of P_i $(i = 1,2)$.

For more general cases we will use the following definition.

Definition 3.4.5. For any compound distance μ the functional

$$\overset{(s)}{\mu}(P_1, P_2) := \sup\{\mu(P) : T_1 P + T_2 P = P_1 + P_2\}$$

will be called the *μ-upper bound with marginal sum fixed.*

Let us consider another possible improvement of Minkowski's inequality (3.4.24). Suppose we need to estimate from above (in the best possible way) the value $\mathcal{L}(X,Y)$ $(p > 0)$ having available only the moments

$$m_p(X) := [Ed^p(X,\mathbf{0})]^{p'}, \quad p' := \min(1, 1/p) \tag{3.4.41}$$

and $m_p(Y)$. Then the problem consists in evaluating the quantity

$$\psi_p(a_1,a_2) := \sup\left\{\mathcal{L}_p(P) : P \in \mathcal{P}_2(U), \left(\int_U d^p(x,\mathbf{0}) T_i P(\mathrm{d}x)\right)^{p'} = a_i, i=1,2\right\},$$
$$p' = \min(1, 1/p),$$

for each $a_i \geq 0$ and $a_2 \geq 0$.

Obviously, ψ_p is a moment function. Subsequently (Sect. 9.2), we will obtain an explicit representation of $\psi_p(a_1,a_2)$.

Definition 3.4.6. For any p. distance μ the function

$$\overset{(m,p)}{\mu}(a_1,a_2) := \sup\left\{\mu(P) : P \in \mathcal{P}_2(U), \left(\int_U d^p(x,\mathbf{0}) T_i P(\mathrm{d}x)\right)^{p'} = a_i, i=1,2\right\},$$

where $a_1 \geq 0$, $a_2 \geq 0$, $p > 0$, is said to be the *μ-upper bound with fixed pth marginal moments* a_1 and a_2.

Hence, $\overset{(m,1)}{\mathcal{L}}(a_1,a_2)$ is the best possible improvement of the triangle inequality (3.4.28) when we know only the "marginal" moments

$$a_1 = Ed(X, \mathbf{0}), \quad a_2 = Ed(Y, \mathbf{0}).$$

We will investigate improvements of inequalities of the type

$$Ed(X, \mathbf{0}) - Ed(Y, \mathbf{0}) \leq Ed(X, Y) \leq Ed(X, \mathbf{0}) + Ed(Y, \mathbf{0})$$

for dependent RVs X and Y. We set down the following definition.

Definition 3.4.7. For any p. distance μ

(i) The functional

$$\underset{(m,p)}{\mu}(a_1, a_2) := \inf\left\{\mu(P) : P \in \mathcal{P}_2(U), \left[\int_U d^p(x, \mathbf{0}) T_i P(\mathrm{d}x)\right]^{p'} = a_i, i=1,2\right\},$$

where $a_1 \geq 0$, $a_2 \geq 0$, $p > 0$, is said to be the *μ-lower bound with fixed marginal pth moments* a_1 and a_2;

(ii) The functional

$$\overline{\mu}(a_1 + a_2; m, p) := \sup\left\{\mu(P) : P \in \mathcal{P}_2(U), \left[\int_U d^p(x, \mathbf{0}) T_1 P(\mathrm{d}x)\right]^{p'} + \left[\int_U d^p(x, \mathbf{0}) T_2 P(\mathrm{d}x)\right]^{p'} = a_1 + a_2\right\},$$

where $a_1 \geq 0, a_2 \geq 0$, $p > 0$, is said to be the *μ-upper bound with fixed sum of marginal pth moments* $a_1 + a_2$;

(iii) The functional

$$\underline{\mu}(a_1 - a_2; m, p) := \inf\left\{\mu(P) : P \in \mathcal{P}_2(U), \left[\int_U d^p(x, \mathbf{0}) T_1 P(\mathrm{d}x)\right]^{p'} - \left[\int_U d^p(x, \mathbf{0}) T_2 P(\mathrm{d}x)\right]^{p'} = a_1 - a_2\right\},$$

where $a_1 \geq 0$, $a_2 \geq 0$, $p > 0$, is said to be the *μ-lower bound with fixed difference* of marginal p. moments $a_1 - a_2$.

Knowing explicit formulae for $\overset{(m,p)}{\mu}$ and $\underset{(m,p)}{\mu}$ (Sect. 9.2), we can easily determine $\overline{\mu}(a_1 + a_2; m, p)$ and $\overline{\mu}(a_1 - a_2; m, p)$ using the representations

$$\overline{\mu}(a;m,p) = \sup\left\{ \overset{(m,p)}{\mu}(a_1,a_2) : a_1 \geq 0, a_2 \geq 0, a_1 + a_2 = a \right\}$$

and

$$\underline{\mu}(a;m,p) = \inf\left\{ \underset{(m,p)}{\mu}(a_1,a_2) : a_1 \geq 0, a_2 \geq 0, a_1 - a_2 = a \right\}.$$

Let us summarize the bounds for μ we have obtained up to now. For any compound distance μ (Fig. 3.3) the maximal distance $\breve{\mu}$ (Definition 3.4.4) is not greater than the moment distance

$$\overset{(m,p)}{\mu}(a_1,a_2) := \sup\left\{ \mu(P_1,P_2) : \left[\int_U d^p(x,\mathbf{0})P_i(\mathrm{d}x)\right]^{p'} = a_i, i = 1,2 \right\}. \tag{3.4.42}$$

As we have seen, all compound distances μ can be estimated from above by means of $\breve{\mu}$, $\overset{(s)}{\mu}$, $\overset{(m,p)}{\mu}$, and $\mu(\cdot;m,p)$; in addition, the following inequality holds:

$$\mu \leq \breve{\mu} \leq \overset{(s)}{\mu} \leq \overline{\mu}(\cdot;m,p), \quad \breve{\mu} \leq \overset{(m,p)}{\mu}. \tag{3.4.43}$$

The p. distance μ can be estimated from below by means of the minimal metric $\widehat{\mu}$ (Definition 3.3.2), the cominimal metric $\mu\nu$ (Definition 3.3.3), and the primary minimal distance $\widetilde{\mu}_h$ (Definition 3.2.2), as well as for such μ as $\mu = \mu_c$ [see (3.3.40)] by means of minimal norms $\overset{\circ}{\mu}$ (Definition 3.3.4).

Thus

$$\underline{\mu}(\cdot;m,p) \leq \widetilde{\mu}_h \leq \widehat{\mu} \leq \mu\nu \leq \mu, \quad \overset{\circ}{\mu}_c \leq \mu_c, \tag{3.4.44}$$

and, moreover, we can compute the values of $\widetilde{\mu}_h$ using the values of the minimal distances μ since

$$\widetilde{\mu}_h(a_1,a_2) = (\widetilde{\widehat{\mu}})_h(a_1,a_2) := \inf\{\widehat{\mu}(P_1,P_2) : hP_i = a_i, i = 1,2\}. \tag{3.4.45}$$

Also, if $c(x,y) = H(d(x,y))$, $H \in \mathcal{H}$, then μ_c is a p. distance and

$$\overset{\circ}{\mu}_c \leq \widehat{\mu}_c \leq \mu. \tag{3.4.46}$$

Inequalities (3.4.42)–(3.4.46) are represented in the scheme in Fig. 3.3.

The functionals $\mu(\cdot;m,p)$, $\overset{(s)}{\mu}$, $\breve{\mu}$, μ, $\mu\nu$, $\widehat{\mu}$, $\overset{\circ}{\mu}$, and $\underline{\mu}(\cdot;m,p)$, are listed in order of numerical size in Fig. 3.3. The double arrows are interpreted in the following way. The functional $\overset{(s)}{\mu}$ dominates $\breve{\mu}$, but $\overset{(s)}{\mu}$ and $\overset{(m,p)}{\mu}$ are not generally comparable.

As an example illustrating the list of bounds in Fig. 3.3, let us consider the case $p = 1$ and $\mu(X,Y) = Ed(X,Y)$. Then, for a fixed point $\mathbf{0} \in U$,

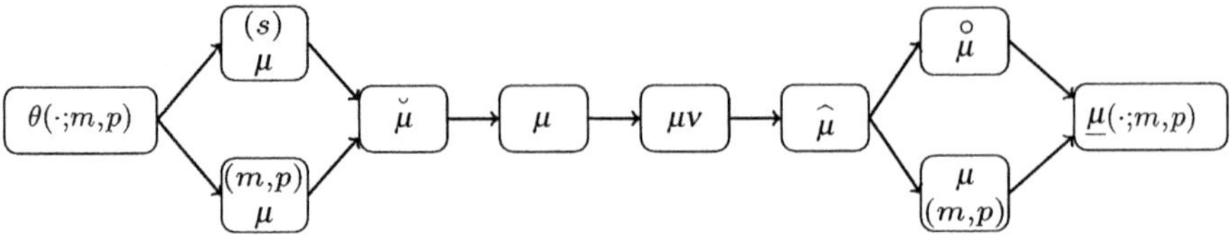

Fig. 3.3 Lower and upper bounds for $\mu(P)$ $(P \in \mathcal{P}_2)$ of a compound distance μ when different kinds of marginal characteristics of P are fixed. The *arrow* $\rightarrow$ indicates an inequality of the type $\leq$

(a)

$$\mu(a_1 + a_2; m, 1) = \sup\{Ed(X,Y) : Ed(X,\mathbf{0}) + Ed(Y,\mathbf{0}) = a_1 + a_2\},$$
$$a_1 + a_2 \geq 0; \tag{3.4.47}$$

(b)

$$\overset{(m,1)}{\mu}(a_1, a_2) = \sup\{Ed(X,Y) : Ed(X,\mathbf{0}) = a_1, Ed(Y,\mathbf{0}) = a_2\},$$
$$a_1 \geq 0, \quad a_2 \geq 0; \tag{3.4.48}$$

(c)

$$\overset{(s)}{\mu}(P_1 + P_2) = \sup\{Ed(X,Y) : \Pr_X + \Pr_Y = P_1 + P_2\},$$
$$P_1, P_2 \in \mathcal{P}_1; \tag{3.4.49}$$

(d)

$$\breve{\mu}(P_1, P_2) = \sup\{Ed(X,Y) : \Pr_X = P_1, \Pr_Y = P_2\},$$
$$P_1, P_2 \in \mathcal{P}_1; \tag{3.4.50}$$

and each of these functionals gives the best possible refinement of the inequality

$$Ed(X,Y) \leq Ed(X,\mathbf{0}) + Ed(Y,\mathbf{0})$$

under the respective conditions

(a)

$$Ed(X,\mathbf{0}) + Ed(Y,\mathbf{0}) = a_1 + a_2,$$

(b)

$$Ed(X,\mathbf{0}) = a_1, \quad Ed(Y,\mathbf{0}) = a_2,$$

(c)

$$\Pr_X + \Pr_Y = P_1 + P_2,$$

(d)

$$\Pr_X = P_1, \quad Pr_Y = P_2.$$

Analogously, the functionals

1.

$$\underline{\mu}(a_1 - a_2; m, 1) = \inf\{Ed(X,Y) : Ed(X,\mathbf{0}) - Ed(Y,\mathbf{0}) = a_1 - a_2\},$$
$$a_1, a_2 \in \mathbb{R}, \tag{3.4.51}$$

2.

$$\underset{(m,1)}{\mu}(a_1, a_2) = \inf\{Ed(X,Y) : Ed(X,\mathbf{0}) = a_1, Ed(Y,\mathbf{0}) = a_2\},$$
$$a_1 \geq 0, \quad a_2 \geq 0, \tag{3.4.52}$$

3.

$$\overset{\circ}{\mu}(P_1, P_2) = \inf\{\alpha Ed(X,Y) : \text{ for some } \alpha > 0, X \in \mathfrak{X}, Y \in \mathfrak{X}$$
$$\text{such that } \alpha(\Pr_X - \Pr_Y) = P_1 - P_2, \quad P_1, P_2 \in \mathcal{P}_1, \tag{3.4.53}$$

4.

$$\widehat{\mu}(P_1, P_2) = \inf\{Ed(X,Y) : \Pr_X = P_1, \Pr_Y = P_2\},$$
$$P_1, P_2 \in \mathcal{P}_1, \tag{3.4.54}$$

5.

$$\mu\nu(P_1, P_2) = \inf\{Ed(X,Y) : \Pr_X = P_1, \Pr_Y = P_2, \nu(X,Y) < \alpha\},$$
$$[P_1, P_2 \in \mathcal{P}_1, \nu \text{ is a p. distance in } \mathfrak{X}(U)] \tag{3.4.55}$$

describe the best possible refinement of the inequality

$$Ed(X,Y) \geq Ed(X,\mathbf{0}) - Ed(Y,\mathbf{0})$$

under the respective conditions

1. $Ed(X,\mathbf{0}) - Ed(Y,\mathbf{0}) = a_1 - a_2$,
2. $Ed(X,\mathbf{0}) = a_1 \;\; Ed(Y,\mathbf{0}) = a_2$,
3. $\alpha(Pr_X - \Pr_Y) = P_1 - P_2$ for some $\alpha > 0$,
4. $\Pr_X = P_1 \;\; \Pr_Y = P_2$,
5. $\Pr_X = P_1 \;\; \Pr_Y = P_2 \;\; \nu(X,Y) < \alpha$.

Remark 3.4.6. If $\mu(X,Y) = Ed(X,Y)$, then $\overset{\circ}{\mu} = \widehat{\mu}$ (Theorem 6.2.1); hence, in this case,

$$\overset{\circ}{\mu}(P_1, P_2) = \inf\{Ed(X,Y) : \Pr_X - \Pr_Y = P_1 - P_2\}. \tag{3.4.56}$$

References

Billingsley P (1999) Convergence of probability measures, 2nd edn. Wiley, New York
Dudley RM (1976) Probabilities and metrics: convergence of laws on metric spaces, with a view to statistical testing. Aarhus university mathematics institute lecture notes series no. 45, Aarhus
Dudley RM (2002) Real analysis and probability, 2nd edn. Cambridge University Press, New York
Kantorovich LV (1942) On the transfer of masses. Dokl Akad Nauk USSR 37:227–229
Kantorovich LV (1948) On a problem of Monge. Usp Mat Nauk 3:225–226 (in Russian)
Kantorovich LV, Akilov GP (1984) Functional analysis. Nauka, Moscow (in Russian)
Kemperman JHB (1983) On the role of duality in the theory of moments. Semi-infinite programming and applications. In: Lecture notes economic mathematical system, vol 215. Springer, Berlin, pp 63–92
Levin VL, Rachev ST (1990) New duality theorems for marginal problems with some applications in stochastics. In: Lecture notes mathematics, vol 1412. Springer, Berlin, pp 137–171
Prokhorov YuV (1956) Convergence of random processes and limit theorems in probability theory. Theor Prob Appl 1:157–214
Zolotarev VM (1976) The effect of stability for characterization of probability distributions. Zap Nauchn Sem LOMI 61:38–55 (in Russian)

Chapter 4
A Structural Classification of Probability Distances

The goals of this chapter are to:

- Introduce and motivate three classifications of probability metrics according to their metric structure,
- Provide examples of probability metrics belonging to a particular structural group,
- Discuss the generic properties of the structural groups and the links between them.

Notation introduced in this chapter:

Notation	Description
$\mathbf{L}_\lambda$	Parametric version of Lévy metric
$\mathbf{W}$	Uniform metric between generalized inverse functions
r_λ	Hausdorff metric with parameter λ
$\widetilde{r}_\lambda$	Hausdorff semimetric between functions
$h_{\lambda,\phi,\mathcal{B}_0}$	Hausdorff representation of a probability semidistance
$\mathcal{F}^n = \mathcal{F}(\mathbb{R}^n)$	Space of distribution functions on $\mathbb{R}^n$
$\mathbf{e}$	Unit vector $(1, 1, \ldots, 1)$ in $\mathbb{R}^n$
$\mathbf{L}_{\lambda,H}$	Lévy probability distance
$\mathbf{H}_\lambda$	Hausdorff metric in $\mathcal{F}(\mathbb{R}^n)$
$\widetilde{\mathbf{W}}$	Limit of $\lambda\mathbf{H}_\lambda$ as $\lambda \to \infty$
$\rho_1 \overset{\text{top}}{\leq} \rho_2$	ρ_2-convergence implies ρ_1-convergence
$\rho_1 \overset{\text{top}}{<} \rho_2$	$\rho_1 \overset{\text{top}}{\leq} \rho_2$ but not $\rho_2 \overset{\text{top}}{\leq} \rho_1$
$\rho_1 \overset{\text{top}}{\sim} \rho_2$	$\rho_1 \overset{\text{top}}{\leq} \rho_2$ and $\rho_2 \overset{\text{top}}{\leq} \rho_1$
$\mathbf{SB}$	Skorokhod–Billingsley metric
ω'_F, ω''_F	Moduli of continuity in the space of distribution functions
$\pi\mathbf{H}_\lambda$	Metric with a Hausdorff structure satisfying the property $\pi_\lambda \leq \pi\mathbf{H}_\lambda \leq \sigma$
$\Lambda_{\lambda,\nu}$	Λ-structure of a probability semidistance
$C^b(U)$	Set of bounded continuous functions on U

S.T. Rachev et al., *The Methods of Distances in the Theory of Probability and Statistics*, DOI 10.1007/978-1-4614-4869-3_4, © Springer Science+Business Media, LLC 2013

Notation	Description
$\zeta(\cdot,\cdot;\mathcal{G}^p)$	Fortet–Mourier metric
$\boldsymbol{\beta}$	Dudley metric
$\boldsymbol{\kappa}_Q$	Q-difference pseudomoment
$\boldsymbol{\tau}_Q$	Compound Q-difference pseudomoment
$\mathbf{AS}_p$	Szulga metric

4.1 Introduction

Chapter 3 was devoted to a classification of probability [semidistances $\mu(P)$ $(P \in \mathcal{P}_2)$] with respect to various partitionings of the set $\mathcal{P}_2$ into classes $\mathcal{PC}$ such that $\mu(P)$ takes a constant value on each $\mathcal{PC}$. For instance, if $\mathcal{PC} := \mathcal{PC}(P_1, P_2) := \{P \in \mathcal{P}_2 : T_1 P = P_1, T_2 P = P_2\}$, $P_1, P_2 \in \mathcal{P}_1$, and $\mu(P') = \mu(P'')$ for each P', $P'' \in \mathcal{PC}$, then μ was said to be a simple semidistance. Analogously, if

$$\mathcal{PC} := \mathcal{PC}(\overline{a}_1, \overline{a}_2) := \{P \in \mathcal{P}_2 : h(T_1 P) = \overline{a}_1, h(T_2 P) = \overline{a}_2\}$$

[see (3.2.2) and Definition 3.2.1 in Chap. 3] and $\mu(P') = \mu(P'')$ as P', $P'' \in \mathcal{PC}(\overline{a}_1, \overline{a}_2)$, then μ was said to be a primary distance.

In the present chapter, we classify the probability semidistances (p. semidistances) on the basis of their metric structure. For example, a p. metric that admits a representation as a Hausdorff metric [see (2.6.1) of Chap. 2] will be called a metric with a Hausdorff structure. See, for instance, the H-metric introduced in Sect. 2.4.

Some probability metrics are more naturally defined in the following form:

$$\Lambda_{\lambda,\nu}(X,Y) := \inf\{\varepsilon > 0 : \nu(X,Y;\lambda\varepsilon) < \varepsilon\},$$

where the functional $\nu(X,Y;t)$ has a particular axiomatic structure. Examples include the Lévy metric $\mathbf{L}$ (2.2.3), the Prokhorov metric $\boldsymbol{\pi}$ (3.3.18), and the Ky Fan metric $\mathbf{K}_\lambda$ (2.2.5).

Finally, some simple probability distances can be represented as $\zeta_{\mathcal{F}}$-metrics, namely,

$$\mu(P_1, P_2) = \zeta_{\mathcal{F}}(P_1, P_2) := \sup_{f \in \mathcal{F}} \left| \int f \mathrm{d}(P_1 - P_2) \right|, \quad P_i \in \mathcal{P} \subset \mathcal{P}(U),$$

where $\mathcal{F}$ is a class of functions on an s.m.s. U that are P-integrable for any $P \in \mathcal{P}$. In this case, μ is said to be a probability metric with a ζ-structure. Examples of such μ are the Kantorovich metric ℓ_1 (3.3.12), the total variation metric $\boldsymbol{\sigma}$ (3.3.13), the Kolmogorov metric $\boldsymbol{\rho}$ (2.2.2), and the $\boldsymbol{\theta}$-metric (Remark 2.2.2).

From a general perspective, a single probability metric can enjoy all three representations. In this case, the representation chosen depends on the particular problem at hand. Three sections are devoted to these three structural classifications. We begin with the Hausdorff structure, then we continue with the Λ-structure, and finally we discuss the ζ-structure.

4.2 Hausdorff Structure of Probability Semidistances

The definition of a Hausdorff p. semidistance structure (henceforth simply *h*-structure) is based on the notion of a *Hausdorff semimetric* in the space of all subsets of a given metric space (S, ρ):

$$r(A, B) = \inf\{\varepsilon > 0 : A^{\varepsilon} \supseteq B, B^{\varepsilon} \supseteq A\}$$
$$= \max\{\inf\{\varepsilon > 0 : A^{\varepsilon} \supseteq B\}, \inf\{\varepsilon > 0 : B^{\varepsilon} \supset A\}\}, \quad (4.2.1)$$

where A^{ε} is the open ε-neighborhood of A.

From definition (4.2.1) the second Hausdorff semidistance representation follows immediately:

$$r(A, B) := \max(r', r''), \quad (4.2.2)$$

where

$$r' := \sup_{x\in A} \inf_{y\in B} \rho(x, y)$$

and

$$r'' := \sup_{y\in B} \inf_{x\in A} \rho(x, y).$$

As an example of a probability metric with a representation close to that of equality (4.2.2), let us consider the following *parametric version of the Lévy metric for* $\lambda > 0$, $X, Y \in \mathfrak{X}(\mathbb{R})$ (Fig. 4.1):

$$\mathbf{L}_{\lambda}(X, Y) := \mathbf{L}_{\lambda}(F_X, F_Y) := \inf\{\varepsilon > 0 : F_X(x - \lambda\varepsilon) - \varepsilon \le F_Y(x)$$
$$\le F_X(x + \lambda\varepsilon) + \varepsilon \quad \forall x \in \mathbb{R}\}. \quad (4.2.3)$$

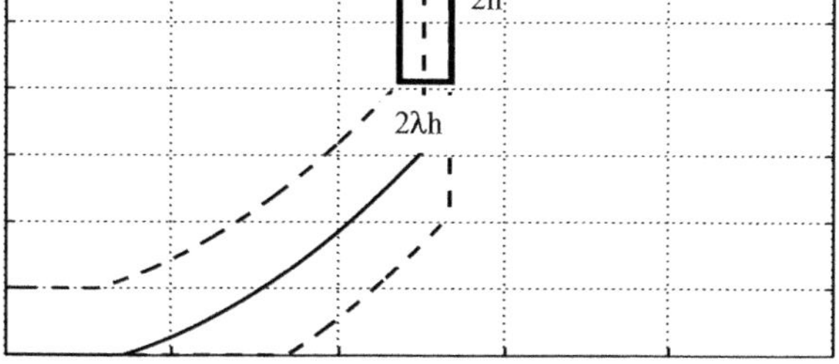

Fig. 4.1 $\mathrm{St}(F_X, h)$ is the strip in which the graph of F_Y must be positioned in order for the inequality $\mathbf{L}_{\lambda}(X, Y) \le h$ to hold

Obviously, $\mathbf{L}_\lambda$ is a simple metric in $\mathfrak{X}(\mathbb{R})$ for any $\lambda > 0$, and $\mathbf{L} := \mathbf{L}_1$ is the usual Lévy metric [see (2.2.3)]. Moreover, it is not difficult to verify that $\mathbf{L}_\lambda(F, G)$ is a metric in the space $\mathcal{F}$ of all distribution functions (DFs). Considering $\mathbf{L}_\lambda$ as a function of λ, we see that $\mathbf{L}_\lambda$ is nonincreasing on $(0, \infty)$, and the following limit relations hold:

$$\lim_{\lambda \to 0} \mathbf{L}_\lambda(F, G) = \boldsymbol{\rho}(F, G), \quad F, G \in \mathcal{F}, \tag{4.2.4}$$

and

$$\lim_{\lambda \to 0} \lambda \mathbf{L}_\lambda(F, G) = \mathbf{W}(F, G). \tag{4.2.5}$$

In equality (4.2.4), $\boldsymbol{\rho}$ is the *Kolmogorov metric* [see (2.2.2)] in $\mathcal{F}$

$$\boldsymbol{\rho}(F, G) := \sup_{x \in \mathbb{R}} |F(x) - G(x)|. \tag{4.2.6}$$

In equality (4.2.5), $\mathbf{W}(F, G)$ is the *uniform metric between the inverse functions* F^{-1}, G^{-1}

$$\mathbf{W}(F, G) := \sup_{0<t<1} |F^{-1}(t) - G^{-1}(t)|, \tag{4.2.7}$$

where F^{-1} is the generalized inverse of F

$$F^{-1}(t) := \sup\{x : F(x) < t\}. \tag{4.2.8}$$

Equality (4.2.4) follows from (4.2.3) (Fig. 4.1). Likewise, (4.2.5) is handled by the equalities

$$\begin{aligned}\lim_{\lambda \to \infty} \lambda \mathbf{L}_\lambda(F, G) &= \inf\{\delta > 0 : F(x) \le G(x + \delta), G(x) \le F(x + \delta) \quad \forall x \in \mathbb{R}\} \\ &= \mathbf{W}(F, G).\end{aligned}$$

Another way to prove (4.2.5) is to use the representation of $\mathbf{L}_\lambda(F, G)$ in terms of the inverse functions F^{-1} and G^{-1}:

$$\begin{aligned}\mathbf{L}_\lambda(F, G) &= \inf\{\varepsilon > 0 : F_X^{-1}(t - \varepsilon) - \lambda\varepsilon \le F_Y^{-1}(t), \\ &\qquad F_Y^{-1}(t - \varepsilon) - \lambda\varepsilon \le F_X^{-1}(t) \forall \varepsilon \le t \le 1\} \\ &= \frac{1}{\lambda} \inf\left\{\delta > 0 : F_X^{-1}\left(t - \frac{1}{\lambda}\delta\right) - \delta \le F_Y^{-1}(t),\right. \\ &\qquad \left. F_Y^{-1}\left(t - \frac{1}{\lambda}\delta\right) - \delta \le F_X^{-1}(t) \forall \frac{1}{\lambda}\delta \le t \le 1\right\}.\end{aligned}$$

We will prove subsequently [Corollaries 7.4.1 and (7.5.15)] that $\mathbf{W}$ coincides with the ℓ_∞-metric

$$\ell_\infty(F_1, F_2) := \ell_\infty(P_1, P_2) := \inf\{\varepsilon > 0 : P_1(A) \le P_2(A^\varepsilon), \quad \forall A \subset \mathbb{R}\},$$

where P_i is the law determined by F_i. The equality $\mathbf{W} = \ell_\infty$ illustrates – together with equality (4.2.5) – the main relationship between the Lévy metric and ℓ_∞.

Let us define the Hausdorff metric between two bounded functions on the real line $\mathbb{R}$. Let dm_λ ($\lambda > 0$) be the Minkowski metric on the plane $\mathbb{R}^2$; that is, for each $A = (x_1, y_1)$ and $B = (x_2, y_2)$ we have $dm_\lambda(A, B) := \max\{(1/\lambda)|x_1 - x_2|, |y_1 - y_2|\}$. *The Hausdorff metric* r_λ ($\lambda > 0$) *in the set* $C(\mathbb{R}^2)$ (of all closed nonempty sets $G \subset \mathbb{R}^2$) is defined as follows: for $G_1 \subseteq \mathbb{R}^2$ and $G_2 \subseteq \mathbb{R}^2$

$$r_\lambda(G_1, G_2) := \max\left\{\sup_{A \in G_1} \inf_{B \in G_2} dm_\lambda(A, B), \sup_{B \in G_2} \inf_{A \in G_1} dm_\lambda(A, B)\right\}. \quad (4.2.9)$$

We will say that r_λ is generated by the metric dm_λ just as the Hausdorff distance r was generated by ρ in equality (4.2.2). Let $f \in D(\mathbb{R})$ be the set of all bounded right-continuous functions on $\mathbb{R}$ having limits $f(x-)$ from the left. The set

$$\overline{f} = \{(x, y) : x \in \mathbb{R} \text{ and either } f(x-) \le y \le f(x) \text{ or } f(x) \le y \le f(x-)\}$$

is called the *completed graph* of the function f.

Remark 4.2.1. Obviously, the completed graph $\overline{F}$ of a DF $F \in \mathcal{F}$ is given by

$$\overline{F} := \{(x, y) : x \in \mathbb{R}, F(x-) \le y \le F(x)\}. \quad (4.2.10)$$

Using equality (4.2.9), we define the Hausdorff metric $r_\lambda = r_\lambda(\overline{f}, \overline{g})$ in the space of completed graphs of bounded, right-continuous functions.

Definition 4.2.1. The metric

$$r_\lambda(f, g) := r_\lambda(\overline{f}, \overline{g}), \quad f, g \in D(\mathbb{R}), \quad (4.2.11)$$

is said to be the *Hausdorff metric in* $D(\mathbb{R})$.

Lemma 4.2.1 (Sendov 1969). *For any* $f, g \in D(\mathbb{R})$

$$r_\lambda(f, g) = \max\left\{\sup_{x \in \mathbb{R}} \inf_{(x_2, y_2) \in \overline{g}} dm_\lambda((x, f(x)), (x_2, y_2)),\right.$$
$$\left.\sup_{x \in \mathbb{R}} \inf_{(x_1, y_1) \in \overline{f}} dm_\lambda((x_1, y_1), (x, g(x)))\right\}.$$

Proof. It is sufficient to prove that if for each $x_0 \in \mathbb{R}$ there exist points $(x_1, y_1) \in \overline{f}$, $(x_2, y_2) \in \overline{g}$ such that $\max\{(1/\lambda)|x_0 - x_1|, |g(x_0) - y_1|\} \le \delta$, $\max\{(1/\lambda)|x_0 - x_2], |f(x_0) - y_2|\} \le \delta$, then $r_\lambda(f, g) \le \delta$. Suppose the contrary is true. Then there exists a point (x_0, y_0) in the completed graph of one of the two functions, say $f(x)$, such that in the rectangle $|x - x_0| \le \lambda\delta$, $|y - y_0| \le \delta$, there is no point of the completed graph $\overline{g}$. Writing

$$y_0' = \min_{(x_0,y)\in\overline{f}} y, \qquad y_0'' = \max_{(x_0,y)\in\overline{f}} y,$$

we then have $y_0' \le y_0 < y_0''$. From the definition of (x_0, y_0') and (x_0, y_0'') it follows that there exist two sequences $\{x_n'\}$ and $\{x_n''\}$ in $\mathbb{R}$, converging to x_0, such that $\lim_{n\to\infty} f(x_n') = y_0'$, $\lim_{n\to\infty} f(x_n'') = y_0''$. Then from the hypothesis and the fact that $\overline{g}$ is a closed set it follows that there exist two points (x_1, y_1), $(x_2, y_2) \in \overline{g}$ for which $x_1, x_2 \in [x_0 - \lambda\delta, x_0 + \lambda\delta]$, $y_1 \le y_0'$, $y_2 \ge y_0''$. This contradicts our assumptions since by the definition of the completed graph $\overline{g}$, there exists $\widetilde{x}_0 \in [x_0 - \lambda\delta, x_0 + \lambda\delta]$ such that $(\widetilde{x}_0, y_0) \in \overline{g}$. □

Remark 4.2.2. Before proceeding to the proof of the fact that the Lévy metric is a special case of the Hausdorff metric (Theorem 4.2.1), we will mention the following two properties of the metric $r_\lambda(f, g)$ that can be considered as generalizations of well-known properties of the Lévy metric.

Property 4.2.1. Let ρ be the uniform distance in $D(\mathbb{R})$, i.e., $\rho(f,g) := \sup_{u\in\mathbb{R}} |f(u) - g(u)|$, and let $\omega_f(\delta) := \sup\{|f(u) - f(u')| : |u - u'| < \delta\}$, $f \in C_b(\mathbb{R})$, $\delta > 0$, be the modulus of f-continuity. Then

$$r_\lambda(f,g) \le \rho(f,g) \le r_\lambda(f,g) + \min(\omega_f(\lambda r_\lambda(f,g)), \omega_g(\lambda r_\lambda(f,g))). \tag{4.2.12}$$

Proof. If $r_\lambda(f,g) = \sup_{a\in\overline{f}} \inf_{b\in\overline{g}} dm_\lambda(a,b)$, then following the proof of Lemma 4.2.1 we have

$$\begin{aligned} r_\lambda(f,g) &= \sup_{x\in\mathbb{R}} \inf_{(x_2,y_2)\in\overline{g}} dm_\lambda(x, f(x)), (x_2, y_2)) \\ &\le \sup_{x\in\mathbb{R}} \inf_{y\in\mathbb{R}} \max\left\{\frac{1}{\lambda}|x-y|, |f(x) - g(y)|\right\} \le \rho(f,g). \end{aligned}$$

For any $x \in \mathbb{R}$ there exists $(y_0, z_0) \in \overline{g}$ such that

$$r_\lambda(f,g) \ge \inf_{(y,z)\in\overline{g}} dm_\lambda((x, f(x)), (y,z)) = \max\left(\frac{1}{\lambda}|x - y_0|, |f(x) - z_0|\right).$$

Hence

$$\begin{aligned} |f(x) - g(x)| &\le |f(x) - z_0| + |g(x) - z_0| \\ &\le r(f,g) + \max(|g(x) - g(y_0-)|, |g(x) - g(y_0)|) \\ &\le r(f,g) + \omega_g(\lambda r_\lambda(f,g)). \end{aligned}$$

□

As a consequence of inequalities (4.2.12), we obtain the following property.

Property 4.2.2. Let $\{f_n(x), n = 1, 2, \dots\}$ be a sequence in $D(\mathbb{R})$, and let $f(x)$ be a continuous-bounded function on the line. The sequence $\{f_n\}$ converges uniformly on $\mathbb{R}$ to $f(x)$ if and only if $\lim_{n\to\infty} r_\lambda(f_n, f) = 0$.

Theorem 4.2.1. *For all $F, G \in \mathcal{F}$ and $\lambda > 0$*

$$\mathbf{L}_\lambda(F, G) = r_\lambda(F, G). \tag{4.2.13}$$

Proof. Consider the completed graphs $\overline{F}$ and $\overline{G}$ of the DFs F and G and denote by P and Q the points where they intersect the line $(1/\lambda)x + y = u$, where u can be any real number. Then

$$\mathbf{L}_\lambda(F, G) = \max_{u \in \mathbb{R}} |PQ|(1 + \lambda^2)^{-1/2}, \tag{4.2.14}$$

where $|PQ|$ is the length of the segment joining the points P and Q.[1] We will show that $r_\lambda(F, G) \leq \mathbf{L}_\lambda(F, G)$ by applying Lemma 4.2.1.

Choose a point $x_0 \in \mathbb{R}$. The line $(1/\lambda)x + y = (1/\lambda)x_0 y + F(x_0)$ intersects $\overline{F}$ and $\overline{G}$ at the points $P(x_0, F(x_o))$ and $Q(x_1, y_1)$. It follows from (4.2.14) that $|F(x_0) - y_1| \leq \mathbf{L}_\lambda(F, G)$ and $(1/\lambda)|x_0 - x_1| \leq \mathbf{L}_\lambda(F, G)$. Permuting F and G, we find that for some $(x_2, y_2) \in \overline{F}$

$$\max\left[\frac{1}{\lambda}|x_0 - x_2|, |G(x_0) - y_2|\right] \leq \mathbf{L}_\lambda(F, G).$$

By Lemma 4.2.1, this means that $r_\lambda(F, G) \leq \mathbf{L}_\lambda(F, G)$.

Now let us show the reverse inequality. Assume otherwise, i.e., assume $\mathbf{L}_\lambda(F, G) > r_\lambda(F, G)$. Let $P_0(x', y')$ and $Q_0(x'', y'')$ be points such that

$$\mathbf{L}_\lambda(F, G) = \frac{|P_0 Q_0|}{(1 + \lambda^2)^{1/2}} > r_\lambda(F, G).$$

Suppose that $x' < x''$. Since the points P_0 and Q_0 lie on some $(1/\lambda)x + y = u_0$, and, say, $u_0 > 0$, we have $y' > y''$. By the definition of the metric $r_\lambda(F, G)$ and our assumptions, it follows that

$$\frac{|P_0 Q_0|}{(1 + \lambda^2)^{1/2}} > \max_{A \in \overline{F}} \min_{B \in \overline{G}} dm_\lambda(A, B)$$

[see (4.2.9)]. Since $P_0 \in \overline{F}$, there exists a point $B_0(x^*, y^*) \in \overline{G}$ such that

$$\frac{|P_0 Q_0|}{(1 + \lambda^2)^{1/2}} > \min_{B \in \overline{G}} dm_\lambda(P_0, B) = dm_\lambda(P_0, B_0).$$

Thus,

$$dm_\lambda(P_0, B_0) = \max\left[\frac{1}{\lambda}|x' - x^*|, |y' - y^*|\right] < |P_0 Q_0|(1 + \lambda^2)^{-1/2}. \tag{4.2.15}$$

---

[1]The proof of (4.2.14) is quite analogous to that given in Hennequin and Tortrat (1965, Chap. 19), for the case $\lambda = 1$.

Suppose that $x' \geq x^*$. Then $x^* \leq x' < x''$. The function G is nondecreasing, so $y^* \leq y'$, i.e.,

$$y' - y^* \geq y' - y'' = \frac{|P_0 Q_0|}{(1+\lambda^2)^{1/2}},$$

which is impossible by virtue of (4.2.15). If $x' < x^*$, then

$$0 < \frac{1}{\lambda}(x^* - x') < \frac{|P_0 Q_0|}{(1+\lambda^2)^{1/2}} = \frac{1}{\lambda}(x'' - x').$$

Then $x^* < x''$ and $y^* \leq y''$, which, as we have proved, is impossible. Thus, $\mathbf{L}_\lambda(F,G) \leq r_\lambda(F,G)$. □

To cover other probability metrics by means of the Hausdorff metric structure, the following generalization of the notion of Hausdorff metric r is needed. Let $\mathcal{FS}$ be the space of all real-valued functions $F_A : A \to \mathbb{R}$, where A is a subset of the metric space (S,ρ).

Definition 4.2.2. Let $f = f_A$ and $g = g_B$ be elements of $\mathcal{FS}$. The quantity

$$\widetilde{r}_\lambda(f,g) := \max(\widetilde{r}'_\lambda(f,g), \widetilde{r}'_\lambda(g,f)), \tag{4.2.16}$$

where

$$\widetilde{r}'_\lambda(f,g) := \sup_{x\in A} \inf_{y\in B} \max\left\{\frac{1}{\lambda}\rho(x,y), f(x) - g(y)\right\},$$

is called the *Hausdorff semimetric* between the functions f_A and g_B.

Obviously, if $f(x) = g(y) =$ constant for all $x \in A$, $y \in B$, then $\widetilde{r}_\lambda(f,g) = r(A,B)$ [see (4.2.2)]. Note that $\widetilde{r}_\lambda$ is a metric in the space of all upper semicontinuous functions with closed domains.

The next two theorems are straightforward consequences of the more general Theorem 4.3.1.

Theorem 4.2.2. *The Lévy metric* $\mathbf{L}_\lambda$ (4.2.3) *admits the following representation in terms of metric* $\widetilde{r}$ *[*(4.2.16)*]:*

$$\mathbf{L}_\lambda(X,Y) = \widetilde{r}_\lambda(f_A, g_B), \tag{4.2.17}$$

where $f_A = F_X$, $g_B = F_Y$, $A \equiv B \equiv \mathbb{R}$, $\rho(x,y) = |x-y|$.

Thus, the Lévy metric $\mathbf{L}_\lambda$ has two representations in terms of r_λ and in terms of $\widetilde{r}_\lambda$. Concerning the Prokhorov metric $\boldsymbol{\pi}_\lambda$ (3.3.22), only a representation in terms of $\widetilde{r}_\lambda$ is known. That is, let $\mathcal{S} = \mathcal{C}((U,d))$ be the space of all closed nonempty subsets of a metric space (U,d), and let r be the Hausdorff distance (4.2.1) in $\mathcal{S}$. Any law $P \in \mathcal{P}_1(U)$ can be considered as a function on the metric space $(\mathcal{S}, r)$ because P is determined uniquely on $\mathcal{S}$, that is,

$$P(A) := \sup\{P(C) : C \in \mathcal{S}, C \subseteq A\} \quad \text{for any } A \in \mathcal{B}_1.$$

Define a metric $\widetilde{r}_\lambda(P_1, P_2)$ $(P_1, P_2 \in \mathcal{P}_1(U))$ by setting $A = B = \mathcal{S}$ and $\rho = r$ in equality (4.2.16).

Theorem 4.2.3. *For any $\lambda > 0$ the Prokhorov metric π_λ takes the form*

$$\pi_\lambda(P_1, P_2) = \widetilde{r}_\lambda(P_1, P_2) \qquad (P_1, P_2 \in \mathcal{P}_1(U)),$$

where $U = (U, d)$ is assumed to be an arbitrary metric space.

Remark 4.2.3. By Theorem 4.2.3, for all $P_1, P_2 \in \mathcal{P}_1$ we have the following Hausdorff representation of the Prokhorov metric π_λ, $\lambda > 0$:

$$\pi_\lambda(P_1, P_2) := \max\left\{ \sup_{A \in \mathcal{B}_1} \inf_{B \in \mathcal{B}_1} \max\left[\frac{1}{\lambda} r(A, B), P_1(A) - P_2(B)\right], \right.$$
$$\left. \sup_{B \in \mathcal{B}_1} \inf_{A \in \mathcal{B}_1} \max\left[\frac{1}{\lambda} r(A, B), P_2(B) - P_1(A)\right]\right\}. \qquad (4.2.18)$$

Problem 4.2.1. Is it possible to represent the Prokhorov metric π_λ by means of r_λ or to find a probability metric with a r_λ-structure that metrizes the weak convergence in $\mathcal{P}(U)$ for an s.m.s. U?

Remark 4.2.4. We can use the Hausdorff representation (4.2.18) of $\pi = \pi_1$ to extend the definition of the Prokhorov metric over the set $\Phi(U)$ that strictly contains the set $\mathcal{P}(U)$ of all probability laws on an arbitrary metric space (U, d). Specifically, let $\Phi(U)$ be the family of all set functions $\phi : (\mathcal{S}, r) \to [0, 1]$ that are continuous from above, i.e., for any sequence $\{C_n\}_{n \geq 0}$ of closed subsets of U

$$r(C_n C_0) \to 0 \quad \Rightarrow \quad \overline{\lim_{n \to \infty}} \phi(C_n) \leq \phi(C_0).$$

Clearly, each law $P \in \Phi(U)$. We extend the Prokhorov metric over $\Phi(U)$ by simply setting

$$\pi(\phi_1, \phi_2) = \max\left\{ \sup_{C_1 \in \mathcal{S}} \inf_{C_2 \in \mathcal{S}} \max[r(C_1, C_2), \phi_1(C_1) - \phi_2(C_2)], \right.$$
$$\left. \sup_{C_2 \in \mathcal{S}} \inf_{C_1 \in \mathcal{S}} \max[r(C_1, C_2), \phi_2(C_2) - \phi_1(C_1)]\right\}.$$

For $\phi_i = P_i \in \mathcal{P}(U)$ the preceding formula gives

$$\pi(P_1, P_2) = \inf\{\varepsilon > 0 : P_1(C) < P_2(C^\varepsilon) + \varepsilon, P_2(C) \leq P_1(C^\varepsilon) + \varepsilon, \ \forall C \in \mathcal{S}\},$$

i.e., the usual Prokhorov metric (see Theorem 4.3.1 for details).

The next step is to extend the notion of weak convergence. We will use the analog of the Hausdorff topological convergence of sequences of sets. For a sequence $\{\phi_n\} \subset \Phi(U)$, define the *upper topological limit* $\overline{\phi} = \overline{\ell t}\phi_n$ by

$$\overline{\phi}(C) := \sup\left\{\overline{\lim_{n\to\infty}}\phi_n(C_n) : C_n \in \mathcal{S}, r(C_n, C) \to 0\right\}.$$

Analogously, define the *lower topological limit* $\underline{\phi} = \underline{\ell t}\phi_n$ by

$$\underline{\phi}(C) := \sup\left\{\underline{\lim_{n\to\infty}}\phi_n(C_n) : C_n \in \mathcal{S}, r(C_n, C) \to 0\right\}.$$

If $\overline{\ell t}\phi_n = \underline{\ell t}\phi_n$, then $\{\phi_n\}$ is said to be *topologically convergent* and $\phi := \ell t\phi_n := \overline{\ell t}\phi_n$ is said to be the *topological limit* of $\{\phi_n\}$. One can see that $\phi = \ell t\phi_n \in \Phi(U)$.

For any metric space (U, d) the following conditions hold:

(a) Suppose P_n and P are laws on U. If $P = \ell t P_n$, then $P_n \xrightarrow{\text{w}} P$. Conversely, if (U, d) is an s.m.s., then the weak convergence $P_n \xrightarrow{\text{w}} P$ yields the topological convergence $P = \ell t P_n$.
(b) If $\pi(\phi_n, \phi) \to 0$ for $\{\phi_n\} \subset \Phi(U)$, then $\phi = \ell t\phi_n$.
(c) If $\{\phi_n\}$ is fundamental (Cauchy) with respect to π, then ϕ_n is topologically convergent.
(d) If (U, d) is a compact set, then the π-convergence and the topological convergence coincide in $\Phi(U)$.
(e) If (U, d) is a complete metric space, then the metric space $(\phi(U), \pi)$ is also complete.
(f) If (U, d) is totally bounded, then $(\Phi(U), \pi)$ is also totally bounded.
(g) If (U, d) is a compact metric space, then $(\Phi(U), \pi)$ is also a compact metric space.

The extension $\Phi(U)$ of the set of laws $\mathcal{P}(U)$ seems to enjoy properties that are basic in the application of the notions of weak convergence and Prokhorov metric. Note also that in an s.m.s. (U, d), if $\{P_n\} \subset \mathcal{P}(U)$ is π-fundamental, then clearly $\{P_n\}$ may not be weakly convergent; however, by (c), $\{P_n\}$ has a topological limit, $\phi = \ell t P_n \in \Phi(U)$.

Next, taking into account Definition 4.2.2, we will define the Hausdorff structure of p. semidistances.

Without loss of generality (Sect. 2.7), we assume that any p. semidistance $\mu(P)$, $P \in \mathcal{P}_2(U)$, has a representation in terms of pairs of U-valued random variables $X, Y \in \mathfrak{X} := \mathfrak{X}(U)$:

$$\mu(P) = \mu(\Pr_{X,Y}) = \mu(X, Y).$$

Let $\mathcal{B}_0 \subseteq \mathcal{B}(U)$ and let the function $\phi : \mathfrak{X}^2 \times \mathcal{B}_0^2 \to [0, \infty]$ satisfy the following relations:

(a) If $\Pr(X = Y) = 1$, then $\phi(X, Y; A, B) = 0$ for all $A, B \in \mathcal{B}_0$.
(b) There exists a constant $K_\phi \geq 1$ such that for all $A, B, C \in \mathcal{B}_0$ and RV X, Y, Z

$$\phi(X, Z; A, B) \leq K_\phi[\phi(X, Y; A, C) + \phi(Y, Z, C, B)].$$

Definition 4.2.3. Let μ be a p. semidistance. The representation of μ in the form

$$\mu(X,Y) = h_{\lambda,\phi,\mathcal{B}_0}(X,Y) := \max\{h'_{\lambda,\phi,\mathcal{B}_0}(X,Y), h'_{\lambda,\phi,\mathcal{B}_0}(Y,X)\}, \qquad (4.2.19)$$

where

$$h'_{\lambda,\phi,\mathcal{B}_0}(X,Y) = \sup_{A\in\mathcal{B}_0}\inf_{B\in\mathcal{B}_0}\max\left\{\frac{1}{\lambda}r(A,B), \phi(X,Y;A,B)\right\}, \qquad (4.2.20)$$

is called the *Hausdorff structure* of μ, or simply *h-structure.*

In (4.2.20), $r(A,B)$ is the Hausdorff semimetric in the Borel σ-algebra $\mathcal{B}((U,d))$ [see (4.2.1) with $\rho \equiv d$], λ is a positive number. $B_0 \subseteq \mathcal{B}(U)$, and ϕ satisfies the foregoing conditions (a) and (b).

Using conditions (a) and (b) we easily obtain the following lemma.

Lemma 4.2.2. *Each μ in the form* (4.2.19) *is a p. semidistance in $\mathfrak{X}$ with a parameter $\mathbb{K}_\mu = K_\phi$.*

In the limit cases $\lambda \to 0$, $\lambda \to \infty$, the Hausdorff structure turns into a "uniform" structure. More precisely, the following limit relations hold.

Lemma 4.2.3. *Let μ have Hausdorff structure* (4.2.19)*; then, as $\lambda \to 0$, $\mu(X,Y) = h_{\lambda,\phi,\mathcal{B}_0}(X,Y)$ has a limit defined to be*

$$h_{0,\phi,\mathcal{B}_0}(X,Y) = \max\left\{\sup_{A\in\mathcal{B}_0}\inf_{B\in\mathcal{B}_0}\phi(X,Y;A,B), \sup_{A\in\mathcal{B}_0}\inf_{B\in\mathcal{B}_0}\phi(Y,X;A,B)\right\}.$$

As $\lambda \to \infty$, the limit

$$\lim_{\lambda\to\infty}\lambda h_{\lambda,\phi,\mathcal{B}_0}(X,Y) = h_{\infty,\phi,\mathcal{B}_0}(X,Y) \qquad (4.2.21)$$

exists and is defined to be

$$\max\left\{\sup_{A\in\mathcal{B}_0}\inf_{B\in\mathcal{B}_0,\phi(X,Y;A,B)=0} r(A,B), \sup_{A\in\mathcal{B}_0}\inf_{B\in\mathcal{B}_0,\phi(Y,X;A,B)=0} r(A,B)\right\}.$$

Remark 4.2.5. Since $\lim_{\lambda\to\infty} h_{\lambda,\phi,\mathcal{B}_0}(X,Y) = 0$, we normalized the quantity $h_{\lambda,\phi,\mathcal{B}_0}(X,Y)$, multiplying it by λ, so that $\lambda \to \infty$ yields a nontrivial limit $h_{\infty,\phi,\mathcal{B}_0}(X,Y)$.

Proof. We will prove equality (4.2.21) only. That is, for each $X, Y \in \mathfrak{X}$

$$\lim_{\lambda\to\infty}\lambda h'_{\lambda,\phi,\mathcal{B}_0}(X,Y)$$

$$= \lim_{\lambda\to 0}\sup_{A\in\mathcal{B}_0}\inf_{B\in\mathcal{B}_0}\max\left\{r(A,b)\frac{1}{\lambda}\phi(X,Y;A,B)\right\}$$

$$= \lim_{\lambda\to 0}\inf\left\{\varepsilon > 0 : \inf_{B\in\mathcal{B}_0, r(A,B)<\varepsilon}\frac{1}{\lambda}\phi(X,Y;A,B) < \varepsilon \text{ for all } A \in \mathcal{B}_0\right\}$$

$$= \inf\left\{\varepsilon > 0 : \inf_{B\in\mathcal{B}_0, r(A,B)<\varepsilon} \phi(X,Y;A,B) = 0 \text{ for all } A \in \mathcal{B}_0\right\}$$

$$= \sup_{A\in\mathcal{B}_0} \inf_{B\in\mathcal{B}_0, \phi(X,Y;A,B)=0} r(A,B).$$

Now, by equality (4.2.19), we claim equality (4.2.21). □

Let us consider some examples of probability semidistances with a Hausdorff structure.

Example 4.2.1 (Universal Hausdorff representation). Each p. semidistance μ has the trivial form $h_{\lambda,\phi,\mathcal{B}_0} = \mu$, where the set $\mathcal{B}_0$ is a singleton, say, $\mathcal{B}_0 \equiv \{A_0\}$, and $\phi(X,Y;A_0,A_0) = \mu(X,Y)$.

Example 4.2.2 (Hausdorff structure of Prokhorov metric π_λ). The Prokhorov metric (3.3.22) admits a Hausdorff structure representation $h_{\lambda,\phi,\mathcal{B}_0} = \mu$ [see representations (4.2.18) and (4.2.19)], where $\mathcal{B}_0$ is either the class $\mathcal{C}$ of all nonempty closed subsets of U or $\mathcal{B}_0 \equiv \mathcal{B}(U)$ and $\phi(X,Y;A,B) = \Pr(X \in A) - \Pr(Y \in B)$, $A, B \in \mathcal{B}(U)$. As $\lambda \to 0$ and $\lambda \to \infty$ (Lemma 3.3.1), we obtain the limits

$$h_{0,\phi,\mathcal{B}_0} = \boldsymbol{\sigma} \quad \text{(distance in variation)}$$

and

$$h_{\infty,\phi,\mathcal{B}_0} = \ell_\infty.$$

Example 4.2.3 (Lévy metric $\mathbf{L}_\lambda$, $\lambda > 0$, in the space $\mathcal{P}(\mathbb{R}^n)$). Let $\mathcal{F}(\mathbb{R}^n)$ be the space of all right-continuous DFs F on $\mathbb{R}^n$. We extend the definition of the Lévy metric $(\mathbf{L}_\lambda, \lambda > 0)$ in $\mathcal{F}(\mathbb{R}^1)$ [see definition (4.2.3)] considering the multivariate case $\mathbf{L}_\lambda$ in $\mathcal{F}(\mathbb{R}^n)$:

$$\begin{aligned}\mathbf{L}_\lambda(P_1,P_2) &:= \mathbf{L}_\lambda(F_1,F_2) := \inf\{\varepsilon > 0 : F_1(x-\lambda\varepsilon\mathbf{e}) - \varepsilon \le F_2(x)\\ &\le F_1(x+\lambda\varepsilon\mathbf{e}) + \varepsilon \quad \forall x \in \mathbb{R}^n\},\end{aligned} \tag{4.2.22}$$

where F_i is the DF of P_i ($i = 1,2$) and $\mathbf{e} = (1,1,\ldots,1)$ is the unit vector in $\mathbb{R}^n$.

The Hausdorff representation of $\mathbf{L}_\lambda$ is handled by representation (4.2.19), where $\mathcal{B}_0$ is the set of all multivariate intervals $(-\infty, x]$ ($x \in \mathbb{R}^n$) and

$$\phi(X,Y;(-\infty,x],(-\infty,y]) := F_1(x) - F_2(y),$$

i.e., for RVs X and Y with DFs F_1 and F_2, respectively,

$$\begin{aligned}\mathbf{L}_\lambda(X,Y) = \mathbf{L}_\lambda(F_1,F_2) := \max\Bigg\{ &\sup_{x\in\mathbb{R}^n}\inf_{y\in\mathbb{R}^n} \max\left[\frac{1}{\lambda}\|x-y\|_\infty, F_1(x)-F_2(y)\right],\\ &\sup_{y\in\mathbb{R}^n}\inf_{x\in\mathbb{R}^n} \max\left[\frac{1}{\lambda}\|x-y\|_\infty, F_2(y)-F_1(x)\right]\Bigg\}\end{aligned} \tag{4.2.23}$$

for all $F_1, F_2 \in \mathcal{F}(\mathbb{R}^n)$, where $\|\cdot\|$ stands for the Minkowski norm in $\mathbb{R}^n$, $\|(x_1,\ldots,x_n)\|_\infty := \max_{1\le i\le n}|x_i|$. Letting $\lambda \to 0$ in Definition (4.2.23) we get the *Kolmogorov distance* in $\mathcal{F}(\mathbb{R}^n)$:

$$\lim_{\lambda\to 0}\mathbf{L}_\lambda(F_1,F_2) = \boldsymbol{\rho}(F_1,F_2) := \sup_{x\in\mathbb{R}^n}|F_1(x)-F_2(x)|. \tag{4.2.24}$$

The limit of $\lambda\mathbf{L}_\lambda$ as $\lambda\to\infty$ is given by (4.2.21), that is,

$$\begin{aligned}\lim_{\lambda\to\infty}\lambda_\lambda(F_1,F_2) &= \inf\{\varepsilon>0 : \inf[F_1(x)-F_2(y) : y\in\mathbb{R}^n, \|x-y\|_\infty\le\varepsilon] = 0,\\ &\quad \inf[F_2(x)-F_1(y) : x\in\mathbb{R}^n, \|x-y\|_\infty\le\varepsilon] = 0 \quad \forall x\in\mathbb{R}^n\}\\ &= \mathbf{W}(F_1,F_2) := \inf\{\varepsilon>0 : F_1(x)\le F_2(x+\varepsilon\mathbf{e}), F_2(x)\le F_1(x+\varepsilon\mathbf{e})\\ &\qquad \forall x\in\mathbb{R}^n\}.\end{aligned} \tag{4.2.25}$$

Problem 4.2.2. If $n = 1$, then

$$\lim_{\lambda\to\infty}\lambda\mathbf{L}_\lambda(P_1,P_2) = \ell_\infty(P_1,P_2),\ P_1,P_2\in\mathcal{P}(\mathbb{R}^n), \tag{4.2.26}$$

where $\ell_\infty(P_1,P_2) := \inf\{\varepsilon>0 : P_1(A)\le P_2(A^\varepsilon)$ for all Borel subsets of $\mathbb{R}^n\}$.[2] Let us see if it is true that equality (4.2.26) is valid for any integer n.

Example 4.2.4 (Lévy p. distance $\mathbf{L}_{\lambda,H}$, $\lambda>0$, $H\in\mathcal{H}$). The Lévy metric $\mathbf{L}_\lambda$ (4.2.22) can be rewritten in the form

$$\begin{aligned}\mathbf{L}_\lambda(F_1,F_2) := \inf\{\varepsilon>0 : &(F_1(x)-F_2(x-\lambda\varepsilon\mathbf{e}))_+ < \varepsilon,\\ &(F_2(x)-F_1(x-\lambda\varepsilon\mathbf{e}))_+ < \varepsilon \quad \forall x\in\mathbb{R}^n\}, \quad (\cdot)_+ := \max(\cdot,0),\end{aligned}$$

which can be viewed as a special case $[H(t)=t]$ of the *Lev́y p. distance* $\mathbf{L}_{\lambda,H}$ ($\lambda>0$, $H\in\mathcal{H}$) defined as

$$\begin{aligned}\mathbf{L}_{\lambda,H}(F_1,F_2) := \inf\{\varepsilon>0 : &\widetilde{H}(F_1(x)-F_2(x+\lambda\varepsilon\mathbf{e})) < \varepsilon,\\ &\widetilde{H}(F_2(x)-F_1(x+\lambda\varepsilon\mathbf{e})) < \varepsilon, \quad \forall x\in\mathbb{R}^n\},\end{aligned} \tag{4.2.27}$$

where

$$\widetilde{H}(t) := \begin{cases} H(t), & t\ge 0,\\ 0, & t\le 0.\end{cases}$$

---

[2] See (4.2.5) and subsequently Corollary 7.4.2 and (7.5.15) in Chap. 7.

$\mathbf{sL}_{\lambda,H}$ admits a Hausdorff representation of the following type:

$$\mathbf{L}_{\lambda,H}(F_1,F_2) = \max\left\{\sup_{x\in\mathbb{R}^n}\inf_{y\in\mathbb{R}^n}\max\left[\frac{1}{\lambda}\|x-y\|,\widetilde{H}(F_1(x)-F_2(y))\right],\right.$$

$$\left.\sup_{y\in\mathbb{R}^n}\inf_{x\in\mathbb{R}^n}\max\left[\frac{1}{\lambda}\|x-y\|,\widetilde{H}(F_2(y)-F_1(x))\right]\right\}. \quad (4.2.28)$$

The last representation of $\mathbf{L}_{\lambda,H}$ shows that $\mathbf{L}_{\lambda,H}$ is a simple distance with parameter $\mathbb{K}_{\mathbf{L}_{\lambda,H}} := K_H$ [see (2.4.3)]. Also, from (4.2.28) as $\lambda \to 0$ we get the *Kolmogorov p. distance*

$$\lim_{\lambda\to 0}\mathbf{L}_{\lambda,H}(F_1,F_2) = H(\rho(F_1,F_2)) = \boldsymbol{\rho}_H(F_1,F_2) := \sup_{x\in\mathbb{R}^n} H(|F_1(x)-F_2(x)|). \quad (4.2.29)$$

Analogously, letting $\lambda \to \infty$ in (4.2.28), we have

$$\lim_{\lambda\to\infty}\lambda\mathbf{L}_{\lambda,H}(F_1,F_2) = \mathbf{W}(F_1,F_2). \quad (4.2.30)$$

We prove equality (4.2.30) by arguments provided in the limit relation (4.2.25).

Example 4.2.5 (Hausdorff metric on $\mathcal{F}(\mathbb{R})$ and $\mathcal{P}(U)$). The Lévy metric in $\mathcal{F} := \mathcal{F}(\mathbb{R})$ (4.2.22) has a Hausdorff structure [see (4.2.23)]; however, the function

$$\widetilde{D}((x,F_1(x)),(y,F_2(y))) := \max\left\{\frac{1}{\lambda}|x-y|, F_1(x)-F_2(y)\right\}$$

is not a metric in the space $\mathbb{R}\times[0,1]$, and hence (4.2.23) is not a "pure" Hausdorff metric [see (4.2.2)]. In the next definition we will replace the semimetric $\widetilde{D}$ with the Minkowski metric dm_λ in $\mathbb{R}\times[0,1]$:

$$dm_\lambda((x,F_1(x)),(y,F_2(y))) := \max\left\{\frac{1}{\lambda}|x-y|, |F_1(x)-F_2(y)|\right\}. \quad (4.2.31)$$

By means of equality (4.2.31), we define the Hausdorff metric in $\mathcal{F}(\mathbb{R}^n)$ as follows.

Definition 4.2.4. The metric

$$\mathbf{H}_\lambda(F,G) := \max\left\{\sup_{x\in\mathbb{R}^n}\inf_{y\in\mathbb{R}^n} dm_\lambda((x,F(x)),(y,G(y))),\right.$$

$$\left.\sup_{y\in\mathbb{R}^n}\inf_{x\in\mathbb{R}^n} dm_\lambda((x,F(x)),(y,G(y)))\right\}, \quad F,G\in\mathcal{F}^n, \quad (4.2.32)$$

is said to be a *Hausdorff metric with parameter* λ (or simply H_λ-metric) in DF space $\mathcal{F}$.

Lemma 4.2.4. *(a) For any $\lambda > 0$, H_λ is a metric in $\mathcal{F}$.*
(b) H_λ is a nonincreasing function of λ, and the following relation hold:

$$\lim_{\lambda\to 0} \mathbf{H}_\lambda(F,G) = \boldsymbol{\rho}(F,G) \tag{4.2.33}$$

and

$$\lim_{\lambda\to\infty} \lambda\mathbf{H}_\lambda(F,G) = \widetilde{\mathbf{W}}(F,G)$$

$$:= \inf\{\varepsilon > 0 : (F_1(x) - F_2(x+\varepsilon))_+ = 0,$$
$$(F_2(x-\varepsilon) - F_1(x))_+ = 0 \quad \forall x \in \mathbb{R}\}. \tag{4.2.34}$$

(c) If F and G are continuous DFs, then $\mathbf{H}_\lambda(F,G) = \mathbf{L}_\lambda(F,G)$.

Proof. (a) By means of the Minkowski metric

$$dm_\lambda((x_1,y_1),(x_2,y_2)) := \max\left\{\frac{1}{\lambda}|x_1 - x_2|, |y_1 - y_2|\right\}$$

in the space $D := \mathbb{R} \times [0,1]$, define the Hausdorff semimetric in the space 2^D of all subsets $B \subseteq D$:

$$h_\lambda(B_1,B_2) := \max\left\{\sup_{b_1\in B_1}\inf_{b_2\in B_2} dm_\lambda(b_1,b_2),\ \sup_{b_2\in B_2}\inf_{b_1\in B_1} dm_\lambda(b_1,b_2)\right\}.$$

In the Hausdorff representation (4.2.11) of the Lévy metric, the main role was played by the notion of the completed graph $\overline{F}$ of a DF F. Here, we need the notion of the closed graph Γ_F of a DF F defined as follows:

$$\Gamma_F := \left(\bigcup_{x\in\mathbb{R}}(x,F(x))\right) \cup \left(\bigcup_{x\in\mathbb{R}}(x,F(x-0))\right), \tag{4.2.35}$$

i.e., the closed graph Γ_F is handled by adding the points $(x, F(x-))$ to the graph of F, where x denotes points of F-discontinuity (Figs. 4.1 and 4.2).

Obviously, $H_\lambda(F,G) = h_\lambda(\Gamma_F,\Gamma_G)$. Moreover, if the closed graphs of F and G coincide, then $F(x) = G(x)$ for all continuity points x of F and G. Since F and G are right-continuous, then $\Gamma_F \equiv \Gamma_G \iff F \equiv G$.

(b) The limit relation (4.2.33) is a consequence of (4.2.24) and

$$\mathbf{L}_\lambda(F_1,F_2) \le \mathbf{H}_\lambda(F_1,F_2) \le \boldsymbol{\rho}(F_1,F_2), \qquad F_1,F_2 \in \mathcal{F}. \tag{4.2.36}$$

Analogously to (4.2.25), we claim that

$$\lim_{\lambda\to 0} \lambda\mathbf{H}_\lambda(F,G) = \inf\{\varepsilon > 0 : \inf\{|F_1(x) - F_2(y)| : y \in \mathbb{R}, |x-y| \le \varepsilon\} = 0$$
$$\inf\{|F_2(x) - F_1(x)| : y \in \mathbb{R}, |x-y| \le \varepsilon\} = 0 \quad \forall x \in \mathbb{R}\}$$
$$= \widetilde{\mathbf{W}}(F,G).$$

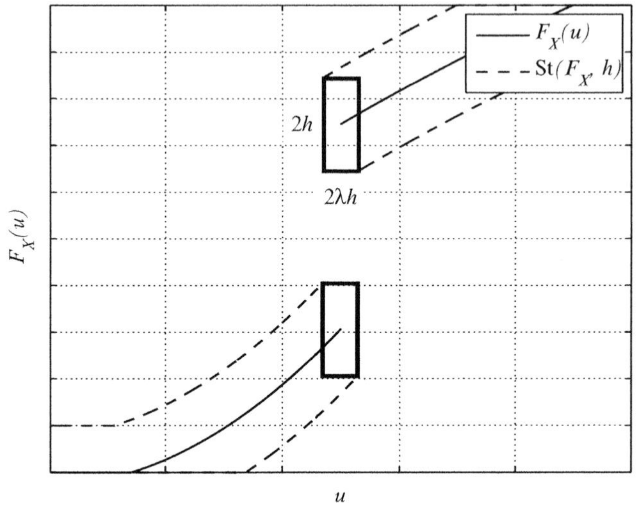

Fig. 4.2 St(F, h) is the strip into which the graph of the DF G has to be located so that $H_\lambda(F, G) \le h$ for $F, G \in \mathcal{F}'$

(c) See Figs. 4.1 and 4.2. □

Remark 4.2.6. Further, we need the following notations. For two metrics ρ_1 and ρ_2 on a set S, $\rho_1 \overset{\text{top}}{\le} \rho_2$ means that ρ_2-convergence implies ρ_1-convergence, and $\rho_1 \overset{\text{top}}{<} \rho_2$ means $\rho_1 \overset{\text{top}}{\le} \rho_2$ but not $\rho_2 \overset{\text{top}}{\le} \rho_1$. Finally, $\rho_1 \overset{\text{top}}{\sim} \rho_2$ means that $\rho_1 \overset{\text{top}}{\le} \rho_2$ and $\rho_2 \overset{\text{top}}{\le} \rho_1$. By (4.2.36) it follows that

$$\mathbf{L}_\lambda \overset{\text{top}}{\le} \mathbf{H}_\lambda \overset{\text{top}}{\le} \boldsymbol{\rho}. \tag{4.2.37}$$

Moreover, the following simple examples show that

$$\mathbf{L}_\lambda \overset{\text{top}}{<} \mathbf{H}_\lambda \overset{\text{top}}{<} \boldsymbol{\rho}.$$

Example 4.2.6. Let

$$F_n(x) = \begin{cases} 0, \ x < \dfrac{1}{n}, \\ 1, \ x \le \dfrac{1}{n}, \end{cases} \qquad F_0(x) = \begin{cases} 0, \ x < 0, \\ 1, \ x \ge 1. \end{cases}$$

Then $\boldsymbol{\rho}(F_n, F) = 1$, $\mathbf{H}_\lambda(F_n, F) = 1/\lambda n \to 0$ as $n \to \infty$.

Example 4.2.7. Let

$$\phi_n(x) = \begin{cases} 0, \ x < 0, \\ \frac{1}{2}, \ 0 \le x < \frac{1}{n}, \\ 1, \ x \ge \frac{1}{n}. \end{cases}$$

Then

$$\mathbf{L}_\lambda(\phi_n, F_0) = \min\left(1, \frac{1}{\lambda}\right) n^{-1} \to 0 \ \text{as} \ n \to \infty,$$

but

$$H_\lambda(\phi_n, F_0) \ge \inf_{y\in\mathbb{R}} \max\left\{\frac{1}{\lambda}\left|\frac{1}{2n} - y\right|, \left|\phi_n\left(\frac{1}{2n}\right) - F_0(y)\right|\right\} \ge \frac{1}{2}$$

for any $n = 1, 2, \ldots$.

Remark 4.2.7. For any $0 < \lambda < \infty$, $\mathbf{H}_\lambda$ metrizes one and the same topology. We characterize the $\mathbf{H}$-topology ($\mathbf{H} := \mathbf{H}_1$) by the following compactness criterion. Recall that a subset $\mathcal{A}$ of a metric space (S, ρ) is said to be *ρ-relatively compact* if any sequence in $\mathcal{A}$ has a ρ-convergent subsequence. Define the Skorokhod–Billingsley metric in the space $\mathcal{F}$ of distribution functions on $\mathbb{R}$

$$\mathbf{SB}(F, G) = \inf_{\lambda\in\Lambda} \max\left\{\sup_{s\neq t}\left|\log\frac{\lambda(s) - \lambda(t)}{s - t}\right|,\right.$$

$$\left.\sup_{t\in\mathbb{R}}|\lambda(t) - t|, \sup_{t\in\mathbb{R}}|F(t) - G(\lambda(t))|\right\},$$

where Λ is the class of all strictly increasing continuous functions λ from $\mathbb{R}$ onto $\mathbb{R}$. The metrics $\mathbf{H}$ and $\mathbf{SB}$ generate the same exact topology in $\mathcal{F}$; the metric space $(\mathcal{F}, \mathbf{H})$ is not complete, whereas $(\mathcal{F}, \mathbf{SB})$ is complete. To show that $\mathbf{H}$ is not a complete metric, observe that ϕ_n, introduced in Example 4.2.7, is $\mathbf{H}$-fundamental but not $\mathbf{H}$-convergent. The proof that $(\mathcal{F}, \mathbf{SB})$ is complete is the same as the proof that $D[0, 1]$ is complete with the Skorokhod–Billingsley metric d_0.[3] The equivalence of $\mathbf{H}$ and $\mathbf{SB}$ topologies is a consequence of the compactness criterion given below. Consider the following moduli of $\mathbf{H}$-continuity:

1.
$$\omega'_F(\delta) := \inf_{\{t_0,\ldots,t_r\}} \max_{0\le i\le r} [F(t_i-) - F(t_{i-1})], \quad F \in \mathcal{F}, \ \delta \in (0, 1),$$

where the infimum is taken over all $\{t_0, t_1, \ldots, t_r\}$ satisfying the conditions: $-\infty = t_0 < t_1 < \cdots < t_r = \infty$, $t_i - t_{i-1} > \delta$, $i = 1, \ldots, r$.

---

[3]See Billingsley (1999, Theorem 14.2).

2.

$$\omega''_F := \sup_{x\in\mathbb{R}} \min\{F(x+\delta/2) - F(x), F(x) - F(x-\delta/2)\}, \qquad F \in \mathcal{F},\ \delta \in (0,1).$$

For any $f \in \mathcal{F}$, $\lim_{\delta\to\infty} \omega'_F(\delta) = 0$ and $\omega''_F(\delta) \le \omega'_F(2\delta)$.[4] Let $\mathcal{A} \subset \mathcal{F}$. Then the following are equivalent[5]:

(a) $\mathcal{A}$ is **H**-relatively compact.
(b) $\mathcal{A}$ is **SB**-relatively compact.
(c) $\lim_{\delta\to\infty} \sup_{F\in\mathcal{A}} \omega'_F(\delta) = 0$.
(d) $\mathcal{A}$ is weakly compact (i.e., **L**-relatively compact) and $\lim_{\delta\to\infty} \sup_{F\in\mathcal{A}} \omega''_F(\delta) = 0$.

Moreover, for $F, G \in \mathcal{F}$, and $\delta > 0$ the following relations hold:

$$\mathbf{H}(F,G) \le \mathbf{SB}(F,G),$$
$$\omega'_G(\delta) \le \omega'_F(\delta + 2\mathbf{H}(F,G)) + 4\mathbf{H}(F,G),$$
$$\mathbf{H}(F,G) \le \max\{\omega''_F(4\mathbf{L}(F,G)), \omega''_G(4\mathbf{L}(G,G))\}\mathbf{L}(F,G).$$

Next, let (U,d) be a metric space and define the following analog of **H**-metrics:

$$\boldsymbol{\pi}\mathbf{H}_\lambda(P_1,P_2) := \max\left\{ \sup_{A\in\mathcal{B}_1} \inf_{B\in\mathcal{B}_1} \max\left[\frac{1}{\lambda} r(A,B), |P_1(A) - P_2(B)|\right] \right.$$
$$\left. \sup_{B\in\mathcal{B}_1} \inf_{A\in\mathcal{B}_1} \max\left[\frac{1}{\lambda} r(A,B), |P_1(A) - P_2(B)|\right]\right\} \tag{4.2.38}$$

for any laws $P_1, P_2 \in \mathcal{P}(U)$.

Lemma 4.2.5. *The following statements hold:*

(a) For any $\lambda > 0$ the functional $\boldsymbol{\pi}\mathbf{H}_\lambda$ on $\mathcal{P}_1 \times \mathcal{P}_1$ is a metric in $\mathcal{P}_1 = \mathcal{P}(U)$.
(b) $\boldsymbol{\pi}\mathbf{H}_\lambda$ is a nonincreasing function of λ, and the following relation holds:

$$\lim_{\lambda\to 0} \boldsymbol{\pi}\mathbf{H}_\lambda(P_1,P_2) = \boldsymbol{\sigma}(P_1,P_2) := \sup_{A\in\mathcal{B}_1} |P_1(A) - P_2(A)|, \qquad P_1,P_2 \in \mathcal{P}_1, \tag{4.2.39}$$

$$\lim_{\lambda\to\infty} \lambda\boldsymbol{\pi}\mathbf{H}_\lambda(P_1,P_2) = \boldsymbol{\pi}\mathbf{H}_\infty(P_1,P_2)$$
$$:= \inf\{\varepsilon > 0 : \inf[|P_1(A) - P_2(B)| : B \in \mathcal{B}_1, r(A,B) < \varepsilon] = 0,$$
$$\inf[|P_2(A) - P_1(B)| : B \in \mathcal{B}_1, r(A,B) < \varepsilon] = 0\ \forall A \in \mathcal{B}_1\}. \tag{4.2.40}$$

---

[4] See Billingsley (1999, Sect. 12).

[5] See Rachev (1984) and Kakosyan et al. (1988, Sect. 2.5).

(c) $\boldsymbol{\pi}\mathbf{H}_\lambda$ *is "between" the Prokhorov metric* $\boldsymbol{\pi}_\lambda$ (4.2.18) *and the total variation metric* $\boldsymbol{\sigma}$, *i.e.,*

$$\boldsymbol{\pi}_\lambda \le \boldsymbol{\pi}\mathbf{H}_\lambda \le \boldsymbol{\sigma} \tag{4.2.41}$$

and

$$\boldsymbol{\pi}_\lambda \overset{\text{top}}{<} \boldsymbol{\pi}\mathbf{H}_\lambda \overset{\text{top}}{<} \boldsymbol{\sigma}. \tag{4.2.42}$$

Proof. Let us prove only (4.2.40). We have

$$\begin{aligned}\boldsymbol{\pi}\mathbf{H}_\lambda(P_1, P_2) = \inf\{\varepsilon > 0 : &\inf[|P_1(A) - P_2(B)| : B \in \mathcal{B}_1, r(A, B) < \lambda\varepsilon] < \varepsilon,\\ &\inf[|P_2(A) - P_1(B)| : B \in \mathcal{B}_1, r(A, B) < \lambda\varepsilon] < \varepsilon \ \forall A \in \mathcal{B}_1\}.\end{aligned} \tag{4.2.43}$$

Further multiplying the two sides of (4.2.43) by λ, and letting $\lambda \to \infty$, we get (4.2.40). □

4.3 Λ-Structure of Probability Semidistances

The p. semidistance structure Λ in $\mathcal{X} = \mathcal{X}(U)$ is defined by means of a nonnegative function ν on $\mathcal{X} \times \mathcal{X} \times [0, \infty)$ that satisfies the following relationships for all $X, Y, Z \in \mathfrak{X}$:

(a) If $\Pr(X = Y) = 1$, then $\nu(X, Y; t) = 0 \forall t \ge 0$.
(b) $\nu(X, Y; t) = \nu(Y, X; t)$.
(c) If $t' < t''$, then $\nu(X, Y; t') \ge \nu(X, Y; t'')$.
(d) For some $K_\nu > 1$, $\nu(X, Z; t' + t'') \le K_\nu[\nu(X, Y; t') + \nu(Y, Z, t'')]$.

If $\nu(X, Y; t)$ is completely determined by the marginals $P_1 = \Pr_X$, $P_2 = \Pr_Y$, then we will use the notation $\nu(P_1, P_2; t)$ instead of $\nu(X, Y; t)$. For the case $K_\nu = 1$, the following definition is due to Zolotarev (1976).

Definition 4.3.1. The p. semidistance μ has a Λ-structure if it admits a Λ-representation, i.e.,

$$\mu(X, Y) = \Lambda_{\lambda,\nu}(X, Y) := \inf\{\varepsilon > 0 : \nu(X, Y; \lambda\varepsilon) < \varepsilon\} \tag{4.3.1}$$

for some $\lambda > 0$ and ν satisfying (a)–(d).

Obviously, if μ has a Λ-representation (4.3.1), then μ is a p. semidistance with $\mathbb{K}_\mu = K_\nu$. In Example 4.2.1 it was shown that each p. semidistance has a Hausdorff representation $h_{\lambda,\phi,\mathcal{B}_0}$. In the next theorem we will prove that each p. semidistance μ with a Hausdorff structure (Definition 4.2.3) also has a Λ-representation. Hence, in particular, each p. semidistance has a Λ-structure as well as a Hausdorff structure.

Theorem 4.3.1. *Suppose a p. semidistance μ admits the Hausdorff representation $\mu = h_{\lambda,\phi,\mathcal{B}_0}$ [see (4.2.19)]. Then μ enjoys also a Λ-representation*

$$h_{\lambda,\phi,\mathcal{B}_0}(X,Y) = \Lambda_{\lambda,\nu}(X,Y), \tag{4.3.2}$$

where

$$\nu(X,Y;t) := \max\left\{\sup_{A\in\mathcal{B}_0}\inf_{B\in A(t)}\phi(X,Y;A,B),\ \sup_{A\in\mathcal{B}_0}\inf_{B\in A(t)}\phi(Y,X;A,B)\right\},$$

and $A(t)$ is the collection of all elements B of $\mathcal{B}_0$ such that the Hausdorff semimetric $r(A,B)$ is not greater than t.

Proof. Let $\Lambda_{\lambda,\nu}(X,Y) < \varepsilon$. Then for each $A \in \mathcal{B}_0$ there exists a set $B \in A(\lambda\varepsilon)$ such that $\phi(X,Y;A,B) < \varepsilon$, i.e.,

$$\sup_{A\in\mathcal{B}_0}\inf_{B\in\mathcal{B}_0}\max\left\{\frac{1}{\lambda}r(A,B),\phi(X,Y;A,B)\right\} < \varepsilon.$$

By symmetry, it follows that $h_{\lambda,\phi,\mathcal{B}_0}(X,Y) < \varepsilon$. If, conversely, $h_{\lambda,\phi,\mathcal{B}_0}(X,Y) < \varepsilon$, then for each $A \in \mathcal{B}_0$ there exists $B \in \mathcal{B}_0$ such that $r(A,B) < \lambda\varepsilon$ and $\phi(X,Y;A,B) < \varepsilon$. Thus

$$\sup_{A\in\mathcal{B}_0}\inf_{B\in A(\lambda\varepsilon)}\phi(X,Y;A,B) < \varepsilon.$$ □

Example 4.3.1 (Λ-structure of the Lévy metric and the Lévy distance). Recall the definition of the Lévy metric in $\mathcal{P}(\mathbb{R}^n)$ [see (4.2.22)]:

$$\mathbf{L}_\lambda(P_1,P_2) := \inf\left\{\varepsilon > 0 : \sup_{x\in\mathbb{R}^n}(F_1(x) - F_2(x+\lambda\varepsilon\mathbf{e})) \le \varepsilon \right.$$
$$\left. \text{and } \sup_{x\in\mathbb{R}^n}(F_2(x) - F_1(x+\lambda\varepsilon\mathbf{e})) \le \varepsilon\right\},$$

where obviously F_i is the DF of P_i. By Definition 4.3.1, $\mathbf{L}_\lambda$ has a Λ-representation

$$\mathbf{L}_\lambda(P_1,P_2) = \Lambda_{\lambda,\nu}(P_1,P_2), \quad \lambda > 0,$$

where

$$\nu(P_1,P_2;t) := \sup_{x\in\mathbb{R}^n}\max\{(F_1(x) - F_2(x+\lambda t\mathbf{e})), (F_2(x) - F_1(x+\lambda t\mathbf{e}))\}$$

and F_i is the DF of P_i. With an appeal to Theorem 4.3.1, for any $F_1, F_2 \in \mathcal{F}(\mathbb{R}^n)$, we have that the metric h defined below admits a Λ-representation:

$$h(F_1, F_2) := \max\left\{\sup_{x\in\mathbb{R}^n}\inf_{y\in\mathbb{R}^n}\max\left\{\frac{1}{\lambda}\|x-y\|_\infty, F_1(x)-F_2(y)\right\},\right.$$

$$\left.\sup_{x\in\mathbb{R}^n}\inf_{y\in\mathbb{R}^n}\max\left\{\frac{1}{\lambda}\|x-y\|_\infty, F_2(x)-F_1(y)\right\}\right\}$$

$$= \Lambda_{\lambda,\nu}(P_1, P_2),$$

where

$$\nu(P_1, P_2; t) = \max\left\{\sup_{x\in\mathbb{R}^n}\inf_{y:\|x-y\|_\infty\le t}(F_1(x)-F_2(y)),\right.$$

$$\left.\sup_{x\in\mathbb{R}^n}\inf_{y:\|x-y\|_\infty\le t}(F_2(x)-F_1(y))\right\}.$$

By virtue of the Λ-representation of the $\mathbf{L}_\lambda$, we conclude that $h(F_1, F_2) = \mathbf{L}_\lambda(F_1, F_2)$, which proves (4.2.23) and Theorem 4.2.2.

Analogously, consider the Lévy distance $\mathbf{L}_{\lambda,H}$ (4.2.27) and apply Theorem 4.3.1 with

$$\begin{aligned}\nu(X, Y; \lambda t) &= \nu(P_1, P_2; \lambda t)\\ &:= H\left(\sup_{x\in\mathbb{R}^n}\max\{F_1(x)-F_2(x+\lambda t\mathbf{e}), \{F_2(x)-F_1(x+\lambda t\mathbf{e})\right)\end{aligned}$$

to prove the Hausdorff representation of $\mathbf{L}_{\lambda,H}$ (4.2.28).

Example 4.3.2 (Λ-structure of the Prokhorov metric $\boldsymbol{\pi}_\lambda$). [6] Let

$$\begin{aligned}\nu(P_1, P_2; \varepsilon) :=& \sup_{A\in\mathcal{B}(U)}\max\{P_1(A)-P_2(A^\varepsilon), P_2(A)-P_1(A^\varepsilon)\}\\ =& \sup_{A\in\mathcal{B}(U)}\{P_1(A)-P_2(A^\varepsilon)\}.\end{aligned}$$

Then $\Lambda_{\lambda,\nu}$ is the Λ-representation of the Prokhorov metric $\boldsymbol{\pi}_\lambda(P_1, P_2)$ [see (3.3.22)]. In this way, Theorem 4.2.3 and equality (4.2.18) are corollaries of Theorem 4.3.1.

For each $\lambda > 0$ the Prokhorov metric $\boldsymbol{\pi}_\lambda$ induces a weak convergence in $\mathcal{P}_1$; thus,

$$\boldsymbol{\pi}_\lambda(P_n, P) \to 0 \quad \Longleftrightarrow \quad P_n \xrightarrow{\text{w}} P.$$

Remark 4.3.1. As is well known, the weak convergence $P_n \xrightarrow{\text{w}} P$ means that

$$\int_U f\,\mathrm{d}P_n \to \int_U f\,\mathrm{d}P \tag{4.3.3}$$

[6] See Dudley (1976, Theorem 8.1).

for each continuous and bounded function f on (U,d). The Prokhorov metric π (3.3.20) metrizes the weak convergence in $\mathcal{P}(U)$, where U is an s.m.s.[7] The next definition was essentially used by Dudley (1966), Ranga (1962), and Bhattacharya and Ranga Rao (1976).

Definition 4.3.2. Let G be a nonnegative continuous function on U and $\mathcal{P}_G$ be the set of laws P such that $\int_U G\mathrm{d}P < \infty$. The joint convergence

$$P_n \xrightarrow{\mathrm{w}} P \int_U -G\mathrm{d}P_n \to \int_U G\mathrm{d}P \quad (P_n, P \in \mathcal{P}_G) \tag{4.3.4}$$

will be called a *G-weak convergence in* $\mathcal{P}_G$.

As in Prokhorov (1956), one can show that the *G-weighted Prokhorov metric*

$$\pi_{\lambda,G}(P_1,P_2) := \inf\{\varepsilon > 0 : \lambda_1(A) \le \lambda_2(A^{\lambda\varepsilon}) + \varepsilon, \lambda_2(A) \le \lambda_1(A^{\lambda\varepsilon}) + \varepsilon \forall A \in \mathcal{B}(U)\}, \tag{4.3.5}$$

where $\lambda_i(A) := \int_A (1 + G(x))P_i(\mathrm{d}x)$, metrizes the G-weak convergence in $\mathcal{P}_G$, where U is an s.m.s. (see Theorem 11.2.2 subsequently for details).

The metric $\pi_{\lambda,G}$ admits a Λ-representation with

$$\nu(P_1,P_2;\varepsilon) := \sup_{A\in\mathcal{B}(U)} \max\{\lambda_1(A) - \lambda_2(A^{\varepsilon}), \lambda_2(A) - \lambda_1(A^{\varepsilon})\}.$$

Example 4.3.3 (Λ-structure of the Ky Fan metric and Ky Fan distance). The Λ-structure of the Ky Fan metric $\mathbf{K}_\lambda$ [see (3.4.10)] and the Ky Fan distance $\mathbf{KF}_H$ [see (3.4.9)] is handled by assuming that in (4.3.1), $\nu(X,Y;\lambda t) := \Pr(d(X,Y) > \lambda t)$ and $\nu(X,Y;t) := \Pr(H(d(X,Y)) > t)$, respectively.

4.4 ζ-Structure of Probability Semidistances

In Example 3.3.6 we considered the notion of a minimal norm $\overset{\circ}{\mu}_c$

$$\overset{\circ}{\mu}_c(P_1,P_2) := \inf\left\{\int_{U^2} c\mathrm{d}m : m \in \mathcal{M}_2, T_1m - T_2m = P_1 - P_2\right\}, \tag{4.4.1}$$

where $U = (U,d)$ is an s.m.s. and c is a nonnegative, continuous symmetric function on U^2.

Let $\mathcal{F}_{c,1}$ be the space of all bounded $(c,1)$-Lipschitz functions $f : U \to \mathbb{R}$, i.e.,

$$\|f\|_{cL} := \sup_{c(x,y)\neq 0} \frac{|f(x) - f(y)|}{c(x,y)} \le 1. \tag{4.4.2}$$

Remark 4.4.1. If c is a metric in U, then $\mathcal{F}_{c,1}$ is the space of all functions with Lipschitz constant ≤ 1, w.r.t. c. Note that, if c is not a metric, then the set $\mathcal{F}_{c,1}$

[7] See Prokhorov (1956) and Dudley (2002, Theorem 11.3.3).

might be a very "poor" one. For instance, if $U = \mathbb{R}$, $c(x,y) = |x-y|^p$ $(p > 1)$, then $\mathcal{F}_{c,1}$ contains only constant functions.

By (4.4.2), we have that for each nonnegative measure m on U^2 whose marginals $T_i m$, $i = 1, 2$, satisfy $T_1 m - T_2 m = P_1 - P_2$, and for each $f \in \mathcal{F}_{c,1}$ the following inequalities hold:

$$\left|\int_U f(x)(P_1 - P_2)(\mathrm{d}x)\right| = \left|\int_{U^2} (f(x) - f(y))m(\mathrm{d}x, \mathrm{d}y)\right|$$
$$\leq \|f\|_{cL} \int_{U^2} c(x,y)m(\mathrm{d}x, \mathrm{d}y)$$
$$\leq \int_{U^2} c(x,y)m(\mathrm{d}x, \mathrm{d}y).$$

The minimal norm $\mathring{\mu}_c$ then has the following estimate from below:

$$\zeta(P_1, P_2; \mathcal{F}_c) \leq \mathring{\mu}_c(P_1, P_2), \tag{4.4.3}$$

where

$$\zeta(P_1, P_2; \mathcal{F}_{c,1}) := \sup\left\{\left|\int_{U^2} f \mathrm{d}(P_1 - P_2)\right| : f \in \mathcal{F}_{c,1}\right\}. \tag{4.4.4}$$

Further, in Sect. 5.4 in Chap. 5 and Sect. 6.2 in Chap. 6, we will prove that for some c (as, for example, $c = d$) we have equality (4.4.3).

Let $C^b(U)$ be the set of all bounded continuous functions on U. Then for each subset $\mathfrak{F}$ of $C^b(U)$ the functional

$$\zeta_{\mathfrak{F}}(P_1, P_2) := \zeta(P_1, P_2; \mathfrak{F}) := \sup_{f \in \mathfrak{F}} \left|\int_U f \mathrm{d}(P_1 - P_2)\right| \tag{4.4.5}$$

on $\mathcal{P}_1 \times \mathcal{P}_1$ defines a simple p. semimetric in $\mathcal{P}_1$. The metric $\zeta_{\mathfrak{F}}$ was introduced by Zolotarev (1976) and is called the *Zolotarev $\zeta_{\mathfrak{F}}$-metric* (or simply *$\zeta_{\mathfrak{F}}$-metric*).

Definition 4.4.1. A simple semimetric μ having the *$\zeta_{\mathfrak{F}}$-representation*

$$\mu(P_1, P_2) = \zeta_{\mathfrak{F}}(P_1, P_2) \tag{4.4.6}$$

for some $\mathcal{F} \subseteq C^b(U)$ is called semimetric with a *ζ-structure*.

Remark 4.4.2. In the space $\mathfrak{X} = \mathfrak{X}(U)$ of all U-valued RVs, the $\zeta_{\mathfrak{F}}$-metric ($\mathfrak{F} \subseteq C^b(U)$) is defined by

$$\zeta_{\mathfrak{F}}(X, Y) := \zeta_{\mathfrak{F}}(Pr_X, Pr_Y) := \sup_{f \in \mathfrak{F}} |Ef(X) - Ef(Y)|. \tag{4.4.7}$$

Simple metrics with a ζ-structure are well known in probability theory. Let us consider some examples of such metrics.

Example 4.4.1 (Engineer metric). Let $U = \mathbb{R}$ and $\mathfrak{X}^{(1)}$ be the set of all real-valued RVs X with finite first absolute moment, i.e., $E|X| < \infty$. In the set $\mathfrak{X}^{(1)}$, the engineer metric $\mathbf{EN}(X,Y) := |EX - EY|$ admits a ζ-representation, where $\mathcal{F}$ is a collection of functions

$$f_N(x) = \begin{cases} -N, & x < N, \\ x, & |x| \le N, \\ N, & x > N, N = 1,2,\ldots. \end{cases} \tag{4.4.8}$$

Example 4.4.2 (Kolmogorov metric and $\mathcal{L}_p$-metric in distribution function space). Let $\mathcal{F} = \mathcal{F}(R)$ be the space of all DFs on $\mathbb{R}$. The Kolmogorov metric $\boldsymbol{\rho}(F_1, F_2) := \sup_{x\in\mathbb{R}} |F_1(x) - F_2(x)|$ in $\mathcal{F}$ has $\zeta_{\mathfrak{F}}$-structure. In fact

$$\boldsymbol{\rho}(F_1, F_2) = \|f_1 - f_2\|_\infty = \sup\left\{\left|\int_{-\infty}^{\infty} u(x)(F_1(x) - F_2(x))\mathrm{d}x\right| : \|u\|_1 \le 1\right\}. \tag{4.4.9}$$

Here and subsequently $\|\cdot\|_p$ $(1 \le p < \infty)$ stands for the $\mathcal{L}^p$-norm

$$\|u\|_p := \left\{\int_{-\infty}^{\infty} |u(x)|^p \mathrm{d}x\right\}^{1/p}, \qquad 1 \le p < \infty,$$

$$\|u\|_\infty := \operatorname*{ess\,sup}_{x\in\mathbb{R}} |u(x)|.$$

Further, let us denote by $\mathfrak{F}(p)$ the space of all (Lebesgue) almost everywhere (a.e.) differentiable functions f such that f' has $\mathcal{L}^p$-norm $\|f'\|_p \le 1$. Hence, integrating by parts the right-hand side of (4.4.9) we obtain a ζ-representation of the uniform metric $\boldsymbol{\rho}$:

$$\boldsymbol{\rho}(F_1, F_2) := \sup_{f\in\mathfrak{F}(1)} \left|\int_{-\infty}^{\infty} f(x)\mathrm{d}(F_1(x) - F_2(x))\right| = \zeta(F_1, F_2; \mathfrak{F}(1)). \tag{4.4.10}$$

Analogously, we have a $\zeta_{\mathfrak{F}(q)}$-representation for $\boldsymbol{\theta}_p$-metric $(p \ge 1)$ [see (3.3.28)]:

$$\begin{aligned} \boldsymbol{\theta}_p(F_1, F_2) &:= \|F_1 - F_2\|_p \\ &= \sup\left\{\left|\int_{-\infty}^{\infty} u(x)(F_1(x) - F_2(x))\mathrm{d}x\right| : \|u\|_q \le 1\right\} \\ &= \zeta(F_1, F_2; \mathcal{F}(q)). \end{aligned} \tag{4.4.11}$$

Next, we will examine some n-dimensional analogs of (4.4.9) and (4.4.10) by investigating the ζ-structure of (weighted) mean and uniform metrics in the space $\mathcal{F}^n = \mathcal{F}(\mathbb{R}^n)$ of all DFs $F(x)$, $x \in \mathbb{R}^n$.

Let $g(x)$ be a positive continuous function on $\mathbb{R}^n$ and let $p \in [1,\infty]$. Define the distances

$$\boldsymbol{\theta}_p(F,G;g) = \left(\int_{\mathbb{R}^n} |F(x) - G(x)|^p g(x)^p \mathrm{d}x\right)^{1/p}, \quad p \in [1,\infty], \tag{4.4.12}$$

$$\boldsymbol{\theta}_{\infty}(F, G; g) = \sup\{g(x)|F(x) - G(x)| : x \in \mathbb{R}^n\}. \tag{4.4.13}$$

Remark 4.4.3. In (4.4.12), for $n \geq 2$, the weight function $g(x)$ must vanish for all x with $\|x\|_{\infty} = \max_{1 \leq i \leq n} |x_i| = \infty$ in order to provide finite values of $\boldsymbol{\theta}_p$.

Let $A_{n,p}$ be the class of real functions f on $\mathbb{R}^n$ having a.e. the derivatives $\mathrm{D}^n f$, where

$$(\mathrm{D}^k f)(x) := \frac{\mathrm{d}^k f}{\mathrm{d}x_1 \cdots \mathrm{d}x_k}, \; x = (x_1, \ldots, x_n) \in \mathbb{R}^n, \; k = 1, 2, \ldots, n, \tag{4.4.14}$$

and

$$\int_{\mathbb{R}^n} \left| \frac{\mathrm{D}^n f(x)}{g(x)} \right|^q \mathrm{d}x \leq 1, \quad \frac{1}{p} + \frac{1}{q} = 1 \quad \text{if } p > 1, \tag{4.4.15}$$

and

$$|\mathrm{D}^n f(x)| \leq g(x) \text{ a.e. if } p = 1.$$

Denote by $g^*(x)$ a continuous function on $\mathbb{R}^n$ such that for some point $a = (a_1, \ldots, a_n)$ the function $g^*(x)$ is nondecreasing (resp. nonincreasing) in the variables x_i if $x_i \geq a_i$ (resp. $x_i \leq a_i$), $i = 1, \ldots, n$, and $g^* \geq g$.

Theorem 4.4.1. *Suppose that $p \in [1, \infty]$ and the functions $F, G \in \mathcal{F}^n$ satisfy the following conditions:*

(1) $\boldsymbol{\theta}_p(F, G; g) < \infty$.
(2) The derivative $\mathrm{D}^{n-1}(F - G)$ exists a.e., and for any $k = 1, \ldots, n$ the limit relation

$$\lim_{x_k \to \pm\infty} |x_k|^{1/p} g^*(x) |\mathrm{D}^{k-1}(F \quad G)(x)| = 0, \quad x = (x_1, \ldots, x_n) \tag{4.4.16}$$

holds a.e. for $x_j \in \mathbb{R}^1$, $j \neq k$, $j = 1, \ldots, n$. Then

$$\boldsymbol{\theta}_p(F, G; g) = \zeta(F, G; A_{n,p}). \tag{4.4.17}$$

Proof. As in equalities (4.4.9)–(4.4.11) we use the duality between $\mathcal{L}^p$ and $\mathcal{L}^q$ spaces. Integrating by parts and using the tail condition (4.4.16) we get (4.4.17). □

In the case $n = 1$, we get the following ζ-representation for the mean and uniform metrics with a weight.

Corollary 4.4.1. *If $p \in [1, \infty]$, $F, G \in \mathcal{F}^1$, and*

$$\lim_{x \to \pm\infty} |x|^{1/p} g^*(x) |F(x) - G(x)| = 0, \tag{4.4.18}$$

then

$$\boldsymbol{\theta}_p(F, G; g) = \zeta(F, G; A_{1,p}). \tag{4.4.19}$$

As a consequence of Theorem 4.4.1, we will subsequently investigate estimates of some classes of ζ-metrics with the help of metrics of type $\boldsymbol{\theta}_p(\cdot,\cdot;g)$. This is connected with the problem of characterizing *uniform classes* with respect to $\boldsymbol{\theta}_p(\cdot,\cdot;g)$-convergence.

Definition 4.4.2. If μ is a metric on $\mathcal{F}^n$, then a class A of measurable functions on $\mathbb{R}^n$ is called a *uniform class with respect to μ-convergence* (or simply a μ-u.c.) if for any F_n $(n = 1,2,\dots)$ and $F \in \mathcal{F}^n$ the condition $\mu(F_n, F) \to 0$ $(n \to \infty)$ implies that $\zeta_A(F_n, F) \to 0$ $(n \to \infty)$.

Bhattacharya and Ranga Rao (1976), Kantorovich and Rubinshtein (1958), Billingsley (1999), and Dudley (1976) have studied uniform classes w.r.t. weak convergence. It is clear that $A_{n,p}$ is a $\boldsymbol{\theta}(\cdot,\cdot;g)$-u.c. in the set of distribution functions satisfying (1) and (2) of Theorem 4.4.1.

Let $\mathcal{G}_{n,p}$ be the class of all functions in $A_{n,p}$ such that for any tuple $I = (1,\dots,k)$, $1 \le k \le n-1$, we have

$$\mathrm{D}_I^k f(x^I) = 0 \quad \text{a.e.} \quad x^I \in \mathbb{R}^n, \quad x_i^I = \begin{cases} x_i, & \text{if } i \in I, \\ +\infty, & \text{if } i \notin I. \end{cases}$$

Any function in $A_{n,p}$ constant outside a compact set obviously belongs to the class $\mathcal{G}_{n,p}$. Now we can omit the restriction (4.4.16) to get

Corollary 4.4.2. *For any $F, G \in \mathcal{F}^n$*

$$\zeta(F,G;\mathcal{G}_{n,p}) \le \boldsymbol{\theta}_p(F,G;g), \qquad p \in [1,\infty]. \tag{4.4.20}$$

In the case of the uniform metric

$$\boldsymbol{\rho}_n(F,G) := \sup_{x\in\mathbb{R}^n} |F(x) - G(x)| = \boldsymbol{\theta}_\infty(F,G;1), \tag{4.4.21}$$

we get the following refinement of Corollary 4.4.2. Denote by B_n the set of all real functions on $\mathbb{R}^n$ having a.e. the derivatives $\mathrm{D}^n f$ such that for any $I = (i_1,\dots,i_k)$, $1 \le k \le n$, $1 \le i_1 < \dots < i_k \le n$,

$$\int_{\mathbb{R}^k} |\mathrm{D}_I^k f(x^I)| \mathrm{d}x_{i_1} \dots \mathrm{d}x_{i_k} \le 1.$$

Denote by $F_I(x_1,\dots,x_i) = F(x^I)$ the marginal distribution of $F \in \mathcal{F}^n$ on the first k coordinates.

Corollary 4.4.3. *For any $F, G \in \mathcal{F}^n$*

$$\zeta(F,G;B_n) \le \sum_{\substack{I=(i,\dots,k)\\ 1\le k\le n}} \boldsymbol{\rho}_k(F_I,G_I). \tag{4.4.22}$$

The obvious inequality [see (4.4.22)]

$$\zeta(F,G;B_n) \le n\boldsymbol{\rho}_n(F,G) \tag{4.4.23}$$

implies that B_n is ρ_n-u.c.

Open Problem 4.4.1. Investigating the uniform estimates of the rate of convergence in the multidimensional central limit theorem, several authors[8] consider the following metric:

$$\boldsymbol{\rho}(P,Q;\mathcal{CB}) = \sup\{|P(A) - Q(A)| : A \in \mathcal{CB},\ P,Q \in \mathcal{P}(\mathbb{R}^n)\}, \tag{4.4.24}$$

where $\mathcal{CB}$ denotes the set of all convex Borel subsets of $\mathbb{R}^n$. The metric $\boldsymbol{\rho}(\cdot,\cdot;\mathcal{CB})$ may be viewed as a generalization of the notion of uniform metric $\boldsymbol{\rho}$ on $\mathcal{P}(\mathbb{R}^1)$; that is why $\boldsymbol{\rho}(\cdot,\cdot;\mathcal{CB})$ is called the *uniform metric* in $\mathcal{P}(\mathbb{R}^n)$. However, using the ζ-representation (4.4.10) of the Kolmogorov metric $\boldsymbol{\rho}$ on $\mathcal{P}(\mathbb{R}^1)$, it is possible to extend the notion of uniform metric in a way that is different from (4.4.24). That is, define the *uniform $\boldsymbol{\rho}\zeta$-metric* in $\mathcal{P}(\mathbb{R}^n)$ as follows:

$$\boldsymbol{\rho}\zeta(P,Q) := \zeta(P,Q;A_{n,1}(1)), \tag{4.4.25}$$

where $A_{n,1}(1)$ is the class of real functions f on $\mathbb{R}^n$ having a.e. the derivatives $\mathrm{D}^n f$ and

$$\int_{\mathbb{R}^n} |\mathrm{D}^n f(x)|\mathrm{d}x \le 1. \tag{4.4.26}$$

What kind of quantitative relationships exist between the metrics $\boldsymbol{\rho}_n$, $\boldsymbol{\rho}(\cdot,\cdot;\mathcal{CB})$, and $\boldsymbol{\rho}\zeta$ [see (4.4.21), (4.4.24), and (4.4.25)]? Such relationships would yield the rate of convergence for the central limit theorem in terms of $\boldsymbol{\rho}\zeta$.

Example 4.4.3 (ζ-metrics that metrize G-weak convergence). In Example 4.3.2 we considered a Λ-metric that metrizes G-weak convergence in $\mathcal{P}_G \subseteq \mathcal{P}(U)$ [see Definition 4.3.2 and (4.3.5)]. Now we will be interested in ζ-metrics generating G-weak convergence in $\mathcal{P}_G$. Let $\mathbb{F} = \mathbb{F}(G)$ be the class of real-valued functions f on an s.m.s. U such that the following conditions hold:

(i) $\mathbb{F}$ is an equicontinuous class, i.e.,

$$\lim_{d(x,y)\to 0} \sup_{f\in\mathbb{F}} |f(x) - f(y)| = 0;$$

(ii)

$$\sup_{f\in\mathbb{F}} |f(x)| \le G(x) \quad \forall x \in U;$$

(iii) $\alpha G \in \mathbb{F}$ for some constant $\alpha \ne 0$;

[8]See, for instance, Sazonov (1981) and Senatov (1980).

(iv) For each nonempty closed set $C \subseteq U$ and for each integer k, the function

$$f_{k,C}(x) := \max\{0, 1/k - d(x, C)\}$$

belongs to $\mathbb{F}$.

Note that if $\mathbb{F}$ satisfies (i) and (ii) only, then $\mathbb{F}$ is $\pi_{\lambda,G}$-u.c. [see Definition 4.4.2 and (4.3.5)], i.e., G-weak convergence implies $\zeta_{\mathbb{F}}$-convergence.[9] The next theorem determines the cases in which $\zeta_{\mathbb{F}}$-convergence is equivalent to G-weak convergence.

Theorem 4.4.2. *If $\mathbb{F} = \mathbb{F}(G)$ satisfies (i)–(iv), then $\zeta_{\mathbb{F}}$ metrizes the G-weak convergence in $\mathcal{P}_G$.*

In fact, we will prove a more general result (see further Sect. 11.2, Theorem 11.2.2 in Chap. 11).

Let us consider some particular cases of the classes $\mathbb{F}(G)$.

Case A. Let c be a fixed point of U, a and b be positive constants, and $h : [0, \infty] \to [0, \infty]$ be a nondecreasing function, $h(0) = 0$, $h(\infty) \leq \infty$. Define the class $S = S(a, b, h)$ of all functions $f : U \to \mathbb{R}$ such that

$$\|f\|_\infty := \sup_{x \in U} |f(x)| \leq a \tag{4.4.27}$$

and

$$\mathrm{Lip}_h(f) := \sup_{x \neq y,\ x,y \in U} \frac{|f(x) - f(y)|}{d(x, y)\max\{1, h(d(x, c)), h(d(y, c))\}} \leq b. \tag{4.4.28}$$

Corollary 4.4.4. *(a) If $0 < a < \infty$, $0 < b < \infty$, then $\zeta_{S(a,b,h)}$ metrizes the weak convergence in $\mathcal{P}(U)$.*

(b) If $a = \infty$, $b < \infty$ and

$$\sup_{t \neq s} \frac{|t\max\{1, h(t)\} - s\max\{1, h(s)\}|}{|t - s|\max\{1, h(t), h(s)\}} < \infty, \tag{4.4.29}$$

then $\zeta_{S(a,b,h)}$ metrizes the G-weak convergence with

$$G(x) = d(x, c)\max\{1, h(d(x, c))\}.$$

Case B. Fortet and Mourier (1953) investigated the following two $\zeta_{\mathbb{F}}$-metrics.

(a) $\zeta(\cdot, \cdot; \mathcal{G}^p)$ $(p \geq 1)$, where the class $\mathcal{G}^p$ is defined as follows. For each function $f : U \to \mathbb{R}$ let

[9] See Bhattacharya and Ranga Rao (1976) and Ranga (1962).

$$L(f,t) := \sup\left\{\frac{f(x)-f(y)}{d(x,y)} : x \neq y, d(x,c) \leq t, d(y,c) \leq t\right\} \quad (4.4.30)$$

and

$$M(f) := \sup \frac{L(f,t)}{\max(1,t^{p-1})}. \quad (4.4.31)$$

Then

$$\mathcal{G}^p := \{f : U \to \mathbb{R},\ M(f) \leq 1\}. \quad (4.4.32)$$

(b) $\zeta(\cdot,\cdot;\overline{\mathcal{G}}^p)$, where

$$\overline{\mathcal{G}}^p := \{f \in \mathcal{G}^p, \|f\|_\infty \leq 1\}. \quad (4.4.33)$$

Lemma 4.4.1. *Let $h_p(t) = t^{p-1}$ $(p > 1, t \geq 0)$. Then*

$$\zeta(P,Q;\mathcal{G}^p) = \zeta(P,Q;S(\infty,1,h_p)) \quad (4.4.34)$$

and

$$\zeta(P,Q;\overline{\mathcal{G}}^p) = \zeta(P,Q;S(1,1,h_p)). \quad (4.4.35)$$

Proof. It is enough to check that $\mathrm{Lip}_{h_p}(f) = M(f)$. Actually, let $x \neq y$ and $t_0 := \max\{d(x,c), d(y,c)\}$. Then $t_0 > 0$ and $|f(x)-f(y)| \leq L(f,t_0)\, d(x,y) \leq M(f)\max(1,t_0^{p-1})d(x,y)$; hence, $\mathrm{Lip}_{h_p}(f) < M(f)$. Conversely, for each $t > 0$ $L(f,t) \leq \mathrm{Lip}_{h_p}(f)\max(1,t^{p-1})$, and thus $M(f) \leq \mathrm{Lip}_{h_p}(f)$. □

Corollary 4.4.4 and Lemma 4.4.1 imply the following corollary.

Corollary 4.4.5. *Let (U,d) be an s.m.s. Then,*

(i) $\zeta(\cdot,\cdot;\overline{\mathcal{G}}^p)$ metrizes the weak convergence in $\mathcal{P}(U)$;
(ii) In the set

$$\mathcal{P}^{(p)}(U) := \left\{P \in \mathcal{P}(U), \int_U d^p(x,c)P(\mathrm{d}x) < \infty\right\}, \quad (4.4.36)$$

the $\zeta(\cdot,\cdot;\mathcal{G}^p)$-convergence is equivalent to the G-weak convergence with $G(x) = d^p(x,c)$.

Case C. Dudley (1966, 1976) considered $\boldsymbol{\beta}$-metric in $\mathcal{P}(U)$, which is defined as $\zeta_{\mathbb{F}}$-metric with

$$\mathbb{F} := \left\{f : U \to \mathbb{R}, \|f\|_\infty + \sup_{x,y\in U, x\neq y} \frac{|f(x)-f(y)|}{d(x,y)} \leq 1\right\}. \quad (4.4.37)$$

Corollary 4.4.6. *The Dudley metric* $\boldsymbol{\beta} := \zeta_{\mathbb{F}}$ *defined by* (4.4.7) *and* (4.4.37) *metrizes the weak convergence in* $\mathcal{P}(U)$.[10]

Proof. Using Corollary 4.4.5(i) with $p = 1$ and the inequality

$$\frac{1}{2}\zeta(P, Q; \overline{\mathcal{G}}^1) \le \boldsymbol{\beta}(P, Q) \le \zeta(P, Q; \overline{\mathcal{G}}^1), \qquad P, Q \in \mathcal{P}(U), \tag{4.4.38}$$

we claim that $\boldsymbol{\beta}$ induces weak convergence in $\mathcal{P}(U)$. □

Case D. The Kantorovich metric ℓ_1 [see (3.3.12) and (3.3.17)] admits the ζ-representation $\zeta(\cdot, \cdot; \mathcal{G}^1)$, and ℓ_p $(0 < p \le 1)$ [see (3.3.12)] has the form

$$\ell_p(P_1, P_2) = \zeta(P_1, P_2; \overline{\mathcal{G}}^1), \qquad P_1, P_2 \in \mathcal{P}(U), \quad U = (U, d^p). \tag{4.4.39}$$

On the right-hand side of (4.4.39), U is an s.m.s. with the metric d^p, i.e., in the definition of $\zeta(\cdot, \cdot; \mathcal{G}^1)$ [see (4.4.30), (4.4.33)], we replace the metric d with d^p.

Now let us touch on some special cases of (4.4.39).

(a) Let U be a separable normed space with norm $\|\cdot\|$ and $Q : U \to U$ be a function on U such that the metric $d_Q(x, y) = \|Q(x) - Q(y)\|$ metrizes the space U as an s.m.s. For instance, if Q is a homeomorphism of U, i.e., Q is a one-to-one function and both Q and Q^{-1} are continuous, then (U, d_Q) is an s.m.s. Further, let $p = 1$ and $d = d_Q$ in (4.4.39). Then

$$\boldsymbol{\kappa}_Q(P_1, P_2) := \ell_1(P_1, P_2) = \sup\left\{\left|\int_U f \mathrm{d}(P_1 - P_2)\right| : f : U \to \mathbb{R}, \right.$$
$$\left. |f(x) - f(y)| \le d_Q(x, y) \quad \forall x, y \in U\right\} \tag{4.4.40}$$

is called a *Q-difference pseudomoment in* $\mathcal{P}(U)$.

If U is a separable normed space and Q is a homeomorphism of U, then (noting our earlier discussions in Theorem 2.7.1 and Example 3.3.2), in the space $\mathfrak{X}(U)$ of U-valued RVs, $\kappa_Q(X, Y) := \kappa_Q(\Pr_X, \Pr_Y)$ is the minimal metric w.r.t. the *compound Q-difference pseudomoment*

$$\tau_Q(X, Y) := E d_Q(X, Y) \tag{4.4.41}$$

and

$$\boldsymbol{\kappa}_Q(X, Y) = \widehat{\tau}_Q(X, Y) = \sup\{|E[f(Q(X)) - f(Q(Y))]| : f : U \to \mathbb{R},$$
$$|f(x) - f(y)| \le \|x - y\| \quad \forall x, y \in U\}. \tag{4.4.42}$$

---

[10]See Dudley (1966).

In the particular case $U = \mathbb{R}$, $\|x\| = |x|$,

$$Q(x) := \int_0^x q(u)\mathrm{d}u \quad q(u) \geq 0, u \in \mathbb{R}, \quad x \in \mathbb{R},$$

the metric κ_Q has the following explicit representation:

$$\kappa_Q(P_1, P_2) := \kappa_Q(F_1, F_2) := \int_{-\infty}^{\infty} q(x)|F_1(x) - F_2(x)|\mathrm{d}x. \tag{4.4.43}$$

If, in (4.4.40), $Q(x) = x\|x\|^{s-1}$ for some $s > 0$, then $x_s := x_Q$ is called an *s-difference pseudomoment*.[11]

(b) By (4.4.39), we have that

$$\ell_p(P_1, P_2) := \sup\left\{\left|\int_U f\mathrm{d}(P_1 - P_2)\right| : f : U \to \mathbb{R}, \right.$$

$$\left. |f(x) - f(y)| \leq d^p(x, y), x, y \in U\right\} \tag{4.4.44}$$

for any $p \in (0, 1)$. Hence, letting $p \to 0$ and defining the indicator metric

$$i(x, y) = \begin{cases} 1, x \neq y, \\ 0, x = y, \end{cases}$$

we get

$$\lim_{p \to 0} \ell_p(P_1, P_2)$$

$$= \sup\left\{\left|\int_U f\mathrm{d}(P_1 - P_2)\right| : f : U \to \mathbb{R}, |f(x) - f(y)| \right.$$

$$\left. \leq i(x, y) \forall x, y \in U\right\}$$

$$= \sigma(P_1, P_2) = \ell_0(P_1, P_2), \tag{4.4.45}$$

where σ (resp. ℓ_0) is the total variation metric [see (3.3.13)].

Examples 4.4.1–4.4.3 show that the ζ-structure encompasses the simple metrics ℓ_p that are minimal with respect to the compound metric $\mathcal{L}_p$ {see (3.4.18) for $p \in [0, 1]$}. If, however, $p > 1$, then $\ell_p = \widehat{\mathcal{L}}_p$ [see equalities (3.4.18), (3.4.3), and (3.3.11)] has a form different from the ζ-representation, namely,

---

[11]See Zolotarev (1976, 1977, 1978) and Hall (1981).

$$\ell_p(P_1, P_2) = \sup\left\{\left[\int_U f\mathrm{d}P_1 + \int_U g\mathrm{d}P_2\right]^{1/p} : (f, g) \in \mathcal{G}_p\right\}, \tag{4.4.46}$$

where $\mathcal{G}_p$ is the class of all pairs (f, g) of Lipschitz bounded functions $f, g \in \mathrm{Lip}^b(U)$ [see (3.3.8)] that satisfy the inequality

$$f(x) + g(y) \leq d^p(x, y), \quad x, y \in U. \tag{4.4.47}$$

The following lemma shows that $\ell_p = \widehat{\mathcal{L}}_p$ $(p > 1)$ has no ζ-representation.

Lemma 4.4.2. *If an s.m.s.* (U, d) *has more than one point and the minimal metric* $\widehat{\mathcal{L}}_p$ $(p > 1)$ *has a* ζ*-representation* (4.4.5)*, then* $p = 1$.[12]

Proof. Assuming that $\widehat{\mathcal{L}}_p$ has a $\zeta_{\mathbb{F}}$-representation for a certain class $\mathbb{F} \subseteq C^b(U)$, then

$$\sup_{f\in\mathbb{F}}\left\{\left|\int_U f\mathrm{d}(P_1 - P_2)\right|\right\} = \widehat{\mathcal{L}}_p(P_1, P_2), \quad \forall P_1, P_2 \in \mathcal{P}_1(U). \tag{4.4.48}$$

If in (4.4.48) the law P_1 is concentrated at the point x and P_2 is concentrated at y, then $\sup\{|f(x) - f(y)| : f \in \mathbb{F}\} \leq d(x, y)$. Thus, $\mathbb{F}$ is contained in the Lipschitz class

$$\begin{aligned}\mathrm{Lip}^b_{1,1} &= \mathrm{Lip}^b_{1,1}(U)\\ &:= \{f : U \to \mathbb{R}, f \text{ bounded}, |f(x) - f(y)| \leq d(x, y)\ \forall x, y \in U\}.\end{aligned} \tag{4.4.49}$$

For each law $P \in \mathcal{P}_2$ with marginals P_1 and P_2

$$\begin{aligned}\widehat{\mathcal{L}}_p(P_1, P_2) &\leq \sup_{f\in\mathrm{Lip}^b_{1,1}}\left|\int_U f\mathrm{d}(P_1 - P_2)\right|\\ &\leq \sup_{f\in\mathrm{Lip}^b_{1,1}}\int_{U^2}|f(x) - f(y)|P(\mathrm{d}x, \mathrm{d}y) \leq \mathcal{L}_1(P).\end{aligned}$$

Next, we can pass to the minimal metric $\widehat{\mathcal{L}}_1$ on the right-hand side of the preceding inequality and then claim $\widehat{\mathcal{L}}_p = \widehat{\mathcal{L}}_1$. In particular, by the Minkowski inequality we have

$$\left\{\int_U d^p(x, a)P_1(\mathrm{d}x)\right\}^{1/p} - \left\{\int_U d^p(x, a)P_1(\mathrm{d}x)\right\}^{1/p} \leq \widehat{\mathcal{L}}_p(P_1, P_2) = \widehat{\mathcal{L}}_1(P_1, P_2). \tag{4.4.50}$$

[12] See Neveu and Dudley (1980).

Assuming that there exists $b \in U$ such that $d(a,b) > 0$, let us consider the laws P_1, P_2 with $P_1(\{a\}) = r \in (0,1)$, $P_1(\{b\}) = 1-r$, $P_2(\{a\}) = 1$; then the ζ-representation of $\widehat{\mathcal{L}}_1 = \ell_1$ [see (3.3.12), (3.4.18)],

$$\widehat{\mathcal{L}}_1(P_1, P_2) = \sup_{f \in \mathrm{Lip}^b_{1,1}} |rf(a) + (1-r)f(b) - f(a)|$$
$$= (1-r) \sup_{f \in \mathrm{Lip}^b_{1,1}} |f(a) - f(b)| \leq (1-r)d(a,b)$$

and hence

$$(1-r)d(a,b) \geq \widehat{\mathcal{L}}_1(P_1, P_2) \geq \{d^p(b,a)(1-r)\}^{1/p} = (1-r)^{1/p}d(a,b),$$

i.e., $p = 1$. □

Remark 4.4.4. Szulga (1982) made a conjecture that $\widehat{\mathcal{L}}_p$ $(p > 1)$ has a dual form close to that of the ζ-metric, namely,

$$\widehat{\mathcal{L}}_p(P_1, P_2) = \mathbf{AS}_p(P_1, P_2), \qquad P_1, P_2 \in \mathcal{P}^{(p)}(U). \tag{4.4.51}$$

In (4.4.49), the class $\mathcal{P}^{(p)}(U)$ consists of all laws P with finite "pth moment," $\int d^p(x,a)P(\mathrm{d}x) < \infty$ and

$$\mathbf{AS}_p(P_1, P_2) := \sup_{f \in \mathrm{Lip}^b_{1,1}} \left| \left\{ \int_U |f|^p \mathrm{d}P_1 \right\}^{1/p} - \left\{ \int_U |f|^p \mathrm{d}P_2 \right\}^{1/p} \right|. \tag{4.4.52}$$

By the Minkowski inequality it follows easily that

$$\mathbf{AS}_p \leq \widehat{\mathcal{L}}_p. \tag{4.4.53}$$

Rachev and Schief (1992) construct an example illustrating that the conjecture is wrong. However, the following lemma shows that Szulga's conjecture is partially true in the sense that $\widehat{\mathcal{L}}_p \overset{\mathrm{top}}{\sim} \mathbf{AS}_p$.

Lemma 4.4.3. *In the space $\mathcal{P}^{(p)}(U)$, the metrics $\mathbf{AS}_p$ and $\widehat{\mathcal{L}}_p$ generate the same exact topology.*

Proof. It is known that (see further Sect. 8.3, Corollary 8.3.1) $\widehat{\mathcal{L}}_p$ metrizes G_p-weak convergence in $\mathcal{P}^{(p)}(U)$ (Definition 4.3.2), where $G_p(x) = d^p(x,a)$. Hence, by (4.4.51) it is sufficient to prove that $\mathbf{AS}_p$-convergence implies G_p-weak convergence. In fact, since $G_1 \in \mathrm{Lip}_{1,1}$, then

$$\mathbf{AS}_p(P_n, P) \to 0 \Rightarrow \int_U d^p(x,a)P_n(\mathrm{d}x) \to \int_U d^p(x,a)P(\mathrm{d}x). \tag{4.4.54}$$

Further, for each closed nonempty set C and $\varepsilon > 0$ let

$$f_C := \max\left(0, 1 - \frac{1}{\varepsilon}d(x, C)\right).$$

Then $f_C \in \mathrm{Lip}_{1/\varepsilon,1}(U)$ [see (3.3.6)] and

$$\begin{aligned} P_n^{1/p}(C) &\leq \left\{\int_U f_C^p \mathrm{d}P_n\right\}^{1/p} \\ &\leq \left\{\int_U f_C^p \mathrm{d}P\right\}^{1/p} + \frac{1}{\varepsilon}\mathbf{AS}_p(P_n, P) \\ &\leq \{P(C^\varepsilon)\}^{1/p} + \frac{1}{\varepsilon}\mathbf{AS}_p(P_n, P), \end{aligned}$$

which implies

$$\mathbf{AS}_p(P_n, P) \to 0 \quad \Rightarrow \quad P_n \xrightarrow{\mathrm{w}} P, \tag{4.4.55}$$

as desired. □

By Lemma 4.4.2 it follows, in particular, that there exist simple metrics that have no $\zeta_{\mathbb{F}}$-representation. In the case of a $\widehat{\mathcal{L}}_p$-metric, however, we can find a $\zeta_{\mathbb{F}}$-metric that is topologically equivalent to $\widehat{\mathcal{L}}_p$, i.e.,

$$\widehat{\mathcal{L}}_p \overset{\mathrm{top}}{\sim} \zeta_{\mathcal{G}}^p \tag{4.4.56}$$

[see (4.4.6), (4.4.34), and Corollary 4.4.5(ii)]. Also, it is not difficult to see that the Prokhorov metric π [see (3.3.20)] has no $\zeta_{\mathbb{F}}$-representation, even in the case where $U = \mathbb{R}$, $d(x, y) = |x - y|$. In fact, assume that

$$\pi(P, Q) = \zeta_{\mathbb{F}}(P, Q), \quad \forall P, Q \in \mathcal{P}(\mathbb{R}). \tag{4.4.57}$$

Denoting the measure concentrated at the point x by P_x we have

$$\pi(P_x, P_y) = \min(1, |x - y|) \leq |x - y|. \tag{4.4.58}$$

Hence, by (4.4.56),

$$|x - y| \geq \pi(P_x, P_y) = \sup_{f \in \mathbb{F}} |f(x) - f(y)|;$$

hence,

$$\begin{aligned} \pi(P, Q) &\leq \sup\left\{\left|\int f \mathrm{d}(F - G)\right| : f : U \to \mathbb{R}, f - \text{bounded}, \right. \\ &\qquad \left. |f(x) - f(y)| \leq |x - y|, x, y \in \mathbb{R}\right\} \\ &\leq \int_{-\infty}^{\infty} |F(x) - G(x)| \mathrm{d}x =: \kappa(F, G), \end{aligned}$$

where F is the DF of P and G is the DF of Q. Obviously, $\pi(P, Q) > \mathbf{L}(F, G)$, where $\mathbf{L}$ is the Lévy metric in the distribution function space $\mathcal{F}$ [see (4.2.3)]. Hence, the equality (4.4.57) implies

$$\mathbf{L}(F, G) \leq \kappa(F, G), \quad \forall F, G \in \mathcal{F}. \tag{4.4.59}$$

Let $1 > \varepsilon > 0$ and

$$F_\varepsilon(x) = \begin{cases} 0, & x \leq 0, \\ 1 - \varepsilon, & 0 < x \leq \varepsilon, \\ 1, & x > \varepsilon, \end{cases} \qquad G_\varepsilon = \begin{cases} 0, & x \leq 0, \\ 1, & x > 0. \end{cases}$$

Then the equalities

$$\kappa(F, G) = \varepsilon^2 = \mathbf{L}^2(F, G)$$

contradict (4.4.59); hence, π does not admit a ζ-representation.

Although there is no ζ-representation for the Prokhorov metric π, nevertheless π is topologically equivalent to various ζ-metrics. To see this, simply note that both π and certain ζ-metrics metrize weak convergence [Corollary 4.4.4(a)]. Therefore, the following question arises: is there a simple metric μ such that

$$\mu \overset{\text{top}}{\sim} \zeta_{\mathbb{F}}$$

fails for any set $\mathbb{F} \subseteq C^b(U)$? The following lemma gives an affirmative answer to this question, where $\mu = \pi\mathbf{H}$ [see (4.2.38) and (4.2.43)], and if $U = \mathbb{R}$, $d(x, y) = |x - y|$, then one can take $\mu = \mathbf{H}$ [see (4.2.33) and Fig. 4.2].

Lemma 4.4.4. *Let $\lambda > 0$ and let (U, d) be a metric space containing a nonconstant sequence $a_1, a_2, \ldots \to a \in U$.*

(i) If (U, d) is an s.m.s., then there is no set $\mathbb{F} \subseteq C^b(U)$ such that $\pi\mathbf{H}_\lambda \overset{\text{top}}{\sim} \zeta_{\mathbb{F}}$.
(ii) If $U = \mathbb{R}$, $d(x, y) = |x - y|$, then there is no set $\mathbb{F} \subseteq C^b(U)$ such that $\mathbf{H}_\lambda \overset{\text{top}}{\sim} \zeta_{\mathbb{F}}$.

Proof. We will consider only case (i) with $\lambda = 1$. Choose the laws P_n and P as follows: $P(\{a\}) = 1$, $P_n(\{a\}) = P_n(\{a_n\}) = \frac{1}{2}$. Then for each $B \in \mathcal{B}_1$ the measure P takes a value 0 or 1, and thus

$$\pi\mathbf{H}_1(P_n, P) \geq \inf_{B \in \mathcal{B}_1} \max\{d(a_n, B), |P_n(a_n) - P(B)|\} \geq \frac{1}{2}. \tag{4.4.60}$$

Assuming that $\pi\mathbf{H}_1 \overset{\text{top}}{\sim} \zeta_{\mathbb{F}}$ we have, by (4.4.60), that

$$0 < \lim_{n \to \infty} \sup \zeta_{\mathbb{F}}(P_n, P)$$

$$= \lim_{n\to\infty} \sup \left| \frac{1}{2} f(a) + \frac{1}{2} f(a_n) - f(a) \right|$$

$$= \frac{1}{2} \lim_{n\to\infty} \sup |f(a) - f(a_n)|. \tag{4.4.61}$$

Further, let $Q_n(\{a_n\}) = 1$. Then $\pi\mathbf{H}_1(Q_n, P) \to 0$, and hence

$$0 = \lim_{n\to\infty} \sup \zeta_{\mathbb{F}}(Q_n, P)$$

$$= \lim_{n\to\infty} \sup |f(a) - f(a_n)|. \tag{4.4.62}$$

Relationships (4.4.61) and (4.4.62) give the necessary contradiction. □

Lemma 4.4.4 claims that the ζ-structure of simple metrics does not describe all possible topologies arising from simple metrics. Next, we will extend the notion of ζ-structure to encompass all simple p. semidistances as well as all compound p. semidistances. To this end, note first that for the compound metric $\mathcal{L}_p(X, Y)$ $[p \geq 1,\ X, Y \in \mathfrak{X}(\mathbb{R})]$ [see (3.4.3) with $d(x, y) = |x - y|$, $U = \mathbb{R}$] we have the following dual representation as shown by Neveu and Dudley (1980):

$$\mathcal{L}_p(X, Y) = \sup\{|E(XZ - YZ)| : Z \in \mathfrak{X}(\mathbb{R}), \mathcal{L}_q(Z, 0) \leq 1\},$$
$$1 \leq p \leq \infty\ 1/p + 1/q = 1. \tag{4.4.63}$$

The next definition generalizes the notion of the ζ-structure as well as the metric structure of $\widehat{\mathcal{L}}_H$-distances [see (3.3.10) and (3.4.17)] and $\mathcal{L}_p$-metrics [see (3.4.3)].

Definition 4.4.3. We say that a p. semidistance μ admits a *ζ-structure* if μ can be written in the following way:

$$\mu(X, Y) = \overline{\zeta}(X, Y; \overline{\mathbb{F}}(X, Y)) = \sup_{f \in \overline{\mathbb{F}}(X,Y)} Ef, \tag{4.4.64}$$

where $\mathbb{F}(X, Y)$ is a class of integrable functions $f : \Omega \to (\mathbb{R}, \mathcal{B}(\mathbb{R}))$ given on a probability space $(\Omega, \mathcal{A}, \Pr)$.

In general, $\overline{\zeta}$ is not a p. semidistance, but each p. semidistance has a $\overline{\zeta}$-representation. Actually, for each p. semidistance μ equality (4.4.64) is valid where $\overline{\mathbb{F}}(X, Y)$ contains only a constant function $\mu(X, Y)$.

Let us consider some examples of $\overline{\zeta}$-structures of p. semidistances.

Example 4.4.4 (see (3.4.1)*).* $\mathcal{L}_H$ has a trivial $\overline{\zeta}$-representation, where $\overline{\mathbb{F}}(X, Y)$ contains only the function $H(d(X, Y))$.

Example 4.4.5. $\mathcal{L}_p$ on $\mathfrak{X}(\mathbb{R})$ [see (4.4.63)] enjoys a nontrivial $\overline{\zeta}$-representation, where

$$\overline{\mathbb{F}}(X, Y) = \{f_Z(X, Y) = XZ - YZ : \mathcal{L}_q(Z, 0) \leq 1\}.$$

Example 4.4.6. The simple distance ℓ_H [see (3.3.10)] has a $\overline{\zeta}$-representation, where

$$\overline{\mathbb{F}}(X,Y) = \{f(X,Y) = f_1(X) + f_2(Y), (f_1, f_2) \in \mathcal{G}_H(U)\}$$

for each $X, Y \in \mathfrak{X}$.

Example 4.4.7. A $\zeta_{\mathbb{F}}$-structure of simple metrics is a particular case of a $\overline{\zeta}$-structure with

$$\overline{\mathbb{F}}(Z,Y) = \{f(X,Y) = f(X) - f(Y) : f \in \mathbb{F}\}$$
$$\cup \{f(X,Y) = f(Y) - f(X) : f \in \mathbb{F}\}.$$

Additional examples and applications of metrics with ζ-structures are discussed in Sriperumbudur et al. (2010). A variety of ζ-representations with applications in various central limit theorems are discussed in Boutsikas and Vaggelatou (2002). Kantorovich-type metrics are applied by Koeppl et al. (2010) in the area of stochastic chemical kinetics and by Rachev and Römisch (2002) to the problem of the stability of stochastic programming and convergence of empirical processes. Other applications in the area of stochastic programming are provided by Rachev and Römisch (2002), Dupacová et al. (2003), and Stockbridge and Güzin (2012). An extension of the Prokhorov metric to fuzzy sets is provided by Repovš et al. (2011), and other applications are provided in Graf and Luschgy (2009). A metric with a ζ-structure based on the Trotter operator is applied to the convergence rate problem in moment central limit theorems by Hung (2007). Other applications of probability metrics include Rüschendorf et al. (1996), Toscani and Villani (1999), Greven et al. (2009), Sriperumbudur et al. (2009), Bouchitté et al. (2011), and Hutchinson and Rüschendorf (2000).

We have completed the investigation of the three universal metric structures (h, Λ, and $\overline{\zeta}$). The reason we call them universal is that each p. semidistance μ has h-, Λ-, and $\overline{\zeta}$-representations simultaneously. Thus, depending on the specific problem under consideration, one can use one or another p. semidistance representation.

References

Bhattacharya RM, Ranga Rao R (1976) Normal approximation and asymptotic expansions. Wiley, New York

Billingsley P (1999) Convergence of probability measures, 2nd edn. Wiley, New York

Bouchitté G, Jimenez C, Mahadevan R (2011) Asymptotic analysis of a class of optimal location problems. J de Mathématiques Pures et Appliquées 96(4):382–419

Boutsikas M, Vaggelatou E (2002) On distance between convex ordered random variables, with applications. Adv Appl Probab 34:349–374

Dudley RM (1966) Convergence of baire measures. Studia Math XXVII:251–268

Dudley RM (1976) Probabilities and metrics: convergence of laws on metric spaces, with a view to statistical testing. Aarhus university mathematics institute lecture notes series no. 45, Aarhus.

Dudley RM (2002) Real analysis and probability, 2nd edn. Cambridge University Press, New York
Dupacová J, Gröwe-Kuska N, Römisch W (2003) Scenario reduction in stochastic programming: an approach using probability metrics. Math Program 95(3):493–511
Fortet R, Mourier B (1953) Convergence de la réparation empirique vers la répétition theorétique. Ann Sci Ecole Norm 70:267–285
Graf S, Luschgy H (2009) Quantization for probability measures in the prokhorov metric. Theor Probab Appl 53:216–241
Greven A, Pfaffelhuber P, Winter A (2009) Convergence in distribution of random metric measure spaces: (lambda-coalescent measure trees). Probab Theor Relat Fields 145(1):285–322
Hall P (1981) Two-sided bounds on the rate of convergence to a stable law. Z Wahrsch Verw Geb 57:349–364
Hennequin PL, Tortrat A (1965) Théorie des probabilités et quelques applications. Masson, Paris
Hung TL (2007) On a probability metric based on the Trotter operator. Vietnam J Math 35(1):21–32
Hutchinson J, Rüschendorf L (2000) Random fractals and probability metrics. Adv Appl Probab 32(4):925–947
Kakosyan AB, Klebanov L, Rachev ST (1988) Quantitative criteria for convergence of probability measures. Ayastan Press, Yerevan (in Russian)
Kantorovich LV, Rubinshtein GSh (1958) On the space of completely additive functions. Vestnik LGU Ser Mat Mekh i Astron 7:52–59
Koeppl H, Setti G, Pelet S, Mangia M, Petrov T, Peter M (2010) Probability metrics to calibrate stochastic chemical kinetics. In: Circuits and systems (ISCAS), proceedings of 2010 IEEE international symposium, pp 541–544
Neveu J, Dudley RM (1980) On Kantorovich–Rubinstein theorems (transcript). Forthcoming Math Program DOI 10.1007/S10107-012-0563-6
Prokhorov YuV (1956) Convergence of random processes and limit theorems in probability theory. Theor Prob Appl 1:157–214
Rachev S, Schief A (1992) On L_p-minimal metrics. Probab Math Statist 13:311–320
Rachev ST (1984) Hausdorff metric construction in the probability measure space. Pliska Studia Math Bulgarica 7:152–162
Rachev S, Römisch W (2002) Quantitative stability in stochastic programming: the method of probability metrics. Math Oper Res 27(4):792–818
Ranga RR (1962) Relations between weak and uniform convergence of measures with applications. Ann Math Stat 33:659–680
Repovš D, Savchenko A, Zarichnyi M (2011) Fuzzy Prokhorov metric on the set of probability measures. Fuzzy Sets Syst 175:96–104
Rüschendorf L, Schweizer B, Taylor MD (1996) Distributions with fixed marginals and related topics. In: Lecture notes monograph series, Institute of Mathematical Statistics, vol 2–8, Hayward, CA
Sazonov VV (1981) Normal approximation – some recent advances. In: Lecture Notes in Mathematics, vol 879. Springer, New York
Senatov VV (1980) Uniform estimates of the rate of convergence in the multi-dimensional central limit theorem. Theor Prob Appl 25:745–759
Sendov B (1969) Some problems of the theory of approximations of functions and sets in the Hausdorff metrics. Usp Mat Nauk 24/5:141–173
Sriperumbudur B, Gretton A, Fukumizu K, Schölkopf B (2010) Hilbert space embeddings and metrics on probability measures. J Mach Learn Res 11:1517–1561
Sriperumbudur B, Fukumizu K, Gretton A, Schölkopf B, Lanckriet G (2009) On integral probability metrics, ϕ-divergences and binary classification. arXiv:0901.2698v4
Stockbridge R, Güzin B (2012) A probability metrics approach for reducing the bias of optimality gap estimators in two-stage stochastic programming. Forthcoming Mathematical Programming DOI 10.1007/S10107-012-0563-6.
Szulga A (1982) On minimal metrics in the space of random variables. Theor Prob Appl 27:424–430

Toscani G, Villani C (1999) Probability metrics and uniqueness of the solution to the Boltzmann equation for a Maxwell gas. J Stat Phys 94:619–637

Zolotarev VM (1976) Metric distances in spaces of random variables and their distributions. Math USSR Sb 30:373–401

Zolotarev VM (1977) Ideal metrics in the problem of approximating distributions of sums of independent random variables. Theor Prob Appl 22:433–449

Zolotarev VM (1978) On pseudomoments. Theor Prob Appl 23:269–278

Part II
Relations Between Compound, Simple, and Primary Distances

Chapter 5
Monge–Kantorovich Mass Transference Problem, Minimal Distances and Minimal Norms

The goals of this chapter are to:

- Introduce the Kantorovich and Kantorovich–Rubinstein problems in one-dimensional and multidimensional settings;
- Provide examples illustrating applications of the abstract problems;
- Discuss the multivariate Kantorovich and Kantorovich–Rubinstein theorems, which provide dual representations of certain types of minimal distances and norms;
- Discuss a particular application leading to an explicit representation for a class of minimal norms.

Notation introduced in this chapter:

5.1 Introduction

The Kantorovich and Kantorovich–Rubinstein problems, also known respectively as the mass transportation and mass transshipment problems, represent abstract formulations of optimization problems of high practical importance. They can be regarded as infinite-dimensional versions of the well-known transportation and transshipment problems in mathematical programming. An extensive treatment of both the theory and application of mass-transportation problems is provided by Rachev and Rüschendorf (1998, 1999). More recent discussions of applications of mass-transportation problems include Talagrand (1996), Levin (1997, 1999), Evans and Gangbo (1999), Ambrosio (2002, 2003) Feldman and McCann (2002), Carlier (2003), Angenent et al. (2003), Villani (2003), Brenier (2003), Feyel and Üstünel (2004), Barrett and Prigozhin (2009), Chartrand et al. (2009), Zhang (2011), Gabriel et al. (2010), Igbida et al. (2011), and Léonard (2012). More recently, an international conference on the Monge–Kantorovich optimal transportation problem, transport metrics, and their applications organized by the St. Petersburg

DOI 10.1007/978-1-4614-4869-3_5,

Notation	Description
$\mathcal{A}_c$	Kantorovich functional
$\mathcal{P}^{(P_1,P_2)}$	Space of all laws on $U \times U$ with marginals P_1 and P_2 or, alternatively, the space of all translocations of masses without transits permitted
P^*	Optimal transference plan
$c(x, y)$	Cost of transferring mass from x to y
$\mathcal{Q}^{(P_1,P_2)}$	Space of all translocations of masses with transits permitted
$\mathbf{D}_{n,\alpha}$	Ornstein-type metric
$\widetilde{P}$	Vector of probability measures $P_1, \ldots, P_N$
$\mathfrak{P}(\widetilde{P})$	Space of laws on U^N with fixed one-dimensional marginals
$A_c(\widetilde{P})$	Multidimensional version of Kantorovich functional $\mathcal{A}_c(P_1, P_2)$
$\mathcal{H}^*$	All convex functions in $\mathcal{H}$
$\mathcal{P} = \mathcal{P}_U = \mathcal{P}(U)$	Space of all laws on U
$\mathcal{P}^{\mathcal{H}}$	Space of all laws on (U, d) with a finite $H(d(\cdot, a))$-moment
$\mathcal{D}(x)$	$\|(d(x_1, x_2), d(x_1, x_3)), \ldots, d(x_{N-1}, x_N))\|$
$D(x)$	$H(\mathcal{D}(x))$
$\mathbb{K}(\widetilde{P}; \mathfrak{A})$	Dual form of $A_c(\widetilde{P})$
$\mathfrak{K}_H$	Multivariate analog of Kantorovich distance ℓ_H
$\overset{\circ}{\mu}_c$	Kantorovich–Rubinstein functional (minimal norm w.r.t. μ_c)
$m = m^+ + m^-$	Jordan decomposition of signed measure m
$\|\cdot\|_{\mathrm{w}}$	Kantorovich–Rubinstein or Wasserstein norm
$m_1 \times m_2$	Product measure

branch of the V. A. Steklov Mathematics Institute and the Euler Institute was held in St. Petersburg, Russia in June 2012 marking 100 years since the birth of L. V. Kantorovich.[1]

Despite the theoretical and practical significance of a direct application of the Kantorovich and the Kantorovich-Rubinstein problems, this chapter is devoted to them because of their link to the theory of probability metrics.[2] In fact, the Kantorovich problem and the dual theory behind it provide insights into the structure of some minimal probability distances such as the Kantorovich distance ℓ_H and the ℓ_p metric, respectively [see (3.3.11)]. Likewise, the Kantorovich–Rubinstein functional has normlike properties and can be regarded as a minimal norm (see discussion in Example 3.3.6).

We begin with an introduction to the Kantorovich and Kantorovich–Rubinstein problems and provide examples illustrating their application in different areas such as job assignments, classification problems, and best allocation policy. Then we continue with the dual theory, which leads to alternative representations of some minimal probability distances. Finally, we discuss an explicit representation of a class of minimal norms that define probability semimetrics.

[1]The program of the conference and related materials are available online at http://www.mccme.ru/~ansobol/otarie/MK2012conf.html.

[2]See Rachev (1991), Rachev and Taksar (1992), Rachev and Hanin (1995a,b), Cuesta et al. (1996), and Rachev and Rüschendorf (1999).

5.2 Statement of Monge–Kantorovich Problem

This section should be viewed as an introduction to the Monge–Kantorovich problem (MKP) and its related probability semidistances. There are six known versions of the MKP.

1. *Monge transportation problem.* In 1781, the French mathematician and engineer Gaspard Monge formulated the following problem in studying the most efficient way of transporting soil:

   > Split two equally large volumes into infinitely small particles and then associate them with each other so that the sum of products of these paths of the particles to a volume is least. Along what paths must the particles be transported and what is the lowest transportation cost?

   In other words, two sets S_1 and S_2 are the supports of two masses μ_1 and μ_2 with equal total weight $\mu_1(S_1) = \mu_2(S_2)$. The *initial* mass μ_1 is to be transported from S_1 to S_2 so that the result is the *final* mass μ_2. The transportation should be realized in such a way as to minimize the total labor involved.
2. *Kantorovich's mass transference problem.* In the Monge problem, let A and B be initial and final volumes. For any set $a \subset A$ and $b \subset B$, let $P(a,b)$ be the fraction of volume of A that was transferred from a to b. Note that $P(a,B)$ is equal to the ratio of volumes of a and A and $P(A,b)$ is equal to the ratio of volumes of b and B, respectively.

   In general we need not assume that A and B are of equal volumes; rather, they are bodies with equal masses though not necessarily uniform densities. Let $P_1(\cdot)$ and $P_2(\cdot)$ be the probability measures on a space U, respectively describing the masses of A and B. Then a shipping plan would be a probability measure P on $U \times U$ such that its projections on the first and second coordinates are P_1 and P_2, respectively. The amount of mass shipped from a neighborhood $\mathrm{d}x$ of x into the neighborhood $\mathrm{d}y$ of y is then proportional to $P(\mathrm{d}x, \mathrm{d}y)$. If the unit cost of shipment from x to y is $c(x,y)$, then the total cost is

$$\int_{U\times U} c(x,y)P(\mathrm{d}x,\mathrm{d}y). \tag{5.2.1}$$

   Thus we see that minimization of transportation costs can be formulated in terms of finding a distribution of $U \times U$ whose marginals are fixed and such that the double integral of the cost function is minimal. This is the so-called Kantorovich formulation of the Monge problem, which in abstract form is as follows:

   > Suppose that P_1 and P_2 are two Borel probability measures given on a separable metric space (s.m.s.) (U,d), and $\mathcal{P}^{(P_1,P_2)}$ is the space of all Borel probability measures P on $U \times U$ with fixed marginals $P_1(\cdot) = P(\cdot \times U)$ and $P_2(\cdot) = P_2(U \times \cdot)$. Evaluate the functional

$$\mathcal{A}_c(P_1,P_2) = \inf\left\{\int_{U\times U} c(x,y)P(\mathrm{d}x,\mathrm{d}y) : P \in \mathcal{P}^{(P_1,P_2)}\right\}, \tag{5.2.2}$$

   > where $c(x,y)$ is a given continuous nonnegative function on $U \times U$.

We will call the functional (5.2.2) the *Kantorovich functional* (*Kantorovich metric*) if $c = d$ [see Example 3.3.2, (3.4.18), and (3.4.54)].

The measures P_1 and P_2 may be viewed as the initial and final distributions of mass and $\mathcal{P}^{(P_1,P_2)}$ as the space of admissible transference plans. If the infimum in (5.2.2) is realized for some measure $P^* \in \mathcal{P}^{(P_1,P_2)}$, then P^* is said to be the *optimal transference plan*. The function $c(x, y)$ can be interpreted as the cost of transferring the mass from x to y.

Remark 5.2.1. Kantorovich's formulation differs from the Monge problem in that the class $\mathcal{P}^{(P_1,P_2)}$ is broader than the class of one-to-one transference plans in Monge's sense. Sudakov (1976) showed that if measures P_1 and P_2 are given on a bounded subset of a finite-dimensional Banach space and are absolutely continuous with respect to the Lebesgue measure, then there exists an optimal one-to-one transference plan.

Remark 5.2.2. Another example of the MKP is assigning army recruits to jobs to be filled. The flock of recruits has a certain distribution of parameters such as education, previous training, and physical conditions. The distribution of parameters that are necessary to fill all the jobs might not necessarily coincide with one of the contingents. There is a certain cost involved in training an individual for a specific job depending on the job requirements and individual parameters; thus the problem of assigning recruits to the job and training them so that the total cost is minimal can be viewed as a particular case of the MKP.

Comparing the definition of $\mathcal{A}_c(P_1, P_1)$ with Definition 3.3.2 [see (3.3.2)] of minimal distance $\widehat{\mu}$ we see that

$$\mathcal{A}_c = \widehat{\mu} \tag{5.2.3}$$

for any compound distance μ of the form

$$\mu(P) = \mu_c(P) = \int_{U\times U} c(x, y) P(dx, dy), \qquad P \in \mathcal{P}_2. \tag{5.2.4}$$

(Recall that $\mathcal{P}_k$ is the set of all Borel probability measures on the Cartesian product U^k.) If $\mu(P) = \mathcal{L}_H := \int H(d(x, y)) P(dx, dy)$, $H \in \mathcal{H}$, $P \in \mathcal{P}_2$, is the H-average compound distance [see (3.4.1)], then $\mathcal{A}_c = \widehat{\mathcal{L}}_H$. This example seems to be the most important one from the point of view of the theory of probability metrics. For this reason we will devote special attention to the mass transportation problem with cost function $c(x, y) = H(d(x, y))$.

3. *Kantorovich–Rubinstein–Kemperman problem of multistaged shipping*. In 1957, Kantorovich and Rubinstein studied the problem of transferring masses in cases where transits are permitted. Rather than shipping a mass from a certain subset of U to another subset of U in one step, the shipment is made in n stages. That is, we ship $A = A_1$ to volume A_2, then A_2 to $A_3, \ldots, A_{n-1}$ to $A_n = B$. Let $\gamma_n(a_1, a_2, a_3, \ldots, a_n)$ be a measure equal to the total mass that was removed

from the set a_1 and on its way to a_n passed the sets $a_2, a_3, \ldots, a_{n-1}$. If $c(x,y)$ is the unit cost of transportation from x to y, then the total cost under such a transportation plan is

$$\begin{aligned}&\int_{U\times U} c(x,y)\gamma_n(\mathrm{d}x\times \mathrm{d}y\times U^{n-2})\\ &\quad+\sum_{i=2}^{n-2}\int_{U\times U} c(x,y)\gamma_n(U^{i-1}\times \mathrm{d}x\times \mathrm{d}y\times U^{n-i-1})\\ &\quad+\int_{U\times U} c(x,y)\gamma_n(U^{n-2}\times \mathrm{d}x\times \mathrm{d}y)\\ =:&\int_{U\times U} c(x,y)\Gamma_n(\mathrm{d}x\times \mathrm{d}y). \qquad (5.2.5)\end{aligned}$$

A more sophisticated plan consists of a sequence of transportation subplans $\gamma_n, n = 2, 3, \ldots$, due to Kemperman (1983). Each subplan γ_n need not transfer the whole mass from A to B, rather only a certain part of it. However, combined the subplans complete the transshipment of mass, that is,

$$P_1(A) = \sum_{n=2}^{\infty} \gamma_n(A\times U^{n-1}) \qquad (5.2.6)$$

and

$$P_2(B) = \sum_{n=2}^{\infty} \gamma_n(U^{n-1}\times B). \qquad (5.2.7)$$

The total cost of transshipment under this sequential transportation plan will be the sum of costs of each subplan and is equal to

$$\int_{U\times U} c(x,y)Q(\mathrm{d}x,\mathrm{d}y), \qquad (5.2.8)$$

where

$$Q(A\times B) = \sum_{n=2}^{\infty} \Gamma_n(A\times B) \qquad (5.2.9)$$

and Γ_n is defined by (5.2.5):

$$\begin{aligned}\Gamma_n(A,B) :=\ &\gamma_n(A\times B\times U^{n-2})\\ &+\sum_{i=2}^{n-2}\gamma_n(U^{i-1}+A\times B\times U^{n-i-1})+\gamma_n(U^{n-2}\times A\times B).\end{aligned}$$

Note that now Q is not necessarily a probability measure. The marginals of Q are equal to

$$Q_1(A) = \sum_{n=2}^{\infty}\left(\gamma_n(A \times U^{n-1}) + \sum_{i=1}^{n-2}\gamma_n(U^i \times A \times U^{n-i-1})\right) \tag{5.2.10}$$

and

$$Q_2(B) = \sum_{n=2}^{\infty}\left(\gamma_n(U^{n-1} \times B) + \sum_{i=1}^{n-2}\gamma_n(U^i \times B \times U^{n-i-1})\right), \tag{5.2.11}$$

respectively. Combining equalities (5.2.6), (5.2.7) and (5.2.10), (5.2.11), we obtain

$$Q_1(A) - P_1(A) = Q_2(A) - P_2(A) = \sum_{n=3}^{\infty}\sum_{i=1}^{n-2}\gamma_n(U^i \times A \times U^{n-1-i}) \tag{5.2.12}$$

for any $A \in \mathcal{B}(U)$. Denote the space of all translocations of masses (without transits permitted) by $\mathcal{P}^{(P_1,P_2)}$ [see (5.2.2)]. Under the *translocations of masses with transits permitted* we will understand the finite Borel measure Q on $\mathcal{B}(U \times U)$ such that

$$Q(A \times U) - Q(U \times A) = P_1(A) - P_2(A) \tag{5.2.13}$$

for any $A \in \mathcal{B}(U)$. Denote the space of all Q satisfying (5.2.13) by $\mathcal{Q}^{(P_1,P_2)}$. Let a continuous nonnegative function $c(x, y)$ be given that represents the cost of transferring a unit mass from x to y. The total cost of transferring the given mass distributions P_1 and P_2 is given by

$$\mu_c(P) := \int_{U\times U} c(x, y)P(\mathrm{d}x, \mathrm{d}y), \quad \text{if } P \in \mathcal{P}^{(P_1,P_2)} \tag{5.2.14}$$

[see (5.2.2)] or

$$\mu_c(Q) := \int_{U\times U} c(x, y)Q(\mathrm{d}x, \mathrm{d}y), \quad \text{if } Q \in \mathcal{Q}^{(P_1,P_2)}. \tag{5.2.15}$$

Hence, if μ_c is a probability distance, then the minimal distance

$$\widehat{\mu}_c(P_1, P_2) = \inf\{\mu_c(P) : P \in \mathcal{P}^{(P_1,P_2)}\} \tag{5.2.16}$$

may be viewed as the minimal translocation cost, while the minimal norm (Definition 3.3.4)

$$\mathring{\mu}_c(P_1, P_2) = \inf\{\mu_c(Q) : Q \in \mathcal{Q}^{(P_1,P_2)}\} \tag{5.2.17}$$

may be viewed as the minimal translocation cost in the case of transits permitted.

The problem of calculating the exact value of $\widehat{\mu}_c$ (for general c) is known as the *Kantorovich problem*, and $\widehat{\mu}$ is called the *Kantorovich functional* [see equality (5.2.2)]. Similarly, the problem of evaluating $\overset{\circ}{\mu}_c$ is known as the *Kantorovich–Rubinstein problem*, and $\overset{\circ}{\mu}_c$ is said to be the *Kantorovich–Rubinstein functional*. Some authors refer to $\overset{\circ}{\mu}_c$ as the *Wasserstein norm* if $c = d$. In Example 3.3.6 in Chap. 3 we defined $\overset{\circ}{\mu}_c$ as the *minimal norm*.

The functional $\overset{\circ}{\mu}_c$ is frequently used in mathematical-economical models but is not applied in probability theory.[3] Observe, however, the following relationship between the Fortet–Mourier metric

$$\zeta(P, Q; \mathcal{G}^p) = \sup\left\{ \int_U f\mathrm{d}(P - Q) : f : U \to \mathbb{R}, \quad \text{and} |f(x) - f(y)| \right.$$
$$\left. \le d(x, y)\max[1, d(x, a)^{p-1}, d(y, a)^{p-1}] \quad \forall x, y \in U \right\}$$

[see Lemma 4.4.1, (4.4.35)] and the minimal norm $\overset{\circ}{\mu}_c$:

$$\zeta(P, Q; \mathcal{G}^p) = \overset{\circ}{\mu}_c(P, Q),$$

where the cost function is given by $c(x, y) = d(x, y)\max[1, d(x, a), d^{p-1}(y, a)]$, $p \ge 1$ (see further Theorem 5.4.3).

Open Problem 5.2.1. The last equality provides a representation of the Fortet–Mourier metric in terms of the minimal norm $\overset{\circ}{\mu}_c$. It is interesting to find a similar representation but in terms of a minimal metric $\widehat{\mu}$. On the real line ($U = \mathbb{R}$, $d(x, y) = |x - y|$) one can solve this problem as follows:

$$\zeta(P, Q; \mathcal{G}^p) = \int_{\mathbb{R}} \max(1, |x - a|^{p-1})(P - Q)(-\infty, x]|\mathrm{d}x$$
$$= \inf\left\{ \int_{\mathbb{R}} (\Pr(X \le t < Y) + \Pr(Y \le t < X) \max(1, |t - a|^{p-1})\mathrm{d}t, \right.$$
$$\left. X, Y \in \mathfrak{X}(\mathbb{R}) : \Pr_X = P, \Pr_Y = Q \right\}$$

(see further Theorems 5.5.1 and 6.6.1). Thus, in this particular case, $\zeta(P, Q; \mathcal{G}^p) = \widehat{\mu}_c(P, Q)$, where the cost function c is given by

---

[3] See, for example, Bazaraa and Jarvis (2005).

$$c(x,y) = \int (I\{x \le t < y\} + I\{y \le t < x\}) \max(1, |t-a|^{p-1}) \mathrm{d}t.$$

However, if U is an s.m.s., then the problem of determining a minimal metric $\widehat{\mu}$ such that $\zeta(\cdot,\cdot;\mathcal{G}^p) = \widehat{\mu}$ is still open. Note that we can define a minimal metric, namely, $\widehat{\mathcal{L}}_p = (\widehat{\mu}_{d^p})^{1/p}$ [see (4.4.54) in Chap. 4], that metrizes the same topology as $\zeta(\cdot,\cdot;\mathcal{G}^p)$.

Example 5.2.1. Kantorovich functionals and the problem of classification. In multivariate statistical analysis, the problem of classification is well known [see, for example, Anderson (2003)]. Let us give one popular example of an alternative problem of classification.

Army recruits are given a battery of tests to determine their fitness for different jobs: the scores are a set of measurements $x \in U$, where (U,d) is an s.m.s., for example, $U = \mathbb{R}^k$, $d(x,y) = \|y-x\|$. The distribution of scores is given by the measure P_1,

$$P_1(A) = \frac{\text{number of recruits with scores in } A}{\text{Total number of recruits}}.$$

On the other hand, the army's needs can be expressed by a probability measure P_2 on U that represents the desired distribution of scores for the jobs needed to be filled. The problem is to choose an optimal classification (or assignment) of recruits to jobs. A classification can be specified by choosing a bounded measure Q on $\mathcal{B}(U \times U)$. If a classification satisfies the balancing conditions

$$Q(A \times U) = P_1(A), \quad Q(U \times B) = P_2(B), \tag{5.2.18}$$

then we view the quantity of recruits with scores $x \in A$ that are classified as satisfying (after retraining) the requirements of jobs that call for scores $y \in B$. If we think that the training procedure might be a multistaged one, in which the same individual gradually changes his scores (and fitness for different jobs respectively) in a sequence of n retraining stages, then the measure Q satisfies the balancing conditions

$$Q(A \times U) - Q(U \times A) = P_1(A) - P_2(A). \tag{5.2.19}$$

The interpretation of $Q(A \times B)$ is the combined number of GIs at all stages who had scores x in A and who were trained to fit the jobs that require scores y in B. Let $c_0(x,y)$ be the cost of training a person with a score x to fit a job that requires score y. Consider the joint cost $c(x,y) = c_0(x,y) + c_0(y,x)$. (Nonsymmetric cost functions will be considered in Sect. 7.4; see Theorem 7.4.2.) The obvious assumption on c is that

$$c(x,x) = 0. \tag{5.2.20}$$

Moreover, we can assume that

$$d(x', y') \leq d(x'', y'') \Rightarrow c(x', y') \leq c(x'', y''), \qquad (5.2.21)$$

i.e., the cost $c(x, y)$ increases with $d(x, y)$. In particular, (5.2.21) implies

$$d(x', a) < d(x'', a) \Rightarrow c(x', a) < c(x'', a), \qquad (5.2.22)$$

$$d(a, y') < d(a, y'') \Rightarrow c(a, y') < c(a, y''), \qquad (5.2.23)$$

for a fixed point $a \in U$, which one can consider as the "center" of recruitment possibilities and the army's needs. Implications (5.2.21)–(5.2.23) suggest that one reasonable form of c is given by

$$c(x, y) = d(x, y) \max(h(d(x, a)), h(d(y, a))), \qquad (5.2.24)$$

where h is a continuous nondecreasing function on $[0, \infty)$, $h(0) \geq 0$, $h(x) > 0$, for $x > 0$. Another natural choice of c might be

$$c(x, y) = H(d(x, y)), \qquad (5.2.25)$$

where $H \in \mathcal{H}$ (Examples 2.4.1 and 3.4.1). Fixing the cost function c we conclude that the total cost involved in using the classification Q is calculated by the integral

$$TC(Q) = \int_{U \times U} c(x, y) Q(\mathrm{d}x, \mathrm{d}y). \qquad (5.2.26)$$

The following problems therefore arise.

Problem 5.2.1. Considering the set of classifications $\mathcal{P}^{(P_1, P_2)}$ we seek to characterize the optimal $P^* \in \mathcal{P}^{(P_1, P_2)}$ (if P^* exists) for which

$$TC(P^*) = \inf\{TC(P) : P \in \mathcal{P}^{(P_1, P_2)}\} \qquad (5.2.27)$$

and to evaluate the bound

$$\widehat{\mu}_c(P_1, P_2) = \inf\{TC(P) : P \in \mathcal{P}^{(P_1, P_2)}\}. \qquad (5.2.28)$$

Problem 5.2.2. Considering the set of classifications $\mathcal{Q}^{(P_1, P_2)}$ we seek to characterize the optimal $Q^* \in \mathcal{Q}^{(P_1, P_2)}$ (if Q^* exists) for which

$$TC(Q^*) = \inf\{TC(Q) : Q \in \mathcal{Q}^{(P_1, P_2)}\} \qquad (5.2.29)$$

and to evaluate the bound

$$\overset{\circ}{\mu}_c(P_1, P_2) = \inf\{TC(Q2) : Q \in \mathcal{Q}^{(P_1, P_2)}\}.$$

Problem 5.2.3. What kind of quantitative relationships exist between $\widehat{\mu}_c$ and $\overset{\circ}{\mu}_c$?

In the next three sections, we will attempt to provide some answers to Problems 5.2.1–5.2.3.

Example 5.2.2. Kantorovich functionals and the problem of the best allocation policy. Karatzas (1984) considers d medical treatments (or projects or investigations) with the state of the jth of them (at time $t \geq 0$) denoted by $x_j(t)$.[4] At each instant of time t, it is allowed to use only one medical treatment, denoted by $i(t)$, which then evolves according to some Markovian rule; meanwhile, the states of all other projects remain frozen.

Now we will consider the situation where one is allowed to use a combination of different medical treatments (say, for brevity, medicines) denoted by $M_1, \ldots, M_d$. Let $d = 2$ and (U, d) be an s.m.s. The space U may be viewed as the space of a patient's parameters. Assume that for $i = 1, 2$ and for any Borel set $A \in \mathcal{B}(U)$ the exact quantity $P_1(A)$ of medicine M (which should be prescribed to the patient with parameters A) is known. Normalizing the total quantity $P_i(U)$ that can be prescribed by 1, we can consider P_i as a probability measure on $\mathcal{B}(U)$. Our aim is to handle an optimal policy of treatments with medicines M_1, M_2. Such a treatment should be a combination of medicines M_1 and M_2 varying on different sets $A \subset U$.

A policy can be specified by choosing a bounded measure Q on $\mathcal{B}(U \times U)$ and the quantity of medicine M_i in the case of *patient with parameters*, $i = 1, 2$, by following policy Q. The policy may satisfy the balancing condition

$$Q(A \times U) = P_1(A), \qquad Q(U \times A) = P_2(A), \qquad A \in \mathcal{B}(U), \tag{5.2.30}$$

i.e., $Q \in \mathcal{P}^{(P_1, P_2)}$ or (in the case of a multistage treatment)

$$Q(A \times U) - Q(U \times A) = P_1(A) - P_2(A), \qquad A \in \mathcal{B}S(U), \tag{5.2.31}$$

i.e., $Q \in \mathcal{Q}^{(P_1, P_2)}$. Let $c(x_1, x_2)$ be the cost of treating the patient with instant parameters x_i with medicines M_i, $i = 1, 2$. The $\widehat{\mu}$ and $\overset{\circ}{\mu}$ [see (5.2.16) and (5.2.17)] represent the minimal total costs under the balancing conditions (5.2.30) and (5.2.31), respectively. In this context, Problems 5.2.1–5.2.3 are of interest.

4. *Gini's index of dissimilarity.* Already at the beginning of this century, the following question arose among probabilists: What is the proper way to measure the degree of difference between two random quantities [see the review article by Kruskal (1958)]? Specific contributions to the solution of this problem, which is closely related to Kantorovich's problem 5.2.2, were made by Gini, Hoeffding, Frechet, and their successors. In 1914, Gini introduced the concept of a *simple index of dissimilarity*, which coincides with Kantorovich's metric $\mathcal{A}_d = \mathbb{R}^1$,

---

[4] See also the general discussion in Whittle (1982, p. 210–211).

$d(x, y) = |x - y|$). That is, Gini studied the functional

$$\mathcal{K}(F_1, F_2) = \inf\left\{\int_{\mathbb{R}^2} |x - y| \mathrm{d}F(x, y) : F \in \mathcal{F}(F_1, F_2)\right\} \tag{5.2.32}$$

in the space $\mathcal{F}$ of one-dimensional distribution functions (DF) F_1 and F_2. In (5.2.32), $\mathcal{F}(F_1, F_2)$ is the class of all bivariate DFs F with fixed marginal distributions $F_1(x) = F(x, \infty)$ and $F_2(x) = F(\infty, x)$, $x \in \mathbb{R}^1$ [see (3.4.54)]. Gini and his students devoted a great deal of effort to studying the properties of the sample measure of discrepancy, Glivenko's theorem, and goodness-of-fit tests in terms of $\mathcal{K}$. Of special importance in these investigations was the question of finding explicit expressions for this measure of discrepancy and its generalizations. Thus in 1943, Salvemini showed that

$$\mathcal{K}(F_1, F_2) = \int_{-\infty}^{\infty} |F_1(x) - F_2(x)| \mathrm{d}x \tag{5.2.33}$$

in the class of discrete DFs and in 1956 Dall'Aglio extended it to all of $\mathcal{F}$. This formula was proved and generalized in many ways [see Example 3.4.3, Eq. (3.4.19), and Sect. 7.4].

2. *Ornstein metric*. Let (U, d) be an s.m.s., and let $d_{n,\alpha}$, $\alpha \in [0, \infty]$, be the analog of the Hamming metric on U^n,[5] namely,

$$d_{n,\alpha}(x, y) = \frac{1}{n}\left(\sum_{i=1}^{n} d^{\alpha}(x_i, y_i)\right)^{\alpha'}, \qquad x = (x_1, \ldots, x_n) \in U^n,$$

$$y = (y_1, \ldots, y_n) \in U^n, \quad 0 < \alpha < \infty, \quad \alpha' = \min(1, 1/\alpha),$$

$$d_{n,0}(x, y) = \frac{1}{n}\sum_{i=1}^{n} I\{x_i \neq y_i\},$$

$$d_{n,\infty}(x, y) = \frac{1}{n}\max\{d(x_i, y_i) : i = 1, \ldots, n\}.$$

For any Borel probability measures P and Q on U^n define the following analog of the Kantorovich metric:

$$\mathbf{D}_{n,\alpha}(P, Q) = \inf\left\{\int d_{n,\alpha} \mathrm{d}\widehat{P} : \widehat{P} \in \mathcal{P}^{(P,Q)}\right\}. \tag{5.2.34}$$

The simple probability metric $\mathbf{D}_{n,0}$ is known among specialists in the theory of dynamical systems and coding theory as Ornstein's $\bar{d}$-metric, while $\mathbf{D}_{n,1}$ is called

[5] See Gray (1988, p. 48).

the $\overline{\rho}$-distance. In information theory, the Kantorovich metric $\mathbf{D}_{1,1}$ is known as the Wasserstein (sometimes Lévy–Wasserstein) metric. We will show that

$$\mathbf{D}_{n,\alpha}(P,Q) = \sup\left\{\left|\int f\mathrm{d}(P-Q)\right| : f : u^n \to \mathbb{R}^1,\ L_{n,\alpha}(f) \le 1\right\},$$

$$L_{n,\alpha}(f) = \sup\{|f(x) - f(y)|/d_{n,\alpha}(x,y),\ x \ne y,\ y \in U^n\}, \quad (5.2.35)$$

for all $\alpha \in [0,\infty)$ (see Corollary 6.2.1 for the case where $0 < \alpha < \infty$ and Corollary 7.5.2 for the case where $\alpha = 0$).

3. *Multidimensional Kantorovich problem.* We now generalize the preceding problems as follows.

   Let $\widetilde{P} = \{P_i, i = 1,\ldots,N\}$ be a set of probability measures given on an s.m.s. (U,d), and let $\mathfrak{P}(\widetilde{P})$ be the space of all Borel probability measures P on the direct product U^N with fixed projections P_i on the ith coordinates, $i = 1,\ldots,N$. Evaluate the functional

$$A_c(\widetilde{P}) = \inf\left\{\int_{U^N} c\mathrm{d}P : P \in \mathfrak{P}(\widetilde{P})\right\}, \quad (5.2.36)$$

   where c is a given continuous function on U^N.

This transportation problem of infinite-dimensional linear programming is of interest in its own right in problems of stability of stochastic models.[6] This is related to the fact that if $\{P_1^{(i)},\ldots,P_N^{(i)}\}$, $i = 1,2$, are two sets of probability measures on (U,d) and $P^{(i)} := P_1^{(i)} \times \cdots \times P_N^{(i)}$ are their products, then the value of the Kantorovich functional

$$\mathcal{A}_{c^*}(P^{(1)},P^{(2)}) = \inf\left\{\int_{U^{2N}} c^*\mathrm{d}\widehat{P} : \widehat{P} \in \mathcal{P}^{(P^{(1)},P^{(2)})}\right\} \quad (5.2.37)$$

with cost function c^* given by

$$c^*(x_1,\ldots,x_N,y_1,\ldots,y_N) := \phi(c_1(x_1,y_1),\ldots,c_N(x_N,y_N)), \quad (5.2.38)$$

$$x_i, y_i \in U, \quad i = 1,\ldots,N,$$

where ϕ is some nondecreasing, nonnegative continuous function on $\mathbb{R}^n$, coincides with

[6] See Kalashnikov and Rachev (1988, Chaps. 3 and 6).

$$A_{c^*}(P_1^{(1)},\dots,P_N^{(1)},P_1^{(2)},\dots,P_N^{(2)})$$
$$= \inf\left\{\int_{U^{2N}} c^* \mathrm{d}P : P \in \mathfrak{P}(P_1^{(1)},\dots,P_N^{(1)},P_1^{(2)},\dots,P_N^{(2)})\right\}. \qquad (5.2.39)$$

See further Theorem 7.2.3 of Chap. 7.

5.3 Multidimensional Kantorovich Theorem

In this section, we will prove the duality theorem for the multidimensional Kantorovich problem [see (5.2.36)].

For brevity, $\mathcal{P}$ will denote the space $\mathcal{P}_U$ of all Borel probability measures on an s.m.s. (U,d). Let $N = 2,3,\dots$ and let $\|\mathbf{b}\|$ ($\mathbf{b} \in \mathbb{R}^m$, $m = \binom{N}{2}$) be a *monotone seminorm* $\|\cdot\|$, i.e., $\|\cdot\|$ is a seminorm in $\mathbb{R}^m$ with the following property: if $0 < b'_i, \le b''_i$, $i = 1,\dots,m$, then $\|\mathbf{b}'\| \le \|\mathbf{b}''\|$. For example,

$$\|\mathbf{b}\|_p := \left(\sum_{i=1}^m |b_i|^p\right)^{1/p}, \qquad \|\mathbf{b}\|_\infty := \max\{|b_i| : i = 1,\dots,m\},$$

$$\|\mathbf{b}\| := \left|\sum_{i=1}^m b_i\right| \quad \text{and} \quad \|\mathbf{b}\| := \left(\left|\sum_{i=1}^k b_i\right|^p + \left|\sum_{i=k+1}^n b_i\right|^p\right)^{1/p}, \quad p \ge 1.$$

For any $x = (x_1,\dots,x_N) \in U^N$ let

$$\mathcal{D}(x) = \|d(x_1,x_2), d(x_1,x_3),\dots,d(x_1,x_N), d(x_2,x_3),\dots,d(x_{N-1},x_N)\|.$$

Let $\widetilde{P} = (P_1,\dots,P_N)$ be a finite set of measures in $\mathcal{P}$, and let

$$A_D(\tilde{P}) := \inf\left\{\int_{U^N} D\mathrm{d}P : P \in \mathfrak{P}(\tilde{P}\right\}, \qquad (5.3.1)$$

where $D(x) := H(\mathcal{D}(x))$, $x \in U^N$, and $H \in \mathcal{H}^* = \{H \in \mathcal{H},\ H \text{ convex}\}$ (see Example 2.4.1).

Let $\mathcal{P}^H$ be the space of all measures in $\mathcal{P}$ for which $\int_U H(d(x,a))P(\mathrm{d}x) < \infty$, $a \in U$. For any $U_0 \subseteq U$ define the class $\mathrm{Lip}(U_0) := \bigcup_{\alpha>0} \mathrm{Lip}_{1,\alpha}(U_0)$, where

$$\mathrm{Lip}_{1,\alpha}(U_0) := \{f : U \to \mathbb{R}^1 : |f(x) - f(y)| \le \alpha d(x,y), \quad \forall x,y \in U_0,$$
$$\text{and} \quad \sup\{|f(x)| : x \in U_0\} < \infty\}.$$

Define the class

$$\mathfrak{G}(U_0) = \Big\{\mathbf{f} = (f_1, \dots, f_n) : \sum_{i=1}^{n} f_i(x_i) \le D(x_1, \dots, x_N)$$

$$\text{for} \quad x_i \in U_0, f_i \in \mathrm{Lip}(U_0), i = 1, \dots, N\Big\},$$

and for any class $\mathfrak{U}$ of vectors $\mathbf{f} = (f_1, \dots, f_N)$ of measurable functions let

$$\mathbb{K}(\widetilde{P}; \mathfrak{A}) = \sup\left\{\sum_{i=1}^{N} \int_U f_i \mathrm{d}P : f \in \mathfrak{A}\right\}, \tag{5.3.2}$$

assuming that $P_i \in \mathcal{P}^H$ and f_i is P-integrable.

Lemma 5.3.1.

$$A_D(\widetilde{P}) \ge \mathbb{K}(\widetilde{P}; \mathfrak{G}(U)). \tag{5.3.3}$$

Proof. Let $\mathbf{f} = (f_1, \dots, f_N) \in \mathfrak{G}(U)$ and $P \in \mathfrak{B}(\widetilde{P})$, where, as in (5.2.36), $\mathcal{B}(\widetilde{P})$ is the set of all laws on U^N with fixed projections P_i on the ith coordinates, $i = 1, \dots, N$. Then

$$\sum_{i=1}^{N} \int_U f_i(x_i) P(\mathrm{d}x_i) = \int_{U^N} \sum_{i=1}^{N} f_i(x_i) P(\mathrm{d}x_1, \dots, \mathrm{d}x_N)$$
$$\le \int_{U^N} D \mathrm{d}P.$$

The last inequality, together with (5.3.1) and (5.3.2), completes the proof (5.3.3). □

The next theorem [an extension of Kantorovich's (1940) theorem to the multidimensional case] shows that exact equality holds in (5.3.3).

Theorem 5.3.1. *For any s.m.s.* (U, d) *and for any set* $\widetilde{P} = (P_1, \dots, P_N)$, $P_i \in \mathcal{P}^H$, $i = 1, \dots, N$,

$$A_D(\widetilde{P}) = \mathbb{K}(\widetilde{P}; \mathfrak{G}(U)). \tag{5.3.4}$$

If the set P *consists of tight measures, then the infimum is attained in* (5.3.1).

Proof. 1. Suppose first that d is a bounded metric in U, and let

$$\rho_i(x_i, y_i) = \sup\{|D(x_1, \dots, x_N) - D(y_1, \dots, y_N)| : x_j = y_j \in U,$$
$$j = 1, \dots, N, \ j \ne i\}, \tag{5.3.5}$$

for $x_i, y_i \in U$, $i = 1, \dots, N$. Since H is a convex function and d is bounded, $\rho_1, \dots, \rho_N$ are bounded metrics. Let $U_0 \subseteq U$ and let $\mathfrak{G}'(U_0)$ be the space of all collections $\mathbf{f} = (f_1, \dots, f_N)$ of measurable functions on U_0 such that $f_1(x_1) + \dots + f_N(x_N) < D(x_1, \dots, x_N)$, $x_1, \dots, x_N \in U_0$. Let $\mathfrak{G}''(U_0)$ be a subset of

$\mathfrak{G}'(U_0)$ of vectors $\mathbf{f}$ for which $|f_i(x) - f_i(y)| \le \rho_i(x, y)$, $x, y \in U_0$, $i = 1, \dots, N$. Observe that $\mathfrak{G}'' \subset \mathfrak{G} \subset \mathfrak{G}'$. We wish to show that if $P_i(U_0) = 1$, $i = 1, \dots, N$, then

$$\mathbb{K}(\widetilde{P}; \mathfrak{G}'(U_0)) = \mathbb{K}(\widetilde{P}; \mathfrak{G}''(U)). \tag{5.3.6}$$

Let $\mathbf{f} \in \mathfrak{G}''(U_0)$. We define sequentially the functions

$$\begin{aligned}
f_1^*(x_1) &= \inf\{D(x_1, \dots, x_N) - f_2(x_2) - \dots - f_N(x_N) : \\
&\qquad x_2, \dots, x_N \in U_0\}, \quad x_1 \in U, \\
f_2^*(x_2) &= \inf\{D(x_1, \dots, x_N) - f_1^*(x_1) - f_3(x_3) - \dots - f_N(x_N) : \\
&\qquad x_1 \in U, x_3, \dots, x_N \in U_0\}, \quad x_2 \in U, \dots, \\
f_N^*(x_N) &= \inf\{D(x_1, \dots, x_N) - f_1^*(x_1) - \dots - f_{N-1}^*(x_{N-1}) : \\
&\qquad x_1, \dots, x_{N-1} \in U\}, \quad x_N \in U.
\end{aligned}$$

Since D is continuous, it follows that f_j^* are upper semicontinuous and, hence, Borel measurable. Also, $f_1^*(x_1) + \dots + f_N^*(x_N) \le D(x_1, \dots, x_N)$ $\forall x_1, \dots, x_N \in U$. Furthermore, for any $x_1, y_1 \in U$

$$\begin{aligned}
f_1^*(x_1) - f_1^*(y_1) &= \inf\{D(x_1, \dots, y_N) - f_2(x_2) - \dots - f_N(x_N) : \\
&\qquad x_2, \dots, x_N \in U_0\} \\
&\quad + \sup\{f_2(y_2) + \dots + f_n(y_N) - D(y_1, \dots, y_N) : \\
&\qquad y_2, \dots, y_N \in U_0\} \\
&\le \sup\{D(x_1, y_2, \dots, y_N) - D(y_1, \dots, y_N) : \\
&\qquad y_2, \dots, y_N \in U_0\} \\
&\le \rho_1(x_1, y_1).
\end{aligned}$$

A similar argument proves that the collection $\mathbf{f}^* = (f_1^*, \dots, f_N^*)$ belongs to the set $\mathfrak{G}''(U)$. Given $x_1 \in U_0$, we have $f(x_1) \le D(x_1, x_2, \dots, x_N) - f_2(x_2) - \dots - f_N(x_N)$ for all $x_2, \dots, x_N \in U_0$. Thus, $f(x_1) \le f^*(x_1)$. Also, if $x_2 \in U_0$, then

$$\begin{aligned}
f_2^*(x_2) = &\inf_{x_1 \in U, x_3, \dots, x_N \in U_0} \{D(x_1, \dots, x_N) \\
&- \inf_{y_2, \dots, y_N \in U_0} [D(x_1, y_2, \dots, y_N) - f_2(x_2) - \dots - f_N(y_N)] \\
&- f_3(x_3) - \dots - f_N(x_N)\} \\
\ge &\inf_{x_1 \in U, x_3, \dots, x_N \in U_0} \{D(x_1, \dots, x_N) - D(x_1, x_2, \dots, x_N) + f_2(x_2) \\
&+ \dots + f_N(x_N) - f_3(x_3) - \dots - f_N(x_N)\} \\
= &\ f_2(x_2).
\end{aligned}$$

Similarly, $f_i^*(x_j) \geq f_i(x_i)$ for all $i = 1, \ldots, N$ and $x_i \in U_0$. Hence,

$$\sum_{i=1}^{N} \int f_i \mathrm{d}P_i \leq \sum_{i=1}^{N} \int f_i^* \mathrm{d}P_i,$$

which implies the inequality

$$\mathbb{K}(\widetilde{P}; \mathfrak{G}'(U_0) < \mathbb{K}(\widetilde{P}; \mathfrak{G}''(U)), \tag{5.3.7}$$

from which (5.3.6) clearly follows.

Case 1. Let U be a finite space with the elements $u_1, \ldots, u_n$. From the duality principle in linear programming, we have[7]

$$\begin{aligned} A_D(\widetilde{P}) = \inf & \left\{ \sum_{i_1=1}^{n} \cdots \sum_{i_N=1}^{n} D(u_{i_1}, \ldots, u_{i_N}) \pi(i_1, \ldots, i_N) : \right. \\ & \left. \pi(i_1, \ldots, i_N) \geq 0, \sum_{i_j : j \neq k} \pi(i_1, \ldots, i_N) = P_k(u_{i_k}), k = 1, \ldots, N \right\} \\ = \sup & \left\{ \sum_{i=1}^{n} \sum_{j=1}^{N} f_j(u_j) P_j(u_i) : \sum_{j=1}^{N} f_j(\tilde{u}_j) \leq D(\tilde{u})i, \ldots, \tilde{u}_N), \right. \\ & \left. \tilde{u}_1, \ldots, \tilde{u}_N \in U \right\} \\ = & \mathbb{K}(\widetilde{P}; \mathfrak{G}'(U)). \end{aligned}$$

Therefore, (5.3.7) implies the chain of inequalities

$$\mathbb{K}(\widetilde{P}; \mathfrak{G}(U)) \geq \mathbb{K}(\widetilde{P}; \mathfrak{G}''(U)) \geq \mathbb{K}(\widetilde{P}; \mathfrak{G}'(U)) \geq A_D(\widetilde{P}),$$

from which (5.3.4) follows by virtue of (5.3.3).

Case 2. Let U be a compact set. For any $n = 1, 2, \ldots$, choose disjoint nonempty Borel sets $A_1, \ldots, A_{m_n}$ of diameter less than $1/n$ whose union is U. Define a mapping $h_h : U \to U_n = \{u_1, \ldots, u_{m_n}\}$ such that $h_n(A_i) = u_i$, $i = 1, \ldots, m_n$. According to (5.3.6) we have for the collection $\widetilde{P}_n = (P_1 \circ h_n^{-1}, \ldots, P_N \circ h_n^{-1})$ the relation

$$\mathbb{K}(\widetilde{P}_n; \mathfrak{G}'(U_n)) = \sup \left\{ \sum_{i=1}^{N} \int_U f_i(h_n(u)) P_i(\mathrm{d}u) : \mathbf{f} \in \mathfrak{G}'(u) \right\}. \tag{5.3.8}$$

[7] See, for example, Bazaraa and Jarvis (2005).

If $\mathbf{f} \in \mathfrak{G}'(U_n)$, then $\sum_{i=1}^{N} f_i(h_n(\tilde{u}_i)) \leq D(h_n(\tilde{u}_1), \dots, h_n(\tilde{u}_N)) \leq D(\tilde{u}_1, \dots, \tilde{u}_N) + K/n$, where the constant K is independent of n and $\tilde{u}_1, \dots, \tilde{u}_N \in U$. Hence, from (5.3.8) we have

$$\mathbb{K}(\widetilde{P}_n; \mathfrak{G}'(U_n)) \leq \mathbb{K}(\widetilde{P}; \mathfrak{G}'(U)) + K/n. \tag{5.3.9}$$

According to Case 1, there exists a measure $P^{(n)} \in \mathfrak{P}(\widetilde{P}_n)$ such that

$$\int_{U^N} D \mathrm{d}P^{(n)} \leq \mathbb{K}(\widetilde{P}_n; \mathfrak{G}'(U_n)). \tag{5.3.10}$$

Since $P_i \circ h_n^{-1}$ converges weakly to $P_i, i = 1, \dots, N$, the sequence $\{P^{(n)}, n = 1, 2, \dots\}$ is weakly compact (Billingsley 1999, Sect. 6).
Let P^* be a weak limit of $P^{(n)}$. From estimate (5.3.9) and equality (5.3.10) it follows that

$$\int_{U^N} D \mathrm{d}P^* \leq \mathbb{K}(\widetilde{P}; \mathfrak{G}'(U)),$$

which together with Lemma 5.3.1 implies (5.3.4).

Case 3. Let (U, d) be a bounded s.m.s. Since $\int_U H(d(x, a)) P_i(\mathrm{d}x) < \infty$, the convexity of H and (5.3.5) imply that $\int_U \rho_i(x, a) P_i(\mathrm{d}x) < \infty, i = 1, \dots, N$. Let the P_i be tight measures (Definition 2.6.1). Then for each $n = 1, 2, \dots$ there exists a compact set K_n such that

$$\sup_{1 \leq i \leq N} \int_{U \setminus K_n} (1 + \rho_i(x, a)) P_i(\mathrm{d}x) < \frac{1}{n}. \tag{5.3.11}$$

For any $A \in \mathfrak{B}(U)$ set

$$P_{i,n}(A) := P_i(A \cap K_n) + P_i(U \setminus K_n) \delta_\alpha(A), \quad \widetilde{P}_n := (P_{1,n}, \dots, P_{N,n}),$$

where

$$\delta_\alpha(A) := \begin{cases} 1, & \alpha \in A, \\ 0, & \alpha \notin A. \end{cases}$$

By (5.3.6),

$$\begin{aligned} &\mathbb{K}(\widetilde{P}_n; \mathfrak{G}'(K_n \cup \{a\})) = \mathbb{K}(\widetilde{P}; \mathfrak{G}''(U)) \\ &\leq \sup \left\{ \sum_{i=1}^{N} \int_U f_i(x) P_i(\mathrm{d}x) + \int_{U \setminus K_n} \rho_i(x, a) P_i(\mathrm{d}x) : \mathbf{f} \in \mathfrak{G}(u) \right\} \\ &\leq \mathbb{K}(\widetilde{P}; \mathfrak{G}(U)) + N/n. \end{aligned} \tag{5.3.12}$$

According to Case 2, there exists a measure $P^{(n)} \in \mathfrak{P}(\widetilde{P})$ such that

$$\int_{U^n} D \mathrm{d} P^{(n)} \leq \mathbb{K}(\widetilde{P}_n; \mathfrak{G}'(K_n \cup \{a\})). \tag{5.3.13}$$

Similarly to Case 2, we then obtain (5.3.4) from relations (5.3.12) and (5.3.13).

Now let $P_1, \dots, P_N$ be measures that are not necessarily tight. Let $\overline{U}$ be the completion of U. To any positive ε choose the largest set A such that $d(x, y) \geq \varepsilon/2$ $\forall x, y \in A$. The set A is countable: $A = \{x_1, x_2, \dots\}$. Let $\overline{A}_n = \{x \in \overline{U} : d(x, x_n) < \varepsilon/2 \leq d(x, x_j)\ \forall j < n\}$, and let $A_n = \overline{A}_n \cap U$. Then $\overline{A}_n$, $n = 1, 2, \dots$, are disjoint Borel sets in $\overline{U}$ and A_n, $n = 1, 2, \dots$, are disjoint sets in U of diameter less than ε. Let $\overline{P}_i$ be a measure generated on $\overline{U}$ by P_i, $i = 1, \dots, N$. Then for $\mathbb{Q} = (\overline{P}_1, \dots, \overline{P}_N)$ there exists a measure $\overline{\mu} \in \mathfrak{P}(\mathbb{Q})$ such that

$$\int_{\overline{U}^N} D \mathrm{d}\overline{\mu} = \mathbb{K}(\mathbb{Q}; \mathfrak{G}(U)).$$

Let $P_{i,m}(B) = P_i(B \cap A_m)$ for all $B \in \mathfrak{B}(U)$, $i = 1, \dots, N$. To any multiple index $\mathbf{m} = (m_1, \dots, m_n)$, $m_i = 1, 2, \dots, i = 1, \dots, N$, define the measure

$$\mu_{\mathbf{m}} = c_{\mathbf{m}} P_{1,m_1} \times \cdots \times P_{N,m_N},$$

where the constant $c_{\mathbf{m}}$ is chosen such that

$$\mu_{\mathbf{m}}(A_{m_1} \times \cdots \times A_{m_N}) = \overline{\mu}_{\mathbf{m}}(A_{m_1} \times \cdots \times A_{m_N}).$$

Let $\mu_\varepsilon = \sum_{\mathbf{m}} \mu_{\mathbf{m}}$. Then for any $B \in \mathcal{B}(U)$

$$\begin{aligned}
\mu_\varepsilon(B \times U^{N-1}) &= \sum_{\mathbf{m}} c_{\mathbf{m}} P_{1,m_1}(B) P_{2,m_2}(U) \cdots P_{N,m_N}(U) \\
&= \sum_{\mathbf{m}} c_{\mathbf{m}} P_1(B \cap A_{m_1}) P_2(A_{m_2}) \cdots P_N(A_{m_N}) \\
&= \sum_{\mathbf{m}}{}' \frac{\overline{\mu}(\overline{A}_{m_1} \times \cdots \times \overline{A}_{m_N})}{P_{1,m_1}(A_{m_1}) \dots P_{N,m_N}(A_{m_N})} \\
&\quad \times P_1(B \cap A_{m_1}) P_2(A_{m_2}) \dots P_N(A_{m_N}),
\end{aligned}$$

where $\sum_{\mathbf{m}}{}'$ indicates summation over all $\mathbf{m}$ such that $P_{j,m_j}(A_{m_j}) > 0$ for all $j = 1, \dots, m_N$. Note that if $P_{1,m_1}(A_{m_1}) > 0$, then we have

$$\begin{aligned}
\sum_{m_2, \dots, m_N} \frac{\overline{\mu}(\overline{A}_{m_1} \times \cdots \times \overline{A}_{m_N})}{P_{1,m_1}(A_{m_1})} &= \overline{\mu}(\overline{A}_{m_1} \times U^{N-1}) / P_{1,m_1}(A_{m_1}) \\
&= \overline{P}_1(\overline{A}_{m_1}) / P_{1,m_1}(A_{m_1}) = 1.
\end{aligned}$$

This, together with analogous calculations for $\mu_\varepsilon(U^k \times B \times U^{N-k-1})$, $k = 1, 2, \ldots, N-1$, shows that $\mu_\varepsilon \in \mathfrak{P}(\widetilde{P})$; hence, to each positive ε,

$$\begin{aligned}\mu_\varepsilon(\mathcal{D}(y_1, \ldots, y_n) > \alpha + 2\varepsilon\|\mathbf{e}\|) &\\ \leq \sum\{\mu_m(A_{m_1} \times \cdots \times A_{m_n}) &: \mathcal{D}(x_1, \ldots, x_N) > \alpha + \varepsilon\|\mathbf{e}\|\}\\ \leq \overline{\mu}(\mathcal{D}(y_1, \ldots, y_n) > \alpha),&\end{aligned}$$

where $\mathbf{e}$ is a unit vector in $\mathbb{R}^m$. Since $H(t)$ is strictly increasing and $D(\mathbf{x}) = H(D(\mathbf{x}))$,

$$\begin{aligned}\int_N D(\mathbf{x})\mu_\varepsilon(\mathrm{d}\mathbf{x}) &= \int_0^\infty \mu_\varepsilon(\mathcal{D}(\mathbf{x}) > t)\mathrm{d}H(t)\\ &\leq \int_0^\infty \overline{\mu}(\mathcal{D}(\mathbf{x}) > t)\mathrm{d}H(t + 2\varepsilon\|\mathbf{e}\|) + H(2\varepsilon\|\mathbf{e}\|)\\ &\leq \int_{\overline{U}^N} D(\mathbf{x})\overline{\mu}(\mathrm{d}\mathbf{x}) + \int_{\overline{U}^N} (H(\mathcal{D}(\mathbf{x}) + 2\varepsilon\|\mathbf{e}\|)\\ &\quad -D(\mathbf{x}))\overline{\mu}(\mathrm{d}\mathbf{x}) + H(2\varepsilon\|\mathbf{e}\|).\end{aligned}$$

From the Orlicz condition it follows that for any positive p the inequality

$$\begin{aligned}\int_{U^N} (H(D(\mathbf{x}) + 2\varepsilon\|\mathbf{e}\|) - D(\mathbf{x}))\overline{\mu}(\mathrm{d}\mathbf{x})&\\ \leq \sup\{H(t + 2\varepsilon\|\mathbf{e}\|) - H(t) : t \in [0, 2p\|\mathbf{e}\|]\}&\\ +c_1 \sum_{i=1}^N \int_U H(d(x,a))I\{d(x,a) > p/N\}P_i(\mathrm{d}x)&\end{aligned}$$

holds, where c_1 is a constant independent of ε and p. As $\varepsilon \to 0$ and $p \to \infty$, we obtain

$$\limsup_{\varepsilon\to 0} \int_{U^N} D\mathrm{d}\mu_\varepsilon \leq \int_{\overline{U}^N} D\mathrm{d}\overline{\mu} = \mathbb{K}(\mathbb{Q}; \mathfrak{G}(\overline{U})) = \mathbb{K}(\widetilde{P}; \mathfrak{G}(U)).$$

2. Let U be any s.m.s. Suppose that $P_1, \ldots, P_N$ are tight measures. For any $n = 1, 2, \ldots$, define the bounded metric $d_n = \min(n, d)$. Write $D_n(x_1, \ldots, x_N) = H(\|d_n(x_1, x_2), \ldots, d_n(x_1, x_N), d_n(x_2, x_3), \ldots, d_n(x_{N-1}, x_N)\|)$. According to Part 1 of the proof, there exists a measure $P^{(n)} \in \mathfrak{P}(\widetilde{P})$ such that

$$\int_{U^N} D_n\mathrm{d}P^{(n)} = \mathbb{K}(\widetilde{P}; \mathfrak{G}(U, d_n)). \tag{5.3.14}$$

Since $P^{(n)}$, $n = 1, 2, \dots$, is a uniformly tight sequence, passing on to a subsequence if necessary, we may assume that $P^{(n)}$ converges weakly to $P^{(0)} \in \mathfrak{P}(\widetilde{P})$. By the Skorokhod–Dudley theorem [see Dudley (2002, Theorem 11.7.1)], there exist a probability space (Ω, μ) and a sequence $\{X_k, k = 0, 1, \dots\}$ of N-dimensional random vectors defined on (Ω, μ) and assuming values on U^N. Moreover, for any $k = 0, 1, \dots$, the vector X_k has distribution $P^{(k)}$ and the sequence $X_1, X_2, \dots$ converges μ-almost everywhere to X_0. According to (5.3.14) and the Fatou lemma,

$$\begin{aligned}\liminf_{n\to\infty} \mathbb{K}(\widetilde{P}; \mathfrak{G}(U, d_n)) &= \liminf_{n\to\infty} E_\mu D_n(X_n) \geq E_\mu \liminf_{n\to\infty} D_n(X_n) \\ &\geq E_\mu D(X_0) - E_\mu \limsup_{n\to\infty} |D_n(X_n) - D(X_0)|,\end{aligned}$$

where

$$|D_n(X_n) - D(X_0)| \leq |D_n(X_n) - D_n(X_0)| + |D_n(X_0) - D(X_0)| \to 0$$
$$\mu\text{-a.e. as } n \to \infty$$

and

$$E_\mu \limsup_{n\to\infty} (D_n(X_n) + D(X_0)) < \text{const} \times \sum_{i=1}^{N} \int_U H(d(x, a)) P_i(\mathrm{d}x) < \infty.$$

Hence

$$\mathbb{K}(\widetilde{P}; \mathfrak{G}(U)) \geq \lim_{k\to\infty} \mathbb{K}(\widetilde{P}; \mathfrak{G}(U, d_k)) \geq A_D(\widetilde{P}),$$

which by virtue of (5.3.3) implies (5.3.4). If $P_1, \dots, P_N$ are not necessarily tight, then one can use arguments similar to those in Case 3 of Part 1 and prove (5.3.4), which completes the proof of the theorem. □

As already mentioned, the multidimensional Kantorovich theorem can be interpreted naturally as a criterion for the closeness of n-dimensional sets of probability measures. Let (U_i, d_i) be an s.m.s., and $P_i, Q_i \in \mathcal{P}_{U_i}$, $i = 1, \dots, n$. Write $\widetilde{P} = (P_1, \dots, P_n)$, $\widetilde{Q} = (Q_1, \dots, Q_n)$, $P_i, Q_i \in \mathcal{P}_{U_i}$, and $\Delta(\mathbf{x}, \mathbf{y}) = H(\|d_1(x_1, y_1), \dots, d_n(x_n, y_n)\|)$, where $\mathbf{x} = (x_1, \dots, x_n)$ and $\mathbf{y} = (y_1, \dots, y_n) \in U_1 \times \dots \times U_n = \mathfrak{A}$ and $\|\cdot\|_n$ is a monotone seminorm in $\mathbb{R}^n$. The analog of the Kantorovich distance in $\widetilde{\mathcal{P}} = \mathcal{P}_{U_1} \times \dots \times \mathcal{P}_{U_n}$ is defined as follows:

$$\mathfrak{K}_H(\widetilde{P}, \widetilde{Q}) = \inf\left\{\int_{\mathfrak{A}\times\mathfrak{A}} \Delta(\mathbf{x}, \mathbf{y}) P(\mathrm{d}\mathbf{x}, \mathrm{d}\mathbf{y}) : P \in \mathfrak{P}(\widetilde{P}, \widetilde{Q})\right\}, \qquad (5.3.15)$$

where $\mathfrak{P}(\widetilde{P}, \widetilde{Q})$ is the space of all probability measures on $\mathfrak{A} \times \mathfrak{A}$ with fixed one-dimensional marginal distributions $P_1, \dots, P_n, Q_1, \dots, Q_n$. Subsequently (Chap. 7) we will consider more examples of minimal functionals of the type (5.3.15) (the so-called K-minimal metrics).

Case $N = 2$. Dual representation of the Kantorovich functional $\mathcal{A}_c(P_1, P_2)$. $\widehat{\mathcal{L}}_H = \ell_H$. Let $\mathfrak{C}$ be the class of all functions $c(x, y) = H(d(x, y))$, $x, y \in U$, where the function H belongs to the class $\mathcal{H}$ of all nondecreasing continuous functions on $[0, \infty)$ for which $H(0) = 0$ and that satisfy Orlicz' condition

$$K_H = \sup\{H(2t)/H(t) : t > 0\} < \infty. \tag{5.3.16}$$

We also recall that $\mathcal{H}^*$ is the subset of all convex functions in $\mathcal{H}$ and let $\mathfrak{C}^*$ be the set of all $c(x, y) = H(d(x, y))$, $H \in \mathcal{H}^*$.

Corollary 5.3.1. *Let (U, d) be an s.m.s. and P_1, P_2 be Borel probability measures on U. Let $c \in \mathfrak{C}^*$ and $\mathcal{A}_c(P_1, P_2)$ be given by (5.2.2). Let* $\mathrm{Lip}_{1,\alpha}(U) := \{f : U \to \mathbb{R} : |f(x) - f(y)| \leq \alpha d(x, y), x, y \in U\}$,

$$\mathrm{Lip}^c(U) = \left\{(f, g) \in \bigcup_{\alpha > 0} [\mathrm{Lip}_{1,\alpha}(U)]^{\times 2};\ f(x) + g(y) \leq c(x, y), x, y \in U\right\}$$

and

$$\mathcal{B}_c(P_1, P_2) = \sup\left\{\int_U f \mathrm{d}P_1 + \int_U g \mathrm{d}P_2 : (f, g) \in \mathrm{Lip}^c(U)\right\}.$$

If $\int_U c(x, a)(P_1 + P_2)(\mathrm{d}x) < \infty$ for some $a \in U$, then

$$\mathcal{A}_c(P_1, P_2) = \mathcal{B}_c(P_2, P_2).$$

Moreover, if P_1 and P_2 are tight measures, then there exists an optimal measure $P^* \in \mathcal{P}^{(P_1, P_2)}$ for which the infimum in (5.2.2) is attained.

The corollary implies that if $\mathfrak{A}$ is a class of pairs (f, g) of P_1-integrable (resp. P_2-integrable) functions satisfying $f(x) + g(y) \leq c(x, y)$ for all $x, y \in U$ and $\mathfrak{A} \supset [\mathrm{Lip}^c(U)]^{\times 2}$, then the Kantorovich functional (5.2.2) admits the following dual representation:

$$\mathcal{A}_c(P_1, P_2) = \sup\left\{\int_U f \mathrm{d}P_1 + \int_U g \mathrm{d}P_2 : (f, g) \in \mathfrak{A}\right\}.$$

The equality $\mathcal{A}_c = \mathcal{B}_c$ furnishes the main relationship between the *H-average distance* $\mathcal{L}_H(X, Y) = EH(d(X, Y))$ [(3.4.1)], [resp. *p-average metric* $\mathcal{L}_p(X, Y) = [Ed^p(X, Y)]^{1/p}$, $p \in (1, \infty)$, (3.4.3)] and the *Kantorovich distance* ℓ_H [resp. *ℓ_p-metric*; see (3.3.11)].

Corollary 5.3.2. *(i) If (U, d) is an s.m.s., $H \in \mathcal{H}^*$, and*

$$P_1, P_2 \in \mathcal{P}^H(U) := \left\{P \in \mathcal{P}(U) : \int_U H(d(x, a)) P(\mathrm{d}x) < \infty\right\},$$

then

$$\ell_H(P_1, P_2) = \widehat{\mathcal{L}}_H(P_1, P_2) := \inf\{\mathcal{L}_H(X_1, X_2) : X_i \in \mathfrak{X}(U), \\ \Pr_X = P_i, i = 1, 2\}. \tag{5.3.17}$$

Moreover, if U is a u.m.s.m.s., then ℓ_H is a simple distance in $\mathcal{P}^H(U)$ with parameter $\mathbb{K}_{\ell_H} = K_H$, i.e., for any P_1, P_2, and $P_3 \in \mathcal{P}^H(U)$, $\ell_H(P_1, P_2) \leq K_H(\ell_H(P_1, P_3) + \ell_H(P_3, P_2))$. In this case, the infimum in (5.3.17) is attained.

(ii) *If $1 < p < \infty$ (U, d) is an s.m.s. and*

$$P_1, P_2 \in \mathcal{P}^{(p)}(U) := \left\{P \in \mathcal{P}(U : \int_U d^p(x, a) P(\mathrm{d}x) < \infty\right\},$$

then

$$\ell_p(P_1, P_2) = \widehat{\mathcal{L}}_p(P_1, P_2). \tag{5.3.18}$$

In the space $\mathcal{P}^p(U)$, ℓ_p is a simple metric, provided U is a u.m.s.m.s.

Proof. See Theorem 3.3.1, Corollary 5.3.1, and Remark 2.7.1. □

5.4 Dual Representation of Minimal Norms $\overset{\circ}{\mu}_c$: Generalization of Kantorovich–Rubinstein Theorem

The Kantorovich–Rubinstein duality theorem has a long and colorful history, originating in the 1958 work of Kantorovich and Rubinstein on the mass transport problem. For a detailed survey, see Kemperman (1983). Given probabilities P_1 and P_2 on a space U and a measurable cost function $c(x, y)$ on $U \times U$ satisfying some integrability conditions, let us consider the *Kantorovich–Rubinstein functional*

$$\overset{\circ}{\mu}_c(P_1, P_2) := \inf \int c(x, y) \mathrm{d}b(x, y), \tag{5.4.1}$$

where the infimum is over all finite measures b on $U \times U$ with marginal difference $b_1 - b_2 = P_1 - P_2$, where $b_i = T_i b$ is the ith projection of b [see (5.2.17)]. ($\overset{\circ}{\mu}_c$ is sometimes called the Wasserstein functional; in Example 3.3.6, we defined $\overset{\circ}{\mu}_c$ as a *minimal norm*.)

The duality theorem for $\overset{\circ}{\mu}_c$ is of the general form

$$\overset{\circ}{\mu}_c(P_1, P_2) = \sup \int_U f \mathrm{d}(P_1 - P_2), \tag{5.4.2}$$

with the supremum taken over a class of $f : U \to \mathbb{R}$ satisfying the *Lipschitz condition* $f(x) - f(y) < c(x, y)$. When the probabilities in question have a finite support, this becomes a dual representation of the minimal cost in a network flow problem.[8]

The results for (5.4.2) were obtained by Kantorovich and Rubinstein (1958) with cost function $c(x, y) = d(x, y)$, where (U, d) is a compact metric space. Levin and Milyutin (1979) proved the dual relation (5.4.2) for a compact space U and for an arbitrary continuous cost function $c(x, y)$. Dudley (1976, Theorem 20.1) proved (5.4.2) for s.m.s. U and $c = d$. Following the proofs of Kantorovich and Rubinstein (1958) and Dudley (1976), we will show (5.4.2) for cost functions $c(x, y)$, which are not necessarily metrics. The supremum in (5.4.2) is shown to be attained for some optimal function f.

Let (U, d) be a separable metric space. Suppose that $c : U \times U \to [0, \infty)$ and $\lambda : U \to [0, \infty)$ are measurable functions such that

(C1) $c(x, y) = 0$ iff $x = y$;
(C2) $c(x, y) = c(y, x)$ for x, y in U;
(C3) $c(x, y) \le \lambda(x) + \lambda(y)$ for $x, y \in U$;
(C4) λ maps bounded sets to bounded sets;
(C5) $\sup\{c(x, y) : x, y \in B(a; R), d(x, y) \le \delta\}$ tends to 0 as $\delta \to 0$ for each $a \in U$ and $R > 0$. Here, $B(a; R) := \{x \in U : d(x, a) < R\}$.

We give two examples of function c satisfying (C1)–(C5), which are related to our discussion in Sect. 5.2 (Examples 5.2.1 and 5.2.2):

1. $c(x, y) = H(d(x, y))$, $H \in \mathcal{H}$ [see (5.3.16)].
2. $c(x, y) = d(x, y) \max(1, h(d(x, a)), h(d(y, a)))$, where $h : [0, \infty) \to [0, \infty)$ is a continuous nondecreasing function.

Given a real-valued function $f : U \to \mathbb{R}$, we define

$$\|f\|_c := \sup\{|f(x) - f(y)|/c(x, y) : x \neq y\} \tag{5.4.3}$$

and set

$$\mathbb{L} := \{f : \|f\|_c < +\infty\}. \tag{5.4.4}$$

It is easy to see that $\|\cdot\|_c$ is a seminorm on the linear space $\mathbb{L}$. Notice that for $f \in \mathbb{L}$ we have $|f(x) - f(y)| \le \|f\|_c c(x, y)$ $\forall x, y \in U$. It follows from Condition (C5) on c that each function in L is continuous and, hence, measurable. Note also that $\|f\|_c = 0$ if and only if f is constant. Define $\mathbb{L}_0$ as the quotient of $\mathbb{L}$ modulo the constant functions. Then $\|\cdot\|_c$ is naturally defined on $\mathbb{L}_0$, and $(\mathbb{L}_0, \|\cdot\|)$ is a normed linear space.[9]

---

[8] See Bazaraa and Jarvis (2005) and Berge and Chouila-Houri (1965, Sect. 9.8).

[9] See Fortet and Mourier (1953).

Now suppose that $\mathcal{M} = \mathcal{M}_\lambda(U)$ denotes the linear space of all finite signed measures m on U such that

$$m(U) = 0 \quad \text{and} \quad \int \lambda \mathrm{d}|m| \leq \infty. \tag{5.4.5}$$

Here $|m| := m^+ + m^-$, where $m = m^+ - m^-$ is the Jordan decomposition of m.

For each $m \in \mathcal{M}$ let $\mathbb{B}(m)$ be the set of all finite measures b on $U \times U$ such that

$$b(A \times U) - b(U \times A) = m(A) \tag{5.4.6}$$

for each Borel $A \subseteq U$. Note that $\mathbb{B}(m)$ is always nonempty since it contains $(m^+ \times m^-)/m^+(U)$. Here, $m^+ \times m^-$ denotes the product measure $m^+ \times m^-(A) = m^+(A)m^-(A)$. $A \in \mathcal{B}(U)$. Define a function $m \to \|m\|_{\mathrm{w}}$ on $\mathcal{M}$ by

$$\|m\|_{\mathrm{w}} := \inf\left\{\int c(x,y)b(\mathrm{d}x,\mathrm{d}y) : b \in \mathbb{B}(m)\right\}. \tag{5.4.7}$$

We have

$$\begin{aligned}\|m\|_{\mathrm{w}} &\leq \int c(x,y)(m^+ \times m^-)(\mathrm{d}x,\mathrm{d}y)/m^+(U)\\ &\leq \int \lambda(x)m^+(\mathrm{d}x) + \int \lambda(y)m^-(\mathrm{d}y)\\ &= \int \lambda \mathrm{d}|m| < \infty.\end{aligned} \tag{5.4.8}$$

For $c(x,y) = d(x,y)$, $\|m\|_{\mathrm{w}}$ is sometimes called the *Kantorovich–Rubinstein* or *Wasserstein norm* of m [see also (3.3.38) and Definition 3.3.4].

We will demonstrate that for probabilities P and Q on U with $P - Q \in \mathcal{M}$ we have

$$\|P - Q\|_{\mathrm{w}} = \sup\left\{\int f \mathrm{d}(P - Q) : \|f\|_c \leq 1\right\}, \tag{5.4.9}$$

which furnished (5.4.2) with cost function c satisfying (Cl)–(C5). When $c(x,y) = d(x,y)$ and $\lambda(x) = d(x,a)$, a being some fixed point of U, this is a straightforward generalization of the classic Kantorovich–Rubinstein duality theorem [see Dudley (1976, Lecture 20)].

First note that $\|\cdot\|_{\mathrm{w}}$ is a seminorm on $\mathcal{M}$: (Lemma 3.3.2). Now given $m \in \mathcal{M}$, $f \in Ł$, and a fixed $a \in U$, we have

$$\begin{aligned}|f(x)| &\leq |f(x) - f(a)| + |f(a)| \leq \|f\|_c c(x,a) + |f(a)|\\ &\leq \|f\|_c(\lambda(x) + \lambda(a)) + |f(a)| = K_1\lambda(x) + K_2, \quad \forall x \in U,\end{aligned}$$

for constants $K_1, K_2 \geq 0$. Thus, each $f \in Ł$ is $|m|$-integrable and induces a linear form $\phi_f : \mathcal{L} \to \mathbb{R}$ defined by

$$\phi_f(m) := \int f\mathrm{d}m. \tag{5.4.10}$$

Note that if f and g differ by a constant, then $\phi_f = \phi_g$. Given $b \in \mathbb{B}(m)$, we have

$$\begin{aligned}|\phi_f(m)| &= \left|\int f\mathrm{d}m\right| = \left|\int (f(x) - f(y))b(\mathrm{d}x, \mathrm{d}y)\right| \\ &\leq \int |f(x) - f(y)|b(\mathrm{d}x, \mathrm{d}y) \leq \|f\|_c \int c(x, y)b(\mathrm{d}x, \mathrm{d}y).\end{aligned}$$

Taking the infimum over all $b \in \mathbb{B}(m)$, this yields $|\phi_f(m)| \leq \|f\|_c\|m\|_{\mathrm{w}}$, so that ϕ_f is a continuous linear functional with dual norm $\|\phi_f\|_{\mathrm{w}}^*$ such that

$$\|\phi_f\|_{\mathrm{w}}^* \leq \|f\|_c. \tag{5.4.11}$$

Thus, we may define a continuous linear transformation

$$(\mathbb{L}_0, \|\cdot\|_c) \xrightarrow{D} (\mathcal{M}^*, \|\cdot\|_{\mathrm{w}}^*) \tag{5.4.12}$$

by $D(f) = \phi_f$.

Lemma 5.4.1. *The map D is an isometry, i.e.,* $\|f\|_c = \|\phi_f\|_{\mathrm{w}}^*$.

Proof. Given $x \in U$, denote the point mass at x by δ_x. Note first that if $m_{xy} := \delta_x - \delta_y$ for some $x, y \in U$, then

$$\|m_{xy}\|_{\mathrm{w}} \leq \int c(u, t)(\delta_x \times \delta_y)(\mathrm{d}u, \mathrm{d}t) = c(x, y).$$

Then for each $f \in \mathbb{L}$,

$$\begin{aligned}\|f\|_c &= \sup\{|f(x) - f(y)|/c(x, y) : x \neq y\} \\ &= \sup\{|\phi_f(m_{xy})|/c(x, y) : x \neq y\} \\ &\leq \|\phi_f\|_{\mathrm{w}}^* \sup\{\|m_{xy}\|_{\mathrm{w}}/c(x, y) : x \neq y\} \leq \|\phi_f\|_{\mathrm{w}}^*,\end{aligned}$$

so that $\|f\|_c = \|\phi_f\|_{\mathrm{w}}^*$ by (5.4.11). □

We now set about proving that the map D is subjective and, hence, an isometric isomorphism of Banach spaces. Recall that an isometric isomorphism between two normed linear spaces $\mathbb{A}_1$ and $\mathbb{A}_2$ is a one-to-one continuous linear map $T : \mathbb{A}_1 \to \mathbb{A}_2$ with $T\mathbb{A}_1 = \mathbb{A}_2$ and $\|Tx\|_{\mathbb{A}_2} = \|x\|_{\mathbb{A}_1}$.[10]

We need some preliminary facts. Let $\mathcal{M}_0$ be the set of signed measures of the form $m = m_1 - m_2$, where m_1 and m_2 are finite measures on U with bounded support such that $m_1(U) = m_2(U)$. Condition (C4) on λ implies that $\mathcal{M}_0 \subseteq \mathcal{M}$.

---

[10] See Dunford and Schwartz (1988, p. 65).

Lemma 5.4.2. *$\mathcal{M}_0$ is a dense subspace of $(\mathcal{M}, \|\cdot\|_{\mathrm{w}})$.*

Proof. Given $m \in \mathcal{M}$ $(m \neq 0)$, fix $a \in U$ and set

$$B_n = B(a,n) := \{x \in U : d(x,a) < n\}$$

for $n = 1, 2, \ldots$. For all sufficiently large n, we have $m^+(B_n)m^-(B_n) > 0$. For such n, let us denote

$$m_n(A) := m^+(U)\left[\frac{m^+(A \cap B_n)}{m^+(B_n)} - \frac{m^-(A \cap B_n)}{m^-(B_n)}\right],$$

$$\delta_n := \frac{m^-(U)}{m^-(B_n)} - 1, \qquad \varepsilon_n := \frac{m^+(U)}{m^+(B_n)} - 1.$$

Then $\delta_n, \varepsilon_n \to 0$ as $n \to \infty$. Also,

$$(m - m_n)(A) = m(A \setminus B_n) - \varepsilon_n m^+(A \cap B_n) + \delta_n m^-(A \cap B_n).$$

Define finite measures μ_n and ν_n on U by

$$\mu_n(A) := m^+(A \setminus B_n) + \delta_n m^-(A \cap B_n),$$
$$\nu_n(A) := m^-(A \setminus B_n) + \varepsilon_n m^+(A \cap B_n).$$

Then, $m - m_n = \mu_n - \nu_n$. Moreover, μ_n and ν_n are absolutely continuous with respect to $|m|$. Letting P and N be the supports of m^+ and m^- in the Jordan–Hahn decomposition for m, we determine the Radon–Nikodym derivatives

$$\frac{\mathrm{d}\mu_n}{\mathrm{d}|m|}(x) = \begin{cases} 1, & x \in P \setminus B_n, \\ \delta_n, & x \in N \cap B_n, \\ 0, & \text{otherwise}, \end{cases} \qquad \frac{\mathrm{d}\nu_n}{\mathrm{d}|m|}(y) = \begin{cases} 1, & y \in N \setminus B_n, \\ \varepsilon_n, & y \in P \cap B_n, \\ 0, & \text{otherwise}. \end{cases}$$

Then the measure $b_n = (\mu_n \times \nu_m)/\mu_n(U)$ belongs to $\mathbb{B}(m - m_n)$. Noting that

$$\begin{aligned} \nu_n(U) = \mu_n(U) &= m^+(U \setminus B_n) + \delta_n m^-(B_n) \\ &= m^+(U \setminus B_n) + (m^-(U) - m^-(B_n)) \\ &= |m|(U \setminus B_n) = |m|(B_n^c), \end{aligned}$$

we write the Radon–Nikodym derivative

$$f_n(x,y) := \frac{\mathrm{d}b_n}{\mathrm{d}(|m| \times |m|)}(x,y) := \frac{1}{|m|(B_n^c)}\frac{\mathrm{d}\mu_n}{\mathrm{d}|m|}(x)\frac{\mathrm{d}\nu_n}{\mathrm{d}|m|}(y).$$

Then we make the following claim.

Claim. The function $g(x, y) = \sup_n f_n(x, y)c(x, y)$ is $|m| \times |m|$-integrable.

Proof of Claim. We show that g is integrable over various subsets of $U \times U$.

(i) g is integrable over $P \times N$: we suppose that $x \in P$ and $y \in N$. Then

$$g(x, y) \leq \sum_{n=1}^{\infty} \frac{c(x, y)}{|m|(B_n^c)} I_{C_n}(x, y),$$

where $C_n = (B_n^c \times B_n^c) - (B_{n+1}^c \times B_{n+1}^c)$ and $I_{(\cdot)}$ is the indicator of $(\cdot)$. Thus,

$$\begin{aligned}
\int_{P\times N} g\mathrm{d}|m| \times |m| &\leq \sum_{n=1}^{\infty} \frac{1}{|m|(B_n^c)} \int_{C_n} (\lambda(x) + \lambda(y))|m| \times |m|(\mathrm{d}x, \mathrm{d}y) \\
&\leq \sum_{n=1}^{\infty} \frac{2}{|m|(B_n^c)} \int_{(B_n^c - B_{n+1}^c)\times B_n^c} \lambda(x)|m| \times |m|(\mathrm{d}x, \mathrm{d}y) \\
&= 2\sum_{n=1}^{\infty} \int_{B_n^c - B_{n+1}^c} \lambda(x)|m|\mathrm{d}x \\
&= 2\int_{B_1} \lambda \mathrm{d}|m| < +\infty.
\end{aligned}$$

(ii) $g(x, y) < Kc(x, y)$ for some $K \geq 0$ on $P \times P$: we suppose $x, y \in P$. Then

$$\begin{aligned}
g(x, y) \leq \sup_n \frac{\varepsilon_n c(x, y)}{|m|(B_n^c)} &= \sup_n \frac{c(x, y)}{m^+(B_n)} (m^+(U) - m^+(B_n)) \frac{1}{|m|(B_n^c)} \\
&= \sup_n \frac{m + (B_n^c)}{|m|(B_n^c)} \frac{c(x, y)}{m^+(B_n)} \leq \frac{c(x, y)}{m^+(B_1)}.
\end{aligned}$$

Very similar arguments serve to demonstrate

(iii) $g(x, y) \leq Kc(x, y)$ for some $K \geq 0$ on $N \times N$;

(iv) $g(x, y) \leq Kc(x, y)$ for some $K \geq 0$ on $N \times P$.

Combining (i)–(iv) establishes the claim.

Now $f_n(x, y) \to 0$ as $n \to \infty$ $\forall x, y \in U$. In view of the claim, Lebesgue's dominated convergence theorem implies that

$$\begin{aligned}
\|m - m_n\|_{\mathrm{w}} &\leq \int c(x, y)(\mathrm{d}x, \mathrm{d}y) \\
&= \int c(x, y) f_n(x, y)(|m| \times |m|)(\mathrm{d}x, \mathrm{d}y) \to 0
\end{aligned}$$

as $n \to \infty$. □

Call a signed measure on U *simple* if it is a finite linear combination of signed measures of the form $\delta_x - \delta_y$. $\mathcal{M}$ contains all the simple measures. In the next lemma we will use the Strassen–Dudley theorem.

Theorem 5.4.1. *Suppose that (U, d) is an s.m.s. and that $P_n \to P$ weakly in $\mathcal{P}(U)$. Then for each $\varepsilon, \delta > 0$ there is some N such that whenever $n \geq N$, there is some law b_n on $U \times U$ with marginals P_n and P such that*

$$b_n\{(x, y) : d(x, y) > \delta\} < \varepsilon. \tag{5.4.13}$$

Proof. Further (Corollary 7.5.2),[11] we will prove that the Prokhorov metric π is minimal with respect to the Ky Fan metric $\mathbf{K}$. In other words,

$$\pi(P_1, P_2) = \inf\{\mathbf{K}(P) : P \in \mathcal{P}(U \times U), P(\cdot \times U) = P_1(\cdot), P(U \times \cdot) = P_2(\cdot)\},$$

where $\mathbf{K}(P) = \inf\{\varepsilon > 0 : P((x, y) : d(x, y) > \varepsilon) < \varepsilon\}$. Since π metrizes the weak topology in $\mathcal{P}(U)$ (Dudley 2002), the preceding equality yields (5.4.13). □

Lemma 5.4.3. *The simple measures are dense in $(\mathcal{M}, \|\cdot\|_{\mathrm{w}})$.*

Proof. In view of Lemmas 3.3.2 and 5.4.2, there is no loss of generality to assume that $m = P - Q$, where P and Q are laws on U supported on a bounded set $U_0 \subseteq U$. Then there are laws $P_n \xrightarrow{\mathrm{w}} P$, $Q_n \xrightarrow{\mathrm{w}} Q$ such that for each n, we have $P_n(U_0) = Q_n(U_0) = 1$ and $P_n - Q_n$ is simple [see, for example, the Glivenko-Cantelli–Varadarajan theorem (Dudley 2002)]. To prove the lemma, it is enough to show that $\|P_n - P\|_{\mathrm{w}} \to 0$ as $n \to \infty$.

Given $\varepsilon > 0$, use the boundedness of U_0 and Condition (C5) on c to find $\delta > 0$ such that $c(x, y) < \varepsilon/2$ whenever $x, y \in U_0$ with $d(x, y) \leq \delta$. Set $K = \sup\{\lambda(x) : x \in U_0\}$. By Theorem 5.4.1, for all large n, there is a law b_n on $U \times U$ with marginals P_n and P such that $b_n\{(x, y) : d(x, y) > \delta\} < \delta\} < \varepsilon/4K$. Set $A = \{(x, y) : d(x, y) > \delta\}$. Then

$$\begin{aligned}
\|P_n - P\|_{\mathrm{w}} &\leq \int c(x, y) b_n(\mathrm{d}x, \mathrm{d}y) \\
&= \int_A c(x, y) b_n(\mathrm{d}x, \mathrm{d}y) + \int_{U/A} c(x, y) b_n(\mathrm{d}x, \mathrm{d}y) \\
&\leq \int_A (\lambda(x) + \lambda(y) b_n(\mathrm{d}x, \mathrm{d}y) + \varepsilon/2 \leq 2K b_n(A) + \varepsilon/2 < \varepsilon
\end{aligned}$$

for all large n. □

Lemma 5.4.4. *The linear transformation D (5.4.12) is an isometric isomorphism of $(\mathbb{L}_0, \|\cdot\|_c)$ onto $(\mathcal{M}^*, \|\cdot\|^*_{\mathrm{w}})$.*

[11] See also Dudley (2002, Theorem 11.6.2).

Proof. Suppose that $\phi : \mathcal{M} \to \mathbb{R}$ is a continuous linear functional on $\mathcal{M}$. Fix $a \in U$ and define $f : U \to \mathbb{R}$ by $f(x) = \phi(\delta_x - \delta_a)$. For any $x, y \in U$

$$|f(x) - f(y)| = |\phi(\delta_x - \delta_y)| \leq \|\phi\|_{\mathrm{w}}^* \|\delta_x - \delta_y\|_{\mathrm{w}} \leq \|\phi\|_{\mathrm{w}}^* c(x, y),$$

so that $\|f\|_c \leq \|\phi\|_{\mathrm{w}}^* < \infty$. We see that $\phi(m) = \phi_f(m)$ for $m = \delta_x - \delta_y$ and, hence, for all simple $m \in \mathcal{M}$. Lemma 5.4.3 implies that $\phi(m) = \phi_f(m)$ for all $m \in \mathcal{M}$. Thus $\phi = D(f)$.

We have shown that D is subjective. Earlier results now apply to complete the argument. □

Now we consider the adjoint of the transformation D. As usual, the Hahn–Banach theorem applies to show that $(\mathbb{L}_0^*, \|\cdot\|_c^*) \overset{D^*}{\longleftarrow} (\mathcal{M}^{**}, \|\cdot\|_{\mathrm{w}}^*)$ is an isometric isomorphism; see Dunford and Schwartz (1988, Theorem II 3.19). Let $(\mathcal{M}^{**}, \|\cdot\|_{\mathrm{w}}^{**}) \overset{T}{\longleftarrow} (\mathcal{M}, \|\cdot\|_{\mathrm{w}})$ be the natural isometric isomorphism of $\mathcal{M}$ into its second conjugate $\mathcal{M}^{**}$. Then $(L_0^*, \|\cdot\|_c^* \overset{D^* \circ T}{\longleftarrow} (\mathcal{M}, \|\cdot\|_{\mathrm{w}})$ is an isometry. A routine diagram shows that $\|m\|_{\mathrm{w}} = \sup\{\int f \mathrm{d}m : \|f\|_c \leq 1\}$.

We summarize by stating the following, the main result of this section.

Theorem 5.4.2. *Let m be a measure in $\mathcal{M}$. Then*

$$\|m\|_{\mathrm{w}} = \sup\left\{\int f \mathrm{d}m : f(x) - f(y) \leq c(x, y)\right\}.$$

We now show that the supremum in Theorem 5.4.2 is attained for some optimal f.

Theorem 5.4.3. *Let m be a measure in $\mathcal{M}$. Then there is some $f \in \mathbb{L}$ with $\|f\|_c = 1$ such that $\|m\|_{\mathrm{w}} = \int f \mathrm{d}m$.*

Proof. Using the Hahn–Banach theorem, choose a linear functional ϕ in $\mathcal{M}^*$ with $\|\phi\|^* = 1$ and such that $\phi(m) = \|m\|_{\mathrm{w}}$. By Lemma 5.4.4, we have $\phi = \phi_f$ for some $f \in \mathbb{L}$ with $\|f\|_c = \|\phi\|^* = 1$. □

Given probability measures P_1 and P_2 on U, define the *minimal norm*

$$\overset{\circ}{\mu}_c(P_1, P_2) = \inf\left\{\int c(x, y) b(\mathrm{d}x, \mathrm{d}y) : b \in \mathbb{B}(P_1 - P_2)\right\} \qquad (5.4.14)$$

[see (5.4.6) and Example 3.3.6]. Let $\mathcal{P}(U)$ be the set of all laws P on U such that λ is P-integrable. Then $\overset{\circ}{\mu}_c(P_1, P_2)$ defines a semimetric on $\mathcal{P}_\lambda(U)$ (Remark 3.3.1). In the next section, we will analyze the explicit representations and the topological properties of these semimetrics.

It should also be noted that if X and Y are random variables taking values in U, then it is natural to define

$$\overset{\circ}{\mu}_c(X,Y) = \overset{\circ}{\mu}_c(\Pr_X, \Pr_Y),$$

where $\Pr_X$ is the law of X. We will freely use both notations in the next section.

Example 5.4.1. Suppose that $c(x,y) = d(x,y)$ and set $\lambda(x) = d(x,a)$ for some $a \in U$. Then Conditions (C1)–(C5) are satisfied, and Theorem 5.4.2 yields

$$\inf\left\{\int d(x,y)b(\mathrm{d}x,\mathrm{d}y) : b \in \mathbb{B}(P_1 - P_2)\right\}$$
$$= \sup\left\{\int f\mathrm{d}(P_1 - P_2) : \|f\|_L \leq 1\right\}, \qquad (5.4.15)$$

where $P_1, P_2 \in \mathcal{P}_\lambda(U)$ and $\|f\|_L$ is the Lipschitz norm of f. In this case, $\overset{\circ}{\mu}_c(P_1, P_2)$ is a metric in $\mathcal{P}_\lambda(U)$. This classic situation has been much studied; see Kantorovich and Rubinstein (1958) and Dudley (1976, Lecture 20). In particular, (5.4.15) gives us the dual representations of $\overset{\circ}{\mu}_c(P_1, P_2)$ given by [(3.4.53)]

$$\overset{\circ}{\mu}(P_1, P_2) = \inf\{\alpha E d(X,Y) : \text{for some } \alpha > 0, X \in \mathfrak{X}(U), Y \in \mathfrak{X}(U),$$
$$\text{such that } \alpha(\Pr_X - \Pr_Y) = P_1 - P_2\}$$
$$= \sup\left\{\left|\int_U f\mathrm{d}(P_1 - P_2)\right| ; \|f\|_L \leq 1\right\}. \qquad (5.4.16)$$

5.5 Application: Explicit Representations for a Class of Minimal Norms

Throughout this section, we take $U = \mathbb{R}$, $d(x,y) = |x - y|$, and we also define $c : \mathbb{R} \times \mathbb{R} \to [0, +\infty)$ by

$$c(x,y) = |x - y| \max(h(|x - a|), h(|y - a|)), \qquad (5.5.1)$$

where a is a fixed point of $\mathbb{R}$ and $h : [0,\infty) \to [0,\infty)$ is a continuous nondecreasing function such that $h(x) > 0$ for $x > 0$. Note that the cost function in Example 5.2.1 [see (5.2.24)] has precisely the same form as (5.5.1). Define $\lambda : \mathbb{R} \to [0,\infty)$ by

$$\lambda(x) = 2|x|h(|x - a|).$$

It is not difficult to verify that c and λ satisfy Conditions (C1)–(C5) specified in Sect. 5.4. As in Sect. 5.4, the normed space $(\mathbb{L}_0, \|\cdot\|_c)$ and the set $\mathcal{M}$, comprising all finite signed measures m on $\mathbb{R}$ such that $m(U) = 0$ and $\int \lambda \mathrm{d}|m| < +\infty$, are to be investigated.

We consider random variables X and Y in $\mathfrak{X} = \mathfrak{X}(\mathbb{R})$ with $E(\lambda(X)) + E(\lambda(Y)) < \infty$. Then $m = \Pr_X - Pr_Y$ is an element of $\mathcal{M}$, and Theorem 5.4.2 implies the dual representation of $\overset{\circ}{\mu}_c$:

$$\overset{\circ}{\mu}_c(X,Y) = \inf\{\alpha E(c(X',Y')) : X',Y' \in \mathfrak{X}, \alpha > 0, \alpha(\Pr_{X'} - \Pr_{Y'}) = m\}$$
$$= \sup\left\{\left|\int_{\mathbb{R}} f\mathrm{d}m\right| : |f(x) - f(y)| < c(x,y), \forall x,y \in \mathbb{R}\right\}. \quad (5.5.2)$$

An explicit representation is given in the following theorem.

Theorem 5.5.1. *Suppose c is given by* (5.5.1) *and $X, Y \in \mathfrak{X}$ with $E(\lambda(X)) + E(\lambda(Y)) < \infty$; then*

$$\overset{\circ}{\mu}_c(X,Y) = \int_{-\infty}^{\infty} h(|x-a|)|F_X(x) - F_Y(x)|\mathrm{d}x. \quad (5.5.3)$$

Proof. We begin by proving the theorem in the special case where X and Y are bounded. Suppose that $|X| \le N$ and $|Y| \le N$ for some N. Application of Theorem 5.4.2 with $U := U_N := [-N, N]$ yields $\overset{\circ}{\mu}_c(X,Y) = \sup\left\{\left|\int f\mathrm{d}m\right| : f : U_n \to \mathbb{R}, |f(x) - f(y)| < c(x,y), \forall x,y \in U_N\right\}$, where $m = \Pr_X - \Pr_Y$. It is easy to check that if $|f(x) - f(y)| \le c(x,y)$ as previously, then f is absolutely continuous on any compact interval. Thus, f is differentiable a.e. on $[-N, N]$, and $|f'(x)| \le h(|x-a|)$ wherever f' exists. Therefore

$$\overset{\circ}{\mu}_c(X,Y) \le \left\{\left|\int_{-\infty}^{\infty}(F_X(x) - F_Y x))f'(x)\mathrm{d}x\right| : f : U_N \to \mathbb{R},\right.$$
$$\left.|f'(x)| \le h(|x-a|) \text{ a.e}\right\}$$
$$\le \int_{-\infty}^{\infty} h(|x-a|)|F_X(x) - F_Y(y)|\mathrm{d}x$$

using integration by parts.

On the other hand, if f is absolutely continuous with $|f'(x)| \le h(|x-a|)$ a.e., then $|f(x) - f(y)| = \left|\int_x^y f'(t)\mathrm{d}t\right| \le |x-y| \max(h(|x-a|), h(|y-a|)) = c(x,y)$. Define $f_* : \mathbb{R} \to \mathbb{R}$ by

$$f_*' = h(|x-a|)\mathrm{sgn}(F_X(x) - F_Y(x)) \text{ a.e.}$$

Then

$$\overset{\circ}{\mu}_c(X,Y) = \sup\left\{\left|\int F_X(x) - F_Y(x))f'(x)\mathrm{d}x\right| : |f'(x)| \leq h(|x-a|) \text{ a.e.}\right\}$$
$$\leq \left|\int (F_X(x) - F_Y(x))f'_*(x)\mathrm{d}x\right|$$
$$= \int h(|x-a|)|F_X(x) - F_Y(x)|\mathrm{d}x.$$

We have shown that whenever X and Y are bounded random variables, (5.5.3) holds. Now define $H : \mathbb{R} \to \mathbb{R}$ by

$$H(t) = \int_0^t h(|x-a|)\mathrm{d}x. \tag{5.5.4}$$

For $t \geq 0$, $H(t) \leq h(|a|)|a| + |t-a|h(|t-a|)$, so that $E(\lambda(X)) + E(\lambda(Y)) < \infty$ implies that $E|H(X)| + E|H(Y)| < \infty$. Under this assumption, integrating by parts we obtain

$$E|H(X)| = \int_0^\infty h(|x-a|)(1 - F_X(x))\mathrm{d}x + \int_{-\infty}^0 h(|x-a|)F_X(x)\mathrm{d}x.$$

An analogous equality holds for the variable Y. These imply that

$$\int_{-\infty}^\infty h(|x-a|)|F_X(x) - F_Y(x)|\mathrm{d}x < \infty.$$

For $n \geq 1$, define random variables X_n, Y_n by

$$X_n = \begin{cases} n, & \text{if } X > n, \\ X, & \text{if } -n \leq X \leq n, \\ -n, & \text{if } X < -n, \end{cases} \qquad Y_n = \begin{cases} n, & \text{if } Y > n, \\ Y, & \text{if } -n \leq Y \leq n, \\ -n, & \text{if } Y < -n. \end{cases}$$

Then $X_n \to X$, $Y_n \to Y$ in distribution, and for $n \geq |a|$

$$\overset{\circ}{\mu}_c(X_n, X) \leq Ec(X_n, X) \leq E(|X|I\{|X| \geq n\}h(|X-a|)),$$

which tends to 0 as $n \to \infty$ $[E(\lambda(X)) < \infty]$. Similarly, $\overset{\circ}{\mu}_c(X_n, Y) \to 0$. Then $\overset{\circ}{\mu}_c(X_n, Y_n) \to \overset{\circ}{\mu}_c(X, Y)$ as $n \to \infty$. Also, we have

$$|F_{X_n}(x) - F_{Y_n}(x)| = \begin{cases} |F_X(x) - F_Y(x)|, & \text{for } -n \leq x < n, \\ 0, & \text{otherwise.} \end{cases}$$

Applying dominated convergence, we see that as $n \to \infty$,

$$\int h(|x-a|)|F_{X_n}(x) - F_{Y_n}(x)|\mathrm{d}x \to \int h(|x-a|)|F_X(x) - F_Y(x)|\mathrm{d}x.$$

Combining this with $\overset{\circ}{\mu}_c(X_n, Y_n) \to \overset{\circ}{\mu}_c(X, Y)$ and the result for bounded random variables yields

$$\overset{\circ}{\mu}_c(X, Y) = \int_{-\infty}^{\infty} h(|x-a|)|F_X(x) - F_Y(x)|\mathrm{d}x. \qquad \square$$

For $h(x) = 1$, this yields a well-known formula presented in Dudley (1976, Theorem 20.10). We also note the following formulation, which is not hard to derive from the strict monotonicity of H [see (5.5.4)].

Corollary 5.5.1. *Suppose c is given by* (5.5.1) *and $X, Y \in \mathfrak{X}$ with $E(\lambda(X)) + E(\lambda(Y)) < \infty$, and set $P = \mathrm{Pr}_X$, $Q = \mathrm{Pr}_Y$. Then*

$$\overset{\circ}{\mu}_c(P, Q) = \int_{-\infty}^{\infty} |F_{H(X)}(x) - F_{H(Y)}(x)|\mathrm{d}x, \tag{5.5.5}$$

where H is given by (5.5.4).

For $h(x) = 1$ we see that $H(t) = t$ and that $\overset{\circ}{\mu}$ gives the Kantorovich metric.[12]

Corollary 5.5.2. *In this context, $\overset{\circ}{\mu}_c(P_1, P_2)$ defines a metric on $\mathcal{P}_\lambda(\mathbb{R}) := \{P : \int_{\mathbb{R}} \lambda \mathrm{d}P < \infty\}$.*

References

Ambrosio L (2002) Optimal transport maps in the Monge-Kantorovich problem. Lect Notes ICM III:131–140

Ambrosio L (2003) Lecture notes on optimal transport problems. In: Mathematical aspects of evolving interfaces, lecture notes in mathematics, vol 1812. Springer, Berlin/Heidelberg, pp 1–52

Anderson TW (2003) An introduction to multivariate statistical analysis, 3rd edn. Wiley, Hoboken

Angenent S, Haker S, Tannenbaum A (2003) Minimizing flows for the Monge-Kantorovich problem. SIAM J Math Anal 35(1):61–97

Barrett JW, Prigozhin L (2009) L1 Monge-Kantorovich problem: variational formulation and numerical approximation. Interfaces and Free Boundaries 11:201–238

Bazaraa MS, Jarvis JJ (2005) Linear programming and network flows, 3rd edn. Wiley, Hoboken

Berge C, Chouila-Houri A (1965) Programming, games, and transportation networks. Wiley, New York

[12] See Sect. 2.2 of Chap. 2.

Billingsley P (1999) Convergence of probability measures, 2nd edn. Wiley, New York
Brenier Y (2003) Extended Monge-Kantorovich theory. In: Optimal transportation and applications. Lecture notes in mathematics, vol 1813. Springer, Berlin, pp 91–121
Carlier G (2003) Duality and existence for a class of mass transportation problems and economic applications. Adv Math Econ 5:1–21
Chartrand R, Vixie KR, Wohlberg B, Bollt EM (2009) A gradient descent solution to the Monge-Kantorovich problem. Appl Math Sci 3(22):1071–1080
Cuesta J, Matran C, Rachev S, Rüschendorf L (1996) Mass transportation problems in probability theory. Math Sci 21:34–72
Dudley RM (1976) Probabilities and metrics: convergence of laws on metric spaces, with a view to statistical testing. In: Aarhus University Mathematics Institute lecture notes, series no. 45, Aarhus
Dudley RM (2002) Real analysis and probability, 2nd edn. Cambridge University Press, New York
Dunford N, Schwartz J (1988) Linear operators, vol 1. Wiley, New York
Evans LC, Gangbo W (1999) Differential equations methods for the Monge-Kantorovich mass transfer problem. Mem Am Math Soc 137:1–66
Feldman M, McCann RJ (2002) Monge's transport problem on a Riemannian manifold. Trans Am Math Soc 354:1667–1697
Feyel D, Üstünel A (2004) Monge-Kantorovitch measure transportation and Monge-Ampère equation on Wiener space. Probab Theor Relat Fields 128:347–385
Fortet R, Mourier B (1953) Convergence de la réparation empirique vers la répétition theorétique. Ann Sci Ecole Norm 70:267–285
Gabriel J, Gonzlez-Hernndez J, Lpez-Martnez R (2010) Numerical approximations to the mass transfer problem on compact spaces. J Numer Anal 30(4):1121–1136
Gray RM (1988) Probability, random processes, and ergodic properties. Springer, New York
Igbida N, Mazón JM, Rossi JD, Toledo J (2011) A Monge-Kantorovich mass transport problem for a discrete distance. J Funct Anal 260:494–3534
Kalashnikov VV, Rachev ST (1988) Mathematical methods for construction of stochastic queueing models. Nauka, Moscow (in Russian). [English transl. (1990) Wadsworth, Brooks—Cole, Pacific Grove, CA]
Kantorovich LV (1940) On one effective method of solving certain classes of extremal problems. Dokl. Akad. Nauk, USSR 28:212–215
Kantorovich LV, Rubinshtein GSh (1958) On the space of completely additive functions. Vestnik LGU Ser Mat, Mekh i Astron 7:52–59
Karatzas I (1984) Gittins indices in the dynamic allocation problem for diffusion processes. Ann Prob 12:173–192
Kemperman JHB (1983) On the role of duality in the theory of moments. Semi-infinite programming and applications. In: Lecture notes economic mathematical system, vol 215. Springer, Berlin, pp 63–92
Kruskal WM (1958) Ordinal measures of association. J Am Stat Assoc 53:814–861
Léonard C (2012) From the Schrödinger problem to the Monge-Kantorovich problem. J Funct Anal 262:1879–1920
Levin V (1997) Reduced cost functions and their applications. J Math Econ 28(2):155–186
Levin V (1999) Abstract cyclical monotonicity and Monge solutions for the general Monge-Kantorovich problem. Set-Valued Anal 7(1):7–32
Levin VL, Milyutin AA (1979) The mass transfer problem with discontinuous cost function and a mass setting for the problem of duality of convex extremum problems. Usp Mat Nauk 34:3–68 (in Russian). (English transl. (1979) Russian Math Surveys **34**, 1–78]
Rachev S (1991) Mass transshipment problems and ideal metrics. Numer Funct Anal Optim 12(5&6):563–573
Rachev S, Hanin L (1995a) An extension of the kantorovich–Rubinstein mass-transshipment problem. Numer Funct Anal Optim 16:701–735
Rachev S, Hanin L (1995b) Mass transshipment problems and ideal metrics. J Comput Appl Math 56:114–150

Rachev S, Rüschendorf L (1998) Mass transportation problems, vol I: Theory. Springer, New York

Rachev S, Rüschendorf L (1999) Mass transportation problems: applications, vol II. Springer, New York

Rachev S, Taksar M (1992) Kantorovich's functionals in space of measures. In: Applied stochastic analysis, proceedings of the US-French workshop. Lecture notes in control and information science, vol 177, pp 248–261

Sudakov VN (1976) Geometric problems in the theory of infinite-dimensional probability distributions. Tr Mat Inst V A Steklov, Akad Nauk, SSSR, vol 141 (in Russian). (English transl. (1979) Proc Steklov Inst Math (1979). no 2, 141)

Talagrand M (1996) Transportation cost for Gaussian and other product measures. Geom Funct Anal 6:587–600

Villani C (2003) Topics in Optimal transportation, graduate studies in mathematics, vol 58. AMS, Providence

Whittle P (1982) Optimization over time: dynamic programming and stochastic control. Wiley, Chichester, UK

Zhang X (2011) Stochastic Monge-Kantorovich problem and its duality. Forthcoming in stochastics an international journal of probability and stochastic processes DOI: 10.1080/17442508.2011.624627

Chapter 6
Quantitative Relationships Between Minimal Distances and Minimal Norms

The goals of this chapter are to:

- Explore the conditions under which there is equality between the Kantorovich and the Kantorovich–Rubinstein functionals;
- Provide inequalities between the Kantorovich and Kantorovich–Rubinstein functionals;
- Provide criteria for convergence, compactness, and completeness of probability measures in probability spaces involving the Kantorovich and Kantorovich–Rubinstein functionals;
- Analyze the problem of uniformity between the two functionals.

Notation introduced in this chapter:

Notation	Description
$\mathcal{P}_\lambda = \mathcal{P}_\lambda(U)$	Space of laws with a finite λ-moment
$\widehat{\Lambda}$	Generalized Kantorovich functional
$\overset{\circ}{\Lambda}$	Generalized Kantorovich–Rubinstein functional

6.1 Introduction

In Chap. 5, we discussed the Kantorovich and Kantorovich–Rubinstein functionals. They generate minimal distances, $\widehat{\mu}_c$, and minimal norms, $\overset{\circ}{\mu}_c$, respectively, and we considered the problem of evaluating these functionals. The similarities between the two functionals indicate there can be quantitative relationships between them.

In this chapter, we begin by exploring the conditions under which $\widehat{\mu}_c = \overset{\circ}{\mu}_c$. It turns out that equality holds if and only if the cost function $c(x, y)$ is a metric

S.T. Rachev et al., *The Methods of Distances in the Theory of Probability and Statistics*, DOI 10.1007/978-1-4614-4869-3_6, © Springer Science+Business Media, LLC 2013

itself. Under more general conditions, certain inequalities hold involving $\widehat{\mu}_c$, $\mathring{\mu}_c$, and other probability metrics. These inequalities imply criteria for convergence, compactness, and uniformity in the spaces of probability measures $(\mathcal{P}(U), \widehat{\mu}_c)$ and $(\mathcal{P}(U), \mathring{\mu}_c)$. Finally, we conclude with a generalization of the Kantorovich and Kantorovich–Rubinstein functionals.

6.2 Equivalence Between Kantorovich Metric and Kantorovich–Rubinstein Norm

Levin (1975) proved that if U is a compact, $c(x,x) = 0$, $c(x,y) > 0$, and $c(x,y) + c(y,x) > 0$ for $x \neq y$, then $\widehat{\mu}_c = \mathring{\mu}_c$ if and only if $c(x,y) + c(y,x)$ is a metric in U. In the case of an s.m.s. U, we have the following version of Levin's result.

Theorem 6.2.1 (Neveu and Dudley 1980). *Suppose U is an s.m.s. and $c \in \mathfrak{C}^*$ (Corollary 5.3.1). Then*

$$\widehat{\mu}_c(P_1, P_2) = \mathring{\mu}_c(P_1, P_2) \tag{6.2.1}$$

for all P_1 and P_2 with

$$\int_U c(x,a)(P_1 + P_2)(\mathrm{d}x) < \infty \tag{6.2.2}$$

if and only if c is a metric.

Proof. Suppose (6.2.1) holds and set $P_1 = \delta_x$ and $P_2 = \delta_y$ for $x, y \in U$. Then the set $\mathcal{P}^{(P_1,P_2)}$ of all laws in $U \times U$ with marginals P_1 and P_2 contains only $P_1 \times P_2 = \delta_{(x,y)}$, and by Theorem 5.4.2,

$$\begin{aligned}\widehat{\mu}_c(P_1, P_2) = c(x,y) = \mathring{\mu}_c(P_1, P_2) &= \sup\left\{\left|\int f\mathrm{d}(P_1 - P_2) : \|f\|_c \leq 1\right|\right\} \\ &= \sup\{|f(x) - f(y)| : \|f\|_c \leq 1\} \\ &\leq \sup\{|f(x) - f(z)| + |f(z) - f(y)| : \|f\|_c \leq 1\} \\ &\leq c(x,z) + c(z,y).\end{aligned}$$

By assumption, $c \in \mathfrak{C}^*$, and therefore the triangle inequality implies that c is a metric in U.

Now define $\mathcal{G}(U)$ as the set of all pairs (f, g) of continuous functions $f : U \to \mathbb{R}$ and $g : U \to \mathbb{R}$ such that $f(x) + g(y) < c(x,y)$ $\forall x, y \in U$. Let $\mathcal{G}_B(U)$ be the set of all pairs $(f, g) \in \mathcal{G}(U)$ with f and g bounded.

Now suppose that $c(x,y)$ is a metric and that $(f, g) \in \mathcal{G}_B(U)$. Define $h(x) = \inf\{c(x,y) - g(y) : y \in U\}$. As the infimum of a family of continuous functions, h is upper semicontinuous. For each $x \in U$ we have $f(x) \leq h(x) \leq -g(x)$. Then

$$\begin{aligned} h(x) - h(x') &= \inf_u (c(x,u) - g(u)) - \inf_v (c(x',v) - g(v)) \\ &< \sup_v (g(v) - c(x',v) + c(x,v) - g(v)) \\ &= \sup_v (c(x,v) - c(x',v)) \leq c(x,x'), \end{aligned}$$

so that $\|h\|_c \leq 1$. Then for P_1, P_2 satisfying (6.2.2) we have

$$\int f \mathrm{d}P_1 + \int g \mathrm{d}P_2 \leq \int h \mathrm{d}(P_1 - P_2),$$

so that (according to Corollary 5.3.1 and Theorem 5.4.2 of Chap. 5) we have

$$\begin{aligned} \widehat{\mu}_c(P_1, P_2) &= \sup\left\{\int f \mathrm{d}P_1 + \int g \mathrm{d}P_2 : (f, g) \in \mathcal{G}_B(U)\right\} \\ &\leq \sup\left\{\int h \mathrm{d}(P_1 - P_2) : \|h\|_c \leq 1\right\} = \mathring{\mu}_c(P_1, P_2). \end{aligned}$$

Thus $\widehat{\mu}_c(P_1, P_2) = \mathring{\mu}_c(P_1, P_2)$. □

Corollary 6.2.1. *Let (U, d) be an s.m.s. and $a \in U$. Then*

$$\widehat{\mu}_d(P_1, P_2) = \mathring{\mu}_d(P_1, P_2) = \sup\left\{\int f \mathrm{d}(P_1 - P_2) : \|f\|_L \leq 1\right\} \tag{6.2.3}$$

whenever

$$\int d(x, a) P_i(\mathrm{d}x) < \infty, \qquad i = 1, 2. \tag{6.2.4}$$

The supremum is attained for some optimal f_0 with $\|f_0\|_L := \sup_{x \neq y}\{|f(x) - f(y)|/d(x,y)\}$.

If P_1 and P_2 are tight, there are some $b_0 \in \mathcal{P}^{(P_1,P_2)}$ and $f_0 : U \to \mathbb{R}$ with $\|f_0\|_L \leq 1$ such that

$$\widehat{\mu}_d(P_1, P_2) = \int d(x,y) b_0(\mathrm{d}x, \mathrm{d}y) = \int f_0 \mathrm{d}(P_1 - P_2),$$

where $f_0(x) - f_0(y) = d(x, y)$ for b_0-a.e. (x, y) in $U \times U$.

Proof. Set $c(x, y) = d(x, y)$. Application of the theorem proves the first statement. The second (existence of f_0) follows from Theorem 5.4.3.

For each $n \geq 1$ choose $b_n \in \mathcal{P}^{(P_1,P_2)}$ with

$$\int d(x,y) b_n(\mathrm{d}x, \mathrm{d}y) < \widehat{\mu}_d(P_1, P_2) + \frac{1}{n}.$$

If P_1 and P_2 are tight, then by Corollary 5.3.1 there exists $b_0 \in \mathcal{P}^{(P_1,P_2)}$ such that

$$\widehat{\mu}_d(P_1, P_2) = \int d(x,y) b_0(\mathrm{d}x, \mathrm{d}y),$$

i.e., that b_0 is optimal. Integrating both sides of $f_0(x) - f_0(y) \le d(x,y)$ with respect to b_0 yields $\int f_0 \mathrm{d}(P_1 - P_2) \le \int d(x,y) b_0(\mathrm{d}x, \mathrm{d}y)$. However, we know that we have equality of these integrals. This implies that $f_0(x) - f_0(y) = d(x,y)$ b_0-a.e. □

6.3 Inequalities Between $\widehat{\mu}_c$ and $\overset{\circ}{\mu}_c$

In the previous section we looked at conditions under which $\widehat{\mu}_c = \overset{\circ}{\mu}_c$. In general, $\widehat{\mu}_c \neq \overset{\circ}{\mu}_c$. For example, if $U = \mathbb{R}$, $d(x,y) = |x - y|$,

$$c(x,y) = d(x,y) \max(1, d^{p-1}(x,a), d^{p-1}(y,a)), \quad p \ge 1, \tag{6.3.1}$$

then for any laws P_i $(i = 1,2)$ on $\mathcal{B}(R)$ with distribution functions (DFs) F_i we have the following explicit expressions:

$$\widehat{\mu}_c(P_1, P_2) = \int_0^1 c(F_1^{-1}(t), F_2^{-1}(t)) \mathrm{d}t, \tag{6.3.2}$$

where F_i^{-1} is the function inverse to the DF F_i (see Theorem 7.4.2 in Chap. 7). On the other hand,

$$\overset{\circ}{\mu}_c(P_1, P_2) = \int_{-\infty}^{\infty} |F_1(x) - F_2(x)| \max(1, |x - a|^{p-1}) \mathrm{d}x \tag{6.3.3}$$

(see Theorem 5.5.1 in Chap. 5). However, in the space $\mathcal{M}_p = \mathcal{M}_p(U)$ [$U = (U,d)$ is an s.m.s.] of all Borel probability measures P with finite $\int d^p(x,a) P(\mathrm{d}x)$, the functionals $\widehat{\mu}_c$ and $\overset{\circ}{\mu}_c$ [where c is given by (6.3.1)] metrize the same exact topology, that is, the following $\widehat{\mu}_c$- and $\overset{\circ}{\mu}_c$-convergence criteria will be proved.

Theorem 6.3.1. *Let (U,d) be an s.m.s., let c be given by (6.3.1), and let $P, P_n \in \mathcal{M}_p$ $(n = 1,2,\dots)$. Then the following relations are equivalent:*

(I)

$$\widehat{\mu}_c(P_n, P) \to 0;$$

(II)

$$\overset{\circ}{\mu}_c(P_n, P) \to 0;$$

(III)

$$P_n \text{ converges weakly to } P\ (P_n \xrightarrow{w} P) \text{ and}$$

$$\lim_{N\to\infty} \sup_n \int d^p(x,a) I\{d(x,a) > N\} P_n(\mathrm{d}x) = 0;$$

(IV)

$$P_n \xrightarrow{w} P \text{ and } \int d^p(x,a) P_n(\mathrm{d}x) \to \int d^p(x,a) P(\mathrm{d}x).$$

(The assertion of the theorem is an immediate consequence of Theorems 6.3.2–6.3.5 below and the more general Theorem 6.4.1).

Theorem 6.3.1 is a qualitative $\widehat{\mu}_c$ ($\overset{\circ}{\mu}_c$)-convergence criterion. One can rewrite (III) as

$$\pi(P_n, P) \to 0 \text{ and } \lim_{\varepsilon\to 0} \sup_n \omega(\varepsilon; P_n; \lambda) = 0,$$

where π is the Prokhorov metric[1]

$$\pi(P, Q) := \inf\{\varepsilon > 0 : P(A) \le Q(A^\varepsilon) + \varepsilon \quad \forall A \in \mathcal{B}(U)\}$$
$$(A^\varepsilon := \{x : d(x, A) < \varepsilon\}) \tag{6.3.4}$$

and $\omega(\varepsilon; P; \lambda)$ is the following modulus of λ-integrability:

$$\omega(\varepsilon; P; \lambda) := \int \lambda(x) I\left\{d(x,a) > \frac{1}{\varepsilon}\right\} P(\mathrm{d}x), \tag{6.3.5}$$

where $\lambda(x) := \max(d(x,a), d^p(x,a))$. Analogously, (IV) is equivalent to

(IV^*)

$$\pi(P_n, P) \to 0 \text{ and } D(P_n, P; \lambda) \to 0,$$

where

$$D(P, Q; \lambda) := \left| \int \lambda(x)(P - Q)(\mathrm{d}x) \right|. \tag{6.3.6}$$

In this section we investigate quantitative relationships between $\widehat{\mu}_c$, $\overset{\circ}{\mu}_c$, π, ω, and D in terms of inequalities between these functionals. These relationships yield convergence and compactness criteria in the space of measures w.r.t. the Kantorovich-type functionals $\widehat{\mu}_c$ and $\overset{\circ}{\mu}_c$ (see Examples 3.3.2 and 3.3.6 in Chap. 3) as well as the $\overset{\circ}{\mu}_c$-completeness of the space of measures.

[1]See Examples 3.3.3 and 4.3.2 in Chaps. 3 and 4, respectively.

In what follows, we assume that the cost function c has the form considered in Example 5.2.1:

$$c(x,y) = d(x,y)k_0(d(x,a),d(y,a)) \quad x,y \in U, \tag{6.3.7}$$

where $k_0(t,s)$ is a symmetric continuous function nondecreasing on both arguments $t \geq 0$, $s \geq 0$, and satisfying the following conditions:

(C1)

$$\alpha := \sup_{s \neq t} \frac{|K(t) - K(s)|}{|t-s|k_0(t,s)} < \infty,$$

where $K(t) := tk_0(t,t)$, $t \neq 0$;

(C2)

$$\beta := k(0) > 0,$$

where $k(t) = k_0(t,t)$ $t \geq 0$; and

(C3)

$$\gamma := \sup_{t \geq 0, s \geq 0} \frac{k_0(2t,2s)}{k_0(t,s)} < \infty.$$

If c is given by (6.3.1), then c admits the form (6.3.7) with $k_0(t,s) = \max(1, t^{p-1}, s^{p-1})$, and in this case $\alpha = p$, $\beta = 1$, $\gamma = 2^{p-1}$. Further, let $\mathcal{P}_\lambda = \mathcal{P}_\lambda(U)$ be the space of all probability measures on the s.m.s. (U,d) with finite λ-moment

$$\mathcal{P}_\lambda(U) = \left\{ P \in \mathcal{P}(U) : \int_U \lambda(x) P(\mathrm{d}x) < \infty \right\}, \tag{6.3.8}$$

where $\lambda(x) = K(d(x,a))$ and a is a fixed point of U.

In Theorems 6.3.2–6.3.5 we assume that $P_1 \in \mathcal{P}_\lambda$, $P_2 \in \mathcal{P}_\lambda$, $\varepsilon > 0$, and denote $\widehat{\mu}_c := \widehat{\mu}_c(P_1,P_2)$ [see (5.2.16)], $\overset{\circ}{\mu}_c := \overset{\circ}{\mu}_c(P_1,P_2)$ [see (5.2.17)], $\boldsymbol{\pi} := \boldsymbol{\pi}(P_1,P_2)$,

$$\omega_i(\varepsilon) := \omega(\varepsilon; P_i; \lambda) := \int \lambda(x) I\{d(x,a) > 1/\varepsilon\} P_i(\mathrm{d}x), \quad P_i \in \mathcal{P}_\lambda$$

$$D := D(P_1,P_2;\lambda) := \left| \int \lambda \mathrm{d}(P_1 - P_2) \right|,$$

and the function c satisfies conditions (C1)–(C3). We begin with an estimate of $\widehat{\mu}_c$ from above in terms of $\boldsymbol{\pi}$ and $\omega_i(\varepsilon)$.

Theorem 6.3.2.

$$\widehat{\mu}_c \leq \boldsymbol{\pi}[4K(1/\varepsilon) + \omega_1(1) + \omega_2(1) + 2k(1)] + 5\omega_1(\varepsilon) + 5\omega_2(\varepsilon). \tag{6.3.9}$$

Proof. Recall that $\mathcal{P}^{(P_1,P_2)}$ is the space of all laws P on $U \times U$ with prescribed marginals P_1 and P_2. Let $\mathbf{K} = \mathbf{K}_1$ be the Ky Fan metric with parameter 1 (see Example 3.4.2 in Chap. 3)

$$\mathbf{K}(P) := \inf\{\delta > 0 : P(d(x,y) > \delta) < \delta\} \quad P \in \mathcal{P}_\lambda(U). \tag{6.3.10}$$

Claim 1. For any $N > 0$ and for any measure P on U^2 with marginals P_1 and P_2, i.e., $P \in \mathcal{P}^{(P_1,P_2)}$, we have

$$\int_{U\times U} c(x,y)P(\mathrm{d}x,\mathrm{d}y) \le \mathbf{K}(P)\left[4K(N) + \int_U k(d(x,a))(P_1+P_2)(\mathrm{d}x)\right] + 5\omega_1(1/N) + 5\omega_2(1/N). \tag{6.3.11}$$

Proof of Claim 1. Suppose $\mathbf{K}(P) < \sigma \le 1$, $P \in \mathcal{P}^{(P_1,P_2)}$. Then by (6.3.7) and (C3),

$$\int c(x,y)P(\mathrm{d}x,\mathrm{d}y) \le \int d(x,y)k(\max\{d(x,a),d(y,a)\})P(\mathrm{d}x,\mathrm{d}y) \le I_1 + I_2,$$

where

$$I_1 := \int_{U\times U} d(x,y)k(d(x,a))P(\mathrm{d}x,\mathrm{d}y)$$

and

$$I_2 := \int_{U\times U} d(x,y)k(d(y,a))P(\mathrm{d}x,\mathrm{d}y).$$

Let us estimate I_1:

$$\begin{aligned}
I_1 &:= \int d(x,y)k(d(x,a))[I\{d(x,y) < \delta\} + I\{d(x,y) \ge \delta\}]P(\mathrm{d}x,\mathrm{d}y) \\
&\le \delta \int k(d(x,a))P(\mathrm{d}x,\mathrm{d}y) \\
&\quad + \int d(x,y)k(d(x,a))I\{d(x,y) \ge \delta\}P(\mathrm{d}x,\mathrm{d}y) \\
&\le I_{11} + I_{12} + I_{13},
\end{aligned} \tag{6.3.12}$$

where

$$I_{11} := \delta \int_U k(d(x,a))[I\{d(x,a) \ge 1\} + I\{d(x,a) \le 1\}]P_1(\mathrm{d}x),$$

$$I_{12} := \int_{U\times U} d(x,a)k(d(x,a))I\{d(x,y) \ge \delta\}P(\mathrm{d}x,\mathrm{d}y), \quad \text{and}$$

$$I_{13} := \int_{U\times U} d(y,a)k(d(x,a))I\{d(x,y) \ge \delta\}P(\mathrm{d}x,\mathrm{d}y).$$

Obviously, by $\lambda(x) := K(d(x,a))$, $I_{11} \leq \delta \int k(d(x,a))I\{d(x,a) \geq 1\}P_1(\mathrm{d}x) + \delta k(1) \leq \delta\omega_1(1) + \delta k(1)$. Further,

$$\begin{aligned}
I_{12} &= \int K(d(x,a))I\{d(x,y) \geq \delta\}[I\{d(x,a) > N\} + I\{d(x,a) \leq N\}]P(\mathrm{d}x,\mathrm{d}y) \\
&\leq \int_U \lambda(x)I\{d(x,a) > N\}P_1(\mathrm{d}x) + K(N)\int_{U\times U} I\{d(x,y) \geq \delta\}P(\mathrm{d}x,\mathrm{d}y) \\
&\leq \omega_1(1/N) + K(N)\delta.
\end{aligned}$$

Now let us estimate the last term in estimate (6.3.12):

$$\begin{aligned}
I_{13} &= \int_{U\times U} d(y,a)k(d(x,a))I\{d(x,y) \geq \delta\}[I\{d(x,a) \geq d(y,a) > N\} \\
&\quad + I\{d(y,a) > d(x,a) > N\} + I\{d(x,a) > N, d(y,a) \leq N\} \\
&\quad + I\{d(x,a) \leq N, d(y,a) > N\} + I\{d(x,a) \leq N, d(y,a) \leq N\}]P(\mathrm{d}x,\mathrm{d}y) \\
&\leq \int_{U\times U} \lambda(x)I\{d(x,a) > d(y,a) > N\}P(\mathrm{d}x,\mathrm{d}y) \\
&\quad + \int_{U\times U} \lambda(y)I\{d(y,a) \geq d(x,a) \geq N\}P(\mathrm{d}x,\mathrm{d}y) \\
&\quad + \int_U \lambda(x)I\{d(x,a) > N\}P_1(\mathrm{d}x) + \int_U \lambda(y)I\{d(y,a) > N\}P_2(\mathrm{d}y) \\
&\quad + K(N)\int_{U\times U} I\{d(x,y) \geq \delta\}P(\mathrm{d}x,\mathrm{d}y) \\
&\leq 2\omega_1(1/N) + 2\omega_2(1/N) + K(N)\delta.
\end{aligned}$$

Summarizing the preceding estimates we obtain $I_1 \leq \delta\omega_1(1) + \delta k(1) + 3\omega_1(1/N) + 2\omega_2(1/N) + 2K(N)\delta$. By symmetry we have $I_2 \leq \delta\omega_2(1) + \delta k(1) + 3\omega_2(1/N) + 2\omega_1(1/N) + 2K(N)\delta$. Therefore, the last two estimates imply

$$\begin{aligned}
\int c(x,y)P(\mathrm{d}x,\mathrm{d}y) &\leq I_1 + I_2 \\
&\leq \delta(\omega_1(1) + \omega_2(1) + 2k(1) + 4K(N)) \\
&\quad + 5\omega_1(1/N) + 5\omega_2(1/N).
\end{aligned}$$

Letting $\delta \to \mathbf{K}(P)$ we obtain (6.3.11), which proves the claim.

Claim 2 (Strassen–Dudley Theorem).

$$\inf\{\mathbf{K}(P) : P \in \mathcal{P}^{(P_1,P_2)}\} = \pi(P_1, P_2). \tag{6.3.13}$$

Proof of Claim 2. See Dudley (2002) (see also further Corollary 7.5.2 in Chap. 7).

Claims 1 and 2 complete the proof of the theorem. □

The next theorem shows that $\widehat{\mu}_c$-convergence and $\overset{\circ}{\mu}_c$-convergence imply the weak convergence of measures.

Theorem 6.3.3.

$$\beta\pi^2 \le \overset{\circ}{\mu}_c \le \widehat{\mu}_c. \tag{6.3.14}$$

Proof. Obviously, for any continuous nonnegative function c,

$$\overset{\circ}{\mu}_c \le \widehat{\mu}_c \tag{6.3.15}$$

and

$$\overset{\circ}{\mu}_c \ge \boldsymbol{\zeta}_c, \tag{6.3.16}$$

where $\boldsymbol{\zeta}_c$ is the Zolatarev simple metric with a ζ-structure (Definition 4.4.1)

$$\begin{aligned}\boldsymbol{\zeta}_c &:= \boldsymbol{\zeta}_c(P_1, P_2)\\ &:= \sup\left\{\left|\int_U f\,\mathrm{d}(P_1 - P_2)\right| : f : U \to \mathbb{R}, |f(x) - f(y)| \le c(x,y) \forall x, y \in U\right\}.\end{aligned} \tag{6.3.17}$$

Now, using assumption (C2) we have that $c(x,y) \ge \beta d(x,y)$ and, hence, $\boldsymbol{\zeta}_c \ge \beta\boldsymbol{\zeta}_d$. Thus, by (6.3.16),

$$\overset{\circ}{\mu}_c \ge \beta\boldsymbol{\zeta}_d. \tag{6.3.18}$$

Claim 3.

$$\boldsymbol{\zeta}_d \ge \boldsymbol{\pi}^2. \tag{6.3.19}$$

Proof of Claim 3. Using the dual representation of $\widehat{\mu}_d$ [see (6.2.3)] we are led to

$$\widehat{\mu}_d = \boldsymbol{\zeta}_d, \tag{6.3.20}$$

which in view of the inequality

$$\int d(x,y) P(\mathrm{d}x, \mathrm{d}y) \ge \mathbf{K}^2(P) \ \text{ for any } \ P \in \mathcal{P}^{(P_1,P_2)} \tag{6.3.21}$$

establishes (6.3.19). The proof of the claim is now completed.

The desired inequalities (6.3.14) are the consequence of (6.3.15), (6.3.16), (6.3.18), and Claim 3. □

The next theorem establishes the uniform λ-integrability

$$\lim_{\varepsilon\to 0} \sup_n \omega(\varepsilon, P_n, \lambda) = 0$$

of the sequence of measures $P_n \in \mathcal{P}_\lambda$ $\overset{\circ}{\mu}_c$-converging to a measure $P \in \mathcal{P}_\lambda$.

Theorem 6.3.4.

$$\omega_1(\varepsilon/2) \le \alpha(2\gamma+1)\overset{\circ}{\mu}_c + 2(\gamma+1)\omega_2(\varepsilon). \tag{6.3.22}$$

Proof. For any $N > 0$, by the triangle inequality, we have

$$\omega_1(1/2N) := \int \lambda(x) I\{d(x,a) > 2N\} P_1(\mathrm{d}x) \le \mathcal{T}_1 + \mathcal{T}_2, \tag{6.3.23}$$

where

$$\mathcal{T}_1 := \left| \int \lambda(x) I\{d(x,a) > 2N\}(P_1 - P_2)(\mathrm{d}x) \right|$$

and

$$\mathcal{T}_2 := \int \lambda(x) I\{d(x,a) > N\} P_2(\mathrm{d}x) = \omega_2(1/N).$$

Claim 4.

$$\mathcal{T}_1 \le \alpha\overset{\circ}{\mu}_c + K(2N) \int I\{d(x,a) > 2N\}(P_1 + P_2)(\mathrm{d}x). \tag{6.3.24}$$

Proof of Claim 4. Denote $f_N(x) := (1/\alpha)\max(\lambda(x), K(2N))$. Since $\lambda(x) = K(d(x,a)) = d(x,a)k_0(d(x,a), d(x,a))$, then by (C1),

$$\begin{aligned}|f_N(x) - f_N(y)| &\le (1/\alpha)|\lambda(x) - \lambda(y)| \\ &\le |d(x,a) - d(y,a)| k_0(d(x,a), d(y,a)) \le c(x,y)\end{aligned}$$

for any $x, y \in U$. Thus the inequalities

$$\left| \int_U f_N(x)(P_1 - P_2)(\mathrm{d}x) \right| \le \zeta_c(P_1, P_2) \le \overset{\circ}{\mu}_c(P_1, P_2) \tag{6.3.25}$$

follow from (6.3.16) and (6.3.17). Since $\alpha f_N(x) = \max(K(d(x,a)), K(2N))$ and (6.3.25) holds, then

$$\begin{aligned}\mathcal{T}_1 < &\left| \int_U K(d(x,a)) I\{d(x,a) > 2N\}(P_1 - P_2)(\mathrm{d}x) \right. \\ &\left. - \int_U K(2N) I\{d(x,a) \le 2N\}(P_1 - P_2)(\mathrm{d}x) \right| \\ &+ K(2N) \left| \int_U I\{d(x,a) \le 2N\}(P_1 - P_2)(\mathrm{d}x) \right|\end{aligned}$$

$$= \left|\int_U \alpha f_N(x)(P_1 - P_2)(\mathrm{d}x)\right| + K(2N)\left|\int_U I\{d(x,a) > 2N\}(P_1 - P_2)(\mathrm{d}x)\right|$$

$$\leq \alpha\mathring{\mu}_c + K(2N)\int I\{d(x,a) > 2N\}(P_1 + P_2)(\mathrm{d}x),$$

which proves the claim.

Claim 5.

$$A(P_1) := K(2N)\int_U I\{d(x,a) > 2N\}P_1(\mathrm{d}x) \leq 2\alpha\gamma\mathring{\mu}_c + 2\gamma\omega_2(1/N). \quad (6.3.26)$$

Proof of Claim 5. As in the proof of Claim 4, we choose an appropriate Lipschitz function. That is, write

$$g_N(x) = (1/(2\alpha\gamma))\min\{K(2N), K(2d(x, O(a,N)))\},$$

where $O(a,N) := \{x : d(x,a) < N\}$. Using (C1) and (C3),

$$|g_N(x) - g_N(y)| \leq (1/(2\alpha\gamma))|K(2d(x, O(a,N))) - K\{2d(y, O(a,N)))|$$

by (C1)

$$\leq (1/\gamma)|d(x, O(a,N))$$
$$-d(y, O(a,N))|k_0(2d(x, O(a,N)), 2d(y, O(a,N)))$$

by (C3)

$$\leq d(x,y)k_0(d(x, O(a,N)), d(y, O(a,N))) \leq c(x,y).$$

Hence

$$\left|\int g_N(P_1 - P_2)(\mathrm{d}x)\right| \leq \zeta_c \leq \mathring{\mu}_c. \quad (6.3.27)$$

Using (6.3.27) and the implications

$$d(x,a) > 2N \Rightarrow d(x, O(a,N)) > N \Rightarrow K(2d(x, O(a,N))) \geq K(2N)$$

we obtain the following chain of inequalities:

$$A(P_1) \leq 2\alpha\gamma\int g_N(x)P_1(\mathrm{d}x)$$

$$\leq 2\alpha\gamma\left|\int g_N(x)(P_1 - P_2)(\mathrm{d}x)\right| + 2\alpha\gamma\int_U g_N(x)P_2(\mathrm{d}x)$$

$$\leq 2\alpha\mathring{\mu}_c + \int K(2d(x, O(a,N)))I\{d(x,a) \geq N\}P_2(\mathrm{d}x),$$

$$\left(\text{by C3, } \frac{K(2t)}{K(t)} = \frac{2tk_0(2t,2t)}{tk_0(t,t)} \le 2\gamma\right)$$

$$\le 2\alpha\gamma\mathring{\mu}_c + 2\gamma\int K(d(x,O(a,N)))I\{d(x,a)\ge N\}P_2(\mathrm{d}x)$$

$$\le 2\alpha\gamma\mathring{\mu}_c + 2\gamma\omega_2(1/N), \tag{6.3.28}$$

which proves the claim.

For $A(P_2)$ [see (6.3.26)] we have the following estimate:

$$A(P_2) \le \int_U K(d(x,a))I\{d(x,a) > 2N\}P_2(\mathrm{d}x) \le \omega_2(1/N). \tag{6.3.29}$$

Summarizing (6.3.23), (6.3.24), (6.3.26), and (6.3.29) we obtain

$$\omega_1(1/2N) \le \alpha\mathring{\mu}_c + A(P_1) + A(P_2) + \omega_2(1/N)$$
$$\le (\alpha + 2\alpha\gamma)\mathring{\mu}_c + (2\gamma + 2)\omega_2(1/N)$$

for any $N > 0$, as desired. □

The next theorem shows that $\mathring{\mu}_c$-convergence implies convergence of the λ-moments.

Theorem 6.3.5.

$$D \le \alpha\mathring{\mu}_c. \tag{6.3.30}$$

Proof. By (C1), for any finite nonnegative measure Q with marginals P_1 and P_2 we have

$$D := \left|\int_U \lambda(x)(P_1 - P_2)(\mathrm{d}x)\right| = \left|\int_{U\times U} \lambda(x) - \lambda(y)Q(\mathrm{d}x,\mathrm{d}y)\right|$$
$$\le \int_{U\times U} \alpha|d(x,a) - d(y,a)|k_0(d(x,a),d(y,a))Q(\mathrm{d}x,\mathrm{d}y)$$
$$\le \alpha\int_{U\times U} c(x,y)Q(\mathrm{d}x,\mathrm{d}y)$$

which completes the proof of (6.3.30). □

Inequalities (6.3.9), (6.3.14), (6.3.22), and (6.3.30), described in Theorems 6.3.2–6.3.5, imply criteria for convergence, compactness, and uniformity in the spaces of probability measures $(\mathcal{P}(U),\widehat{\mu}_c)$ and $(\mathcal{P}(U),\mathring{\mu}_c)$ (see also the next section). Moreover, the estimates obtained for $\widehat{\mu}_c$ and $\mathring{\mu}_c$ may be viewed as quantitative results demanding conditions that are necessary and sufficient for

$\widehat{\mu}_c$-convergence and $\mathring{\mu}_c$-convergence. Note that, in general, quantitative results require assumptions additional to the set of necessary and sufficient conditions implying the qualitative results. The classic example is the central limit theorem, where the uniform convergence of the normalized sum of i.i.d. RVs can be at any low rate assuming only the existence of the second moment.

6.4 Convergence, Compactness, and Completeness in $(\mathcal{P}(U), \widehat{\mu}_c)$ and $(\mathcal{P}(U), \mathring{\mu}_c)$

In this section, we assume that the cost function c satisfies conditions (C1)–(C3) in the previous section and $\lambda(x) = K(d(x,a))$. We begin with the criterion for $\widehat{\mu}_c$- and $\mathring{\mu}_c$-convergence.

Theorem 6.4.1. *If P_n, and $P \in \mathcal{P}_\lambda(U)$, then the following statements are equivalent*

(A)

$$\widehat{\mu}_c(P_n, P) \to 0;$$

(B)

$$\mathring{\mu}_c(P_n, P) \to 0;$$

(C)

$$P_n \xrightarrow{w} P \ (P_n \text{ converges weakly to } P) \text{ and } \int \lambda \mathrm{d}(P_n - P) \to 0 \text{ as } n \to \infty;$$

(D)

$$P_n \xrightarrow{w} P \ \text{ and } \lim_{\varepsilon\to 0} \sup_n \omega_n(\varepsilon) = 0,$$

where $\omega_n(\varepsilon) := \omega(\varepsilon; P_n; \lambda) = \int \lambda(x)\{d(x,a) > 1/\varepsilon\} P_n(\mathrm{d}x)$.

Proof. From inequality (6.3.14) it is apparent that $A \Rightarrow B$ and $B \Rightarrow P_n \xrightarrow{w} P$. Using (6.3.30) we obtain that B implies $\int \lambda \mathrm{d}(P_n - P) \to 0$, and thus $B \Rightarrow C$. Now, let C hold.

Claim 6. $C \Rightarrow D$.

Proof of Claim 6. Choose a sequence $\varepsilon_1 > \varepsilon_2 > \cdots \to 0$ such that $P(d(x,a) = 1/\varepsilon_n) = 0$ for any $n = 1, 2, \ldots$. Then for fixed n

$$\int \lambda(x) I\{d(x,a) \leq 1/\varepsilon_n\}(P_k - P)(\mathrm{d}x) \to 0 \ \text{ as } \ k \to \infty$$

by Billingsley (1999, Theorem 5.1). Since $P \in \mathcal{P}_\lambda$, $\omega(\varepsilon_n) := \omega(\varepsilon_n; P; c) \to 0$ as $n \to \infty$, and hence

$$\begin{aligned}
\limsup_{k\to\infty} \omega_k(\varepsilon_n) &\le \limsup_{k\to\infty} \left| \int \lambda(x)\{d(x,a) > 1/\varepsilon_n\}(P_k - P)(\mathrm{d}x) \right| + \omega(\varepsilon_n) \\
&\le \limsup_{k\to\infty} \left| \int \lambda(x)(P_k - P)(\mathrm{d}x) \right| \\
&\quad + \limsup_{k\to\infty} \left| \int \lambda(x) I\{d(x,a) \le 1/\varepsilon_n\}(P_k - P)(\mathrm{d}x) \right| \\
&\quad + \omega(\varepsilon_n) \to 0 \text{ as } n \to \infty.
\end{aligned}$$

The last inequality and $P_k \in \mathcal{P}_\lambda$ imply $\lim_{\varepsilon\to 0} \sup_n \omega_n(\varepsilon) = 0$, and hence D holds.

The claim is proved.

Claim 7. $D \Rightarrow A$.

Proof of Claim 7. By Theorem 6.3.2,

$$\widehat{\mu}_c(P_n, P) \le \pi(P_n, P)[4K(1/\varepsilon_n) + \omega_n(1) + \omega(1) + 2k(1)] + 5\omega_n(\varepsilon_n) + 5\omega(\varepsilon_n),$$

where ω_n and ω are defined as in Claim 6 and, moreover, $\varepsilon_n > 0$ is such that

$$4K(1/\varepsilon_n) + \sup_{n\ge 1} \omega_n(1) + \omega(1) + 2k(1) \le (\pi(P_n, P))^{-1/2}.$$

Hence, using the last two inequalities we obtain

$$\widehat{\mu}_c(P_n, P) \le \sqrt{\pi(P_n, P)} + 5 \sup_{n\ge 1} \omega_n(\varepsilon_n) + 5\omega(\varepsilon_n),$$

and hence $D \Rightarrow A$, as we claimed. □

The Kantorovich–Rubinstein functional $\overset{\circ}{\mu}_c$ is a metric in $\mathcal{P}_\lambda(U)$, while $\widehat{\mu}_c$ is not a metric except for the case $c = d$ (see the discussion in the previous section). The next theorem establishes a criterion for $\overset{\circ}{\mu}_c$-relative compactness of sets of measures. Recall that a set $\mathcal{A} \subset \mathcal{P}_\lambda$ is said to be *$\overset{\circ}{\mu}_c$-relatively compact* if any sequence of measures in $\mathcal{A}$ has a $\overset{\circ}{\mu}_c$-convergent subsequence and the limit belongs to $\mathcal{P}_\lambda$. Recall that the set $\mathcal{A} \subset \mathcal{P}(U)$ is *weakly compact* if $\mathcal{A}$ is π-relatively compact, i.e., any sequence of measures in $\mathcal{A}$ has a weakly (π-) convergent subsequence.

Theorem 6.4.2. *The set $\mathcal{A} \subset \mathcal{P}_\lambda$ is $\overset{\circ}{\mu}_c$-relatively compact if and only if $\mathcal{A}$ is weakly compact and*

$$\lim_{\varepsilon\to 0} \sup_{P\in\mathcal{A}} \omega(\varepsilon; P; \lambda) = 0. \tag{6.4.1}$$

Proof. *"If" part*: If $\mathcal{A}$ is weakly compact, (6.4.1) holds and $\{P_n\}_{n\geq 1} \subset \mathcal{A}$, then we can choose a subsequence $\{P_{n'}\} \subset \{P_n\}$ that converges weakly to a probability measure P.

Claim 8. $P \in \mathcal{P}_\lambda$.

Proof of Claim 8. Let $0 < \alpha_1 < \alpha_2 < \cdots$, $\lim \alpha_n = \infty$ be such a sequence that $P(d(x,a) = \alpha_n) = 0$ for any $n \geq 1$. Then, by Billingsley (1999, Theorem 5.1) and (6.4.1),

$$\int \lambda(x) I\{d(x,a) \leq \alpha_{n'}\} P(\mathrm{d}x) = \lim_{n\to\infty} \int \lambda(x) I\{d(x,a) \leq \alpha_{n'}\} P_{n'}(\mathrm{d}x)$$
$$\leq \liminf_{n\to\infty} \int \lambda(x) P_{n'}(\mathrm{d}x) < \infty,$$

which proves the claim.

Claim 9.

$$\overset{\circ}{\mu}_c(P_{n'}, P) \to 0.$$

Proof of Claim 9. Using Theorem 6.3.2, Claim 8, and (6.4.1) we have, for any $\delta > 0$,

$$\overset{\circ}{\mu}_c(P_{n'}, P) \leq \widehat{\mu}_c(P_{n'}, P) \leq \pi(P_{n'}, P)[4K(1/\varepsilon) + \omega_1(1) + \omega_2(1) + 2K(1)]$$
$$+5 \sup_{n'} \omega(P_{n'}; \varepsilon; \lambda) + \omega(P; \varepsilon; \lambda)$$
$$\leq \pi(P_{n'}, P)[4K(1/\varepsilon) + \omega_1(1) + \omega_2(1) + 2K(1)] + \delta$$

if $\varepsilon = \varepsilon(\delta)$ is small enough. Hence, by $\pi(P_{n'}, P) \to 0$, we can choose $N = N(\delta)$ such that $\overset{\circ}{\mu}_c(P_{n'}, P) < 2\delta$ for any $n' \geq N$, as desired.

Claims 8 and 9 establish the "if" part of the theorem.

"Only if" part: If $\mathcal{A}$ is $\overset{\circ}{\mu}_c$-relatively compact and $\{P_n\} \subset \mathcal{A}$, then there exists a subsequence $\{P_{n'}\} \subset \{P_n\}$ that is convergent w.r.t. $\overset{\circ}{\mu}_c$, and let P be the limit. Hence, by Theorem 6.3.3, $\overset{\circ}{\mu}_c(P_n, P) \geq \beta\pi^2(P_n, P) \to 0$, which demonstrates that the set $\mathcal{A}$ is weakly compact.

Further, if (6.4.1) is not valid, then there exists $\delta > 0$ and a sequence $\{P_n\}$ such that

$$\omega(1/n; P_n; \lambda) > \delta \quad \forall n \geq 1. \tag{6.4.2}$$

Let $\{P_{n'}\}$ be a $\overset{\circ}{\mu}_c$-convergent subsequence of $\{P_n\}$, and let $P \in \mathcal{P}_\lambda$ be the corresponding limit. By Theorem 6.3.4, $\omega(1/n'; P_{n'}; \lambda) \geq (2\gamma + 2)(\alpha \overset{\circ}{\mu}_c(P_{n'}, P) + \omega(1/n'; P; \lambda)) \to 0$ as $n' \to \infty$, which is in contradiction with (6.4.2). □

In the light of Theorem 6.4.1, we can now interpret Theorem 6.4.2 as a criterion for $\overset{\circ}{\mu}_c$-relative compactness of sets of measures in $\mathcal{P}$ by simply changing $\overset{\circ}{\mu}_c$ with $\widehat{\mu}_c$ in the formation of the last theorem.

The well-known Prokhorov theorem says that (U,d) *is a complete s.m.s; then the set of all laws on* U *is complete w.r.t. the Prokhorov metric* $\boldsymbol{\pi}$.[2] The next theorem is an analog of the Prokhorov theorem for the metric space $\mathcal{P}_\lambda, \overset{\circ}{\mu}_c)$.

Theorem 6.4.3. *If* (U,d) *is a complete s.m.s., then* $(\mathcal{P}_\lambda(U), \overset{\circ}{\mu}_c)$ *is also complete.*

Proof. If $\{P_n\}$ is a $\overset{\circ}{\mu}_c$-fundamental sequence, then by Theorem 6.3.3, $\{P_n\}$ is also π-fundamental, and hence there exists the weak limit $P \in \mathcal{P}(U)$.

Claim 10. $P \in \mathcal{P}_\lambda$.

Proof of Claim 10. Let $\varepsilon > 0$ and $\overset{\circ}{\mu}_c(P_n, P_m) \le \varepsilon$ for any $n, m \ge n_\varepsilon$. Then, by Theorem 6.3.5, $\left|\int \lambda(x)(P_n - P_{n_\varepsilon})(\mathrm{d}x)\right| < \alpha\varepsilon$ for any $n > n_\varepsilon$; hence,

$$\sup_{n \ge n_\varepsilon} \int \lambda(x) P_n(\mathrm{d}x) < \alpha\varepsilon + \int \lambda(x) P_{n_\varepsilon}(\mathrm{d}x) < \infty.$$

Choose the sequence $0 < \alpha_1 < \alpha_2 < \cdots$, $\lim_{k\to\infty} \alpha_k = \infty$, such that $P(d(x,a) = \alpha_k) = 0$ for any $k > 1$. Then

$$\begin{aligned}\int \lambda(x) I\{d(x,a) \le \alpha_k\} P(\mathrm{d}x) &= \lim_{n\to\infty} \int \lambda(x) I\{d(x,a) \le \alpha_k\} P_n(\mathrm{d}x)\\ &\le \liminf_{n\to\infty} \int \lambda(x) P_n(\mathrm{d}x)\\ &\le \sup_{n \ge n_\varepsilon} \int_U \lambda(x) P_n(\mathrm{d}x) < \infty.\end{aligned}$$

Letting $k \to \infty$ the assertion follows.

Claim 11.

$$\overset{\circ}{\mu}_c(P_n, P) \to 0.$$

Proof of Claim 11. Since $\overset{\circ}{\mu}_c(P_n, P_{n_\varepsilon}) \le \varepsilon$ for any $n \ge n_\varepsilon$, then, by Theorem 6.3.4,

$$\sup_{n \ge n_\varepsilon} \omega(\delta; P_n; \lambda) \le 2(\gamma + 1)(\alpha\varepsilon + \omega(2\delta; P_{n_\varepsilon}; \lambda))$$

for any $\delta > 0$. The last inequality and Theorem 6.3.2 yield

$$\overset{\circ}{\mu}_c(P_n, P) \le \widehat{\mu}_c(P_n, P) \le \pi(P_n, P)[4K(1/\delta)$$

[2]See, for example, Hennequin and Tortrat (1965) and Dudley (2002, Theorem 11.5.5).

$$+ \sup_{n \geq n_\varepsilon} \omega(1; P_n; \lambda) + \omega(1; P; \lambda) + 2K(1)]$$

$$+10(\gamma + 1)(\alpha\varepsilon + \omega(2\delta; P_{n_\varepsilon}; \lambda) + 5\omega(\delta; P_{n_\varepsilon}; \lambda)) \quad (6.4.3)$$

for any $n \geq n_\varepsilon$ and $\delta > 0$. Next, choose $\delta_n = \delta_{n,\varepsilon} > 0$ such that $\delta_n \to 0$ as $n \to \infty$ and

$$4K(1/\delta_n) + \sup_{n \geq n_\varepsilon} \omega(1; P_n; \lambda) + \omega(1; P; \lambda) + 2k(1) \leq \frac{1}{(\pi(P_n, P))^{1/2}}. \quad (6.4.4)$$

Combining (6.4.3) and (6.4.4) we have that $\overset{\circ}{\mu}_c(P_n, P) \leq$ const. ε for n large enough, which proves the claim. □

6.5 $\overset{\circ}{\mu}_c$- and $\widehat{\mu}_c$-Uniformity

In the previous section, we saw that $\overset{\circ}{\mu}_c$ and $\widehat{\mu}_c$ induce the same exact convergence in $\mathcal{P}_\lambda$. Here we would like to analyze the uniformity of $\overset{\circ}{\mu}_c$ and $\widehat{\mu}_c$-convergence. Namely, if for any $P_n, Q_n \in \mathcal{P}_\lambda$, the equivalence

$$\overset{\circ}{\mu}_c(P_n, Q_n) \iff \widehat{\mu}_c(P_n, Q_n) \to 0 \quad n \to \infty \quad (6.5.1)$$

holds. Obviously, $\Leftarrow$, by $\overset{\circ}{\mu}_c(P_n, Q_n) \leq \widehat{\mu}_c$. So, if

$$\widehat{\mu}_c(P, Q) \leq \phi(\overset{\circ}{\mu}_c(P, Q)) \qquad P, Q \in \mathcal{P}_\lambda \quad (6.5.2)$$

for a continuous nondecreasing function, $\phi(0) = 0$, then (6.5.1) holds.

Remark 6.5.1. Given two metrics, say μ and ν, in the space of measures, the equivalence of μ- and ν-convergence does not imply the existence of a continuous nondecreasing function ϕ vanishing at 0 and such that $\mu \leq \phi(\nu)$. For example, both the Lévy metric $\mathbf{L}$ [see (4.2.3)] and the Prokhorov metric π [see (3.3.18)] metrize the weak convergence in the space $\mathcal{P}(\mathbb{R})$. Suppose there exists ϕ such that

$$\pi(X, Y) \leq \phi(\mathbf{L}(X, Y)) \quad (6.5.3)$$

for any real-valued r.v.s X and Y. (Recall our notation $\mu(X, Y) := \mu(\Pr_X, \Pr_Y)$ for any metric μ in the space of measures.) Then, by (4.2.4) and (3.3.23),

$$\mathbf{L}(X/\lambda, Y/\lambda) = \mathbf{L}_\lambda(X, Y) \to \rho(X, Y) \quad \text{as} \quad \lambda \to 0 \quad (6.5.4)$$

and

$$\pi(X/\lambda, Y/\lambda) = \pi_\lambda(X, Y) \to \sigma(X, Y) \quad \text{as} \quad \lambda \to 0, \tag{6.5.5}$$

where $\boldsymbol{\rho}$ is the Kolmogorov metric [see (4.2.6)] and $\boldsymbol{\sigma}$ is the total variation metric [see (3.3.13)]. Thus, (6.5.3)–(6.5.5) imply that $\boldsymbol{\sigma}(X, Y) \leq \phi(\boldsymbol{\rho}(X, Y))$. The last inequality simply is, however, not true because in general $\boldsymbol{\rho}$-convergence does not yield $\boldsymbol{\sigma}$-convergence. [For example, if X_n is a random variable taking values k/n, $k = 1, \ldots, n$ with probability $1/n$, then $\boldsymbol{\rho}(X_n, Y) \to 0$ where Y is a $(0, 1)$-uniformly distributed random variable. On the other hand, $\boldsymbol{\sigma}(X_n, Y) = 1$.]

We are going to prove (6.5.2) for the special but important case where $\mathring{\mu}_c$ is the Fortet–Mourier metric on $\mathcal{P}_\lambda(\mathbb{R})$, i.e., $\mathring{\mu}_c(P, Q) = \zeta(P, Q; \mathcal{G}^p)$ [see (4.4.34)]; in other words, for any $P, Q \in \mathcal{P}_\lambda$,

$$\mathring{\mu}_c(P, Q) = \sup\left\{\int f \mathrm{d}(P - Q) : f : \mathbb{R} \to \mathbb{R}, |f(x) - f(y)| \leq c(x, y) \forall x, y \in \mathbb{R}\right\},$$

where

$$c(x, y) = |x - y| \max(1, |x|^{p-1}, |y|^{p-1}) \quad p \geq 1. \tag{6.5.6}$$

Since $\lambda(x) := 2\max(|x|, |x|^p)$, then $\mathcal{P}_\lambda(\mathbb{R})$ is the space of all laws on $\mathbb{R}$, with finite pth absolute moment.

Theorem 6.5.1. *If c is given by* (6.5.6)*, then*

$$\widehat{\mu}_c(P, Q) \leq p \mathring{\mu}_c(P, Q) \quad \forall P, Q \in \mathcal{P}_\lambda(\mathbb{R}). \tag{6.5.7}$$

Proof. Denote $h(t) = \max(1, |t|^{p-1})$, $t \in \mathbb{R}$, and $H(x) = \int_0^x h(t)\mathrm{d}t$, $x \in \mathbb{R}$. Let X and Y be real-valued RVs on a nonatomic probability space $(\Omega, \mathcal{A}, \Pr)$ with distributions P and Q, respectively. Theorem 5.5.1 gives us explicit representation of $\mathring{\mu}_c$, namely,

$$\mathring{\mu}_c(P, Q) = \int_{-\infty}^{\infty} h(t)|F_X(t) - F_Y(t)|\mathrm{d}t, \tag{6.5.8}$$

and thus

$$\mathring{\mu}_c(P, Q) = \int_{-\infty}^{\infty} |F_{H(X)}(x) - F_{H(Y)}(x)|\mathrm{d}x. \tag{6.5.9}$$

Claim 12. Let X and Y be real-valued RVs with distributions P and Q, respectively. Then

$$\mathring{\mu}_c(P, Q) = \inf\{E|H(\widetilde{X}) - H(\widetilde{Y})| : F_{\widetilde{X}} = F_X, F_{\widetilde{Y}} = F_Y\}. \tag{6.5.10}$$

Proof of Claim 12. Using the equality $\widehat{\mu}_d = \mathring{\mu}_d$ [see (6.2.3) and (5.5.5)] with $H(t) = t$ we have that

$$\overset{\circ}{\mu}_d(F,G) = \widehat{\mu}_d(F,G) = \inf\{E|X'-Y'| : F_{X'} = F, F_{Y'} = G\}$$
$$= \int_{-\infty}^{\infty} |F(x) - G(x)| \mathrm{d}x \tag{6.5.11}$$

for any DFs F and G. Hence, by (6.5.9)

$$\overset{\circ}{\mu}_c(P,Q) = \inf\{E|X'-Y'| : F_{X'} = F_{H(X)}, F_{Y'} = F_{H(Y)}\}$$
$$= \inf\{E|H(\widetilde{X}) - H(\widetilde{Y})| : F_{\widetilde{X}} = F_X, F_{\widetilde{Y}} = F_Y\}$$

which proves the claim.

Next we use Theorem 2.7.2, which claims that on a nonatomic probability space, the class of all joint distributions $\Pr_{X,Y}$ coincides with the class of all probability Borel measures on $\mathbb{R}^2$. This implies

$$\widehat{\mu}_c(P,Q) = \inf\{Ec(\widetilde{X},\widetilde{Y}) : F_{\widetilde{X}} = F_X, F_{\widetilde{Y}} = F_Y\}. \tag{6.5.12}$$

Claim 13. For any $x, y \in \mathbb{R}$, $c(x,y) \leq p|H(x) - H(y)|$.

Proof of Claim 13.

(a) Let $y > x > 0$. Then

$$c(x,y) = (y-x)h(y) = yh(y) - xh(y) \leq yh(y) - xh(x)$$
$$\leq (H(y) - H(x)) \sup_{y>x>0} \frac{yh(y) - xh(x)}{H(y) - H(x)}.$$

Since $H(t)$ is a strictly increasing continuous function,

$$B := \sup_{y>x>0} \frac{yh(y) - xh(x)}{H(y) - H(x)} = \sup_{t>s>0} \frac{f(t) - f(s)}{t - s},$$

where $f(t) := H^{-1}(t)h(H^{-1}(t))$ and H^{-1} is a function inverse to H; hence, $B = \operatorname{ess\,sup}_t |f'(t)| \leq p$.

(b) Let $y > 0 > x > -y$. Then $c(x,y) = |x - y|h(y) = (y + (-x))h(y) = yh(y) + (-x)h(|x|) + ((-x)h(y) - (-x)h(|x|)) \leq yh(y) + (-x)h(|x|)$. Since

$$th(t) = \begin{cases} t & \text{if } t \leq 1, \\ t^p & \text{if } t \geq 1, \end{cases} \qquad H(t) = \begin{cases} t & \text{if } 0 < t \leq 1, \\ \dfrac{p-1}{p} + \dfrac{1}{p}t^p & \text{if } t \geq 1, \end{cases}$$

then $yh(y) + (-x)h(|x|) \leq p(H(y) + H(-x)) = p(H(y) - H(x))$. By symmetry, the other cases are reduced to (a) or (b). The claim is shown. Now, (6.5.7) is a consequence of Claims 12, 13, and (6.5.12). □

6.6 Generalized Kantorovich and Kantorovich–Rubinstein Functionals

In this section, we consider a generalization of the Kantorovich-type functionals $\widehat{\mu}_c$ and $\overset{\circ}{\mu}_c$ [see (5.2.16) and (5.2.17)].

Let $U = (U, d)$ be an s.m.s. and $\mathcal{M}(U \times U)$ the space of all nonnegative Borel measures on the Cartesian product $U \times U$. For any probability measures P_1 and P_2 define the sets $\mathcal{P}^{(P_1,P_2)}$ and $\mathcal{Q}^{(P_1,P_2)}$ as in Sect. 5.2 [see (5.2.2) and (5.2.13)].

Let $\Lambda : \mathcal{M}(U \times U) \to [0, \infty]$ satisfy the conditions

1. $\Lambda(\alpha P) = \alpha\Lambda(P) \quad \forall \alpha \geq 0$,
2. $\Lambda(P + Q) \leq \Lambda(P) + \Lambda(Q) \quad \forall P$ and Q in $\mathcal{M}(U \times U)$.

We introduce the *generalized Kantorovich functional*

$$\widehat{\Lambda}(P_1, P_2) := \inf\{\Lambda(P) : P \in \mathcal{P}^{(P_1,P_2)}\} \tag{6.6.1}$$

and the *generalized Kantorovich–Rubinstein functional*

$$\overset{\circ}{\Lambda}(P_1, P_2) := \inf\{\Lambda(P) : P \in \mathcal{Q}^{(P_1,P_2)}\}. \tag{6.6.2}$$

Example 6.6.1. The Kantorovich metric[3]

$$\ell_1(P_1, P_2) := \sup\left\{\left|\int f \mathrm{d}(P_1 - P_2)\right| : f : U \to \mathbb{R}, |f(x) - f(y)| \leq d(x, y), x, y \in U\right\}$$

in the space of measures P with finite "first moment," $\int d(x,a)P(\mathrm{d}x) < \infty$, has the dual representations $\ell_1(P_1, P_2) = \overset{\circ}{\Lambda}(P_1, P_2) = \widehat{\Lambda}(P_1, P_2)$, where

$$\Lambda(P) = \Lambda_1(P) := \int_{U\times U} d(x, y)P(\mathrm{d}x, \mathrm{d}y). \tag{6.6.3}$$

Example 6.6.2. Let $U = \mathbb{R}$, $d(x, y) = |x - y|$. Then

$$\ell_1(P_1, P_2) = \int_{\mathbb{R}} |F_1(t) - F_2(t)|\mathrm{d}t,$$

[3]See Example 3.3.2 in Chap. 3.

where F_i is the DF of P_i and

$$\begin{aligned}\Lambda_1(P) &= \int_{\mathbb{R}} (\Pr(X \le t < Y) + \Pr(Y \le t < X))\mathrm{d}t \\ &= \int_{\mathbb{R}} \Pr(X \le t) + \Pr(Y \le t) - 2\Pr(\max(X,Y) \le t)\mathrm{d}t \\ &= E(2\max(X,Y) - X - Y) = E|X - Y|\end{aligned}$$

for RVs X and Y with $\Pr_{X,Y} = P$. We generalize (6.6.3) as follows: for any $1 \le p \le \infty$, define

$$\Lambda(P) := \Lambda_p(P) := \begin{cases} \left\{\int_{\mathbb{R}} \left[\int_{\mathbb{R}} c_t(x,y)P(\mathrm{d}x,\mathrm{d}y)\right]^p \lambda(\mathrm{d}t)\right\}^{1/p} & 1 \le p < \infty \\ \operatorname{ess\,sup}_\lambda \int_{\mathbb{R}^2} c_t(x,y)P(\mathrm{d}x,\mathrm{d}y) & \\ := \inf\left\{\varepsilon > 0 : \lambda\left\{t : \int_{\mathbb{R}^2} c_t\mathrm{d}P > \varepsilon\right\} = 0\right\} & p = \infty, \end{cases} \tag{6.6.4}$$

where c_t $(t \in \mathbb{R})$ is the following semimetric in $\mathbb{R}$

$$c_t(x,y) := I\{x \le t \le y\} + I\{y \le t \le x\} \forall x,y \in \mathbb{R}, \tag{6.6.5}$$

and $\lambda(\cdot)$ is a nonnegative measure on $\mathbb{R}$. In the space $\mathfrak{X} = \mathfrak{X}(\mathbb{R})$ of all real-valued RVs on a nonatomic probability space $(\Omega, \mathcal{A}, \Pr)$, the minimal metric w.r.t. Λ is given by

$$\widehat{\Lambda}_p(P_1,P_2) = \begin{cases} \inf\left\{\left[\int_{\mathbb{R}} \phi_t^p(X,Y)\lambda(\mathrm{d}t)\right]^{1/p} : X,Y \in \mathfrak{X}, \Pr_X = P_1, \Pr_Y = P_2\right\} \\ \qquad\qquad 1 \le p < \infty \\ \inf\left\{\sup_{t\in\mathbb{R}} \phi_t(X,Y) : X,Y \in \mathfrak{X}, \Pr_X = P_1, \Pr_Y = P_2\right\} \\ \qquad\qquad p = \infty. \end{cases} \tag{6.6.6}$$

Similarly, the minimal norm with respect to Λ is

$$\overset{\circ}{\Lambda}_p(P_1,P_2) = \begin{cases} \inf\left\{\alpha\left[\int_{\mathbb{R}} \phi_t^p(X,Y)\lambda(\mathrm{d}t)\right]^{1/p} : \alpha>0, \quad X,Y\in\mathfrak{X}, \right. \\ \qquad\qquad \left. \alpha(\mathrm{Pr}_X - \mathrm{Pr}_Y) = P_1 - P_2\right\} & \text{if } p<\infty \\ \inf\left\{\alpha \sup_\lambda \phi_t(X,Y) : \alpha>0,\ X,Y\in\mathfrak{X}, \right. \\ \qquad\qquad \left. \alpha(\mathrm{Pr}_X - \mathrm{Pr}_Y) = P_1 - P_2\right\} & \text{if } p=\infty, \end{cases} \tag{6.6.7}$$

where in (6.6.6) and (6.6.7)

$$\phi_t(X,Y) := \mathrm{Pr}(X\le t<Y) + \mathrm{Pr}(Y\le t<X). \tag{6.6.8}$$

The next theorem gives the explicit form of $\widehat{\Lambda}_p$ and $\overset{\circ}{\Lambda}_p$.

Theorem 6.6.1. *Let F_i be the DF of P_i ($i=1,2$). Then*

$$\widehat{\Lambda}_p(P_1,P_2) = \overset{\circ}{\Lambda}_p(P_1,P_2) = \boldsymbol{\lambda}_p(F_1,F_2), \tag{6.6.9}$$

where

$$\boldsymbol{\lambda}_p(F_1,F_2) = \begin{cases} \left(\int_{\mathbb{R}} |F_1(t)-F_2(t)|^p\lambda(\mathrm{d}t)\right)^{1/p} & 1\le p<\infty \\ \operatorname{ess\,sup}_\lambda|F_1-F_2| = \inf\{\varepsilon>0 : \lambda(t : |F_1(t)-F_2(t)|>\varepsilon)=0\} & p=\infty. \end{cases} \tag{6.6.10}$$

Claim 14. $\boldsymbol{\lambda}_p(F_1,F_2) \le \overset{\circ}{\Lambda}_p(P_1,P_2)$.

Proof of Claim 14. Let $P\in\mathcal{Q}^{(P_1,P_2)}$. Then in view of Remark 2.7.2 in Chap. 2, there exist $\alpha>0$, $X\in\mathfrak{X}$, $Y\in\mathfrak{X}$, such that $\alpha\,\mathrm{Pr}_{X,Y} = P$ and $\alpha(F_X - F_Y) = F_1 - F_2$; thus

$$\begin{aligned} |F_1(x)-F_2(x)| &= \alpha|F_X(t)-F_Y(t)| \\ &= \alpha[\max(F_X(t)-F_Y(t),0) + \max(F_Y(t)-F_X(t),0)] \\ &\le \alpha\phi_t(X,Y). \end{aligned} \tag{6.6.11}$$

By (6.6.7) and (6.6.11), it follows that $\boldsymbol{\lambda}_p(F_1,F_2)\le\overset{\circ}{\Lambda}_p(P_1,P_2)$, as desired.

Further

$$\overset{\circ}{\Lambda}_p(P_1, P_2) \le \widehat{\Lambda}_p(P_1, P_2) \tag{6.6.12}$$

by the representations (6.6.6) and (6.6.7).

Claim 15.

$$\widehat{\Lambda}_p(P_1, P_2) \le \boldsymbol{\lambda}_p(F_1, F_2).$$

Proof of claim 15. Let $\widetilde{X} := F_1^{-1}(V), \widetilde{Y} := F_2^{-1}(V)$, where F_i^{-1} is the generalized inverse to the DF F_i [see (3.3.16) in Chap. 3] and V is a $(0, 1)$-uniformly distributed RV. Then $F_{\widetilde{X},\widetilde{Y}}(t, s) = \min(F_1(t), F_2(s))$ for all $t, s \in \mathbb{R}$. Hence, $\phi_t(\widetilde{X}, \widetilde{Y}) = |F_1(t) - F_2(t)|$, which proves the claim by using (6.6.6) and (6.6.7).

Combining Claims 14, 15, and (6.6.12) we obtain (6.6.9). □

Problem 6.6.1. In general, dual and explicit solutions of $\widehat{\Lambda}_p$ and $\overset{\circ}{\Lambda}_p$ in (6.6.1) and (6.6.2) are not known.

References

Billingsley P (1999) Convergence of probability measures, 2nd edn. Wiley, New York
Dudley RM (2002) Real analysis and probability, 2nd edn. Cambridge University Press, New York
Hennequin PL, Tortrat A (1965) Théorie des probabilités et quelques applications. Masson, Paris
Levin VL (1975) On the problem of mass transfer. Sov Math Dokl 16:1349–1353
Neveu J, Dudley RM (1980) On Kantorovich–Rubinstein theorems (transcript)

Chapter 7
K-Minimal Metrics

The goals of this chapter are to:

- Define the notion of K-minimal metrics and describe their general properties;
- Provide representations of the K-minimal metrics with respect to several particular metrics such as the Lévy metric, Kolmogorov metric, and p-average metric;
- Consider K-minimal metrics when probability measures are defined on a general separable metric space;
- Provide relations between the multidimensional Kantorovich and Strassen theorems.

Notation introduced in this chapter:

Notation	Description
$\overset{\wedge}{\mu}$	K-minimal metric w.r.t. a probability distance μ on $(\mathcal{P}_2(U^n)$
ρ_α	Metric on the Cartesian product U^n
E	A (0, 1)-distributed random variable
$X_E = (X_1, \ldots, X_n)_E$	Random vector with components $F_{X_i}(E)$, $i = 1, \ldots, n$
$\widetilde{\mu}$	For a given compound distance μ, $\widetilde{\mu}(X, Y) = \mu(X_E, Y_E)$
$\mathbf{L}(X, Y; \alpha)$	Lévy distance in space of random vectors on $(\mathbb{R}^2, \rho_\alpha)$
$\mathbf{W}(X, Y; \alpha)$	Limit $\lambda \mathbf{L}(X/\lambda, Y/\lambda; \alpha)$ as $\lambda \to \infty$
$\boldsymbol{\delta}$	Discrete metric in space of distribution functions on $\mathbb{R}^n$
$\mathcal{KF}_\alpha$	Ky Fan functional in $\mathfrak{X}(U^N)$
$\Pi_\lambda(\widetilde{P})$	Prokhorov functional in $(\mathcal{P}(U))^N$
$[x]$	Integer part of x

S.T. Rachev et al., *The Methods of Distances in the Theory of Probability and Statistics*, DOI 10.1007/978-1-4614-4869-3_7, © Springer Science+Business Media, LLC 2013

7.1 Introduction

As we saw in the previous two chapters, the notion of minimal distance

$$\widehat{\mu}(P_1, P_2) = \inf\{\mu(P) : P \in \mathcal{P}(U^2), T_i P = P_i, i = 1,2\} \quad P_1, P_2 \in \mathcal{P}(U) \tag{7.1.1}$$

represents the main relationship between compound and simple distances (see the general discussion in Sect. 3.3). In view of the multidimensional Kantorovich problem (Sect. 5.2, VI), we have been interested in the n-dimensional analog of the notion of minimal metrics, that is, we have defined the following distance between n-dimensional vectors of probability measures [see (5.3.15)]

$$\mathfrak{R}(\widetilde{P}, \widetilde{Q}) = \inf\left\{\int_{U^n \times U^n} \Delta(x, y) P(\mathrm{d}x, \mathrm{d}y) : P \in \mathfrak{P}(\widetilde{P}, \widetilde{Q})\right\}, \tag{7.1.2}$$

where $\widetilde{P} = (P_1 i, \dots, P_n)$, $\widetilde{Q} = (Q_1, \dots, Q_n)$, $P_i, Q_i \in \mathcal{P}(U)$, $\Delta(x, y)$ is a distance in the Cartesian product U^n, and $\mathfrak{P}(\widetilde{P}, \widetilde{Q})$ is the space of all probability measures on U^{2n} with fixed one-dimensional marginals $P_1, \dots, P_n, Q_1, \dots, Q_n$.

In the 1960s, H. G. Kellerer investigated the multidimensional marginal problem. His results on this topic were the major source for the famous Strassen (1965) work on minimal probabilistic functionals. In this chapter, we study the properties of metrics in the space of vectors $\widetilde{\mathcal{P}}$ that have representation similar to that of $\mathfrak{R}$.

7.2 Definition and General Properties

In this section, we define K-minimal distances and provide some general properties.

Definition 7.2.1. Let μ be a probability distance (p. distance) in $\mathcal{P}_2(U^n)$ (U is an s.m.s.). For any two vectors $\widetilde{P}_i = (P_i^{(1)}, \dots, P_i^{(n)})$, $i = 1, 2$, of probability measures $P_i^{(j)} \in \mathcal{P}_1(U)$ define the *K-minimal distance*

$$\overset{n}{\mu}(\widetilde{P}_1, \widetilde{P}_2) = \inf\{\mu(P) : P \in \mathfrak{P}(\widetilde{P}_1, \widetilde{P}_2)\}, \tag{7.2.1}$$

where $\mathfrak{P}(\widetilde{P}_1, \widetilde{P}_2) = \{P \in \mathcal{P}_2(U^n) : T_j P = P_1^{(j)}, T_{j+n} P = P_2^{(j)}, j = 1, \dots, n\}$.

Obviously, $\overset{1}{\mu} = \widehat{\mu}$. One of the main reasons to study K-minimal metrics is based on the simple observation that in most cases the minimal metric between the product measures $\widehat{\mu}(P_1^{(1)} \times \cdots \times P_1^{(n)}, P_2^{(1)} \times \cdots \times P_2^{(n)})$ coincides with $\overset{n}{\mu}(\widetilde{P}_1, \widetilde{P}_2)$. Surprisingly, it is much easier to find explicit representations for $\overset{n}{\mu}(\widetilde{P}_1, \widetilde{P}_2)$

$\widetilde{P}_i \in \mathcal{P}(U^n)$ than for $\widehat{\mu}(P_1, P_2)$ [$P_i \in \mathcal{P}(U)$]. Some general relations between compound, minimal, and K-minimal distances are given in the next four theorems. Recall that for any $P \in \mathcal{P}(U^k)$, $k \geq 2$, the law $T_{\alpha_1,\dots,\alpha_m} P \in \mathcal{P}(U^m)$ $(1 \leq m \leq k)$ is the marginal distribution of P on the coordinates $\alpha_1 < \alpha_2 < \cdots < \alpha_m$.

Theorem 7.2.1. *Let ψ be a right semicontinuous (RSC) function on $(0, \infty)$ and $\phi(t_1, \dots, t_n)$ a nondecreasing function in each argument $t_i \geq 0$, $i = 1, \dots, n$. Suppose that a p. distance μ on $\mathcal{P}(U^n)$ and p. distances $\mu_1, \dots, \mu_n$ on $\mathcal{P}_2(U)$ satisfy the following inequality: for any $P \in \mathfrak{P}(\widetilde{P}_1, \widetilde{P}_2)$*

$$\psi(\mu(P)) \geq \phi(\mu(T_{1,n+1}P), \mu_2(T_{2,n+2}P), \dots, \mu_n(T_{n,2n}P)). \tag{7.2.2}$$

Then

$$\psi\left(\overset{\,n}{\widehat{\mu}}(\widetilde{P}_1, \widetilde{P}_2)\right) \geq \phi\left(\widehat{\mu}\left(P_1^{(1)}, P_2^{(1)}\right), \dots, \widehat{\mu}\left(P_1^{(n)}, P_2^{(n)}\right)\right).$$

Proof. Given $\varepsilon > 0$, there exists $P^{(\varepsilon)} \in \mathfrak{P}(\widetilde{P}_1, \widetilde{P}_2)$ such that $|\mathcal{D}_\varepsilon| < \varepsilon$, where $\mathcal{D}_\varepsilon = \psi\left(\overset{\,n}{\widehat{\mu}}(\widetilde{P}_1, \widetilde{P}_2)) - \psi(\mu(P^{(\varepsilon)})\right)$. Thus, by (7.2.2),

$$\psi\left(\overset{\,n}{\widehat{\mu}}(\widetilde{P}_1, \widetilde{P}_2)\right) = \psi\left(\mu\left(P^{(\varepsilon)}\right)\right) + \mathcal{D}_\varepsilon \geq \phi\left(\widehat{\mu}_1\left(P_1^{(1)}, P_2^{(2)}\right), \dots, \widehat{\mu}\left(P_1^{(n)}, P_2^{(n)}\right)\right) - \varepsilon.$$

□

Theorem 7.2.2. *Let $\mu_1, \dots, \mu_k, \nu_1, \dots, \nu_k$ be probability distances on $\mathcal{P}_2(U^n)$ and suppose*

$$\psi(\mu_1(P_1), \dots, \mu_k(P_k)) \geq \phi(\nu_1(P_1), \dots, \nu_k(P_k)), \quad P_i \in \mathcal{P}_2(U^n),$$

where ϕ is nondecreasing in each argument and ψ is an RSC function on $\mathbb{R}^n$. Then

$$\psi(\overset{\,n}{\widehat{\mu}}_1, \dots, \overset{\,n}{\widehat{\mu}}_k) \geq \phi(\overset{\,n}{\widehat{\nu}}_1, \dots, \overset{\,n}{\widehat{\nu}}_k).$$

The proof is straightforward.

In what follows, $P_1 \times \cdots \times P_n$ denotes the product measure generated by $P_1, \dots, P_n$. The next theorem describes conditions providing an equality between $\overset{\,n}{\widehat{\mu}}(\widetilde{P}_1, \widetilde{P}_2)$ and $\mu(P_1^{(1)} \times \cdots \times P_1^{(n)}, P_2^{(1)} \times \cdots \times P_2^{(n)})$.

Theorem 7.2.3. *Suppose that a p. distance μ on $\mathcal{P}_2(U^n)$ and p. distances $\mu_1, \dots, \mu_n$ on $\mathcal{P}_2(U)$ satisfy the equality*

$$\mu(P) = \phi(\mu_1(T_{1,n+1}P), \dots, \mu_n(T_{n,2n}P)), \tag{7.2.3}$$

where ϕ is an RSC function, nondecreasing in each argument. Then for any vectors of measures $\widetilde{P}_1, \widetilde{P}_2 \in \mathcal{P}_1(U)^n$

$$\overset{n}{\hat{\mu}}\left(\widetilde{P}_1, \widetilde{P}_2\right) = \widehat{\mu}\left(P_1^{(1)} \times \cdots \times P_1^{(n)}, P_2^{(1)} \times \cdots \times P_2^{(n)}\right)$$

$$= \phi\left(\widehat{\mu}_1\left(P_1^{(1)}, P_2^{(1)}\right), \dots, \widehat{\mu}\left(P_1^{(n)}, P_2^{(n)}\right)\right). \tag{7.2.4}$$

Proof. Given $\varepsilon > 0$, choose $\delta_\varepsilon \in (0, \varepsilon)$ and $P^{(\varepsilon)} \in \mathfrak{P}(\widetilde{P}_1, \widetilde{P}_2)$ such that

$$\overset{n}{\hat{\mu}}(\widetilde{P}_1, \widetilde{P}_2) = \mu(P^{(\varepsilon)}) - \delta_\varepsilon = \phi(\mu_1(T_{1,n+1}P^{(\varepsilon)}, \dots, \mu_n(T_{n,2n}P^{(\varepsilon)})) - \delta_\varepsilon. \tag{7.2.5}$$

Take

$$Q^{(\varepsilon)} = T_{1,n+1}P^{(\varepsilon)} \times \cdots \times T_{n,2n}P^{(\varepsilon)}.$$

Then

$$T_{1,\dots,n}Q^{(\varepsilon)} = P_1^{(1)} \times \cdots \times P_1^{(n)}, T_{n+1,\dots,2n}Q^{(\varepsilon)} = P_2^{(1)} \times \cdots \times P_2^{(n)},$$

and by (7.2.3), $\mu(P^{(\varepsilon)}) = \mu(Q^{(\varepsilon)})$, which, together with (7.2.5), implies

$$\overset{n}{\hat{\mu}}(\widetilde{P}_1, \widetilde{P}_2) = \mu(Q^{(\varepsilon)}) - \delta_\varepsilon \geq \widehat{\mu}(P_1^{(1)} \times \cdots \times P_1^{(n)}, P_2^{(1)} \times \cdots \times P_2^{(n)}) - \delta_\varepsilon$$

and

$$\overset{n}{\hat{\mu}}(\widetilde{P}_1, \widetilde{P}_2) \geq \phi\left(\widehat{\mu}_1(P_1^{(1)}, P_2^{(1)}), \dots, \widehat{\mu}_n(P_1^{(n)}, P_2^{(n)})\right) - \delta_\varepsilon.$$

On the other hand, $\overset{n}{\hat{\mu}}(\widetilde{P}_1, \widetilde{P}_2) \leq \widehat{\mu}(P_1^{(1)} \times \cdots \times P_1^{(n)}, P_2^{(1)} \times \cdots \times P_2^{(n)})$, and if

$$D_\varepsilon := \phi\left(\widehat{\mu}_1(P_1^{(1)}, P_2^{(1)}), \dots, \widehat{\mu}_n(P_1^{(n)}, P_2^{(n)})\right) - \phi\left(\mu_1(T_{1,n+1}P^{(\varepsilon)}, \dots, T_{n,2n}P^{(\varepsilon)})\right),$$

then, taking into account (7.2.3), we get

$$\phi(\widehat{\mu}_1(P_1^{(1)}, P_2^{(1)}), \dots, \widehat{\mu}_n(P_1^{(n)}, P_2^{(n)})) = \mu(P^{(\varepsilon)}) + D_\varepsilon \geq \overset{n}{\hat{\mu}}(\widetilde{P}_1, \widetilde{P}_2) + D_\varepsilon,$$

where $D_\varepsilon \to 0$ as $\varepsilon \to 0$. □

In terms of distributions of random variables (RVs), the last theorem can be rewritten as follows. Let $X_i = (X_i^{(1)}, \dots, X_i^{(n)})$ $(i = 1, 2)$ be two vectors in $\mathfrak{X}(U^n)$ with independent components, and suppose that the compound metric μ in $\mathfrak{X}(U^n)$ has the following representation:

$$\mu(X, Y) = \phi\left(\mu_1(X^{(1)}, Y^{(1)}), \dots, \mu_n(X^{(n)}, Y^{(n)})\right) \; X, Y \in \mathfrak{X}(U^n), \tag{7.2.6}$$

where ϕ is defined as in Theorem 7.2.3. Then

$$\hat{\hat{\mu}}(X_1, X_2) = \widehat{\mu}(X_1, X_2) = \phi\left(\widehat{\mu}_1(X_1^{(1)}, X_2^{(1)}), \ldots, \widehat{\mu}_n(X_1^{(n)}, X_2^{(n)})\right). \qquad (7.2.7)$$

Remark 7.2.1. The implication (7.2.6) $\Rightarrow$ (7.2.7) is often used in problems of estimating the closeness between two RVs with independent components. In many cases, working with compound distances is more convenient than working with simple ones. That is, when we are seeking inequalities, estimators, and so on, then, considering all RVs on a common probability space, we are dealing with simple operations (for example, sums and maximums) in the space of RVs. However, considering inequalities between simple metrics and distances, we must evaluate functionals in the space of distributions involving, e.g., convolutions or product of distribution functions (DFs). Among many specialists, this simple idea is referred to as the "method of one probability space."

A particular case of Theorem 7.2.3 asserts that the equality

$$\mu(X_1, X_2) = \phi(\mu_1(X_1, X_2)) \quad X_1, X_2 \in \mathfrak{X}(U) \qquad (7.2.8)$$

yields

$$\widehat{\mu}(X_1, X_2) = \phi(\widehat{\mu}_1(X_1, X_2)) \qquad (7.2.9)$$

for any RSC nondecreasing function ϕ on $[0, \infty)$. The next theorem is a variant of the implication (7.2.8) $\Rightarrow$ (7.2.9) and essentially says that if

$$\mu_\phi(X_1, X_2) = \mu(\phi(X_1), \phi(X_2)), \qquad (7.2.10)$$

then

$$\widehat{\mu}_\phi = (\widehat{\mu})_\phi \qquad (7.2.11)$$

for any measurable function ϕ. More precisely, let $(U, \mathcal{A})$, $(V, \mathcal{B})$ be measurable spaces and $\phi : U \to V$ be a measurable function. Let μ be a p. distance on $\mathcal{P}(V^2)$; then define

$$\mu_\phi : \mathcal{P}(V^2) \to [0, \infty] \quad \mu_\phi(Q) := \mu(Q_{(\phi,\phi)}) \quad Q \in \mathcal{P}(V^2), \qquad (7.2.12)$$

where $Q_{(\phi,\phi)}$ is the image of Q under the transformation $(\phi, \phi)(x, y) = (\phi(x), \phi(y))$. Similarly, if ν is a simple distance $\nu_\phi(P_1, P_2) = \nu(P_{1\phi}, P_{2\phi})$, where $\mathcal{P}_{i,\phi}(A) = P_i(\phi^{-1}(A))$.

It is easy to see that μ_ϕ defines a probability semidistance on $P(U^2)$. In terms of RVs, the preceding definition can also be written in the following way: $\mu_\phi(X, Y) = \mu(\phi(X), \phi(Y))$.

Definition 7.2.2. A measurable space $(U, \mathcal{A})$ is called a *Borel space* if there exists a Borel subset $B \in \mathcal{B}_1 = \mathcal{B}(\mathbb{R}^1)$ and a Borel isomorphism $\phi : (U, \mathcal{A}) \to (B, B \cap \mathcal{B}_1)$, i.e., if U and B are Borel-isomorphic (see Definition 2.6.6 in Chap. 2).

Theorem 7.2.4. *Let $(U, \mathcal{A})$ be a Borel space, $(V, \mathcal{B})$ a measurable space such that $\{v\} \in \mathcal{B}$ for all $v \in V$, and $\phi : U \to V$ a measurable mapping. Let $\widehat{\mu}$, $\widehat{\mu}_\phi$ denote the minimal distance corresponding to μ, μ_ϕ. Then*

$$\widehat{\mu}_\phi(P_1, P_2) = \widehat{\mu}(P_{1\phi}, P_{2\phi}) \tag{7.2.13}$$

for all $P_1, P_2 \in \mathcal{P}_1(U)$.

Proof. We need an auxiliary result on the construction of RVs. Let $(\Omega, \mathcal{E}, \Pr)$ be a probability space, and let $(S, Z) : \Omega \to V \times \mathbb{R}$ be a pair of independent RVs, where S is a V-valued RV and Z is uniformly distributed on $[0, 1]$. Let P be a probability measure on $(U, \mathcal{A})$ such that $P \circ \phi^{-1}$ coincides with the law of S, $\Pr_S$. □

Lemma 7.2.1. *There exists a U-valued RV X such that*

$$\Pr_X = P \quad \textit{and} \quad \phi(X) = S \textit{ a.e.} \tag{7.2.14}$$

Proof. We start with the special case $(U, \mathcal{A}) = (\mathbb{R}, \mathcal{B}_1)$. Let $I : \mathbb{R} \to \mathbb{R}$ denote the identity, $I(x) = x$, and define the set $(P_s)_{s \in V}$ of regular conditional distributions $P_s := P_{I|\phi=s}$, $s \in V$. Let F_s be the DF of P_s, $s \in V$. Then it is easily verified that

$$F : V \times \mathbb{R} \to [0, 1], \quad F(s, x) := F_s(x) \tag{7.2.15}$$

is product-measurable. For $s \in V$ let $F_s^{-1}(x) := \sup\{y : F_s(y) < x\}$, $x \in (0, 1)$, be the generalized inverse of F_s and define the RV $X := F_S^{-1}(Z)$. For any $A \in \mathcal{A} = \mathcal{B}_1$ we have

$$\Pr(X \in A) = \int_V \Pr_{X|S=s}(A) \Pr_S(\mathrm{d}s).$$

For the regular conditional distributions we obtain, by the independence of S and Z,

$$\Pr_{X|S=s} = \Pr_{F_s^{-1}(Z)|S=s} = \Pr_{F_s^{-1}(Z)}.$$

Since $\Pr_{F_s^{-1}(Z)} = P_s = P_{I|\phi=s}$, then $\Pr(X \in A) = \int P_{I|\phi=s}(A) P \circ \phi^{-1}(\mathrm{d}s) = P(A)$. Thus, the law of X is P. To show that $\phi(X) = S$ a.e., observe that, by $\Pr_S = P \circ \phi^{-1}$ and $\Pr_{X|S=s} = P_{I|\phi=s}$, we have

$$\begin{aligned}\Pr(\phi(X) = S) &= \int_V \Pr_{X|S=s}(x : \phi(x) = s) \Pr_S(\mathrm{d}s) \\ &= \int_V P_{I|\phi=s}(x : \phi(s) = s) P \circ \phi^{-1}(\mathrm{d}s) = 1.\end{aligned}$$

Now let $(U, \mathcal{A})$ be a Borel space. Let $\psi : (U, \mathcal{A}) \to (B : B_n \cap \mathcal{B}_1)$, $B \in \mathcal{B}_1$, be a measure isomorphism, and define $P' := P \circ \psi^{-1}$, $\phi' := \phi \circ \psi^{-1}$. By the first part of this proof, there exists a RV $X' : \Omega \to B$ such that $\Pr_{X'} = P'$ and $\phi' \circ X' = S$ a.e.; thus, $\Pr_X = P$ and $\phi \circ X = S$ a.e., where $X = \psi^{-1} \circ X'$, as desired in (7.2.14). □

Now let $\mathcal{P}^{(P_1,P_2)}$ be the set of all probability measures on $U \times U$ with marginals P_1, P_2. Then

$$\{Q_{(\phi,\phi)} : Q \in \mathcal{P}^{(P_1,P_2)}\} \subset \mathcal{P}^{(P_{1\phi},P_{2\phi})},$$

and hence

$$\begin{aligned}\widehat{\mu}_\phi(P_1, P_2) &= \inf\{\mu(P_{\phi,\phi}) : P \in \mathcal{P}^{(P_1,P_2)}\}\\ &\geq \inf\{\mu(P) : P \in \mathcal{P}^{(P_{1\phi},P_{2\phi})}\} = \widehat{\mu}(P_{1\phi}, P_{2\phi}).\end{aligned}$$

On the other hand, suppose $P \in \mathcal{P}^{(P_{1\phi},P_{2\phi})}$. Let $(\Omega, \mathcal{E}, \Pr)$ be a probability space with V-valued RVs S, S' such that $\Pr_{(S,S')} = P$ and rich enough to contain a further RV $Z : M \to [0,1]$ uniformly distributed on $[0,1]$ and independent of S, S'. By Lemma 7.2.1, there exist U-valued RVs X and Y such that $\Pr_X = P_1$, $\Pr_Y = P_2$ and $\phi(X) = S$, $\phi(Y) = S'$ a.e. Therefore, $\mu(P) = \mu(\phi \circ X, \phi \circ V) = \mu_\phi(X,Y)$, implying that

$$\begin{aligned}\widehat{\mu}_\phi(P_1, P_2) &= \inf\{\mu_\phi(X,Y) : P_X = P_1, P_Y = P_2\}\\ &\leq \inf\{\mu(S,S') : \Pr_S = P_{1\phi}, \Pr_{S'} = P_{2\phi}\} = \widehat{\mu}(P_{1\phi}, P_{2\phi}).\end{aligned}$$

Remark 7.2.2. Theorem 7.2.4 is valid under the alternative condition of U being a u.m.s.m.s. and V being an s.m.s.

Remark 7.2.3. Let $U = V$ be a Banach space, $d_s(x,y) = \|x\|x\|^{s-1} - y\|y\|^{s-1}\|$, $x, y \in U$, where $s \geq 0$ and $x\|x\|^{s-1} = 0$ for $x = 0$. Let $\mu_s(X,Y) = Ed_s(X,Y)$. Then the corresponding minimal metrics $\kappa_s(X,Y) := \widehat{\mu}_s(X,Y)$ are the *absolute pseudomoments of order s* [see (4.4.40)–(4.4.43)]. By Theorem 7.2.4, κ_s can be expressed in terms of the more simple metric κ_1, $\kappa_s(P_1,P_2) = \kappa_1(P_{1\phi}, P_{2\phi})$, where $\phi(x) = x\|x\|^{s-1}$.

7.3 Two Examples of K-Minimal Metrics

Let (U,d) be an s.m.s. with metric d and Borel σ-algebra $\mathcal{B}(U)$. Let U^n be the Cartesian product of n copies of the space U. We consider in U^n the metrics $\rho_\alpha(x,y)$, $\alpha \in [0,\infty]$, $x = (x_1,\dots,x_n)$, $y = (y_1,\dots,y_n) \in U^n$ of the following form:

$$\begin{aligned}\rho_\alpha(x,y) &= \left(\sum_{i=1}^n d^\alpha(x_i,y_i)\right)^{\min(1,1/\alpha)} \qquad \text{for } \alpha \in (0,\infty)\\ \rho_\infty(x,y) &= \max\{d(x_i,y_i); i = 1,\dots,n\}\\ \rho_0(x,y) &= \sum_{i=1}^n I\{(x,y); x_i \neq y_i\}, \end{aligned} \tag{7.3.1}$$

where I is the indicator in U^{2n}. Let $\mathfrak{X}(U^n) = \{X = (X_1, \ldots, X_n)\}$ be the space of all n-dimensional U-valued RVs defined on a probability space $(\Omega, \mathcal{A}, \Pr)$ that is rich enough.[1]

Let μ be a probability semimetric in the space $\mathfrak{X}(U^n)$. For every pair of random vectors $X = (X_1, \ldots, X_n)$, $Y = (Y_1, \ldots, Y_n)$ in $\mathfrak{X}(U^n)$ we define the K-minimal metric

$$\overset{n}{\hat{\mu}}(X, Y) = \inf \mu(X, Y),$$

where the infimum is taken over all joint distributions $\Pr_{X,Y}$ with fixed one-dimensional marginal distributions $\Pr_X$, $\Pr_Y$, $i = 1, \ldots, n$. In the case $n = 1$, $\overset{n}{\hat{\mu}} = \widehat{\mu}$ is the minimal metric with respect to μ. Following the definitions in Sect. 2.5, a semimetric μ in $\mathfrak{X}(U^n)$ is called a simple semimetric if its values $\mu(X, Y)$ are determined by the pair of marginal distributions $\Pr_X$, $\Pr_Y$. A semimetric $\mu(X, Y)$ in $\mathfrak{X}(U^n)$ is called *componentwise simple* (or K-simple) if its values are determined by the one-dimensional marginal distributions $\Pr_{X_i}$, $\Pr_{Y_i}$, $i = 1, \ldots, n$. Obviously, every K-simple semimetric is simple in $\mathfrak{X}(U^n)$.

We give two examples of K-simple semimetrics that will be used frequently in what follows.

Example 7.3.1. Suppose that in $\mathbb{R}^n$ a monotone seminorm $\|x\|$ is given, that is, (a) $\|x\| \geq 0$ for any $x \in \mathbb{R}^n$; (b) $\|\lambda x\| = |\lambda| \cdot \|x\|$ for $\lambda \in \mathbb{R}$, $x \in \mathbb{R}^n$; (c) $\|x + y\| \leq \|x\| + \|y\|$; (d) if $0 < x_i < y_i$, $i = 1, \ldots, n$, then $\|x\| \leq \|y\|$. Examples of monotone seminorms:

1. A monotone norm

$$\|a\|_\alpha = \left(\sum_{i=1}^n |a_i|^\alpha\right)^{1/\alpha} \qquad 1 \leq \alpha < \infty \quad a = (a_1, \ldots, a_n) \in \mathbb{R}^n, \tag{7.3.2}$$

$$\|a\|_\infty = \max\{|a_i|, i = 1, \ldots, n\}; \tag{7.3.3}$$

2. A monotone seminorm

$$\|a\| = \left|\sum_{i=1}^n a_i\right|. \tag{7.3.4}$$

Let $\mu^{(1)}, \ldots, \mu(n)$ be simple metrics in $\mathfrak{X}(U)$. The semimetric $\mu(X, Y) = \|\mu^{(1)}(X_1, Y_1), \ldots, \mu^{(n)}(X_n, Y_n)\|$ is K-simple in $\mathfrak{X}(U^n)$.

Example 7.3.2. Denote by E an RV uniformly distributed on $(0, 1)$, and for every $X = (X_1, \ldots, X_n) \in \mathfrak{X}(\mathbb{R}^n)$ denote by X_E the random vector $X_E = (F_{X_1}^{-1}(E), \ldots, F_{X_n}^{-1}(E))$, where $F_{X_i}^{-1}(t) = \sup\{x : F_{X_i}(x) \leq t\}$. For any p. metric $\mu(X, Y)$ in the space $\mathfrak{X}(\mathbb{R}^n)$

---

[1] See Sect. 2.7 and Remark 2.7.1 in Chap. 2.

$$\widetilde{\mu}(X,Y) = \mu(X_E, Y_E) \tag{7.3.5}$$

is K-simple in $\mathfrak{X}(\mathbb{R}^n)$. Obviously, $\overset{\hat{n}}{\mu} \leq \widetilde{\mu}$.

In the next two sections, for some simple and compound probability metrics, we will find the explicit form of the corresponding K-minimal metrics. We will often use the following obvious assertion.

Theorem 7.3.1. *Let* $\nu = \widehat{\mu}$. *Then* $\overset{\hat{n}}{\mu} = \overset{\hat{n}}{\nu}$.

7.4 K-Minimal Metrics of Given Probability Metrics: The Case of $U = R$

In this section, we will examine the representations of the K-minimal metrics w.r.t. the following probability metrics in $\mathfrak{X}(\mathbb{R}^n)$: Lévy metric, Kolmogorov metric, and the p-average metric $\mathcal{L}_p$.[2]

Let $0 < \alpha < \infty$ and ρ_α be defined by (7.3.1). The expression $x \leq y$ or $x \in (-\infty, y]$ for $x, y \in \mathbb{R}^n$ means that $x_i \leq y_i$ for all $i = 1, \ldots, n$. As a metric d in $U = \mathbb{R}^1$ we take the uniform metric $d(x_1, y_1) = |x_1 - y_1|$ for $x_1, y_1 \in \mathbb{R}$. For every $\alpha \in (0, \infty)$ we define a Lévy metric in $\mathfrak{X}(\mathbb{R}^n)$

$$\begin{aligned}\mathbf{L}(X,Y;\alpha) = \inf\{\varepsilon > 0; &\Pr(X \leq x) \leq \Pr(Y \in (-\infty, x]^\varepsilon_\alpha) + \varepsilon,\\ &\Pr(Y \leq x) \leq \Pr(X \in (-\infty, x]^\varepsilon_\alpha) + \varepsilon, \quad \forall x \in \mathbb{R}^n\},\end{aligned}$$

where $A^\varepsilon_\alpha = \{x : \rho_\alpha(x, A) \leq \varepsilon\}$ for any $A \subset \mathbb{R}^n$. As is well known, $\mathbf{L}(X,Y;\alpha)$, $\alpha \in (0,\infty]$ metrizes the weak convergence in $\mathfrak{X}(\mathbb{R}^n)$. In $\mathfrak{X}(\mathbb{R}^1)$ we define the Lévy metric $\mathbf{L}(X_1, Y_1; \alpha)$ in the foregoing manner. Obviously, $\mathbf{L}(X_1, Y_1; \alpha) = \mathbf{L}(X_1, Y_1; 1)$ for $\alpha \in [1, \infty]$ is the usual Lévy metric (2.2.3) (Fig. 4.1). We recall the uniform metric (Kolmogorov metric) $\boldsymbol{\rho}(X,Y)$ in $\mathfrak{X}(\mathbb{R}^n)$

$$\boldsymbol{\rho}(X,Y) = \sup\{|\Pr(X \leq x) - \Pr(Y \leq x)| : x \in \mathbb{R}^n\}.$$

Denote by $\mathbf{W}$ and $\boldsymbol{\delta}$ the following simple metrics in $\mathfrak{X}(\mathbb{R}^n)$:

$$\begin{aligned}\mathbf{W}(X,Y;\alpha) := \inf\{\varepsilon > 0; &\Pr(X \leq x) \leq \Pr(Y \in (-\infty, x]^\varepsilon_\alpha),\\ &\Pr(Y \leq x) \leq \Pr(X \in (-\infty, x]^\varepsilon_\alpha), \quad \forall x \in \mathbb{R}^n\},\end{aligned}$$

[2]See (3.4.3) in Chap. 3 and (4.2.22) and (4.2.24) in Chap. 4.

and $\boldsymbol{\delta}(X,Y)$ is the *discrete metric*: $\boldsymbol{\delta}(X,Y)=0$ if $F_X = F_Y$ and $\delta(X,Y)=+\infty$ if $F_X \neq F_Y$. The following relations are valid (Example 4.2.3):

$$\mathbf{L}\left(\frac{1}{\lambda}X,\frac{1}{\lambda}Y;\alpha\right) \to \boldsymbol{\rho}(X,Y) \text{ as } \lambda \to 0, \lambda > 0,\ \alpha \in (0,\infty], \tag{7.4.1}$$

$$\lim_{\lambda\to\infty} \lambda\mathbf{L}\left(\frac{1}{\lambda}X,\frac{1}{\lambda}Y;\alpha\right) = \mathbf{W}(X,Y;\alpha),\ \text{for } \alpha \in [1,\infty],$$

$$\lim_{\lambda\to\infty} \lambda\mathbf{L}\left(\frac{1}{\lambda}X,\frac{1}{\lambda}Y;\alpha\right) = \boldsymbol{\delta}(X,Y),\ \text{for } \alpha \in (0,1). \tag{7.4.2}$$

For any $X = (X_1,\dots,X_n) \in \mathfrak{X}(\mathbb{R}^n)$ we denote by $M_X(x) = \min(F_X(x_1), \dots, F_X(x_n)) = \Pr(F_{X_1}^{-1}(E) \le x_1,\dots,F_{X_n}^{-1}(E) \le x_n)$, $x=(x_1,\dots,x_n) \in \mathbb{R}^n$, the maximal DF having fixed one-dimensional marginal distributions $F_{X_i}, i = 1,\dots,n$. For any semimetric $\mu(X,Y)$ in $\mathfrak{X}(\mathbb{R}^n)$ we denote by $\mu(X_i,Y_i)$, $i=1,\dots,n$, the corresponding semimetric in $\mathfrak{X}(\mathbb{R})$.

Theorem 7.4.1. *For any $\alpha \in (0,\infty]$ and $X,Y \in \mathfrak{X}(\mathbb{R}^n)$*

$$\max_{1\le i\le n} \mathbf{L}(X_i,Y_i;\alpha)^{\alpha^*} \le \overset{\wedge}{\mathbf{L}}(X,Y;\alpha) \le \max_{1\le i\le n} \mathbf{L}(n^{1/\alpha}X_i, n^{1/\alpha}Y_i;\alpha)$$
$$\alpha^* := \max(1,1/\alpha). \tag{7.4.3}$$

Proof. The lower estimate for $\overset{\wedge}{\mathbf{L}}$ follows from the inequality

$$\max\{\mathbf{L}(X_i,Y_i;\alpha); i=1,\dots,n\} \le \mathbf{L}(X,Y;\alpha)^{\beta},\ \beta := \min(1,\alpha),$$

for any $X,Y \in \mathfrak{X}(\mathbb{R}^n)$. Let $\max\{\mathbf{L}(n^{1/\alpha}X_i, n^{1/\alpha}Y_i;\alpha); i = 1,\dots,n\} < \varepsilon$ and $x \in \mathbb{R}^n$. Then for any $i=1,\dots,n$, any $x_i \in \mathbb{R}$

$$\Pr(X_i \le x_i) < \Pr(Y_i \le x_i + n^{-1/\alpha}\varepsilon) + \varepsilon,$$

and thus

$$\min_{1\le i\le n} \Pr(X_i \le x_i) \le \min_{1\le i\le n} \Pr(Y_i \le x_i + n^{-1/\alpha}\cdot\varepsilon^{\max(1,1/\alpha)}) + \varepsilon. \tag{7.4.4}$$

Given $X,Y \in \mathfrak{X}(R^n)$, denote $\widetilde{X} = X_E$, $\widetilde{Y} = Y_E$ (see Example 7.3.2 in Sect. 7.3). Then X and Y have DFs M_X and M_Y, respectively. Now, (7.4.4) implies that $M_X(X) = \Pr(\widetilde{X} \le x) \le \Pr(\widetilde{Y} \in (-\infty,x]_{\alpha}^{\varepsilon}) + \varepsilon$. Therefore, $\mathbf{L}(X,Y;\alpha) < \varepsilon$ and thus the upper bound in (7.4.3) is established. □

Letting $\alpha = \infty$ in (7.4.3) we obtain the following corollary immediately.

Corollary 7.4.1. *For any X and $Y \in \mathfrak{X}(\mathbb{R}^n)$*

$$\overset{\wedge}{\mathbf{L}}(X,Y;\infty) = \mathbf{L}(X_E,Y_E;\infty) = \max\{\mathbf{L}(X_i,Y_i,\infty);\ i=1,\dots,n\}. \tag{7.4.5}$$

Corollary 7.4.2. *For any X and Y* $\in \mathfrak{X}(\mathbb{R}^n)$

$$\overset{n}{\widehat{\boldsymbol{\rho}}}(X,Y) = \boldsymbol{\rho}(X_E, Y_E) = \max\{\boldsymbol{\rho}(X_i, Y_i); i = 1, \ldots, n\}. \tag{7.4.6}$$

Proof. One can prove (7.4.6) using the same arguments as in the proof of Theorem 7.4.1. Another way is to use (7.4.5) and (7.4.1). □

Corollary 7.4.3. *For every* $\alpha \in (0, \infty]$ *and* $X, Y \in \mathfrak{X}(\mathbb{R}^n)$

$$\max_{1 \le i \le n} \mathbf{W}(X_i, Y_i; \alpha)^{\alpha^*} \le \overset{n}{\widehat{\mathbf{W}}}(X, Y; \alpha) \le \max_{1 \le i \le n} \mathbf{W}(n^{1/\alpha} X_i, n^{1/\alpha} Y_i; \alpha)$$

$$\overset{n}{\widehat{\mathbf{W}}}(X, Y; \infty) = \sup\{|F_{X_i}^{-1}(t) - F_{Y_i}^{-1}(t)|; t \in [0,1], i = 1, \ldots, n\}. \tag{7.4.7}$$

Proof. The first estimates follow from (7.4.2) and (7.4.3). The representation for $\overset{n}{\widehat{\mathbf{W}}}(X, Y, \infty)$ is a consequence of the preceding estimates. □

Corollary 7.4.4. *For any* $X, Y \in \mathfrak{X}(\mathbb{R}^n)$

$$\overset{n}{\widehat{\boldsymbol{\delta}}}(X,Y) = \boldsymbol{\delta}(X_E, Y_E) = \max\{\boldsymbol{\delta}(X_i, Y_i); i = 1, \ldots, n\}. \tag{7.4.8}$$

Equalities (7.4.5)–(7.4.8) describe the sharp lower bounds of the simple metrics $\mathbf{L}(X,Y)$, $\boldsymbol{\rho}(X,Y)$, $\mathbf{W}(X,Y)$, and $\boldsymbol{\delta}(X,Y)$ in $\mathfrak{X}(\mathbb{R}^n)$ in the case of fixed one-dimensional distributions, F_{X_i}, F_{Y_i}, $(i = 1, \ldots, n)$.

We will next consider the K-minimal metric with respect to the average compound distance

$$\mathcal{L}_H(X,Y) = EH(d(X,Y)), X, Y \in \mathfrak{X}(\mathbb{R}^n) \tag{7.4.9}$$

[see Example 3.4.1 and (3.4.3)], where $d(x,y) = \rho_\alpha(x,y)$ (7.3.2)–(7.3.4) ($\alpha \ge 1$) and H is a convex function on $[0,\infty)$, $H(0) = 0$. We will examine minimal functionals that are more general than $\overset{n}{\widehat{\mathcal{L}}}_H$.

Definition 7.4.1 (Cambanis et al. 1976). A function $\phi : E \subset \mathbb{R}^2 \to \mathbb{R}$ is said to be *quasiantitone* if

$$\phi(x,y) + \phi(x',y') \le \phi(x',y) + \phi(x,y') \tag{7.4.10}$$

for all $x' > x$, $y' > y$, $x, x', y, y' \in E$. We call $\phi : E \subset \mathbb{R}^n \to \mathbb{R}$ quasiantitone if it is a quasiantitone function of any two coordinates considered separately.

Some examples of quasiantitone functions are as follows: $f(x-y)$ where f is a nonnegative convex function on $\mathbb{R}$; $|x-y|^p$ for $p \ge 1$; $\max(x,y)$, $x, y \in \mathbb{R}$; any concave function on $\mathbb{R}^n$; and any DF of a nonpositive measure in $\mathbb{R}^n$.

At first we will find an explicit solution of the multidimensional Kantorovich problem [see Sect. 5.2, VI, and (5.2.36)] in the case of $U = \mathbb{R}$, $d = \rho_\alpha$, and a cost function c being quasiantitone. That is, let $\widetilde{F} = \{F_i, i = 1, \dots, N\}$ be the vector of N DFs $F_1, \dots, F_N$ on $\mathbb{R}$, and let $\mathfrak{P}(\widetilde{F})$ be the set of all DFs F on $\mathbb{R}^N$ with fixed one-dimensional marginal $F_1, \dots, F_N$. The pointwise upper bound of the distributions F in $\mathfrak{P}(\widetilde{F})$ is obtained at the Hoeffding distribution

$$M(x) := \min(F_1(x_1), \dots, F_N(x_N)), \quad x = (x_1, \dots, x_N). \tag{7.4.11}$$

The next theorem shows that the minimal total cost in the multidimensional Kantorovich transportation problem

$$\mathcal{A}_c(\widetilde{F}) = \inf\left\{\int_{\mathbb{R}^N} c \, \mathrm{d}F : F \in \mathfrak{P}(\widetilde{F})\right\} \tag{7.4.12}$$

coincides with the total cost of $\int_{\mathbb{R}^N} c \, \mathrm{d}M$, i.e., M describes the optimal plan of transportation.

Lemma 7.4.1 (Lorentz 1953). *For a p-tuple $(x_1, \dots, x_p)$ let $(\overline{x}_1, \dots, \overline{x}_p)$ denote its rearrangement in increasing order. Then, given N p-tuples $(x_1^{(1)}, \dots, x_p^{(1)}), \dots, (x_1^{(N)}, \dots, x_p^{(N)})$ for any quasiantitone function ϕ, the minimum of $\sum_{i=1}^p \phi(x_i^{(1)}, \dots, x_i^{(N)})$ over all the rearrangements of the p-tuples is attained at $(\overline{x}_1^{(1)}, \dots, \overline{x}_p^{(1)}), \dots, (\overline{x}_1^{(N)}, \dots, \overline{x}_p^{(N)})$.*

Proof. Let $X^{(k)} = (x_1^{(k)}, \dots, x_p^{(k)})$. Further, in inequalities containing values of the function ϕ at different points, we will omit those arrangements that take the same but arbitrary values. For a group I of indices i, $1 \le i \le N$, we denote $U_I := \{u_i\}_{i \in I}$, $U_I' = \{u_i'\}_{i \in I}$, and $U_I + U_I' = \{u_i + u_i'\}_{i \in I}$.

Claim 1. For any two disjoint groups of indices I, J, and h_i, $h_i \ge 0$,

$$\phi(U_I + H_I, U_J + H_J) - \phi(U_I + H_I, U_J) - \phi(U_I, U_J + H_J) + \phi(U_I, U_J) \le 0. \tag{7.4.13}$$

Proof of the Claim 1. Let I' be the group consisting of I and the index k, which belongs to neither I nor J. Then

$$\begin{aligned}
&\phi(U_{I'} + H_{I'}, U_J + H_J) - \phi(U_{I'} + H_{I'}, U_J) - \phi(U_{I'}, U_J + H_J) + \phi(U_{I'}, U_J) \\
&= \{\phi(U_I + H_I, u_k + h_k, U_J + H_J) - \phi(U_I + H_I, u_k + h_k, U_J) \\
&\quad - \phi(U_I, u_k + h_k, U_J + H_J) + \phi(U_I, u_k + h_k, U_J)\} \\
&\quad + \{\phi(U_I, u_k + h_k, U_J + H_J) - \phi(U_I, u_k + h_k, U_J) \\
&\quad - \phi(U_I, u_k, U_J + H_J) + \phi(U_I, u_k, U_J)\}.
\end{aligned}$$

Starting the inductive arguments with the inequality

$$\phi(x', y') - \phi(x', y) - \phi(x, y') + \phi(x, y) \leq 0, \quad x' \geq y,\ y' \geq y,$$

we prove the claim by induction with respect to the number of elements of I and J.

Further, for any $1 \leq s < p$ we consider the following operation, which gives a new set of p-tuples $\widetilde{X}^{(k)}$. We set $\widetilde{x}_i^{(k)} = x_i^{(k)}$ for $i \neq s$, $i \neq s+1$, and $\widetilde{x}_s^{(k)} = \min(x_s^{(k)}, x_{s+1}^{(k)}), \widetilde{x}_{s+1}^{(k)} = \max(x_s^{(k)}, x_{s+1}^{(k)})$. If I consists of indices k for which $x_s^{(k)} \leq x_{s+1}^{(k)}$, J of indices k for which $x_s^{(k)} \geq x_{s+1}^{(k)}$, u_k is the smaller, and $u_k + h_k$ is the larger of the two values, then

$$\sum_{i=1}^{p} \phi(x_i^{(1)}, \ldots, x_i^{(N)}) \geq \sum_{i=1}^{p} \phi(\widetilde{x}_i^{(1)}, \ldots, \widetilde{x}_i^{(N)}) \tag{7.4.14}$$

is exactly inequality (7.4.13). Continuing in the same manner we prove the theorem after a finite number of steps. □

Theorem 7.4.2 (Tchen 1980). *Let $\widetilde{F} = (F_1, \ldots, F_N)$ be a set of N DFs on $\mathbb{R}$ and M be defined by* (7.4.11)*. Given a quasiantitone function $\phi : \mathbb{R}^N \to \mathbb{R}$, suppose that the family $\{\phi(X), X$ distributed as $F \in \mathfrak{P}(\widetilde{F})\}$ is uniformly integrable. Then*

$$\mathcal{A}_\phi(\widetilde{F}) = \int \phi \, \mathrm{d}M. \tag{7.4.15}$$

Remark 7.4.1. For $N = 2$, this theorem is known as the Cambanis et al. (1976) theorem.[3]

Proof. Suppose first that the F_i have compact support. Let $X = (X_1, \ldots, X_N)$ be distributed as $F \in \mathfrak{P}(\widetilde{F})$ and defined on $[0, 1]$ with the Lebesgue measure. By Lemma 7.4.1, if the distribution F is concentrated on p atoms $(x_i^{(1)}, \ldots, x_i^{(N)})$ $(i = 1, \ldots, p)$ of mass $1/p$, then $E\phi(X) \geq E\phi(X_E)$, where $X_E = (F_{X_1}^{-1}(E), \ldots, F_{X_N}^{-1}(E))$, $E(\omega) = \omega$, $\omega \in [0, 1]$ (Sect. 7.3, Example 7.3.2). In the general case, let

$$x_{i,k}^m = 2^m E\{X_i I[k2^{-m} \leq X_i \leq (k+1)2^{-m}]\}$$

and

$$X_i^m(\omega) = \sum_{k=0}^{2^m-1} x_{i,k}^m \cdot I[k2^{-m} \leq \omega \leq (k+1)2^{-m}] \quad i = 1, \ldots, N, \quad \omega \in [0, 1].$$

$X_1^m, X_2^m, \ldots, X_N^m$ are step functions and bounded martingales converging almost surely (a.s.) to $X_1, \ldots, X_N$, respectively; see Breiman (1992).

[3] See Kalashnikov and Rachev (1988, Theorem 7.1.1).

Call $\overline{X_i^m}$, $i = 1, \ldots, N$ the reorderings of X_i^m. $\overline{X_i^m}$ and X_i^m have the same distribution and $\overline{X_i^m} = F_{X_i^m}^{-1}(E)$; hence, $\overline{X_i^m} \to F_{X_i}^{-1}(E)$ a.s., so that in the bounded case the theorem follows by bounded convergence.

Consider the general case. Let $\mathbb{B}_N = (-B, B)^N$, and let F_B be the distribution that is F outside $\mathbb{B}_N$ and $F_B\{A\} = F\{A \cap \mathbb{B}_N^c\} + \overline{F}_B\{A \cap \mathbb{B}_N\}$ for all Borel sets on $\mathbb{R}^N$, where $\overline{F}_B$ is the maximal subprobability with the sub-DF

$$M_B(x) = \min_{1 \le i \le N} F\{(-B, B]^{i-1} \times (-B, x_i] \times (-B, B]^{N-i}\}$$

for

$$x = (x_1, \ldots, x_N) \in \mathbb{B}_N.$$

Clearly, $F_B \in \mathfrak{P}(\widetilde{F})$ and F_B converges weakly to M as $B \to \infty$, which completes the proof of the theorem. □

As a consequence of the explicit solution of the N-dimensional Kantorovich problem, we will find an explicit representation of the following minimal functional:

$$\begin{aligned} \mathcal{L}_{p,q}(\widetilde{F}) := \inf\{ED_{p,q}(X) : X = (X_1, \ldots, X_N) \in \mathfrak{X}(\mathbb{R}^N), \\ F_{X_i} = F_i, \quad i = 1, \ldots, N\}, \end{aligned} \tag{7.4.16}$$

where $D_{p,q}(x) = \left[\sum_{1 \le i \le j \le N} |x_i - x_j|^p\right]^q$, $p \ge 1$, $q \le 1$, and $\widetilde{F} = (F_1, \ldots, F_N)$ is a vector of one-dimensional DFs.

Corollary 7.4.5. *For any $p \ge 1$ and $q \le 1$*

$$\mathcal{L}(\widetilde{F}) = \int_0^\infty D_{p,q}(F_1^{-1}(t), \ldots, F_n^{-1}(t))\mathrm{d}t. \tag{7.4.17}$$

As a special case of Theorem 7.4.2 [$N = 2$, $\phi(x, y) = H(|x - y|]$, H convex on $[0, \infty)$, $H \in \mathcal{H}$ (Example 2.4.1), we obtain the following corollary.

Corollary 7.4.6. *Let H be a convex function from $\mathcal{H}$ and*

$$\mathcal{L}_H(X, Y) = EH(|X - Y|)$$

be the H-average distance on $\mathfrak{X}(\mathbb{R})$ (Example 3.4.1). Then

$$\widehat{\mathcal{L}}_H(X, Y) = \widetilde{\mathcal{L}}_H(X, Y) = \int_0^1 H(|F_X^{-1}(t) - F_Y^{-1}(t)|)\mathrm{d}t. \tag{7.4.18}$$

Further, we will consider other examples of explicit formulae for K-minimal and minimal distances and metrics. Denote by $\mathbf{m}(X, Y)$ the following probability metric:

$$\mathbf{m}(X,Y) = E\left[2\max(X_1,\ldots,X_n,Y_1,\ldots,Y_n) - \frac{1}{n}\sum_{i=1}^{n}(X_i+Y_i)\right]. \quad (7.4.19)$$

Theorem 7.4.3. *Suppose that the set of random vectors X and Y with fixed one-dimensional marginals is uniformly integrable. Then*

$$\overset{\hat{n}}{\mathbf{m}}(X,Y) = \widetilde{\mathbf{m}}(X,Y) = \int_{-\infty}^{\infty} \frac{1}{n}\sum_{i=1}^{n}[F_{X_i}(u) + F_{Y_i}(u)]$$
$$-2\min[F_{X_1}(u),\ldots,F_{X_n}(u),F_{Y_1}(u),\ldots,F_{Y_n}(u)]\mathrm{d}u. \quad (7.4.20)$$

Proof. Suppose $E|X_i| + E|Y_i| < \infty$, $i = 1,\ldots,n$. Then from the representation

$$\mathbf{m}(X,Y) = \int_{-\infty}^{\infty} \frac{1}{n}\sum_{i=1}^{n}[F_{X_i}(u) + F_{Y_i}(u)]$$
$$-2\Pr(\max(X_1,\ldots,X_n,Y_1,\ldots,Y_n) \le u)\mathrm{d}u$$

and the Hoeffding inequality,

$$\Pr(\max(X_1,\ldots,X_n,Y_1,\ldots,Y_n) \le u)$$
$$\le \min(F_{X_1}(u),\ldots,F_{X_n}(u),F_{Y_1}(u),\ldots,F_{Y_n}(u)),$$

we obtain (7.4.20). The weaker regularity condition is obtained as in the previous theorem. □

Consider the special case $n = 1$. We will prove the equality

$$\widehat{\mu}(X,Y) = \widetilde{\mu}(X,Y) := \mu(F_X^{-1}(E), F_Y^{-1}(E)), \quad (7.4.21)$$

where E is uniformly distributed on $(0,1)$ for various compound distances in $\mathfrak{X}(\mathbb{R})$.

In Example 3.4.3 we introduced the Birnbaum–Orlicz compound distances

$$\mathbf{\Theta}_H(X_1,X_2) = \int_{-\infty}^{\infty} H(\Pr(X_1 \le t < X_2) + \Pr(X_2 \le t < X_1))\mathrm{d}t$$
$$H \in \mathcal{H} \quad (7.4.22)$$

$$\mathbf{R}_H(X_1,X_2) = \sup_{t\in\mathbb{R}} H(\Pr(X_1 \le t < X_2) + \Pr(X_2 \le t < X_1))$$

and compound metrics

$$\boldsymbol{\Theta}_p(X_1, X_2) = \left\{ \int_{-\infty}^{\infty} [\Pr(X_1 \le t < X_2) + \Pr(X_2 \le t < X_1)]^p \mathrm{d}t \right\}^{p'},$$
$$p' = \min(1, 1/p),$$
$$\boldsymbol{\Theta}_\infty(X_1, X_2) = \sup_{t \in \mathbb{R}} [\Pr(X_1 \le t < X_2) + \Pr(X_2 \le t < X_1)].$$

Note that $\boldsymbol{\Theta}_1(X_1, X_2) = E|X_1 - X_2|$ for $H(t) = t$. In Example 3.3.4, we consider the corresponding simple Birnbaum–Orlicz distances

$$\boldsymbol{\theta}_H(F_1, F_2) = \int_{-\infty}^{\infty} H(|F_1(x) - F_2(x)|\mathrm{d}x, \quad H \in \mathcal{H},$$
$$\boldsymbol{\theta}_H(F_1, F_2) = \sup_{x \in \mathbb{R}} H(|F_1(x) - F_2(x)|) \tag{7.4.23}$$

and simple metrics

$$\boldsymbol{\theta}_p(F_1, F_2) = \left(\int_{-\infty}^{\infty} |F_1(x) - F_2(x)|^p \mathrm{d}x \right)^{p'},$$
$$\boldsymbol{\theta}_\infty(F_1, F_2) = \boldsymbol{\rho}(F_1, F_2) = \sup_{x \in \mathbb{R}} |F_1(x) - F_2(x)|.$$

Theorem 7.4.4.

$$\boldsymbol{\theta}_H = \widetilde{\boldsymbol{\Theta}}_H = \widehat{\boldsymbol{\Theta}}_H \quad \boldsymbol{\rho}_H = \widetilde{\mathbf{R}}_H = \widehat{\mathbf{R}}_H \quad \boldsymbol{\theta}_p = \widetilde{\boldsymbol{\Theta}}_p = \widehat{\boldsymbol{\Theta}}_p \quad 0 < p \le \infty. \tag{7.4.24}$$

Proof. To prove the first equality in (7.4.24), consider the set of all random pairs (X_1, X_2) with marginal DFs F_1 and F_2. For any such pair

$$\begin{aligned}
\boldsymbol{\Theta}_H(X_1, X_2) &= \int_{-\infty}^{\infty} H(F_1(t) + F_2(t) - 2\Pr(X_1 \vee X_2 \le t))\mathrm{d}t \\
&\ge \widetilde{\boldsymbol{\Theta}}_H(X_1, X_2) = \int_{-\infty}^{\infty} H(F_1(t) + F_2(t) - 2\min(F_1(t), F_2(t))\mathrm{d}t \\
&= \int_{-\infty}^{\infty} H(|F_1(t) - F_2(t)|)\mathrm{d}t = \boldsymbol{\theta}_H(F_1, F_2).
\end{aligned}$$

Thus $\widetilde{\boldsymbol{\Theta}}_H = \widehat{\boldsymbol{\Theta}}_H = \boldsymbol{\theta}_H$. In a similar way one proves the other equalities in (7.4.24). □

Remark 7.4.2. Theorem 7.4.2 for $N = 2$ shows that the infimum of $E\phi(X_1, X_2)$ (ϕ is a quasiantitone function (7.4.10) over $\mathfrak{P}(F_1, F_2)$, the set of all possible joint DF $H = F_{X_1,X_2}$ with fixed marginals $F_{X_i} = F_i$) is attained at the upper Hoeffding–Fréchet bound $\overline{H}(x, y) = \min(F_1(x), F_2(y))$. Similarly,[4]

$$\sup\{E\phi(X_1, X_2) : H \in \mathfrak{P}(F_1, F_2) = \int_0^1 \phi(F_1(t), F_2(1-t))\mathrm{d}t, \qquad (7.4.25)$$

i.e., the supremum of $E\phi(X_1, X_2)$ is attained at the lower Hoeffding–Fréchet bound $\underline{H}(x, y) = \max(0, F_1(x) + F_2(y) - 1)$. The multidimensional analogs of (7.4.25) are not known. Notice that the multivariate lower Hoeffding–Fréchet bound $\underline{H}(x_1, \ldots, x_N) = \max(0, F_1(x_1) + \cdots + F_N(x_N) - N + 1)$ is not a DF on $\mathbb{R}^N$, in contrast to the upper bound $\overline{H}(x_1, \ldots, x_N) = \min(F_1(x_1), \ldots, F_N(x_N))$, which is a DF on $\mathbb{R}^N$. That is why we do not have an analog for Theorem 7.4.2 when the supremum of $E\phi(X_1, \ldots, X_N)$ over the set of N-dimensional DFs with fixed one-dimensional marginals is considered.

Remark 7.4.3. In 1981, Kolmogorov stated the following problem to Makarov: find the infimum and supremum of $\Pr(X + Y < z)$ over $\mathfrak{P}(F_1, F_2)$ for any fixed z. The problem was solved independently by Makarov (1981) and Rüschendorf (1982). Rüschendorf (1982) considered also the multivariate extension. Another solution was given by Frank et al. (1987). Their solution was based on the notion of *copula* linking the multidimensional DFs to their one-dimensional marginals.[5]

7.5 The Case Where U Is a Separable Metric Space

We begin with a multivariate extension of the Strassen theorem, $\pi = \widehat{\mathbf{K}}$, where π is the Prokhorov metric.[6]

The following theorem was proved by Schay (1979) in the case where (U, d) is a complete separable space. We will use the method of Dudley (1976, Theorem 18.1) to extend this result in the case of a separable space.

Denote by $\mathcal{P}(U)$ the space of all Borel probability measures (laws) on an s.m.s. (U, d). Let $N \geq 2$ be an integer, let $\|x\|$, $x \in \mathbb{R}^m$, be a monotone norm (if $0 < x < y$, then $\|x\| < \|y\|$) in $\mathbb{R}^n$, where $m = \binom{N}{2}$, and let

$$\mathcal{D}(x_1, \ldots, x_N) = \|d(x_1, x_2), \ldots, d(x_1, x_n), d(x_2, x_3), \ldots, d(x_{N-1}, x_N)\|. \qquad (7.5.1)$$

[4] See Cambanis et al. (1976) and Tchen (1980).

[5] See Sklar (1959), Schweizer and Sklar (2005), Wolff and Schweizer (1981), and Genest and MacKay (1986).

[6] See Example 3.3.3 and (3.3.18) in Chap. 3.

Theorem 7.5.1. *For any $P_1, \ldots, P_N$ in $\mathcal{P}(U)$, $\alpha \geq 0$, $\beta \geq 0$, the following two assertions are equivalent:*

(I) For any $a > \alpha$ there exists $\mu \in \mathcal{P}(U^N)$ with marginal distributions $P_1, \ldots, P_N$ such that

$$\mu\{\mathcal{D}(x_1, \ldots, x_N) > a\} \leq \beta. \tag{7.5.2}$$

(II) For any Borel sets $B_1, \ldots, B_{N-1} \in \mathcal{B}(U)$

$$P_1(B_1) + \cdots + P_{N-1}(B_{N-1}) \leq P_N B^{(\alpha)} + \beta + N - 2, \tag{7.5.3}$$

where $B^{(\alpha)} = \{x_N \in U : \mathcal{D}(x_1, \ldots, x_N) \leq \alpha$, for some $x_1 \in B_1, \ldots, x_{N-1} \in B_{N-1}\}$. If $P_1, \ldots, P_N$ are tight measures, then $a = \alpha$.

Proof. Assertion (I) implies (II) since

$$\begin{aligned} P_1(B_1) &\leq \mu(\mathcal{D}(x_1, \ldots, x_N) > a) + \mu\left(\bigcap_{i=1}^{N-1}\{x_i \in B_i\}, \mathcal{D}(x_1, \ldots, x_N) \leq a\right) \\ &\quad + \mu\left(x_1 \in B_1, \bigcup_{i=2}^{N-1}\{x_i \notin B_i\}, \mathcal{D}(x_1, \ldots, x_N) \leq a\right) \\ &\leq \beta + \mu(B^{(a)}) + \sum_{i=2}^{N-1}(1 - P_i(B_i)). \end{aligned}$$

As $a \to \alpha$ we obtain (II).

To prove that (II) $\Rightarrow$ (I), suppose first that $P_1, \ldots, P_N$ are tight measures. Let $\{x_i : i = 1, 2, \ldots\}$ be a dense sequence in U, and let $P_{i,n}$ $(i = 1, \ldots, N)$ be probability measures on the set $U_n := \{x_1, \ldots, x_n\}$. We first fix n and prove (II) $\to$ (I) for $a = \alpha$, U_n, and $P_{1,n}, \ldots, P_{N,n}$ in place of U and $P_1, \ldots, P_N$, and then let $n \to \infty$.

For any $I = (i_1, \ldots, i_N) \in \{1, \ldots, n\}^N$ and $X_I = (x_{i_1}, \ldots, x_{i_N})$ define the indicator: $\mathrm{Ind}(X_I) = 1$ if $\mathcal{D}(X_I) \leq \alpha$ and $\mathrm{Ind}(X_I) = 0$ otherwise. To obtain the μ of the theorem, we consider μ_n on U_n^N. We denote

$$\xi_I = \mu_n(\{X_I\}) \quad P_{i_k,j} = P_{j,n}(\{x_{i_k}\}) \quad i_k = 1, \ldots, n, \quad k, j = 1, \ldots, N.$$

Since we want μ_n to have $P_{1,n}, \ldots, P_{N,n}$ as one-dimensional projections, we require the constraints

$$\begin{aligned} &\sum_{i_\ell} \xi_I \leq P_{i_k,j} \quad j = 1, \ldots, N \quad i_k = 1, \ldots, n, \\ &\xi_I \geq 0, \end{aligned} \tag{7.5.4}$$

where in (7.5.4) i_ℓ runs from 1 to n for all $\ell \in \{1, \ldots, k-1, k+1, \ldots, N\}$.

If we denote by μ_n^* the "optimal" μ_n that assigns as much probability as possible to the "diagonal cylinder" C_α in U_n^N given by $\mathcal{D}(X_I) \leq \alpha$, then we will determine $\mu_n^*(C_\alpha)$ by looking at the following linear programming problem of canonical form:

$$\text{maximize } Z = \sum_{I\in(1,\ldots,n)^N} \text{Ind}(X_I)\xi_I \quad \text{subject to (7.5.4).} \tag{7.5.5}$$

The dual of the foregoing problem is easily seen to be

$$\text{minimize } W = \sum_{i_k=1}^{n}\sum_{j=1}^{N} P_{i_k,j}u_{i_k,j}$$

subject to $u_{i_k,j} \geq 0$

$$\sum_{j=1}^{N} u_{i_k,j} \geq \text{Ind}(X_I) \quad \forall i_k = 1,\ldots,n,\ k = 1,\ldots,N,\ j = 1,\ldots,N, \tag{7.5.6}$$

and by the duality theorem,[7] the maximum of Z equals the minimum of W. Let us write $\overline{u}_{i_k,j} = 1 - u_{i_k,j}$. Then (7.5.6) becomes

$$\text{minimize } W = N - 1 - \sum_{i_k=1}^{n}\sum_{j=1}^{N-1} P_{i_k,j}\overline{u}_{i_k,j} + \sum_{i_k=1}^{n} p_{i_k,N}u_{i_k,N}$$

$$\text{subject to } \overline{u}_{i_k,j} \leq 1,\ j = 1,\ldots,N-1,\ u_{i_k,N} \geq 0,$$

$$\text{and } u_{i_k,N} \geq (\text{Ind}(X_I) - N - 1) + \sum_{j=1}^{n-1} \overline{u}_{i_k,j}$$

$$\forall i_k = 1,\ldots,n \quad k = 1,\ldots,N. \tag{7.5.7}$$

We may also assume

$$\overline{u}_{i_k,j} \geq 0 \quad j = 1,\ldots,N-1 \quad u_{i_k,N} \leq 1 \tag{7.5.8}$$

since these additional constraints cannot affect the minimum of W. Now the set of "feasible" solutions $u_{i_k,j}$, $j = 1,\ldots,N-1$, $u_{i_k,N}$, $i_k = 1,\ldots,n$, $k = 1,\ldots,N$ for the dual problem (7.5.7), (7.5.8) is a convex polyhedron contained in the unit cube $[0,1]^{Nn}$, the extreme points of which are the vertices of the cube. Since the minimum of W is attained at one of these extreme points, there exists $\overline{u}_{i_k,j}$, $u_{i_k,N}$ equal to 0 or 1, which minimize W under the constraints in (7.5.7) and (7.5.8). Thus, without loss of generality, we may assume that $\overline{u}_{i_k,j}$ $u_{i_k,N}$ are 0s and 1s.

[7] See, for example, Berge and Chouila-Houri (1965, Sect. 5.2).

Define the sets $F_j \subset U^n$, $j = 1, \ldots, N-1$, such that $\overline{u}_{i_k,j} = 1$ for all j such that $x_{i_k} \in F_j$ and $u_{i_k,j} = 0$ otherwise. Then, by (7.5.7), $u_{i_k,N} = 1$ for all k such that $\mathrm{Ind}(X_I) = 1$ when $\overline{u}_{i_k,j} = 1$, $j = 1, \ldots, N-1$, that is, whenever x_{i_N} satisfies $\mathcal{D}(X_I) \le \alpha$ with $x_{i_j} \in F_j$, $j = 1, \ldots, N-1$. Hence

$$\min W = N - 1 - \max[P_{1,n}(F_1) + \cdots + P_{N-1,n}(F_{N-1}) - P_{N,n}(F_n^{(\alpha)})],$$

where

$$F_n^{(\alpha)} := \{x_{i_N} : D(X_I) \le \alpha \quad \text{for some} \quad x_{i_j} \in F_j,\ j = 1, \ldots, N-1\}.$$

Thus, by the duality theorem in linear programming, maximum $Z =$ minimum W, and then

$$\begin{aligned}\mu_n^*(\mathcal{D}(X_I) > \alpha) &= 1 - \mu_n^*(C_\alpha) \\ &= 2 - N + \max\{[P_{1,n}(F_1) + \cdots + P_{N-1,n}(F_{N-1}) \\ &\quad - P_{N,n}(F_n^{(\alpha)}] : F_1, \ldots, F_{N-1} \subset U_n\}.\end{aligned}$$

The latter inequality is true for any $\alpha > 0$, and therefore

$$\begin{aligned}&\inf\{\alpha : \mu_n^* \mathcal{D}((X_I) \ge \alpha) \le \alpha\} \\ &= \inf\left\{\alpha : \max_{F_1,\ldots,F_{N-1} \subset U_n} [P_{1,n}(F_1) + \cdots + P_{N-1,n}(F_{N-1}) \right. \\ &\qquad \left. - P_{N,n}(F_n^{(\alpha)}] + 2 - N \le \alpha \right\}.\end{aligned}$$

Given P_j ($j = 1, \ldots, N$), one can take $P_{j,n}$ concentrated in finitely many atoms, say in U_n such that the Prokhorov distance $\pi(P_{j,n}, P_j) \le \varepsilon$. The latter follows, for example, by the Glivenko–Cantelli–Varadarajan theorem.[8] As P_j is tight, then $P_{j,n}$ is uniformly tight and thus there is a weakly convergent subsequence $P_{j,n(k)} \to P_j$. The corresponding sequence of optimal measures $\mu_{n(k)}^*$ with marginals $P_{j,n(k)}$ ($j = 1, \ldots, N$) is also uniformly tight. Now the same "tightness" argument implies the existence of a measure μ for which (7.5.2) holds. □

Remark 7.5.1. It is easy to see that (II) is equivalent to (7.5.3) for all closed sets B_j ($j = 1, \ldots, N-1$) and/or $B^{(\alpha)}$ given by $\{x_N \in U : \mathcal{D}(x_1, \ldots, x_N) < \alpha\}$.

Now, suppose that $P_1, \ldots, P_N$ are not tight. Let $\overline{U}$ be a completion of the space U. For a given $a > \alpha$ let $\varepsilon \in (0, (a - \alpha)/2\|e\|)$, where $e = (1, \ldots, 1)$ and A is a

---

[8] See Dudley (2002, Theorem 11.4.1).

maximal subset of U such that $d(x, y) \geq \varepsilon/2$ for $x \neq y$ in A. Then A is countable; $A = \{x_k\}_{k=1}^{\infty}$. Let $\overline{A}_k = \{x \in \overline{U}; d(x, x_k) < \varepsilon/2 \leq d(x, x_j), j = 1, \ldots, k-1\}$ and $A = \overline{A}_k \cap U$. The measure $P_1, \ldots, P_N$ on U determines the probability measures $\overline{P}_1, \ldots, \overline{P}_N$ on $\overline{U}$. Then $\overline{P}_1, \ldots, \overline{P}_N$ are tight, and consequently there exists $\overline{\mu} \in \mathcal{P}(\overline{U}^N)$ with marginal distributions $\overline{P}_1, \ldots, \overline{P}_N$ for which (I) holds for $a = \alpha$. Let $P_{k,m}(B) = P_k(B \cap A_m)$, $k = 1, \ldots, N$, for any $B \in \mathcal{B}(U)$. We define the measure

$$\mu_{m_1,\ldots,m_N} = c_{m_1,\ldots,m_N} P_{m_1} \times \cdots \times P_{m_N},$$

where the number $c_{m_1,\ldots,m_N}$ is chosen such that

$$\mu_{m_1,\ldots,m_N}(A_{m_1} \times \cdots \times A_{m_N}) = \overline{\mu}(\overline{A}_{m_1} \times \cdots \times \overline{A}_{m_N}).$$

We set

$$\mu_\varepsilon = \sum_{m_1,\ldots,m_N} \mu_{m_1,\ldots,m_n}.$$

Then μ_ε has marginal distributions $P_1, \ldots, P_N$ (see the proof of Case 3, Theorem 5.3.1 in Chap. 5) and

$$\begin{aligned} \mu_\varepsilon(\mathcal{D}(y_1, \ldots, y_N) > a) &\leq \sum_{m_1,\ldots,m_N} \mu_{m_1,\ldots,m_N}(\mathcal{D}(y_1, \ldots, y_N) > \alpha + 2\varepsilon\|e\|) \\ &\leq \sum_{m_1,\ldots,m_N} \overline{\mu}\{(\overline{A}_{m_1} \times \cdots \times \overline{A}_{m_N}) : \\ &\qquad \mathcal{D}(x_1, \ldots, x_N) > \alpha + \varepsilon\|e\|\} \\ &\leq \overline{\mu}(\mathcal{D}(y_1, \ldots, y_n) > \alpha) \leq \beta. \end{aligned}$$

Thus (II) $\to$ (I), as desired.

Let us apply Theorem 7.5.1 to the set $\mathfrak{X}(U)$ of RVs defined on a rich enough probability space (Remark 2.7.1), taking values in the s.m.s. (U, d).

Given $\alpha > 0$ and a vector of laws $\widetilde{P} = (P_1, \ldots, P_N) \in (\mathcal{P}(U))^N$, define

$$\begin{aligned} S_1(\widetilde{P}; \alpha) = \inf\{&\Pr(\mathcal{D}(X) > \alpha) : X = (X_1, \ldots, X_N) \in \mathfrak{X}(U^N), \\ &\Pr{}_{X_i} = P_i, \quad i = 1, \ldots, N\} \end{aligned} \tag{7.5.9}$$

and

$$\begin{aligned} S_2(\widetilde{P}; \alpha) = \sup\{&P_1(B_1) + \cdots + P_{N-1}(B_{N-1}) - P_N(B_N^{(\alpha)}) \\ &-N + 2 : B_1, B_2, \ldots, B_{N-1} \in \mathcal{B}(U)\}, \end{aligned} \tag{7.5.10}$$

where $\mathcal{D}(x_1,\dots,x_N) = \|d(x_1,x_2),\dots,d(x_1,x_N),\dots,d(x_{N-1},x_N)\|$, $\|\cdot\|$ is a monotone seminorm and $B_N^{(\alpha)}$ is defined as in Theorem 7.5.1. Then the following duality theorem holds.

Corollary 7.5.1. *For any* $\alpha > 0$

$$S_1(\widetilde{P};\alpha) = S_2(\widetilde{P};\alpha). \tag{7.5.11}$$

If P_i*s are tight measures, then the infimum in* (7.5.9) *is attained.*

In the case $N = 1$, we obtain the Strassen–Dudley theorem.

Corollary 7.5.2. *Let* $\mathbf{K}_\lambda$ *(*$\lambda > 0$*) be the Ky Fan metric [see* (3.4.10)*] and* $\boldsymbol{\pi}_\lambda$ *the Prokhorov metric [see* (3.3.22)*]. Then* $\boldsymbol{\pi}_\lambda$ *is the minimal metric relative to* $\mathbf{K}_\lambda$*, i.e.,*

$$\widehat{\mathbf{K}}_\lambda = \boldsymbol{\pi}_\lambda. \tag{7.5.12}$$

In particular, by the limit relations $\boldsymbol{\pi}_\lambda \underset{\lambda\to 0}{\longrightarrow} \ell_0 = \boldsymbol{\sigma}$ (Lemma 3.3.1) and $\mathbf{K}_\lambda \underset{\lambda\to 0}{\longrightarrow} \mathcal{L}_0$ [see (3.4.11) and (3.4.6)], we have that the minimal metric relative to the indicator metric $\mathcal{L}_0(X,Y) = EI\{X \neq Y\}$ equals the total variation metric

$$\boldsymbol{\sigma}(X,Y) = \sup_{A\in\mathcal{B}(U)} |\Pr(X \in A) - \Pr(Y \in A)|,$$

i.e., (Dobrushin (1970)) $\widehat{\mathcal{L}}_0 = \boldsymbol{\sigma}$.

By the duality Theorem 7.5.1, for any $\lambda > 0$ and $\widetilde{P} = (P_1,\dots,P_N) \in \mathcal{P}(U)^N$,

$$\inf_{\substack{X\in\mathfrak{X}(U^N)\\ \Pr_{X_i}=P_i,\ i=1,\dots,N}} \mathcal{KF}_\lambda(X) = \Pi_\lambda(\widetilde{P}), \tag{7.5.13}$$

where $\mathcal{KF}_\lambda$ is the *Ky Fan functional* in $\mathfrak{X}(U^N)$,

$$\mathcal{KF}_\lambda(X) := \inf\{\varepsilon > 0 : \Pr(\mathcal{D}(X) > \lambda\varepsilon) \le \varepsilon\},$$

and $\Pi_\lambda(\widetilde{P})$ is the *Prokhorov functional in* $(\mathcal{P}(U))^N$ *with parameter* $\lambda > 0$

$$\Pi_\lambda(\widetilde{P}) = \inf\{\varepsilon > 0 : S_2(\widetilde{P},\lambda\varepsilon) \le \varepsilon\}.$$

Letting $\lambda \to 0$ in (7.5.13), we obtain the following multivariate version of the Dobrushin (1970) duality theorem:

$$\inf_{\substack{X\in\mathfrak{X}(U^N)\\ \Pr_{X_i}=P_i,\ i=1,\dots,N}} \Pr(X_i \neq X_j \quad \forall 1 \le i < j \le N)$$

$$= \sup_{B_1,\dots,B_{N-1}\in\mathcal{B}(U)} \left[P_1(B_1) + \cdots + P_{N-1}(B_{N-1}) - P_N\left(\bigcap_{i=1}^{N-1} B_i\right) - N + 2 \right]$$

$$= \sup_{B_1,\dots,B_{N-1}\in\mathcal{B}(U)} \left[P_N\left(\bigcup_{i=1}^{N-1} B_i\right) - P_1(B_1) - \cdots - P_{N-1}(B_{N-1}) \right]. \tag{7.5.14}$$

Note that the preceding quantities are symmetric with respect to any rearrangement of the vector $\widetilde{P}$.

Multiplying both sides of (7.5.13) by λ and then letting $\lambda \to \infty$ [or simply using (7.5.11)] we obtain

$$\inf_{\substack{X\in\mathfrak{X}(U^N) \\ \Pr_{X_i}=P_i,\ i=1,\dots,N}} \operatorname{ess\,sup} \mathcal{D}(X) = \inf\{\varepsilon > 0 : S_2(\widetilde{P};\varepsilon) = 0\}.$$

Using the preceding equality for $N = 2$, we obtain that the minimal metric relative to $\mathcal{L}_\infty(X,Y) = \operatorname{ess\,sup} d(X,Y)$ [see (3.4.5), (3.4.7), (3.4.11)] is equal to ℓ_∞ [see (3.3.14) and Lemma 3.3.1], i.e.,

$$\widehat{\mathcal{L}}_\infty = \ell_\infty. \tag{7.5.15}$$

Suppose that $d_1,\dots,d_n$ are metrics in U and that U is a separable metric space with respect to each d_i, $i = 1,\dots,n$. We introduce in U^n the metric

$$d_\Sigma(x,y) = \sum_{i=1}^n d_i(x_i,y_i), x = (x_1,\dots,x_n), y = (y_1,\dots,y_n) \in U^n. \tag{7.5.16}$$

We consider in $\mathfrak{X}(U^n)$ the metric $\tau_\Sigma(X,Y) := Ed_\Sigma(X,Y)$. Denote by $\kappa(X_i,Y_i;d_i)$ the Kantorovich metric in the space $\mathfrak{X}(U,d_i)$

$$\kappa(X_i,Y_i;d_i) = \sup\left\{ |E[f(X_i) - f(Y_i)]| : \|f\|_L^{(i)} := \sup_{x\neq y} \frac{|f(x)-f(y)|}{d_i(x,y)} \le 1 \right\} \tag{7.5.17}$$

(Example 3.3.2).

Theorem 7.5.2. *Suppose that for $X = (X_1,\dots,X_n)$, $Y = (Y_1,\dots,Y_n) \in \mathfrak{X}(U^n)$, $\tau_\Sigma(X,a) + \tau_\Sigma(Y,a) < +\infty$ for some $a \in U^n$. Then*

$$\hat{\tau}_\Sigma(X,Y) = \sum_{i=1}^n \kappa(X_i,Y_i;d_i). \tag{7.5.18}$$

Proof. By the Kantorovich theorem (Corollary 6.2.1), the minimal metric relative to the metric $\boldsymbol{\tau}(X_i, Y_i; d_i) = Ed_i(X_i, Y_i)$ in $\mathfrak{X}(U)$ is $\boldsymbol{\kappa}(X_i, Y_i; d_i)$. Hence,

$$\overset{\hat{n}}{\boldsymbol{\tau}}_{\Sigma}(X, Y) \geq \sum_{i=1}^{n} \widehat{\boldsymbol{\tau}}(X_i, Y_i; d_i) = \sum_{i=1}^{n} \boldsymbol{\kappa}(X_i, Y_i; d_i). \tag{7.5.19}$$

Conversely, let $\mathfrak{L}_m^{(i)}$, $i = 1, 2, \ldots$, be a sequence of joint distributions of RVs X_i, Y_i such that $\boldsymbol{\kappa}(X_i, Y_i; d_i) = \lim_{m\to\infty} \boldsymbol{\tau}(X_i, Y_i; d_i, \mathfrak{L}_m^{(i)})$, where $\boldsymbol{\tau}(X_i, Y_i; d_i; \mathfrak{L}_m^{(i)})$ is the value of the metric $\boldsymbol{\tau}$ for the joint distribution $\mathfrak{L}_m^{(i)}$. Then $\overset{\hat{n}}{\boldsymbol{\tau}}_{\Sigma}(X, Y) \leq \sum_{i=1}^{n} \boldsymbol{\tau}(X_i, Y_i; d_i; \mathfrak{L}_m^{(i)})$, and as $m \to +\infty$ we get the inequality

$$\overset{\hat{n}}{\boldsymbol{\tau}}(X, Y) \leq \sum_{i=1}^{n} \boldsymbol{\kappa}(X_i, Y_i; d_i). \tag{7.5.20}$$

Inequalities (7.5.19) and (7.5.20) imply equality (7.5.18). □

Corollary 7.5.3. *For any* $\alpha \in [0, 1]$

$$\overset{\hat{n}}{\boldsymbol{\tau}}(X, Y; d^{\alpha}) = \sum_{i=1}^{n} \boldsymbol{\kappa}(X_i, Y_i; d^{\alpha}) \text{ for } 0 < \alpha \leq 1, \tag{7.5.21}$$

$$\overset{\hat{n}}{\boldsymbol{\tau}}(X, Y; d^{0}) = \sum_{i=1}^{n} \boldsymbol{\sigma}(X_i, Y_i). \tag{7.5.22}$$

The proof of (7.5.21) follows from (7.5.18) if we set $d_i = d^{\alpha}$. Equality (7.5.21) follows from equality (7.5.21) as $\alpha \to 0$.

7.6 Relations Between Multidimensional Kantorovich and Strassen Theorems: Convergence of Minimal Metrics and Minimal Distances

Recall the multidimensional Kantorovich theorem [see (5.3.1), (5.3.2), and (5.3.4)]

$$A_D(\widetilde{P}) = \mathbb{K}(\widetilde{P}) := K(\widetilde{P}, \mathfrak{G}(U)), \tag{7.6.1}$$

where $\widetilde{P} = (\widetilde{P}_1, \ldots, \widetilde{P}_N) \in (\mathcal{P}(U))^N$

$$A_D(\widetilde{P}) = \inf\left\{\int_{U^N} D\mathrm{d}P,\ P \in \mathfrak{P}(\widetilde{P})\right\}, D = H(\mathcal{D}). \tag{7.6.2}$$

In the preceding relations, the minimal functional $\mathcal{D}(x)$ is given by

$$\mathcal{D}(x) = \|d(x_1, x_2), d(x_1, x_3), \ldots, d(x_1, x_N), d(x_2, x_3), \ldots, d(x_{N-1}, x_N)\|.$$

$\|\cdot\|$ is a monotone seminorm on $\mathbb{R}^n$, $m = \binom{N}{2}$, and $\mathfrak{B}(\widetilde{P})$ is the space of all Borel probability measures P on U^N with fixed one-dimensional marginals $P_1, \ldots, P_N$ (Sect. 5.3).

Next we turn our attention to the relationship between (7.6.1) and the multidimensional Strassen theorem [see (7.5.13)].

Theorem 7.6.1. *Suppose that* (U, d) *is an s.m.s.,*

$$\mathcal{KF}(P) = \inf\{\alpha > 0 : P(\mathcal{D}(x) > \alpha) < \alpha\} \tag{7.6.3}$$

is the Ky Fan functional *in* $\mathcal{P}(U^N)$*, and*

$$\begin{aligned}\Pi(\widetilde{P}) = \inf\{\alpha > 0 : &\ P_1(B_1) + \cdots + P_{N-1}(B_{N-1}) \\ &\leq P_N(B^{(\alpha)}) + \alpha + N - 2 \\ &\quad \text{for all } B_1, \ldots, B_{N-1}, \text{ Borel subsets of } U\}\end{aligned} \tag{7.6.4}$$

is the Prokhorov functional *in* $(\mathcal{P}(U))^N$*, where* $B^{(\alpha)} = \{x_N \in U : \mathcal{D}(x_1, \ldots, x_N) \leq \alpha$ *for some* $x_1 \in B_1, \ldots, x_{N-1} \in B_{N-1}\}$. *Then*

$$\inf\{\mathcal{KF}(P) : P \in \mathfrak{P}(\widetilde{P})\} = \Pi(\widetilde{P}), \tag{7.6.5}$$

and if $\widetilde{P}$ *is a set of tight measures, then the infimum is attained in* (7.6.3).

The next inequality represents the main relationship between the Kantorovich functional $A_D(\widetilde{P})$ and the Prokhorov functional $\Pi(\widetilde{P})$.

Theorem 7.6.2. *For any* $H \in \mathcal{H}^*$ *(i.e.,* $H \in \mathcal{H}$*, Example 2.4.1, and* H *is convex),* $M > 0$*, and* $a \in U$

$$\begin{aligned}\Pi(\widetilde{P})H(\Pi(\widetilde{P})) \leq \mathbb{K}(\widetilde{P}) \leq &\ H(\Pi(\widetilde{P})) + c_1 H(M)\Pi(\widetilde{P}) \\ &+ c_2 \sum_{i=1}^{N} \int_U H(d(x, a)) I(d(x, a) > M) P_i(\mathrm{d}x),\end{aligned} \tag{7.6.6}$$

where $c_2 := K_H^\ell$ *[see* (2.4.3)*],* $\ell := [\log_2(A_m N^2)] + 1$, $c_1 = Nc_2$, $[x]$ *is the integer part of* x*, and*

$$A_m := \max_{1 \leq j \leq m} \{\|(i_1, \ldots, i_m)\| : i_k = 0, k \neq j, i_j = 1\} \quad m = \binom{N}{2}.$$

Proof. For any probability measure P on U^N and $\varepsilon > 0$ the inequality $\int_{U^N} H(\mathcal{D}(x))P(\mathrm{d}x) < \delta = \varepsilon H(\varepsilon)$ follows from $P(\mathcal{D}(x) > \varepsilon) < \varepsilon$; hence,

$$\mathcal{KF}(P) \cdot H(\mathcal{KF}(P)) \le \int_{U^N} H(\mathcal{D}(x))P(\mathrm{d}x).$$

From (7.6.1), (7.6.2), and (7.6.5) it follows that $\Pi(\widetilde{P})H(\Pi(\widetilde{P})) \le A_D(\widetilde{P})$. We will now prove the right-hand-side inequality in (7.6.6). Given $\mathcal{KF}(P) < \delta$ and $a \in U$, we have

$$\begin{aligned}\int H(\mathcal{D}(x))P(\mathrm{d}x) &= \left(\int_{\mathcal{D}(x)\le\delta} + \int_{\mathcal{D}(x)>\delta}\right) H(\mathcal{D}(x))P(\mathrm{d}x)\\ &\le H(\delta) + \int_{\mathcal{D}(x)>\delta} H\left(A_m \sum_{i<j} d(x_i, x_j)\right) P(\mathrm{d}x)\\ &\le H(\delta) + \int_{\mathcal{D}(x)>\delta} H\left(A_m N^2 \max_{1\le i<\le N} d(x_i, a)\right) P(\mathrm{d}x)\end{aligned}$$

[by (2.4.3), $H(2^k t) \le K_H^k H(t)$]

$$\le H(\delta) + K_H^{\ell} \sum_{i=1}^{N} I_i,$$

where

$$\begin{aligned}I_i &:= \int_{\mathcal{D}(x)>\delta} H(d(x_i, a))P(\mathrm{d}x)\\ &= \left(\int_{\mathcal{D}(x)>\delta, d(x_i,a)>M} + \int_{\mathcal{D}(x)>\delta, d(x_i,a)\le M}\right) H(d(x_i, a))P(\mathrm{d}x)\\ &\le \int_{d(x_i,a)\ge M} H(d(x_i, a))P(\mathrm{d}x) + H(M)\delta.\end{aligned}$$

Hence,

$$\begin{aligned}\int H(\mathcal{D}(x))P(\mathrm{d}x) &\le H(\mathcal{KF}(P)) + c_2 N H(M)\mathcal{KF}(P)\\ &\quad + C_2 \sum_{i=1}^{N} \int_{d(x_i,a)>M} H(d(x_i, a))P_i(\mathrm{d}x)).\end{aligned}$$

Together with (7.6.1) and (7.6.5), the latter inequality yields the required estimate (7.6.6). □

The inequality (7.6.6) provides a "merging" criterion for a sequence of vectors $\widetilde{P}^{(n)} = (P_1^{(n)}, \ldots, P_N^{(n)})$.

As in Diaconis and Freedman (1984), D'Aristotile et al. (1988), and Dudley (2002, Sect. 11.7), we call two sequences $\{P^{(n)}\}_{n\geq 1}, \{Q^{(n)}\}_{n\geq 1} \in \mathcal{P}(U)$, *$\mu$-merging*, where μ is a simple probability metric if $\mu(P^{(n)}, Q^{(n)}) \to 0$ as $n \to \infty$. More generally, we say *the sequence $\{\widetilde{P}^{(n)}\}_{n\geq 1} \subset (\mathcal{P}(U))^N$ is μ-merging* if

$$\mu(P_i^{(n)}, P_j^{(n)}) \to 0 \text{ as } n \to \infty$$

for any $i, j = 1, \ldots, N$.

The next corollary gives criteria for merging it and the minimal distance ℓ_H (3.3.10) with respect to the Prokhorov metric.

Corollary 7.6.1. *Let $\{\widetilde{P}^{(n)}\}_{n\geq 1} \subset (\mathcal{P}(U))^N$. Then the following statements hold:*

(i) $\{\widetilde{P}^{(n)}\}_{n\geq 1}$ is π-merging if and only if

$$\Pi(\widetilde{P}^{(n)}) \to 0 \text{ as } n \to \infty. \tag{7.6.7}$$

(ii) If $H \in \mathcal{H}^*$ and $\int H(d(x,a))P_i(\mathrm{d}x) < \infty$, $i = 1, \ldots, N$, then $\{\widetilde{P}^{(n)}\}_{n\geq 1}$ is ℓ_H-merging if and only if

$$\mathbb{K}(\widetilde{P}^{(n)}) \to 0 \text{ as } n \to \infty.$$

Proof. (i) There exist constants C_1 and C_2 depending on the seminorm $\|\cdot\|$ such that

$$C_1 \sum_{1\leq i\leq j\leq N} \mathbf{K}(T_{ij}P) \leq \mathcal{KF}(P) \leq C_2 \sum_{1\leq i\leq j\leq N} \mathbf{K}(T_{ij}P),$$

where $\mathbf{K}$ is the Ky Fan distance in $\mathcal{P}_i(U)$ (Example 3.4.2). Now Theorem 7.6.1 can be used to yield the assertion.

(ii) The same argument is applied. Here we make use of the multidimensional Kantorovich theorem 7.6.1. □

Theorem 7.6.2 and Corollary 7.6.1 show that ℓ_H-merging implies π-merging. On the other hand, if

$$\lim_{M\to\infty} \max_{n\geq 1, 1\leq i\leq N} \int H(d(x,a))I\{d(x,a) > M\}P_i^{(n)}(\mathrm{d}x) = 0,$$

then ℓ_H-merging and π-merging of $\{\widetilde{P}^{(n)}\}_{n\geq 1}$ are equivalent.

Regarding the K-minimal metric $\overset{\wedge}{\boldsymbol{\tau}}_{\Sigma}$ [see (7.5.16) and (7.5.18)], we have the following criterion for the $\overset{\wedge}{\boldsymbol{\tau}}_{\Sigma}$-convergence.

Corollary 7.6.2. *Given* $X^{(k)} = (X_1^{(k)}, \dots, X_n^{(k)}) \in \mathfrak{X}(U^n)$ *such that* $Ed_j(X_j^{(k)}, a) < \infty$, $j = 1, \dots, n$, $k = 0, 1, \dots,$ *the convergence* $\overset{n}{\boldsymbol{\tau}}_{\Sigma}(X^{(k)}, X^{(0)}) \to 0$ *as* $k \to \infty$ *is equivalent to convergence in distributions,* $X_j^{(k)} \xrightarrow{w} X_j^{(0)}$, *and the moment convergence* $Ed_j(X_j^{(k)}, a) \to Ed_j(X_j^{(0)}, a)$ $\forall j = 1, \dots, n$.

Corollary 7.6.2 is a consequence of Theorems 7.5.2 and 6.4.1 [for $\lambda(x) = d(x, a), c(x, y) = d(x, y)$].

To conclude, we turn our attention to the inequalities between minimal distances $\widehat{\mathcal{L}}_H$, the Kantorovich distance ℓ_H [see (3.3.10), (3.3.15), and (5.3.17)], and the Prokhorov metric π [see (3.3.18)].

Corollary 7.6.3. *(i) For any* $H \in \mathcal{H}$, $M > 0$, $a \in U$, *and* $P_1, P_2 \in \mathcal{P}(U)$ *such that*

$$\int H(d(x, a))(P_1 + P_2)(\mathrm{d}x) < \infty \tag{7.6.8}$$

the following inequality holds:

$$\begin{aligned} H(\pi(P_1, P_2))\pi(P_1, P_2) &\leq \widehat{\mathcal{L}}_H(P_1, P_2) \\ &\leq H(\pi(P_1, P_2)) + K_H\Bigg[2\pi(P_1, P_2)H(M) \\ &\quad + \int_{d(x,a)>M} H(d(x, a))(P_1 + P_2)(\mathrm{d}x)\Bigg]. \end{aligned} \tag{7.6.9}$$

If $H \in \mathcal{H}$ *is a convex function, then one can replace* $\widehat{\mathcal{L}}_H$ *with* ℓ_H *in (7.6.9).*

(ii) Given a sequence $P_0, P_1, \dots \in \mathcal{P}(U)$ *with* $\int H(d(x, a))P_j(\mathrm{d}x) < \infty$ $(j = 0, 1, \dots)$, *the following assertions are equivalent as* $n \to \infty$:

(a) $\widehat{\mathcal{L}}(P_n, P_0) \to 0$,

(b) P_n *converges weakly to* P $(P_n \xrightarrow{w} P)$ *and* $\int H(d(x, a))(P_n - P)(\mathrm{d}x) \to 0$,

(c) $P_n \xrightarrow{w} P$ *and* $\lim_{N\to\infty} \limsup_n \int H(d(x, a))I\{d(x, a) > N\}P_n(\mathrm{d}x) = 0$.

This theorem is a particular case of more general theorems (see further Theorems 8.3.1 and 11.2.1).

References

Berge C, Chouila-Houri A (1965) Programming, games, and transportation networks. Wiley, New York

Breiman L (1992) Probability, society for industrial and applied mathematics. Society for Industrial and Applied Mathematics, Philadelphia

Cambanis S, Simons G, Stout W (1976) Inequalities for $Ek(X, Y)$ when the marginals are fixed. Z Wahrsch Verw Geb 36:285–294

D'Aristotile A, Diaconis P, Freedman D (1988) On a merging of probabilities. Technical Report, vol 301, Department of Statistics, Stanford University, Stanford, CA

Diaconis P, Freedman D (1984) A note on weak star uniformities. Technical report, vol 39, Department of Statistics, University of California, Berkeley, Berkeley, CA

Dobrushin RL (1970) Prescribing a system of random variables by conditional distributions. Theor Prob Appl 15:458–486

Dudley RM (1976) Probabilities and metrics: convergence of laws on metric spaces, with a view to statistical testing. Aarhus University Mathematics Institute lecture notes series no. 45, Aarhus, Denmark

Dudley RM (2002) Real analysis and probability, 2nd edn. Cambridge University Press, New York

Frank MJ, Nelsen RB, Schweizer B (1987) Best-possible bounds for the distribution of a sum: a problem of Kolmogorov. Prob Theor Relat Fields 74:199–211

Genest C, MacKay J (1986) The joy of copulas, bivariate distributions with uniform marginals. Am Statist 40:280–283

Kalashnikov VV, Rachev ST (1988) Mathematical methods for construction of stochastic queueing models. Nauka, Moscow (in Russian). [English transl., (1990) Wadsworth, Brooks–Cole, Pacific Grove, CA]

Lorentz GG (1953) An inequality for rearrangements. Am Math Mon 60:176–179

Makarov GD (1981) Estimates for the distribution function of a sum of two random variables when the marginal distributions are fixed. Theor Prob Appl 26:803–806

Rüschendorf L (1982) Random variables with maximum sums. Adv Appl Prob 14:623–632

Schay G (1979) Optimal joint distributions of several random variables with given marginals. Stud Appl Math LXI:179–183

Schweizer B, Sklar A (2005) Probabilistic metric spaces. Dover, New York

Sklar M (1959) Fonctions de réparation à dimensions et leurs marges. Institute of Statistics, University of Paris 8, Paris, pp 229–231

Strassen V (1965) The existence of probability measures with given marginals. Ann Math Stat 36:423–439

Tchen AH (1980) Inequalities for distributions with given marginals. Ann Prob 8:814–827

Wolff EF, Schweizer B (1981) On nonparametric measures of dependence for random variables. Ann Stat 9:879–885

Chapter 8
Relations Between Minimal and Maximal Distances

The goals of this chapter are to:

- Discuss dual representations of the maximal distances $\check{\mu}_c$ and $\overset{(s)}{\mu}_c$ and to compare them with the corresponding dual representations of the minimal metric $\widehat{\mu}$ and minimal norm $\overset{\circ}{\mu}_c$,
- Provide closed-form expressions for $\check{\mu}_c$ and $\overset{(s)}{\mu}_c$ in some special cases,
- Study the topological structure of minimal distances and minimal norms.

Notation introduced in this chapter:

Notation	Description
$\mathcal{F}(F_1, F_2)$	Set of bivariate distribution functions with fixed marginals F_1 and F_2
$F_-(F_1, F_2)$	Hoeffding–Fréchet lower bound in $\mathcal{F}(F_1, F_2)$
$F_+(F_1, F_2)$	Hoeffding–Fréchet upper bound in $\mathcal{F}(F_1, F_2)$
$\mathbf{D}$	Metric between pth moments

8.1 Introduction

The metric structure of the functionals $\widehat{\mu}_c$, $\overset{\circ}{\mu}$, $\check{\mu}_c$, and $\overset{(s)}{\mu}$ was discussed in Chap. 3 (see, in particular, Fig. 3.3). In Chap. 6, we found dual and explicit representations for the minimal distance $\widehat{\mu}_c$ and minimal norm $\overset{\circ}{\mu}_c$ choosing some special form of the function c. Here we will deal mainly with the following two questions:

1. What are the dual representations and explicit forms of $\check{\mu}_c$, $\overset{(s)}{\mu}$?
2. What are the necessary and sufficient conditions for $\lim_{n\to\infty} \widehat{\mu}_c(P_n, P) = 0$, resp. $\lim_{n\to\infty} \overset{\circ}{\mu}_c(P_n, P) = 0$?

S.T. Rachev et al., *The Methods of Distances in the Theory of Probability and Statistics*, DOI 10.1007/978-1-4614-4869-3_8, © Springer Science+Business Media, LLC 2013

We begin with duality theorems and explicit representations for $\check{\mu}_c$ and $\overset{(s)}{\mu}_c$ and then proceed with a discussion of the topological structure of $\widehat{\mu}_c$ and $\overset{\circ}{\mu}_c$.

8.2 Duality Theorems and Explicit Representations for $\check{\mu}_c$ and $\overset{(s)}{\mu}_c$

Let us begin by considering the dual form for the maximal distance $\check{\mu}_c$ and $\overset{(s)}{\mu}_c$, and let us compare them with the corresponding dual representations for the minimal metric $\widehat{\mu}$ and minimal norm $\overset{\circ}{\mu}_c$ (Definitions 3.3.2, 3.3.4, 3.4.4, and 3.4.5). Recall that

$$\overset{\circ}{\mu}_c(P_1, P_2) \leq \widehat{\mu}_c(P_1, P_2) \leq \check{\mu}_c(P_1, P_2) \leq \overset{(s)}{\mu}_c(P_1, P_2). \tag{8.2.1}$$

Subsequently, we will use the following notation:

$$\begin{aligned}
L_\alpha &= \{f : U \to \mathbb{R}^1; |f(x) - f(y)| \leq \alpha d(x, y), x, y \in U\},\\
\mathrm{Lip} &:= \bigcup_{\alpha>0} L_\alpha,\\
\mathrm{Lip}^b &:= \{f \in \mathrm{Lip} : \sup\{|f(x)| : x \in U\} < \infty\},\\
c(x, y) &:= H(d(x, y)), x, y \in U, H \in \mathcal{H} \text{ (Example 2.4.1)},\\
\mathcal{P}_H &:= \left\{P \in \mathcal{P}(U) : \int c(x, a) P(\mathrm{d}x) < \infty\right\},
\end{aligned}$$

$$\underline{\mathcal{G}}_H := \{(f, g) : f, g \in \mathrm{Lip}^b, f(x) + g(y) \leq c(x, y),\ x, y \in U\}, \tag{8.2.2}$$

$$\overline{\mathcal{G}}_H := \{(f, g) : f, g \in \mathrm{Lip}^b, f(x) \geq 0, g(y) \geq 0, f(x) + g(y) \geq c(x, y),\ x, y \in U\}, \tag{8.2.3}$$

$$h(x, y) := d(x, y) h_0(d(x, a) \vee d(y, a)) \quad x, y \in U, \quad \vee := \max, \tag{8.2.4}$$

where a is a fixed point of U and h_0 is a nonnegative, nondecreasing, continuous function on $[0, \infty)$

$$\begin{aligned}
\mathrm{Lip}_h &:= \{f : U \to \mathbb{R}^1 : |f(x) - f(y)| \leq h(x, y),\ x, y \in U\}\\
\mathcal{H}^* &:= \{\text{convex } H \in \mathcal{H}\}\\
\mathcal{F} &:= \{f \in \mathrm{Lip}^b : f(x) + f(y) \geq c(x, y),\ x, y \in U\}
\end{aligned} \tag{8.2.5}$$

and

$$\mathbb{T}(P_1, P_2; \mathcal{F}) := \inf\left\{\int f \mathrm{d}(P_1 + P_2) : f \in \mathcal{F}\right\}. \tag{8.2.6}$$

Theorem 8.2.1. *Let (U,d) be an s.m.s.*

(i) If $H \in \mathcal{H}^*$ and $P_1, P_2 \in \mathcal{P}_H$, then the minimal distance,

$$\widehat{\mu}_c(P_1,P_2) := \inf\{\mu_c(P) : P \in \mathcal{P}(U\times U), T_iP = P_i, i=1,2\}, \quad (8.2.7)$$

relative to the compound distance,

$$\mu_c(P) = \int_{U\times U} c(x,y)P(\mathrm{d}x,\mathrm{d}y), \quad (8.2.8)$$

admits the dual representation

$$\widehat{\mu}_c(P_1,P_2) = \sup\left\{\int f\mathrm{d}P_1 + \int g\mathrm{d}P_2 : (f,g) \in \mathcal{G}_H\right\}. \quad (8.2.9)$$

If P_1 and P_2 are tight measures, then the infimum in (8.2.7) is attained.

(ii) If $\int h(x,a)(P_1+P_2)(\mathrm{d}x) < \infty$, then the minimal norm

$$\overset{\circ}{\mu}_h(P_1,P_2) := \inf\{\mu_h(m) : m\text{-bounded nonnegative measures with fixed } T_1m - T_2m = P_1 - P_2\} \quad (8.2.10)$$

has a dual form

$$\overset{\circ}{\mu}_h(P_1,P_2) = \sup\left\{\left|\int f\mathrm{d}(P_1-P_2)\right| : f \in \mathrm{Lip}_h\right\}, \quad (8.2.11)$$

and the supremum in (8.2.11) is attained.

(iii) If $H \in \mathcal{H}^*$ and $P_1, P_2 \in \mathcal{P}_H$, then the maximal distance

$$\check{\mu}(P_1,P_2) := \sup\{\mu_c(P_1,P_2) : P \in \mathcal{P}(U\times U),\ T_iP = P_i,\ i=1,2\} \quad (8.2.12)$$

has the dual representation

$$\check{\mu}_c(P_1,P_2) = \inf\left\{\int f\mathrm{d}P_1 + \int g\mathrm{d}P_2 : (f,g) \in \overline{\mathcal{G}}_H\right\}. \quad (8.2.13)$$

If P_1 and P_2 are tight measures, then the supremum in (8.2.12) is attained.

(iv) If $H \in \mathcal{H}^*$ and $P_1, P_2 \in \mathcal{P}_H$, then

$$\overset{(s)}{\mu}_c(P_1,P_2) := \overset{(s)}{\mu}_c(P_1+P_2) := \sup\{\mu(P) : P \in \mathcal{P}(U\times U), T_1P + T_2P = P_1 + P_2\} \quad (8.2.14)$$

has the dual representation

$$\overset{(s)}{\mu}_c(P_1, P_2) = \mathbb{T}(P_1, P_2; \mathcal{F}). \tag{8.2.15}$$

If P_1 *and* P_2 *are tight measures, then the supremum in* (8.2.14) *is attained.*

Proof. (i) This is Corollary 5.3.2 in Chap. 5.

(ii) This is a special case of Theorems 5.4.2 and 5.4.3 with $c(x, y) = h(x, y)$ given by (8.2.9).

(iii) The proof here is quite similar to that of Corollary 5.3.2 and Theorem 5.3.1 and is thus omitted.

(iv) For any probability measures P_1 and P_2 on U, any $P \in \mathcal{P}(U \times U)$ with fixed sum of marginals, $T_1P + T_2P = P_1 + P_2$, and any $f \in \mathcal{F}$ [see (8.2.5)] we have

$$\begin{aligned}\int f\,\mathrm{d}(P_1 + P_2) &= \int f\,\mathrm{d}(T_1P + T_2P) \\ &= \int f(x) + f(y)P(\mathrm{d}x, \mathrm{d}y) \geq \int c(x, y)P(\mathrm{d}x, \mathrm{d}y),\end{aligned}$$

hence

$$\overset{(s)}{\mu}_c(P_1 + P_2) \leq \mathbb{T}(P_1, P_2; \mathcal{F}). \tag{8.2.16}$$

Our next step is to prove the inequality

$$\overset{(s)}{\mu}_c(P_1 + P_2) \geq \mathbb{T}(P_1, P_2, \mathcal{F}), \tag{8.2.17}$$

and here we will use the main idea of the proof of Theorem 5.3.1. To prove (8.2.17), we first treat the following case.

Case A. (U, d) is a bounded s.m.s. For any subset $U_1 \subset U$ define

$$\begin{aligned}&\overline{\mathcal{F}}(U_1) = \{f : U \to \mathbb{R}^1,\ f(x) + f(y) \geq c(x, t) \text{ for all } x, y \in U_1\}, \\ &\mathcal{F}(U_1) = \overline{\mathcal{F}}(U_1) \cap \mathrm{Lip}_\tau(U_1),\end{aligned}$$

where $\mathrm{Lip}_\tau(U_1) := \{f : U \to \mathbb{R}^1 : |f(x) - f(y)| \leq \tau(x, y) \text{ for all } x, y \in U_1\}$ and $\tau(x, y) := \sup\{|c(x, z) - c(y, z)| : z \in U\}$, $x, y \in U$. We need the following equality: if $P_1(U_1) = P_2(U_1) = 1$, then

$$\mathbb{T}(P_1, P_2; \overline{\mathcal{F}}(U_1)) = \mathbb{T}(P_1, P_2; \mathcal{F}(U)). \tag{8.2.18}$$

Let $f \in \overline{\mathcal{F}}(U_1)$. We extend f to a function on the whole U letting $f(x) = \infty$ for $x \notin U_1$, and hence

$$f(x) \geq f^*(x) := \sup\{c(x, y) - f(y) : y \in U\} \qquad \forall x \in U. \tag{8.2.19}$$

Since for any $x, y \in U$

$$f^*(x) - f^*(y) = \sup_{z \in U}\{c(x,z) - f(z)\} - \sup_{w \in U}\{c(y,w) - f(w)\}$$
$$\leq \sup_{z \in U}\{c(x,z) - c(y,z)\} \leq \tau(x,y),$$

then $f^* \in \mathcal{F}(U)$. Moreover, if $P_1(U_1) = P_2(U_1) = 1$, then by (8.2.19),

$$\mathbb{T}(P_1, P_2; \overline{\mathcal{F}}(U_1)) \geq \mathbb{T}(P_1, P_2; \mathcal{F}(U)) \tag{8.2.20}$$

which yields (8.2.18).

Case A1. Let (U, d) be a finite set, say, $U = \{u_1, \ldots, u_n\}$. By (8.2.18) and the duality theorem in the linear programming, we obtain

$$\overset{(s)}{\mu}_c(P_1 + P_2) = \mathbb{T}(P_1, P_2; \overline{\mathcal{F}}(U)) = \mathbb{T}(P_1, P_2; \mathcal{F}(U)), \tag{8.2.21}$$

as desired.

The remaining cases A2 [(U, d) is a compact space], A3 [(U, d) is a bounded s.m.s.], and B [(U, d) is an s.m.s.] are treated in a way quite similar to that in Theorem 5.3.1. □

In the special case $c = d$, one can get more refined duality representations for $\overset{(s)}{\mu}_c$. This is the following corollary.

Corollary 8.2.1. *If (U, d) is an s.m.s. and $P_1, P_2 \in \mathcal{P}(U)$, $\int d(x,a)(P_1 + P_2)(\mathrm{d}x) < \infty$, then*

$$\overset{(s)}{\mu}_d(P_1, P_2) = \inf\left\{\int f \mathrm{d}(P_1 + P_2) : f \in L_1, f(x) + f(y) \geq d(x,y)\ \forall x, y \in U\right\}. \tag{8.2.22}$$

Here the proof is identical to the proof of (iv) in Theorem 8.2.1 with some simplifications due to the fact that $c = d$.

Open Problem 8.2.1. Let us compare the dual forms of $\widehat{\mu}_d$, $\mathring{\mu}_d$, $\check{\mu}_d$, and $\overset{(s)}{\mu}_d$. The Kantorovich metric $\widehat{\mu}_d$ in the space $\mathcal{P}^1$ of all measures P with finite moment $\int d(x,a)P(\mathrm{d}x) < \infty$ has two dual representations:

$$\widehat{\mu}_d(P_1, P_2) = \sup\left\{\int f \mathrm{d}P_1 + \int g \mathrm{d}P_2 : f, g \in L, f(x) + g(y) \leq d(x,y),\right.$$
$$\left. x, y \in U\right\}$$
$$= \sup\left\{\int f \mathrm{d}(P_1 - P_2) : f \in L_1\right\} = \mathring{\mu}_d(P_1, P_2) \tag{8.2.23}$$

[see Sect. 6.2, (5.4.15), and (8.2.9)]. On the other hand, by (8.2.13) and (8.2.16), a dual form of $\check{\mu}_d$ is

$$\check{\mu}(P_1, P_2) = \inf\left\{ \int f \mathrm{d}P_1 + \int g\mathrm{d}P_2 : f, g \in L_1, f(x) + g(y) \geq d(x, y) \right.$$
$$\left. \forall x, y \in U \right\}, \tag{8.2.24}$$

which corresponds to the first expression for $\widehat{\mu}_d$ in (8.2.23), so an open problem is to check whether the equality

$$\check{\mu}_d(P_1, P_2) = \overset{(s)}{\mu}_d(P_1, P_2) \tag{8.2.25}$$

holds [here, $\overset{(s)}{\mu}_d$ is given by (8.2.22)]. In the special case $(U, d) = (\mathbb{R}, |\cdot|)$, equality (8.2.25) is true (see further Remark 8.2.1).

Next we will concern ourselves with the explicit representations for $\widehat{\mu}_c$, $\overset{\circ}{\mu}_c$, $\check{\mu}_d$, and $\overset{(s)}{\mu}_c$ in the case $U = \mathbb{R}$, $d(x, y) = |x - y|$.

Suppose $\phi : \mathbb{R}^2 \to \mathbb{R}$ is a quasiantitone upper-semicontinuous function (Sect. 7.4). Then $\widehat{\mu}_\phi$, $\check{\mu}_\phi$, and $\overset{(s)}{\mu}_\phi$ have the following representations.

Lemma 8.2.1. *Given P_1 and $P_2 \in \mathcal{P}(\mathbb{R})$ with finite moments $\int \phi(x, a)\mathrm{d}P_i(x) < \infty$, $i = 1, 2$, we have:*

(i) (Cambanis–Simons–Stout)

$$\widehat{\mu}_\phi(P_1, P_2) = \int_0^1 \phi(F_1^{-1}(t), F_2^{-1}(t))\mathrm{d}t, \tag{8.2.26}$$

where F_i is the DF of P_i and

$$\check{\mu}_\phi(P_1, P_2) = \int_0^1 \phi(F_1^{-1}(t), F_2^{-1}(1-t))\mathrm{d}t. \tag{8.2.27}$$

(ii) Assuming that $\phi(x, y)$ is symmetric,

$$\overset{(s)}{\mu}(P_1 + P_2) = \int_0^1 \phi(A(t), A(1-t))\mathrm{d}t, \tag{8.2.28}$$

where $A(t) = \frac{1}{2}(F_1(t) + F_2(t))$.

Proof. (i) Equality (8.2.26) follows from Theorem 7.4.2 (with $N = 2$). Analogously, one can prove (8.2.27). That is, let $\mathcal{F}(F_1, F_2)$ be the set of all DFs

F on $\mathbb{R}^2$ with marginals F_1 and F_2. By the well-known Hoeffding–Fréchet inequality, $\mathcal{F}(F_1, F_2)$ has a lower bound

$$F_-(x_1, x_2) := \max(0, F_1(x_1) + F_2(x_2) - 1), \quad F_- \in \mathcal{F}(F_1, F_2), \tag{8.2.29}$$

and an upper bound

$$F_+(x_1, x_2) = \min(F_1(x_1), F_2(x_2)), \quad F_+ \in \mathcal{F}(F_1, F_2). \tag{8.2.30}$$

Consider the space $\mathfrak{X}(\mathbb{R})$ of all random variables (RVs) on a nonatomic probability space (Remark 2.7.2). Then

$$\widehat{\mu}_\phi(P_1, P_2) = \inf\{E\phi(X_1, X_2) : X_i \in \mathfrak{X}(\mathbb{R}),\ F_{X_i} = F_i, i = 1, 2\}, \tag{8.2.31}$$

$$\check{\mu}_\phi(P_1, P_2) = \sup\{E\phi(X_1, X_2) : X_i \in \mathfrak{X}(\mathbb{R}),\ F_{X_i} = F_i, i = 1, 2\}. \tag{8.2.32}$$

If E is a (0,1)-uniformly distributed RV, then $F_-(x_1, x_2) = P(X_1^- \le x_1, X_2^- \le x_2)$, where $X_1^- := F_1^{-1}(E)$, $X_2^- := F_2^{-1}(1 - E)$ and $F_i^{-1}(u) := \inf\{t : F_i(t) \ge u\}$ is the generalized inverse function to F_i. Similarly, $F_+(x_1, x_2) = P(X_1^+ \le x_1, X_2^+ \le x_2)$, where $X_i^+ = F_i^{-1}(E)$, $i = 1, 2$. Thus

$$\check{\mu}_\phi(P_1, P_2) \ge E\phi(X_1^-, X_2^-) = \int_0^1 \phi(F_1^{-1}(t), F_2^{-1}(1 - t))\mathrm{d}t \tag{8.2.33}$$

and

$$\widehat{\mu}_\phi(P_1, P_2) \le E\phi(X_1^+, X_2^+) = \int_0^1 \phi(F_1^{-1}(t), F_2^{-1}(t))\mathrm{d}t. \tag{8.2.34}$$

In Theorem 7.4.2 of Chap. 7 (in the special case $N = 2$), we showed that (8.2.34) is true with an equality sign. Using the same method, one can check that $\check{\mu}_\phi(P_1, P_2) = E\phi(X_1^-, X_2^-)$.[1]

(ii) From the definition of $\overset{(s)}{\mu}(P_1, P_2)$ [see (8.2.14)] it follows that

$$\begin{aligned}\overset{(s)}{\mu}_\phi(P_1 + P_2) &= \overset{(s)}{\mu}_\phi(F_1 + F_2)\\ &:= \sup\{E\phi(X_1, X_2) : X_1, X_2 \in \mathfrak{X}(\mathbb{R}),\\ &\qquad F_{X_1} + F_{X_2} = F_1 + F_2 =: 2A\},\end{aligned}$$

or, in other words,

[1] See Kalashnikov and Rachev (1988, Theorem 7.1.1).

$$\overset{(s)}{\mu}_\phi(F_1+F_2) = \sup\left\{\int_{\mathbb{R}^2}\phi(x,y)\mathrm{d}F(x,y) : F \in \mathcal{F}(F_1,F_2), \frac{1}{2}(F_1+F_2) = A\right\}.$$

For any $F \in \mathcal{F}(F_1,F_2)$ denote $\widetilde{F}(x,y) = \frac{1}{2}[F(x,y)+F(y,x)]$. Then, by the symmetry of $\phi(x,y)$,

$$\overset{(s)}{\mu}_\phi(F_1,F_2) = \sup\left\{\int_{\mathbb{R}^2}\phi(x,y)\mathrm{d}\widetilde{F}(x,y) : \widetilde{F} \in \mathcal{F}(A,A)\right\}$$
$$= \check{\mu}_\phi(A,A) = \int_0^1 \phi(A^{-1}(t), A^{-1}(1-t))\mathrm{d}t. \qquad \square$$

Remark 8.2.1. It is easy to see that for any symmetric cost function c

$$\overset{(s)}{\mu}_\phi(P_1+P_2) = \check{\mu}_c\left(\frac{1}{2}(P_1+P_2), \frac{1}{2}(P_1+P_2)\right), \quad P_i \in \mathcal{P}(U). \tag{8.2.35}$$

On the other hand, in the case $U = \mathbb{R}$, $c(x,y) = |x-y|$, by Lemma 8.2.1,

$$\check{\mu}_c(P_1,P_2) = \int_0^1 |F_1^{-1}(t) - F_2^{-1}(1-t)|\mathrm{d}t$$
$$= \int_{-\infty}^{\infty} |x-a|\mathrm{d}(F_1(x)+F_2(x)), \tag{8.2.36}$$

where a is the point of *intersection* of the graphs of F_1 and $1-F_2$, i.e., $F_1(a-0) \le 1-F_2(a-0)$ but $F_1(a+0) \ge 1-F_2(a+0)$. Hence, by (8.2.1) and (8.2.37),

$$\overset{(s)}{\mu}_c(P_1,P_2) \ge \widehat{\mu}_c(P_1,P_2)$$
$$= \sup\{E|X_1-a| + E|X_2-a| : X_1, X_2 \in \mathfrak{X}(\mathbb{R}), F_{X_1}+F_{X_2} = F_1+F_2\}$$
$$\ge \overset{(s)}{\mu}_c(P_1,P_2),$$

i.e., $\overset{(s)}{\mu}_c = \widehat{\mu}_c$.

By virtue of Lemma 8.2.1 [with $\phi(x,y) = c(x,y) := H(|x-y|)$, H convex on $[0,\infty)$], we obtain the following explicit expressions for $\widehat{\mu}_c$, $\check{\mu}_c$, $\overset{(s)}{\mu}_c$, and $\overset{\circ}{\mu}_h$.

Theorem 8.2.2. *(i) Suppose $P_1, P_2 \in \mathcal{P}(\mathbb{R})$ have finite H-absolute moments, $\int H(|x|)(P_1+P_2)(\mathrm{d}x) < \infty$, where $H \in \mathcal{H}^*$. Then*

$$\widehat{\mu}_c(P_1,P_2) = \int_0^1 c(F_1^{-1}(t), F_2^{-1}(t))\mathrm{d}t, \tag{8.2.37}$$

$$\check{\mu}(P_1,P_2) = \int_0^1 c(F_1^{-1}(t), F_2^{-1}(1-t))\mathrm{d}t, \tag{8.2.38}$$

and

$$\overset{(s)}{\mu}_c(P_1, P_2) = \int_0^1 c(A^{-1}(t), A^{-1}(1-t))\mathrm{d}t, \tag{8.2.39}$$

where F_i is the DF of P_i, F_i^{-1} is the inverse of F_i, and $A = \frac{1}{2}(F_1 + F_2)$.

(ii) Suppose $h : \mathbb{R}^2 \to \mathbb{R}$ is given by (8.2.3)*, where $d(x, y) = |x - y|$ and $h(t) > 0$ for $t > 0$. Then*

$$\overset{\circ}{\mu}_h(P_1, P_2) = \int_{-\infty}^{\infty} h(|x - a|)|F_1(x) - F_2(x)|\mathrm{d}x. \tag{8.2.40}$$

8.3 Convergence of Measures with Respect to Minimal Distances and Minimal Norms

In this section, we investigate the topological structure of minimal distances ($\widehat{\mu}_c$) and minimal norms $\overset{\circ}{\mu}_h$ defined as in Sect. 8.2 in Chap. 8.

First, note that the definition of a simple distance ν (say, $\nu = \widehat{\mu}_c$ or $\nu = \overset{\circ}{\mu}_h$) does not exclude infinite values of ν. Hence, the space $\mathcal{P}_1 = \mathcal{P}(U)$ of all laws P on an s.m.s. (U, d) is divided into the classes $\mathcal{D}(\nu, P_0) := \{P \in \mathcal{P}_1 : \nu(P, P_0) < \infty\}$, $P_0 \in \mathcal{P}_1$ with respect to the equivalence relation $P_1 \sim P_2 \iff \nu(P_1, P_2) < \infty$. In Sects. 6.3, 6.4, and 7.6, the *topological* structure of the Kantorovich distance $\widehat{\mathcal{L}}_H = \widehat{\mu}_c$, where $c(x, y) = H(d(x, y))$, $H \in \mathcal{H}$ [Example 3.3.2 and (5.3.17)], was analyzed *only* in the set $\mathcal{D}(\widehat{\mu}_c, \delta_\alpha)$, $\alpha \in U$, where $\delta_\alpha(\{\alpha\}) = 1$. Here we will consider the $\widehat{\mu}_c$ convergence in the following sets: $\mathcal{D}(\widehat{\mu}_c, P_0)$, $\widetilde{\mathcal{D}}_c(P_0) := \{P \in \mathcal{P}_1 : \mu_c(P \times P_0) := \int_{U\times U} c(x, y)P(\mathrm{d}x)P_0(\mathrm{d}y) < \infty\}$ and $\mathcal{D}(\check{\mu}_c, P_0) := \{P \in \mathcal{P}_1 : \check{\mu}_c(P, P_0) \le \infty\}$, where $\check{\mu}$ is the maximal distance relative to μ_c [see (8.2.12)] and P_0 *is an arbitrary law in* $\mathcal{P}_1$. Obviously, $\mathcal{D}(\check{\mu}_c, P_0) \subset \widetilde{\mathcal{D}}_c(P_0) \subset \mathcal{D}(\widehat{\mu}_c, P_0)$ for any $P_0 \in \mathcal{P}_1$ and $\mathcal{D}(\check{\mu}_c, \delta_\alpha) \equiv \widetilde{\mathcal{D}}_c(\delta_\alpha) \equiv \mathcal{D}(\widehat{\mu}_c; \delta_\alpha)$, $\alpha \in U$.

Let $H_N(t) = H(t)I\{t > N\}$ for $H \in \mathcal{H}$, $t \ge 0$, $N > 0$, and define $c_N(x, y) := H_N(d(x, y))$. μ_{c_N}, $\widehat{\mu}_{c_N}$, $\check{\mu}_{c_N}$ by (8.2.8), (8.2.7), and (8.2.12), respectively. Therefore,

$$\mathcal{D}(\widehat{\mu}_c, P_0) = \left\{P \in \mathcal{P}_1 : \lim_{N\to\infty} \widehat{\mu}_{c_N}(P, P_0) = 0\right\}, \tag{8.3.1}$$

$$\widetilde{\mathcal{D}}_c(P_0) = \left\{P \in \mathcal{P}_1 : \lim_{N\to\infty} \widehat{\mu}_{c_N}(P \times P_0) = 0\right\}, \tag{8.3.2}$$

$$\mathcal{D}(\check{\mu}_c, P_0) \supset \widetilde{\mathcal{D}}(\check{\mu}_c, P_0) := \left\{P \in \mathcal{P}_1 : \lim_{N\to\infty} \check{\mu}_{c_N}(P, P_0) = 0\right\}. \tag{8.3.3}$$

As usual, we denote the weak convergence of laws $\{P_n\}_{n=1}^{\infty}$ to the law P by $P_n \xrightarrow{\mathrm{w}} P$.

Theorem 8.3.1. *Let* (U, d) *be a u.m.s.m.s. (Sect. 2.6),* $H \in \mathcal{H}$ *[*$H(t) > 0$ *for* $t > 0$*], and* P_0 *be a law in* $\mathcal{P}_1$*.*

(i) If $\{P_1, P_2, \dots\} \subset \mathcal{D}(\widehat{\mu}_c, P_0)$ *and* $Q \in \widetilde{\mathcal{D}}(\check{\mu}_c, P_0)$*, then*

$$\lim_{N \to \infty} \widehat{\mu}_c(P_n, Q) = 0 \tag{8.3.4}$$

if and only if the following two conditions are satisfied:

(1)* $P_n \xrightarrow{\text{w}} Q$*;*
(2)* $\lim_{N \to \infty} \sup_n \widehat{\mu}_{c_N}(P_n, P_0) = 0.$

(ii) If $\{Q, P_1, P_2, \dots\} \subset \widetilde{\mathcal{D}}_c(P_0)$*, then* (8.3.4) *holds if and only if the conditions (1*) and*

(3)* $\lim_{N \to \infty} \sup_n \mu_c(P_n \times P_0) = 0$

are fulfilled.

(iii) If $\{P_1, P_2, \dots\} \subset \widetilde{\mathcal{D}}(\check{\mu}_c, P_0)$ *and* $Q \in \mathcal{D}(\widehat{\mu}_c, P_0)$*, then* (8.3.4) *holds if and only if the conditions (1*) and*

(4)* $\lim_{N \to \infty} \sup_n \check{\mu}_c(P_n, P_0) = 0$

are fulfilled.

Theorem 8.3.1 is an immediate corollary of the following lemma. Further, we use the same notation as in (8.2.1)–(8.2.2).

Lemma 8.3.1. *Let* U *be a u.m.s.m.s.,* π *the Prokhorov metric in* $\mathcal{P}$*, and* $H \in \mathcal{H}$*. For any* $P_0, P_1, P_2 \in \mathcal{P}_1$ *and* $N > 0$ *the following inequalities are satisfied:*

$$\check{\mu}_c(P_1, P_2) \le H(\pi(P_1, P_2)) + K_H\{2\pi(P_1, P_2)H(N) + \widehat{\mu}_{c_N}(P_1, P_0) + \check{\mu}_{c_N}(P_2, P_0)\}, \tag{8.3.5}$$

$$\widehat{\mu}_c(P_1, P_2) \le H(\pi(P_1, P_2)) + K_H\{2\pi(P_1, P_2)H(N) + \mu_{c_N}(P_1 \times P_0) + \mu_{c_N}(P_2 \times P_0)\}, \tag{8.3.6}$$

$$\pi(P_1, P_2)H(\pi(P_1, P_2)) \le \widehat{\mu}(P_1, P_2), \tag{8.3.7}$$

$$\widehat{\mu}_{c_N}(P_1, P_0) \le K(\widehat{\mu}_c(P_1, P_2) + \widehat{\mu}_{c_{N/2}}(P_2, P_0)), \tag{8.3.8}$$

$$\mu_{c_N}(P_1 \times P_0) \le K(\widehat{\mu}_c(P_1, P_2) + \mu_{c_{N/2}}(P_2 \times P_0)), \tag{8.3.9}$$

$$\check{\mu}_{c_N}(P_1, P_0) \le K(\widehat{\mu}_c(P_1, P_2) + \check{\mu}_{c_{N/2}}(P_2, P_0)), \tag{8.3.10}$$

where K_H *is given by* (2.4.3) *and* $K = K_H + K_H^2$*.*

Remark 8.3.1. Relationships (8.3.5)–(8.3.10) give us necessary and sufficient conditions for $\widehat{\mu}_c$-convergence as well as quantitative representations of these conditions. Clearly, such treatment of the $\widehat{\mu}_c$-convergence is preferable because it gives not only a qualitative answer when $\widehat{\mu}_c(P_n, Q) \to 0$ but also establishes a quantitative estimate of the convergence $\widehat{\mu}_c(P_n, Q) \to 0$.

Proof of Lemma 8.3.1. To get (8.3.5), we require the following relation between the H-average compound $\mu_c = \mathcal{L}_H$ and the Ky Fan metric $\mathbf{K}$ (Examples 3.4.1 and 3.4.2):

$$\mathcal{L}_H(P) \leq H(\mathbf{K}(P)) \\ + K_H\{2\mathbf{K}(P)H(N) + \mathcal{L}_{H_N}(P') + \mathcal{L}_{H_N}(P'')\}\ \mathcal{L}_{H_N} := \mu_{c_N} \tag{8.3.11}$$

for $N > 0$ and any triplet of laws $(P, P', P'') \in \mathcal{P}_2$ such that there exists a law $Q \in \mathcal{P}_3$ with marginals

$$T_{12}Q = P \quad T_{13}Q = P' \quad T_{23}Q = P''. \tag{8.3.12}$$

If $\mathbf{K}(P) > \delta$, then, by (2.4.3),

$$\int H(d(x,y))P(\mathrm{d}x,\mathrm{d}y) \\ \leq K_H \int [H(d(x,x_0)) + H(d(y,x_0))]I\{d(x,y) > \delta\}Q(\mathrm{d}x,\mathrm{d}y,\mathrm{d}x_0) + H(\delta) \\ \leq H(\delta) + K_H\{2H(N)\delta + \mathcal{L}_{H_N}(P') + \mathcal{L}_{H_N}(P'')\}.$$

Letting $\delta \to \mathbf{K}(P)$ completes the proof of (8.3.11).

For any $\varepsilon > 0$ we choose $P \in \mathcal{P}_2$ with marginals P_1, P_2, and $P' \in \mathcal{P}_2$ with marginals P_1 and P_0 such that

$$\widehat{\mathbf{K}}(P_1, P_2) > \mathbf{K}(P) - \varepsilon, \qquad \widehat{\mathcal{L}}_{H_N}(P_1, P_0) > \mathcal{L}_{H_N}(P') - \varepsilon. \tag{8.3.13}$$

Choosing Q with property (8.3.12) [see (3.3.5)] we obtain

$$\widehat{\mathcal{L}}_H(P_1, P_2) \leq \mathcal{L}(P) \\ \leq H(\widehat{\mathbf{K}}(P_1, P_2) + \varepsilon) + \mathbb{K}_H\{2(\widehat{\mathbf{K}}(P_1, P_2) + \varepsilon)H(N) \\ + \widehat{\mathcal{L}}_{H_N}(P_1, P_0) + \varepsilon + \mathcal{L}_{H_N}(T_{23}Q)\}$$

by (8.3.11) and (8.3.13). The last inequality, together with the Strassen theorem (see Corollary 7.5.2), proves (8.3.5).

If $P_1 \times P_0$ and $P_2 \times P_0$ stand for P' and P'', respectively, then (8.3.11) implies (8.3.6). To prove that (8.3.7)–(8.3.10) hold, we use the following two inequalities: for any $P \in \mathcal{P}_2$ with marginals P_1 and P_2

$$\mathbf{K}(P)H(\mathbf{K}(P)) \leq \mathcal{L}_H(P) \tag{8.3.14}$$

and

$$\mathcal{L}_{H_N}(P') \le K[\mathcal{L}_H(P) + \mathcal{L}_{H_{N/2}}(P'')], \tag{8.3.15}$$

where (P, P', P'') are subject to conditions (8.3.12) and $N > 0$. Using the same arguments as in the proof of (8.3.5), we get (8.3.7)–(8.3.10) by means of (8.3.14) and (8.3.15). □

Given a u.m.s.m.s. (U, d) and an s.m.s. (V, g), let $\phi : U \to V$ be a measurable function. For any probability distance μ on $\mathcal{P}(V^2)$ define the probability distance μ_ϕ on $\mathcal{P}(U^2)$ by (7.2.12). Theorem 7.2.4 states that (Remark 7.2.2)

$$\widehat{\mu}_\phi(P_1, P_2) = \widehat{\mu}(P_{1,\phi}, P_{2,\phi}), \tag{8.3.16}$$

or, in terms of U-valued RVs,

$$\widehat{\mu}_\phi(X_1, X_2) = \widehat{\mu}(\phi(X_1), \phi(X_2)), \quad X_1, X_2 \in \mathfrak{X}(U). \tag{8.3.17}$$

Next we will generalize Theorem 8.3.1, considering criteria for $\widehat{\mu}_{c,\phi}$-convergence. We start with the special but important case of $\mu_c = \mathcal{L}_p^p$ $(p \ge 1)$. Define the $\mathcal{L}_p$-metric in $\mathcal{P}(V^2)$

$$\mathcal{L}_p(Q) := \left(\int_{V\times V} g^p(x, y)(\mathrm{d}x, \mathrm{d}y)\right)^{1/p}, \qquad p \ge 1 \quad Q \in \mathcal{P}(V^2).$$

Then, by (7.2.12), $\mathcal{L}_{p,\phi}$ is a probability metric in $\mathcal{P}(U^2)$ and $\widehat{\mathcal{L}}_{p,\phi}$ is the corresponding minimal metric. In the next corollary, we apply Theorems 7.2.4 and 8.2.1 to get a criterion for $\widehat{\mathcal{L}}_{p,\phi}$-convergence.

Let $Q, P_1, P_2, \ldots$ be probability measures on $\mathcal{P}(U)$. Denote $\pi_{n,\phi} = \pi(P_{n,\phi}, Q_\phi)$, π being the Prokhorov metric in $\mathcal{P}(V)$

$$D_{n,\phi} := \mathbf{D}(P_{n,\phi}, Q_\phi) := \left|\left(\int_V g^p(x, a) P_{n,\phi}(\mathrm{d}x)\right)^{1/p} - \left(\int_V g^p(x, a) Q_\phi(\mathrm{d}x)\right)^{1/p}\right|$$

(a is a fixed point in V),

$$\mathcal{A}(Q_\phi) = \left(p\int_V (g(x, a) + 1)^{p-1} Q_\phi(\mathrm{d}x)\right)^{1/p},$$

$$M(Q_\phi, N) := \left(\int_V g^p(x, a) I\{g(x, a) > N\} Q_\phi(\mathrm{d}x)\right)^{1/p},$$

$$M(Q_\phi) := \left(\int_V g^p(x, a) Q_\phi(\mathrm{d}x)\right)^{1/p}.$$

Corollary 8.3.1. *For all* $n = 1, 2, \ldots$ *let*

$$M(P_{n,\phi}) + M(Q_\phi) < \infty. \tag{8.3.18}$$

Then $\widehat{\mathcal{L}}_{p,\phi}(P_n, Q) \to 0$ *as* $n \to \infty$ *if and only if* $P_{n,\phi}$ *weakly tends to* Q_ϕ *and* $D_{n,\phi} \to 0$ *as* $n \to \infty$*. Moreover, the quantitative estimates*

$$\widehat{\mathcal{L}}_{p,\phi}(P_n, Q) \geq \max(D_{n,\phi}, (\pi_{n,\phi})^{1+1/p}), \tag{8.3.19}$$

$$\widehat{\mathcal{L}}_{p,\phi}(P_n, Q) \leq (1+2N)\pi_{n,\phi} + 5M(Q_\phi, N) + (\pi_{n,\phi})^{1/p}(3\mathcal{A}(Q_\phi) + 2^{2+1/p}N) + D_{n,\phi} \tag{8.3.20}$$

are valid for each positive N.

Proof. The first part of Corollary 8.3.1 follows immediately from (8.3.19), (8.3.20) [for the "if" part set, for instance, $N = (\pi_{n,\phi})^{-1/2p}$]. Relations (8.3.19) and (8.3.20) establish additionally a quantitative estimate of the convergence of $\widehat{\mathcal{L}}_{p,\phi}(P_n, Q)$ to zero. To prove the latter relations, we use (8.3.16) and the following inequalities:

$$\widehat{\mathcal{L}}_p(Q_1, Q_2) \geq \max(\boldsymbol{\pi}(Q_1, Q_2)^{1+1/p}, \mathbf{D}(Q_1, Q_2)), \tag{8.3.21}$$

$$\widehat{\mathcal{L}}_p(Q_1, Q_2) \leq (1+2N)\boldsymbol{\pi}(Q_1, Q_2) + M(Q_1, N) + M(Q_2, N), \tag{8.3.22}$$

and

$$M(Q_1, 2N) \leq \mathbf{D}(Q_1, Q_2) + 4M(Q_2, N) + \boldsymbol{\pi}(Q_1, Q_2)^{1/p}(3\mathcal{A}(Q_2) + 2^{2+1/p}N) \tag{8.3.23}$$

for each positive N and $Q_1, Q_2 \in \mathcal{P}(V)$, where $\mathbf{D}$ is the primary metric given by

$$\mathbf{D}(Q_1, Q_2) = \left| \left(\int_V g^p(x,a)Q_1(\mathrm{d}x)\right)^{1/p} - \left(\int_V g^p(x,a)Q_2(\mathrm{d}x)\right)^{1/p} \right|. \tag{8.3.24}$$

Claim 1. Equation (8.3.21) holds.
For any V-valued RVs X_1 and X_2 with distributions Q_1 and Q_2, respectively,

$$\mathcal{L}_p(X_1, X_2) = [Eg^p(X_1, X_2)]^{1/p} > \mathbf{D}(Q_1, Q_2)$$

by the Minkowski inequality. Thus $\widehat{\mathcal{L}}_p(Q_1, Q_2) \geq \mathbf{D}(Q_1, Q_2)$. Using (8.3.7) with $H(t) = t^p$, we have also that $\widehat{\mathcal{L}}_p \geq \boldsymbol{\pi}^{1+1/p}$.

Claim 2. Equation (8.3.22) holds.
We start with Chebyshev's inequality: for any X_i with laws Q_i

$$\mathcal{L}_p(X_1, X_2) \leq (1+2N)\mathbf{K}(X_1, X_2) + M(Q_1, N) + M(Q_2, N),$$

where $\mathbf{K}$ is the Ky Fan metric in $\mathfrak{X}(V)$ and $M(Q_i) = \left(\int_V g^p(x,a)Q_i(\mathrm{d}x)\right)^{1/p}$. The proof is analogous to that of (8.3.11). By virtue of the Strassen theorem, it follows that $\widehat{\mathbf{K}} = \boldsymbol{\pi}$, and the preceding inequality yields (8.3.22).

Claim 3. Equation (8.3.23) holds.
Observe that

$$\begin{aligned}M(Q_1,2N) &:= \left(\int_V g^p(x,a)I\{g(x,a)>2N\}Q(\mathrm{d}x)\right)^{1/p}\\ &\le \mathbf{D}(Q_1,Q_2)+\left|\int_V g^p(x,a)I\{g(x,a)\le 2N\}(Q_1-Q_2)(\mathrm{d}x)\right|^{1/p}\\ &\quad +M(Q_2,2N).\end{aligned}$$

Denote $f(x) := \min\{g^p(x,a),(2N)^p\}$, $h(x) := \min\{2^p g^p(x,O(a,N)),(2N)^p\}$, where $O(a,N) := \{x\in V : g(x,a)\le N\}$. Then

$$\begin{aligned}I &:= \left|\int_V g^p(x,a)I\{g(x,a)\le 2N\}(Q_1-Q_2)(\mathrm{d}x)\right|^{1/p}\\ &\le \left|\int_V f(x)(Q_1-Q_2)(\mathrm{d}x)\right|^{1/p}+2N\left|\int_V I\{g(x,a)>2N\}(Q_1-Q_2)(\mathrm{d}x)\right|^{1/p}\\ &=: I_1+I_2.\end{aligned}$$

Using the inequality

$$\begin{aligned}|f(x)-f(y)| &\le |g^p(x,a)-g^p(y,a)|\\ &\le p\max(g^{p-1}(x,a),g^{p-1}(y,a))|g(x,a)-g(y,a)|\\ &\le p\max(g^{p-1}(x,a),g^{p-1}(y,a))g(x,y)\quad x,y\in V\end{aligned}$$

we get for any pair (X_1,X_2) of V-valued RVs with marginal distributions Q_1 and Q_2

$$\begin{aligned}I_1^p &:= |E(f(X_1)-f(X_2))|\\ &\le E|f(X_1)-f(X_2)|I\{g(X_1,X_2)\le\gamma\}\\ &\quad +E[|f(X_1)|+|f(X_2)|]I\{g(X_1,X_2)\ge\gamma\}\\ &\le \gamma pE(g(X_2,a)+\gamma)^{p-1}+2(2N)^p\Pr(g(X_1,X_2)\ge\gamma\}\text{ for any }\gamma\in[0,1].\end{aligned}$$

Let $K=\mathbf{K}(X_1,X_2)$ be the Ky Fan metric in $\mathfrak{X}(V)$. Then from the preceding bound

$$I_1\le K^{1/p}[\mathcal{A}(Q_2)^p+2(2N)^p]^{1/p}\le K^{1/p}[\mathcal{A}(Q_2)+2^{1+1/p}N].$$

Now let us estimate the second term in the upper bound for I:

$$\begin{aligned}I_2 &:= \left|\int_V (2N)^p I\{g(x,a)>2N\}(Q_1-Q_2)(\mathrm{d}x)\right|^{1/p}\\ &\le \left(\int_V (2N)^p I\{g(x,a)>2N\}Q_1(\mathrm{d}x)\right)^{1/p}+M(Q_2,2N).\end{aligned}$$

If $g(x,c) > 2N$, then $g(x, O(c,N)) \geq N$, and therefore

$$\left[\int_V (2N)^p I\{g(x,a) > 2N\} Q_1(\mathrm{d}x)\right]^{1/p} \leq [Eh(X_1)]^{1/p}$$
$$\leq |Eh(X_1) - Eh(X_2)|^{1/p}$$
$$+[Eh(X_2)]^{1/p} =: I_1' + I_2'.$$

The inequality

$$|h(x) - h(y)| \leq 2^p |g^p(x, O(a,N)) - g^p(y, O(a,N))|$$
$$\leq 2^p p \max[g^{p-1}(x, O(a,N)), g^{p-1}(y, O(a,N))] g(x,y)$$

implies

$$I_1' \leq [E|h(X_1) - h(X_2)| I\{g(X_1, X_2) \leq \gamma\}]^{1/p}$$
$$+ [E(h(X_1) + h(X_2)) I\{g(X_1, X_2) > \gamma\}]^{1/p}$$
$$\leq 2\{\gamma E p [g(X_2, O(a,N)) + 1]^{p-1}\}^{1/p}$$
$$+ 2(2N)^p \Pr(g(X_1, X_2) > \gamma)^{1/p} \quad \text{for } K < \gamma.$$

On the other hand, by the definition of h,

$$I_2' := [Eh(X_2)]^{1/p} \leq [E(2N)^p I\{g(X_2, a) > N\}]^{1/p} \leq 2M(Q_2, N).$$

Combining the foregoing estimates we get

$$I_2 \leq 3M(Q_2, N) + 2K^{1/p} \mathcal{A}(Q_2) + 2^{1+1/p} N M^{1/p}.$$

Making use of the estimates for I_1 and I_2 and the Strassen theorem we get

$$I \leq I_1 + I_2 \leq 3M(Q_2, N) + \pi(Q_1, Q_2)^{1/p} (3\mathcal{A}(Q_2) + 2^{2+1/p} N).$$

This completes the proof of (8.3.23). □

We can extend Corollary 8.3.1 considering the H-average compound distance

$$\mu_c(Q) := \mathcal{L}_H(Q) = \int_{V^2} c(x,y) Q(\mathrm{d}x, \mathrm{d}y) \qquad Q \in \mathcal{P}(V^2), \tag{8.3.25}$$

where $c(x,y) = H(g(x,y))$ and $H(t)$ is a nondecreasing continuous function on $[0,\infty)$ vanishing at zero (and only there) and satisfying the Orlicz condition

$$K_H := \sup\{H(2t)/H(t); t > 0\} < \infty \tag{8.3.26}$$

[see (3.4.1) and Example 2.4.1].

Corollary 8.3.2. *Assume that* $\int_V c(x,a)(P_{n,\phi} + Q_\phi)(\mathrm{d}x) < \infty$. *Then the convergence* $\widehat{\mu}_{c,\phi}(P_n, Q) \to 0$ *as* $n \to \infty$ *is equivalent to the following relations:* $P_{n,\phi}$ *tends weakly to* Q_ϕ *as* $n \to \infty$, *and for some* $a \in U$

$$\lim_{N\to\infty} \overline{\lim_{n}} \int_V c(x,a) I\{g(x,a) > N\} P_{n,\phi}(\mathrm{d}x) = 0.$$

Proof. See Theorems 8.3.1 and 7.2.4. □

Note that the Orlicz condition (8.3.26) implies a power growth of the function H. To extend the $\widehat{\mu}_{c,\phi}$-convergence criterion in Corollary 8.3.2, we consider the functions H in (8.3.25) with exponential growth. Let RB represent the class of all bounded from above real-valued RVs. Then

$$\xi \in RB \iff \tag{8.3.27}$$

$$\tau(\xi) := \inf\{a > 0 : E \exp \lambda\xi \le \exp \lambda a \ \forall \lambda > 0\} = \sup_{\lambda>0} \frac{1}{\lambda} \ln E \exp(\lambda\xi) < \infty.$$

In fact, clearly, if $\xi \in RB$, then $\tau(\xi) < \infty$. On the other hand, if $F_\xi(x) < 1$ for $x \in \mathbb{R}$, then for any $a > 0$, $E \exp[\lambda(\xi - a)] \ge \exp(\lambda a) \Pr(\xi > 2a) \to \infty$ as $\lambda \to \infty$. By the Holder inequality one gets

$$\tau(\xi + \eta) \le \tau(\xi) + \tau(\eta), \tag{8.3.28}$$

and hence, if $Q \in \mathcal{P}(V^2)$ and (Y_1, Y_2) is a pair of V-valued RVs with joint distribution Q, then

$$\boldsymbol{\tau}(Q) := \tau(g(Y_1, Y_2)) \tag{8.3.29}$$

determines a compound metric on $\mathcal{P}(V^2)$.[2] The next theorem gives us a criterion for $\widehat{\boldsymbol{\tau}}_\phi$-convergence, where $\widehat{\boldsymbol{\tau}}_\phi$ is defined by (8.3.16) and (8.3.17).

Theorem 8.3.2. *Let* X_n, $n = 1, 2, \ldots$, *and* Y *be* U*-valued RVs with distributions* P_n *and* Q, *respectively, and let* $\boldsymbol{\tau}(g(\phi(X_n), a)) + \boldsymbol{\tau}(g(\phi(Y), a)) < \infty$. *Then the convergence* $\widehat{\tau}_\phi(P_n, Q) \to 0$ *as* $n \to \infty$ *is equivalent to the following relations:*

(a) $P_{n,\phi}$ *tends weakly to* Q_ϕ,
(b) $\lim_{N\to\infty} \overline{\lim}_n \boldsymbol{\tau}(g(\phi(X_n), a) I\{g(\phi(X_n), a) > N\}) = 0$.

Proof. As in Corollary 8.3.1, the assertion of the theorem is a consequence of (8.3.16) and the following three claims. Let V-valued RVs Y_1 and Y_2 have distributions Q_1 and Q_2, respectively.

Claim 1.

$$\boldsymbol{\pi}^2(Q_1, Q_2) \le \widehat{\boldsymbol{\tau}}(Q_1, Q_2). \tag{8.3.30}$$

---

[2] See Sect. 2.5 of Chap. 2.

By the Strassen theorem $\widehat{\mathbf{K}} = \pi$ (see Corollary 7.5.2 in Chap. 7), it is enough to prove that $\tau(g(Y_1, Y_2)) \geq \mathbf{K}^2(Y_1, Y_2)$. Let $\xi = g(Y_1, Y_2)$ and $\tau(\xi) < \varepsilon^2 \leq 1$. Then

$$\Pr(\xi > \varepsilon) \leq \frac{E e^{\xi} - 1}{e^{\varepsilon} - 1} \leq \frac{e^{\tau(\xi)} - 1}{e^{\varepsilon} - 1} \leq \frac{e^{\varepsilon^2} - 1}{e^{\varepsilon} - 1} \leq \varepsilon.$$

Letting $\varepsilon^2 \to \tau(\xi)$ we obtain (8.3.30).

Claim 2.

$$\boldsymbol{\tau}(g(Y_1,a)I\{g(Y_1,a) > N\}) \leq 2\widehat{\boldsymbol{\tau}}(Q_1, Q_2) + 2\boldsymbol{\tau}(g(Y_2,c)I\{g(Y_2,a) > N/2\}). \tag{8.3.31}$$

Note that the inequality $\xi \leq \eta$ with probability 1 implies $\tau(\xi) \leq \tau(\eta)$. Hence

$$\begin{aligned}
&\tau(g(Y_1,a)I\{g(Y_1,a) > N\}) \\
&\quad \leq \tau[(g(Y_1,Y_2) + g(Y_2,a))I\{g(Y_2,a) + g(Y_1,Y_2) > N\}] \\
&\quad \leq \tau[(g(Y_1,Y_2) + g(Y_2,a))\max(I\{g(Y_2,a) > N/2\}, I\{g(Y_1,Y_2) > N/2\})] \\
&\quad \leq \tau(g(Y_1,Y_2)I\{g(Y_1,Y_2) > N/2\}) + \tau(g(Y_1,Y_2)I\{g(Y_1,Y_2) \leq N/2\} \\
&\qquad \times I\{g(Y_2,a) > N/2\}) + \tau(g(Y_2,a)\{g(Y_2,a) > N/2\}) + \tau(g(Y_2,a) \\
&\qquad \times I\{g(Y_2,a) > N/2\}I\{g(Y_1,Y_2) > N/2\}) \\
&\quad \leq 2\tau(g(Y_1,Y_2)I\{g(Y_1,Y_2) > N/2\}) + 2\tau(g(Y_2,a)I\{g(Y_2,a) \geq N/2\}) \\
&\quad \leq 2\tau(g(Y_1,Y_2)) + 2\tau(g(Y_2,a)I\{g(Y_2,a) > N/2\}).
\end{aligned}$$

Passing to the minimal metric $\widehat{\boldsymbol{\tau}}$ we get (8.3.31).

Claim 3.

$$\begin{aligned}
\widehat{\boldsymbol{\tau}}(Q_1, Q_2) &\leq \boldsymbol{\pi}(Q_1, Q_2)(1 + 2N) + \tau(g(Y_1,a)I\{g(Y_1,a) > N\}) \\
&\quad + \tau(g(Y_2,a)I\{g(Y_2,a) > N\}), \quad \forall N > 0,\ a \in V.
\end{aligned} \tag{8.3.32}$$

For each δ the following relation holds: $\tau(g(Y_1,Y_2)) \leq \tau(g(Y_1,Y_2)I\{g(Y_1,Y_2) \leq \delta\}) + \tau(g(Y_1,Y_2)I\{g(Y_1,Y_2) > \delta\}) =: I_1 + I_2$. For I_1 we obtain the estimate

$$I_1 = \sup_{\lambda>0} 1/\lambda \ln E\exp(\lambda g(Y_1,Y_2)I\{g(Y_1,Y_2) \leq \delta\}) \leq \sup_{\lambda>0} 1/\lambda E\exp\lambda\delta = \delta.$$

For I_2 we have

$$\begin{aligned}
I_2 &\leq \tau(g(Y_1,a) + g(Y_2,a))I\{g(Y_1,Y_2) > \delta\}) \\
&\leq \tau(g(Y_1,a)I\{g(Y_1,Y_2) \geq \delta\}) \\
&\quad + \tau(g(Y_2,a)I\{g(Y_1,Y_2) \geq \delta\}) =: A_1 + A_2.
\end{aligned}$$

Furthermore,

$$\begin{aligned}A_1 &\le \tau(g(Y_1,a)I\{g(Y_1,Y_2)>\delta\}I\{g(Y_1,a)\le N\})\\ &\quad+\tau(g(Y_1,a)I\{g(Y_1,Y_2)>\delta\}I\{g(Y_1,a)>N\})\\ &\le \tau(NI\{g(Y_1,Y_2)>\delta\})+\tau(g(Y_1,a)I\{g(Y_1,a)\ge N\}).\end{aligned}$$

Hence, if $\mathbf{K}(Y_1,Y_2)<\delta$, then

$$\begin{aligned}\tau(g(Y_1,Y_2)) &\le (1+2N)\delta+\tau(g(Y_1,c)I\{g(Y_1,c)>N\})\\ &\quad+\tau(g(Y_2,a)I\{g(Y_2,a)>N\}).\end{aligned}$$

Letting $\delta\to\mathbf{K}(Y_1,Y_2)$ and passing to the minimal metrics, we obtain (8.3.32). □

In the rest of this section, we look at the topological structure of the minimal norms $\mathring{\mu}_h(P_1,P_2)$, $P_1,P_2\in\mathcal{P}_1$ [see (8.2.10)], where the function $h(x,y)=d(x,y)\,h_o(d(x,a)\vee d(y,a))$, $x,y\in U$, is defined as in (8.2.3).

Theorem 8.3.3. *Let (U,d) be an s.m.s.*

(a) If $g:=d/(1+d)$ and $a_n:=\sup_{t>0}h_0(2t)/h_0(t)<\infty$, then

$$\begin{aligned}\mathring{\mu}_h(P_1,P_2) &\le (1+N)\mathring{\mu}_g(P_1,P_2)\\ &\quad+(2a_h+4)\int h(x,a)I\{d(x,a)>N\}(P_1+P_2)(\mathrm{d}x)\ \textit{for}\ N\ge 1.\end{aligned}$$

(b) If $b_h=\sup_{0<s<1}[(1+t-s)/h_0(t)]^{-1}<\infty$, then $\mathring{\mu}_g(P_1,P_2)\le\mathring{\mu}_h(P_1,P_2)$.
(c) If

$$c_h:=\sup_{0<s<1}[th_0(t)-sh_0(s)]/[(t-s)/h_0(t)]<\infty,$$

then

$$\left|\int h(x,a)(P_1-P_2)(\mathrm{d}x)\right|\le c_h\mathring{\mu}_h(P_1,P_2).$$

(d) If $a_h+b_h+c_h<\infty$ and $\int h(x,a)(P_n+P)(\mathrm{d}x)<\infty$, $n=1,2,\dots$, then

$$\lim_{n\to\infty}\mathring{\mu}_h(P_n,P)=0$$

if and only if $P_n\xrightarrow{\mathrm{w}}P$ and

$$\lim_{n\to\infty}\left|\int h(x,a)(P_n-P)(\mathrm{d}x)\right|=0.$$

The proof of the theorem is similar to that of Theorem 6.4.1 in Chap. 6 and can therefore be omitted. Note that, in contrast to Theorems 6.3.2 and 6.3.3, the preceding bounds are based only on the relationships between minimal norms.

Open Problem 8.3.1. A question of great interest concerning the topological structure of minimal distances is the *necessary and sufficient conditions for the convergence* $\widehat{\mu}_c(P_n, P) \to 0$, *where* $\{P, P_n, n = 1, 2, \ldots\} \subset \mathcal{D}(\widehat{\mu}_c, P_0)$ *and* P_0 *is an arbitrary law of* $\mathcal{P}_1$. Note that in the case $(U, d) = (\mathbb{R}^1, |\cdot|)$, if $\{P, P_n, n = 1, 2, \ldots\} \subset \mathcal{D}(\widehat{\mu}_d, P_0)$, then $\widehat{\mu}_d(P_n, P) = \int_{-\infty}^{\infty} |F_n(x) - F(x)|\mathrm{d}x = \widehat{\mathcal{L}}_1(P_n, P) \to 0$ if and only if $P_n \xrightarrow{\text{w}} P$ and

$$\lim_{N\to\infty} \sup_n \int_{|x|>N} |F_n(x) - F_0(x)|\mathrm{d}x = 0,$$

where F_n is a DF of P_n, $n = 0, 1, \ldots,$ and F is the DF of P.

Reference

Kalashnikov VV, Rachev ST (1988) Mathematical methods for construction of stochastic queueing models. Nauka, Moscow (in Russian). [English transl. (1990) Wadsworth, Brooks–Cole, Pacific Grove, CA]

Chapter 9
Moment Problems Related to the Theory of Probability Metrics: Relations Between Compound and Primary Distances

The goals of this chapter are to:

- Explore the general relations between compound and primary probability distances that are similar to the relations between compound and simple probability distances,
- Study the primary minimal metrics arising from minimal functionals with one pair of marginal moments fixed,
- Extend the setting to minimal functionals with two pairs of marginals and with linear combinations of moments fixed.

9.1 Introduction

In Chaps. 5–8, we investigated the relationships between compound and simple distances. The main method we used was based on the dual and explicit solutions of the following problem:

Marginal problem. For fixed probability measures (laws) P_1 and P_2 on an s.m.s. (U, d) and a continuous function c on the product space $U^2 = U \times U$

$$\text{minimize (maximize)} \int_{U^2} c(x, y) P(\mathrm{d}x, \mathrm{d}y),$$

where the laws P on U^2 have marginals P_1 and P_2, i.e., $T_i P = P_i$, $i = 1, 2$.

Similarly, in this chapter we will study the connection between compound and primary distances (see Sect. 3.2 of Chap. 3) solving the following problem:

Moment problem. For fixed real numbers a_{ij} and real-valued continuous functions f_{ij} $(i = 1, 2, j = 1, \dots, n)$

S.T. Rachev et al., *The Methods of Distances in the Theory of Probability and Statistics*,
DOI 10.1007/978-1-4614-4869-3_9,

$$\text{minimize (maximize)} \int_{U^2} c(x, y) P(\mathrm{d}x, \mathrm{d}y),$$

where the law P on U^2 satisfies the marginal moment conditions

$$\int_U f_{ij} \mathrm{d}P_i = a_{ij}, \qquad i = 1, 2, \ j = 1, \ldots, n.$$

We begin with moment problems in which one pair of marginal moments is fixed; then, we extend the setting to moment problems with two pairs of marginal moments fixed and with linear combinations of marginal moments fixed.

9.2 Primary Minimal Distances: Moment Problems with One Fixed Pair of Marginal Moments

Let U be a separable norm space with norm $\|\cdot\|$, $\mathfrak{X} = \mathfrak{X}(U)$ the space of all U-valued random variables (RVs), μ a compound metric in $\mathfrak{X}(U)$, and $\mathcal{M}$ the class of all strictly increasing continuous functions $f : [0, \infty] \to [0, \infty]$, $f(0) = 0$, $f(\infty) = \infty$. Following the definition of primary distances (see Sect. 3.2 of Chap. 3) let us define the spaces $h(\mathfrak{X}) = \{Eh(\|X\|) : X \in \mathfrak{X}\}$ [see (3.2.3)] for a fixed $h \in \mathcal{M}$ and a primary minimal distance $\widetilde{\mu}_h$ (in $h(\mathfrak{X})$)

$$\widetilde{\mu}(a, b) := \inf\{\mu(X, Y) : X, Y \in \mathfrak{X}, Eh(\|X\|) = a, Eh(\|Y\|) = b\}. \qquad (9.2.1)$$

Given the H-average compound distance

$$\mu(X, Y = \mathcal{L}_H(X, Y) = EH(\|X - Y\|) \quad H \in \mathcal{M} \cap \mathcal{H} \qquad (9.2.2)$$

(see Example 3.4.1) we will treat the explicit representations of the following extremal functional:

$$I(H, h, a, b) := \widetilde{\mu}_h(a, b). \qquad (9.2.3)$$

Moreover, for $\mu(X, Y){:=}\mathcal{L}_H(X, Y)$ we will consider the upper bound

$$S(H, h; a, b) := \sup\{\mu(X, Y) : X, Y \in \mathfrak{X}, Eh(\|X\|) = aEh(\|Y\|) = b\} \qquad (9.2.4)$$

whose explicit form will lead to the expression for the moment functions discussed in Sect. 3.4 (Definition 3.4.6). Denote for all $p \geq 0$, $q \geq 0$ the values

$$I(p, q; a, b) := I(H, h; a, b)(H(t) = t^p, h(t) = t^q), \qquad (9.2.5)$$

$$S(p, q; a, b) := S(H, h; a, b)(H(t) = t^p, h(t) = t^q), \qquad (9.2.6)$$

where here and in the sequel 0^0 means 0, and thus $E\|X-Y\|^0$ means $\Pr(X \neq Y)$. Clearly, $I(p,q;a,b)^{\min(1,1/p)}$ represents the primary h-minimal metric $(\widetilde{\mathcal{L}}_{p,h})$ with respect to the $\mathcal{L}_p$-metric

$$\mathcal{L}_p(X,Y) := \{E\|X-Y\|^p\}^{\min(1,1/p)}, \mathcal{L}_0(X,Y) = \text{ess sup}\,\|X-Y\|,$$

where $hX := E\|X\|^q$, $q \geq 0$, i.e.,[1]

$$I(p,q;a,b)^{\min(1,1/p)} = \widetilde{\mathcal{L}}_{p,h}(a,b) := \inf\{\mathcal{L}_p(X,Y) : hX = a, hY = b\}.$$

Further (Corollary 9.2.1), we will find explicit expressions for $\widetilde{\mathcal{L}}_{p,h}$ for any $p \geq 0$ and any $q \geq 0$.

The scheme of the proofs of all statements here is as follows. First we prove the necessary inequalities that give us the required bounds, and then we construct pairs of random variables that achieve the bounds or approximate them with arbitrary precision.

Let $f, f_1, f_2 \in \mathcal{M}$, and consider the following conditions (in what follows, f^{-1} is the inverse function of $f \in \mathcal{M}$):

A. $(f_1, f_2) : f_1 \circ f_2^{-1}(t)$ $(t \geq 0)$ is convex.
B. (f) : $f^{-1}(Ef(\|X+Y\|)) \leq f^{-1}(Ef(\|X\|)) + f^{-1}(Ef(\|Y\|))$ for any $X, Y \in \mathfrak{X}$.
C. $(f) : Ef(\|X+Y\|) \leq Ef(\|X\|) + Ef(\|Y\|)$ for any $X, Y \in \mathfrak{X}$.
D. $(f_1, f_2) : \lim_{t\to\infty} f_1(t)/f_2(t) = 0$.
E. $(f_1, f_2) : f_1 \circ f_2(t)$ $(t \geq 0)$ is concave.
F. (f_1, f_2) : f_1 is concave and f_2 is convex.
G. $(f_1, f_2) : \lim_{t\to\infty} f_1(t)/f_2(t) = \infty$.

Obviously, if $H(t) = t^p$, $h(t) = t^q$ $(p > 0, q > 0)$, then $A(H,h) \iff p \geq q$, $B(h) \iff q \geq 1$, $C(h) \iff q \leq 1$, $D(H,h) \iff q > p$, $E(H,h) \iff q \geq p$, $F(H,h) \iff p \leq 1 \leq q$, $G(H,h) \iff p > q$, and hence conditions A to G cover all possible values of the pairs (p,q).

Theorem 9.2.1. *For any $a \geq 0$ and $b \geq 0$, $a+b > 0$, the following equalities hold:*

(i)

$$I(H,h;a,b) = \begin{cases} (H(|h^{-1}(a) - h^{-1}(b)|) & \text{if } A(H,h) \text{ and } B(h) \text{ hold,} \\ H \circ h^{-1}(|a-b|) & \text{if } A(H,b) \text{ and } C(h) \text{ hold,} \\ 0 & \text{if } D(H,h) \text{ holds.} \end{cases} \tag{9.2.7}$$

(ii) For any $H \in \mathcal{M}$ and $h \in \mathcal{M}$

$$\inf\{\Pr\{X \neq y\} : Eh(\|X\|) = a, Eh(\|Y\|) = b\} = 0, \tag{9.2.8}$$

[1] See Definition 3.2.2 in Chap. 3.

$$\inf\{EH(\|X-Y\|) : \Pr\{X \neq u\} = a, \Pr\{Y \neq u\} = 0 \ (u \in U,\ a, b \in [0,1]). \quad (9.2.9)$$

(iii)

$$S(H,h;a,b) =$$

$$\begin{cases} H(h^{-1}(a)+h^{-1}(b)) & \text{if } F(H,h) \text{ holds or if } B(h) \text{ and } E(H,h) \text{ hold,} \\ H \circ h^{-1}(\alpha+\beta) & \text{if } C(h) \text{ and } E(H,h) \text{ hold,} \\ \infty & \text{if } G(H,h) \text{ holds.} \end{cases} \quad (9.2.10)$$

(iv) For any $u \in U,\ H \in \mathcal{M},\ h \in \mathcal{M}$

$$\sup\{\Pr\{X+Y\} : Eh(\|X\|) = a, Eh(\|Y\|) = b\} = 1, \quad (9.2.11)$$

$$\sup\{\Pr\{\neq Y\} : \Pr\{X \neq u\} = a, \Pr\{Y \neq u\} = b\} = \min(a+b,1)(a,b \in [0,1]), \quad (9.2.12)$$

$$\sup\{EH(\|X-Y\|) : \Pr(X \neq u) = a, \Pr\{Y \neq u\} = b\} = \infty. \quad (9.2.13)$$

Proof. (i) *Case 1.* Let $A(H,h)$ and $B(h)$ be fulfilled. Denote $\phi(a,b) := H(|h^{-1}(a) - h^{-1}(b)|), a \geq 0, b \geq 0$.

Claim. $I(H,h,a,b) \geq \phi(a,b)$.

By Jensen's inequality and $A(H,h)$,

$$H \circ h^{-1}(EZ) \leq EH \circ h^{-1}(Z). \quad (9.2.14)$$

Taking $Z = h(\|X-Y\|)$ and using $B(h)$ we obtain $H^{-1}(EH(\|X-Y\|)) = H^{-1}(EH \circ h^{-1}(Z)) \geq h^{-1}(Eh(\|X-Y\|)) \geq |h^{-1}(Eh(\|X\|) - h^{-1}(Eh(\|Y\|))|$ for any $X, Y \in \mathfrak{X}$, which proves the claim.

Claim. There exists an optimal pair (X^*, Y^*) of RVs such that $Eh(\|X^*\|) = a$, $Eh(\|Y^*\|) = b$, $EH(\|X^* - Y^*\|) = \phi(a,b)$. (Note that an optimal pair of RVs has this restricted meaning in the chapter.)

Let $\overline{e}$ here and in what follows be a fixed point of U with $\|\overline{e}\| = 1$. Then the required pair (X^*, Y^*) is given by

$$X^* = h^{-1}(a)\overline{e} \quad Y^* = h^{-1}(b)\overline{e}, \quad (9.2.15)$$

which proves the claim.

Case 2. Let $A(H,h)$ and $C(h)$ be fulfilled. Denote $\phi_1(t) := H \circ h^{-1}(t), t \geq 0$. As in Claim 1, we get $I(H,h;a,b) \geq \phi_1(|a-b|)$. Suppose that $a > b$, and for each $\varepsilon > 0$ define a pair $(X_\varepsilon, Y_\varepsilon)$ of RVs as follows: $\Pr\{X_\varepsilon = c_\varepsilon \overline{e}, Y_\varepsilon = \overline{0}\} = p_\varepsilon$, $\Pr\{X_\varepsilon = d_\varepsilon \overline{e}, Y_\varepsilon = d_\varepsilon \overline{e}\} = 1 - p_\varepsilon$, where

$$\overline{0} := 0\overline{e} \quad p_\varepsilon := \frac{a-b}{a-b+\varepsilon} \quad c_\varepsilon := h^{-1}(a-b+\varepsilon) \quad d_\varepsilon := h^{-1}\left(\frac{b}{1-p_\varepsilon}\right). \tag{9.2.16}$$

Then $(X_\varepsilon, Y_\varepsilon)$ enjoys the side conditions in (9.2.1) and $EH(\|X_\varepsilon - Y_\varepsilon\|) = \phi_1(a-b+\varepsilon)(a-b)/(a-b+\varepsilon)$. Letting $\varepsilon \to 0$, we claim (9.2.7).

Case 3. Let $D(H,h)$ be fulfilled. To obtain (9.2.7), it is sufficient to define a sequence (X_n, Y_n) $(n \geq N)$ such that $\lim_{n\to\infty} EH(\|X_n - Y_n\|) = 0$, $Eh(\|X_n\|) = a$, $Eh(\|Y_n\|) = b$. An example of such a sequence is the following one: $\Pr\{X_n = \overline{0}, Y_n = \overline{0}\} = 1 - c_n - d_n$, $\Pr\{X_n = na\overline{e}, Y_n = \overline{0}\} = c_n$, $\Pr\{X_n = \overline{0}, Y_n = nb\overline{e}\} = d_n$, where $c_n = a/h(na)$, $d_n = b/h(nb)$, and N satisfies $c_N + d_N < 1$.

(ii) Define the sequence (X_n, Y_n) $(n = 2, 3, \ldots)$ such that $\Pr\{X_n = h^{-1}(na)\overline{e}, Y_n = h^{-1}(nb)\} = 1/n$, $\Pr\{X_n = 0, Y_n = 0\} = (n-1)/n$. Hence, $Eh(\|X_n\|) = a$, $Eh(\|Y_n\|) = b$, and $\Pr(X_n \neq Y_n) = 1/n$, which proves (9.2.8).

Further, suppose $a \geq b$. Without loss of generality, we may assume that $u = \overline{0}$. Then consider the random pair $(\widetilde{X}_n, \widetilde{Y}_n)$ with the following joint distribution: $\Pr\{\widetilde{X}_n = \overline{0}, \widetilde{Y}_n = \overline{0}\} = 1-a$, $\Pr\{\widetilde{X}_n = (1/n)\overline{e}, \widetilde{Y}_n = \overline{0}\} = a-b$, $\Pr\{\widetilde{X}_n = (1/n)\overline{e}, \widetilde{Y}_n = (1/n)\overline{e}\} = b$. Obviously $(\widetilde{X}_n, \widetilde{Y}_n)$ satisfies the constraints $\Pr(\widetilde{X}_n \neq 0) = a$, $\Pr(\widetilde{Y}_n \neq 0) = b$, and $\lim_{n\to\infty} EH(\|X_n - Y_n\|) = 0$, which proves (9.2.9).

The proofs of (iii) and (iv) are quite analogous to those of (i) and (ii), respectively. □

Remark 9.2.1. If $A(H,h)$ and $B(h)$ hold, then we have constructed an optimal pair (X^*, Y^*) [see (9.2.15)], i.e., (X^*, Y^*) realizes the infimum in $I(H, h; a, b)$. However, if $D(H,h)$ holds and $a \neq b$, then optimal pairs do not exist because $EH(\|X - Y\|) = 0$ implies $a = b$. Note that the latter was not the case when we studied the minimal or maximal distances on a u.m.s.m.s. (U,d) since, by Theorem 8.2.1 of Chap. 8, (X^*, Y^*) and (X^{**}, Y^{**}) exist such that

$$\begin{aligned}\widehat{\mu}_c(X,Y) &:= \inf\{\mu_c(\widetilde{X}, \widetilde{Y}) : X, Y \in \mathfrak{X}(U,d), \Pr_{\widetilde{X}} = \Pr_X, \Pr_{\widetilde{Y}} = \Pr_Y\} \\ &= \mu_c(X^*, Y^*)\end{aligned}$$

and

$$\begin{aligned}\check{\mu}_c(X,Y) &:= \sup\{\mu_c(\widetilde{X}, \widetilde{Y}) : X, Y \in \mathfrak{X}(U,d), \Pr_{\widetilde{X}} = \Pr_X, Pr_{\widetilde{Y}} = \Pr_Y\} \\ &= \mu_c(X^{**}, Y^{**}).\end{aligned}$$

Corollary 9.2.1. *For any $a \geq 0$, $b \geq 0$, $a+b>0$, $p \geq 0$, $q \geq 0$,*

$$I(p,q;a,b) = \begin{cases} |a^{1/q} - b^{1/q}|^p, & \text{if } p \geq q \geq 1, \\ |a-b|^{p/q}, & \text{if } p \geq q \quad 0<q<1, \\ 0, & \text{if } 0 \leq p < q \text{ or } q=0, \ p>0, \\ |a-b|, & \text{if } p=q=0, \end{cases} \tag{9.2.17}$$

and in particular, the primary h-minimal metric, $\widetilde{\mathcal{L}}_{p,h}(hX = E\|X\|^q)$, admits the following representation:

$$\widetilde{\mathcal{L}}_{p,h}(a,b) = \begin{cases} |a^{1/q} - b^{1/q}|, & \text{if } p \geq q \geq 1, \\ |a-b|^{1/q}, & \text{if } p \geq 1, \ 0<q<1, \\ |a-b|^{p/q}, & \text{if } 1 \geq p \geq q > 0, \\ 0, & \text{if } 0 \leq p < q \text{ or } q=0, \ p>0, \\ |a-b|, & \text{if } p=q=0. \end{cases} \tag{9.2.18}$$

One can verify that if μ is a compound or simple probability distance with parameter $\mathbb{K}_H$, then

$$M(P_1,P_2) := \sup\left\{\mu(X_1,X_2) : X_1, X_2 \in \mathfrak{X}, \right.$$
$$\left. Eh(\|X_i\|) = \int_U h(\|x\|)P_i(\mathrm{d}x), i=1,2\right\} \tag{9.2.19}$$

is a moment function with the same parameter $\mathbb{K}_M = \mathbb{K}_\mu$ (see Definition 3.4.2 of Chap. 3). In particular, in (9.2.4), $S(H,h;a,b)$ $(H \in \mathcal{H} \cap \mathcal{M})$, and in (9.2.6), $M_p(P_1,P_2) = S(p,q;a,b)^{\min(1,1/p)}$, $a = \int \|x\|^q P_1(\mathrm{d}x)$, $b = \int \|x\|^q P_2(\mathrm{d}x)$ may be viewed as moment functions with parameters $\mathbb{K}_M = K_H$ [see (2.4.3)] and $\mathbb{K}_M = 1$, respectively.

Corollary 9.2.2. *For any $a \geq 0$, $b \geq 0$, $a+b>0$, $p \geq 0$, $q \geq 0$,*

$$S(p,q;a,b) = \begin{cases} (a^{1/q} + b^{1/q})^q, & \text{if } 0 \leq p \leq q, \ q \geq 1, \\ (a+b)^{p/q}, & \text{if } 0 \leq p \leq q < 1, \ q \neq 0, \\ \infty, & \text{if } p > q \geq 0, \\ \min(a+b,1), & \text{if } p=q=0. \end{cases} \tag{9.2.20}$$

Obviously, if $q=0$ in (9.2.17), (9.2.18), or (9.2.20), then the values of I and S make sense for $a,b \in [0,1]$.

The following theorem is an extension for $p = q = 1$ of Corollaries 9.2.1 and 9.2.2 to a nonnormed space U such as the Skorokhod space $D[0, 1]$.[2]

Theorem 9.2.2. *Let (U, d) be a separable metric space and $\mathfrak{X} = \mathfrak{X}(U)$ the space of all U-valued RVs, and let $u \in U$, $a \geq 0$, $b \geq 0$. Assume that there exists $z \in U$ such that $d(z, u) \geq \max(a, b)$. Then*

$$\min\{Ed(X,Y) : X, Y \in \mathfrak{X}, Ed(X,u) = a, Ed(Y,u) = b\} = |a - b| \quad (9.2.21)$$

and

$$\max\{Ed(X,Y) : X, Y \in \mathfrak{X}, Ed(X,u) = a, Ed(Y,u) = b\} = a + b. \quad (9.2.22)$$

Proof. Let $a \leq b$, $\gamma = d(z, u)$. By the triangle inequality, the minimum in (9.2.21) is greater than $b - a$. On the other hand, if $\Pr(X = u, Y = u) = 1 - b/\gamma$, $\Pr(X = u, Y = z) = (b - a)/\gamma$, $\Pr(X = z, Y = u) = 0$, $\Pr(X = z, Y = z) = a/\gamma$, then $Ed(X,u) = a$, $Ed(Y,u) = b$, $Ed(X,Y) = b - a$, which proves (9.2.21). One proves (9.2.22) analogously. □

From (9.2.21) it follows that the primary h-minimal metric, $\widetilde{\mathcal{L}}_{1,h}(hX, hY)$, with respect to the average metric $\mathcal{L}_1(X,Y) = Ed(X,Y)$ with $hX = Ed(X,u)$, is equal to $|hX - hY|$. The preceding theorem provides the exact values of the bounds (3.4.48) and (3.4.52).

Open Problem 9.2.1. Find the explicit solutions of moment problems with one fixed pair of marginal moments for RVs with values in a separable metric space U. In particular, find the primary h-minimal metric $\widetilde{\mathcal{L}}_{p,h}(hX, hY)$, with respect to $\mathcal{L}_p(X,Y) = \{Ed^p(X,Y)\}^{1/p}$, $p > 1$, with $hX = Ed^q(X,u)$, $q > 0$.

Suppose that $U = \mathbb{R}^n$ and $d(x, y) = \|x - y\|_1$, where $\|x_1, \ldots, x_n)\|_1 = |x_1| + \cdots + |x_n|$. Consider the H-average distance $\mathcal{L}_H(X,Y) := EH(\|X - Y\|_1)$, with convex $H \in \mathcal{H}$, and the L_p-metric $\mathcal{L}_p(X,Y) = \{E\|X - Y\|_1^p\}^{1/p}$. Define the engineer distance $\mathbf{EN}(X, Y; H) = H(\|EX - EY\|_1)$, where $EX = (EX_1, \ldots, EX_n)$ in the space of $\widetilde{\mathfrak{X}}(\mathbb{R}^n)$ of all n-dimensional random vectors with integrable components (Example 3.2.5). Similarly, define L_p*-engineer metric*, $\mathbf{EN}(X, Y, p) = \left(\sum_{i=1}^n |EX_i - EY_i|^p\right)^{1/p}$, $p \geq 1$ [see (3.2.14)]. Let $hX = EX$ for any $X \in \widetilde{\mathfrak{X}}(\mathbb{R}^n)$. Then the following relations between the compound distances $\mathcal{L}_H$, $\mathcal{L}_p$ and the primary distances $\mathbf{EN}(\cdot, \cdot; H)$, $\mathbf{EN}(\cdot, \cdot; p)$ hold.

Corollary 9.2.3. *(i) If H is convex, then*

$$\begin{aligned}\widetilde{\mathcal{L}}_{H,h}(hX, hY) &:= \min\{\mathcal{L}_H(\widetilde{X}, \widetilde{Y}) : h\widetilde{X} = hX, h\widetilde{Y} = hY\}\\ &= \mathbf{EN}(X, Y; H).\end{aligned} \quad (9.2.23)$$

[2] See, for example, Billingsley (1999).

(ii) For any $p \geq 1$

$$\widetilde{\mathcal{L}}_{p,h}(hX, hY) = \mathbf{EN}(X, Y; p). \tag{9.2.24}$$

Proof. Use Jensen's inequality to obtain the necessary lower bounds. The "optimal pair" is $\widetilde{X} = EX, \widetilde{Y} = EY$. □

Combining Theorems 8.2.1, 8.2.2, and 9.2.1, we obtain the following sharp bounds of the extremal functionals $\widehat{\mathcal{L}}_H(P, Q)$ $P, Q \in \mathcal{P}(U)$ and $\check{\mathcal{L}}_H(P, Q)$ (Theorem 8.2.1) in terms of the moments

$$a = \int_U h(x)P(\mathrm{d}x), \qquad b = \int_U h(x)Q(\mathrm{d}x). \tag{9.2.25}$$

Corollary 9.2.4. *Let* $(U, \|\cdot\|)$ *be a separable normed space and* $H \in \mathcal{H}$.

(i) If $A(H, h)$ *and* $B(h)$ *hold, then* $\widehat{\mathcal{L}}_H(P, Q) \geq H(|h^{-1}(a) - h^{-1}(b)|)$.
(ii) If $B(h)$ *and* $E(H, h)$ *hold, then* $\check{\mathcal{L}}_H(P, Q) \leq H(h^{-1}(a) + h^{-1}(b))$.

Moreover, there exist $P_i, Q_i \in \mathcal{P}(U)$, $i = 1, 2$, *with* $a = \int_U h(x)P_i(\mathrm{d}x)$, $b = \int_U h(x)Q_i(\mathrm{d}x)$ *such that* $\widehat{\mathcal{L}}_H(P_1, Q_1) = H(|h^{-1}(a) - h^{-1}(b)|)$ *and* $\check{\mathcal{L}}_H(P_2, Q_2) = H(h^{-1}(a) + h^{-1}(b))$.

9.3 Moment Problems with Two Fixed Pairs of Marginal Moments and with Fixed Linear Combination of Moments

In this section we will consider the explicit representation of the following bounds:

$$I(H, h_1, h_2; a_1, b_1, a_2, b_2) := \inf EH(\|X - Y\|), \tag{9.3.1}$$

$$S(H, h_1, h_2; a_1, b_1, a_2, b_2) := \sup EH(\|X - Y\|), \tag{9.3.2}$$

where H, h_1, $h_2 \in \mathcal{M}$, and the infimum in (9.3.1) and the supremum in (9.3.2) are taken over the set of all pairs of RVs $X, Y \in \mathfrak{X}(U)$, satisfying the moment conditions

$$Eh_i(\|X\|) = a, \quad Eh_i(\|Y\|) = b_i, \quad i = 1, 2, \tag{9.3.3}$$

and U is a separable normed space with norm $\|\cdot\|$. In particular, if $H(t) = t^p$, $h_i(t) = t^{q_i}$, $i = 1, 2$ ($p \geq 0$, $q_2 > q_1 \geq 0$), then we write

$$I(p, q_1, q_2; a_1, b_1, a_2, b_2) := I(H, h_1, h_2; a_1, b_1, a_2, b_2), \tag{9.3.4}$$

$$S(p, q_1, q_2; a_1, b_1, a_2, b_2) := S(H, h_1, h_2; a_1, b_1, a_2, b_2). \tag{9.3.5}$$

If $H \in \mathcal{H}$, then the functional I represents a primary h-minimal distance with respect to $\mathcal{L}_H(X,Y) = EH(\|X - Y\|)$ with $hX = (Eh_1(\|X\|), Eh_2(\|X\|))$.[3] In particular, $I(p, q_1, q_2; a_1, b_1, a_2, b_2)^{\min(1,1/p)}$ is a primary h-minimal metric with respect to $\mathcal{L}_p(X,Y) = \{E\|X - Y\|^p\}^{\min(1,1/p)}$. The functionals (9.3.2) and (9.3.5) may be viewed as moment functions with parameters K_H and $2^{\min(1,p)}$, respectively.[4] A moment problem with two pairs of marginal conditions is considerably more complicated, and in the present section, our results are not as complete as in the previous one. Further, conditions A to G are defined as in the previous section.

Theorem 9.3.1. *Let the conditions $A(h_2, h_1)$ and $G(h_2, h_1)$ hold. Let $a_i \geq 0$, $b_i \geq 0$, $i = 1, 2$, $a_1 + a_2 > 0$, $b_1 + b_2 > 0$, and*

$$h_1^{-1}(a_1) \leq h_2^{-1}(a_2), \quad h_1^{-1}(b_1) \leq h_2^{-1}(b_2). \tag{9.3.6}$$

(i) If $A(H, h_1)$, $B(h_1)$ and $D(H, h_2)$ are fulfilled, then

$$I(H, h_1, h_2; a_1, b_1, a_2, b_2) = I(H, h_1; a_1, b_1) = H(|h_1^{-1}(a_1) - h_1^{-1}(b_1)|). \tag{9.3.7}$$

(ii) Let $D(H, h_2)$ be fulfilled. If $F(H, h_1)$ holds or if $B(h_1)$ and $E(H, h_1)$ hold, then

$$\begin{aligned} S(H, h_1, h_2; a_1, b_1, a_2, b_2) &= S(H, h_1; a_1, b_1) \\ &= H(h_1^{-1}(a_1) + h_1^{-1}(b_1)). \end{aligned} \tag{9.3.8}$$

(iii) If $G(H, h_2)$ is fulfilled and $h_1^{-1}(a_1) \neq h_2^{-1}(a_2)$ or $h_1^{-1}(b_1) \neq h_2^{-1}(b_2)$, then

$$S(H, h_1, h_2; a_1, b_1, a_2, b_2) = S(H, h_1; a_1, b_1) = \infty. \tag{9.3.9}$$

Proof. By Theorem 9.2.1 (i), we have

$$\begin{aligned} I(H, h_1, h_2; fli, a_1, b_1, a_2, b_2) &\geq I(H, h_1; a_1, b_1) = \phi(a_1, b_1) \\ &:= H(|h_1^{-1}(a_1) - h_1^{-1}(b_1)|). \end{aligned} \tag{9.3.10}$$

Further, we will define an appropriate sequence of RVs (X_t, Y_t) that satisfy the side conditions (9.3.3) and $\lim_{t\to\infty} EH(\|X_t - Y_t\|) = \phi(a_1, b_1)$. Let $f(x) = h_2 \circ h_1^{-1}(x)$. Then, by Jensen's inequality and $A(h_2, h_1)$,

$$f(a_1) = f(Eh_1(\|X\|)) \leq Ef \circ h_1(\|X\|) = a_2 \tag{9.3.11}$$

and $f(b_1) < b_2$. Moreover, $\lim_{t\to\infty} f(t)/t = \infty$ by $G(h_1, h_2)$.

---

[3]See Sect. 3.2 in Chap. 3.

[4]See Definition 3.4.2 in Chap. 3.

Case 1. Suppose that $f(a_1) < a_2$, $f(b_1) < b_2$.

Claim. If the convex function $f \in \mathcal{M}$ and the real numbers c_1, c_2 possess the properties

$$f(c_1) < c_2 \quad \lim_{t\to\infty} f(t)/t = \infty, \tag{9.3.12}$$

then there exist a positive t_0 and a function $k(t)$ $(t \geq t_0)$ such that the following relations hold for any $t \geq t_0$:

$$0 < k(t) < c_1, \tag{9.3.13}$$

$$tf(c_1 - k(t)) + k(t)f(c_1 + t) = c_2(k(t) + t), \tag{9.3.14}$$

$$\frac{k(t)}{k(t)+t} \leq \frac{c_2}{f(c_1+t)}, \tag{9.3.15}$$

and

$$\lim_{t\to\infty} k(t) = 0. \tag{9.3.16}$$

Proof of the claim. Let us take t_0 such that $f(c_1 + t)/(c_1 + t) > c_2/c_1$, $t \geq t_0$, and consider the equation

$$F(t, X) = c_2,$$

where $F(t,x) := (f(c_1 - x)t + f(c_1 + t)x)/(x + t)$. For each $t \geq t_0$ we have $F(t, c_1) > c_2$, $F(t, 0) = f(c_1) < c_2$. Hence, for each $t \geq t_0$ there exists $x = k(t)$ such that $k(t) \in (0, c_1)$ and $F(t, k(t)) = c_2$, which proves (9.3.13) and (9.3.14). Further, (9.3.14) implies (9.3.15), and (9.3.13), (9.3.15) imply (9.2.16). The claim is established.

From the claim we see that there exist $t_0 > 0$ and functions $\ell(t)$ and $m(t)$ $(t \geq t_0)$ such that for all $t > t_0$ we have

$$0 < \ell(t) < a_1, \quad 0 < m(t) < b_1, \tag{9.3.17}$$

$$tf(a_1 - \ell(t)) + \ell(t)f(a_1 + t) = a_2(\ell(t) + t), \tag{9.3.18}$$

$$tf(b_1 - m(t)) + m(t)f(b_1 + t) = b_2(m(t) + t), \tag{9.3.19}$$

$$\lim_{t\to\infty} \ell(t) = 0, \qquad \lim_{t\to\infty} m(t) = 0. \tag{9.3.20}$$

Using (9.3.17)–(9.3.20) and the conditions $A(H, h_1)$, $D(H, h_2)$, and $G(h_2, h_1)$ one can obtain that the RVs (X_t, Y_t) $(t > t_0)$ determined by the equalities are

$$\Pr\{X_t = x_i(t), \quad Y_t = y_j(t)\} = p_{ij}(t), \quad i, j = 1, 2,$$

where

$$x_1(t) := h_1^{-1}(a_1 - \ell(t))\overline{e}, \qquad x_2(t) := h_1^{-1}(a_1 + t)\overline{e},$$

$$y_1(t) := h_1^{-1}(b_1 - m(t))\overline{e}, \qquad y_2(t) := h_1^{-1}(b_1 + t)\overline{e},$$

$$p_{11}(t) := \min\{t/(\ell(t)+t), t/(m(t)+t)\}, \quad p_{12}(t) := t/(\ell(t)+t) - p_{11}(t),$$

$$p_{21}(t) := t/(m(t)+t) - p_{11}(t), \; p_{22}(t) := \min\{\ell(t)/(\ell(t)+t), m(t)/(m(t)+t)\}$$

possess all the desired optimal properties. □

Case 2. Suppose $f(a_1) = a_2$ [i.e., $h_1^{-1}(a_1) = h_2^{-1}(a_2)$], $f(b_1) < b_2$. Then we can determine (X_t, Y_t) by the equalities $\Pr\{X_t = h_1^{-1}(a_1), Y_t = y_1(t)\} = t/(m(t)+t)$, $\Pr\{X_t = h_1^{-1}(a_1), Y_t = y_2(t)\} = m(t)/(m(t)+t)$.

Case 3. The cases $(f(a_1) < a_2,\ f(b_1) = b_2)$ and $(f(a_1) = a_2,\ f(b_1) = b_2)$ are considered in the same way as in Case 2.

Parts (ii) and (iii) are proved by analogous arguments. □

Corollary 9.3.1. *Let* $a_1 \geq 0$, $b_i \geq 0$, $a_1 + a_2 > 0$, $b_1 + b_2 > 0$, $a_1^{1/q_1} \leq a_2^{1/q_2}$, $b_1^{1/q_1} \leq b_2^{1/q_2}$.

(i) If $1 \leq q_1 \leq p < q_2$, *then*

$$I(p,q_1,q_2;a_1,b_1,a_2,b_2) = I(p,q_1;a_1,b_1) = (a_1^{1/q_1} - b_1^{1/q_1})^p. \tag{9.3.21}$$

(ii) If $0 < p \leq q_1$, $1 \leq q_1 < q_2$, *then*

$$S(p,q_1,q_2;a_1,b_1,a_2,b_2) = S(p,q_1;a_1,b_1) = (a_1^{1/q_1} - b_1^{1/q_1})^p. \tag{9.3.22}$$

(iii) If $0 < q_1 < q_2 < p$ *and* $a_1^{1/q_1} = a_2^{1/q_2}$ *or* $b_1^{1/q_1} = b_2^{1/q_2}$, *then*

$$S(p,q_1,q_2;a_1,b_1,a_2,b_2) = S(p,q_1;a_1,b_1) = \infty.$$

Corollary 9.3.1 describes situations in which the "additional moment information" $a_2 = E\|X\|^{q_2}$, $b_2 = E\|Y\|^{q_2}$ does not affect the bounds

$$I(p,q_1,q_2;a_1,b_1,a_2,b_2) = I(p,q_1;a_1,a_2),$$

$$S(p,q_1,q_2;a_1,b_1,a_2,b_2) = S(p,q_1;a_1,a_2)$$

(and likewise Theorem 9.3.1).

Open Problem 9.3.1. Find the explicit expression of $I(p,q_1,q_2,a_1,b_1,a_2,b_2)$ and $S(p,q_1,q_2;a_1,b_1,a_2,b_2)$ for all $p \geq 0$, $q_2 > 0$, $q_1 \geq 0$ [see (9.3.4), Corollary 9.3.1, and Theorem 9.3.1]. One could start with the following one-dimensional version of the problem. Let $h_i : [0,\infty) \to \mathbb{R}$ $(i = 1,2)$ and $H : \mathbb{R} \to \mathbb{R}$ be given continuous functions with H symmetric and strictly increasing on $[0,\infty)$. Further, let X and Y be nonnegative RVs having fixed moments $a_i = Eh_i(X)$, $b_i = Eh_i(Y)$, $i = 1,2$. The problem is to evaluate

$$I = \inf EH(X - Y), \quad S = \sup EH(X + Y). \tag{9.3.23}$$

If one desired, one could think of $X = X(t)$ and $Y = Y(t)$ as functions on the unit interval (with Lebesgue measure).[5] The five moments a_1, a_2, b_1, b_2, and $EH(X \pm Y)$ depend only on the joint distribution of the pair (X,Y) and the extremal values in (9.3.23) are realized by a probability measure supported by six points.[6] Thus the problem can also be formulated as a nonlinear programming problem to find

$$I = \inf \sum_{j=1}^{6} p_j H(u_j - v_j), \quad S = \sup \sum_{j=1}^{6} p_j H(u_j + v_j),$$

subject to

$$p_j \geq 0, \quad \sum_{j=1}^{6} p_j = 1, \quad u_j \geq 0, \quad v_j \geq 0, \quad j = 1,\dots,6,$$

$$\sum_{j=1}^{6} p_j h_i(u_j) = a_i, \qquad \sum_{j=1}^{6} p_j h_i(v_j) = b_i, \qquad i = 1,2.$$

Such a problem becomes simpler when the function h_i and the function H on $\mathbb{R}_+$ are convex.[7]

Note that in the case where U is a normed space, the moment problem was easily reduced to the one-dimensional moment problem ($U = \mathbb{R}$). This is no longer possible for general (nonnormed) spaces U, rendering the problem quite different from that considered in Sects. 9.2 and 9.3.

Open Problem 9.3.2. Let $\mu(X,Y)$ be a given compound probability metric in $(U, \|\cdot\|)$, I an arbitrary index set, α_i, β_i $(i \in I)$ positive constants, and $h_i \in \mathcal{M}$, $i \in I$. Find

$$I\{\mu;\alpha_i,\beta_i,i \in I\} = \inf\{\mu(X,Y) : X,Y \in \mathfrak{X}(U),$$
$$Eh_i(\|X\|) = \alpha_i \; Eh_i(\|Y\|) = \beta_i,\ i \in I\}, \tag{9.3.24}$$

and define $S\{\mu;\alpha_i,\beta_i,i \in I\}$ by changing in to sup in (9.3.24). One very special case of the problem is

---

[5]See Karlin and Studden (1966, Chap. 3) and Rogosinky (1958).

[6]See Rogosinky (1958, Theorem 1), Karlin and Studden (1966, Chap. 3), and Kemperman (1983).

[7]See, for example, Karlin and Studden (1966, Chap. 14).

$$\mu(X,Y) = \delta(X,Y) = \begin{cases} 0 \text{ if } \Pr(X = Y) = 1, \\ 1 \text{ if } \Pr(X \neq Y) = 1 \end{cases}$$

(Example 3.2.4). Then one can easily see thats

$$I\{\delta;\alpha_i,\beta_i,i \in I\} = \begin{cases} 0 \text{ if } \alpha_i = \beta_i,\ \forall i \in I, \\ 1 \text{ otherwise,} \end{cases} \tag{9.3.25}$$

and

$$S(\delta;\alpha_i,\beta_i,i \in I) = 1. \tag{9.3.26}$$

In Sect. 3.4 we introduced the μ-upper bound with fixed sum of marginal qth moments

$$\overline{\mu}(c;m,q) := \sup\{\mu(X,Y) : X,Y \in \mathfrak{X}(U), m_q(X) + m_q(Y) = c\}, \tag{9.3.27}$$

where μ is a compound probability distance in $\mathfrak{X}(U)$ and $m_p(X)$ is the "qth moment"

$$m_q(X) := Ed(X,a)^q, \quad q > 0,$$
$$m_0(X) := EI\{d(X,a) \neq 0\} = \Pr(X \neq a).$$

Similarly, we defined the μ-lower bound with fixed difference of marginal qth moments

$$\underline{\mu}(c;m,q) := \inf\{\mu(X,Y) : X,Y \in \mathfrak{X}(U), m_q(X) - m_q(Y) = c\}. \tag{9.3.28}$$

The next theorem gives us explicit expressions for $\underline{\mu}(c;m,q)$ and $\overline{\mu}(c;m,q)$ when μ is the p-average metric (Example 3.4.1), $\mu(X,Y) = \mathcal{L}_p(X,Y) = \{E\|X - Y\|^p\}^{p'}$, $p' = \min(1,1/p)$ $(p > 0)$ or μ is the indicator metric, $\mu(X,Y) = \mathcal{L}_0(X,Y) = E\|X - Y\|^0 = EI\{X \neq Y\}$. [We assume, as before, that (U,d), $d(x,y) := \|x - y\|$, is a separable normed space.]

We will generalize the functionals $\underline{\mu}$ and $\overline{\mu}$ given by (9.3.27) and (9.3.28) in the following way. For any $p \geq 0$, $q \geq 0$, α, β, $c \in \mathbb{R}$, consider

$$I(p,q,c,\alpha,\beta) := \inf\{E\|X - Y\|^p : \alpha m_q + \beta m_q = c\} \tag{9.3.29}$$

and

$$S(p,q,c,\alpha,\beta) := \sup\{E\|X - Y\|^p : \alpha m_q + \beta m_q = c\}. \tag{9.3.30}$$

Theorem 9.3.2. *For any $\alpha > 0$, $\beta > 0$, $c > 0$, $p \geq 0$, $q \geq 0$, the following relations hold:*

$$I(p,q,c,\alpha,\beta) = \begin{cases} 0 & \textit{if } q \neq 0 \textit{ or if } q = 0,\ c \leq \alpha + \beta, \\ +\infty & \textit{if } q = 0,\ c > \alpha + \beta \end{cases} \tag{9.3.31}$$

[the value $+\infty$ means that the infimum in (9.3.29) is taken over an empty set],

$$
\begin{aligned}
I(p,q,c,\alpha,-\beta) &= 0 \text{ if } \beta < \alpha,\, p^2+q^2 \neq 0, \text{ or } 0 \le p < q, \\
&\quad \text{or } q = 0,\, p > 0, \text{ and } c \le \alpha \\
&= [c(\beta^{1/(q-1)} - \alpha^{1/(q-1)})^{q-1}/(\alpha\beta)]^{p/q} \text{ if } \alpha \le \beta,\, p \ge \beta > 1 \\
&= (c/\alpha)^{p/q} \text{ if } \alpha \le \beta,\, p \ge q,\, 0 < q \le 1 \\
&= \max\left(\frac{c-\alpha+\beta}{\alpha}, 0\right) \quad \text{if } p = q = 0,\, \beta < \alpha,\, c \le \alpha \\
&= +\infty \qquad \text{if } q = 0,\, c > \alpha.
\end{aligned}
\tag{9.3.32}
$$

Proof. Clearly, if $c > \alpha + \beta$, then there is no (X, Y) such that $\alpha m_0(X) + \beta m_0(Y) = c$. Suppose $q > 0$. Define the optimal pair (X^*, Y^*) by $X^* = Y^* = (c/(\alpha+\beta))^{1/q}\overline{e}$, where $\|\overline{e}\| = 1$. Then $\mathcal{L}_p(X^*, Y^*) = 0$ for all $0 \le p < \infty$ and clearly $\alpha m_q + \beta m_q = c$, i.e., (9.3.31) holds.

To prove (9.3.32), we will make use of Corollary 9.2.1 [see (9.2.17)]. By the definition of the extremal functional $I(p,q;a,b)$ (9.2.5),

$$I(p,q,c,\alpha,-\beta) = \inf\{I(p,q;d,f) : d \ge 0, f \ge 0, \alpha d - \beta f = c\}, \tag{9.3.33}$$

where $I(p,q;a,b)$ admits the explicit representation (9.2.17). Solving the minimization problem (9.3.33) yields (9.3.32). □

Similarly, we have the following explicit formulae for $S(p,q,c,\alpha,\beta)$ (9.3.30).

Theorem 9.3.3. *For any $\alpha > 0$, $\beta > 0$, $c > 0$, $p \ge 0$, $q \ge 0$,*

$$
S(p,q,c,\alpha,-\beta) = \begin{cases} +\infty \text{ if } p > 0,\, q > 0, \text{ or } p > 0,\, q = 0,\, c \le \alpha \\ 1 \quad \text{if } p = q = 0,\, c \le \alpha, \text{ or } p = 0,\, q > 0 \\ -\infty \text{ if } q = 0,\, c > \alpha, \end{cases}
$$

[the value $-\infty$ means that the supremum in (9.3.30) is taken over an empty set],

$$
\begin{aligned}
S(p,q,c,\alpha,\beta) &= [c(\alpha^{1/(q-1)} + \beta^{1/(q-1)})^{q-1}/(\alpha\beta)]^{p/q} \text{ if } 0 \le p \le q,\, q > 1, \\
&= \left(\frac{1}{c}\min(\alpha,\beta)\right)^{p/q} \qquad \text{if } 0 \le p \le q \le 1,\, q > 0 \\
&= +\infty \qquad \text{if } p > q > 0 \text{ or } p > q = 0, \\
&\qquad\qquad c \le \alpha + \beta, \\
&= \min[1, c/\min(\alpha,\beta)] \qquad \text{if } p = q = 0,\, c \le \alpha+\beta, \\
&= -\infty \qquad \text{if } q = 0,\, c > \alpha + \beta.
\end{aligned}
$$

Using Corollary 9.3.1 one can study similar but more general moment problems:

$$
\begin{array}{l} \text{minimize} \\ \text{(maximize)} \end{array} \{\mathcal{L}_p(X,Y) : F(m_{q_1}(X), m_{q_2}(X), m_{q_1}(Y), m_{q_2}(Y)) = 0\}.
$$

References

Billingsley P (1999) Convergence of probability measures, 2nd edn. Wiley, New York

Karlin S, Studden W (1966) Tchebycheff systems: with applications in analysis and statistics. Wiley, New York

Kemperman JHB (1983) On the role of duality in the theory of moments. Semi-infinite programming and applications. In: Lecture notes economic mathematical system, vol 215. Springer, Berlin, pp 63–92

Rogosinky WW (1958) Moments of non-negative mass. Proc R Soc 245A:1–27

Part III
Applications of Minimal Probability Distances

Chapter 10
Moment Distances

The goals of this chapter are to:

- Discuss convergence criteria in terms of a simple metric between characteristic functions assuming they are analytic,
- Provide estimates of a simple metric between characteristic functions of two distributions in terms of moment-based primary metrics,
- Discuss the inverse problem of estimating moment-based primary metrics in terms of a simple metric between characteristic functions.

Notation introduced in this chapter:

Notation	Description
$\lambda(F, G)$	Simple probability metric calculating the distance between two distributions F and G in terms of their characteristic functions
$\mu_j(F)$	jth moment of distribution F
$d_\alpha(F, G)$	Primary metric based on moments of F and G with parameter $\alpha > 0$
D	Set of all distributions with finite moments of all orders and uniquely determined by them
$\underline{\mathrm{N}}$	Subclass of D defined by distributions having moments that do not increase faster than a given sequence

10.1 Introduction

In this chapter we show that in some cases the investigation of the convergence of a sequence of distributions $\{F_n\}$ to a prescribed distribution function (DF) F (or to a prescribed class $\mathcal{K}$ of DFs) can be replaced by studying the convergence of certain characteristics of F_n to the corresponding characteristics of F (or characteristics of $\mathcal{K}$). For example, if $\underline{I}$ is a functional on a class of DFs for which a function F is

DOI 10.1007/978-1-4614-4869-3_10,

the only minimum point and the problem of minimizing $\underline{I}$ is well posed in the sense that any minimizing $\underline{I}$ sequence of functions converges to F, then

$$F_n \to F \iff \underline{I}(F_n) \to \underline{I}(F).$$

Thus, the convergence of $\{F_n\}$ to F is equivalent to the convergence of $\underline{I}(F_n)$ to $\underline{I}(F)$. Of course, estimating the closeness of F_n to F from that of $\underline{I}(F_n)$ to $\underline{I}(F)$ is interesting by itself. Sometimes it is useful to construct a functional $\underline{I}$ for which F is a minimum point in order to have scalar characteristics whose convergence to the corresponding characteristics of F implies the convergence of the distributions themselves. These problems are considered below for certain special distributions.

We begin the discussion by introducing a simple metric between characteristic functions of probability distributions denoted by $\lambda(F, G)$ and derive bounds for the metric assuming the characteristic functions are analytic. In Sect. 10.3, we introduce a primary metric $d_\alpha(F, G)$ defined through the absolute distance between the corresponding moments of the distributions F and G. Bounds of $\lambda(F, G)$ are derived in terms of $d_\alpha(F, G)$. Finally, we consider the question of estimating the primary metric $d_\alpha(F, G)$ in terms of the simple metric $\lambda(F, G)$. Although not always possible because of the nature of the metrics, we consider the conditions under which an estimate can be provided.

10.2 Criteria for Convergence to a Distribution with an Analytic Characteristic Function

Consider two characteristic functions $f_0(t)$ and $f_1(t)$ of real random variables (RVs). Assume that the function $f_0(t)$ has derivatives of all orders and is uniquely determined by them (i.e., the corresponding random variable has all moments and its distribution is determined by these moments). In this case, the coincidence of $f_0(t)$ and $f_1(t)$ in a neighborhood of $t = 0$ implies their coincidence for all values of t. Therefore, it is natural to think that the convergence on a fixed interval of a sequence of characteristic functions $\{f_n(t)\}$: $\lim_{n\to\infty} f_n(t) = f_0(t)$ for $|t| \le T_0$ (where $T_0 > 0$ is a fixed number) will imply the weak convergence of the sequence $\{F_n\}$ of the corresponding DFs to F_0. To measure the distance between two distributions F and G in terms of their characteristic functions, we employ the following metric:

$$\lambda(F, G) = \min_{T>0} \max \left(\max_{|z|\le T}(|f(z) - g(z)|, 1/T) \right), \tag{10.2.1}$$

where $f(z)$ and $g(z)$ denote the characteristic functions of F and G.

In this section, we will consider this problem for an analytic characteristic function f_0. The first result [see Klebanov and Mkrtchyan (1980)] is formulated for an entire characteristic function f_0.

Theorem 10.2.1. *Suppose that a nondegenerate DF $F(x)$ has the moments $\mu_j = \int_{-\infty}^{\infty} x^j \, dF(x)$ of all orders with $\mu_{2k}^{1/(2k)}/(2k) \to 0$ as $k \to \infty$. For $\{F_n\}_{n=1}^{\infty}$ to converge weakly to F, it is necessary and sufficient that for some $T_0 > 0$,*

$$\sup_{|t| \le T_0} |f_n(t) - f(t)| \to 0, \; n \to \infty, \tag{10.2.2}$$

where f_n and f are the characteristic functions of F_n and F, respectively. Moreover,

$$\lambda(F_n, F) \le C \min_{k=1,2,\dots} \left\{ \mu_2^{1/2} k^{3/2} \varepsilon_n^{1/(4k+1)} + \mu_{2k}^{1/(2k)}/(2k) \right\}, \tag{10.2.3}$$

where $\lambda(F, G)$ is defined in (10.2.1),

$$\varepsilon_n = \sup_{|t| \le T_0} |f_n(t) - f(t)|,$$

and C is a constant depending only on F and T_0.

Proof. Clearly, the weak convergence of F_n to F implies that $\lambda(F_n, F) \to 0$, especially as $\sup_{|t| \le T_0} |f_n(t) - f(t)| \to 0$. Therefore, it is enough to prove (10.2.3). Let $g(t)$ be an arbitrary characteristic function. We write

$$\varepsilon = \sup_{|t| \le T_0} |f(t) - g(t)| \tag{10.2.4}$$

and prove that

$$\lambda(F, G) \le C \min_{k=1,2,\dots} \left\{ \mu_2^{1/2} k^{3/2} \varepsilon^{1/(4k+1)} + \mu_{2k}^{1/(2k)}/(2k) \right\}, \tag{10.2.5}$$

where G is the DF corresponding to the characteristic function $g(t)$. This will lead to (10.2.3) when we take $g(t) = f_n(t)$.

For all real t, relation (10.2.4) can be written as

$$f(t) - g(t) = R(t; \varepsilon), \tag{10.2.6}$$

where $|R(t; \varepsilon)| \le \varepsilon$ for $|t| \le T_0$. Suppose that

$$u(t) = \begin{cases} \exp\{-1/(1+t^2) - 1/(1-t^2)\} & \text{for } t \in (-1, 1), \\ 0 & \text{for } t \notin (-1, 1), \end{cases}$$

and

$$u_\delta(t) = u(t/\delta) \Big/ \left(\delta \int_{-\infty}^{\infty} u(\tau) d\tau \right), \quad \delta > 0.$$

Let

$$\tilde{u}_\delta(z) = \int_{-\infty}^{\infty} u_{\delta/z}(t) u_{\delta/z}(t - z) dt.$$

Clearly, $\tilde{u}_\delta(z) = 0$ for $|z| \geq \delta$ and $\int_{-\infty}^{\infty} \tilde{u}_\delta(z)\mathrm{d}z = 1$. Without much difficulty we can verify that $\tilde{u}_\delta(z)$ is an infinitely differentiable function and

$$\sup_z \left|\tilde{u}_\delta^{(m)}(z)\right| \leq C m^{3m}, \quad m = 1, 2, \ldots, \tag{10.2.7}$$

where $C > 0$ is an absolute constant.

Multiplying both sides of (10.2.6) by $\tilde{u}_\delta(t - z)$ and integrating with respect to t, we obtain

$$f_\delta(z) - g_\delta(z) = R_\delta(z; \varepsilon), \tag{10.2.8}$$

where

$$f_\delta(z) = \int_{-\infty}^{\infty} f(t)\tilde{u}_\delta(t - z)\mathrm{d}t,$$

$$g_\delta(z) = \int_{-\infty}^{\infty} g(t)\tilde{u}_\delta(t - z)\mathrm{d}t,$$

$$R_\delta(z; \varepsilon) = \int_{-\infty}^{\infty} R(t; \varepsilon)\tilde{u}_\delta(t - z)\mathrm{d}t.$$

Clearly, all functions that appear in (10.2.8) are infinitely differentiable, and by (10.2.7), for any integer $n \geq 1$,

$$\left|R_\delta^{(n)}(z; \varepsilon)\right| \leq C^{3n}\varepsilon/\delta^n \quad \text{for } |z| \leq T_0 - \delta. \tag{10.2.9}$$

Differentiating both sides of (10.2.8) k times with respect to z and letting $z = 0$, in view of (10.2.9) we find that

$$\left|f_\delta^{(k)}(0) - g_\delta^{(k)}(0)\right| \leq C_k k^{3k}\varepsilon/\delta^k, \quad k = 1, 2, \ldots. \tag{10.2.10}$$

Note that although f_δ and g_δ are not characteristic functions of probability distributions, by the construction of $\tilde{u}_\delta$, they are the Fourier transforms of positive and finite (but not necessarily normalized) measures. Therefore, they have all the properties of a characteristic function with the exception of the fact that for $z = 0$ their values may be different than unity. Thus, for an arbitrary integer $k \geq 1$ and any $\delta \in (0, T_0)$, taking (10.2.10) into account we have for all real z

$$\begin{aligned} |f_\delta(z) - g_\delta(z)| &\leq \left|\sum_{j=0}^{2k-1} \frac{f_\delta^{(j)}(0) - g_\delta^{(j)}(0)}{j!} z^j\right| + \frac{\left|f_\delta^{(2k)}(0) + g_\delta^{(2k)}(0)\right|}{(2k)!} |z|^{2k} \\ &\leq C\varepsilon \frac{(2k)}{\delta^{2k}} \exp|z| + \frac{2\left|f_\delta^{(2k)}(0)\right|}{(2k)!} |z|^{2k}. \end{aligned} \tag{10.2.11}$$

Since

$$\left|f_\delta^{(2k)}(z)\right| \le \int_{-\infty}^{\infty} \left|f^{(2k)}(t)\right| \tilde{u}_\delta(t-z)\mathrm{d}t \le \mu_{2k},$$

we find from (10.2.11) that

$$|f_\delta(z) - g_\delta(z)| \le C \exp|z|(2k)^{\delta k}/\delta^{2k} + 2\mu_{2k}|z|^{2k}/(2k)!. \tag{10.2.12}$$

In addition,

$$\begin{aligned}|f_\delta(z) - f(z)| &\le \int_{-\infty}^{\infty} |f(t) - f(z)|\tilde{u}_\delta(t-z)\mathrm{d}t \\ &\le \sup_{|t-z|\le\delta} |f(t) - f(z)| \\ &\le \mu_2^{1/2}\delta.\end{aligned} \tag{10.2.13}$$

Next,

$$\begin{aligned}|g_\delta(z) - g(z)| &\le \sup_{|t-z|\le\delta} |g(t) - g(z)| = \sup_{|t-z|\le\delta} \left|\int_{-\infty}^{\infty} \left(\mathrm{e}^{ix(t-z)} - 1\right) \mathrm{e}^{izx} dG(x)\right| \\ &\le \sup_{|t-z|\le\delta} \int_{-\infty}^{\infty} \left|\mathrm{e}^{ix(t-z)} - 1\right| dG(x) \\ &\le \sup_{|t-z|\le\delta} \int_{-\infty}^{\infty} \{1 - \cos(t-z)x^2 + \sin^2(t-z)x\}^{1/2} dG(x) \\ &\le \sup_{|t-z|\le\delta} \sqrt{3} \int_{-\infty}^{\infty} \left(1 - \cos(t-z)x\right)^{1/2} dG(x) \\ &\le \sup_{|t-z|\le\delta} \sqrt{3}\left(\int_{-\infty}^{\infty} (1 - \cos(t-z)x) dG(x)\right)^{1/2} \\ &\le \sqrt{3}\left(\sup_{|\tau|\le\delta} |1 - g(\tau)|\right)^{1/2}.\end{aligned}$$

Since $\delta < T_0$,

$$\begin{aligned}\sup_{|\tau|\le\delta} |1 - g(\tau)| &\le \sup_{|\tau|\le\delta} |g(\tau) - f(\tau)| + \sup_{|\tau|\le\delta} |1 - f(\tau)| \\ &\le \varepsilon + \mu_2^{1/2}\delta.\end{aligned}$$

From this and the preceding inequality we deduce that

$$|g_\delta(z) - g(z)| \le \sqrt{3}\,\varepsilon^{1/2} + \sqrt{3}\,\mu_2^{1/4}\delta^{1/2}. \tag{10.2.14}$$

Relations (10.2.12)–(10.2.14) lead us to

$$|f(z) - g(z)| \le C\varepsilon \frac{(2k)^{\delta k}}{\delta^{2k}} \exp|z| + \frac{2\mu_{2k}}{(2k)!}|z|^{2k} + C\mu_2^{1/2}\delta^{1/2}, \tag{10.2.15}$$

which holds for any real z. Here, we assume that $\varepsilon \le \delta < T_0$, and C is an absolute constant.

Let us find the minimum with respect to δ of the right-hand side of (10.2.15). Without difficulty we can verify that this minimum is attained when

$$\begin{aligned}\delta^{1/2} &= \delta^{1/2}(z)\\ &= C^{1/(4k+1)}\varepsilon^{1/(4k+1)}(4k)^{1/(4k+1)}(2k)^{6k/(4k+1)}\exp\left\{\frac{|z|}{4k+1}\right\}/\mu_2^{1/(8k+2)}\end{aligned} \tag{10.2.16}$$

and is equal to

$$\frac{2\mu_{2k}}{(2k)!}|z|^{2k} + 2C^{\frac{4k+2}{4k+1}}\varepsilon^{\frac{1}{(4k+1)}}(4k)^{\frac{1}{(4k+1)}}(2k)^{\frac{6k}{(4k+1)}}\exp\left\{\frac{|z|}{4k+1)}\right\}/\mu_2^{1/2-1/(8k+2)}. \tag{10.2.17}$$

Since

$$\lambda(F,G) = \min_{T>0}\max\{\max_{|z|\le T}|f(z)-g(z)|,\ 1/T\},$$

we find from (10.2.15) and (10.2.17) that

$$\lambda(F,G) \le \min_{k=1,2,\dots}\ \min_{0<T\le 4k+1}\ \max\left\{\frac{2\mu_{2k}}{(2k)!}T^{2k} + C\varepsilon^{\frac{1}{4k+1}}(2k)^{3/2}\mu_2^{1/2},\ 1/T\right\}. \tag{10.2.18}$$

Here, $\delta = \delta(T)$ determined by (10.2.16) must be less than T_0.

For $T = C_1(2k)!/(2\mu_{2k})^{1/(2k+1)}$, where $C_1 > 0$ is a constant, we have

$$\begin{aligned}&\max\{2\mu_{2k}T^{2k}/(2k)! + C\varepsilon^{\frac{1}{4k+1}}(2k)^{3/2}\mu_2^{1/2}, 1/T\}\\ &\quad\le (2\mu_{2k}/(2k)!)^{1/(2k+1)} + C\varepsilon^{\frac{1}{4k+1}}(2k)^{3/2}\mu_2^{1/2}\\ &\quad\le C\mu_{2k}^{1/(2k)}/k + \varepsilon,\end{aligned}$$

where C is a new absolute constant. We now see from (10.2.18) that

$$\lambda(F,G) \le \min_{k=1,2\dots} C\{\mu_{2k}^{1/(2k)}/k + \varepsilon^{\frac{1}{4k+1}}(2k)^{3/2}\mu_2^{1/2}\} \tag{10.2.19}$$

if only $\delta(T) \le T_0$ holds. However, for $T \le 4k+1$,

$$\delta^{1/2}(T) = \tilde{C}\varepsilon^{\frac{1}{4k+1}}(2k)^{3/2}/\mu_2, \tag{10.2.20}$$

where $\tilde{C} > 0$ is an absolute constant.

Since the moments μ_{2k} cannot decrease faster than a geometric progression, it is clear that C_1 can be chosen in such a way that

$$T = C_1\left((2k)!/(2\mu_{2k})\right)^{\frac{1}{2k+1}} \le 4k + 1.$$

It is easy to see that the minimum on the right-hand side of (10.2.19) is attained for $k = k(\varepsilon)$, satisfying

$$k(\varepsilon) \le C \ln\frac{1}{\varepsilon}/\ln\ln\frac{1}{\varepsilon},$$

and that for this k the right-hand side of (10.2.20) can be made to be less than T_0 for sufficiently small $\varepsilon > 0$. □

Corollary 10.2.1. *Suppose that $F(x)$ is a DF concentrated on a finite interval (a, b) and DF $G(x)$ is such that the characteristic function $g(t)$ satisfies (10.2.4). Then there exist a constant $\varepsilon_0 > 0$, depending only on T_0, a, and b, and a constant $C > 0$, depending only on a and b, for which*

$$\lambda(F, G) \le C\frac{\ln\ln\frac{1}{\varepsilon}}{\ln\frac{1}{\varepsilon}}$$

when $0 < \varepsilon \le \varepsilon_0$.

Corollary 10.2.2. *Suppose that $F(x)$ is the standard normal DF and $G(x)$ is such that its characteristic function $g(t)$ satisfies (10.2.4). Then there exist constants $\varepsilon_0 = \varepsilon_0(T_0) > 0$ and $C > 0$ such that for $\varepsilon \in (0, \varepsilon_0]$,*

$$\lambda(F, G) \le C\left(\frac{\ln\ln\frac{1}{\varepsilon}}{\ln\frac{1}{\varepsilon}}\right)^{1/2}.$$

Let us now turn to the case of an analytic characteristic function f_0. First let us obtain an estimate of the closeness of F and G in λ knowing that $f(t) := f_0(t)$ and $g(t)$ are close in some fixed neighborhood of zero.

Theorem 10.2.2. *Let $F(x)$ be a distribution function whose characteristic function $f(t)$ is analytic in $|t| \le R$. Assume that $G(x)$ is such that its characteristic function $g(t)$ satisfies*

$$|f(t) - g(t)| \le \varepsilon \tag{10.2.21}$$

for real $t \in [-T_0, T_0]$. Then there exist $\varepsilon_0 = \varepsilon_0(T_0) > 0$ and $C > 0$, depending only on F and T_0, such that for $\varepsilon \in (0, \varepsilon_0]$,

$$\lambda(F, G) \leq C \left(\ln \ln \frac{1}{\varepsilon} \right)^{-1}. \tag{10.2.22}$$

The proof of this theorem will require the following lemma, which was obtained in Sapogov (1980).

Lemma 10.2.1. *Suppose that $F(x)$ is an arbitrary distribution function and $f(t)$ its characteristic function. We denote by $\Delta_u^{(2k)}(f;t)$ a symmetric, $2k$th-order finite difference of f with step $u \geq 0$ at $t \in \mathbb{R}^1$. Then*

$$F\left(-\frac{2\pi}{s}\right) + 1 - F\left(\frac{2\pi}{s}\right) \leq \frac{(-1)^k 2\pi}{4^k I_{2k} s} \int_0^s \Delta_u^{(2k)}(f, 0) \mathrm{d}u, \tag{10.2.23}$$

where $s > 0$ is arbitrary, $k = 1, 2, \ldots$, and

$$I_{2k} = \frac{(2k-1)!!}{(2k)!!} \pi.$$

Proof of Lemma 10.2.1. It is well known that for any function $\varphi(t)$

$$\Delta_u^{(2k)}(\varphi; t) = \sum_{l=0}^{2k} (-1)^l \varphi\left(t - (k-1)u\right). \tag{10.2.24}$$

For $\varphi(t) = \exp(itx)$ we have

$$\Delta_u^{(2)}\big(\exp(itx); t\big) = \exp(itx) \left(-4 \sin^2 \frac{ux}{2}\right).$$

Since

$$\Delta_u^{(2k)}(\varphi; t) = \Delta_u^{(2)}\big(\Delta_u^{(2k-2)}(\varphi; t); t\big),$$

for any $k \geq 1$ we obtain

$$\Delta_u^{(2k)}\big(\exp(itx); t\big) = \exp(itx) \left(-4 \sin^2 \frac{ux}{2}\right)^k,$$

$$\Delta_u^{(2k)}\big(\exp(itx); 0\big) = \left(-4 \sin^2 \frac{ux}{2}\right)^k.$$

By the fact that $\Delta_u^{(2k)}(\varphi; t)$ is linear, we find from the last relation that

$$(-1)^k \Delta_u^{(2k)}(f; 0) = 4^k \int_{-\infty}^{\infty} \sin^{2k} \frac{xu}{2} \mathrm{d}F(x). \tag{10.2.25}$$

Integrating this identity with respect to u between $0 \leq u \leq s < \infty$ we obtain

$$
\begin{aligned}
(-1)^k \int_0^s \Delta_u^{(2k)}(f;0)\mathrm{d}u &= 4^k \int_0^s \int_{-\infty}^{\infty} \sin^{2k} \frac{xu}{2} \mathrm{d}F(x)\mathrm{d}u \\
&= 4^k \int_0^s \int_{-\infty}^{\infty} \sin^{2k} \frac{xu}{2} d[F(x) - F(-x)]\mathrm{d}u. \quad (10.2.26)
\end{aligned}
$$

Suppose that for $h = 0, 1, 2, \ldots$ the domains $E_h \subset \mathbb{R}^2$ and $G_h \subset \mathbb{R}^2$ are determined as follows:

$$
\begin{aligned}
E_h &:= \{(x,u) : 0 \le u \le s,\ 2h\pi \le x \le 2(h+1)\pi\}, \\
G_h &:= \{(x,u) : x_h \le x < \infty,\ 2h\pi \le xu \le 2(h+1)\pi\},
\end{aligned}
$$

where

$$
x_h := 2\pi(h+1)/s. \quad (10.2.27)
$$

If $(x,u) \in G_h$, then $2h\pi/x \le u \le 2(h+1)\pi/x$ and according to (10.2.27), $0 \le u \le 2(h+1)\pi/x_h = s$. Consequently, $(x,u) \in E_h$. This means that $G_n \subset E_h$, $h = 0, 1, 2, \ldots$. Now, letting $F_1(x) = F(x) - F(-x)$, we obtain by (10.2.26) that

$$
\begin{aligned}
(-1)^k \int_0^s \Delta_u^{(2k)}(f;0)\mathrm{d}u &= 4^k \sum_{h=0}^{\infty} \int\int_{E_h} \sin^{2k} \frac{xu}{2} \mathrm{d}F_1(x)\mathrm{d}u \\
&\ge 4^k \sum_{h=0}^{\infty} \int\int_{G_h} \sin^{2k} \frac{xu}{2} \mathrm{d}F_1(x)\mathrm{d}u \\
&= 4^k \sum_{h=0}^{\infty} \int_{x_h}^{\infty} \mathrm{d}F_1(x) \int_{2\pi h/x}^{2\pi(h+1)/x} \sin^{2k} \frac{xu}{2} \mathrm{d}u \\
&= 4^k \sum_{h=0}^{\infty} \int_{x_h}^{\infty} \mathrm{d}F_1(x) \frac{x}{2} \int_{\pi h}^{\pi(h+1)} \sin^{2k} y \mathrm{d}y \\
&= 4^k 2 \sum_{h=0}^{\infty} \int_{x_h}^{\infty} \frac{\mathrm{d}F_1(x)}{x} I_2 k \\
&= 4^k 2 I_{2k} \sum_{h=0}^{\infty} \sum_{l=h}^{\infty} \int_{x_i}^{X_{i+1}} \frac{\mathrm{d}F_1(x)}{x} \\
&= 4^k 2 I_{2k} \sum_{h=1}^{\infty} h \int_{x_{h-1}}^{x_h} \frac{\mathrm{d}F_1(x)}{x} \\
&\ge 4^k 2 I_{2k} \sum_{h=1}^{\infty} \frac{h}{x_h} \int_{x_{h-1}}^{x_h} \mathrm{d}F_1(x)
\end{aligned}
$$

$$
\begin{aligned}
&= 4^k 2 I_{2k} \sum_{h=1}^{\infty} \frac{hs}{2\pi(h+1)} \int_{x_{h-1}}^{x_h} \mathrm{d}F_1(x) \\
&\geq 4^k I_{2k} \frac{s}{2\pi} \int_{x_0}^{\infty} \mathrm{d}F_1(x) \\
&= \frac{4^k s}{2\pi} I_{2k} \left[F(-x_0) + 1 - F(x_0)\right].
\end{aligned}
$$

Here we denoted $x_0 = 2\pi/s$ and used

$$
I_{2k} := \frac{(2k-1)!!}{(2k)!!} = \int_0^{\pi} \sin^{2k} y \mathrm{d}y.
$$

This concludes the proof of Lemma 10.2.1. □

Remark 10.2.1. Let us note now that if $f(t)$ has a derivative of order $2k$ at $t = 0$, then

$$
\left|\Delta_u^{(2k)}(f;0)\right| \leq u^{2k} \left|f^{(2k)}(0)\right|. \tag{10.2.28}
$$

Let us now prove the theorem.

Proof of Theorem 10.2.2. Suppose that $g(t)$ satisfies (10.2.21). The distribution function $G(x)$ corresponding to $g(t)$ is truncated at π/s, where s is a positive number. This means that the probability $\int_{|x|\geq\pi/s} dG(x)$ is displaced from $|x| \geq \pi/s$ to $x = 0$ on $\mathbb{R}^1$.

For the corresponding distribution function $G^*(x)$ we have

$$
\int_{-\pi/s}^{\pi/s} \varphi(x) d\left(G^*(x) - G(x)\right) = \varphi(0) \int_{|x|\geq\pi/s} dG(x),
$$

regardless of what the continuous function $\varphi : [-\pi/s, \pi/s] \to \mathbb{R}^1$ is. Therefore, for any integer $k \geq 1$,

$$
\left|g^{*(2k)}(0)\right| = \int_{-\pi/s}^{\pi/s} x^{2k} dG(x), \tag{10.2.29}
$$

$$
\begin{aligned}
|g^*(t) - g(t)| &\leq \left|\int_{-\infty}^{\infty} e^{itx} d\left(G^*(x) - G(x)\right)\right| \\
&\leq 2 \int_{|x|\geq\pi/s} dG(x),
\end{aligned} \tag{10.2.30}
$$

where $t \in \mathbb{R}^1$ and $g^*(t) = \int_{-\infty}^{\infty} \exp(itx) dG^*(x)$. Next, according to (10.2.24), letting $\varphi = f$ and $\varphi = g$, we find, with $0 \leq ku \leq ks \leq T_0$ and (10.2.21) taken into account, that

$$\left|\Delta_u^{(2k)}(f;0) - \Delta_u^{(2k)}(g;0)\right| \le \sum_{l=0}^{2k} \binom{2k}{l} \left|f\big(t-(k-l)u\big) - g\big(t-(k-l)u\big)\right|$$
$$\le \varepsilon 4^k. \tag{10.2.31}$$

Therefore, for $u = s$,

$$\left|\Delta_s^{(2k)}(g;0)\right| \le \left|\Delta_s^{(2k)}(f;0)\right| + \varepsilon 4^k \tag{10.2.32}$$

and

$$\left|\int_0^s \Delta_u^{(2k)}(g;0)\mathrm{d}u\right| \le \left|\int_0^s \Delta_u^{(2k)}(f;0)\right| + \varepsilon s 4^k. \tag{10.2.33}$$

To estimate $\int_{|x|\ge\pi/s} dG(x)$, we apply Lemma 10.2.1 and (10.2.33), substituting s for $2s$:

$$\int_{|x|\ge\pi/s} dG(x) \le \frac{\pi}{4^k l_{2k} s}\left|\int_0^{2s} \Delta_u^{(2k)}(g;0)\mathrm{d}u\right|$$
$$\le \frac{\pi}{4^k I_{2k} s}\left(\left|\int_0^{2s} \Delta_u^{(2k)}(f;0)\mathrm{d}u\right| + 2\varepsilon 4^k\right). \tag{10.2.34}$$

Denote $\mu_{2k} = \int_{-\infty}^{\infty} x^{2k}\mathrm{d}F(x)$. By (10.2.28) and the fact that $\left|f^{(2k)}(0)\right| = \mu_{2k}$, we have

$$\left|\int_0^{2s} \Delta_u^{(2k)}(f;0)\mathrm{d}u\right| \le \mu_{2k}\frac{(2s)^{2k+1}}{2k+1}. \tag{10.2.35}$$

Since $f(t)$ is analytic in $|t| \le R$, for all integers $k \ge 1$

$$\mu_{2k} \le C\frac{(2k)!}{\mathbb{R}^{2k}}. \tag{10.2.36}$$

Here and subsequently in the proof of the theorem, C denotes a constant (possibly different on each occasion) that depends only on f.

Relations (10.2.35) and (10.2.36) and Stirling's formula imply

$$\frac{\pi}{4^k I_{2k} s}\left|\int_0^{2s} \Delta_u^{(2k)}(f;0)\mathrm{d}u\right| \le \frac{\mu_{2k}(2s)^{2k}(2k)!!}{\pi(2k+1)!!}$$
$$\le C\left(\frac{s2k}{eR}\right)^{2k}\sqrt{k}. \tag{10.2.37}$$

From (10.2.21), (10.2.30), (10.2.34), and (10.2.37) we derive, for $-T_0 \le t \le T_0$,

$$|F(t) - G^*(t)| \le \varepsilon + \frac{\pi}{4^k I_{2k} s}\left(\left|\int_0^{2s} \Delta_u^{(2k)}(f;0)\mathrm{d}u\right| + 2\varepsilon s 4^k\right)$$
$$\le C\sqrt{k}\left[\left(\frac{s2k}{eR}\right)^{2k} + \varepsilon\right]. \tag{10.2.38}$$

Passing from $F(x)$ to its truncation $F^*(x)$, we obtain

$$|f(t) - g^*(t)| \le C\sqrt{k}\left[\left(\frac{2ks}{eR}\right)^{2k} + \varepsilon\right] \tag{10.2.39}$$

for $-T_0 \le t \le T_0$.

Let us now choose $\rho > 0$ and consider $g^*(t)$ for complex t with $|\mathrm{Im}(t)| \le \rho$. We assume that ρ is fixed and chosen sufficiently small so that $\rho < \min(R, T_0, 1)$. Subsequently more constraints will be imposed on ρ. We write

$$W(x) := 1 - G^*(x) + G^*(-x).$$

We have

$$|g^*(t)| = \left|\int_{-\pi/s}^{\pi/s} \mathrm{e}^{itx} dG^*(x)\right|$$
$$= \int_{-\pi/s}^{\pi/s} \mathrm{e}^{\rho x} dG^*(x)$$
$$= \sum_{n=0}^{\infty} \frac{\rho^{\pi}}{n!} \int_{-\pi/s}^{\pi/s} |x|^n dG^*(x). \tag{10.2.40}$$

However, for $n \ge 1$,

$$\int_{-\pi/s}^{\pi/s} |x|^n dG^*(x) = -\int_0^{\pi/s} x^n \mathrm{d}W(x)$$
$$= n\int_0^{\pi/s} x^{n-1} W(x)\mathrm{d}x.$$

In addition, because (10.2.34) and (10.2.37) imply

$$W(x) \le C\sqrt{k}\left[\left(\frac{2k\pi}{xRe}\right)^{2k} + \varepsilon\right],$$

therefore, for $\pi/s > 1$ and $n \ge 1$,

$$\int_{-\pi/s}^{\pi/s} |x|^n dG^*(x) = n\int_0^1 x^{n-1}W(x)\mathrm{d}x + n\int_1^{\pi/s} x^{n-1}W(x)\mathrm{d}x$$

$$\leq n\int_0^1 x^{n-1}\mathrm{d}x + n\int_1^{\pi/s} x^{n-1}C\sqrt{k}\left[\left(\frac{2\pi k}{xRe}\right)^{2k} + \varepsilon\right]\mathrm{d}x$$

$$\leq 1 + C\sqrt{k}\varepsilon\left(\frac{\pi}{s}\right)^n + C\sqrt{k}\left(\frac{2\pi k}{Re}\right)^{2k} n\int_1^{\pi/s} x^{n-2k-1}\mathrm{d}x.$$

Thus,

$$\int_{-\pi/s}^{\pi/s} |x|^n dG^*(x) = 1 + C\sqrt{k}\varepsilon\left(\frac{\pi}{s}\right)^n + C\sqrt{k}\left(\frac{2\pi k}{Re}\right)^{2k}\frac{n}{n-2k}\left(\frac{\pi}{s}\right)^{n-2k}$$

for $n \neq 2k$ and

$$\int_{-\pi/s}^{\pi/s} |x|^{2k} dG^*(x) \leq 1 + C\sqrt{k}\varepsilon\left(\frac{\pi}{s}\right)^{2k} + C\sqrt{k}\left(\frac{2\pi k}{Re}\right)^{2k} 2k\ln\frac{\pi}{s}.$$

Substituting the last two estimates into (10.2.34) and applying Stirling's formula, we find that

$$|g^*(t)| \leq 1 + \sum_{\substack{n=1\\ n\neq 2k}} \frac{\rho^n}{n!}\left[1 + C\sqrt{k}\varepsilon\left(\frac{\pi}{s}\right)^n + C\sqrt{k}\left(\frac{2\pi k}{Re}\right)^{2k}\frac{n}{n-2k}\left(\frac{\pi}{s}\right)^{n-2k}\right]$$

$$+\frac{\rho^{2k}}{(2k)!}\left(1 + C\sqrt{k}\varepsilon\left(\frac{\pi}{s}\right)^{2k} + C\sqrt{k}\left(\frac{2\pi k}{Re}\right)^{2k} 2k\ln\frac{\pi}{s}\right)$$

$$\leq \mathrm{e}^{\rho} + \varepsilon C\sqrt{k}\mathrm{e}^{\frac{\rho\pi}{s}} + C\sqrt{k}\left(\frac{2k}{Re}\right)^{2k} s^{2k}\mathrm{e}^{\frac{\rho\pi}{s}} + C\left(\frac{2\pi\rho}{Re}\right)^{2k}\ln\frac{\pi}{s}. \tag{10.2.41}$$

Note that (10.2.39) and (10.2.41) hold for all $s > 0$ and all integers $k \geq 1$. Let us first choose $s = \rho\pi/(2k)$ in these relations. We then have

$$|f^*(t) - g^*(t)| \leq C\sqrt{k}\left[\left(\frac{\rho\pi}{Re}\right)^{2k} + \varepsilon\right] \tag{10.2.42}$$

for real $t \in [-T_0, T_0]$ and

$$|g^*(t)| \leq \mathrm{e}^{\rho} + C\sqrt{k}\varepsilon\mathrm{e}^{2k} + C\left(\frac{\pi\rho}{R}\right)^{2k} + \ln\frac{2k}{\rho} \tag{10.2.43}$$

for complex t with $|\mathrm{Im}(t)| \leq \rho$. Without loss of generality, we can assume that $\rho\pi/R < 1$. Let us choose in (10.2.42) and (10.2.43)

$$k = \left[\alpha \ln \frac{1}{\varepsilon}\right],$$

where $\alpha > 0$ is a sufficiently small number and $[x]$ denotes the integer part of x. We can assume that $\varepsilon_0 = \varepsilon_0(\alpha)$ is chosen sufficiently small so that $k > 0$. Then we obtain from (10.2.42)

$$|f^*(t) - g^*(t)| \leq C\varepsilon^\gamma, \quad t \in [-T_0, T_0], \tag{10.2.44}$$

for some $\gamma = \gamma(\alpha) > 0$, and from (10.2.43) we find that

$$|g^*(t)| \leq M(\rho), \quad |\mathrm{Im}(t)| \leq \rho, \tag{10.2.45}$$

where $M(\rho)$ depends only on ρ and ε_0.

The arguments given in deriving (10.2.45) also apply in estimating the modulus of $f^*(t)$ (the corresponding calculations can even be simplified). That is, we can assume that

$$|f^*(t)| \leq M(\rho), \quad |\mathrm{Im}(t)| \leq \rho. \tag{10.2.46}$$

Next, we obtain a relation similar to (10.2.44) but for complex t. Let us choose an arbitrary integer $n > 1$ and write

$$f^*(t) - g^*(t) = \sum_{j=0}^{2n-1} \frac{f^{*(j)}(0) - g^{*(j)}(0)}{j!} t^j + \frac{f^{*(2n)}(\tau) - g^{*(2n)}(\tau)}{(2n)!} t^{2n}, \tag{10.2.47}$$

where τ satisfies $|\tau| \leq |t|$. We have

$$f^{*(2n)}(\tau) = \int_{-\pi/s}^{\pi/s} (-1)^n x^{2n} e^{i\tau x} dF^*(x),$$

$$g^{*(2n)}(\tau) = \int_{-\pi/s}^{\pi/s} (-1)^n x^{2n} e^{i\tau x} dG^*(x).$$

From this we find that for real τ

$$\begin{aligned} \left|f^{*(2n)}(\tau)\right| &\leq \int_{-\pi/s}^{\pi/s} x^{2n} dF^*(x) \\ &\leq \left(\frac{\pi}{s}\right)^{2n}, \\ \left|g^{*(2n)}(\tau)\right| &\leq \left(\frac{\pi}{s}\right)^{2n}. \end{aligned} \tag{10.2.48}$$

If, however, τ is a complex number with $|\mathrm{Im}(\tau)| < \rho/2$, then

$$|f^{*(2n)}(\tau)| \le \int_{-\pi/s}^{\pi/s} x^{2n} e^{\rho s/2} dF^*(x)$$

$$\le \left(\int_{-\pi/s}^{\pi/s} x^{4n} dF^*(x) \right)^{1/2},$$

$$\left(\int_{-\pi/s}^{\pi/s} e^{\rho x} dF^*(x) \right)^{1/2} \le (M(\rho))^{1/2} \left(\frac{\pi}{s}\right)^{2n}, \tag{10.2.49}$$

and, analogously,

$$|g^{*(2n)}(\tau)| \le (M(\rho))^{1/2} \left(\frac{\pi}{s}\right)^{2n}, \quad |\mathrm{Im}(\tau)| \le \rho/2. \tag{10.2.50}$$

From (10.2.47) and (10.2.48) we obtain for real t such that $|t| \le \min(T_0, 1)$

$$\left| \sum_{j=0}^{2n-1} \frac{f^{*(j)}(0) - g^{*(j)}(0)}{j!} t^j \right| \le C\varepsilon^\gamma + \left(\frac{\pi}{s}\right)^{2n} \frac{1}{(2n)!}.$$

Since we let $s = \rho\pi/(2k)$, for real t such that $|t| \le T_0 := \min(T_0, 1)$ the last inequality produces

$$\left| \sum_{j=0}^{2n-1} \frac{f^{*(j)}(0) - g^{*(j)}(0)}{j!} t^j \right| \le C\varepsilon^\gamma + \left(\frac{2k}{\rho}\right)^{2n} \frac{1}{(2n)!}. \tag{10.2.51}$$

But then it is well known that[1]

$$\left| \sum_{j=0}^{2n-1} \frac{f^{*(j)}(0) - g^{*(j)}(0)}{j!} t^j \right| \le C \left(\varepsilon^\gamma + \left(\frac{ke}{\rho n}\right)^{2n} \frac{1}{\sqrt{n}} \right) \kappa^{2n} \tag{10.2.52}$$

for $\kappa > 1$ and complex t inside an ellipse with foci $\pm T_1$, real semi-axis $T_1 \frac{\kappa + 1/\kappa}{2}$, and imaginary semi-axis $T_1 \frac{\kappa - 1/\kappa}{2}$.

Let us consider complex t such that $|t| \le \rho_1$, where $\rho_1 > 0$ and $\kappa > 1$ are chosen in such a way that

$$\rho_1 < T_1 \frac{\kappa - 1/\kappa}{2} < \rho/2.$$

[1] See Bernstein (1937, Chap. II, Sect. 1).

Then for complex t such that $|t| \le \rho_1$, we obtain from (10.2.52), (10.2.47), (10.2.49), and (10.2.50) that

$$\left|f^{*(j)}(t) - g^{*(j)}(t)\right| \le C\left(\varepsilon^{\gamma} + \left(\frac{ke}{\rho n}\right)^{2n}\frac{1}{\sqrt{n}}\right)\kappa^{2n}$$
$$+M^{1/2}(\rho)C\left(\frac{ke}{\rho n}\right)^{2n}\frac{(\kappa - 1/\kappa)^{2n}}{2^n\sqrt{n}}.$$

If we assume in the last relation that $n = \beta k$, where $e/(\beta\rho) < 1$, then we obtain

$$|f^*(t) - g^*(t)| \le C\varepsilon^{\gamma_1}, \tag{10.2.53}$$

where $\gamma_1 > 0$ is such that $\gamma_1 < \gamma$ and (10.2.53) is fulfilled for complex t with $|t| \le \rho_1$.

Let us now consider a conformal mapping $\zeta = th\left(\frac{\pi t}{4\rho_1}\right)$ of $\{t : |\text{Im}(t)t| \le \rho_1\}$ onto the unit circle $\{\zeta : |\zeta| \le 1\}$. Denote by r the radius of the largest circle with center at $\zeta = 0$ that can be inscribed into the image of $|t| \le \rho_1$ under the mapping

$$\zeta = th\left(\frac{\pi t}{4\rho_1}\right). \tag{10.2.54}$$

Let

$$\varphi(\zeta) = f^*\big(t(\zeta)\big),$$
$$\Psi(\zeta) = g^*\big(t(\zeta)\big).$$

From (10.2.53) we have

$$|\varphi(\zeta) - \Psi(\zeta)| \le C\varepsilon^{\gamma_1}, \quad |\zeta| \le r, \tag{10.2.55}$$

and from (10.2.45) and (10.2.46)

$$|\varphi(\zeta)| \le M(\rho), \quad \Psi(\zeta)| \le M(\rho), \quad |\zeta| \le 1. \tag{10.2.56}$$

Since $\varphi(\zeta)$ and $\Psi(\zeta)$ are analytic in $|\zeta| \le 1$, we can let

$$\varphi(\zeta) = \sum_{j=0}^{\infty} a_j\zeta^j,$$

$$\Psi(\zeta) = \sum_{j=0}^{\infty} b_j\zeta^j.$$

When $|\zeta| \le r_1 < r$, taking (10.2.55) into account and using the Cauchy inequality for coefficients of the expansion of an analytic function into a power series, we obtain

$$\left|\varphi(\zeta) - \Psi(\zeta) - \sum_{j=0}^{m}(a_j - b_j)\zeta^j\right| \le \left|\sum_{j=m+1}^{\infty}(a_j - b_j)\zeta^j\right|$$
$$\le C\varepsilon^{\gamma}\frac{(r_1/r)^{m+1}}{1 - r_1/r}, \qquad (10.2.57)$$

where $m > 1$ is an arbitrary integer. Hence,

$$\left|\sum_{j=0}^{m}(a_j - b_j)\zeta^j\right| \le C\varepsilon^{\gamma_1}\left[1 + \frac{(r_1/r)^{m+1}}{1 - r_1/r}\right] \qquad (10.2.58)$$

for $|\zeta| \le r_1 < r$. According to Bernstein (1937), for $\sigma > 1$ and real

$$\zeta \in \left[-r_1\frac{\sigma + 1/\sigma}{2}, r_1\frac{\sigma + 1/\sigma}{2}\right]$$

we have

$$\left|\sum_{j=0}^{m}(a_j - b_j)\zeta^j\right| \le C\varepsilon^{\gamma_1}\left[1 + \frac{(r_1/r)^{m+1}}{1 - r_1/r}\right]\sigma^{m+2}. \qquad (10.2.59)$$

However, for all complex $\zeta \in \left\{|\zeta| < r_1\frac{\sigma - 1/\sigma}{2}\right\}$

$$\left|\varphi(\zeta) - \Psi(\zeta) - \sum_{j=0}^{m}(a_j - b_j)\zeta^j\right| \le \left|\sum_{j=m+1}^{\infty}(a_j - b_j)\zeta^j\right|$$
$$\le M(\rho)\frac{\left(r_1(\sigma + 1/\sigma)/2\right)^{m+1}}{1 - r_1(\sigma + 1/\sigma)/2}, \qquad (10.2.60)$$

where we used (10.2.56) and the Cauchy inequality for coefficients of the expansion of an analytic function into a power series. From (10.2.59) and (10.2.60) we conclude that

$$|\varphi(\zeta) - \Psi(\zeta)| \le C\varepsilon^{\gamma_1}\left[1 + \frac{(r_1/r)^{m+1}}{1 - r_1/r}\right]\sigma^{m+1} + M(\rho)\frac{\left(r_1(\sigma + 1/\sigma)/2\right)^{m+1}}{1 - r_1(\sigma + 1/\sigma)/2}$$

for real ζ such that $|\zeta| \le r_1(\sigma + 1/\sigma)/2$. Denoting

$$\Theta := 1 - r_1(\sigma + 1/\sigma)/2$$

and taking into account that $0 < r_1 < r$, we can make the last inequality slightly cruder:

$$|\varphi(\zeta) - \Psi(\zeta)| \le C\varepsilon^{\gamma_1}\sigma^{m+2} + M(\rho)(1 - \Theta)^{m+1}/\Theta \qquad (10.2.61)$$

for real ζ such that $|\zeta| \le 1 - \Theta$. Note that (10.2.61) is true for all integers $m \ge 1$ and all real $\Theta \in (0, 1)$. Now, in (10.2.61) let

$$m = \left[\alpha_1 \ln \frac{1}{\varepsilon}\right],$$

$$\Theta = \frac{\ln\ln 1/\varepsilon}{\alpha_2 \ln 1/\varepsilon},$$

where $\alpha_1 > 0$ and $\alpha_2 > 0$ are sufficiently small (but are independent of ε). Then, elementary (though still quite cumbersome) calculations show that

$$\varepsilon^{\gamma_1}\sigma^{m+2} + M(\rho)(1-\Theta)^{m+1}/\Theta \le C(\ln 1/\varepsilon)^{-\gamma_2},$$

where $\gamma_\epsilon > 0$ is a constant. Therefore, (10.2.7) yields

$$|\varphi(\zeta) - \Psi(\zeta)| \le C(\ln 1/\varepsilon)^{-\gamma_2}$$

for real ζ, satisfying the condition

$$|\zeta| \le 1 - \ln\ln \frac{1}{\varepsilon} \Big/ \left(\alpha_1 \ln \frac{1}{\varepsilon}\right).$$

Turning from $\varphi(\zeta)$ and $\Psi(\zeta)$ back to $f(t)$ and $g(t)$ we obtain

$$|f(t) - g(t)| \le C\left(\ln \frac{1}{\varepsilon}\right)^{-\gamma_2}$$

for real t, satisfying

$$|t| \le \frac{4\rho_1}{\pi} ath \left(1 - \frac{\ln\ln \frac{1}{\varepsilon}}{\alpha_1 \ln \frac{1}{\varepsilon}}\right),$$

that is, for

$$|t| \le C \ln\ln \frac{1}{\varepsilon},$$

which concludes the proof of Theorem 10.2.2. □

Corollary 10.2.3. *Suppose that a nondegenerate $F(x)$ satisfies Cramér's condition: there exists a positive constant R such that $\int_{-\infty}^{\infty} \exp(R|x|)\mathrm{d}F(x) < \infty$. A sequence $\{F_n\}_{n=1}^{\infty}$ of DFs converges weakly to $F(x)$ if and only if for some $T_0 > 0$*

$$\varepsilon_n = \sup_{|t| \le T_0} |f_n(t) - f(t)| \to 0, \quad n \to \infty,$$

where $f_n(t), f(t)$ are the characteristic functions of F_n and F, respectively. Moreover,

$$\lambda(F_n, F) \le C \Big/ \left(\ln\ln\frac{1}{\varepsilon_n}\right).$$

To prove this result, it is enough to note that Cramér's condition is equivalent to the analyticity of $f(t)$ in $|t| < R$ and then use Theorem 10.2.2.

10.3 Moment Metrics

Suppose that D is a set of DFs given on the real line $\mathbb{R}^1$ with finite moments of all orders and uniquely determined by them. Below we give a definition of metrics on D in which closeness means closeness (or coincidence) of a certain number of moments of the corresponding distributions.

Assume that $F \in D$. We denote by $\mu_j(F)$ the jth moment of the DF F:

$$\mu_j(F) := \int_{-\infty}^{\infty} x^j \,\mathrm{d}F(x) \quad (j \ge 0 \text{ is an integer}).$$

For each positive number α we introduce in D a metric d_α by setting

$$d_\alpha(F_1, F_2) := \min_{k=0,1,\dots} \max\left\{\frac{1}{k+1}, \alpha|\mu_0(F_1) - \mu_0(F_2)|, \dots, \alpha|\mu_k(F_1) - \mu_k(F_2)|\right\}. \tag{10.3.1}$$

Let us show that d_α is indeed a metric. Clearly, d_α is symmetric in F_1 and F_2, and $0 \le d_\alpha(F_1, F_2) \le 1$. Moreover, $d_\alpha(F_1, F_2) = 0$ implies the coincidence of all moments of F_1 and F_2, and consequently the equality of F_1 and F_2 since $F_i \in D$ $(i = 1, 2)$. It remains to show that d_α satisfies the triangle inequality. To this end, let us clarify the meaning of d_α. Let

$$d_\alpha(F_1, F_2) = d, \tag{10.3.2}$$

where $d > 0$ is a number. If $1/d - 1 = k$ is an integer, then (10.3.2) is equivalent to

$$|\mu_j(F_1) - \mu_j(F_2)| \le d/\alpha \tag{10.3.3}$$

being fulfilled for $j = 0, 1, \dots, k$. If, however, $1/d$ is not an integer, then (10.3.2) is equivalent to (10.3.3) for $j = 0, 1, \dots, [1/d]$. Let $F_1, F_2, F_3 \in D$. For any j

$$\alpha|\mu_j(F_1) - \mu_j(F_2)| \le \alpha|\mu_j(F_1) - \mu_j(F_3)| + \alpha|\mu_j(F_2) - \mu_j(F_3)|. \tag{10.3.4}$$

Let us show that $d_\alpha(F_1, F_2) \le d_\alpha(F_1, F_3) + d_\alpha(F_3, F_2)$. Without loss of generality, we can assume that $d_\alpha(F_1, F_3) \le d_\alpha(F_3, F_2)$.

To prove the triangle inequality, it is enough to show that

$$\alpha|\mu_j(F_1) - \mu_j(F_2)| \le d_\alpha(F_1, F_3) + d_\alpha(F_3, F_2)$$

for

$$j \leq [1/(d_\alpha(F_1, F_3) + d_\alpha(F_3, F_2))].$$

However,

$$[1/(d_\alpha(F_1, F_3) + d_\alpha(F_3, F_2))] \leq [1/d_\alpha(F_1, F_3)]$$

if $1/d_\alpha(F_1, F_3)$ is not an integer, and

$$[1/(d_\alpha(F_1, F_3) + d_\alpha(F_3, F_2))] \leq 1/d_\alpha(F_1, F_3) - 1$$

if $1/d_\alpha(F_1, F_3)$ is an integer. The conclusion now follows from the value of $d_\alpha(F_1, F_3)$, $d_\alpha(F_2, F_3)$, and (10.3.4).

Clearly, for $0 < \alpha < \infty$ the metrics d_α are topologically equivalent to each other.

Let us now introduce in D the metric d_∞ by setting

$$d_\infty(F_1, F_2) = \frac{1}{k+1} \quad (k \geq 0)$$

if all moments of F_1 and F_2 up to and including order k coincide and $\mu_{k+1}(F_1) \neq \mu_{k+1}(F_2)$. It is easy to verify that $d_\infty(F_1, F_2)$ is the limit of $d_\alpha(F_1, F_2)$ as $\alpha \to \infty$. Clearly, d_∞ is a metric on D. Moreover, it satisfies the strengthened version of the triangle inequality:

$$d_\infty(F_1, F_2) \leq \max(d_\infty(F_1, F_3), d_\infty(F_3, F_2))$$

for all $F_1, F_2, F_3 \in D$ (so that d_∞ is an ultrametric on D). It is also clear that for $\alpha_1 < \alpha_2 < \infty$ we have

$$d_{\alpha_1}(F_1, F_2) < d_{\alpha_2}(F_2, F_2) \leq d_\infty(F_1, F_2).$$

It is easy to verify that d_∞ is a stronger metric than any of the d_α with $\alpha < \infty$.

The metrics d_α and d_∞ can be extended to the space of all distribution functions on $\mathbb{R}^1$ by setting

$$d_\alpha(F_1, F_2) = \min_{k=0,1,\dots,m} \max\left\{\frac{1}{k+1}, \alpha|\mu_0(F_1) - \mu_0(F_2)|, \dots, \alpha|\mu_k(F_1) - \mu_k(F_2)|\right\},$$

$$d_\infty(F_1, F_2) = \lim_{\alpha\to\infty} d_\alpha(F_1, F_2).$$

Here m is determined from the condition that moments of F_1 and F_2, up to and including order m, exist (are finite) and at least one of the F_1 and F_2 does not have a finite moment of order $m+1$. Note that under such considerations, d_α ceases to be a metric. Indeed, $d_\alpha(F_1, F_2) = 0$ does not generally imply that $F_1 = F_2$ (this occurs if F_1 and F_2 have coinciding moments of all orders, but the problem of moments is indeterminate for them). However, the loss of this property is inconsequential for our purposes, and we retain the term "metric" for $d_\alpha(F_1, F_2)$ $(0 < \alpha \le \infty)$ even when $F_1, F_2 \notin D$. Note that d_∞ was first introduced in Mkrtchyan (1978) and d_α $(0 < \alpha < \infty)$ in Klebanov and Mkrtchyan (1979).

10.3.1 Estimates of λ by Means of d_∞

Suppose that for two DFs F and G

$$d_\infty(F, G) = \frac{1}{2m+1}, \quad m \ge 2.$$

Then

$$\mu_j(F) = \mu_j(G), \quad j = 0, 1, \ldots, 2m.$$

Let

$$\mu_j = \mu_j(F) = \mu_j(G), \quad j = 0, 1, \ldots, 2m,$$

$$\beta_m = \sum_{j=1}^{m} \mu_{2j}^{-1/(2j)}. \tag{10.3.5}$$

Clearly, β_m is a truncated Carleman's series. Since the divergence of a Carleman's series is a sufficient condition for the problem of moments to be determinate,[2] it is natural to seek an estimate of the closeness of F and G in λ in terms of β_m^{-1}. Note that $m \le [1/d_\infty(F, G) - 1]/2$, that is, a large m corresponds to distributions close in d_∞ and, if Carleman's series $\sum_{j=1}^{\infty} \mu_{2j}^{-1/(2j)}$ diverges, then also a small β_m^{-1}. We start with the following result due to Klebanov and Mkrtchyan (1980).

Theorem 10.3.1. *Let F and G be two DFs for which (10.3.5) holds. Then there exists an absolute constant C such that*

$$\lambda(F, G) \le C\beta_{m-1}^{-1/4}\left(1 + \mu_2^{1/2}\right)^{1/4}, \tag{10.3.6}$$

where

$$m \le \frac{1}{2}[1/d_\infty(F, G) - 1].$$

Proof. We will use some results from Akhiezer (1961), which, for the convenience of the reader, are stated below.

---

[2] See, for example, Akhiezer (1961).

From a sequence of moments $\mu_0 = 1,\ \mu_1, \dots, \mu_{2m}$ we can construct a sequence of determinants

$$D_k = \begin{vmatrix} \mu_0 & \mu_1 & \dots & \mu_k \\ \mu_1 & \mu_2 & \dots & \mu_{k+1} \\ \dots & \dots & \dots & \dots \\ \mu_k & \mu_{k+1} & \dots & \mu_{2k} \end{vmatrix}, \qquad k = 0, 1, \dots, m,$$

a sequence of polynomials

$$P_0(\zeta) = 1,$$

$$P_k(\zeta) = \frac{1}{\sqrt{D_{k-1} D_k}} \begin{vmatrix} \mu_0 & \mu_1 & \dots & \mu_k \\ \mu_1 & \mu_2 & \dots & \mu_{k+1} \\ \dots & \dots & \dots & \dots \\ \mu_k & \mu_{k+1} & \dots & \mu_{2k} \\ 1 & \zeta & \dots & \zeta \end{vmatrix},$$

and numbers

$$\alpha_k = \int_{-\infty}^{\infty} \zeta P_k^2(\zeta) \mathrm{d}F(\zeta),$$

$$b_k = \frac{\sqrt{D_{k-1} D_{k+1}}}{D_k}, \qquad k = 1, 2, \dots, m-1.$$

Here, $P_k(\zeta)$ $(k = 0, 1, \dots, m)$ are solutions of the finite-difference equations

$$b_{k+1} y_{k+1} + a_k y_k + b_{k+1} y_{k+1} = \zeta y_{k+1},$$

the second linearly independent solution of which is denoted by $Q_k(\zeta)$.

The following analog of the Liouville–Ostrogradski formula holds for $P_k(\zeta)$ and $Q_k(\zeta)$ for any complex ζ[3]:

$$P_{k-1}(\zeta) Q_k(\zeta) - P_k(\zeta) Q_{k-1}(\zeta) = \frac{1}{b_{k-1}}, \qquad k = 1, \dots, m.$$

In Akhiezer (1961, pp. 110–111), it is shown that

$$\beta_{m-1} = \sum_{n=1}^{m-1} \mu_{2n}^{-1/(2n)} \le e \sum_{n=0}^{m-2} \frac{1}{b_n}. \tag{10.3.7}$$

[3]For more information on these concepts see Akhiezer (1961, pp. 1–18).

Let us now estimate $\sum_{n=0}^{m-2} 1/b_n$. Using an analog of the Liouville–Ostrogradski formula and the Cauchy–Buniakowsky inequality, we obtain

$$\begin{aligned}\sum_{n=0}^{m-2} \frac{1}{b_n} &\le \sum_{n=0}^{m-2} |P_n(\zeta)Q_{n+1}(\zeta)| + \sum_{n=0}^{m-2} |P_{n+1}(\zeta)Q_n(\zeta)| \\ &\le 2\left(\sum_{n=1}^{m-1} P_n(\zeta)|^2\right)^{1/2} \left(\sum_{n=0}^{m-1} |Q_n(\zeta)|^2\right)^{1/2}. \end{aligned} \tag{10.3.8}$$

Note that (10.3.8) holds for any complex ζ.

If $\mathrm{Im}(\zeta) \neq 0$, then for any $n \geq 1$[4] we have

$$\sum_{k=0}^{n-1} |wP_k(\zeta) + Q_k(\zeta)|^2 \le \frac{w - \bar{w}}{\zeta - \bar{\zeta}}, \tag{10.3.9}$$

where

$$w := w(\zeta) := \int_{-\infty}^{\infty} \frac{\mathrm{d}F(t)}{t - \zeta}, \tag{10.3.10}$$

and the bar denotes complex conjugate. Using (10.3.9) we find that

$$\begin{aligned}\sum_{n=0}^{m-2} |Q_n(\zeta)|^2 &\le 2\sum_{n=0}^{m-2} |wP_n(\zeta) + Q_{n+1}(\zeta)|^2 + 2\sum_{n=0}^{m-2} |w|^2|P_n(\zeta)|^2 \\ &\le 2\frac{w - \bar{w}}{\zeta - \bar{\zeta}} + 2|w|^2 \sum_{n=0}^{m-1} |P_n(\zeta)|^2. \end{aligned}$$

This, together with (10.3.7) and (10.3.8), implies that

$$\beta_{m-1} \le e2\sqrt{2}\left|\frac{w - \bar{w}}{\zeta - \bar{\zeta}}\right| \left(\sum_{n=0}^{m-2} |P_n(\zeta)|^2 + e2\sqrt{2}|w| \sum_{n=0}^{m-1} |P_n(\zeta)|^2\right). \tag{10.3.11}$$

If $\sum_{n=0}^{m-1} |P_n(\zeta)|^2 \ge 1$, then $\sum_{n=0}^{m-1} |P_n(\zeta)|^2 \ge \left(\sum_{n=0}^{m-1} |P_n(\zeta)|^2\right)^{1/2}$, and we derive from (10.3.11) that

$$\sum_{n=0}^{m-1} |P_n(\zeta)|^2 \ge \beta_{m-1} \Big/ \left(2\sqrt{2}e\left(|w| + \left|\frac{w - \bar{w}}{\zeta - \bar{\zeta}}\right|^{1/2}\right)\right).$$

---

[4] See Akhiezer (1961, pp. 25, 46–48).

If, however, $\sum_{n=0}^{m-1} |P_n(\zeta)|^2 < 1$, then we analogously find from (10.3.11) that

$$\left(\sum_{n=0}^{m-1} |P_n(\zeta)|^2\right)^{1/2} \geq \beta_{m-1} / \left(2\sqrt{2}e\left(|w| + \left|\frac{w-\bar{w}}{\zeta-\bar{\zeta}}\right|^{1/2}\right)\right),$$

and consequently,

$$\max\left\{\sum_{n=0}^{m-1} |P_n(\zeta)|^2, \left(\sum_{n=0}^{m-1} |P_n(\zeta)|^2\right)^{1/2}\right\} \geq \beta_{m-1} \left(2\sqrt{2}e\left(|w| + \left|\frac{w-\bar{w}}{\zeta-\bar{\zeta}}\right|^{1/2}\right)\right). \tag{10.3.12}$$

If G has the same moments $\mu_0, \mu_1, \ldots, \mu_{2m}$ as F, then for any ζ satisfying $\mathrm{Im}(\zeta) = 0$ we have[5]

$$\left|\int_{-\infty}^{\infty} \frac{\mathrm{d}F(t)}{t-\zeta} - \int_{-\infty}^{\infty} \frac{dG(t)}{t-\zeta}\right| \leq \frac{1}{t-\zeta} \frac{2}{\sum_{n=0}^{m-1} |P_n(\zeta)|^2}.$$

The last inequality and (10.3.12) yield the following estimate:

$$\left|\int_{-\infty}^{\infty} \frac{\mathrm{d}F(t)}{t-\zeta} - \int_{-\infty}^{\infty} \frac{dG(t)}{t-\zeta}\right| \leq \frac{C(|w| + |(w-\bar{w})/(\zeta-\bar{\zeta})|^{1/2})}{|\zeta-\bar{\zeta}|\beta_{m-1}}, \tag{10.3.13}$$

where C is an absolute constant (below we denote by C possibly different absolute constants).

To transform (10.3.13) into a more convenient form, we estimate $|w|$ and $|(w - \bar{w})/(\zeta - \bar{\zeta})|$. Denoting $\zeta = \xi + i\eta$, we have

$$\begin{aligned} w &= w(\zeta) \\ &= \int_{-\infty}^{\infty} \frac{\mathrm{d}F(t)}{(t-\zeta)} \\ &= \int_{-\infty}^{\infty} \frac{(t-\xi)}{(t-\xi)^2+\eta} \mathrm{d}F(t) + i \int_{-\infty}^{\infty} \frac{\eta}{(t-\xi)^2+\eta^2} \mathrm{d}F(t). \end{aligned}$$

From this it is easy to obtain that

$$|\mathrm{I\,m}(w)| \leq \frac{1}{|\eta|},$$

$$|\mathrm{Re}(w)| \leq \frac{1}{|\eta|},$$

$$|(w - \bar{w}/(\zeta - \bar{\zeta})| = |\mathrm{I\,m}(w)/\eta| \leq 1/\eta^2.$$

[5]See Akhiezer (1961, pp. 22 and 55).

Next we assume that $0 < \eta \le 1$. Then the preceding inequalities and (10.3.13) produce

$$\left| \int_{-\infty}^{\infty} \frac{\mathrm{d}F(t)}{t-\zeta} - \int_{-\infty}^{\infty} \frac{dG(t)}{t-\zeta} \right| \le \frac{C}{\eta^2 \beta_{m-1}},$$

so that

$$\left| \int_{-\infty}^{\infty} \frac{\eta}{(t-\xi)^2+\eta^2} \mathrm{d}F(t) - \int_{-\infty}^{\infty} \frac{\eta}{(t-\xi)^2+\eta^2} dG(t) \right| \le \frac{C}{\eta^2 \beta_{m-1}}. \qquad (10.3.14)$$

Using (10.3.14) it is easy to verify that

$$\begin{aligned} &\left| \int_{-A}^{A} \mathrm{e}^{iu\xi} \Big(\int_{-\infty}^{\infty} \frac{\eta}{(t-\xi)^2+\eta^2} \mathrm{d}F(t) \Big) \mathrm{d}\xi \right. \\ &\quad \left. - \int_{-A}^{A} \mathrm{e}^{iu\xi} \Big(\int_{-\infty}^{\infty} \frac{\eta}{(t-\xi)^2+\eta^2} dG(t) \Big) \mathrm{d}\xi \right| \\ &\le \frac{CA}{\eta^2 \beta_{m-1}} \end{aligned} \qquad (10.3.15)$$

for all $A > 0$. We want to pass to integrals along the entire axis on the left-hand side of (10.3.15). To this end, we estimate

$$\begin{aligned} &\int_{-\infty}^{\infty} \mathrm{d}\xi \int_{-\infty}^{\infty} \frac{\eta \mathrm{e}^{iu\xi}}{(t-\xi)^2+\eta^2} \mathrm{d}F(t) - \int_{-A}^{A} \mathrm{d}\xi \int_{-\infty}^{\infty} \frac{\eta \mathrm{e}^{iu\xi}}{(t-\xi)^2+\eta^2} \mathrm{d}F(t) \\ &\quad = \int_{-\infty}^{A} \mathrm{d}\xi \int_{-\infty}^{\infty} \frac{\eta \mathrm{e}^{iu\xi}}{(t-\xi)^2+\eta^2} \mathrm{d}F(t) + \int_{-\infty}^{\infty} \mathrm{d}\xi \int_{-\infty}^{\infty} \frac{\eta \mathrm{e}^{iu\xi}}{(t-\xi)^2+\eta^2} \mathrm{d}F(t). \end{aligned} \qquad (10.3.16)$$

For this purpose we consider

$$\begin{aligned} I_1 &= \int_{-\infty}^{A} \mathrm{d}\xi \int_{-\infty}^{\infty} \frac{\eta}{(t-\xi)^2+\eta^2} \mathrm{d}F(t) \\ &= \int_{-\infty}^{\infty} \Big(\int_{-\infty}^{-A+t} \frac{\eta}{(\nu^2+\eta^2} \mathrm{d}\nu \Big) \mathrm{d}F(t) \\ &= \int_{-\infty}^{\infty} \Big(\frac{\pi}{2} - \arctan \frac{A-t}{\eta} \Big) \mathrm{d}F(t). \end{aligned}$$

It is easy to verify that for some constant C

$$\frac{\pi}{2} - \arctan z \le \begin{cases} \frac{C}{1+z}, & \text{for } z \ge 0, \\ \pi, & \text{for } z < 0. \end{cases}$$

Then

$$\begin{aligned}
I_1 &= \int_{-\infty}^{A} \Big(\frac{\pi}{2} - \arctan\frac{A-t}{\eta}\Big)\mathrm{d}F(t) \\
&\quad + \int_{A}^{\infty} \Big(\frac{\pi}{2} - \arctan\frac{A-t}{\eta}\Big)\mathrm{d}F(t) \\
&\le \int_{-\infty}^{A} \frac{C\mathrm{d}F(t)}{1+|(A-t)/\eta|} + \pi\int_{A}^{\infty}\mathrm{d}F(t) \\
&\le \int_{-\infty}^{\infty} \frac{C\mathrm{d}F(t)}{1+|(A-t)/\eta|} + \pi\big(1-F(A)\big). \qquad (10.3.17)
\end{aligned}$$

Let us now show that the first term on the right-hand side of (10.3.17) is not greater than $C(1+\mu_2^{1/2})/A$. Indeed,

$$\begin{aligned}
\frac{A}{\eta}\int_{-\infty}^{\infty}\frac{C}{1+|(A-t)/\eta|}\mathrm{d}F(t) &\le C\bigg|\int_{-\infty}^{\infty}\frac{A/\eta - t/\eta}{1+|(A-t)/\eta}\mathrm{d}F(t) \\
&\quad + \int_{-\infty}^{\infty}\frac{t/\eta}{1+|(A-t)/\eta|}\mathrm{d}F(t) \\
&\le C\bigg(\int_{-\infty}^{\infty}\frac{|(A-t)/\eta|}{1+|(A-t)/\eta|}\mathrm{d}F(t) \\
&\quad + \frac{1}{\eta}\int_{-\infty}^{\infty}\frac{|t|}{1+|(A-t)/\eta|}\mathrm{d}F(t)\bigg) \\
&\le C(1+\mu_2^{1/2}/\eta).
\end{aligned}$$

Therefore,

$$\int_{-\infty}^{\infty}\frac{C\mathrm{d}F(t)}{1+|(A-t)/\eta|}\mathrm{d}F(t) \le \frac{C}{A}\big(\eta+\mu_2^{1/2}\big) \le \frac{C}{A}\big(1+\mu_2^{1/2}\big).$$

In addition, it is clear that

$$1-F(A) \le \mu_2^{1/2}/A.$$

Consequently,

$$\bigg|\int_{-\infty}^{-A}\mathrm{d}\xi\int_{-\infty}^{\infty}\frac{\eta e^{iu\xi}}{(\xi-t)^2+\eta^2}\mathrm{d}F(t)\bigg| \le I_1 \le \frac{C(1+\mu_2^{1/2})}{A}.$$

We now see that

$$\bigg|\int_{A}^{\infty}\mathrm{d}\xi\int_{-\infty}^{\infty}\frac{\eta e^{iu\xi}}{(\xi-t)^2+\eta^2}\mathrm{d}F(t)\bigg| \le \frac{C(1+\mu_2^{1/2})}{A}. \qquad (10.3.18)$$

Arguing analogously we obtain

$$\left|\int_A^\infty \mathrm{d}\xi \int_{-\infty}^\infty \frac{\eta e^{iu\xi}}{(\xi-t)^2+\eta}\mathrm{d}F(t)\right| \le \frac{C(1+\mu_2^{1/2})}{A}. \tag{10.3.19}$$

Substituting (10.3.18) and (10.3.19) into (10.3.16), we find that

$$\left|\int_{-\infty}^\infty \mathrm{d}\xi \int_{-\infty}^\infty \frac{\eta e^{iu\xi}}{(\xi-t)^2+\eta^2}\mathrm{d}F(t) - \int_{-A}^A \mathrm{d}\xi \int_{-\infty}^\infty \frac{\eta e^{iu\xi}}{(\xi-t)^2+\eta}\mathrm{d}F(t)\right|$$
$$\le \frac{C(1+\mu_2^{1/2})}{A}. \tag{10.3.20}$$

Using the same arguments for $G(t)$ we obtain an estimate similar to (10.3.20) but with $F(t)$ replaced by $G(t)$. Taking this and (10.3.15) into account, we obtain

$$\left|\int_{-\infty}^\infty \mathrm{d}\xi \int_{-\infty}^\infty \frac{\eta e^{iu\xi}}{(\xi-t)^2+\eta^2}\mathrm{d}F(t) - \int_{-\infty}^\infty \mathrm{d}\xi \int_{-\infty}^\infty \frac{\eta e^{iu\xi}}{(\xi-t)^2+\eta^2}dG(t)\right| \tag{10.3.21}$$

$$\le \frac{CA}{\eta^2\beta_{m-1}} + \frac{C(1+\mu_2^{1/2})}{A}. \tag{10.3.22}$$

Next, suppose that $f(u)$ and $g(u)$ are the characteristic functions of F and G, respectively. Since for any distribution function H we have

$$\int_{-\infty}^\infty e^{iu\xi} \int_{-\infty}^\infty \frac{\eta}{(\xi-t)^2+\eta^2}\mathrm{d}H(t)d(\xi) = \pi e^{\ |\eta u|} h(u),$$

where h is a characteristic function of H, (10.3.21) implies that

$$|f(u)-g(u)| \le C\left[\frac{A}{\eta^2\beta_{m-1}} + \frac{1+\mu_2^{1/2}}{A}\right] e^{\eta|\nu|}. \tag{10.3.23}$$

Letting here

$$|u| \le T,$$
$$\eta = \min(1/T, 1),$$
$$A = \beta_{m-1}^{1/2}\eta(1+\mu_2^{1/2})^{1/2},$$

we obtain

$$|f(u)-g(u)| \le C\beta_{m-1}^{-1/2}(1+\mu_2^{1/2})^{1/2}T. \tag{10.3.24}$$

Therefore,

$$\lambda(F,G) = \min_{t>0} \max\left\{\frac{1}{2}\max_{|u|\le T}|f(u)-g(u)|,\ 1/T\right\}$$
$$\le C\beta_{m-1}^{-1/2}(1+\mu_2^{1/2})^{1/4}.$$

Note that estimate (10.3.6) is not exact. A better estimate can be easily obtained for distributions with slowly increasing moments. It is, however, altogether unfit for distributions with fast increasing moments. □

Theorem 10.3.2. *Let F and G be two DFs for which (10.3.5) holds. Then*

$$\lambda(F,G) \le C\mu_{2m}^{1/(2m+1)}/m, \tag{10.3.25}$$

where $m \le [1/d_\infty(F,G)-1]/2$ and C is an absolute constant.

Proof. Suppose that $f(t)$ and $g(t)$ are the characteristic functions of F and G, respectively. According to Taylor's formula we have

$$f(t)-g(t) = \sum_{k=0}^{2m-1}\frac{f^{(k)}(0)-g^{(k)}(0)}{k!}t^k + \frac{f^{(2m)}(\theta)-g^{(2m)}(\theta)}{(2m)!}t^{(2m)},$$

where θ is a point between 0 and t. Since

$$\mu_j(F) = \mu_j(G) = \mu_j,\ j = 0,1,\dots,2m,$$

we have $f^{(k)}(0) = g^{(k)}(0),\ k = 0,1,\dots,2m$. In addition,

$$|f^{(2m)}(\theta)| \le \mu_{2m},\ |g^{(2m)}(\theta)| \le \mu_{2m}.$$

Thus,

$$|f(t)-g(t)| \le \frac{2\mu_{2m}}{(2m)!}|t|^{2m},$$

and hence

$$\frac{1}{2}\max_{|t|\le T}|f(t)-g(t)| \le \mu_{2m}T^{2m}/(2m)!. \tag{10.3.26}$$

Clearly,

$$\min_{t>0}\max\{\mu_{2m}T^{2m}/(2m)!,\ 1/T\} = \{\mu_{2m}/(2m)!\}^{1/(2m+1)},$$

that is,

$$\lambda(F,G) \le (\mu_{2m}/(2m)!)^{1/(2m+1)}.$$

Applying Stirling's formula, we obtain the desired result. □

Note that if F is a uniform DF on $(0, 1)$ and G satisfies (10.3.5), then (10.3.25) yields

$$\lambda(F, G) \le C/m, \tag{10.3.27}$$

while (10.3.6) yields only

$$\lambda(F, G) \le C/m^{1/4}. \tag{10.3.28}$$

Of course, estimate (10.3.27) is significantly better than (10.3.28). On the other hand, if F is an exponential DF with parameters $\lambda = 1$ and G satisfies (10.3.5), then (10.3.25) yields only the trivial estimate

$$\lambda(F, G) \le 1,$$

while (10.3.6) implies that

$$\lambda(F, G) \le C(\ln m)^{-1/4}.$$

Thus, the precision of (10.3.6) and (10.3.25) varies rather substantially for different classes of distributions. The following result is intermediate between Theorems 10.3.1 and 10.3.2.

Theorem 10.3.3. *Suppose that the characteristic function $f(t)$ of F is analytic in some circle, $\mu_j := \mu_j(F),\ j = 0, 1, \dots,$ and let G be such that (10.3.5) is fulfilled. Then*

$$\lambda(F, G) \le C_F / \ln(m), \tag{10.3.29}$$

where the constant C_F depends only on F (but not on G and m).

Proof. Let g be a characteristic function corresponding to G. As in the proof of Theorem 10.3.2, we obtain

$$|f(t) - g(t)| \le \frac{2\mu_{2m}}{(2m)!}|t|^{2m}$$

for all real t. If $f(t)$ is analytic in a circle of radius R, then it is clear that for $R_1 < R$ and all m

$$\frac{2\mu_{2m}}{(2m)!}R_1^{2m} \le \tilde{C}_F.$$

Therefore, for $|t| \le T_0 < R_1$

$$|f(t) - g(t)| \le \tilde{C}_F (T_0/R_1)^{2m}.$$

We derive the desired result from Theorem 10.2.2 when $\varepsilon := \tilde{C}_F (T_0/R_1)^{2m}$. □

Remark 10.3.1. It is clear that for an exponential distribution estimate (10.3.29) is better than (10.3.6).

10.3.2 Estimates of λ by Means of d_α, $\alpha \in (0, \infty)$

Let us now turn to estimating the closeness of F and G in λ if we know that these distributions are close in d_α, $\alpha \in (0,\infty)$. The problem of constructing estimates of this type is equivalent to the problem of estimating the closeness of distributions in λ from the closeness of their first $2m$ moments.

Assume that F and G have finite moments up to and including order $2m$ and that

$$|\mu_j(F) - \mu_j(G)| \le \delta,\ j = 1, 2, \ldots, 2m, \tag{10.3.30}$$

where $\delta > 0$ is a given number.

Theorem 10.3.4. *Suppose that F and G satisfy (10.3.30), where $0 < \delta \le 1$. Then*

$$\lambda(F,G) = 2/\ln\bigl(1+\delta^{-1/2}\bigr) + \bigl(2\mu_{2m}/(2m)!\bigr)^{1/(2m+1)}. \tag{10.3.31}$$

Proof. If $f(t)$ and $g(t)$ are the characteristic functions of F and G, then for all $t \in [-T, T]$ $(T > 0)$ we have

$$\begin{aligned}
|f(t) - g(t)| &\le \left|\sum_{j=0}^{2m-1} \frac{f^{(j)}(0) - g^{(j)}(0)}{j!} t^j\right| + \frac{2\mu_{2m} + \delta}{(2m)!}|t|^{2m} \\
&\le \sum_{j=0}^{2m-1} \frac{\delta}{j!} T^j + \frac{2\mu_{2m}}{(2m)!} T^{2m} \\
&\le \delta \mathrm{e}^T + \frac{2\mu_{2m}}{(2m)!} T^{2m}.
\end{aligned} \tag{10.3.32}$$

Letting here

$$T = \min\bigl\{\ln(1+\delta^{-1/2}),\ \bigl((2m)!/(2\mu_{2m})\bigr)^{1/(2m-1)}\bigr\},$$

we obtain

$$\begin{aligned}
\lambda(F,G) &\le \max\bigl\{\delta^{1/2} + \delta + \bigl(2\mu_{2m}(2m)!\bigr)^{1/(2m+1)},\ 1/T\bigr\} \\
&\le \max\bigl\{\delta^{1/2} + \delta + \bigl(2\mu_{2m}/(2m)!\bigr)^{1/(2m+1)}, 1/\ln(1+\delta^{-1/2}) \\
&\quad + (2\mu_{2m}/(2m)!)^{1/(2m+1)}\bigr\} \\
&\le 2/\ln(1+\delta^{-1/2}) + \bigl(2\mu_{2m}/(2m)!\bigr)^{1/(2m+1)}.
\end{aligned}$$

□

The following result follows immediately from Theorem 10.3.4.

Corollary 10.3.1. *The following inequality holds:*

$$\lambda(F,G) \le 2/\ln\big(1+(d_\alpha(F,G)/\alpha)^{-1/2}\big) + \big(2\mu_{2s}/(2s)!\big)^{1/(2s+1)},$$

where

$$s := \big[\big(1-d_\alpha(F,G)\big)/\big(2d_\alpha(F,G)\big)\big],$$
$$\mu_{2s} := \mu_{2s}(F) \quad (0<\alpha<\infty).$$

Theorem 10.3.5. *Suppose that the characteristic function f of F is analytic in a circle and that (10.3.30) is fulfilled for F and G. Then for any $q \in (0,1)$ there exists a value C_q, depending only on q and F (but not on G and m), such that*

$$\lambda(F,G) \le C_q/\ln\ln(\delta + 2q^{2m}). \tag{10.3.33}$$

Proof. For any real t [see (10.3.32)] we have

$$|f(t)-g(t)| \le \delta e^{|t|} + \frac{2\mu_{2m}}{(2m)!}|t|^{2m}. \tag{10.3.34}$$

Let R be the radius of the circle of analyticity of $f(t)$. Then, for $|t| \le R_1 < R_2 < R$,

$$\frac{2\mu_{2m}}{(2m)!}|t|^{2m} \le C_{q_1} q_1^{2m},$$
$$\frac{R_1}{R} < q_1 < \frac{R_2}{R},$$

where C_{q_1} depends only on q_1 (and F) but not on m. From (10.3.34) it follows that for $|t| \le R_1$

$$|f(t)-g(t)| \le e\delta + C_{q_1} q_1^{2m}.$$

Applying Theorem 10.2.2 we find

$$\lambda(F,G) \le C_{q_1,F}/\ln\ln\big(e\delta + C_{q_1} q_1^{2m}\big).$$

Thus, by the fact that R_1 and R_2 are arbitrary under the condition that $R_1 < R_2 < R$, we obtain (10.3.33). □

10.3.3 Estimates of d_α by Means of Characteristic Functions

Previously we obtained estimates of the closeness of distributions in the λ metric from their closeness in the metric d_α. It is natural to ask whether it is possible to construct reverse estimates, that is, whether d_α can be estimated by means of λ or a similar metric. Since in general weak convergence does not imply the convergence of the corresponding moments, even in the class D an estimate of d_α by means of λ is impossible. However, if we consider a subclass $\underline{\mathrm{N}}$ of D formed by distributions with moments that do not increase faster than a specified sequence, then such an estimate becomes possible for $\alpha < \infty$. It is then clear that to construct estimates of this kind it is enough to know the order of the closeness of the characteristic functions of the corresponding distributions in some fixed neighborhood of zero.

Suppose that $N_1 \le N_2 \le \cdots \le N_k \le \dots$ is an increasing sequence of positive numbers. Let

$$\underline{\mathrm{N}} := \underline{\mathrm{N}}(N_1, N_2, \dots, N_k, \dots) = \left\{F : \int_{-\infty}^{\infty} |x^j| \mathrm{d}F(x) \le N_j,\ j = 1, 2, \dots\right\}.$$

Theorem 10.3.6. *Suppose that $F, G \in \underline{\mathrm{N}}$ and the corresponding densities f and g satisfy*

$$\sup_{|t| \le T_0} |f(t) - g(t)| \le \varepsilon, \tag{10.3.35}$$

where $T_0 > 0$ is a constant. Then there exists an absolute constant C such that for all integers $k > 0$ with

$$k^3 C^{\frac{1}{k+1}} \varepsilon^{\frac{1}{k+1}} \le N_k^{\frac{1}{k+1}} T_0/2 \tag{10.3.36}$$

we have

$$|\mu_k(F) - \mu_k(G)| \le C N_{k+1} k^3 \varepsilon^{\frac{1}{k+1}}\ . \tag{10.3.37}$$

Proof. Relation (10.3.35) can be written as

$$f(t) - g(t) = R(t; \varepsilon), \tag{10.3.38}$$

where $|R(t; \varepsilon)| \le \varepsilon$ for $|t| \le T_0$. Let

$$\omega(t) = \begin{cases} \exp(-1/(1+t)^2 - 1/(1-t)^2) & \text{for } t \in (-1, 1), \\ 0 & \text{for } t \notin (-1, 1), \end{cases}$$

and

$$\omega_\delta(t) = \frac{1}{\delta}\omega(1/\delta) \Big/ \int_{-1}^{1} \omega(\tau) d\tau, \quad \delta > 0.$$

We can show that for any integer

$$\sup_t \left|\omega^{(n)}(t)\right| \le C N^{3n}, \tag{10.3.39}$$

where C is an absolute constant.

Let us multiply both sides of (10.3.38) by $\omega_\delta(t-z)$ and integrate with respect to t. We then have

$$f_\delta(z) - g_\delta(z) = R_\delta(z;\varepsilon), \tag{10.3.40}$$

where

$$f_\delta(z) = \int_{-\infty}^{\infty} f(t)\omega_\delta(t-z)\mathrm{d}t,$$

$$g_\delta(z) = \int_{-\infty}^{\infty} g(t)\omega_\delta(t-z)\mathrm{d}t,$$

$$R_\delta(z;\varepsilon) = \int_{-\infty}^{\infty} R(t;\varepsilon)\omega_\delta(t-z)\mathrm{d}t.$$

Moreover, $|R_\delta^{(n)}(z;\varepsilon)| \le \varepsilon$ for $|z| \le T_0 - \delta$.

Clearly, $f_\delta(z)$, $g_\delta(z)$, and $R_\delta(z;\varepsilon)$ are infinitely differentiable with respect to z. By (10.3.39) and the definition of ω_δ, it is clear that

$$\left|R_\delta^{(n)}(z;\varepsilon)\right| \le Cn^{3n}\varepsilon/\delta^n, \quad |z| \le T_0 - \delta. \tag{10.3.41}$$

Differentiating both sides of (10.3.40) k times with respect to z and taking (10.3.41) into account, we find that

$$\left|f_\delta^{(k)}(0) - g_\delta^{(k)}(0)\right| \le C\varepsilon k^{3k}/\delta^k. \tag{10.3.42}$$

On the other hand,

$$\begin{aligned}\left|f_\delta^{(k)}(0) - \mu_k(F)\right| &= \left|f_\delta^{(k)}(0) - f_\delta^{(k)}(0)\right|\\ &\le \int_{-\infty}^{\infty} \left|f^{(k)}(t) - f^{(k)}(0)\right|\omega_\delta(t)\mathrm{d}t\\ &\le N_{k+1}\int_{-\infty}^{\infty} |t|\omega_\delta(t)\mathrm{d}t\\ &\le \delta N_{k+1}.\end{aligned}$$

Similarly,

$$\left|g_\delta^{(k)}(0) - \mu_k(G)\right| \le \delta N_{k+1}.$$

The last two inequalities, together with (10.3.42), show that

$$|\mu_k(F) - \mu_k(G)| \le C\varepsilon k^{3k}/\delta^k + 2\delta N_{k+1}$$

for all $k \ge 1$ and all δ for which $T_0 - \delta > 0$. The right-hand side of the last inequality attains a minimum with respect to δ when

$$\delta = \delta_{\min} = \left(Ck^{3k+1}\varepsilon/\left(2/N_{k+1}\right)\right)^{1/(k+1)},$$

and this minimum is equal to

$$4C^{\frac{1}{k+1}}K^{\frac{3k+1}{k+1}}\varepsilon^{\frac{1}{k+1}}/2^{\frac{1}{k+1}}N_{k+1}^{\frac{1}{k+1}}.$$

From this we see that when (10.3.36) is fulfilled, then so is $T_0 - \delta \geq T_0/2$, and (10.3.35) holds. □

References

Akhiezer NI (1961) The classical moment problem. Gosudar lzdat Fiz-Mat Lit (in Russian), Moscow

Bernstein SN (1937) Extremal properties of polynomials. Leningrad-Moscow (in Russian)

Klebanov LB, Mkrtchyan ST (1979) Characteristic in wide-sense properties of normal distribution in connection with a problem of stability of characterizations. Theor Prob Appl 24:434–435

Klebanov LB, Mkrtchyan ST (1980) Estimate of the closeness of distributions in terms of coinciding moments, problems of stability of stochastic models, Institute of system investigations, pp 64–72

Mkrtchyan ST (1978) Stability of characterization of distributions and some wide-sense characteristic properties of normal distribution. Doklady Arm SSR 67:129–131

Sapogov NA (1980) Problem of stability of the theorem on uniqueness of a characteristic function analytic in neighborhood of the origin. In: Problems of stability of stochastic models. VNIISI, Moscow, pp 88–94

Chapter 11
Uniformity in Weak and Vague Convergence

The goals of this chapter are to:

- Extend the notion of uniformity,
- Study the metrization of weak convergence,
- Describe the notion of vague convergence,
- Consider the question of its metrization.

Notation introduced in this chapter:

Notation	*Description*
$\mathfrak{M}$	Space of bounded nonnegative measures
$\eta_{\mathbb{K}}$	Generalization of ℓ_H on $\mathfrak{M}$
$\pi_{\lambda,G}$	G-weighted Prokhorov metric
$\mathfrak{N}$	Space of nonnegative measures finite on any bounded set
$\xrightarrow{\text{v}}$	Vague convergence
$\mathbf{K}(\nu', \nu'')$	Kantorovich-type metric in $\mathfrak{N}$
$\mathbf{\Pi}(\nu', \nu'')$	Prokhorov metric in $\mathfrak{N}$

11.1 Introduction

In this chapter, we consider ρ-uniform classes in a general setting in order to study uniformity in weak and vague convergence. In the next section, we begin with a few definitions and then proceed to the case of weak convergence. Finally, we introduce the notion of vague convergence and consider the question of its metrization.

S.T. Rachev et al., *The Methods of Distances in the Theory of Probability and Statistics*, DOI 10.1007/978-1-4614-4869-3_11,

11.2 ζ-Metrics and Uniformity Classes

Let (U, d) be a separable metric space (s.m.s) with Borel σ-algebra $\mathfrak{B}$. Let $\mathfrak{M}$ denote the set of all bounded nonnegative measures on $\mathfrak{B}$ and $\mathcal{P}_1 = \mathcal{P}(U)$ the subset of probability measures. Let $\mathfrak{M}' \subset \mathfrak{M}$. For each class $\mathcal{F}$ of μ-integrable functions f on U ($\mu \in \mathfrak{M}'$), define on $\mathfrak{M}'$ the semimetric

$$\zeta_{\mathcal{F}}(\mu', \mu'') = \sup\left\{\left|\int f \mathrm{d}(\mu' - \mu'')\right| : f \in \mathcal{F}\right\}, \tag{11.2.1}$$

with a ζ-structure.[1] There is a special interest in finding, for a given semimetric ρ on $\mathfrak{M}'$, a semimetric $\zeta_{\mathcal{F}}$ that is topologically equivalent to ρ. Note that this is not always possible (see Lemma 4.4.4 in Chap. 4).

Definition 11.2.1. The class $\mathcal{F}$ is said to be *ρ-uniform* if $\zeta_{\mathcal{F}}(\mu_n, \mu) \to 0$ as $n \to \infty$ for any sequence $\{\mu_1, \mu_2, \ldots\} \subset \mathfrak{M}'$ ρ-convergent to $\mu \in \mathfrak{M}'$.

Such ρ-uniform classes were studied in Sect. 4.4 of Chap. 4. Here we investigate ρ-uniform classes in a more general setting. We generalize the notion of ρ-uniform class as follows. Let $\mathbb{K}$ be the class of pairs (f, g) of real measurable functions on U that are μ-integrable for any $\mu \in \mathfrak{M}' \subset \mathfrak{M}$. Consider the functional

$$\eta_{\mathbb{K}}(\mu', \mu'') = \sup\left\{\int f \mathrm{d}\mu' + \int g \mathrm{d}\mu'' : (f, g) \in \mathbb{K}\right\}, \quad \mu', \mu'' \in \mathbb{M}. \tag{11.2.2}$$

The functional $\eta_{\mathbb{K}}$ may provide dual and explicit expressions for minimal distances. For example, define for any measures μ', μ'' with $\mu'(U) = \mu''(U)$ the class $\mathfrak{A}(\mu', \mu'')$ of all Borel measures $\widetilde{\mu}$ on the direct product $U \times U$ with fixed marginals $\mu'(A) = \widetilde{\mu}(A \times U)$, $\mu''(A) = \widetilde{\mu}(U \times A)$, $A \in \mathfrak{B}$. Then (see Corollary 5.3.2 in Chap. 5), for $1 \leq p < \infty$, if $\int d^p(x, a)(\mu' + \mu'')(\mathrm{d}x) < \infty$, then we have that (11.2.2) gives the dual form of the p-average metric, i.e.,

$$\eta_{\mathbb{K}(p)}(\mu', \mu'') = \inf\left\{\int d^p(x, y)\widetilde{\mu}(\mathrm{d}x \times \mathrm{d}y) : \widetilde{\mu} \in \mathfrak{A}(\mu', \mu'')\right\}, \tag{11.2.3}$$

where $\mathbb{K}(p)$ is the set of all pairs (f, g) for which $f(x) + g(y) \leq d^p(x, y)$, $x, y \in U$.

Definition 11.2.2. We call the class $\mathbb{K}$ a *ρ-uniform class* (*in a broad sense*) if for any sequence $\{\mu_1, \mu_2, \ldots\} \subset \mathfrak{M}' \subset \mathfrak{M}$ the ρ-convergence to $\mu \in \mathfrak{M}'$ implies $\lim_{n \to \infty} \eta_{\mathbb{K}}(\mu_n, \mu) = 0$.

---

[1]See Definition 4.4.1 in Chap. 4.

The notation $\mu_n \xrightarrow{\mathrm{w}} \mu$ denotes, as usual, the weak convergence of the sequence $\{\mu_1, \mu_2, \ldots\} \subset \mathfrak{M}$ to $\mu \in \mathfrak{M}$.

Theorem 11.2.1. *Let $\mu, \mu_1, \mu_2, \ldots$ be a sequence of measures in $\mathfrak{M}$ and $\mu(U) = \mu_n(U)$, $n = 1, 2, \ldots$. Let $B(t)$, $t \geq 0$, be a convex nonnegative function, $B(0) = 0$, satisfying the Orlicz condition:* $\sup\{B(2t)/B(t) : t > 0\} < \infty$. *If*

$$\int B(d(x,a))(\mu_n + \mu)(\mathrm{d}x) < \infty,$$

then the joint convergence

$$\mu_n \xrightarrow{\mathrm{w}} \mu \qquad \int B(d(x,a))(\mu_n - \mu)(\mathrm{d}x) \to 0 \tag{11.2.4}$$

is equivalent to the convergence $\eta_{\mathbb{B}}(\mu_n, \mu) \to 0$, where $\mathbb{B}$ is the class of pairs (f, g) such that $f(x) + g(y) < B(d(x, y))$, $x, y \in U$.

Proof. Let π be the Prokhorov metric in $\mathfrak{M}$, i.e.,[2]

$$\pi(\mu', \mu'') = \inf\{\varepsilon > 0 : \mu'(A) \leq \mu''(A^{\varepsilon}) + \varepsilon, \mu''(A) \leq \mu'(A^{\varepsilon}) + \varepsilon \text{ for any closed set } A \subset U\}. \tag{11.2.5}$$

Then, as in Lemma 8.3.1,[3] we conclude that

$$B(\pi(\mu', \mu''))\pi(\mu', \mu'') \leq \eta_{\mathbb{B}}(\mu', \mu'') \leq B(\pi(\mu', \mu'')) + K_B\left[2\pi(\mu', \mu'')B(M) + \int_{d(x,a)>M} B(d(x,a)(\mu' + \mu'')(\mathrm{d}x)\right] \tag{11.2.6}$$

for any $\mu', \mu'' \in \mathfrak{M}$, $M > 0$, $a \in U$, and $K_B := \sup\{B(2t)/B(t); t > 0\}$. Hence, (11.2.4) provides $\eta_{\mathbb{B}}(\mu_n, \mu) \to 0$.

To prove (11.2.4) provided that $\eta_{\mathbb{B}}(\mu_n, \mu) \to 0$, we use the following inequality: for any μ', $\mu'' \in \mathfrak{M}$ with $\mu'(U) = \mu''(U)$ and $\int B(d(x,a))(\mu' + \mu'')(\mathrm{d}x) < \infty$, and for any $M > 0$ and $a \in U$, we have

$$\int B(d(x,a))I\{d(x,a) > M\}\mu'(\mathrm{d}x) \leq (K_B + K_B^2)(\widehat{\mathcal{L}}_B(\mu', \mu'') + \int B(d(x,a))I\{d(x,a) > M/2\}\mu''(\mathrm{d}x). \tag{11.2.7}$$

---

[2]See, for example, Hennequin and Tortrat (1965).

[3]See (8.3.5)–(8.3.7) in Chap. 8.

In the preceding inequality, $\widehat{\mathcal{L}}_B(\mu', \mu'') := \inf\{\mathcal{L}_B(\widetilde{\mu}) : \widetilde{\mu} \in \mathfrak{A}(\mu', \mu'')\}$ is the minimal distance relative to $\mathcal{L}_B(\widetilde{\mu}) := \int B(d(x, y))\widetilde{\mu}(\mathrm{d}x, \mathrm{d}y)$. To prove (11.2.7), observe that for any $\mu \in \mathfrak{A}(\mu', \mu'')$ we have

$$\int B(d(x,a))I\{d(x,a) > M\}\mu'(\mathrm{d}x)$$
$$\leq K_B \int B(d(y,a))I\{d(x,a) > M\}\widetilde{\mu}(\mathrm{d}x, \mathrm{d}y) + K_B\mathcal{L}_B(\widetilde{\mu}),$$

where

$$\int B(d(y,a))I\{d(x,a) > M\}\widetilde{\mu}(\mathrm{d}x, \mathrm{d}y)$$
$$\leq B(M)\mu'(d(x,a) > M) + \int B(d(y,a))I\{d(y,a) > M\}\mu''(\mathrm{d}y)$$

and

$$\mu'(d(x,a){>}M){\leq}\frac{1}{B(M/2)}\left(\mathcal{L}_B(\widetilde{\mu}){+}\int B(d(y,a))I\{d(y,a) > M/2\}\mu''(\mathrm{d}y)\right).$$

Combining the last three inequalities we obtain

$$\int B(d(x,a))I\{d(x,a) > M\}\mu'(\mathrm{d}x)$$
$$\leq K_B\mathcal{L}_B(\widetilde{\mu}) + K_B^2\mathcal{L}_B(\widetilde{\mu}) + K_B \int B(d(y,a))I\{d(y,a) > M\}\mu''(\mathrm{d}y)$$
$$+K_B^2 \int B(d(y,a))I\{d(y,a) > M/2\}\mu''(\mathrm{d}y).$$

Passing to the minimal distances $\widehat{\mathcal{L}}_B$ in the last estimate yields the required (11.2.7). Then $\eta_{\mathbb{B}}(\mu_n, \mu) \to 0$, together with (11.2.6) and (11.2.7), implies

$$\mu_m \to \mu \qquad \lim_{M\to\infty} \sup_n \int B(d(x,a))I\{d(x,a) > M\}\mu_n(\mathrm{d}x) = 0.$$

The preceding limit relations complete the proof of (11.2.4).[4] □

Recall that if $G(x)$ is a nonnegative continuous function on U and $\{\mu_0, \mu_1, \ldots\} \subset \mathfrak{M}$, $\int G\mathrm{d}\mu_n < \infty$, $n = 0, 1, \ldots$, then the joint convergence

[4]See Billingsley (1999, Sect. 5).

$$\mu_n \xrightarrow{\text{w}} \mu_0 \qquad \int G\mathrm{d}(\mu_n - \mu_0) \to 0 \quad n \to \infty \tag{11.2.8}$$

is called a *G-weak convergence* (Definition 4.3.2).

Theorem 11.2.2. *The G-weak convergence* (11.2.8) *in*

$$\mathfrak{M}_G := \left\{\mu \in \mathfrak{M} : \int G\mathrm{d}\mu < \infty\right\}$$

is equivalent to the weak convergence $\lambda_n \xrightarrow{\text{w}} \lambda_0$, *where*

$$\lambda_n(A) = \int_A (1 + G(x))\mu_n(\mathrm{d}x), \quad n = 0, 1, \dots, \quad A \in \mathfrak{B}. \tag{11.2.9}$$

Proof. Suppose (11.2.8) holds; then define the measures $\nu_i(B) := \int_B G\mathrm{d}\mu_i$ $(i = 0, 1, \dots)$ on $A_G \cap \mathfrak{B}$, where $A_G := \{x : G(x) > 0\}$. For any continuous and bounded function f

$$\left|\int f\mathrm{d}(\nu_n - \nu_0)\right| \le \left|\int f(1 + G)I\{G \le N\}\mathrm{d}(\mu_n - \mu)\right| + \int \|f\|(1 + G)I\{G > N\}\mathrm{d}(\mu_n + \mu), \tag{11.2.10}$$

where $\|f\| := \sup\{|f(x)| : x \in U\}$ and $N > 0$. For any N with $\mu_0(G(x) = N) = 0$, by the weak convergence $\mu_n \xrightarrow{\text{w}} \mu_0$, we have that the first integral on the right-hand side of (11.2.10) converges to zero, and hence (11.2.10) and (11.2.8) imply $\lambda_n \xrightarrow{\text{w}} \lambda_0$.

Conversely, if $\lambda_n \xrightarrow{\text{w}} \lambda_0$, then for any continuous and bounded function f and $g = f/(1 + G)$ we have $\int g\mathrm{d}\lambda_n \to \int g\mathrm{d}\lambda_0$ since g is also continuous and bounded (i.e., $\int f\mathrm{d}\mu_n \to \int f\mathrm{d}\mu_0$). Finally, by $\int_U \mathrm{d}\lambda_n \to \int_U \mathrm{d}\lambda_0$, we have $\mu_n(U) + \int G\mathrm{d}\mu_n \to \mu_0(U) + \int G\mathrm{d}\mu_0$, and thus (11.2.8) holds. □

Recall the *G-weighted Prokhorov metric* [see (4.3.5)]

$$\pi_{\lambda,G}(\mu_1, \mu_2) = \inf\{\varepsilon > 0 : \lambda_1(A) \le \lambda_1(A^{\lambda\varepsilon}) + \varepsilon \quad \lambda_2(A) \le \lambda_1(A^{\lambda\varepsilon}) + \varepsilon \quad \forall A \in \mathcal{B}\}, \tag{11.2.11}$$

where λ_i is defined by (11.2.9) and $\lambda > 0$.

Corollary 11.2.1. $\pi_{\lambda,G}$ *metrizes the G-weak convergence in* $\mathfrak{M}_G$.

Proof. For any $\mu_0, \mu_n \in \mathfrak{M}_G$, $\pi_{\lambda,G}(\mu_n, \mu_0) \to 0$ if and only if $\pi_{1,G}(\mu_n, \mu_0) \to 0$, which by the Prokhorov (1956) theorem is equivalent to $\lambda_n \xrightarrow{\text{w}} \lambda_0$. An appeal to Theorem 11.2.2 proves the corollary. □

In the next theorem, Theorem 11.2.3, we will omit the basic restriction in Theorem 11.2.1, $\mu_n(U) = \mu(U), n = 1, 2, \ldots$. Define the class $\mathcal{OR}$ of continuous nonnegative functions $B(t)$, $t \geq 0$, $\overline{\lim}_{t\to 0} \sup_{0\leq s\leq t} B(s) = 0$, satisfying the following condition: there exist a point to $t \geq 0$ and a nondecreasing continuous function $B_0(t)$, $t \geq 0$, $K_{B_0} := \sup\{B_0(2t)/B_0(t) : t > 0\} < \infty$, $B_0(0) = 0$, such that $B(t) = B_0(t)$ for $t \geq t_0$.

Lemma 11.2.1. *Let $B \in \mathcal{OR}$ and $\mu, \mu_1, \mu_2, \ldots$ be a sequence of measures in $\mathfrak{M}$ satisfying* (11.2.4)*, $\mu(U) = \mu_n(U)$, $\int B(d(x,a))(\mu_n+\mu)(\mathrm{d}x) < \infty, n = 1, 2, \ldots$. Then $\eta_{\mathbb{B}}(\mu_n, \mu) \to 0$ as $n \to \infty$, where $\mathbb{B}$ is defined as in Theorem 11.2.1.*

Proof. One can easily see that the joint convergence (11.2.4) is equivalent to

$$\mu_n \to \mu, \quad \lim_{M\to\infty} \sup_n \int B_0(d(x,a)) I\{d(x,a) > M\}\mu_n(\mathrm{d}x) = 0 \qquad (11.2.12)$$

[see Billingsley (1999, Sect. 5) and the proof of Theorem 6.4.1 in Chap. 6 in this book]. Then, as in the proof of (8.3.6), we conclude that for any $M \geq t_0$

$$\sup\left\{\int f \mathrm{d}\mu_n + \int g \mathrm{d}\mu : f(x) + g(y) \leq B(d(x,y)) \ \forall x, y \in U\right\}$$

$$\leq \inf\left\{\int B(d(x,y))\widetilde{\mu}(\mathrm{d}x, \mathrm{d}y) : \widetilde{\mu} \in \mathfrak{A}(\mu, \mu_n)\right\}$$

$$\leq \widetilde{B}(\pi(\mu_n, \mu)) + K_{B_0}\Bigg[2\pi(\mu_n, \mu)B_0(M)$$

$$+ \int B_0(d(x,a))I\{d(x,a) > M\}(\mu_n + \mu)(\mathrm{d}x)\Bigg],$$

where $\widetilde{B}(t) = \sup\{B(s) : 0 \leq s \leq t\}$. The last inequality implies $\eta_{\mathbb{B}}(\mu_n, \mu) \to 0$ [see (11.2.2)]. □

Theorem 11.2.3. *Let $B \in \mathcal{OR}$ and $\mu, \mu_1, \mu_2, \ldots$ be a sequence of measures in $\mathfrak{M}$ satisfying* (11.2.4) *and $\int B(d(x,a))(\mu_n + \mu)(\mathrm{d}x) < \infty$. Then*

$$\lim_{n\to\infty} \eta_{\mathbb{K}_1}(\mu_n, \mu) = 0, \qquad (11.2.13)$$

where $\mathbb{K}_1 = \{(f, g) : f(x) + g(y) \leq B(d(x,y)), |g(x)| \leq B(d(x,b)), x, y \in U\}$, and b is an arbitrary point in U.

Proof. As $B \in \mathcal{OR}$, it is enough to prove (11.2.13) for $b = a$. Setting $c_n = \mu(U)/\mu_n(U)$ we have $\lim_{n\to\infty} \eta_{\mathbb{B}}(c_n\mu_n, \mu) = 0$ by Lemma 11.2.1. Hence, as $n \to \infty$ $0 \leq \eta_{\mathbb{K}_1}(\mu_n, \mu) \leq 1/c_n \eta_{\mathbb{B}}(c_n\mu_n, \mu) + |1/c_n - 1| \int B(d(x,b))\mu(\mathrm{d}x) \to 0$. □

In the next theorem we omit the condition $B \in \mathcal{OR}$ but will assume that the class $\mathcal{G} = \{g : (f, g) \in \mathbb{K}_1\}$ is *equicontinuous*, i.e.,

$$\lim_{y \to x} \sup\{|g(x) - g(y)| : g \in \mathcal{G}\} = 0 \quad x \in U. \tag{11.2.14}$$

Theorem 11.2.4 (See Ranga 1962). *Let G be a nonnegative continuous function on U and h a nonnegative function on $U \times U$. Let $\mathbb{K}$ be the class of pairs (f, g) of measurable functions on U such that $(0, 0) \in \mathbb{K}$ and $f(x) + g(y) \le h(x, y)$, $x, y \in U$. Then $\mathbb{K}$ is a π_G-uniform class (see Definition 11.2.2 with $\mathfrak{M}' = \mathfrak{M}_G$) if at least one of the following conditions holds:*

(a) $\lim_{y \to x} h(x, y) = h(x, x) = 0$ for all $x \in U$, the class $\mathcal{F} = \{f : (f, g) \in \mathbb{K}\}$ is equicontinuous, and $|f(x)| \le G(x)$ for all $x \in U$, $f \in \mathcal{F}$.
(b) $\lim_{y \to x} h(y, x) = h(x, x) = 0$ for all $x \in U$ and the class $\mathcal{G} = \{g : (f, g) \in \mathbb{K}\}$ is equicontinuous, $|g(x)| \le G(x)$ for all $x \in U$, $g \in \mathcal{G}$.

Proof. Suppose that $G \equiv 1$. Let $\varepsilon > 0$, and let (a) hold. For any $z \in U$ there is $\delta = \delta(z) > 0$ such that if $B(z) := \{x : d(x, z) < \delta\}$, then

$$\sup_{x \in B(z)} h(z, x) \le \varepsilon/2 \quad \sup_{f \in \mathcal{F}} \sup_{x \in B(z)} |f(x) - f(z)| < \varepsilon/2. \tag{11.2.15}$$

Without loss of generality, we assume that $\mu(\overset{\circ}{B}(z)) = 0$ ($\overset{\circ}{B}$ is the boundary of B). As U is an s.m.s., there exists $z_1, z_2, \dots$ such that $\cup_{j=1}^{\infty} B(z_j)$. Setting $A_1 := B(z_1)$, $A_j := B(z_j) \setminus \cup_{k=1}^{j-1} B(z_k)$, $j = 2, 3, \dots$, we have $f(x) + g(y) = f(x) - f(z_j) + f(z_j) + g(y) \le \varepsilon/2 + h(z_j, y) \le \varepsilon$ for any $x, y \in A_j$. Let $x_j \in A_j$, $j = 1, 2, \dots$. Then, by

$$f(x) + g(y) \le \varepsilon \quad \forall x, y \in A_j, \; j = 1, 2, \dots, \tag{11.2.16}$$

it follows that

$$\sum_{j=1}^{\infty} f(x_j)\mu(A_j) + \int g \mathrm{d}\mu = \sum_{j=1}^{\infty} \int_{A_j} (f(x_j) + g(x))\mu(\mathrm{d}x) \le \varepsilon\mu(U)$$
$$\text{for any } (f, g) \in \mathbb{K}. \tag{11.2.17}$$

Also,

$$\left| \int f(x)\mu_n(\mathrm{d}x) - \sum_{j=1}^{\infty} f(x_j)\mu_n(A_j) \right| \le \varepsilon\mu_n(U) \tag{11.2.18}$$

and

$$\sum_{j=1}^{\infty} |f(x_j)(\mu_n(A_j) - \mu(A_j))| \le \sum_{j=1}^{\infty} |\mu_n(A_j) - \mu(A_j)| \to 0 \text{ as } n \to \infty \tag{11.2.19}$$

by (11.2.15) and $\mu_n(A_j) \to \mu(A_j)$, respectively. Combining relations (11.2.17)–(11.2.19) and taking into account that $\mu_n(U) \to \mu(U)$, we have that $0 \leq \eta_{\mathbb{K}}(\mu_n, \mu) \to 0$.

In the general case, let $A_G := \{x : G(x) > 0\}$. Define measures ν_n and ν on $\mathfrak{B}$ by $\nu_n(B) = \int_B G\mathrm{d}\mu_n$ and $\nu(B) = \int_B G\mathrm{d}\mu$, respectively. The convergence $\pi_G(\mu_n, \mu) \to 0$ implies $\nu_n \xrightarrow{\mathrm{w}} \nu$ as $n \to \infty$. To reduce the general case to the case $G \equiv 1$, denote $f_1(x) := f(x)/G(x)$, $g_1(x) := g(x)/G(x)$ for $x \in A_G$, $\mathbb{K}_1 := \{(f_1, g_1) : (f, g) \in \mathbb{K}\}$, $\mathcal{F}_1 := \{f_1 : f \in \mathcal{F}\}$, and

$$h_1(x, y) = \frac{h(x, y)}{G(y)} + \left|1 - \frac{G(x)}{G(y)}\right|.$$

Then

$$f_1(x) + g_1(y) = \frac{f(x) + g(y)}{G(y)} + \frac{f(x)}{G(x)} - \frac{f(x)}{G(y)} \leq h_1(x, y),$$

and thus $\eta_{\mathbb{K}}(\mu_n, \mu) = \eta_{\mathbb{K}_1}(\nu_n, \nu) \to 0$ as $n \to \infty$. By symmetry, condition (b) also implies $\eta_{\mathbb{K}_1}(\nu_n, \nu) \to 0$. □

For any continuous nonnegative function $b(t)$, $t \geq 0$, $b(0) = 0$, we define the class $A_b = A_b(c)$, $c \in U$, of all real functions f on U with $f(c) = 0$ and norm

$$\mathrm{Lip}_b(f) = \sup\{|f(x) - f(y)|/D(x, y) : x \neq y, x, y \in U\} \leq 1,$$

where $D(x, y) = d(x, y)\{1 + b(d(x, c)) + b(d(y, c))\}$.

Let $C(t) = t(1 + b(t))$, $t \geq 0$, and $p(x, y)$ be a nonnegative function on $U \times U$ continuous in each argument, $p(x, x) = 0$, $x \in U$, and let $\mathfrak{C}$ be the set of pairs $(f, g) \in A_b \times A_b$ for which $f(x) + g(y) \leq p(x, y)$, $x, y \in U$.

Corollary 11.2.2 (See Fortet and Mourier 1953). *Let*

$$\int C(d(x, c))(\mu_n + \mu)(\mathrm{d}x) < \infty, \quad n = 1, 2, \ldots.$$

Then

(a) If

$$\mu_n \xrightarrow{\mathrm{w}} \mu \qquad \int C(d(x, c))(\mu_n - \mu)(\mathrm{d}x) \to 0, \tag{11.2.20}$$

then

$$\eta_{\mathfrak{C}}(\mu_n, \mu) \to 0. \tag{11.2.21}$$

(b) If $p(x, y) \geq D(x, y)$, $x, y \in U$ *and*

$$K := \sup\{|C(s) - C(t)|/[(s - t)(1 + b(s) + b(t))] : s > t \geq 0\} < \infty, \tag{11.2.22}$$

then (11.2.21) *implies* (11.2.20).

Proof. (a) For any $x \in U$ and $f \in A_b$, $|f(x)| \le C(d(x,c))$. The class A_b is clearly equicontinuous, and thus (11.2.21) follows from Theorem 11.2.4.
(b) As $p(x,y) \ge D(x,y)$, $x, y \in U$, it follows that

$$\eta_{\mathfrak{C}}(\mu', \mu'') \ge \zeta_{A_b}(\mu', \mu'') \quad \mu', \mu'' \in \mathfrak{M}. \tag{11.2.23}$$

Applying Theorem 11.2.4 with $g = -f$ and $h = D$ we see that ζ_{A_b}-convergence yields $\mu_n \xrightarrow{\text{w}} \mu$. As $K < \infty$ in (11.2.22), the function $(1/K)C(d(x,c))$, $x \in U$, belongs to the class A_b, and hence (11.2.21) implies $\int C(d(x,c))(\mu_n - \mu)(\mathrm{d}x) \to 0$. □

11.3 Metrization of the Vague Convergence

In this section we will study ρ-uniform classes in the space $\mathfrak{N}$ of all Borel measures $\nu : \mathfrak{B} \to [0,\infty]$ finite on the ring $\mathfrak{B}_0$ of all bounded Borel subsets of (U,d). In particular, this will give two types of metrics metrizing the vague convergence in $\mathfrak{N}$.

Definition 11.3.1. The sequence of measures $\{\nu_1, \nu_2, \ldots\} \subset \mathfrak{N}$ *vaguely converges* to $\nu \in \mathfrak{N}$ ($\nu_n \xrightarrow{\text{v}} \nu$) if

$$\int f \mathrm{d}\nu_n \to \int f \mathrm{d}\nu \quad \text{for} \quad f \in \bigcup_{m=1}^{\infty} \mathcal{F}_m, \tag{11.3.1}$$

where $\mathcal{F}_m$, $m = 1, 2, \ldots$, is the set of all bounded continuous functions on U equal to zero on $S_m = \{x : d(x,a) < m\}$.[5]

Theorem 11.3.1. *Let h be a nonnegative function on $U \times U$, $\lim_{y \to x} h(x,y) = h(x,x) = 0$. Let $\mathbb{K}_m$ be the class of pairs (f,g) of measurable functions such that $(0,0) \in \mathbb{K}_m$, $f(x) + g(y) \le h(x,y)$, $x, y \in U$, $f(x) = g(x) = 0$, $x \notin S_m$, and let the class $\Phi_m = \{f : (f,g) \in \mathbb{K}_m\}$ be equicontinuous and uniformly bounded. Then, if for the sequence $\{\nu_0, \nu_1, \ldots\}$, $\nu_n \xrightarrow{\text{w}} \nu_0$, then $\lim_{n\to\infty} \eta_{\mathbb{K}_m}(\nu_n, \nu_0) = 0$, where $\eta_{\mathbb{K}_m}$ is given by (11.2.2), with $\mathfrak{M}'$ replaced by $\mathfrak{N}$.*

Proof. Let $\theta(x) := \max(0, 1 - d(x, S_m))$, $x \in U$ and $\mu_n(A) := \int_A \theta \mathrm{d}\nu_n$, $A \in \mathfrak{B}$, $n = 0, 1, \ldots$. Then, by $\nu_n \xrightarrow{\text{v}} \nu_0$, we have $\mu_n \xrightarrow{\text{w}} \mu_0$. By virtue of Theorem 11.2.4, we obtain $\eta_{\mathbb{K}_m}(\mu_n, \mu_0) = \eta_{\mathbb{K}_m}(\nu_n, \nu_0) \to 0$ as $n \to \infty$. □

Now we will look into the question of metrization of vague convergence. Known methods of metrization[6] are too complicated from the viewpoint of the structure of the introduced metrics or use additional restrictions on the space $\mathfrak{N}$.

---

[5]See Kallenberg (1975) and Kerstan et al. (1978).

[6]See Kallenberg (1975), Szasz (1975), and Kerstan et al. (1978).

Let $\mathcal{FL}_m = \{f : U \to \mathbb{R}, |f(x) - f(y)| \leq d(x,y), x, y \in U, f(x) = 0, x \notin S_m\}$, $m = 1, 2, \ldots$. Set $\mathbf{K}_m$ to be the following ζ-metric [see (11.2.1)]:

$$\mathbf{K}_m(\nu', \nu'') = \zeta_{\mathcal{FL}_m}(\nu', \nu''), \nu', \nu'' \in \mathfrak{N}, \quad m = 1, 2, \ldots, \tag{11.3.2}$$

and define the metric

$$\mathbf{K}(\nu', \nu'') = \sum_{m=1}^{\infty} 2^{-m} \mathbf{K}_m(\nu', \nu'')/[1 + \mathbf{K}_m(\nu', \nu'')] \quad \nu', \nu'' \in \mathfrak{N}. \tag{11.3.3}$$

Clearly, in the subspace $\mathfrak{M}_0$ of all Borel nonnegative measures with common bounded support the metric $\mathbf{K}$ is topologically equivalent to the Kantorovich metric[7]

$$\ell_1(\nu', \nu'') := \sup\left\{\left|\int f \mathrm{d}(\nu' - \nu'')\right| : f : U \to R, \text{ bounded}, \right.$$
$$\left. |f(x) - f(y)| \leq d(x,y), x, y \in U \right\}. \tag{11.3.4}$$

Corollary 11.3.1. *For any s.m.s. (U, d) the metric $\mathbf{K}$ metrizes the vague convergence in $\mathfrak{N}$.*

Proof. For any metric space (U, d) a necessary and sufficient condition for $\nu_n \xrightarrow{\mathrm{v}} \nu$ is

$$\int f \mathrm{d}\nu_n \to \int f \mathrm{d}\nu \text{ for any } f \in \mathcal{FL} := \bigcup_m \mathcal{FL}_m. \tag{11.3.5}$$

Actually, if (11.3.5) holds, then for any $\varepsilon > 0$, $B \in \mathfrak{B}_0$ (i.e., B is a bounded Borel set) we have

$$\int f_{\varepsilon,B} \mathrm{d}\nu_n \to \int f_{\varepsilon,B} \mathrm{d}\nu, \tag{11.3.6}$$

where $f_{\varepsilon,B}(x) := \max(0, 1 - d(x,B)/\varepsilon)$. For any $\varepsilon > 0$, $B \in \mathfrak{B}$, define the sets $B^\varepsilon := \{x : d(x,B) < \varepsilon\}$, $B_{-\varepsilon} := \{x : d(x, U \setminus B) \geq \varepsilon\}$, $\mathbb{B}^\varepsilon B := B^\varepsilon \setminus B_{-\varepsilon}$. For any $\mu', \mu'' \in \mathfrak{N}$, and $B \in \mathfrak{B}_0$

$$\mu'(B) \leq \int f_{\varepsilon,B} \mathrm{d}\mu' \leq \int f_{\varepsilon,B} \mathrm{d}(\mu' - \mu'') + \mu''(B^\varepsilon),$$

and hence

[7] See Example 3.3.2 in Chap. 3.

$$\mu'(B) \le \mu'(B_{-\varepsilon}) + \mu'(\mathbb{B}^{\varepsilon} B) \le \int f_{\varepsilon, B_{-\varepsilon}} \mathrm{d}(\mu' - \mu'') + \mu''(B) + \mu'(\mathbb{B}^{\varepsilon} B)$$

and

$$\mu'(B) \le \int f_{\varepsilon, B} \mathrm{d}(\mu' - \mu'') + \mu''(B^{\varepsilon}) \le \int f_{\varepsilon, B} \mathrm{d}(\mu' - \mu'') + \mu''(B) + \mu''(\mathbb{B}^{\varepsilon} B).$$

By symmetry,

$$|\mu'(B) - \mu''(B)| \le \left| \int f_{\varepsilon, B_{-\varepsilon}} \mathrm{d}(\mu' - \mu'') \right| + \left| \int f_{\varepsilon, B} \mathrm{d}(\mu' - \mu'') \right| + \min(\mu'(\mathbb{B}^{\varepsilon} B), \mu''(\mathbb{B}^{\varepsilon} B)).$$

Hence, $\limsup_{n\to\infty} |\mu_n(B) - \mu(B)| \le \mu(\mathbb{B}^{\varepsilon} B)$, and thus $\nu_n \xrightarrow{\mathrm{v}} \nu$.

In particular, from (11.3.5) it follows that the convergence $\mathbf{K}(\nu_n, \nu) \to 0$ implies $\nu_n \xrightarrow{\mathrm{v}} \nu$.

Conversely, suppose $\nu_n \xrightarrow{\mathrm{v}} \nu$. By virtue of Theorem 11.3.1, if Θ_m is a class of equicontinuous and uniformly bounded functions $f(x)$, $x \in U$ such that $f(x) = 0$ for $x \notin S_m$, then $\sup\{|\int f \mathrm{d}(\nu_n - \nu) f \in \Theta_m\} \to 0$ as $n \to \infty$. Setting $\Phi_m = \mathcal{F}\mathcal{L}_m$, $m = 1, 2, \ldots$, we get $\mathbf{K}(\nu_n, \nu) \to 0$. □

For all $m = 1, 2, \ldots, l$ define

$$\pi_m(\nu', \nu'') := \inf\{\varepsilon > 0 : \nu'(B) \le \nu''(B^{\varepsilon}) + \varepsilon, \nu''(B) \le \nu'(B^{\varepsilon}) + \varepsilon, \forall B \in \mathfrak{B}, B \subset S_m\} \quad \nu', \nu'' \in \mathfrak{N}$$

and the *Prokhorov metric* in $\mathfrak{N}$

$$\boldsymbol{\pi}(\nu', \nu'') = \sum_{m=1}^{\infty} 2^{-m} \boldsymbol{\pi}_m(\nu', \nu'')/[1 + \boldsymbol{\pi}_m(\nu', \nu'')]. \tag{11.3.7}$$

Obviously, the metric $\boldsymbol{\pi}$ does not change if we replace $\mathfrak{B}$ by the set of all closed subsets of U or if we replace $B^{\varepsilon} = \{x : d(x, B) < \varepsilon\}$ by its closure. In $\mathfrak{M}_0$ (the space of Borel nonnegative measures with common bounded support) the metric $\boldsymbol{\pi}$ is equivalent to $\boldsymbol{\pi}$. We find from Corollary 11.3.1 that $\boldsymbol{\pi}$ metrizes the vague convergence in $\mathfrak{N}$. If (U, d) is a complete s.m.s., then $(\mathfrak{N}, \mathbf{K})$ and $(\mathfrak{N}, \boldsymbol{\pi})$ are also complete separable metric spaces. Here we refer to Hennequin and Tortrat (1965) for the similar problem (the Prokhorov completeness theorem) concerning the metric space $\mathfrak{M} = \mathfrak{M}(U)$ of all bounded nonnegative measures with the Prokhorov metric

$$\boldsymbol{\pi}(\mu, \nu) = \sup\{\varepsilon > 0 : \mu(F) \le \nu(F^{\varepsilon}) + \varepsilon, \nu(F) \le \mu(F^{\varepsilon}) + \varepsilon \ \forall \text{ closed } F \subset A\}. \tag{11.3.8}$$

References

Billingsley P (1999) Convergence of probability measures, 2nd edn. Wiley, New York
Fortet R, B Mourier (1953) Convergence de la réparation empirique vers la répétition theorétique. Ann Sci Ecole Norm 70:267–285
Hennequin PL, Tortrat A (1965) Théorie des probabilités et quelques applications. Masson, Paris
Kallenberg O (1975) Random measures. Akademie, Berlin
Kerstan J, Matthes K, Mecke J (1978) Infinitely divisible point processes. Wiley, New York
Prokhorov YuV (1956) Convergence of random processes and limit theorems in probability theory. Theor Prob Appl 1:157–214
Ranga RR (1962) Relations between weak and uniform convergence of measures with applications. Ann Math Statist 33:659–680
Szasz D (1975) On a metrization of the vague convergence. Studio Sci Math Hungarica 9:219–222

Chapter 12
Glivenko–Cantelli Theorem and Bernstein–Kantorovich Invariance Principle

The goals of this chapter are to:

- Provide convergence criteria for the classical Glivenko–Cantelli problem in terms of the Kantorovich functional $\mathcal{A}_c$,
- Consider generalizations of the Glivenko–Cantelli theorem and provide convergence criteria in terms of $\mathcal{A}_c$,
- Estimate the rate of convergence in the classic Glivenko–Cantelli theorem through $\mathcal{A}_c$,
- Provide convergence criteria in the functional central limit theorem in terms of $\mathcal{A}_c$,
- Consider the Bernstein–Kantorovich invariance principle and provide examples with the ℓ_p metric.

Notation introduced in this chapter:

Notation	Description
$\mathbb{C}[0,1]$	Space of continuous functions on $[0,1]$
W	Wiener measure
$D[0,1]$	Skorokhod space

12.1 Introduction

This chapter begins with an application of the theory of probability metrics to the problem of convergence of the empirical probability measure. Convergence theorems are provided in terms of the Kantorovich functional $\mathcal{A}_c$ described in Chap. 5 for the classic Glivenko–Cantelli theorem but also for extensions such as

S.T. Rachev et al., *The Methods of Distances in the Theory of Probability and Statistics*, DOI 10.1007/978-1-4614-4869-3_12,

the Wellner and the generalized Wellner theorems. The approach of the theory of probability metrics allows for estimating the convergence rate in limit theorems, which for the Glivenko–Cantelli theorem is illustrated through $\mathcal{A}_c$.

As a next application, we provide a convergence criterion in terms of $\mathcal{A}_c$ for the functional central limit theorem. We consider the Bernstein–Kantorovich invariance principle and provide examples with the ℓ_p metric.

12.2 Fortet–Mourier, Varadarajan, and Wellner Theorems

Let (U, d) be an s.m.s., and let $\mathcal{P}(U)$ be the set of all probability measures on U. Let $X_1, X_2, \ldots$ be a sequence of RVs with values in U and corresponding distributions $P_1, P_2, \ldots \in \mathcal{P}(U)$. For any $n \geq 1$ define the *empirical measure*

$$\mu_n = (\delta_{X_1} + \cdots + \delta_{X_n})/n$$

and the *average measure*

$$\overline{P}_n = (P_1 + \cdots + P_n)/n.$$

Let $\mathcal{A}_c$ be the Kantorovich functional (5.2.2),

$$\mathcal{A}_c(P_1, P_2) = \inf\left\{\int_{U\times U} c(x,y)P(\mathrm{d}x,\mathrm{d}y) : P \in \mathcal{P}^{(P_1,P_2)}\right\}, \tag{12.2.1}$$

where $c \in \mathfrak{C}$. Recall that $\mathcal{P}^{(P_1,P_2)}$ is the set of all laws on $U \times U$ with fixed marginals P_1 and P_2, and $\mathfrak{C}$ is the class of all functions $c(x,y) = H(d(x,y))$, $x, y \in U$, where the function H belongs to the class $\mathcal{H}$ of all nondecreasing functions on $[0,\infty)$ for which $H(0) = 0$ and that satisfy the Orlicz condition

$$K_H = \sup\{H(2t)/H(t) : t > 0\} < \infty \tag{12.2.2}$$

(see Example 2.4.1 in Chap. 2).

We now state the well-known theorems of Fortet and Mourier (1953), Varadarajan (1958), and Wellner (1981) in terms of $\mathcal{A}_c$, relying on the following criterion for the μ-convergence of measures (see Theorem 11.2.1 in Chap. 11).

Theorem 12.2.1. *Let $c \in \mathfrak{C}$ and $\int_U c(x,a)P_n(\mathrm{d}x) < \infty$, $n = 0, 1, \ldots$. Then*

$$\lim_{n\to\infty} \mathcal{A}_c(P_n, P_0) = 0 \ \textit{if and only if}\ P_n \xrightarrow{\mathrm{w}} P_0,$$

$$\lim_{n\to\infty} \int_U c(x,b)(P_n - P_0)(\mathrm{d}x) = 0 \tag{12.2.3}$$

for some (and therefore for any) $b \in U$.

Theorem 12.2.2 (Fortet and Mourier 1953). *If $P_1 = P_2 = \cdots = \mu$ and $c_0(x,y) = d(x,y)/(1+d(x,y))$, then $\mathcal{A}_c(\mu_n,\mu) \to 0$ almost surely (a.s.) as $n \to \infty$.*

Theorem 12.2.3 (Varadarajan 1958). *If $P_1 = P_2 = \cdots = \mu$ and c ($c \in \mathfrak{C}$) is a bounded function, then $\mathcal{A}_c(\mu_n,\mu) \to 0$ a.s. as $n \to \infty$.*

Theorem 12.2.4 (Wellner 1981). *If $\overline{P}_1, \overline{P}_2, \ldots$ is a tight sequence, then $\mathcal{A}_{c_0}(\mu_n, \overline{P}_n) \to 0$ a.s. as $n \to \infty$.*

Proof. We follow the proof of the original Wellner's theorem [see Wellner (1981) and Dudley (1969, Theorem 8.3)]. By the strong law of large numbers,

$$\int_U f \mathrm{d}(\mu_n - \overline{P}_n) \to 0 \text{ a.s, as } n \to \infty \tag{12.2.4}$$

for any bounded continuous function on U. Since $\{\overline{P}_n\}_{n\geq 1}$ is a tight sequence, then for any $\varepsilon > 0$ there exists a compact K_ε such that $\overline{P}_n(K_\varepsilon) \geq 1-\varepsilon$ for all $n = 1, 2, \ldots$. Denote

$$\mathrm{Lip}_{c_0}(U) = \{f : U \to \mathbb{R} : |f(x) - f(y)| \leq c_0(x,y),\ \forall x, y \in U\}. \tag{12.2.5}$$

Thus, for some finite m there are $f_1, f_2, \ldots, f_m \in \mathrm{Lip}_{c_0}(U)$ such that

$$\sup_{f\in \mathrm{Lip}_{c_0}(u)} \inf_{1\leq k\leq m} \sup_{x\in K_\varepsilon} |f(x) - f_k(x)| < \varepsilon;$$

consequently

$$\sup_{f\in \mathrm{Lip}_{c_0}(u)} \inf_{1\leq k\leq m} \sup_{x\in K_\varepsilon^\varepsilon} |f(x) - f_k(x)| < 3\varepsilon, \tag{12.2.6}$$

where $K_\varepsilon^\varepsilon$ means the ε-neighborhood of K_ε with respect to the metric c_0. Let $g(x) := \max(0, 1 - d(x, K_\varepsilon)/\varepsilon)$. Then, by (12.2.4) and $\overline{P}_n(K^\varepsilon) \geq \int g \mathrm{d}\overline{P}_n \geq \overline{P}_n(K) \geq 1-\varepsilon$, we have

$$\mu_n(K^\varepsilon) \geq \int g\mathrm{d}\mu_n \geq \int g\mathrm{d}(\mu_n - \overline{P}_n) + 1 - \varepsilon \geq 1 - 2\varepsilon \quad \text{a.s.} \tag{12.2.7}$$

for n large enough. Inequalities (12.2.6) and (12.2.7) imply that

$$\sup\left\{\left|\int_U f\mathrm{d}(\overline{\mu}_n - \overline{P}_n)\right| : f \in \mathrm{Lip}_{c_0}(U)\right\} \leq 10\varepsilon \quad \text{a.s.} \tag{12.2.8}$$

for n large enough. Note that the left-hand side of (12.2.8) is equal to the minimal norm $\mathring{\mu}_{c_0}(\overline{\mu}_n, \overline{P}_n)$ and thus coincides with $\widehat{\mu}_{c_0}(\overline{\mu}_n, \overline{P}_n)$ (see Theorem 6.2.1 in Chap. 6). □

The following theorem extends the results of Fortet–Mourier, Varadarajan, and Wellner to the case of an arbitrary functional $\mathcal{A}_c$, $c \in \mathfrak{C}$.

Theorem 12.2.5 (A generalized Wellner theorem). *Suppose that $s_1, s_2, \ldots$ is a sequence of operators in U, and denote*

$$D_i = \sup\{d(s_i x, x) : x \in U\}$$
$$L_i = \sup\{d(s_i x, s_i y)/d(x,y) : x \neq y, x, y \in U\}$$
$$\Theta_i = \min[D_i, (L_i + 1)\mathcal{A}_{c_0}(\delta_{X_i}, P_i), 1], i = 1, 2, \ldots.$$

Let $Y_i = s_i(X_i)$, Q_i be the distribution of Y_i, $\overline{Q}_n = (Q_1 + \cdots + Q_n)/n$ and $\nu_n = (\delta_{Y_1} + \cdots + \delta_{Y_n})/n$. If $\overline{Q}_1, \overline{Q}_2, \ldots$ is a tight sequence

$$\overline{\Theta}_n = (\Theta_1 + \cdots + \Theta_n)/n \to 0 \text{ a.s. } n \to \infty \tag{12.2.9}$$

$c \in \mathfrak{C}$ and for some $a \in U$

$$\lim_{M\to\infty} \sup_n \int_U c(x,a) I\{d(x,a) > M\}(\mu_n + \overline{P}_n)(\mathrm{d}x) = 0 \quad \text{a.s.}, \tag{12.2.10}$$

then $\mathcal{A}_c(\mu_n, \overline{P}_n) \to 0$ a.s. as $n \to \infty$.

Proof. From Wellner's theorem it follows that $\lim_n \mathcal{A}_{c_0}(\nu_n, \overline{Q}_n) = 0$ a.s. We next estimate $\mathcal{A}_{c_0}(\mu_n, \overline{P}_n)$ obtaining

$$\mathcal{A}_{c_0}(\mu_n, \overline{P}_n) \leq \mathcal{A}_{c_0}(\nu_n, \overline{Q}_n) + (B_1 + \cdots + B_n)/n, \tag{12.2.11}$$

where

$$B_i = \sup\left\{\left|\int_U [f(s_i x) - f(x)](\delta_{X_i} - P_i)(\mathrm{d}x)\right| : f \in \mathrm{Lip}_{c_0}(U)\right\}.$$

In fact, by the duality representation of $\mathcal{A}_{c_0}$ (see Corollary 6.2.1 of Chap. 6)

$$\mathcal{A}_{c_0}(\mu_n, \overline{P}_n) = \sup\left\{\left|\frac{1}{n}\sum_{i=1}^n \int_U f(x)(\delta_{X_i} - P_i)(\mathrm{d}x)\right| : f \in \mathrm{Lip}_{c_0}(U)\right\},$$

and thus

$$\begin{aligned}
&\mathcal{A}_{c_0}(\mu_n, \overline{P}_n) \\
&\leq \mathcal{A}_{c_0}(\nu_n, \overline{Q}_n) + \sup_{f \in \mathrm{Lip}_{c_0}(U)} \left|\frac{1}{n}\int f(x)(\delta_{Y_i} - Q_i)(\mathrm{d}x) - \frac{1}{n}\int f(x)(\delta_{X_i} - P_i)(\mathrm{d}x)\right| \\
&= \mathcal{A}_{c_0}(\nu_n, \overline{Q}_n) + \sup_{f \in \mathrm{Lip}_{c_0}(U)} \left|\frac{1}{n}\sum_{i=1}^n (f(s_i X_i) - Ef(s_i X_i) - f(X_i) + Ef(X_i))\right| \\
&= \mathcal{A}_{c_0}(\nu_n, \overline{Q}_n) + (B_1 + \cdots + B_n)/n.
\end{aligned}$$

We estimate B_i as follows:

$$B_i \leq \sup_{f \in \mathrm{Lip}_{c_0}(U)} \int |f(s_i X_i) - f(x)|(\delta_{X_i} + P_i)(\mathrm{d}x)$$

$$\leq \sup_{x \in U} \frac{d(s_i x, x)}{1 + d(s_i x, x)} \int (\delta_{X_i} + P_i)(\mathrm{d}x) \leq 2 \min(D_i, 1),$$

and, moreover, since for $g(x) := f(s_k x) - f(x)$, $f \in \mathrm{Lip}_{c_0}(U)$, we have

$$|g(x) - g(y)| \leq d(s_i x, s_i y) + d(x, y) \leq (L_i + 1) d(x, y),$$

$$|g(x) - g(y)| \leq 2 \frac{d(x, y)}{1 + d(x, y)} (L_i + 1) \text{ if } d(x, y) \leq 1,$$

$$\|g\|_\infty := \sup\{|g(x)| : x \in U\} \leq 2,$$

$$\tfrac{1}{4}|g(x) - g(y)| \leq \tfrac{1}{4}\{|g(x)| + |g(y)|\}$$

$$\leq 2 \frac{d(x, y)}{1 + d(x, y)} \text{ if } d(x, y) > 1,$$

and thus

$$B_i \leq \sup\left\{ \left| \int_U g(x)(\delta_{X_i} - P_i)(\mathrm{d}x) \right| : g : U \to \mathbb{R}, \right.$$

$$\left. |g(x) - g(y)| \leq 8(L_i + 1) c_0(x, y) \right\}$$

$$\leq 8(L_i + 1) \mathcal{A}_{c_0}(\delta_{X_i}, P_i).$$

Using the preceding estimates for B_i and assumption (12.2.9) we obtain that $(B_1 + \cdots + B_n)/n \to 0$. According to (12.2.11),

$$\mathcal{A}_{c_0}(\mu_n, \overline{P}_n) \to 0 \text{ a.s. as } n \to \infty. \tag{12.2.12}$$

If $\mathbf{K}$ is the Ky Fan metric (see Example 3.4.2 in Chap. 3) and μ_{c_0} is the probability metric

$$\mu_{c_0}(P) := \int_{U \times U} c_0(x, y) P(\mathrm{d}x, \mathrm{d}y), \quad P \in \mathcal{P}_2(U),$$

then by Chebyshev's inequality we have

$$\frac{\mathbf{K}^2}{1 + \mathbf{K}} \leq \mu_{c_0} \leq \mathbf{K} + \frac{\mathbf{K}}{1 + \mathbf{K}}.$$

Passing to the minimal metrics in the last inequality and using the Strassen–Dudley theorem (see Corollary 7.5.2 in Chap. 7) we get

$$\frac{\pi^2}{1+\pi} \leq \mathcal{A}_{c_0} \leq \pi + \frac{\pi}{1+\pi}, \tag{12.2.13}$$

where π is the Prokhorov metric in $\mathcal{P}(U)$. Applying (12.2.13) and (7.6.9) (see also Lemma 8.3.1) we have, for any positive M,

$$\frac{\pi^2(\mu_n, \overline{P}_n)}{1+\pi(\mu_n, \overline{P}_n)} \leq \mathcal{A}_{c_0}(\mu_n, \overline{P}_n) \leq \pi(\mu_n, \overline{P}_n) + \frac{\pi(\mu_n, \overline{P}_n)}{1+\pi(\mu_n, \overline{P}_n)} \tag{12.2.14}$$

and

$$\begin{aligned}\mathcal{A}_c(\mu_n, \overline{P}_n) &\leq H(\pi(\mu_n, \overline{P}_n)) + 2K_H \pi(\mu_n, \overline{P}_n) H(M) \\ &\quad + K_H \int_U c(x,a) I\{d(x,a) > M\}(\mu_n + P_n)(\mathrm{d}x).\end{aligned} \tag{12.2.15}$$

From (12.2.12), (12.2.14), (12.2.15), and (12.2.10) it follows that $\mathcal{A}_c(\mu_n, \overline{P}_n) \to 0$ a.s. as $n \to \infty$. □

Corollary 12.2.1. *If c ($c \in \mathfrak{C}$) is a bounded function and $\Theta_n \to 0$ a.s., then $\mathcal{A}_c(\mu_n, \overline{P}_n) \to 0$ a.s. as $n \to \infty$.*

Corollary 12.2.1 is a consequence of the preceding theorem when $s_i(x) = x$, $x \in U$, and clearly is a generalization of the Varadarajan theorem 12.2.3. It is also clear that Theorem 12.2.3 implies Theorem 12.2.2. The following example shows that the conditions imposed in Corollary 12.2.1 are actually weaker as compared to the conditions of Wellner's theorem 12.2.4.

Example 12.2.1. Let $(U, \|\cdot\|)$ be a separable normed space. Let x_k be $k\overline{e}$, where $\overline{e}$ is the unit vector in U, and let $X_k = x_k$ a.s. Set $s_k(x) := x - x_k$; then $\overline{Q}_n = \delta_0$ and $\Theta_n = 0$ a.s. Clearly, $\mathcal{A}_c(\mu_n, \overline{P}_n) = 0$ a.s., but $\overline{P}_n$ is not a tight sequence.

In what follows, we will assume that $P_1 = P_2 = \cdots = \mu$. In this case, the Glivenko–Cantelli theorem can be stated as follows in terms of $\mathcal{A}_c$ and the minimal metric $\ell_p = \mathcal{L}_p$ $(0 < p < \infty)$ [see definitions (3.3.11) and (3.3.12), representations (3.4.18) and (5.4.16), and Theorem 8.2.1].

Corollary 12.2.2 (Generalized Glivenko–Cantelli–Varadarajan theorem). *Let $c \in \mathfrak{C}$ and $\int_U c(x,a)\mu(\mathrm{d}x) < \infty$. Then $\mathcal{A}_c(\mu_n, \mu) \to 0$ a.s. as $n \to \infty$. In particular, if*

$$\int_U d^p(x,a)\mu(\mathrm{d}x) < \infty, \qquad 0 < p < \infty, \tag{12.2.16}$$

then $\ell_p(\mu_n, \mu) \to 0$ a.s. as $n \to \infty$.

According to Theorem 6.3.1 and Corollary 7.6.3, the minimal norm $\overset{\circ}{\mu}_{c_p}$ with

$$c_p(x,y) = d(x,y)\max(1, d^{p-1}(x,a), d^{p-1}(y,a)), \quad p \geq 1,$$

and the minimal metric ℓ_p metrize the same exact convergence in the space $\mathcal{P}_p(U) = \{P \in \mathcal{P}(U) : \int_U d^p(x,a)P(\mathrm{d}x) < \infty\}$, namely,

$$\ell_p(P_n, P) \to 0 \iff \overset{\circ}{\mu}_{c_p}(P_n, P) \to 0 \iff \begin{cases} P_n \xrightarrow{\mathrm{w}} P \text{ and} \\ \int_U d^p(x,a)(P_n - P)(\mathrm{d}x) \to 0 \end{cases} \tag{12.2.17}$$

Thus Corollary 12.2.2 implies the following theorem stated by Fortet and Mourier (1953).

Corollary 12.2.3. *If* (12.2.16) *holds, then*

$$\overset{\circ}{\mu}_{c_p}(\mu_n, \mu) \to 0 \text{ a.s. as } n \to \infty.$$

Remark 12.2.1. One could generalize Corollaries 12.2.2 and 12.2.3 by means of Theorem 6.4.1; see also Ranga (1962) for extensions of the original Fortet–Mourier result. We write $\mathcal{H}^*$ for the subset of all convex functions in $\mathcal{H}$ and $\mathfrak{C}^*$ for the set $\{H \circ d : H \in \mathcal{H}^*\}$. Theorem 8.2.2 gives an explicit representation of the functionals $\mathcal{A}_c$, $c \in \mathfrak{C}^*$, when $U = \mathbb{R}^1$ [see (8.2.38)]. Corollary 12.2.2 may be formulated in this case as follows.

Corollary 12.2.4. *Let $c \in \mathfrak{C}^*$, $U = \mathbb{R}^1$, and $d(x,y) = |x - y|$. Let $F_n(x)$ be the empirical distribution function corresponding to the distribution function $F(x)$ with $\int c(x,0)\mathrm{d}F(x)$ finite. Then*

$$\int_0^1 c(F_n^{-1}(x), F^{-1}(x))\mathrm{d}x \to 0 \quad a.s. \tag{12.2.18}$$

In particular, if

$$\int |x|^p \mathrm{d}F(x) < \infty \quad p \geq 1, \tag{12.2.19}$$

then

$$\ell_p^p(F_n, F) = \int_0^1 |F_n^{-1}(x) - F^{-1}(x)|^p \mathrm{d}x \to 0 \quad a.s. \tag{12.2.20}$$

and

$$\overset{\circ}{\mu}_{c_p}(F_n, F) = \int_{-\infty}^{\infty} \max(1, |x|^{p-1})|F_n(x) - F(x)|\mathrm{d}x \to 0 \quad a.s. \tag{12.2.21}$$

Remark 12.2.2. In the case of $p = 1$, Corollary 12.2.4 was proved by Fortet and Mourier (1953). The case $p = 2$ when $F(x)$ is a continuous strictly increasing function was proved by Samuel and Bachi (1964).

We study next the estimation of the convergence speed in the Glivenko–Cantelli theorem in terms of $\mathcal{A}_c$. Estimates of this sort are useful if one has to estimate not only the speed of convergence of the distribution μ_n to μ in weak metrics but also the speed of convergence of their moments. Thus, for example, if $E\ell_p(\mu_n, \mu) = O(\phi(n))$, $n \to \infty$, for some $p \in (0, \infty)$, then Lemma 8.3.1 implies that $(E(\pi(\mu_n, \mu))^{(p+1)/p'} = O(\phi(n))$, $n \to \infty$, where $p' = \max(1, p)$ [see (8.3.7)], and by Minkowski's inequality it follows that

$$E\left|\left[\int_U d^p(x,a)\mu_n(\mathrm{d}x)\right]^{1/p'} - \left[\int_U d^p(x,a)\mu(\mathrm{d}x)\right]^{1/p'}\right| = O(\phi(n))$$

for any point $a \in U$.

We will estimate $E\mathcal{A}_c(\mu_n, \mu)$ in terms of the ε-entropy of the measure μ, as was originally suggested by Dudley (1969). Let $N(\mu, \varepsilon, \delta)$ be the smallest number of sets of diameter at most 2ε whose union covers U except for a set A_0 with $\mu(A_0) \leq \delta$. Using Kolmogorov's definition of the ε-entropy of a set U, we call $\log N(\mu, \varepsilon, \varepsilon)$ the *ε-entropy of the measure μ*. The next theorem was proved by Dudley (1969) for $c = c_0$.

Theorem 12.2.6 (Dudley 1969). *Let $c = H \circ d \in \mathfrak{C}$ and $H(t) = t^\alpha h(t)$, where $0 < \alpha < 1$ and $h(t)$ is a nondecreasing function on $[0, \infty)$. Let $\beta_r = \int_U c^r(x,a)\mu(\mathrm{d}x) < \infty$ for some $r > 1$ and $a \in U$.*

(a) If there exist numbers $k \geq 2$ and $K < \infty$ such that

$$N(\mu, \varepsilon^{1/\alpha}, \varepsilon^{k/(k-2)}) \leq K\varepsilon^{-k}, \tag{12.2.22}$$

then

$$E\mathcal{A}_c(\mu_n, \mu) \leq Cn^{-(1-1/r)/k},$$

where C is a constant depending just on α, k, and K.

(b) If $h(0) > 0$ and, for some positive c_1 and δ

$$N(\mu, \varepsilon^{1/\alpha}, 1/2) \geq c_1\varepsilon^{-k}, \tag{12.2.23}$$

then there exists a $c_2 = c_2(\mu)$ such that

$$E\mathcal{A}_c(\mu_n, \mu) \geq c_2 n^{-1/k}. \tag{12.2.24}$$

The proof of Theorem 12.2.6 is based on Dudley (1969) and the inequality

$$\mathcal{A}_c(\mu, \nu) \leq 2H(N)\mathcal{A}_{c_\alpha}(\mu, \nu) + 2c_H \int c(x,a)\{d(x,a) > N/2\}(\mu + \nu)(\mathrm{d}x), \tag{12.2.25}$$

where $c_\alpha = d^\alpha/(1+d^\alpha)$, $N > 0$, and μ and ν are arbitrary measures on $\mathcal{P}(U)$. The detailed proof is given in Kalashnikov and Rachev (1988, Theorem 9.7, p. 147–150), where the constant C is bounded from above by $\frac{4}{3}(\sqrt{k}3^{2k+1})$.

If $(U, d) = (\mathbb{R}^d, \|\cdot\|)$, $m_\gamma = \int \|x\|^\gamma \mu(\mathrm{d}x) < \infty$, where $\gamma = k\alpha d/[k\alpha - d)(k-2)]$, $k\alpha > d$, $k > 2$, then requirement (12.2.22) is satisfied. If $(U, d) = (\mathbb{R}^{k\alpha}, \|\cdot\|)$, where $k\alpha$ is an integer and μ is an absolutely continuous distribution, then condition (12.2.23) is satisfied. The estimate $E\mathcal{A}_c(\mu_n, \mu) \leq cn^{-1/k}$ has exact exponent $(1/k)$ when $k\alpha$ is an integer, $U = \mathbb{R}^{k\alpha}$, and μ is an absolutely continuous distribution having uniformly bounded moments β_r, $r > 1$, and m_γ, $\gamma > 1$.

Open Problem 12.2.1. What is the exact order of n as $\mathcal{A}_c(\mu_n, \mu) \to 0$ a.s.? For the case where μ is uniform in [0, 1] and

$$c(x, y) = c_0(x, y) = \frac{|x-y|}{1+|x-y|}$$

it follows immediately from a result of Yukich (1989) that there exist constants c and C such that

$$\lim_{n\to\infty} \Pr\left\{ c \leq \left(\frac{n}{\log n}\right)^{1/2} \mathcal{A}_{c_0}(\mu_n, \mu) \leq C \right\} = 1. \tag{12.2.26}$$

12.3 Functional Central Limit and Bernstein–Kantorovich Invariance Principle

Let $\xi_{n1}, \xi_{n2}, \ldots, \xi_{nk_n}$, $n = 1, 2, \ldots$, be an array of independent RVs with distribution functions (DFs) F_{nk}, $k = 1, \ldots, k_n$, obeying the condition of limiting negligibility

$$\lim_n \max_{1\leq k\leq k_n} \Pr(|\xi_{nk}| > \varepsilon) = 0 \tag{12.3.1}$$

and the conditions

$$E\xi_{nk} = 0, \quad E\xi_{nk}^2 = \sigma_{nk}^2 > 0, \quad \sum_{k=1}^{k_n} \sigma_{nk}^2 = 1. \tag{12.3.2}$$

Let $\zeta_{n0} = 0$ and $\zeta_{nk} = \xi_{n1} + \cdots + \xi_{nk}$, $1 \leq k \leq k_n$, and form a random polygonal line $\zeta_n(t)$ with vertices $(E\zeta_{nk}^2, \zeta_{nk})$.[1] Let P_n, from the space of laws on $\mathbb{C}[0, 1]$ with the supremum norm $\|x\| = \sup\{|x(t)| : t \in [0, 1]\}$), be the distribution of $\zeta_n(t)$, and let W be a Wiener measure in $\mathbb{C}[0, 1]$. On the basis of Theorem 8.3.1, we have

[1]See Prokhorov (1956).

the following $\mathcal{A}_c$-convergence criterion:

$$\mathcal{A}_c(P_n, W) \to 0 \iff \begin{cases} P_n \xrightarrow{\mathrm{w}} W \\ \int_{\mathbb{C}[0,1]} c(x,0)(P_n - W)(\mathrm{d}x) \to 0 \end{cases} \tag{12.3.3}$$

for any $c \in \mathfrak{C} = \{c(x,y) = H(\|x-y\|),\ H \in \mathcal{H}$ [see (12.2.2)].

The limit relation (12.3.3) implies the following version of the classic Donsker–Prokhorov theorem.[2]

Theorem 12.3.1 (Bernstein–Kantorovich functional limit theorem). *Suppose that conditions* (12.3.1) *and* (12.3.2) *hold and that* $EH(|\xi_{nk}|) < \infty$, $k = 1,2,\dots,k_n$, $n = 1,2,\dots$, $H \in \mathcal{H}$. *Then the convergence* $\mathcal{A}_c(P_n, W) \to 0$, $n \to \infty$, *is equivalent to the fulfillment of the Lindeberg condition*

$$\lim_{n\to\infty} \sum_{k=1}^{k_n} \int_{|x|>\varepsilon} x^2 \mathrm{d}F_{nk}(x) = 0, \qquad \varepsilon > 0, \tag{12.3.4}$$

and the Bernstein condition

$$\lim_{N\to\infty} \limsup_{n\to\infty} \sum_{k=1}^{k_n} \int_{|x|>N} H(|x|)\mathrm{d}F_{nk}(x) = 0. \tag{12.3.5}$$

Proof. By the well-known theorem by Prokhorov (1956), the necessity of (12.3.4) is a straightforward consequence of $P_n \xrightarrow{\mathrm{w}} W$. Let us prove the necessity of (12.3.5). Define the functional $b : \mathbb{C}[0,1] \to \mathbb{R}$ by $b(x) = x(1)$. For any $N > 2\sqrt{2}$,

$$\begin{aligned} \int_N^\infty \Pr(\|\zeta_n\| > t)\mathrm{d}H(t) &< 2\int_N^\infty \Pr(|\zeta_{n,k_n}| \ge t - \sqrt{2})\mathrm{d}H(t) \\ &\le 2\int_{N/2}^\infty \Pr(|\zeta_{n,k_n}| > t)\mathrm{d}H(2t) \\ &\le 2K_H \int_{M(N)} \Pr(|\zeta_{n,k_n}| \ge t)\mathrm{d}H(t), \end{aligned}$$

where $M(N)$ increases to infinity with $N \uparrow \infty$.[3] From the last inequality it follows that $EH(\|\zeta_n\|) < \infty$ for all $n = 1,2,\dots$. By Theorem 12.2.1 and $\mathcal{A}_c(P_n, W) \to 0$, the relations $P_n \xrightarrow{\mathrm{w}} W$ and

$$\int H(\|x\|)(P_n - W)(\mathrm{d}x) \to 0$$

---

[2] See, for example, Billingsley (1999, Theorem 10.1).

[3] See, for example, Billingsley (1999).

hold as $n \to \infty$, and since for any N

$$EH(|b(\zeta_n)|)I\{|b(\zeta_n)| > N\} \le EH(\|\zeta_n\|)I\{\|\zeta_n\| > N\}$$
$$\le 2\int_{M_1(N)}^{\infty} Pr(\|\zeta_n\| > t)\mathrm{d}H(t),$$

where $M_1(N) \uparrow \infty$ together with $N \uparrow \infty$, we have (i) $P_n \circ b^{-1} \xrightarrow{\mathrm{w}} W \circ b^{-1}$ and (ii) $\int h(\|x\|)(P_n \circ b^{-1} - W \circ b^{-1})(\mathrm{d}x) \to 0$ as $n \to \infty$.

The necessity of condition (12.3.5) is proved by virtue of Kruglov's moment limit theorem.[4] The sufficiency of (12.3.4) and (12.3.5) is proved in a similar way. □

Next we state a functional limit theorem that is based on the Bernstein central limit theorem.[5] We formulate the result in terms of the minimal metric ℓ_p (3.3.11), (3.4.18), and (12.2.17).

Corollary 12.3.1. *Let $\xi_1, \xi_2, \ldots$ be a sequence of independent RVs such that $E\xi_i^2 = b_i$ and $E|\xi_i|^p < \infty$, $i = 1, 2, \ldots$, $p > 2$. Let $B_n = b_1 + \cdots + b_n$, $\zeta_n = \xi_1 + \cdots + \xi_n$, and let the sequence $B_n^{-1/2}/\xi_j$, $j = 1, 2, \ldots$, satisfy the limiting negligibility condition. Let $X_n(t)$ be a random polygonal line with vertices $(B_k/B_n, B_n^{-1/2}/\zeta_k)$, and let P_n be its distribution. Then the convergence*

$$\ell_p(P_n, W) \to 0, \qquad n \to \infty, \tag{12.3.6}$$

is equivalent to the fulfillment of the condition

$$\lim_{n\to\infty} B_n^{-p/2} \sum_{i=1}^{n} E|\xi|^p = 0. \tag{12.3.7}$$

Proof. The proof is analogous to that of Theorem 12.3.1. Here, conditions (12.3.4) and (12.3.5) are equivalent to (12.3.7).[6] □

Corollary 12.3.2 (Bernstein–Kantorovich invariance principle). *Suppose that c, $c \in \mathfrak{C}$, the array $\{\xi_{nk}\}$ satisfies the conditions of Theorem 12.3.1, and conditions* (12.3.4) *and* (12.3.5) *hold. Then $\mathcal{A}_c(P_n \circ b^{-1}, W \circ b^{-1}) \to 0$ as $n \to \infty$ for any functional on $\mathbb{C}[0,1]$ for which*

$$N(b; c, c') = \sup\{c'(b(x), b(y))/c(x, y) : x \ne y, x, y \in \mathbb{C}[0,1]\} < \infty.$$

Proof. Observe that $\mathcal{A}_c(P_n, W) \to 0$ implies $\mathcal{A}_{c'}(P_n \circ b^{-1}, W \circ b^{-1}) \to 0$ as $n \to \infty$, provided $N(b; c, c') < \infty$. Now apply Theorem 12.3.1. □

---

[4] Given (i), then (ii) is equivalent to (12.3.5); see Kruglov (1973, Theorem 1).

[5] See Bernstein (1964, p. 358).

[6] See Bernstein (1964), Kruglov (1973), and de Acosta and Gine (1979).

Let $c'(t,s) = H'(|t-s|)$ and $t,s \in \mathbb{R}$. Consider the following examples of functionals b with finite $N(b;c,c')$.

(a) If $H = H'$ and b has a finite Lipschitz norm,

$$\|b\|_L = \sup\{|b(x)-b(y)|/\|x-y\| : x \neq y, x, y \in \mathbb{C}[0,1]\} < \infty, \quad (12.3.8)$$

then $N(b;c,c') < \infty$. Functionals such as these are $b_1(x) = x(a)$, $a \in [0,1]$; $b_2(x) = \max\{x(t) : t \in [0,1]\}$; $b_3(x) = \|x\|$, and $b_4(x) = \int_0^1 \phi(x(t))dt$, where $\|\phi\|_L := \sup\{|\phi(x)-\phi(y)|/|x-y| : x,y \in [0,1]\} < 1$.

(b) Let $H(t) = t^p$ and $H'(t) = t^{p'}$, $0 < p < p'$. Then $N(b_3^{p/p'};c,c') < \infty$ and $N(b_4;c,c') < \infty$ if

$$|\phi(x)-\phi(y)| \le |x-y|^{p/p'}, \qquad x,y \in [0,1]. \quad (12.3.9)$$

Further, as an example of Corollary 12.3.2 we will consider the functional b_4 and the following moment limit theorem.

Corollary 12.3.3. *Suppose $\xi_1, \xi_2, \ldots$ are independent random variables with $E\xi_i = 0$, $E\xi_i^2 = \sigma^2 > 0$, and*

$$\lim_{n\to\infty} n^{-p/2} \sum_{j=1}^{n} E|\xi_j|^p = 0 \text{ for some } p > 2. \quad (12.3.10)$$

Suppose also that $\phi : [0,1] \to \mathbb{R}$ has a finite Lipschitz seminorm $\|\phi\|_L$. Then

$$\ell_p\left(\frac{1}{n}\sum_{k=1}^{n}\phi\left(\frac{\xi_1+\cdots+\xi_k}{\sigma\sqrt{n}}\right), \int_0^1 \phi(w(t))dt\right) \to 0 \text{ as } n \to \infty, \quad (12.3.11)$$

where the law of w is W.

Proof. Let $X_n(\cdot)$ be a random polygon line with vertices $(k/n, S_k/\sigma\sqrt{n})$, where $S_0 = 0$, $S_k = \xi_1 + \cdots + \xi_k$. From Corollaries 12.3.1 and 12.3.2 it follows that

$$\lim_{n\to\infty} \ell_p\left(\int_0^1 \phi(X_n(t))dt, \int_0^1 \phi(w(t))dt\right) = 0. \quad (12.3.12)$$

Readily,[7] we have

$$\mathbf{K}\left(\left|\int_0^1 \phi(X_n(t))dt - \frac{1}{n}\sum_{k=1}^{n}\phi\left(\frac{S_k}{\sigma\sqrt{n}}\right)\right|, 0\right) \to 0, \quad (12.3.13)$$

[7] See Gikhman and Skorokhod (1971, p. 491, or p. 416 of the English edition).

where **K** is the Ky Fan metric. By virtue of the maximal inequality[8]

$$\int_0^\infty \Pr\left\{\left|\int_0^1 \phi(X_n(t))\mathrm{d}t - \frac{1}{n}\sum_{i=1}^n \left(\frac{S_k}{\sigma\sqrt{n}}\right)\right| > u\right\} \mathrm{d}u^p$$

$$\leq \int_N^\infty \Pr\left\{\frac{|S_n|}{\sigma\sqrt{n}} > \frac{t}{2\|\phi\|_L} - \sqrt{2}\right\} \mathrm{d}t^p. \tag{12.3.14}$$

Corollary 12.3.1 and (12.3.10) imply that the right-hand side of (12.3.14) goes to zero uniformly on n as $N \to \infty$. From (12.3.13) and (12.3.14) it follows that

$$E\left|\int_0^1 \phi(X_n(t))\mathrm{d}t - \frac{1}{n}\sum_{k=1}^n \phi\left(\frac{S_k}{\sigma\sqrt{n}}\right)\right|^p \to 0. \tag{12.3.15}$$

Finally, (12.3.15) and (12.3.13) imply

$$\ell_p\left(\int_0^1 \phi(X_n(t))\mathrm{d}t, \frac{1}{n}\sum_{k=1}^n \phi\left(\frac{S_k}{\sigma\sqrt{n}}\right)\right) \to 0 \text{ as } n \to \infty,$$

which, together with (12.3.12), completes the proof of (12.3.11). □

We state one further consequence of Theorem 12.3.1. Let the series scheme $\{\xi_{nk}\}$ satisfy the conditions of Theorem 12.3.1, and let $\eta_n(t) = \zeta_{nk}$ for $t \in (t_{n(k-1)}, t_{nk})$, $t_{nk} := E\zeta_{nk}^2$, $k = 1, \ldots, k_n$, $\eta(0) = 0$. Let $\widehat{P}_n$ be the distribution of η_n. The distribution $\widehat{P}_n$ belongs to the space of probability measures defined on the Skorokhod space $D[0,1]$.[9]

Corollary 12.3.4. *The convergence* $\mathcal{A}_c(\widehat{P}_n, W) \to 0$ *as* $n \to \infty$ *is equivalent to the fulfillment of* (12.3.4) *and* (12.3.5).

[8] See Billingsley (1999).

[9] See Billingsley (1999, Chap. 3).

References

Bernstein SN (1964) Collected works, vol 4. Nauka, Moscow (in Russian)
Billingsley P (1999) Convergence of probability measures, 2nd edn. Wiley, New York
de Acosta A, Gine E (1979) Convergence of moments and related functionals in the general central limit theorem in Banach spaces. Z Wahrsch Verw Geb 48:213–231
Dudley RM (1969) The speed of mean Glivenko-Cantelli convergence. Ann Math Statist 40:40–50
Fortet R, Mourier B (1953) Convergence de la réparation empirique vers la répétition theorétique. Ann Sci Ecole Norm 70:267–285
Gikhman II, Skorokhod AV (1971) The theory of stochastic processes. Nauka, Moscow (in Russian). [Engl. transl. (1976) Springer, Berlin]
Kalashnikov VV, Rachev ST (1988) Mathematical methods for construction of stochastic queueing models. Nauka, Moscow (in Russian). [Engl. transl., (1990) Wadsworth, Brooks–Cole, Pacific Grove, CA]
Kruglov VM (1973) Convergence of numerical characteristics of independent random variables with values in a Hilbert space. Theor Prob Appl 18:694–712
Prokhorov YuV (1956) Convergence of random processes and limit theorems in probability theory. Theor Prob Appl 1:157–214
Ranga RR (1962) Relations between weak and uniform convergence of measures with applications. Ann Math Statist 33:659–680
Samuel E, Bachi R (1964) Measures of the distance of distribution functions and some applications. Metron XXIII:83–122
Varadarajan VS (1958) Weak convergence of measures on separable metric space. Sankhya 19:15–22
Wellner Jon A (1981) A Glivenko-Cantelli theorem for empirical measures of independent but nonidentically distributed random variables. Stoch Process Appl 11:309–312
Yukich JE (1989) Optimal matching and empirical measures. Proc Am Math Soc 107:1051–1059

Chapter 13
Stability of Queueing Systems

The goals of this chapter are to:

- Explore the question of stability of a sequence of stochastic models in the context of general queueing systems by means of the Kantorovich functional $\mathcal{A}_c$,
- Consider the case of queueing systems with independent interarrival and service times,
- Consider the special case of approximating a stochastic queueing system by means of a deterministic model.

Notation introduced in this chapter:

Notation	Description
$G\|G\|1\|\infty$	Single-server queue with general flows of interarrival times and service times, and infinitely large "waiting room"
$\mathbf{e} = (e_0, e_1, \ldots)$	"Input" flow of interarrival times
$\mathbf{s} = (s_0, s_1, \ldots)$	Flow of service times
$\mathbf{w} = (w_0, w_1, \ldots)$	Flow of waiting times
$GI\|GI\|1\|\infty$	Special case of a $G\|G\|1\|\infty$-system in which $s_n - e_n$ are independent identically distributed random variables
$D\|G\|1\|\infty$	$G\|G\|1\|\infty$-system with a deterministic input flow
$D\|D\|1\|\infty$	Deterministic single-server queueing model
$\mathrm{IND}(X)$	Deviation of $\Pr_X$ from product measure $\Pr_{X_1} \times \cdots \times \Pr_{X_n}$, $X = (X_1, \ldots, X_n)$

13.1 Introduction

The subject of this chapter is the fundamental problem of the stability of a sequence of stochastic models that can be interpreted as approximations or perturbations

S.T. Rachev et al., *The Methods of Distances in the Theory of Probability and Statistics*, DOI 10.1007/978-1-4614-4869-3_13, © Springer Science+Business Media, LLC 2013

of a given initial model. We consider queueing systems and study their stability properties with respect to the Kantorovich functional $\mathcal{A}_c$. We begin with a general one-server queueing system with no assumptions on the interarrival times and then proceed to the special cases of independent interarrival times and independent service times. Finally, we consider deterministic queueing systems as approximations to a stochastic queuing model.

13.2 Stability of $G|G|1|\infty$-Systems

As a model example of the applicability of Kantorovich's theorem in the stability problem for queueing systems, we consider the stability of the system $G|G|1|\infty$.[1] The notation $G|G|1|\infty$ means that we consider a single-server queue with "input flow" $\{e_n\}_{n=0}^{\infty}$ and "service flow" $\{s_n\}_{n=0}^{\infty}$ consisting of dependent nonidentically distributed components. Here, $\{e_n\}_{n=0}^{\infty}$ and $\{s_n\}_{n=0}^{\infty}$ are treated as sequences of the lengths of the time intervals between the nth and $(n+1)$th arrivals and the service times of the nth arrival, respectively.

Define the recursive sequence

$$w_0 = 0, \qquad w_{n+1} = \max(w_n + s_n - e_n, 0), \qquad n = 1, 2, \ldots. \tag{13.2.1}$$

The quantity w_n may be viewed as the waiting time for the nth arrival to begin to be served. We introduce the following notation: $\mathbf{e}_{j,k} = (e_j \ldots, e_k)$, $\mathbf{s}_{j,k} = (s_j, \ldots, s_k)$, $k > j$, $\mathbf{e} = (e_0, e_1, \ldots)$, and $\mathbf{s} = (s_0, s_1, \ldots)$. Along with the model defined by relations (13.2.1), we consider a sequence of analogous models by indexing it with the letter r ($r \geq 1$). That is, all quantities pertaining to the rth model will be designated in the same way as model (13.2.1) but with superscript r: $e_n^{(r)}$, $s_n^{(r)}$, $w_n^{(r)}$, and so on. It is convenient to regard the value $r = \infty$ (which can be omitted) as corresponding to the original model. All of the random variables are assumed to be defined on the same probability space. For brevity, functionals Φ depending just on the distributions of the RVs X and Y will be denoted by $\Phi(X, Y)$.

For the system $G|G|1|\infty$ in question, define for $k \geq 1$ nonnegative functions ϕ_k on $(\mathbb{R}^k, \|x\|)$, $\|(x_1, \ldots, x_k)\| = |x_1| + \cdots + |x_k|$, as follows:

$$\begin{aligned}\phi_k(\xi_1, \ldots, \xi_k, \eta_1, \ldots, \eta_k) := \max[0, \eta_k - \xi_k, (\eta_k - \xi_k) \\ + (\eta_{k-1} - \xi_{k-1}), \ldots, (\eta_k - \xi_k) + \cdots + (\eta_1 - \xi_1)].\end{aligned}$$

It is not hard to see that $\phi(\mathbf{e}_{n-k,n-1}, \mathbf{s}_{n-k,n-1})$ is the waiting time for the nth arrival under the condition that $w_{n-k} = 0$.

[1] Kalashnikov and Rachev (1988) provide a detailed discussion of this problem.

Let $c \in \mathfrak{C} = \{c(x,y) = H(d(x,y)), H \in \mathcal{H}\}$ [see (12.2.2)]. *The system $G|G|1|\infty$ is uniformly stable with respect to the functional $\mathcal{A}_c$ finite time intervals if, for every positive T, the following limit relation holds: as $r \to \infty$,*

$$\delta_{(r)}(T;\mathcal{A}_c) := \sup_{n\geq 0}\max_{1\leq k\leq T} \mathcal{A}_c(\phi_k(\mathbf{e}_{n,n+k-1},\mathbf{s}_{n,n+k-1}),$$

$$\phi_k(\mathbf{e}^{(r)}_{n,n+k-1},\mathbf{s}^{(r)}n,n+k-1)) \to 0, \tag{13.2.2}$$

where $\mathcal{A}_c$ is the Kantorovich functional on $\mathfrak{X}(\mathbb{R}^k)$

$$\mathcal{A}_c(X,Y) = \inf\{Ec(\widetilde{X},\widetilde{Y}) : \widetilde{X} \stackrel{\mathrm{d}}{=} X, \widetilde{Y} \stackrel{\mathrm{d}}{=} Y\} \tag{13.2.3}$$

[see (12.2.1)].

Similarly, we define $\delta^{(r)}(T;\ell_p)$, where $\ell_p = \mathcal{L}_p$ $(0 < p < \infty)$ is the minimal metric w.r.t. the $\mathcal{L}_p$-distance.[2] Relation (13.2.2) means that the largest deviation between the variables w_{n+k} and $w^{(r)}_{n+k}$, $k = 1,\ldots,T$, converges to zero as $r \to \infty$ if at time n both compared systems are free of "customers," and for any positive T this convergence is uniform in n.

Theorem 13.2.1. *If for each $r = 1,2,\ldots,\infty$ the sequences $\mathbf{e}^{(r)}$ and $\mathbf{s}^{(r)}$ are independent, then*

$$\delta^{(r)}_c(T;\mathcal{A}_c) \leq K_H \sup_{n\geq 0} \mathcal{A}_c(\mathbf{e}_{n,n+T-1},\mathbf{e}^{(r)}_{n,n+T-1})$$

$$+K_H \sup_{n\geq 0} \mathcal{A}_c(\mathbf{s}_{n,n+T-1},\mathbf{s}^{(r)}_{n,n+T-1}), \tag{13.2.4}$$

where K_H is given by (12.2.2)*. In particular,*

$$\delta^{(r)}_c(T;\ell_p) \leq \sup_{n\geq 0} \ell_p(\mathbf{e}_{n,n+T-1},\mathbf{e}^{(r)}_{n,n+T-1})$$

$$+\sup_{n\geq 0} \ell_p(\mathbf{s}_{n,n+T-1},\mathbf{s}^{(r)}_{n,n+T-1}). \tag{13.2.5}$$

Proof. We will prove (13.2.4) only. The proof of (13.2.3) is carried out in a similar way. For any $1 \leq k \leq T$ we have the triangle inequality

$$\mathcal{L}_p(\phi_k(\mathbf{e}_{n,n+k-1},\mathbf{s}_{n,n+k-1}),\phi_k(\mathbf{e}^{(r)}_{n,n+k-1},\mathbf{s}^{(r)}_{n,n+k-1}))$$

$$\leq \mathcal{L}_p(\phi_k(\mathbf{e}_{n,n+k-1},\mathbf{s}_{n,n+k-1}),\phi_k(\mathbf{e}^{(r)}_{n,n+k-1},\mathbf{s}_{n,n+k-1}))$$

$$+\mathcal{L}_p(\phi_k(\mathbf{e}_{n,n+k-1},\mathbf{s}_{n,n+k-1}),\phi_k(\mathbf{e}_{n,n+k-1},\mathbf{s}^{(r)}_{n,n+k-1}))$$

[2] See (3.3.11), (3.3.12), (3.4.18), (5.4.16), and Theorem 6.2.1.

$$\leq \mathcal{L}_p(\phi_k(\mathbf{e}_{n,n+T-1}, \mathbf{s}_{n,n+T-1}), \phi_k(\mathbf{e}^{(r)}_{n,n+T-1}, \mathbf{s}_{n,n+T-1}))$$
$$+ \mathcal{L}_p(\phi_k(\mathbf{e}_{n,n+T-1}, \mathbf{s}_{n,n+T-1}), \phi_k(\mathbf{e}_{n,n+T-1}, \mathbf{s}^{(r)}_{n,n+T-1})).$$

Changing over to minimal metric ℓ_p and using the assumption that $\mathbf{e}^{(r)}$ and $\mathbf{s}^{(r)}$ are independent ($r = 1, \ldots, \infty$) we have that

$$\begin{aligned} &\inf\{\mathcal{L}_p(\phi_k(\mathbf{e}_{n,n+k-1}, \mathbf{s}_{n,n+k-1}), \phi_k(\mathbf{e}^{(r)}_{n,n+k-1}, \mathbf{s}^{(r)}_{n,n+k-1}))\} \\ &\quad \leq \ell_p(\mathbf{e}_{n,n+T-1}, \mathbf{e}^{(r)}_{n,n+T-1}), \ell_p(\mathbf{s}_{n,n+T-1}, \mathbf{s}^{(r)}_{n,n+T-1}). \end{aligned} \tag{13.2.6}$$

The infimum in the last inequality is taken over all joint distributions

$$F(x, y, \xi, \eta) = \Pr(\mathbf{e}_{n,n+k-1} < x, \mathbf{e}^{(r)}_{n,n+k-1} < y) \Pr(\mathbf{s}_{n,n+k-1} < \xi, \mathbf{s}^{(r)}_{n,n+k-1} < \eta),$$
$$x, y, \xi, \eta \in \mathbb{R}^k,$$

with fixed marginal distributions

$$F_1(x, \xi) = \Pr(\mathbf{e}_{n,n+k-1} < x, \mathbf{s}_{n,n+k-1} < \xi),$$
$$F_2(y, \eta) = \Pr(\mathbf{e}^{(r)}_{n,n+k-1} < x, \mathbf{s}^{(r)}_{n,n+k-1} < \xi),$$

and thus the left-hand side of (13.2.5) is not greater than

$$\ell_p(\phi_k(\mathbf{e}_{n,n+k-1}, \mathbf{s}_{n,n+k-1}), \phi(\mathbf{e}^{(r)}_{n,n+k-1} < x, \mathbf{s}^{(r)}_{n,n+k-1} < \xi)),$$

which proves (13.2.4). □

From (13.2.3) and (13.2.4) it is possible to derive an estimate of the stability of the system $G|G|1|\infty$ in the sense of (13.2.2). It can be expressed in terms of the deviations of the vectors $\mathbf{e}^{(r)}_{n,n+T-1}$ and $\mathbf{s}^{(r)}_{n,n+T-1}$ from $\mathbf{e}_{n,n+T-1}$ and $\mathbf{s}_{n,n+T-1}$, respectively. Such deviations are easy to estimate if we impose additional restrictions on $\mathbf{e}^{(r)}$ and $\mathbf{s}^{(r)}$, $r = 1, 2, \ldots$. For example, when the terms of the sequences are independent, the following estimates hold:

$$\mathcal{A}_c(\mathbf{e}_{n,n+T-1}, \mathbf{e}^{(r)}_{n,n+T-1}) \leq K_H^q \sum_{j=n}^{n+T-1} \mathcal{A}_c(e_j, e_j^{(r)}), \quad q = [\log_2 T] + 1, \tag{13.2.7}$$

$$\ell_p(\mathbf{e}_{n,n+T-1}, \mathbf{e}^{(r)}_{n,n+T-1}) \leq \sum_{j=n}^{n+T-1} \ell_p(e_j, e_j^{(r)}) \quad \text{for } 0 \leq p \leq \infty. \tag{13.2.8}$$

Let us check (13.2.8). One gets (13.2.7) by a similar argument. By the minimality of ℓ_p, for any vectors $X = (X_1, \ldots, X_T)$, $Y = (Y_1, \ldots, Y_T) \in \mathfrak{X}(\mathbb{R}^T)$ with independent components, we have that the Minkowski inequality

$$\mathcal{L}_p(X, Y) = [E\|X - Y\|^p]^{1/p'} \leq \sum_{i=1}^{T} \mathcal{L}_p(X_i, Y_i), \quad p' = \max(1, p), \tag{13.2.9}$$

implies

$$\ell_p(X,Y) \le \sum_{i=1}^{T} \ell_p(X_i,Y_i), \tag{13.2.10}$$

i.e., (13.2.8) holds.

Estimates (13.2.7) and (13.2.8) can be even further simplified when the terms of these sequences are identically distributed. On the basis of (13.2.3) and (13.2.4), it is possible to construct stability estimates for the system that are uniform over the entire time axis.[3]

13.3 Stability of $GI|GI|1|\infty$-System

The system $GI|GI|1|\infty$ is a special case of $G|G|1|\infty$. For this model the RVs $\zeta_n = s_n - e_n$ are i.i.d., and we assume that $E\zeta_1 < 0$. Then the one-dimensional stationary distribution of the waiting time coincides with the distribution of the following maximum:

$$w = \sup_{k\ge 0} Y_k, \qquad Y_k = \sum_{j=-k}^{-1} \zeta_j, \qquad Y_0 = 0, \qquad \zeta_{-j} \stackrel{\mathrm{d}}{=} \zeta_j. \tag{13.3.1}$$

The perturbed model [i.e., $e_k^{(r)}$, $s_k^{(r)}$, $Y_k^{(r)}$] is assumed to be also of the type $GI|GI|1|\infty$.[4] Borovkov (1984, p. 239) noticed that one of the aims of the stability theorems is to estimate the closeness of $Ef^{(r)}(W^{(r)})$ and $Ef(W)$ for various kinds of functions f, $f^{(r)}$. Borovkov (1984, p. 239–240) proposed considering the case

$$f^{(r)}(x) - f(y) \le A|x-y|, \quad \forall x,y \in \mathbb{R}. \tag{13.3.2}$$

Borovkov (1984, p. 270) proved that

$$D = \sup\{|Ef(w^{(r)}) - Ef(w)| : |f(x)-f(y)| \le A|x-y|, x,y\in\mathbb{R}\} < c\varepsilon, \tag{13.3.3}$$

assuming that $|\zeta_1^{(r)} - \zeta_1| \le \varepsilon$ a.s. Here and in what follows, c stands for an absolute constant that may be different in different places.

By (3.3.12), (3.4.18), (5.4.16), and Theorem 6.2.1, we have for the minimal metric $\ell_1 = \widehat{\mathcal{L}}_1$

$$A\ell_1(w^{(r)},w) = \sup\{Ef^{(r)}(w^{(r)}) - Ef(w) : (f^{(r)},f) \text{ satisfy (13.3.2)}\} = D, \tag{13.3.4}$$

[3]See Kalashnikov and Rachev (1988, Chap. 5).

[4]For a discussion of these problems, see Gnedenko (1970), Kennedy (1972), Iglehart (1973), Borovkov (1984, Chap. IV), Whitt (2010), Baccelli and Bremaud (2010), and Kalashnikov (2010).

provided that $E|w^{(r)}| + E|w| < \infty$. Thus, the estimate in (13.3.3) essentially says that

$$\ell_1(w^{(r)}, w) \leq c\ell_\infty(\zeta_1^{(r)}, \zeta_1), \tag{13.3.5}$$

where for any $X, Y \in \mathfrak{X}(\mathbb{R})$

$$\ell_\infty(X, Y) = \widehat{\mathcal{L}}_\infty(X, Y) = \sup_{0\leq t\leq 1} |F_X^{-1}(t) - F_Y^{-1}(t)| \tag{13.3.6}$$

[see (2.5.4), (3.3.14), (3.4.18), (7.5.15), and Corollary 7.4.2]. Actually, using (7.4.18) with $H(t) = t^p$ we have

$$\ell_p(X, Y) = \widehat{\mathcal{L}}_p(X, Y) = \left(\int_0^1 |F_X^{-1}(t) - F_Y^{-1}(t)|^p \mathrm{d}t\right)^{1/p}, \tag{13.3.7}$$

where F_X^{-1} is the generalized inverse of the DF F_X

$$F_X^{-1}(t) := \sup\{x : F_X(x) \leq t\}. \tag{13.3.8}$$

Letting $p \to \infty$ we obtain (13.3.6).

The estimate in (13.3.5) needs strong assumptions on the disturbances in order to conclude stability. We will refine the bound (13.3.5) considering bounds of

$$A\ell_p^p(w^{(r)}, w) = \sup\{Ef^{(r)}(w^{(r)}) - Ef(w) : f^{(r)}(x) - f(y) \leq A|x - y|^p,$$
$$\forall x, y \in \mathbb{R}^1\}, \qquad 0 < p < \infty, \tag{13.3.9}$$

assuming that $E|w^{(r)}|^p + E|w|^p < \infty$. The next lemma considers the closeness of the prestationary distributions of $w_n = \max(0, w_{n-1} + \zeta_{n-1})$, $w_0 = 0$, and of $w_n^{(r)}$ [see (13.2.1)].

Lemma 13.3.1. *For any* $0 < p < \infty$ *and* $E\zeta_1 = E\zeta_1^{(r)}$, *the following inequality holds:*

$$\ell_p(w_n^{(r)}, w_n) \leq A_p, \tag{13.3.10}$$

$$A_p := \min\left(\frac{n(n+1)}{2}\varepsilon_p, c \min_{1/p-1<\delta<2/p-1} n^{1/(1+\delta)}\varepsilon_{p(1+\delta)}^p\right), \quad \textit{for } p \in (0, 1],$$

where

$$A_p := cn^{1/p}\varepsilon_p \quad \textit{for } 1 < p \leq 2,$$
$$A_p := cn^{1/2}\varepsilon_p \quad \textit{for } p > 2,$$

and

$$\varepsilon_p := \ell_p(\zeta_1, \zeta_1^{(r)}).$$

Remark 13.3.1. The condition $E\zeta_1 = E\zeta_1^{(r)}$ means that we know the mean of $\zeta_1^{(r)}$ for the perturbed "real" model and we chose an "ideal" model with ζ_1 having the same mean.

Proof. The distributions of the waiting times w_n and w_n^* can be determined as follows:

$$w_n = \max(0, \zeta_{n-1}, \zeta_{n-1} + \zeta_{n-2}, \ldots, \zeta_{n-1} + \cdots + \zeta_1) \stackrel{\mathrm{d}}{=} \max_{0\le j\le n} Y_j,$$

$$w_n^{(r)} = \max(0, \zeta_{n-1}^{(r)}, \zeta_{n-1}^{(r)} + \zeta_{n-2}^{(r)}, \ldots, \zeta_{n-1}^{(r)} + \cdots + \zeta_1^{(r)}) \stackrel{\mathrm{d}}{=} \max_{0\le j\le n} Y_j^{(r)}.$$

Further [see (19.4.41), Theorem 19.4.6], we will prove the following estimates of the closeness between $w_n^{(r)}$ and w_n:

$$\ell_p(w_n^{(r)}, w_n) \le \frac{n(n+1)}{2}\ell_p(\zeta_1, \zeta_1^{(r)}) \quad \text{if } 0 < p \le 1 \tag{13.3.11}$$

and

$$\ell_p(w_n^{(r)}, w_n) \le \frac{p}{p-1} B_p n^{1/p}\varepsilon_p \quad \text{if } 1 < p \le 2, \tag{13.3.12}$$

where $B_1 = 1$, $B_p = 18p^{3/2}/(p-1)^{1/2}$ for $1 < p \le 2$. From (13.3.12) and $\ell_p \le \ell_{p(1+\delta)}$ for any $0 < p < 1$ and $(1/p) - 1 < \delta \le 2/p - 1$ we have $1 \le p(1+\delta) \le 2$ and

$$\ell_p(w_n^{(r)}, w_n) \le cn^{1/(1+\delta)}\varepsilon_{p(1+\delta)}^p. \tag{13.3.13}$$

For $p \ge 2$ we have

$$\begin{aligned}\mathcal{L}_p^p(w_n, w_n^{(r)}) &= E\left|\bigvee_{k=1}^n Y_k - \bigvee_{k=1}^n Y_k^{(r)}\right|^p \\ &\le \frac{p}{p-1}E|Y_n - Y_n^{(r)}|^p \le cn^{p/2}\mathcal{L}_p(\zeta_1, \zeta_1^{(r)})^p. \end{aligned}\tag{13.3.14}$$

This last inequality is a consequence of the Marcinkiewicz–Zygmund inequality.[5] Passing to the minimal metrics $\ell_p = \widehat{\mathcal{L}}_p$ in (13.3.14) we get (13.3.10). □

Remark 13.3.2. (a) The estimates in (13.3.10) are of the right order, as can be seen by examples. If, for example, $p \ge 2$, consider $\zeta_i \stackrel{\mathrm{d}}{=} N(0,1)$ and $\zeta_i^{(r)} = 0$; then $\ell_p(w_n^{(r)}, w_n) = cn^{1/2}$.

(b) If $p = \infty$, then $\ell_\infty(w_n^{(r)}, w_n) \le n\varepsilon_\infty$.

[5] See Chow and Teicher (1997, p. 384).

Define the stopping times

$$\theta = \inf\left\{k : w_k = \max_{0\le j\le k} Y_j = w = \sup_{j\ge 0} Y_j\right\},$$

$$\theta^{(r)} = \inf\{k : w_k^{(r)} = w^{(r)}\}. \tag{13.3.15}$$

From Lemma 13.3.1 we now obtain estimates for $\ell_p(w^{(r)}, w)$ in terms of the distributions of θ, $\theta^{(r)}$. Define $G(n) := \Pr(\max(\theta^{(r)}, \theta) = n) < \Pr(\theta^{(r)} = n) + \Pr(\theta = n)$.

Theorem 13.3.1. *If* $1 < p \le 2$, λ, $\mu \ge 1$ *with* $(1/\lambda) + (1/\mu) = 1$ *and* $E\zeta_1 = E\zeta_1^{(r)} < 0$, *then*

$$\ell_p^p(w^{(r)}, w) \le c\varepsilon_{p\lambda} \sum_{n=1}^{\infty} n^{1/\lambda} G(n)^{1/\mu}. \tag{13.3.16}$$

Proof.

$$\mathcal{L}_p^p(w^{(r)}, w) = E|w^{(r)} - w|^p = \sum_{n=0}^{\infty} E|w^{(r)} - w|^p I\{\max(\theta^{(r)}, \theta) = n\}$$

$$= \sum_{n=0}^{\infty} E|w_n - w_n^{(r)}|^p I\{\max(\theta^{(r)}, \theta) = n\}$$

$$\le \sum_{n=0}^{\infty} (E|w_n - w_n^{(r)}|^{p\lambda})^{1/\lambda} G(n)^{1/\mu},$$

and thus, by (13.3.10),

$$\ell_p^p(w^{(r)}, w) \le \sum_{n=0}^{\infty} A_{p\lambda}^p G(n)^{1/\mu} = \sum_{n=0}^{\infty} cn^{1/\lambda}\varepsilon_{p\lambda} G(n)^{1/\mu}.$$ □

Remark 13.3.3. (a) If

$$G(n) \le cn^{-\mu(1/\lambda+1+\varepsilon)} \tag{13.3.17}$$

for some $\varepsilon > 0$, then $\sum_{n=1}^{\infty} n^{1/\lambda} G(n)^{1/\mu} \le c\sum_{n=1}^{\infty} n^{-1/(1+\varepsilon)} \le \infty$. For conditions on ζ_1, ζ_1^* ensuring (13.3.17), compare Borovkov (1984, pp. 229, 230, 240).

(b) For $0 < p \le 1$ and $p > 2$, in the same way we get from Lemma 13.3.1 corresponding estimates for $\ell_p(w^{(r)}, w)$.

(c) Note that $\ell_1(w^{(r)}, w) \leq \ell_p(w^{(r)}, w)$, i.e., ℓ_p-metric represents more functions w.r.t. the deviation [see the side conditions in (13.3.9)] than ℓ_1. Moreover, $\varepsilon_{p\lambda} = \ell_{p\lambda}(\zeta_1^{(r)}, \zeta_1) \leq \ell_\infty(\zeta_1^{(r)}, \zeta_1)$. Therefore, Theorem 13.3.1 is a refinement of the estimates given by Borovkov (1984, p. 270).

13.4 Approximation of a Random Queue by Means of Deterministic Queueing Models

The conceptually simplest class of queueing models are those of the deterministic type. Such models are usually explored under the assumption that the underlying (real) queueing system is close (in some sense) to a deterministic system. It is common practice to change the random variables governing the queueing model with constants in the neighborhood of their mean values. In this section we evaluate the possible error involved by approximating a random queueing model with a deterministic one. To get precise estimates, we explore relationships between distances in the space of random sequences, precise moment inequalities, and the Kemperman (1968, 1987) geometric approach to a certain trigonometric moment problem.

More precisely, as in Sect. 13.2, we consider a single-channel queueing system $G|G|1|\infty$ with sequences $\mathbf{e} = (e_0, e_1, \dots)$ and $\mathbf{s} = (s_0, s_1, \dots)$ of interarrival times and service times, respectively, assuming that $\{e_j\}_{j\geq 1}$ and $\{s_j\}_{j\geq 1}$ are *dependent and nonidentically distributed RVs*. We denote by $\boldsymbol{\zeta} = (f_0, \zeta_1, \dots)$ the difference $\mathbf{s} - \mathbf{e}$ and let $\mathbf{w} = (w_0, w_1, \dots)$ be the sequence of waiting times, determined by (13.2.1).

Along with the queueing model $G|G|1|\infty$ defined by the input random characteristics $\mathbf{e}$, $\mathbf{s}$, ζ and the output characteristic $\mathbf{w}$, we consider an approximating model with corresponding inputs $\mathbf{e}^*$, $\mathbf{s}^*$, $\boldsymbol{\zeta}^*$ and output $\mathbf{w}^*$,

$$w_0^* = 0, \qquad w_{n+1}^* = (w_n^* + S_n^* - e_n^*)_+, \qquad n = 1, 2, \dots, \tag{13.4.1}$$

where $(\cdot)_+ = \max(0, \cdot)$. The latter model has a simpler structure, namely, we assume that $\mathbf{e}^*$ or $\mathbf{s}^*$ is deterministic. We also assume that estimates of the deviations between certain moments of e_j and e_j^* (resp. s_j and s_j^* or ζ_j and ζ_j^*) are given.

We will consider two types of approximating models:

(a) $D|G|1|\infty$ (i.e., e_j^* are constants and in general, $e_j^* \neq e_i^*$ for $i \neq j$) and
(b) $D|D|1|\infty$ (i.e., e_j^* and s_j^* are constants).

The next theorem provides a bound for the deviation between the sequences $\mathbf{w} = (w_0, w_1, \dots)$ and $\mathbf{w}^* = (w_1^*, w_2^*, \dots)$ in terms of the Prokhorov metric π.[6] We

[6] See Example 3.3.3 and (3.3.18).

denote by $U = \mathbb{R}^\infty$ the space of all sequences with the metric

$$d(\overline{x}, \overline{y}) = \sum_{i=0}^{\infty} 2^{-i} |x_i - y_i| \quad [\overline{x} := (x_0, x_1, \dots), \ \overline{y} := (y_0, y_1, \dots)],$$

which may take infinite values. Let $\mathfrak{X}^\infty = \mathfrak{X}(\mathbb{R}^\infty)$ be the space of all random sequences defined on a "rich enough" probability space $(\Omega, \mathcal{A}, \Pr)$; see Remark 2.7.2. Then the Prokhorov metric in $\mathfrak{X}^\infty$ is given by

$$\pi(X, Y) := \inf\{\varepsilon > 0 : \Pr(X \in A) \le \Pr(Y \in A^\varepsilon) + \varepsilon, \quad \forall \text{ Borel sets } A \subset \mathbb{R}^\infty\}, \tag{13.4.2}$$

where A^ε is the open ε-neighborhood of A. Recall the Strassen–Dudley theorem (see Corollary 7.5.2 of Chap. 7):

$$\pi(X, Y) = \widehat{\mathbf{K}}(X, Y) := \inf\{\mathbf{K}(\overline{X}, \overline{Y}) : \overline{X}, \overline{Y} \in \mathfrak{X}^\infty, \overline{X} \stackrel{\mathrm{d}}{=} X, \overline{Y} \stackrel{\mathrm{d}}{=} Y\}, \tag{13.4.3}$$

where $\mathbf{K}$ is the Ky Fan metric

$$\mathbf{K}(X, Y) := \inf\{\varepsilon > 0 : \Pr(d(X, Y) > \varepsilon) < \varepsilon\}, \quad X, Y \in \mathfrak{X}^\infty \tag{13.4.4}$$

(Example 3.4.2).

In stability problems for characterizations of ε-independence the following concept is frequently used.[7] Let $\varepsilon > 0$ and $X = (X_0, X_1, \dots) \in \mathfrak{X}^\infty$. The components of X are said to be *ε-independent* if

$$\mathrm{IND}(X) = \pi(X, \underline{X}) \le \varepsilon,$$

where the components $\underline{X}_i$ of $\underline{X}$ are independent and $\underline{X}_i \stackrel{\mathrm{d}}{=} X_i$ $(i \ge 0)$. The Strassen–Dudley theorem gives upper bounds for $\mathrm{IND}(X)$ in terms of the Ky Fan metric $\mathbf{K}(X, \underline{X})$.

Lemma 13.4.1. *Let the approximating model be of the type $D|G|1|\infty$. Assume that the sequences $\mathbf{e}$ and $\mathbf{s}$ of the queueing model $G|G|1|\infty$ are independent. Then*

$$\pi(\mathbf{w}, \mathbf{w}^*) \le \mathrm{IND}(\mathbf{s}) + \mathrm{IND}(\mathbf{s}^*) + \sum_{j=1}^{\infty} (\pi(e_j, e_j^*) + \pi(s_j, s_j^*)). \tag{13.4.5}$$

Proof. By (13.2.1) and (13.4.1),

$$\begin{aligned} d(\mathbf{w}, \mathbf{w}^*) &= \sum_{n=1}^{\infty} 2^{-n} |w_n - w_n^*| \\ &= \sum_{n=1}^{\infty} 2^{-n} | \max(0, s_{n-1} - e_{n-1}, \dots, (s_{n-1} - e_{n-1}) + \dots + (s_0 - e_0)) \end{aligned}$$

[7] See Kalashnikov and Rachev (1988, Chap. 4).

$$
\begin{aligned}
&- \max(0, s^*_{n-1} - e^*_{n-1}, \ldots, (s^*_{n-1} - e^*_{n-1}) + \cdots + (s^*_0 - e^*_0))| \\
&\le \sum_{n=1}^{\infty} 2^{-n} |\max(0, s_{n-1} - e_{n-1}, \ldots, (s_{n-1} - e_{n-1}) + \cdots + (s_0 - e_0)) \\
&- \max(0, s_{n-1} - e^*_{n-1}, \ldots, (s_{n-1} - e^*_{n-1}) + \cdots + (s_0 - e^*_0))| \\
&+ \sum_{n=1}^{\infty} 2^{-n} |\max(0, s_{n-1} - e^*_{n-1}, \ldots, (s_{n-1} - e^*_{n-1}) + \cdots + (s_0 - e^*_0)) \\
&- \max(0, s^*_{n-1} - e^*_{n-1}, \ldots, (s^*_{n-1} - e^*_{n-1}) + \cdots + (s^*_0 - e^*_0))| \\
&\le \sum_{n=1}^{\infty} 2^{-n} \max(|e_{n-1} - e^*_{n-1}|, \ldots, |e_{n-1} - e^*_{n-1}| + \cdots + |e_0 - e^*_0|) \\
&+ \sum_{n=1}^{\infty} 2^{-n} \max(|s_{n-1} - s^*_{n-1}|, \ldots, |s_{n-1} - s^*_{n-1}| + \cdots + |s_0 - s^*_0|) \\
&\le \sum_{n=1}^{\infty} 2^{-n} \sum_{j=0}^{n-1} (|e_j - e^*_j| + |s_j - s^*_j|) \\
&\le d(\mathbf{e}, \mathbf{e}^*) + d(\mathbf{s}, \mathbf{s}^*).
\end{aligned}
$$

Hence, by the definition of the Ky Fan metric (13.4.4), we obtain $\mathbf{K}(\mathbf{w}, \mathbf{w}^*) \le \mathbf{K}(\mathbf{e}, \mathbf{e}^*) + \mathbf{K}(\mathbf{s}, \mathbf{s}^*)$. Next, using representation (13.4.3) let us choose independent pairs $(\mathbf{e}_\varepsilon, \mathbf{e}^*_\varepsilon)$, $(\mathbf{s}_\varepsilon, \mathbf{s}^*_\varepsilon)$ $(\varepsilon > 0)$ such that $\pi(\mathbf{e}, \mathbf{e}^*) > \mathbf{K}(\mathbf{e}_\varepsilon, \mathbf{e}^*_\varepsilon) - \varepsilon$, $\pi(\mathbf{s}, \mathbf{s}^*) > \mathbf{K}(\mathbf{s}_\varepsilon, \mathbf{s}^*_\varepsilon) - \varepsilon$, and $\mathbf{e} \overset{d}{=} \mathbf{e}_\varepsilon$, $\mathbf{e}^* \overset{d}{=} \mathbf{e}^*_\varepsilon$, $\mathbf{s} \overset{d}{=} \mathbf{s}_\varepsilon$, $\mathbf{s}^* \overset{d}{=} \mathbf{s}^*_\varepsilon$. Then by the independence of $\mathbf{e}$ and $\mathbf{s}$ (resp. $\mathbf{e}^*$ and $\mathbf{s}^*$), we have

$$
\begin{aligned}
\pi(\mathbf{w}, \mathbf{w}^*) &= \inf\{\mathbf{K}(\mathbf{w}_0, \mathbf{w}^*_0) : \mathbf{w}_0 \overset{d}{=} \mathbf{w}, \mathbf{w}^*_0 \overset{d}{=} \mathbf{w}^*\} \\
&\le \inf\{\mathbf{K}(\mathbf{e}_0, \mathbf{e}^*_0) + \mathbf{K}(\mathbf{s}_0, \mathbf{s}^*_0) : (\mathbf{e}_0, \mathbf{s}_0) \overset{d}{=} (\mathbf{e}, \mathbf{s}), (\mathbf{e}^*_0, \mathbf{s}^*_0) \overset{d}{=} (\mathbf{e}^*, \mathbf{s}^*), \\
&\qquad \mathbf{e}_0 \text{ is independent of } \mathbf{s}_0, \mathbf{e} \text{ is independent of } \mathbf{s}, \\
&\qquad \mathbf{e}^*_0 \text{ is independent of } \mathbf{s}^*_0, \mathbf{e}^* \text{ is independent of } \mathbf{s}^*\} \\
&\le \mathbf{K}(\mathbf{e}_\varepsilon, \mathbf{e}^*_\varepsilon) + \mathbf{K}(\mathbf{s}_\varepsilon, \mathbf{s}^*_\varepsilon) \le \pi(\mathbf{e}, \mathbf{e}^*) + \pi(\mathbf{s}, \mathbf{s}^*) + 2\varepsilon,
\end{aligned}
$$

which proves that

$$\pi(\mathbf{w}, \mathbf{w}^*) \le \pi(\mathbf{e}, \mathbf{e}^*) + \pi(\mathbf{s}, \mathbf{s}^*). \tag{13.4.6}$$

Next let us estimate $\pi(\mathbf{e}, \mathbf{e}^*)$ in the preceding inequality. Observe that

$$\mathbf{K}(X, Y) \le \sum_{i=0}^{\infty} \mathbf{K}(X_i, Y_i) \tag{13.4.7}$$

for any $X, Y \in \mathfrak{X}^\infty$. In fact, if $\mathbf{K}(X_i, Y_i) \le \varepsilon_i$ and $1 > \varepsilon = \sum_{i=0}^\infty \varepsilon_i$, then

$$\varepsilon > \sum_{i=0}^{\infty} \Pr(|X_i - Y_i| > \varepsilon_i) \ge \sum_{i=0}^{\infty} \Pr(2^{-i}|X_i - Y_i| > \varepsilon_i)$$
$$\ge \Pr\left(\sum_{i=0}^{\infty} 2^{-i}|X_i - Y_i| > \varepsilon\right).$$

Letting $\varepsilon_i \to \mathbf{K}(X_i, Y_i)$ we obtain (13.4.7). By (13.4.7) and $\boldsymbol{\pi}(\mathbf{e}, \mathbf{e}^*) = \mathbf{K}(\mathbf{e}, \mathbf{e}^*)$, we have

$$\boldsymbol{\pi}(\mathbf{e}, \mathbf{e}^*) \le \sum_{i=0}^{\infty} \mathbf{K}(e_i, e_i^*) = \sum_{i=0}^{\infty} \boldsymbol{\pi}(e_i, e_i^*). \tag{13.4.8}$$

Next we will estimate $\boldsymbol{\pi}(\mathbf{s}, \mathbf{s}^*)$ on the right-hand side of (13.4.6). By the triangle inequality for the metric $\boldsymbol{\pi}$, we have

$$\boldsymbol{\pi}(\mathbf{s}, \mathbf{s}^*) \le \mathrm{IND}(\mathbf{s}) + \mathrm{IND}(\mathbf{s}^*) + \boldsymbol{\pi}(\underline{\mathbf{s}}, \underline{\mathbf{s}}^*), \tag{13.4.9}$$

where the sequence $\underline{\mathbf{s}}$ (resp. $\underline{\mathbf{s}}^*$) in the last inequality consists of independent components such that $\underline{s}_j \stackrel{\mathrm{d}}{=} s_j$ (resp. $\underline{s}_j^* \stackrel{\mathrm{d}}{=} s_j^*$). We now need the "regularity" property of the Prokhorov metric,

$$\boldsymbol{\pi}(X + Z, Y + Z) \le \boldsymbol{\pi}(X, Y), \tag{13.4.10}$$

for any Z independent of X and Y in $\mathfrak{X}^\infty$. In fact, (13.4.10) follows from the Strassen–Dudley theorem (13.4.3) and the corresponding inequality for the Ky Fan metric

$$\mathbf{K}(X + Z, Y + Z) \le \mathbf{K}(X, Y) \tag{13.4.11}$$

for all X, Y, and Z in $\mathfrak{X}^\infty$. By the triangle inequality and (13.4.10), we have

$$\boldsymbol{\pi}\left(\sum_{i=0}^{\infty} X_i, \sum_{i=0}^{\infty} Y_i\right) \le \sum_{i=0}^{\infty} \boldsymbol{\pi}(X_i, Y_i) \tag{13.4.12}$$

for all $X, Y \in \mathfrak{X}^\infty$, $X = (X_0, X_1, \dots)$ and $Y = (Y_0, Y_1, \dots)$ with independent components. Thus $\boldsymbol{\pi}(\underline{\mathbf{s}}, \underline{\mathbf{s}}^*) \le \sum_{j=0}^\infty \boldsymbol{\pi}(s_j, s_j^*)$, which together with (13.4.6), (13.4.8), and (13.4.9) complete the proof of (13.4.5). □

In the next theorem we will omit the restriction that $\mathbf{e}$ and $\mathbf{s}$ are independent, but we will assume that the approximation model is of a completely deterministic type $D|D|1|\infty$. (Note that for this approximation model e_j^* and s_j^* can be different constants for different j.)

Lemma 13.4.2. *Under the preceding assumptions, we have the following estimates:*

$$\pi(\mathbf{w},\mathbf{w}^*) = \mathbf{K}(\mathbf{w},\mathbf{w}^*) \leq \pi(\zeta,\zeta^*) \leq \sum_{j=0}^{\infty} \pi(\zeta_j,\zeta_j^*) = \sum_{j=0}^{\infty} \mathbf{K}(\zeta_j,\zeta_j^*), \quad (13.4.13)$$

$$\pi(\mathbf{w},\mathbf{w}^*) \leq \sum_{j=0}^{\infty} (\pi(e_j,e_j^*) + \pi(s_j,s_j^*)) = \sum_{j=0}^{\infty} (\mathbf{K}(e_j,e_j^*) + \mathbf{K}(s_j,s_j^*)). \quad (13.4.14)$$

The proof is similar to that of the previous theorem.

Lemmas 13.4.1 and 13.4.2 transfer our original problem of estimating the deviation between **w** and **w*** to a problem of obtaining sharp or nearly sharp upper bounds for $\mathbf{K}(e_j,e_j^*) = \pi(e_j,e_j^*)$ [resp. $\mathbf{K}(\zeta_j,\zeta_j^*)$], assuming that certain moment characteristics of e_j (resp. ζ_j) are given. The problem of estimating $\pi(s_j,s_j^*)$ in (13.4.5)[8] reduces to estimating the terms IND(**s**), IND(**s***), and $\pi(e_j,e_j^*)$. IND(**s**) and IND(**s***) can be estimated using the Strassen–Dudley theorem and the Chebyshev inequalities. The estimates for $\pi(e_j,e_j^*)$, $\pi(\zeta_j,\zeta_j^*)$ e_j^*, ζ_j^* being constants, are given in the next Lemmas 13.4.3–13.4.8

Lemma 13.4.3. *Let* $\alpha > 0$, $\delta \in [0,1]$, *and* ϕ *be a nondecreasing continuous function on* $[0,\infty)$. *Then the Ky Fan radius (with fixed moment* ϕ*)*

$$R = R(\alpha,\delta,\phi) := \max\{\mathbf{K}(X,\alpha) : E\phi(|X-\alpha|) \leq \delta\} \quad (13.4.15)$$

is equal to $\min(1,\psi(\delta))$, *where* ψ *is the inverse function of* $t\phi(t)$, $t \geq 0$.

Proof. By Chebyshev's inequality, $\mathbf{K}(X,\alpha) \leq \psi(\delta)$ if $E\phi(|X-\alpha|) \leq \delta$, and thus $R \leq \min(1,\psi(\delta))$. Moreover, $\psi(\delta) < 1$ (otherwise, we have trivially that $R = 1$), then by letting $X = X_0 + \alpha$, where X_0 takes the values $-\varepsilon$, 0, $\varepsilon := \psi(\delta)$ with probabilities $\varepsilon/2, 1-\varepsilon, \varepsilon/2$, respectively, we obtain $\mathbf{K}(X,\alpha) = \psi(\delta)$, as is required. □

Using Lemma 13.4.3 we obtain a sharp estimate of $\mathbf{K}(\zeta_j,\zeta_j^*)$ (ζ_j^* constant) if it is known that $E\phi(|\zeta_j - \zeta_j^*|) \leq \delta$. However, the problem becomes more difficult if one assumes that

$$\zeta_j \in S_{\zeta_j^*}(\varepsilon_{1j},\varepsilon_{2j},f_j,g_j), \quad (13.4.16)$$

where for fixed constants $\alpha \in \mathbb{R}$, $\varepsilon_i \geq 0$, and $\varepsilon_2 > 0$

$$S_\alpha(\varepsilon_1,\varepsilon_2,f,g) := \{X \in \widetilde{\mathfrak{X}} : |Ef(X) - f(\alpha)| \leq \varepsilon_1, |Eg(X) - g(\alpha)| \leq \varepsilon_2\}, \quad (13.4.17)$$

and $\widetilde{\mathfrak{X}}$ is the set of real-valued RVs for which $Ef(X)$ and $Eg(X)$ exist.

---

[8]The problem was considered by Kalashnikov and Rachev (1988, Chap. 4) under different assumptions such as s_j^* being exponentially distributed and s_j possessing certain "aging" or "lack of memory" properties.

Suppose that the only information we have on hand concerns estimates of the deviations $|Ef(\zeta_j) - f(\zeta_j^*)|$ and $|Eg(\zeta_j) - g(\zeta_j^*)|$. Here, the main problem is the evaluation of the *Ky Fan radius*

$$D = D_\alpha(\varepsilon_1, \varepsilon_2, f, g) = \sup_{X \in S_\alpha(\varepsilon_1, \varepsilon_2, f, g)} \mathbf{K}(X, \alpha) = \sup_{X \in S_\alpha(\varepsilon_1, \varepsilon_2, f, g)} \pi(X, \alpha). \tag{13.4.18}$$

The next theorem deals with an estimate of $D_\alpha(\varepsilon_1, \varepsilon_2, f, g)$ for the "classic" case

$$f(x) = x, \quad g(x) = x^2. \tag{13.4.19}$$

Lemma 13.4.4. *If $f(x) = x$, $g(x) = x^2$ then*

$$\varepsilon_2^{1/3} \leq D_\alpha(\varepsilon_1, \varepsilon_2, f, g) \leq \min(1, \gamma), \tag{13.4.20}$$

where $\gamma = (\varepsilon_2 + 2|\alpha|\varepsilon_1)^{1/3}$.

Proof. By Chebyshev's inequality for any $X \in S_\alpha(\varepsilon_1, \varepsilon_2, f, g)$, we have $\mathbf{K}^3(X, \alpha) \leq EX^2 - 2\alpha EX + \alpha^2 := I$. We consider two cases:

If $\alpha > 0$ then $I \leq \alpha^2 + \varepsilon_2 - 2\alpha(\alpha - \varepsilon_1) + \alpha^2 = \gamma^3$.

If $\alpha \leq 0$ then $I \leq 2\alpha^2 + \varepsilon_2 - 2\alpha(\alpha + \varepsilon_1) = \gamma^3$.

Hence the upper bound of D (13.4.20) is established.

Consider the RV X, which takes the values $\alpha - \varepsilon$, α, $\alpha + \varepsilon$ with probabilities p, q, p, $(2p + q = 1)$, respectively. Then $EX = \alpha$, so that $|EX - \alpha| = 0 \leq \varepsilon_1$. Further, $EX^2 = \alpha^2 + 2\varepsilon^2 p = \varepsilon_2 + \alpha^2$ if we choose $\varepsilon = \varepsilon_2^{1/3}$, $p = \varepsilon_2^{1/3}/2$. Then $F_X(\alpha + \varepsilon - 0) - F_X(\alpha - \varepsilon) = q = 1 - \varepsilon_2^{1/3}$, and thus $\mathbf{K}(X, \alpha) \geq \varepsilon_2^{1/3}$, which proves the lower bound of D in (13.4.20). □

Using Lemma 13.4.4 we can easily obtain estimates for $D_\alpha(\varepsilon_1, \varepsilon_2, f, g)$, where

$$f(x) := \lambda + \mu x + \zeta x^2 \qquad x, \lambda, \mu, \zeta \in \mathbb{R}$$

and

$$g(x) := a + bx + cx^2 \qquad x, a, b, c \in \mathbb{R}$$

are polynomials of degree two. That is, assuming $c \neq 0$, we may represent f as follows: $f(x) = A + Bx + Cg(x)$, where $A = \lambda - \zeta a/c$, $B = \mu - \zeta b/c$, $C = \zeta/c$.

Lemma 13.4.5. *Let f and g be defined as previously. Assume $c \neq 0$, and $B \neq 0$. Then*

$$D_\alpha(\varepsilon_1, \varepsilon_2, f, g) \leq D_\alpha(\widetilde{\varepsilon_1}, \widetilde{\varepsilon_2}, \widetilde{f}, \widetilde{g}),$$

where

$$\widetilde{\varepsilon_1} := \frac{1}{|B|}(|C|\varepsilon_2 + \varepsilon_1), \qquad \widetilde{\varepsilon_2} := \frac{1}{|c|}\left[\left|\frac{b}{B}\right|(|C|\varepsilon_2 + \varepsilon_1) + \varepsilon_2\right],$$

$$\widetilde{f}(x) = x, \qquad \widetilde{g}(x) = x^2.$$

In particular, $D_\alpha(\varepsilon_1, \varepsilon_2, f, g) \le (\widetilde{\varepsilon}_2 + 2|\alpha|\widetilde{\varepsilon}_1)^{1/3} = (c_1\varepsilon_2 + c_2\varepsilon_1)^{1/3}$, *where*

$$c_1 = \frac{1}{|c||\mu - \zeta b|}(|b\zeta| + |\mu - \zeta b| + 2|\alpha||\zeta c|)$$

and

$$c_2 = \left|\frac{b}{\mu - \zeta b}\right| + 2|\alpha|.$$

Proof. First we consider the special case $f(x) = x$ and $g(x) = a + bx + cx^2$, $x \in \mathbb{R}$, where $a, b, c \neq 0$ are real constants. We prove first that

$$D_\alpha(\varepsilon_1, \varepsilon_2, f, g) \le D_\alpha(\varepsilon_1, \widetilde{\varepsilon}_2, f, \widetilde{g}), \tag{13.4.21}$$

where $\widetilde{\varepsilon}_2 := (1/|c|)(|b|\varepsilon_1 + \varepsilon_2)$ and $\widetilde{g}(x) = x^2$. Thus, by (13.4.20), we get

$$D_\alpha(\varepsilon_1, \varepsilon_2, f, g) \le (\widetilde{\varepsilon}_2 + 2|\alpha|\varepsilon_1)^{1/3}. \tag{13.4.22}$$

Since $|Ef(X) - f(\alpha)| = |EX - \alpha| \le \varepsilon_1$ and $|Eg(X) - g(\alpha)| = |b(EX - \alpha) + c(EX^2 - \alpha^2)| \le \varepsilon_2$, we have that $|c||EX^2 - \alpha^2| \le |b||EX - \alpha| + \varepsilon_2 \le |b|\varepsilon_1 + \varepsilon_2$. That is, $|EX^2 - \alpha^2| \le \widetilde{\varepsilon}_2$, which establishes the required estimate (13.4.21).

Now we consider the general case of $f(x) = \lambda + \mu x + \zeta x^2$. From $f(x) = A + Bx + Cg(x)$ and the assumptions that $|Ef(X) - f(\alpha)| \le \varepsilon_1$ and $|Eg(X) - g(\alpha)| \le \varepsilon_2$, we have $|B||EX - \alpha| \le |Ef(X) - f(\alpha)| + |C||Eg(X) - g(\alpha)| \le \varepsilon_1 + |C|\varepsilon_2$, that is, $|EX - \alpha| \le \widetilde{\varepsilon}_1$. Therefore, $D_\alpha(\varepsilon_1, \varepsilon_2, f, g) \le D_\alpha(\widetilde{\varepsilon}_1, \varepsilon_2, \underset{\sim}{f}, g)$, where $\underset{\sim}{f}(x) = x$. Using (13.4.22) we have that $D_\alpha(\widetilde{\varepsilon}_1, \varepsilon_2, \underset{\sim}{f}, g) \le D_\alpha(\widetilde{\varepsilon}_1, \widetilde{\varepsilon}_2, \underset{\sim}{f}, \widetilde{g})$, where

$$\widetilde{\varepsilon}_2 = \frac{1}{|c|}(|b|\widetilde{\varepsilon}_1 + \varepsilon_2),$$

which by means of Lemma 13.4.4 completes the proof of Lemma 13.4.5. □

The main assumption in Lemmas 13.4.3–13.4.5 was the monotonicity of ϕ, f, and g, which allows us to use the Chebyshev inequality. More difficult is the problem of finding $D_\alpha(\varepsilon_1, \varepsilon_2, f, g)$ when f and g are not polynomials of degree two. The case of

$$f(x) = \cos x \quad \text{and} \quad g(x) = \sin x,$$

where $x \in [0, 2\pi]$, is particularly difficult.

Remark 13.4.1. In fact, we will investigate the rate of the convergence of $\mathbf{K}(X_n, \alpha) \to 0$ $(0 \le X_n \le 2\pi)$ as $n \to \infty$, provided that $E\cos X_n \to \cos\alpha$ and $E\sin X_n \to \sin\alpha$. In the next lemma, we show Berry–Essen-type bounds for the implication

$$E\exp(iX_n) \to \exp(i\alpha) \Rightarrow \mathbf{K}(X_n, \alpha) = \boldsymbol{\pi}(X_n, \alpha) \to 0.$$

In what follows, we consider probability measures μ on $[0, 2\pi]$ and let

$$M(\varepsilon) = \left\{\mu : \left|\int \cos t \, d\mu - \cos\alpha\right| \leq \varepsilon, \left|\int \sin t \, d\mu - \sin\alpha\right| \leq \varepsilon\right\}. \tag{13.4.23}$$

We would like to evaluate the *trigonometric Ky Fan (or Prokhorov) radius for* $M(\varepsilon)$ defined by

$$D = \sup\{\boldsymbol{\pi}(\mu, \delta_\alpha) : \mu \in M(\varepsilon)\}, \tag{13.4.24}$$

where δ_α is the point mass at α and $\boldsymbol{\pi}(\mu, \delta_\alpha)$ is the Ky Fan (or Prokhorov) metric

$$\boldsymbol{\pi}(\mu, \delta_\alpha) = \inf\{r > 0 : \mu([\alpha - r, \alpha + r]) \geq 1 - r\}. \tag{13.4.25}$$

Our main result is as follows.

Lemma 13.4.6. *Let fixed* $\alpha \in [1, 2\pi - 1]$ *and* $\varepsilon \in (0, (1/\sqrt{2})(1 - \cos 1))$. *We get* D *as the unique solution of*

$$D - D\cos D = \varepsilon(|\cos\alpha| + |\sin\alpha|). \tag{13.4.26}$$

Here we have that $D \in (0, 1)$.

Remark 13.4.2. By (13.4.24), one obtains

$$D \leq [2\varepsilon(|\cos\alpha| + |\sin\alpha|)]^{1/3}. \tag{13.4.27}$$

From (13.4.26), (13.4.27) [and see also (13.4.28)] we have that $D \to 0$ as $\varepsilon \to 0$. The latter implies that $\boldsymbol{\pi}(\mu, \delta_\alpha) \to 0$, which in turn gives that $\mu \xrightarrow{w} \delta_\alpha$, where δ_α is the point mass at α. In fact, D converges to zero quantitatively through (13.4.24) and (13.4.27), that is, the knowledge of D gives the rate of weak convergence of μ to δ_α (see also Lemma 13.4.7).

The proofs of Lemmas 13.4.6 and 13.4.7, while based on the solution of certain moment problems (see Chap. 9), need more facts on the Kemperman geometric approach for the solution of the general moment problem[9] and therefore will be omitted. For the necessary proofs see Anastassiou and Rachev (1992).

Lemma 13.4.7. *Let* $f(x) = \cos x$, $g(x) = \sin x$; $\alpha \in [0, 1)$ *or* $\alpha \in (2\pi - 1, 2\pi)$. *Define*

$$\begin{aligned} D &= D_\alpha(\varepsilon, f, g) \\ &= \sup\{\mathbf{K}(X, \alpha) : |E\cos X - \cos\alpha| \leq \varepsilon, |E\sin X - \sin\alpha| \leq \varepsilon\}. \end{aligned}$$

Let $\beta = \alpha + 1$ *if* $\alpha \in [0, 1)$, *and let* $\beta = \alpha - 1$ *if* $\alpha \in (2\pi - 1, 2\pi)$. *Then*

[9] See Kemperman (1968, 1987).

$$D_\alpha(\varepsilon, f, g) \leq D_\beta(\varepsilon(\cos 1 + \sin 1), f, g).$$

In particular, by (13.4.27),

$$D_\alpha(\varepsilon, f, g) \leq [2\varepsilon(\cos 1 + \sin 1)(|\cos\alpha| + |\sin\alpha|)^{1/3} \qquad (13.4.28)$$

for any $0 \leq \alpha < 2\pi$ *and* $\varepsilon \in (0, (1/\sqrt{2})(1 - \cos 1))$.

Further, we are going to use (13.4.28) to obtain estimates for $D_\alpha(\varepsilon, f, g)$, where $f(x) = \lambda + \mu\cos x + \zeta\sin x$, $x \in [0, 2\pi]$, $\lambda, \mu, \zeta \in \mathbb{R}$, and $g(x) = a + b\cos x + c\sin x$, $x \in [0, 2\pi]$, $a, b, c \in \mathbb{R}$. Assuming $c \neq 0$ we have $f(x) = A + B\cos x + Cg(x)$, where $A = \lambda - \zeta a/c$, $B = \mu - \zeta b/c$, $C = \zeta/c$.

Lemma 13.4.8. *Let the trigonometric polynomials* f *and* g *be defined as previously. Assume* $c \neq 0$ *and* $B \neq 0$*. Then* $D_\alpha(\varepsilon, f, g) \leq D_\alpha(\varepsilon\tau\eta, \widetilde{f}, \widetilde{g})$ *for any* $0 \leq \alpha < 2\pi$*, where*

$$\tau = \max\left(1, \frac{1}{|c|}(|b| + 1|\right)$$

and

$$\eta = \max\left(1, \frac{1}{|B|}(|C| + 1)\right)$$

$\widetilde{f}(x) = \cos x$, $\widetilde{g}(x) = \sin x$*. If*

$$0 \leq \varepsilon \leq \frac{1}{\tau\eta\sqrt{2}}(1 - \cos 1),$$

then we obtain

$$D_\alpha(\varepsilon, f, g) \leq [2\varepsilon\tau\eta(\cos 1 + \sin 1)(|\cos\alpha| + |\sin\alpha|)]^{1/3} \qquad (13.4.29)$$

for any $0 \leq \alpha < 2\pi$.

The proof is similar to that of Lemma 13.4.5.

Now we can state the main result determining the deviation between the waiting times of a deterministic and a random queueing model.

Theorem 13.4.1. *(i) Let the approximating queueing model be of type* $D|G|1|\infty$*. Assume that the sequences* **e** *and* **s** *of the "real" queue of type* $G|G|1|\infty$ *are independent. Then the Prokhorov metric between the sequences of waiting times of* $D|G|1|\infty$ *queue and* $G|G|1|\infty$ *queue is estimated as follows:*

$$\pi(\mathbf{w}, \mathbf{w}^*) \leq \text{IND}(\mathbf{s}) + \text{IND}(\mathbf{s}*) + \sum_{j=1}^{\infty}(\pi(e_j, e_j^*) + \pi(s_j, s_j^*)). \qquad (13.4.30)$$

(ii) *Assume that the approximating model is of type $D|D|1|\infty$ and the "real" queue is of type $G|G|1|\infty$. Then*

$$\pi(\mathbf{w}, \mathbf{w}^*) \leq 2\sum_{j=1}^{\infty} \pi(\zeta_j, \zeta_j^*) \tag{13.4.31}$$

and

$$\pi(\mathbf{w}, \mathbf{w}^*) \leq 2\sum_{j=1}^{\infty} (\pi(e_j, e_j^*) + \pi(s_j, s_j^*)). \tag{13.4.32}$$

(iii) *The right-hand sides of* (13.4.30)–(13.4.32) *can be estimated as follows: let $\pi(X, X^*)$ denote $\pi(e_j, e_j^*)$ in* (13.4.30) *or $\pi(\zeta_j, \zeta_j^*)$ in* (13.4.31) *or $\pi(e_j, e_j^*)$ ($\pi(s_j, s_j^*)$) in* (13.4.32) *(note that X^* is a constant). Then:*

(a) *If the function ϕ is nondecreasing on $[0, \infty)$ and continuous on $[0, 1]$ and satisfies*

$$E\phi(|X - X^*|) \leq \delta \leq 1, \tag{13.4.33}$$

then

$$\pi(X, X^*) \leq \min(1, \psi(\delta)), \tag{13.4.34}$$

where ψ is the inverse function of $t\phi(t)$.

(b) *If $|Ef(X) - f(X^*)| \leq \varepsilon_1$, $|Eg(X) - g(X^*)| \leq \varepsilon_2$, where*

$$f(x) = \lambda + \mu x + \zeta x^2, \quad x, \lambda, \mu, \zeta \in \mathbb{R},$$
$$g(x) = \alpha + bx + cx^2, \quad x, a, b, c \in \mathbb{R},$$

$c \neq 0$, $\mu \neq \zeta b/c$, then for any $\varepsilon_1 > 0$ and $\varepsilon_2 > 0$

$$\pi(X, X^*) \leq (\widetilde{\varepsilon}_2 + 2|X^*|\widetilde{\varepsilon}_1)^{1/3},$$

where $\widetilde{\varepsilon}_1$ and $\widetilde{\varepsilon}_2$ are linear combinations of ε_1 and ε_2 defined as in Lemma 13.4.5.

(c) *If $X \in [0, 2\pi]$ a.e. and $Ef(X) - f(X^*)| \leq \varepsilon$, $|Eg(X) - g(X^*)| \leq \varepsilon$, where $f(x) = \lambda + \mu \cos x + \zeta \sin x$, and $g(x) = a + b\cos x + c \sin x$ for $x \in [0, 2\pi]$, a, b, c, λ, μ, $\zeta \in \mathbb{R}$, $c \neq 0$, $\mu \neq \zeta b/c$, then*

$$\mathbf{K}(X, X^*) \leq [2\varepsilon\tau\eta(\cos 1 + \sin 1)(|\cos X^*| + |\sin X^*|)]^{1/3},$$

where the constants τ and η are defined as in Lemma 13.4.8.

Open Problem 13.4.1. First, one can easily combine the results of this section with those of Kalashnikov and Rachev (1988, Chap. 5), to obtain estimates between the outputs of general multichannel and multistage models and approximating queueing models of types $G|D|1|\infty$ and $D|G|1|\infty$. However, it is much more interesting and difficult to obtain *sharp* estimates for $\pi(\mathbf{e}, \mathbf{e}^*)$, assuming that $\mathbf{e}$ and $\mathbf{e}^*$ are random sequences satisfying

$$|E(e_j - e_j^*)| \leq \varepsilon_{1j}, \qquad |Ef_j(|e_j|) - Ef_j(|e_j^*|) \leq \varepsilon_{2j}.$$

Here, even the case $f_j(x) = x^2$ is open (Chap. 9).

Open Problem 13.4.2. It is interesting to obtain estimates for $\ell_p(\mathbf{w}, \mathbf{w}^*)$, $(0 < p \leq \infty)$, where $\ell_p = \widehat{\mathcal{L}}_p$ (Sects. 13.2 and 13.3).

References

Anastassiou GA, Rachev ST (1992) Moment problems and their applications to the stability of queueing models. Comput Math Appl 24(8/9):229–246

Baccelli F, Bremaud P (2010) Elements of queueing theory: palm Martingale calculus and stochastic recurrencies, 2nd edn. Springer, New York

Borovkov AA (1984) Asymptotic methods in queueing theory. Wiley, New York

Chow YS, Teicher H (1997) Probability theory: independence, interchangeability, Martingales, 3rd edn. Wiley, New York

Gnedenko BV (1970) On some unsolved problems in queueing theory. In: Sixth international telegraphic conference, Munich (in Russian)

Iglehart DL (1973) Weak convergence in queueing theory. Adv Appl Prob 5:570–594

Kalashnikov VV (2010) Mathematical methods in queuing theory. Kluwer, Dordrecht, the Netherlands

Kalashnikov VV, Rachev ST (1988) Mathematical methods for construction of stochastic queueing models. Nauka, Moscow (in Russian). (English transl., (1990) Wadsworth, Brooks–Cole, Pacific Grove, California)

Kemperman JHB (1968) The general moment problem, a geometric approach. Ann Math Statist 39:93–122

Kemperman JHB (1987) Geometry of the moment problem, a geometric approach. Proc SIAM 5:16–53

Kennedy D (1972) The continuity of the single server queue. J Appl Prob 9:370–381

Whitt W (2010) Stochastic-process limits: an introduction to stochastic process limits and their application to queues. Springer, New York

Chapter 14
Optimal Quality Usage

The goals of this chapter are to:

- Discuss the problem of optimal quality usage in terms of a multidimensional Monge–Kantorovich problem,
- Provide conditions for optimality and weak optimality in the multidimensional case,
- Derive an upper bound for the minimal total losses when they can be represented in terms of the ℓ_1 metric.

Notation introduced in this chapter:

Notation	Description
$\Theta(\mu, \nu)$	Collection of admissible plans
$\tau_\phi(\theta)$	Total loss of consumption quality
$\overline{\Theta}(\mu, \nu)$	Collection of weakly admissible plans
$\partial f(x)$	Subdifferential of f at x
$\nu(P_1, P_2)$	First absolute pseudomoment
$\boldsymbol{\kappa}_s(P_1, P_2)$	sth difference pseudomoment

14.1 Introduction

In this chapter, we discuss the problem of optimal quality usage as a multidimensional Monge–Kantorovich problem. We begin by stating and interpreting the one-dimensional and the multidimensional problems. We provide conditions for optimality and weak optimality in the multivariate case for particular choices of the cost function. Finally, we derive an upper bound for the minimal total losses for a special choice of the cost function and compare it to the upper bound involving the first difference pseudomoment.

S.T. Rachev et al., *The Methods of Distances in the Theory of Probability and Statistics*, DOI 10.1007/978-1-4614-4869-3_14,

14.2 Optimality of Quality Usage and the Monge–Kantorovich Problem

The quality of a product is usually described by a collection of its characteristics $x = (x_1, \dots, x_m)$, where m is a required number of quality characteristics and x_i is the real value of the ith characteristic. The quality of all produced items of a given type is described by a probability measure $\mu(A)$, $A \in \mathfrak{B}^m$, where, as before, $\mathfrak{B}^m$ is the Borel σ-algebra sets in $\mathbb{R}^m$. The measure $\mu(A)$ represents the proportion of items with quality x satisfying $x \in A$. On the other hand, the usage (consumption) of all produced items can be represented by another probability measure $\nu(B)$, $B \in \mathfrak{B}^m$, where $\nu(B)$ describes the necessary consumption product for which the quality characteristics satisfy $x \in B$. We call $\mu(A)$ the *production quality measure* and $\nu(B)$ the *consumption quality measure* $A, B \in \mathfrak{B}^m$, and assume that $\mu(\mathbb{R}^m) = \nu(\mathbb{R}^m) = 1$. Clearly, it happens often that $\mu(A) \neq \nu(A)$ at least for some $A \in \mathfrak{B}^m$.

Following the formulation of the Monge–Kantorovich problem discussed in Sect. 5.2 in Chap. 5, we introduce the loss function $\phi(x, y)$ defined for all $x \in \mathbb{R}^m$ and $y \in \mathbb{R}^m$ and taking positive values whenever an item with quality x is used in place of an item with required quality y. Finally, we propose the notion of a distribution plan for production quality [with given measure $\mu(A)$] to satisfy the demand for consumption [with given measure $\nu(B)$]. We define for any distribution plan (or *plan* for short) a nonnegative Borel measure $\theta(A, B)$ on the direct product $\mathbb{R}^m \times \mathbb{R}^m = \mathbb{R}^{2m}$. The measure $\theta(A, B)$ indicates that part of produced items with quality $x \in A$ that is intended to satisfy a required level of consumption of items with quality $y \in B$. The plan $\theta(A, B)$ is called admissible if it satisfies the balance equation

$$\theta(A, \mathbb{R}^m) = \mu(A), \quad \theta(\mathbb{R}^m, B) = \nu(B), \quad \forall A, B \in \mathfrak{B}^m. \tag{14.2.1}$$

In reality the balance equations express the fact that any produced item will be consumed and any demand for an item will be satisfied.

Denote by $\Theta(\mu, \nu)$ the collection of all admissible plans. For a given plan $\theta \in \Theta(\mu, \nu)$ the total loss of consumption quality is defined by the following integral:

$$\tau(\theta) := \tau_\phi(\theta) := \int_{\mathbb{R}^{2m}} \phi(x, y)\theta(\mathrm{d}x, \mathrm{d}y). \tag{14.2.2}$$

θ^* is said to be the *optimal plan for consumption quality* if it satisfies the relationship

$$\tau_\phi(\theta^*) = \widehat{\tau}_\phi(\mu, \nu) := \inf_{\theta \in \Theta(\mu,\nu)} \tau(\theta). \tag{14.2.3}$$

Relations (14.2.1) express the balances between the production quality measure $\mu(A)$, the consumption quality measure $\nu(B)$, and the distribution plan $\theta(A, B)$. It assumes that complete information on the marginals μ and ν is available when the plan is constructed. In most practical cases, the information about production and consumption quality concerns only the set of distributions of $x_i s$ $(i = 1, \dots, m)$.

In this case, it is assumed that the balance equations can be expressed in terms of the corresponding one-dimensional marginal measures. This leads to the formulation of the multidimensional Kantorovich problem.[1] If we denote the ith marginal measure of production quality by $\mu_i(A_i)$ and the jth marginal measure of the consumption quality by $\nu_j(B_j)$, then the following relations hold:

$$\mu_i(A_i) = \mu(\mathbb{R}^{i-1} \times A_i \times R^{m-i}), \qquad A_i \in \mathfrak{B}^1,$$

$$\nu_j(B_j) = \nu(\mathbb{R}^{j-1} \times B_j \times R^{m-j}), \qquad B_j \in \mathfrak{B}^1.$$

We say a distribution plan $\theta(A, B)$ is *weakly admissible* when it satisfies the conditions

$$\theta(\mathbb{R}^{i-1} \times A_i \times R^{m-i}, \mathbb{R}^m) = \mu_i(A_i), \qquad i = 1, \ldots, m, \tag{14.2.4}$$

$$\theta(\mathbb{R}^m, \mathbb{R}^{j-1} \times B_j \times \mathbb{R}^{m-j}) = \nu_j(B_j), \qquad j = 1, \ldots, m. \tag{14.2.5}$$

Denote by $\overline{\Theta}(\mu_1, \ldots, \mu_m; \nu_1, \ldots, \nu_m)$ the collection of all weakly admissible plans. Obviously,

$$\Theta(\mu, \nu) \subseteq \overline{\Theta}(\mu_1, \ldots, \mu_m; \nu_1, \ldots, \nu_m). \tag{14.2.6}$$

A distribution plan Θ^o is called *weakly optimal* if it satisfies the relation

$$\tau(\theta^o) = \inf_{\theta \in \overline{\Theta}} \tau(\theta), \tag{14.2.7}$$

where $\tau(\theta)$ is defined by (14.2.2) for a given loss function $\phi(x, y)$. The inclusion in (14.2.6) means that $\tau(\theta^o) \leq \tau(\theta^*)$, where θ^o and θ^* are determined by (14.2.3) and (14.2.7). Therefore, $\tau(\theta^o)$ is an essential lower bound on the minimal total losses.

First we will evaluate $\tau(\theta^*)$ and determine θ^*. We consider two types of loss functions, $\phi(x, y)$, when the item with quality x is used instead of an item with required quality y.

The first type has the following form:

$$\phi(x, y) = a(x) + b(x, y), \tag{14.2.8}$$

where $a(x)$ is the production cost of an item with quality x and $b(x, y) = b_o(x, y) + b_o(y, x)$, in which $b_o(x, y)$ is the consumer's expenses resulting from replacing the required item with quality y by a product with quality x. We can assume that $b(x, y) = 0$ for all $x = y$ and $b(x, y) \geq 0, a(x) \geq 0, \forall x \in \mathbb{R}^m$.

---

[1] See version (VI) of the Monge–Kantorovich problem in Sect. 5.2 and, in particular, (5.2.36).

From (14.2.3) and (14.2.8)

$$\tau_\phi(\theta^*) := \inf_{\theta\in\Theta(\mu,\nu)} \left\{ \int_{\mathbb{R}^{2m}} [a(x) + b(x,y)]\theta(\mathrm{d}x,\mathrm{d}y) \right\}$$
$$= \int_{\mathbb{R}^m} a(x)\mu(\mathrm{d}x) + \inf_{\theta\in\Theta(\mu,\nu)} \int_{\mathbb{R}^{2m}} b(x,y)\theta(\mathrm{d}x,\mathrm{d}y) =: I_1 + I_2. \tag{14.2.9}$$

Here

$$I_1 := \int_{\mathbb{R}^m} a(x)\mu(\mathrm{d}x) \tag{14.2.10}$$

represents the expected (complete) production price of items with quality measure μ, whereas

$$I_2 := \inf_{\theta\in\Theta(\mu,\nu)} \int_{\mathbb{R}^{2m}} b(x,y)\theta(\mathrm{d}x,\mathrm{d}y) \tag{14.2.11}$$

represents the minimal (expected) means of a consumer's expenses from exchanging the required product for consumption with quality ν by a produced item with quality μ, under its optimal distribution among consumers, according to plan θ^*. Since I_1 in (14.2.10) is completely determined by the measure μ, the only problem is the evaluation of I_2.

The second type of loss function that is of interest has the form

$$\phi(x,y) = H(d(x,y)), \tag{14.2.12}$$

where $H(t)$ is a nondecreasing function and d is a metric in $\mathbb{R}^m$, characterizing the deviation between the production quality x and the required consumption quality y. The function $H(t)$ is defined for all $t \geq 0$ and represents the user's expenses as a function of the deviation $d(x,y)$. Notice that the function $b(x,y)$ in (14.2.11) may also be written in the form (14.2.12), so without loss of generality we may assume that ϕ has the form (14.2.12).

The dual representation for $\widehat{\tau}_\phi$ (14.2.3) is given by Corollary 5.3.1, i.e., if the loss function $\phi(x,y)$ is given by (14.2.12) where H is convex and $K_H := \sup_{t<\infty}[H(2t)/H(t)] < \infty$ [see (2.4.3)], then $\widehat{\tau}_\phi$ is a minimal distance with dual representation

$$\widehat{\tau}_\phi(\mu,\nu) = \sup\left\{ \int_{\mathbb{R}^m} f\,\mathrm{d}\mu + \int_{\mathbb{R}^m} g\mathrm{d}\nu : f,g \in \operatorname{Lip}\mathbb{R}^m, \right.$$
$$\left. f(x) + g(y) \leq H(d(x,y)); x,y \in \mathbb{R}^m \right\}, \tag{14.2.13}$$

where

$$\mathrm{Lip}\,\mathbb{R}^m := \left\{ f : \mathbb{R}^m \to \mathbb{R}^1 : \|f\|_\infty = \sup_{x\in\mathbb{R}^m} |f(x)| < \infty \right.$$

$$\left. \sup_{x,y\in\mathbb{R}^m} |f(x) - f(y)|/d(x,y) < \infty \right\}.$$

By the Cambanis–Simons–Stout formula [see (8.2.26) in Chap. 8 in the case of $m = 1$ and $d(x,y) = |x - y|$], the minimal total losses can be expressed by

$$\widehat{\tau}_\phi(\mu,\nu) = \tau_\phi(\theta^*) = \int_0^1 H(|F^{-1}(x) - G^{-1}(x)|)\mathrm{d}x, \tag{14.2.14}$$

where $F(x) = \mu((-\infty,x])$ and $G(x) = \nu((-\infty,x])$ are the distribution functions of the production quality and the required quality characteristics for usage, respectively. The functions $F^{-1}(x)$ and $G^{-1}(x)$ are their generalized inverses defined by $F^{-1}(x) := \sup\{t : F(t) \le x\}$. Furthermore, the optimal distribution plan is given by

$$\theta^*((-\infty,x] \times (-\infty,y]) = \min(F(x), G(y)). \tag{14.2.15}$$

Equality (14.2.15) essentially means that if $F(x)$ is a continuous DF, then the optimal correspondence between the item of quality x and the item with required quality y is given by

$$y = G^{-1}(F(x)). \tag{14.2.16}$$

The last formula follows immediately from (14.2.14), (14.2.15) since the minimal distance

$$\tau_\phi(\theta^*) = \inf\{EH(|X - Y|) : F_X = F, F_Y = G\} \tag{14.2.17}$$

is equal to $EH(|X^* - Y^*|)$, where $Y^* = G^{-1}(F(X^*))$ and the joint distribution of X^*, Y^* is given by θ^*. Thus, the case of $m = 1$ is solved for any ϕ given by (14.2.2). However, (14.2.16) holds in a more general situation when ϕ is a quasiantitone function (see Definition 7.4.1, Theorem 7.4.2, and Remark 7.4.1 in Chap. 7).

The next theorem deals with the special case where $\phi(x,y)$ is $\|x - y\|^2$ and $\|\cdot\|$ is the Euclidean distance in $\mathbb{R}^m$. Let μ and ν be two probability measures on $\mathfrak{B}^m$ such that

$$\int_{\mathbb{R}^m} \|x\|^2(\mu + \nu)(\mathrm{d}x) < \infty.$$

Recall that the pair of m-dimensional vectors (X^*, Y^*) with joint distribution θ^* and marginal distributions μ and ν is *optimal* if

$$\tau_\phi(\theta^*) = E\|X^* - Y^*\|^2 = \inf\{E\|X - Y\|^2 : \mathrm{Pr}_X = \mu, \mathrm{Pr}_Y = \nu\}. \tag{14.2.18}$$

In the next theorem, we describe the necessary and sufficient condition for a pair (X^*, Y^*) to be optimal. To this end, we recall the definition of a subdifferential.[2] For a lower semicontinuous convex (LSC) function f on $\mathbb{R}^m$, let f^* denote the conjugate function

$$f^*(y) := \sup_{x \in \mathbb{R}^m} \{\langle x, y\rangle - f(x)\}, \tag{14.2.19}$$

where $\langle x, y\rangle := \sum_{i=1}^m x_i y_i$, $x = (x_1, \ldots, x_m)$, $y = (y_1, \ldots, y_m)$, and denote the *subdifferential* of f in x by

$$\partial f(x) = \{y \in \mathbb{R}^m : f(z) - f(x) \geq \langle z - x, y\rangle, z \in \mathbb{R}^m\}. \tag{14.2.20}$$

The elements of $\partial f(x)$ are called subgradients of f at x. Then it holds that for all x, y

$$f(x) + f^*(y) \geq \langle x, y\rangle \tag{14.2.21}$$

with equality if and only if $y \in \partial f(x)$.

Theorem 14.2.1. *(X^*, Y^*) is optimal if for some LSC function f*

$$Y^* \in \partial f(X^*) \quad (\text{Pr } a.s.). \tag{14.2.22}$$

Remark 14.2.1. Note that we can consider only the case where the means $m_\mu := \{\int_{\mathbb{R}^m} x_i \mu(dx), i = 1, \ldots, m\}$ and m_ν are zero vectors with no loss of generality. Simply note that if X and Y are $\mathbb{R}^m$-valued RVs with distributions $\Pr_X = \mu$, $\Pr_Y = \nu$, $\xi = X - m_\mu$, and $\eta = Y - m_\nu$, then

$$E\|X - Y\|^2 = E\|\xi - \eta\|^2 + \|m_\mu - m_\nu\|^2. \tag{14.2.23}$$

Proof. We begin with

$$E\|X - Y\|^2 = E\|X\|^2 + E\|Y\|^2 - 2E\langle X, Y\rangle. \tag{14.2.24}$$

Therefore, problem (14.2.18) is equivalent to finding (X^*, Y^*) such that

$$E\langle X^*, Y^*\rangle = \sup\{E\langle X, Y\rangle : \Pr_X = \mu, \Pr_Y = \nu\}. \tag{14.2.25}$$

By the duality theorem,[3] it follows that

$$\begin{aligned}
&\sup \; \{E\langle X, Y\rangle : \Pr_X = \mu, \Pr_Y = \nu\} \\
&\quad = \int \|x\|^2(\mu + \nu)(dx) - \inf\{E\|X - Y\| : \Pr_X = \mu, \Pr_Y = \nu\}
\end{aligned}$$

---

[2] See, for example, Rockafellar (1970) and Borwein and Lewis (2010).

[3] See (14.2.13), (14.2.24), and Theorem 8.2.1 in Chap. 8.

$$= \int \|x\|^2(\mu+\nu)(\mathrm{d}x) - \sup\left\{\int g\mathrm{d}\mu + \int h\mathrm{d}\nu : g,h \in \mathrm{Lip}\,\mathbb{R}^m,\right.$$

$$\left.\text{and } \forall x,y \in \mathbb{R}^m, g(x)+h(y) \le \|x-y\|^2\right\}$$

$$\ge \inf\left\{\int \widetilde{g}\mathrm{d}\mu + \int \widetilde{h}\mathrm{d}\nu : \int |\widetilde{g}|\mathrm{d}\mu < \infty, \int |\widetilde{h}|\mathrm{d}\nu < \infty,\right.$$

$$\left.\widetilde{g}(x) + \widetilde{h}(y) \ge \langle x,y\rangle\right\}$$

$$\ge \sup\{E\langle X,Y\rangle : \Pr_X = \mu, \Pr_Y = \nu\}. \tag{14.2.26}$$

Here, the last inequality follows from the "trivial" part of the duality theorem, and therefore the last two inequalities are valid with equality signs.

Now, let $\Pr_{X^*} = \mu$, $\Pr_{Y^*} = \nu$, and assume that $Y^* \in \partial f(X^*)$ (Pr-a.s.) for an LSC function f. Then for any other RVs X and Y with distributions μ and ν we have

$$E\langle\widetilde{X},\widetilde{Y}\rangle \le E(f(\widetilde{X}) + f^*(\widetilde{Y})) = E(f(X^*) + f^*(X^*)) = E\langle X^*,Y^*\rangle.$$

Therefore, (14.2.25) holds. □

Remark 14.2.2. Condition (14.2.18) is also necessary.

Sketch of the proof. Let, conversely, $\langle X^*,Y^*\rangle$ be a solution of (14.2.25). Then, by (14.2.26),

$$\sup\{E\langle X,Y\rangle : \Pr_X = \mu, \Pr_Y = \nu\}$$

$$= \inf\left\{\int g\mathrm{d}\mu + \int h\mathrm{d}\nu : g(x) + h(y)\right.$$

$$\left.\ge \langle x,y\rangle, \int |g|\mathrm{d}\mu + \int |h|\mathrm{d}\nu < \infty\rangle\right\}. \tag{14.2.27}$$

Note that the supremum in (14.2.27) is attained [see Corollary 5.3.1 and (14.2.26)]. Moreover, one could see that the infimum in (14.2.26) is also attained (see proof of Theorem 5.3.1).[4] Suppose $f(x)$ and $g(y)$ are *optimal*, i.e., $E\langle X^*,Y^*\rangle = \int g\mathrm{d}\mu + \int h\mathrm{d}\nu$ and $g(x)+h(y) \ge \langle x,y\rangle$. Then $g^*(y) = \sup_x\{\langle x,y\rangle - g(x)\} \le h(y)$, and thus (g,g^*) is also optimal. In the same way, defining $f = g^{**}$ we see that $\langle x,y\rangle \le f(x) + f^*(y)$ and also f is an LSC function. This implies that $\langle X^*,Y^*\rangle = f(X^*) + f^*(Y^*)$ (Pr-a.s.) and therefore, by (14.2.21), that $Y^* \in \partial f(X^*)$ (Pr-a.s.).

[4] See also Kellerer (1984, Theorem 2.21) and Knott and Smith (1984, Theorem 3.2).

Remark 14.2.3. If $m = 1$ and F, G are DFs of μ and ν, then, as we saw by (14.2.14) [with $H(t) = t^2$], the optimal pair X^*, Y^* is given by $X^* = F^{-1}(V)$, $Y^* = G^{-1}(V)$, where V is uniform on $(0, 1)$. Defining $\theta(x) := G^{-1} \circ F(x)$ and $f(x) = \int_0^x \theta(y)\mathrm{d}y$, f is convex and $Y = G^{-1}(V) \in \partial(F^{-1}(V))$. Thus, (14.2.14) is a consequence of Theorem 14.2.1.

Remark 14.2.4. For a symmetric positive semidefinite $(m \times m)$ matrix T, define $f(x) = \frac{1}{2}\langle x, Tx\rangle$ and $g(y) = \langle \frac{1}{2}y, T^{-1}y\rangle$. Then $f(x) + g(Tx) = \langle x, Tx\rangle$. Therefore, if $\nu = \mu \circ T^{-1}$, where T^{-1} denotes the More–Penrose inverse, then the pair (X^*, TX^*) is optimal. This leads to the explicit expression for $\tau_\phi(\theta^*)$ (14.2.14) when μ and ν are Gaussian measures on $\mathbb{R}^m$ with means m_μ and m_ν and nonsingular covariance matrices Σ_μ and Σ_ν.

Corollary 14.2.1 (Olkin and Pukelheim 1982). *In the Gaussian case, where μ and ν are normal laws with means m_μ and m_ν and covariance matrices Σ_μ and Σ_ν,*

$$\tau_\phi(\theta^*) = \|m_\mu - m_\nu\|^2 + \mathrm{tr}(\Sigma_\mu) + \mathrm{tr}(\Sigma_\nu) - 2\,\mathrm{tr}(\Sigma_\mu^{1/2}\Sigma_\nu\Sigma_\mu^{1/2})^{1/2}. \qquad (14.2.28)$$

Proof. We can assume that $m_\mu = m_\nu = 0$ [see (14.2.22)]. Applying Remark 14.2.4, we have that the pair (X^*, TX^*), with

$$T = \Sigma_\nu^{1/2}(\Sigma_\nu^{1/2}\Sigma_\mu\Sigma_\nu^{1/2})^{-1/2}, \Sigma_\nu^{1/2} \qquad (14.2.29)$$

is optimal. Hence, by (14.2.29), $E\langle X^*, TX\rangle = \mathrm{tr}(\Sigma_\mu^{1/2}\Sigma_\nu\Sigma_\mu^{1/2})^{1/2}$ is the maximal possible value for $E\langle X, Y\rangle$ with $\mathrm{Pr}_X = \mu$ and $\mathrm{Pr}_Y = \nu$. □

Thus, if both production quality measure μ and the consumption quality measure ν are Gaussian, then the optimal plan for consumption quality θ^* is determined by the joint distribution of (X^*, TX^*), where T is given by (14.2.29).

To determine θ^*, we need to have complete information on the measures μ and ν. It is much more likely that we can have only the one-dimensional distributions μ_i and ν_j [see (14.2.4), (14.2.5)], i.e., we are dealing with the set of weakly admissible plans $\overline{\theta}(\mu_1, \ldots, \mu_m; \nu_1, \ldots, \nu_m)$ and would like to determine the weakly optimal plan θ^o and evaluate $\tau(\theta^o)$ [see (14.2.7)].

We make use of the multidimensional Kantorovich theorem (Sect. 5.3) to obtain a dual representation for $\tau(\theta^o)$. As in (14.2.13), suppose the cost function ϕ is given by (14.2.12), where H is convex and $K_H < \infty$. Then, by Theorem 5.3.1, there exists a weakly optimal plan θ^o for which the minimal value of the total loss function is

$$\tau_\phi(\theta^o) = \sup_{f_i, g_j \in C_\phi} \left(\sum_{i=1}^m \int_{\mathbb{R}} f_i(x)\mu_i(\mathrm{d}x) + \sum_{j=1}^m \int_{\mathbb{R}} g_j(y)\nu_j(\mathrm{d}y) \right), \qquad (14.2.30)$$

where C_ϕ denotes the collection of all functions $f_i(x_i)$, $g_j(y_j)$ on $\mathbb{R}$ satisfying the constraints

$$\mathrm{Lip}(f_i) := \sup|f_i(x) - f_i(y)|/|x-y| < \infty, \quad \mathrm{Lip}(g_j) < \infty, \tag{14.2.31}$$

and

$$\sum_{i,j=1}^{m} [f_i(x_i) + g_j(y_j)] < \phi(x,y), \quad \forall x, y \in \mathbb{R}^m. \tag{14.2.32}$$

Moreover, by Theorem 7.4.2, we can obtain explicit representations for θ^o and $\tau_\phi(\theta^o)$ for any cost function ϕ that is quasiantitone (Definition 7.4.1). Let F_i denote the DFs of μ_i and G_i the DFs of ν_i. Define the random variables $\overset{\circ}{X}_i = F_i^{-1}(V)$; $\overset{\circ}{Y}_j = G_j^{-1}(V)$, $i,j = 1,\dots,m$, and the random vectors $\overset{\circ}{X} = (\overset{\circ}{X}_1,\dots,\overset{\circ}{X}_m)$; $\overset{\circ}{Y} = (\overset{\circ}{Y}_1,\dots,\overset{\circ}{Y}_m)$, where $F_i^{-1}(x)$, $G_j^{-1}(x)$ are the inverses of the distribution functions $F_i(x)$, $G_j(x)$, respectively, and V is uniform on $[0,1]$.

Theorem 14.2.2. *For any cost function $\phi : \mathbb{R}^{2m} \to \mathbb{R}$ that is quasiantitone, the weak distribution plan θ^o with DF F_o given by*

$$F_o(x_1,\dots,x_m;y_1,\dots,y_m) = \min(F_1(x_1),\dots,F_m(x_m),G_1(y_1),\dots,G_m(y_m)) \tag{14.2.33}$$

is optimal. Moreover, in this case the minimal total cost is given by

$$\tau_\phi(\theta^o) = E\phi(\overset{\circ}{X},\overset{\circ}{Y}) = \int_0^1 \phi(F_1^{-1}(t)),\dots,F_m^{-1}(t),G_1^{-1}(t),\dots,G_m^{-1}(t))dt. \tag{14.2.34}$$

For example, let ϕ be the following metric in $\mathbb{R}^m$ for $x = (x_1,\dots,x_m)$, $y = (y_1,\dots,y_m)$:

$$\phi(x,y) = 2\max(x_1,\dots,x_m;y_1,\dots,y_m) - \frac{1}{m}\sum_{i=1}^{m}(x_i+y_i)$$

[see also (7.4.19)]. Then, by Theorem 14.2.2 and (14.2.30), θ^o with DF F_o is an optimal plan and

$$\tau_\phi(\theta^o) = \int_{-\infty}^{\infty} \frac{1}{n}\sum_{i=1}^{n}(F_i(u)+G_i(u))$$
$$-2\min[F_1(u),\dots,F_n(u),G_1(u),\dots,G_n(u)]du.$$

14.3 Estimates of Minimal Total Losses $\tau_\phi(\phi^*)$

Consider the multidimensional case where the quality vector $x = (x_1,\dots,x_m)$ has $m > 1$ one-dimensional characteristics. We derive an upper bound for $\tau_\phi(\theta^*)$ [see (14.2.3)] in the special case where the loss function has the form

$$\phi(x, y) = K \sum_{i=1}^{m} |x_i - y_i|, x, y \in \mathbb{R}^m, \; x := (x_1, \dots, x_m), \; y := (y_1, \dots, y_m). \tag{14.3.1}$$

Remark 14.3.1. In this particular case,

$$\tau_\phi(\theta^*) = K\ell_1(\mu, \nu) := K \sup \left\{ \left| \int_{\mathbb{R}^m} u \mathrm{d}(\mu - \nu) \right| : u \in \mathrm{Lip}_{1.1}^b(\mathbb{R}^m) \right\}, \tag{14.3.2}$$

where ℓ_1 is the minimal metric with respect to the $\mathcal{L}_1$-distance

$$\mathcal{L}_1(X, Y) = E\|X - Y\|_1, \quad X, Y \in \mathfrak{X}(\mathbb{R}^m), \tag{14.3.3}$$

in which $\|x - y\|_1 := \sum_{i=1}^m |x_i - y_i|, \; x, y \in \mathbb{R}^m$. See (3.3.2), (3.4.3), and (5.3.18) for additional details.

Remark 14.3.2. Dobrushin (1970) called ℓ_1 the Vasershtein (Wasserstein) distance. In our terminology, ℓ_1 is the Kantorovich metric (Example 3.3.2). The problem of bounding ℓ_1 from above also arises in connection with the sufficient conditions implying the uniqueness of the Gibbs random fields; see Dobrushin (1970, Sects. 4 and 5).

By (14.3.2), we need to find precise estimates for ℓ_1 in the space $\mathcal{P}(\mathbb{R}^m)$ of all laws on $(\mathbb{R}^m, \|\cdot\|_1)$. The next two theorems provide such estimates and in certain cases even explicit representations of ℓ_1.

We suppose that $P_1, P_2 \in \mathcal{P}(\mathbb{R}^m)$ have densities p_1 and p_2, respectively.

Theorem 14.3.1. *(i) The following inequality holds:*

$$\ell_1(P_1, P_2) \leq \alpha_1(P_1, P_2), \tag{14.3.4}$$

with

$$\alpha_1(P_1, P_2) := \int_{\mathbb{R}^m} \|x\|_1 \left| \int_0^1 t^{-m-1}(p_1 - p_2)(x/t)\mathrm{d}t \right| \mathrm{d}x.$$

(ii) If

$$\int_{\mathbb{R}^m} \|x\|_1 d(P_1 + P_2) < \infty, \tag{14.3.5}$$

and if a continuous function $g : \mathbb{R}^m \to \mathbb{R}^1$ exists with derivatives $\partial g/\partial x_i$, $i = 1, \dots, m$, defined almost everywhere (a.e.) and satisfying

$$\frac{\partial g}{\partial x_i}(x) = \mathrm{sgn}\left[x_i \int_0^1 t^{-m-1}(p_1 - p_2)(x/t)\mathrm{d}t \right] \; a.e. \; i = 1, \dots, m, \tag{14.3.6}$$

then (14.3.4) *holds with the equality sign.*

Proof. (i) It is easy to see that the constraint set for

$$\ell_1(P_1, P_2) = \sup\left\{\left|\int_{\mathbb{R}^m} u\mathrm{d}(P_1 - P_2)\right| : u : \mathbb{R}^m \to \mathbb{R},\ \text{bounded}\right.$$

$$\left. |u(x) - u(y)| \le \|x - y\|, x, y \in \mathbb{R}^m\right\} \tag{14.3.7}$$

coincides with the class of continuous bounded functions u [$u \in C_b(U)$] that have partial derivatives u_i' defined a.e. and satisfying the inequalities $|u_i'(x)| \le 1$ a.e., $i = 1, \ldots, m$. Now, using the identity

$$u(x) = u(0) + \sum_{i=1}^{m} x_i \int_0^1 u_i'(tx)\mathrm{d}t,$$

passing on from the coordinates t, x to the coordinates $t' = t$, $x' = tx$, and denoting these new coordinates again by t, x, one obtains

$$\ell_1(P_1, P_2) = \sup\left\{\left|\int_{\mathbb{R}^m} \sum_{i=1}^{m} u_i'(x)x_i \left(\int_0^1 t^{-m-1}(p_1 - p_2)(x/t)\mathrm{d}t\right)\mathrm{d}x\right| :\right.$$

$$\left. u \in C_b(\mathbb{R}^m), |u_i'| \le 1, \ldots, |u_m'| \le 1 \text{ a.e.}\right\}. \tag{14.3.8}$$

The estimate (14.3.4) follows obviously from here.

(ii) If the moment condition (14.3.5) holds, then, by Corollary 6.2.1,

$$\ell_1(P_1, P_2) = \sup\left\{\left|\int_{\mathbb{R}^m} u\mathrm{d}(P_1 - P_2)\right| : u \in C(\mathbb{R}^m),\right.$$

$$\left. |u(x) - u(y)| \le \|x - y\| \ \forall x, y \in \mathbb{R}^m\right\}$$

$$= \sup\left\{\left|\int_{\mathbb{R}^m} u\mathrm{d}(P_1 - P_2)\right| :\right.$$

$$\left. u \in C(\mathbb{R}^m), |u_i'| \le 1 \text{ a.e.}, i = 1, \ldots, m\right\}, \tag{14.3.9}$$

where $C(\mathbb{R}^m)$ is the space of all continuous functions on $\mathbb{R}^m$.

Then, in (14.3.8), $C_b(\mathbb{R}^m)$ may also be replaced by $C(\mathbb{R}^m)$. It follows from (14.3.6) and (14.3.8) that the supremum in (14.3.8) is attained by $u = g$, and hence

$$\ell_1(P_1, P_2) = \alpha(P_1, P_2). \tag{14.3.10}$$

The proof is complete. □

Next we give simple sufficient conditions assuring equality (14.3.10). Denote

$$J(P_1, P_2; x) := \int_0^1 t^{-m-1}(p_1 - p_2)(x/t)\mathrm{d}t.$$

Corollary 14.3.1. *If the moment condition* (14.3.5) *holds and* $J(P_1, P_2; x) \geq 0$ *a.e. or* $J(P_1, P_2; x) \leq 0$ *a.e., then equality* (14.3.10) *takes place.*

Proof. Indeed, one can take $g(x) := \|x\|_1$ in Theorem 3.3.1 (ii) if $J(P_1, P_2; x) \geq 0$ a.e. and $g(x) = -\|x\|_1$ if $J(P_1, P_2; x) \leq 0$ a.e. □

Remark 14.3.3. The inequality $J(P_1, P_2; x) \geq 0$ a.e. holds, for example, in the following cases:

(a) $0 < \underline{\lambda} \leq \overline{\lambda}$: $p_1(x) = \mathrm{Weib}_{\underline{\lambda}}(x) := \prod_{i=1}^m \alpha_i \underline{\lambda}(\underline{\lambda} x_i)^{\alpha_i - 1} \exp(-(\underline{\lambda} x_i)^{\alpha_i})$, $\alpha_i > 0$, and $p_2(x) = \mathrm{Weib}_{\overline{\lambda}}(x)$ are constructed assuming the vector components are independent and follow a Weibull distribution.

(b) $0 < \underline{\lambda} \leq \overline{\lambda}$: $p_1(x) = \mathrm{Gam}_{\underline{\lambda}}(x) := \prod_{i=1}^m \underline{\lambda}^{\alpha_i} x_i^{\alpha_i - 1}(\Gamma(\alpha_i))^{-1} \exp(-\underline{\lambda} x_i)$, $\alpha_i > 0$, and $p_2(x) = \mathrm{Gam}_{\overline{\lambda}}(x)$ are constructed assuming the vector components are independent and follow a gamma distribution.

(c) $\overline{\lambda} \geq \underline{\lambda} > 0$: $p_1(x) = \mathrm{Norm}_{\overline{\lambda}}(x) := \prod_{i=1}^m (1/\overline{\lambda}\sqrt{2\pi}) \exp[-(x_i^2/2\overline{\lambda}^2)]$ and $p_2(x) = \mathrm{Norm}_{\underline{\lambda}}(x)$ are constructed assuming the vector components are independent and follow a normal distribution.

Theorem 14.3.2. *(i) The inequality*

$$\ell_1(P_1, P_2) \leq \alpha_2(P_1, P_2) \tag{14.3.11}$$

holds with

$$\begin{aligned} \alpha_2(P_1, P_2) := & \int_{-\infty}^{\infty} \left| \int_{-\infty}^{t} q_1(x_{(1)})\mathrm{d}x_1 \right| \mathrm{d}t \\ & + \sum_{i=2}^m \int_{\mathbb{R}^{i-1}} \left(\int_{-\infty}^0 \left| \int_{-\infty}^t q_i(x_{(i)})\mathrm{d}x_i \right| \mathrm{d}t \right. \\ & \left. + \int_0^{\infty} \left| \int_t^{\infty} q_i(x_{(i)})\mathrm{d}x_i \right| \mathrm{d}t \right) \mathrm{d}x_1 \cdots \mathrm{d}x_{i-1}, \end{aligned} \tag{14.3.12}$$

where

$$x_{(i)} := (x_1, \ldots, x_i),$$

$$q_i(x_{(i)}) := \int_{\mathbb{R}^{m-i}} (p_1 - p_2)(x_1, \ldots, x_m)\mathrm{d}x_1, \ldots, \mathrm{d}x_m,\ i = 1, \ldots, m-1,$$

$$q_m(x_{(m)}) := (p_1 - p_2)(x_1, \ldots, x_m).$$

(ii) If (14.3.5) *holds and if a continuous function* $h : \mathbb{R}^m \to \mathbb{R}^1$ *exists with derivatives* h_i', $i = 1, \ldots, m$, *defined a.e. and satisfying the conditions*

$$h_1'(t,0,\ldots,0) = \mathrm{sgn}[F_{11}(t) - F_{21}(t)],$$

$$h_2'(x_1,t,0,\ldots,0) = \begin{cases} \mathrm{sgn}\displaystyle\int_{-\infty}^{t} q_2(x_{(2)})\mathrm{d}x_2, & \text{if } t \in (-\infty,0],\ x_1 \in \mathbb{R}^1, \\ -\,\mathrm{sgn}\displaystyle\int_{t}^{\infty} q_2(x_{(2)})\mathrm{d}x_2, & \text{if } t \in (0,+\infty),\ x_1 \in \mathbb{R}^1, \end{cases}$$

$$h_m'(x_1,\ldots,x_{m-1},t) = \begin{cases} \mathrm{sgn}\displaystyle\int_{-\infty}^{t} q_m(x_{(m)})\mathrm{d}x_m, & \text{if } t \in (-\infty,0], \\ & x_1,\ldots,x_{m-1} \in \mathbb{R}^1, \\ -\,\mathrm{sgn}\displaystyle\int_{t}^{\infty} q_m(x_{(m)})\mathrm{d}x_m, & \text{if } t \in (0,+\infty), \\ & x_1,\ldots,x_{m-1} \in \mathbb{R}^1, \end{cases}$$

then (14.3.11) *holds with the equality sign. Here* F_{ji} *stands for the DF of the projection* $(T_i P_j)$ *of* P_j *over the* i*th coordinate.*

Proof. (i) Using the formulae

$$q_i(x_{(i)}) = \int_{-\infty}^{\infty} q_{i+1}(x_{(i+1)})\mathrm{d}x_{i+1}, \quad i = 1,\ldots,m-1,$$

$$\int_{-\infty}^{\infty} q_1(x_{(1)})\mathrm{d}x_1 = \int_{\mathbb{R}^m} (p_1 - p_2)(x)\mathrm{d}x = 0,$$

and applying repeatedly the identity

$$\int_{-\infty}^{\infty} a(t)b(t)\mathrm{d}t = \int_{-\infty}^{\infty} a(0)b(t)\mathrm{d}t - \int_{-\infty}^{0} a'(t)\left(\int_{-\infty}^{t} b(s)\mathrm{d}s\right)\mathrm{d}t$$
$$+ \int_{0}^{\infty} a'(t)\left(\int_{t}^{\infty} b(s)\mathrm{d}s\right)\mathrm{d}t$$

for $a(t){=}u(x_1,\ldots,x_{i-1},t,\ldots,0)$, $b(t){=}q_i(x_1,\ldots,x_{i-1},t)$, $i{=}1,\ldots,m$, one obtains

$$\ell_1(P_1,P_2) = \sup\left\{\left| -\int_{-\infty}^{\infty} u_1'(t,0,\ldots,0)\int_{-\infty}^{t} q_1(x_{(1)})\mathrm{d}x_1\mathrm{d}t \right.\right.$$
$$+ \sum_{i=2}^{m}\int_{\mathbb{R}^{i-1}}\left(-\int_{-\infty}^{0} u_i'(x_1\ldots,x_{i-1},t,\ldots,0)\int_{-\infty}^{t} q_i(x_{(i)})\mathrm{d}x_i\mathrm{d}t\right.$$

$$+\int_0^\infty u_i'(x_1,\ldots,x_{i-1},t,\ldots,0)\int_t^\infty q_i(x_{(i)})\mathrm{d}x_i\,\mathrm{d}t\Bigg)\mathrm{d}x_1\ldots\mathrm{d}x_{i-1}\Bigg|$$

$$: u\in C_b(\mathbb{R}^m),\, |u_1'|\le 1,\ldots,|u_m'|\le 1 \text{ a.e.}\Bigg\}, \tag{14.3.13}$$

which obviously implies (14.3.11).

(ii) In view of (14.3.5), $C_b(\mathbb{R}^m)$ in (14.3.13) may be replaced by $C(\mathbb{R}^m)$. Then the function $u = h$ yields the supremum on the right-hand side of (14.3.13), and hence

$$\ell_1(P_1, P_2) = \alpha_2(P_1, P_2). \tag{14.3.14}$$

□

Remark 14.3.4. The bounds (14.3.4) and (14.3.14) are of interest by themselves. They give two improvements of the following bound[5]:

$$\ell_1(P_1, P_2) \le \nu(P_1, P_2),$$

where

$$\nu(P_1, P_2) := \int_{\mathbb{R}^m} \|x\|_1 |p_1(x) - p_2(x)|\mathrm{d}x$$

is the first absolute pseudomoment. Indeed, one can easily check that

$$\alpha_i(P_1, P_2) \le \nu(P_1, P_2), \quad i = 1, 2.$$

Remark 14.3.5. Consider the sth-difference pseudomoment[6]

$$\kappa_s(P_1, P_2) = \sup\Bigg\{\Bigg|\int_{\mathbb{R}^m} u\mathrm{d}(P_1 - P_2)\Bigg| : u : \mathbb{R}^m \to \mathbb{R}^1,$$

$$|u(x) - u(y)| \le d_s(x, y)\Bigg\}, \quad s > 0, \tag{14.3.15}$$

where

$$d_s(x, y) := \|Q_s(x) - Q_s(y)\|, \quad Q_s : \mathbb{R}^m \to \mathbb{R}^m, \tag{14.3.16}$$

$$Q_s(t) := t\|t\|^{s-1}.$$

Since

[5] See Zolotarev (1986, Sect. 1.5).

[6] See Case D in Sect. 4.4.

$$\kappa(P_1, P_2) = \ell_1(P_1 \circ Q_s^{-1}, P_2 \circ Q_s^{-1}), \tag{14.3.17}$$

then by (14.3.4) and (14.3.11) we obtain the bounds

$$\kappa_s(P_1, P_2) \le \alpha_i(P_1 \circ Q_s^{-1}, P_2 \circ Q_s^{-1}), \quad i = 1, 2, \tag{14.3.18}$$

which are better than the following one[7]

$$\kappa_s(P_1, P_2) \le \nu(P_1 \circ Q_s^{-1}, P_2 \circ Q_s^{-1}). \tag{14.3.19}$$

References

Borwein JM, Lewis AS (2010) Convex analysis and nonlinear optimization: theory and examples, 2nd edn. Springer Science, New York

Dobrushin RL (1970) Prescribing a system of random variables by conditional distributions. Theor Prob Appl 15:458–486

Kellerer HG (1984) Duality theorems for marginal problems. Z Wahrsch Verw Geb 67:399–432

Knott M, Smith CS (1984) On the optimal mapping of distributions. J Optim Theor Appl 43:39–49

Olkin I, Pukelheim F (1982) The distance between two random vectors with given dispersion matrices. Linear Algebra Appl 48:257–263

Rockafellar RT (1970) Convex analysis. Princeton University Press, Princeton

Zolotarev VM (1986) Contemporary theory of summation of independent random variables. Nauka, Moscow (in Russian)

[7] See Zolotarev (1986, Sect. 1.5, Eq. 1.5.37).

Part IV
Ideal Metrics

Chapter 15
Ideal Metrics with Respect to Summation Scheme for i.i.d. Random Variables

The goals of this chapter are to:

- Discuss the question of stability in the χ^2 test of exponentially under different contamination mechanisms,
- Describe the notion of ideal probability metrics for summation of independent and identically distributed random variables,
- Provide examples of ideal probability metrics and discuss weak convergence criteria,
- Derive the rate of convergence in the general central limit theorem in terms of metrics with uniform structure.

Notation introduced in this chapter:

Notation	Description
ζ_r	Zolotarev ideal metric of order r
Var	Total variation metric, $\mathbf{Var} = 2\sigma$
ℓ	Uniform metric between densities
$\boldsymbol{\chi}$	Uniform metric between characteristic functions
$\boldsymbol{\chi}_r$	Weighted version of $\boldsymbol{\chi}$-metric
$\boldsymbol{\zeta}_{m,p}$	L^p-version of $\boldsymbol{\zeta}_m$
$\boldsymbol{\mu}_{\theta,r}$	Smoothing version of ℓ
$\boldsymbol{\nu}_{\theta,r}$	Smoothing version of **Var**
$\boldsymbol{\mu}_r$	Special version of $\boldsymbol{\mu}_{\theta,r}$ with θ having α-stable distribution
$\boldsymbol{\nu}_r$	Special version of $\boldsymbol{\nu}_{\theta,r}$ with θ having α-stable distribution

S.T. Rachev et al., *The Methods of Distances in the Theory of Probability and Statistics*,
DOI 10.1007/978-1-4614-4869-3_15, © Springer Science+Business Media, LLC 2013

15.1 Introduction

The subject of this chapter is the application of the theory of probability metrics to limit theorems arising from summing independent and identically distributed (i.i.d.) random variables (RVs). We describe the notion of an ideal metric – a metric endowed with certain properties that make it suitable for studying a particular problem, in this case, the rate of convergence in the corresponding limit theorems.

We begin this chapter with a section on the robustness of the χ^2 test of exponentiality, which serves as an introduction to the general topic. The question of stability is discussed in the context of different contamination mechanisms.

The section on ideal metrics for sums of independent RVs defines axiomatically ideal properties and then introduces various metrics satisfying them. The section also describes relationships between those metrics, conditions under which the metrics are finite, and proves convergence criteria under weak convergence of probability measures.

Finally, we discuss rates of convergence in the central limit theorem (CLT) in terms of metrics with uniform structure. Ideal convolution metrics play a central role in the proofs. Rates of convergence are provided in terms of **Var**, $\boldsymbol{\chi}$, ℓ, and $\boldsymbol{\rho}$.

15.2 Robustness of χ^2 Test of Exponentiality

Suppose that Y is exponentially distributed with density (PDF) $f_Y(x) = (1/a)\exp(x/a)$, $(x \geq 0;\ a > 0)$. To perform hypothesis tests on a, one makes use of the fact that, if $Y_1, Y_2, \ldots, Y_n$ are n i.i.d. RVs, each with PDF f_Y, then $2\sum_{i=1}^n Y_i/a \approx \chi^2_{2n}$. In practice, the assumption of exponentiality is only an approximation; it is therefore of interest to enquire how well the χ^2_{2n} distribution approximates that of $2\sum_{i=1}^n X_i/a$, where $X_1, X_2, \ldots, X_n$ are i.i.d. nonnegative RVs with common mean a, representing the "perturbation," in some sense, of an exponential RV with the same mean. The usual approach requires one either to make an assumption concerning the class of RVs representing the possible perturbations of the exponential distribution or to identify the nature of the mechanism causing the perturbation.

(A) *The case where the X belong to an aging distribution class.* A nonnegative RV X with DF F is said to be *harmonic new better than used in expectation* (HNBUE) if $\int_x^\infty \overline{F}(u)\mathrm{d}u \leq a\exp(-x/a)$ for all $x \geq 0$, where $a = E(X)$ and $\overline{F} = 1 - F$. It is easily seen that if X is HNBUE, then moments of all orders exist. Similarly, X is said to be *harmonic new worse than used in expectation* (HNWUE) if $\int_x^\infty \overline{F}(u)\mathrm{d}u \geq a\exp(-x/a)$ for all $x \geq 0$, assuming that a is finite. The class of HNBUE (HNWUE) distributions include all the standard "aging" ("antiaging") classes – IFR, IFRA, NBU, and NBUE (DFR, DFRA, NWU, and NWUE).[1]

It is well known that *if* X *is* HNBUE *with* $a = EX$ *and* $\sigma^2 = \operatorname{var} X$, *then* X *is exponentially distributed if and only if* $a = \sigma$. To investigate the stability of this characterization, we must select a metric $\mu(X, Y) = \mu(F_X, F_Y)$ in the DF space $\mathcal{F}(\mathbb{R})$ such that

(a) μ *guarantees the convergence in distribution plus convergence of the first two moments;*
(b) μ *satisfies the inequalities*

$$\phi_1(|a - \sigma|) \leq \mu(X, E(a)) \leq \phi_2(|a - \sigma|),$$

where $X \in$ HNBUE, $EX = a$, $\sigma^2 = \operatorname{var} X$, ϕ_1, *and* ϕ_2 *are some continuous, increasing functions with* $\phi_i(0) = 0$, $i = 1, 2$, and $E(a)$ denotes an exponential variable with a mean of a.

Clearly, the most appropriate metric μ should satisfy (a) and (b) with $\phi_1 \equiv \phi_2$. Such a metric is the so-called Zolotarev $\boldsymbol{\zeta}_2$-metric

$$\boldsymbol{\zeta}_2(X, Y) := \boldsymbol{\zeta}_2(F_X, F_Y) = \sup_{f \in \mathbb{F}_2} |E(f(X) - f(Y))|, \tag{15.2.1}$$

where $EX^2 < \infty$, $EY^2 < \infty$, and $\mathbb{F}_2$ is the class of all functions f having almost everywhere (a.e.) the second derivative f'' and $|f''| \leq 1$ a.e. To check (a) and (b) for $\mu = \boldsymbol{\zeta}_2$, first notice that the finiteness of $\boldsymbol{\zeta}_2$ implies $\infty > \boldsymbol{\zeta}_2(X, Y) \geq \sup_{a>0} |E(aX) - E(aY)|$, i.e., $EX = EY$. Secondly, if $EX = EY$, then $\boldsymbol{\zeta}_2(X, Y)$ admits a second representation:

$$\boldsymbol{\zeta}_2(X, Y) = \int_{-\infty}^{\infty} \left| \int_{-\infty}^{x} (F_X(t) - F_Y(t))\mathrm{d}t \right| \mathrm{d}x. \tag{15.2.2}$$

In fact, by Taylor's theorem or integrating by parts,

$$\begin{aligned}
\boldsymbol{\zeta}_2(X, Y) &= \sup_{f \in \mathbb{F}_2} \left| \int_{-\infty}^{\infty} f(t)\mathrm{d}(F_X(t) - F_Y(t)) \right| \\
&= \sup_{f \in \mathbb{F}_2} \left| \int_{-\infty}^{\infty} f''(x) \int_{-\infty}^{x} (F_X(t) - F_Y(t))\mathrm{d}t \right| \mathrm{d}x.
\end{aligned}$$

Now use the isometric isomorphism between L_1^*- and L_∞-spaces to obtain equality (15.2.2).[2]

---

[1]See Barlow and Proschan (1975, Chap. 4) and Kalashnikov and Rachev (1988, Chap. 4) for the necessary definitions.

[2]See, for example, Dunford and Schwartz (1988, Theorem IV.8.3.5).

Using both representations for ζ_2, in the next lemma we show that $\mu = \zeta_2$ satisfies (a).

Lemma 15.2.1. *(i) In the space $\mathfrak{X}^2(\mathbb{R})$ of all square integrable RVs,*

$$\tfrac{1}{2}|EX^2 - EY^2| \leq \zeta_2(X,Y) \tag{15.2.3}$$

and

$$\mathbf{L}(X,Y) \leq [4\zeta_2(X,Y)]^{1/3}, \tag{15.2.4}$$

where $\mathbf{L}$ *is the Lévy metric* (2.2.3) *in Chap. 2. In particular, if X_n, $X \in \mathfrak{X}^2(\mathbb{R})$, then*

$$\zeta_2(X_n, X) \to 0 \Rightarrow \begin{cases} X_n \to X \text{ in distribution} \\ EX_n^2 \to EX^2. \end{cases} \tag{15.2.5}$$

(ii) Given $X_0 \in \mathfrak{X}^2(\mathbb{R})$, let $\mathfrak{X}^2(\mathbb{R}, X_0)$ be the space of all $X \in \mathfrak{X}^2(\mathbb{R})$ with $EX = EX_0$. Then for any $X, Y \in \mathfrak{X}^2(\mathbb{R}, X_0)$

$$2\zeta_2(X,Y) \leq \boldsymbol{\kappa}_2(X,Y), \tag{15.2.6}$$

where $\boldsymbol{\kappa}$ *is the second pseudomoment*

$$\boldsymbol{\kappa}_2(X,Y) = 2\int_{-\infty}^{\infty} |x||F_X(x) - F_Y(x)|\mathrm{d}x. \tag{15.2.7}$$

In particular, for $X_n, X \in \mathfrak{X}^2(\mathbb{R}, X_0)$

$$\begin{cases} X_n \to X \text{ in distribution} \\ EX_n^2 \to EX^2 \end{cases} \Rightarrow \zeta_2(X_n, X) \to 0. \tag{15.2.8}$$

Proof. (i) Clearly, representation (15.2.1) implies (15.2.3). To prove (15.2.4), let $\mathbf{L}(X,Y) > \varepsilon > 0$; then there exists $z \in \mathbb{R}$ such that either

$$F_X(z) - F_Y(z+\varepsilon) > \varepsilon \tag{15.2.9}$$

or $F_Y(z) - F_X(z+\varepsilon) > \varepsilon$. Suppose (15.2.9) holds; then set

$$f_0\left(z + \frac{\varepsilon}{2} + h\right) := \left\{\left[\left(1 - \frac{2|h|}{\varepsilon}\right)_+\right]^2 - 1\right\} \operatorname{sgn} h, \tag{15.2.10}$$

where $(\cdot)_+ = \max(0,\cdot)$. Then $f_0(x) = 1$ for $x \leq z$, $f_0(x) = -1$ for $x > z + \varepsilon$ and $|f_0| \leq 1$. Since $\|f_0''\|_\infty := \operatorname{ess\,sup}|f''(x)| = 8\varepsilon^{-2}$, we have

$$\begin{aligned}\zeta_2(X,Y) &\geq \|f_0''\|_\infty^{-1}\left|\int (f_0(x)+1)\mathrm{d}(F_X(x) - F_Y(x))\right| \\ &\geq (\varepsilon^2/8)\left(\int_{-\infty}^{z} (f_0(x)+1)\mathrm{d}F_X(x) - \int_{z+\varepsilon}^{\infty} (f_0)x)+1)\mathrm{d}F_Y(x)\right) \geq \varepsilon^3/4.\end{aligned}$$

Letting $\varepsilon \to \mathbf{L}(X,Y)$ implies (15.2.4).

(ii) Using representation (15.2.2) and $EX = EY$ one obtains (15.2.6). Clearly,

$$\kappa_2(X,Y) = \ell_1(X|X|, Y|Y|), \tag{15.2.11}$$

where ℓ_1 is the Kantorovich metric $\ell_1(X,Y) = \int_{-\infty}^{\infty} |F_X(x) - F_Y(x)| \mathrm{d}x$ [see also (4.4.39) and (14.3.17)]. For any X_n and X with $E|X_n| + E|X| < \infty$ we have, by Theorem 6.3.1, that

$$\ell_1(X_n, X) \to 0 \iff \begin{cases} X_n \to X \text{ in distribution,} \\ E|X_n| \to E|X|, \end{cases} \tag{15.2.12}$$

which, together with (15.2.11), completes the proof of (ii). □

Thus ζ_2-convergence preserves the convergence in distribution plus convergence of the second moments, and so requirement (a) holds. Concerning property (b), we use the second representation of ζ_2, (15.2.2), to get

$$\begin{aligned} \zeta_2(X,Y) &= \int_0^\infty \left| \int_x^\infty \overline{F}_X(t)\mathrm{d}t - a\exp(-x/a) \right| \mathrm{d}x \\ &= \int_0^\infty \left(a\exp(-x/a) - \int_x^\infty \overline{F}_X(t)\mathrm{d}t \right) \mathrm{d}x \\ &= \frac{1}{2}(a^2 - \sigma^2) \quad \text{for } X \text{ being HNBUE, } Y := E(a). \end{aligned} \tag{15.2.13}$$

Now if one studies the stability of the preceding characterization in terms of a "traditional" metric as the uniform one

$$\boldsymbol{\rho}(X,Y) := \sup_{x\in\mathbb{R}} |F_X(x) - F_Y(x)|, \tag{15.2.14}$$

then one simply compares ζ_2 with $\boldsymbol{\rho}$. That is, by the well-known inequality between the Lévy distance $\mathbf{L}$ and the Kolmogorov distance $\boldsymbol{\rho}$, we have

$$\boldsymbol{\rho}(X,Y) \le \left[1 + \sup_x f_X(x)\right] \mathbf{L}(X,Y) \tag{15.2.15}$$

if $f_X = F_X'$ exists. Thus, by (15.2.4) and (15.2.15),

$$\begin{aligned} \boldsymbol{\rho}(X,Y) &\le \left[1 + \sup_t f_{cX}(t)\right] [4\zeta_2(cX, cY)]^{1/3} \\ &= (c^{2/3} + M_X c^{-1/3})[4\zeta_2(X,Y)^{1/3} \quad \text{for any } c > 0, \end{aligned}$$

where $M_X = \sup_t f_X(t)$. Minimizing the right-hand side of the last inequality with respect to c we obtain

$$\boldsymbol{\rho}(X,Y) \le 3M_X^{2/3}(\zeta_2(X,Y))^{1/3}. \tag{15.2.16}$$

Thus, for any $X \in$ HNBUE with $EX = a$, var $Y = \sigma^2$

$$\rho(X, E(a)) \leq 3(\alpha/2)^{1/3}, \quad \alpha = 1 - \sigma^2/a^2. \tag{15.2.17}$$

Remark 15.2.1. Note that the order 1/3 of α is precise [see Daley (1988) for an appropriate example].

Next, using the "natural" metric ζ_2, we derive a bound on the uniform distance between the χ^2_{2n} distribution and the distribution of $2\sum_{i=1}^n X_i/a$, assuming that X is HNBUE. Define $\overline{X}_i = (X_i - a)/a$ and $\overline{Y}_i = (Y_i - a)/a$ $(i = 1, 2, \ldots, n)$, and write $W_n = 2\sum_{i=1}^n X_i/a$, $\overline{W}_n = \sum_{i=1}^n \overline{X}_i/\sqrt{n}$, $Z_n = 2\sum_{i=1}^n Y_i/a$, and $\overline{Z}_n = \sum_{i=1}^n \overline{Y}_i/\sqrt{n}$. Let $f_{\overline{Z}_n}$ denote the PDF of $\overline{Z}_n$, and let $M_n = \sup_x f_{\overline{Z}_n}(x)$. Then by (15.2.16),

$$\rho(\overline{W}_n, \overline{Z}_n) \leq 3M_n^{2/3}[\zeta_2(\overline{W}_n, \overline{Z}_n)]^{1/3}.$$

Now we use the fact that ζ_2 is the *ideal metric of order* 2 (see further Sect. 15.3), i.e., *for any vectors* $\{X_i\}_{i=1}^n$ and $\{Y_i\}_{i=1}^n$ *with independent components and constants* $c_1, \ldots, c_n$

$$\zeta_2\left(\sum_{i=1}^n c_i X_i, \sum_{i=1}^n c_i Y_i\right) \leq \sum_{i=1}^n |c_i|^2 \zeta_2(X_i, Y_i). \tag{15.2.18}$$

Remark 15.2.2. Since ζ_2 is a simple metric, without loss of generality, we may assume that $\{X_i\}$ and $\{Y_i\}$ are independent. Then (15.2.18) follows from single induction arguments, the triangle inequality, and the following two properties: for any independent X, Y, and Z and any $c \in \mathbb{R}$

$$\mu(X + Z, Y + Z) \leq \mu(X, Y) \quad \text{(regularity)} \tag{15.2.19}$$

and

$$\mu(cX, cY) = c^2\mu(X, Y) \quad \text{(homogeneity of order 2).} \tag{15.2.20}$$

(See Definition 15.3.1 subsequently.)

Thus, by (15.2.18), $\zeta_2(\overline{W}_n, \overline{Z}_n) \leq \zeta(X, Y)/a^2$, and finally the required estimate is

$$\rho(W_n, Z_n) \leq \frac{3}{2^{1/3}} M_n^{2/3}[1 - (\sigma/a)^2]^{1/3} \tag{15.2.21}$$

and it is a straightforward matter to show that

$$M_n = \frac{\sqrt{n}(n-1)^{n-1} - \exp[-(n-1)]}{(n-1)!}. \tag{15.2.22}$$

Expression (15.2.22) may be simplified by using the Robbins–Stirling inequality[3]

$$n^n e^{-n}(2\pi n)^{1/2}\exp[1/(12n+1)] < n! < n^n e^{-n}(2\pi n)^{1/2}\exp(1/12n)$$

to give the following simple bound:

$$M^n < \left(\frac{n}{2\pi(n-1)}\right)^{1/2}\exp[-1/(12n-11)]. \qquad (15.2.23)$$

Further, if X is HNWUE, a similar calculation shows that $\zeta_2(X,Y) = \frac{1}{2}(\sigma^2 - a^2)$, assuming that σ^2 is finite. In summary, we have shown that if X is HNBUE or HNWUE, then

$$\rho(W_n, Z_n) \le \frac{3}{2^{1/3}} M_n^{2/3}[1-(\sigma/a)^2]^{1/3}, \qquad (15.2.24)$$

where M_n can be estimated by (15.2.23).

It follows from (15.2.22) that if X is HNBUE or HNWUE, and if the coefficient of variation of X is close to unity, then the distribution of $2\sum_{i=1}^n X_i/a$ is uniformly close to the χ^2_{2n} distribution.

(B) *The case where X is arbitrary: contamination by mixture.* In practice, a "perturbation" of an exponential RV does not necessarily yield an HNBUE or HNWUE variable, in which case the bound (15.2.24) will not hold. If we make no assumptions concerning X, then it is necessary to make an assumption concerning the "mechanism" by which the exponential distribution is "perturbed." Further, we will deduce bounds for the three most common possible "mechanisms": contamination by mixture, contamination by an additive error, and right-censoring.

Suppose that an exponential RV is contaminated by an arbitrary nonnegative RV with distribution function H, i.e., $\overline{F}_X(t) = (1-\varepsilon)\exp(-t/\lambda) + \varepsilon\overline{H}(t)$. Then $a = (1-\varepsilon)\lambda + \varepsilon h$, where $h = \int_0^\infty t\mathrm{d}H(t)$. It is assumed that $\varepsilon > 0$ is small. Now, since $Y = E(a)$,

$$\begin{aligned}
\zeta_2(X,Y) &= \int_0^\infty \left|\int_x^\infty [\overline{F}_X(t) - \exp(-t/a)]\mathrm{d}t\right| \mathrm{d}x \le \int_0^\infty t|\overline{F}_X(t) - \varepsilon(-t/a)|\mathrm{d}t \\
&\le (1-\varepsilon)\int_0^\infty t|\exp(-t/\lambda) - \exp(-t/a)|\mathrm{d}t \\
&\quad + \varepsilon\int_0^\infty t|\overline{H}(t) - \exp(-t/a)|\mathrm{d}t \\
&\le (1-\varepsilon)|\lambda_2 - a^2| + \varepsilon(b/2 + a^2), \quad \left(\text{where } b = \int_0^\infty t^2\mathrm{d}H(t)\right) \\
&= \varepsilon[|h-\lambda|(\lambda + a) + b/2 + a^2].
\end{aligned}$$

---

[3] See, for example, Erdös and Spencer (1974, p. 17).

Thus, $\zeta_2(X,Y) = O(\varepsilon)$, and so from (15.2.16) it follows that $\rho(W_n, Z_n) = O(\varepsilon^{1/3})$.

(C) *The case where X is arbitrary: contamination by additive error.* Suppose now that an exponential RV is contaminated by an arbitrary additive error, i.e., $X \stackrel{d}{=} Y_\lambda + V$, V is an arbitrary RV, and Y_λ is an exponential RV independent of V with mean $\lambda = a - E(V)$. Consider the metric $\boldsymbol{\kappa}_2$ (15.2.7). For any $N > 0$ we simply estimate $\boldsymbol{\kappa}_2$ by the Kantorovich metric ℓ_1,

$$\begin{aligned}\tfrac{1}{2}\boldsymbol{\kappa}_2(X,Y) &= \int |t|\,|F_X(t) - F_Y(t)|\mathrm{d}t \\ &\leq N\ell_1(X,Y) + N^{-\delta}[E(|X|^{2+\delta}) + E(|Y|^{2+\delta})],\end{aligned}$$

and hence the least upper bound of $\boldsymbol{\kappa}_2(X,Y)$ obtained by varying N is

$$\boldsymbol{\kappa}_2(X,Y) \leq 2(1+1/\delta)[\ell_1(X,Y)]^{\delta/(1+\delta)}(\delta\beta)^{1/(1+\delta)}, \tag{15.2.25}$$

where $\beta = E(|X|^{2+\delta}) + E(|Y|^{2+\delta})$. By the triangle inequality,

$$\begin{aligned}\ell_1(X,Y) &= \ell_1(Y_\lambda + V, Y_a) \leq \ell_1(Y_\lambda, Y_a) \\ &\leq \ell_1(V,0) + \int_0^\infty |\exp(-x/\lambda) - \exp(-x/a)|\mathrm{d}x \\ &= E|V| + |EV| \leq 2E|V|.\end{aligned} \tag{15.2.26}$$

It follows from (15.2.25) and (15.2.26) that

$$\boldsymbol{\kappa}_2(X,V) \leq 2(1+1/\delta)[2E(|V|)]^{\delta/(1+\delta)}(\delta\beta)^{1/(1+\delta)}. \tag{15.2.27}$$

Clearly, from (15.2.27) we see that if $E|V|$ is close to zero, then $\boldsymbol{\kappa}_2(X,Y)$ is small. But $\boldsymbol{\kappa}_2(X,Y) \geq 2\zeta_2(X,Y)$ [see (15.2.6)], and so from (15.2.16) it follows that if $E|V|$ is small, then the uniform distance between the distribution of $2\sum_{i=1}^n X_i/a$ and the χ^2_{2n} distribution is small.

(D) *The case where X is arbitrary: right-censoring.* Finally, suppose that $X = Y_\lambda \wedge N$, where N is a nonnegative RV independent of $Y_\lambda \stackrel{d}{=} E(\lambda)$, so that $a = E(Y_\lambda \wedge N)$. Now, for $\eta > 0$

$$\begin{aligned}\zeta_2(X,Y) &\leq \tfrac{1}{2}\boldsymbol{\kappa}_2(X,Y) = \tfrac{1}{2}\boldsymbol{\kappa}_2(Y_\lambda \wedge N, Y_a) \\ &= \int_0^\eta t|\exp(-t/a) - \exp(-t/\lambda)\overline{F}_N(t)|\mathrm{d}t \\ &\quad + \int_\eta^\infty t|\exp(-t/a) - \exp(-t/\lambda)\overline{F}_N(t)|\mathrm{d}t.\end{aligned}$$

It can easily be shown that

$$\int_0^\eta t|\exp(-t/a) - \exp(-t/\lambda)\overline{F}_N(t)|\mathrm{d}t \le |\lambda^2 - a^2| + \lambda^2 F_N(\eta)$$

and also that

$$\int_\eta^\infty t|\exp(-t/a) - \exp(-t/\lambda)\overline{F}_N(t)|\mathrm{d}t$$
$$\le a(\eta + a)\exp(-\eta/a) + \lambda(\eta + \lambda)\exp(-\eta/\lambda).$$

Hence, for any $\eta > 0$

$$2\zeta_2(X,Y) \le |\lambda^2 - a^2| + \lambda^2 F_N(\eta) + 2\gamma(\eta + \gamma)\exp(-\eta/\gamma),\ \gamma = \max(a,\lambda). \tag{15.2.28}$$

For fixed η the value of $F_N(\eta)$ is small if N is big enough. Thus (15.2.28), together with (15.2.16), gives an estimate of $\zeta_2(X,Y)$ as $N \xrightarrow{\mathrm{d}} \infty$. Finally, by $\zeta_2(\overline{W}_n, \overline{Z}_n) \le \zeta_2(X,Y)/a^2$ and

$$\rho(W_n, Z_n) = \rho(\overline{W}_n, \overline{Z}_n) \le 3M_n^{2/3}[\zeta_2(\overline{W}_n, \overline{Z}_n)]^{1/3},$$

it follows that the distribution of $2\sum_{i=1}^n X_i/a$ is uniformly close to the χ^2_{2n} distribution.

The derivation of the estimates for $\rho(W_n, Z_n)$ is just an illustrative example of how one can use the theory of probability metrics. Clearly, in this simple case one can obtain similar results by traditional methods. However, to study the stability of the characterization of multivariate distributions, the rate of convergence in the multivariate CLT, and other stochastic problems of approximation type, one should use the general relationships between probability distances, which will considerably simplify the task.

15.3 Ideal Metrics for Sums of Independent Random Variables

Let $(U, \|\cdot\|)$ be a complete separable Banach space equipped with the usual algebra of Borel sets $\mathcal{B}(U)$, and let $\mathfrak{X} := \mathfrak{X}(U)$ be the vector space of all RVs defined on a probability space $(\Omega, \mathcal{A}, \Pr)$ and taking values in U. We will choose to work with simple probability metrics on the space $\mathfrak{X}$ instead of the space $\mathcal{P}(U)$.[4] We will show that certain *convolution* metrics on $\mathfrak{X}$ may be used to provide exact rates of convergence of normalized sums to a stable limit law. They will play the role of *ideal metrics* for the approximation problems under consideration. *Traditional* metrics for the rate of convergence in the CLT are uniform-type metrics. Having

[4]See Sect. 2.5 in Chap. 2 and Sect. 3.3 in Chap. 3.

exact estimates in terms of the *ideal* metrics we will pass to the uniform estimates using the Bergström convolution method. The rates of convergence, which hold uniformly in n, will be expressed in terms of a variety of uniform metrics on $\mathfrak{X}$.

Definition 15.3.1 (Zolotarev). A p. semimetric $\mu : \mathfrak{X} \times \mathfrak{X} \to [0, \infty]$ is called an *ideal (probability) metric* of order $r \in \mathbb{R}$ if for any RVs $X_1, X_2, Z \in \mathfrak{X}$, and any nonzero constant c the following two properties are satisfied:

(i) *Regularity:* $\mu(X_1 + Z, X_2 + Z) \le \mu(X_1, X_2)$, and
(ii) *Homogeneity of order r:* $\mu(cX_1, cX_2) = |c|^r \mu(X_1, X_2)$.

When μ is a simple metric (see Sect. 3.3 in Chap. 3), i.e., its values are determined by the marginal distributions of the RVs being compared; then it is assumed in addition that the RV Z is independent of X_1 and X_2 in condition (i). All metrics μ in this section are simple.[5]

Remark 15.3.1. Zolotarev (1976a,b)[6] showed the existence of an ideal metric of a given order $r \ge 0$, and he defined the ideal metric

$$\zeta_r(X_1, X_2) := \sup\{|E(f(X_1) - f(X_2))| : \\ |f^{(m)}(x) - f^{(m)}(y)| \le \|x - y\|^{\beta}\}, \tag{15.3.1}$$

where $m = 0, 1, \dots$ and $\beta \in (0, 1]$ satisfy $m + \beta = r$, and $f^{(m)}$ denotes the mth Fréchet derivative of f for $m \ge 0$ and $f^{(0)} = f(x)$. He also obtained an upper bound for ζ_r (r integer) in terms of the *difference pseudomoment* κ_r, where for $r > 0$

$$\kappa_r(X_1, X_2) \\ := \sup\{|E(f(X_1) - f(X_2))| : |f(x) - f(y)| \le \|x\|x\|^{r-1} - y\|y\|^{r-1}\|\}$$

[see (4.4.40) and (4.4.42)]. If $U = \mathbb{R}$, $\|x\| = |x|$, then [see (4.4.43)]

$$\kappa_r(X_1, X_2) := r \int |x|^{r-1} |F_{X_1}(x) - F_{X_2}(x)| \mathrm{d}x, \quad r > 0, \tag{15.3.2}$$

where F_X denotes the DF for X.

In this section, we introduce and study two ideal metrics of a convolution type on the space $\mathfrak{X}$. These ideal metrics will be used to provide exact convergence rates for convergence to an α-stable RV in the Banach space setting. Moreover, the rates will hold with respect to a variety of uniform metrics on $\mathfrak{X}$.

Remark 15.3.2. Further, in this and the next section, for each $X_1, X_2 \in \mathfrak{X}$ we write $X_1 + X_2$ to mean the sum of independent RVs with laws $\Pr_{X_1}$ and $\Pr_{X_2}$, respectively.

---

[5]Recent publications on applications include Hein et al. (2004) and Sençimen and Pehlivan (2009).
[6]See Zolotarev (1986, Chap. 1).

For any $X \in \mathfrak{X}$, p_X denotes the density of X, if it exists. We reserve the letter Y_α (or Y) to denote a *symmetric stable RV* with parameter $\alpha \in (0, 2]$, i.e., $Y_\alpha \stackrel{d}{=} -Y_\alpha$, and for any $n = 1, 2, \ldots$, $X_1' + \cdots X_n' \stackrel{d}{=} n^{1/\alpha} Y_\alpha$, where $X_1', X_2', \ldots, X_n'$ are i.i.d. RVs with the same distribution as Y_α. If $Y_\alpha \in \mathfrak{X}(\mathbb{R})$, then we assume that Y_α has the characteristic function

$$\phi_Y(t) = \exp\{-|t|^\alpha\}, \qquad t \in \mathbb{R}.$$

For any $f : U \to \mathbb{R}$

$$\|f\|_L := \sup_{x \neq y} \frac{|f(x) - f(y)|}{\|x - y\|}$$

denotes the Lipschitz norm of f, $\|f\|_\infty$ the essential supremum of f, and when $U = \mathbb{R}^k$, $\|f\|_p$ denotes the L^p norm,

$$\|f\|_p^p := \int_{\mathbb{R}^k} |f(x)|^p \mathrm{d}x, \qquad p \geq 1.$$

Letting X, X_1, $X_2, \ldots$ denote i.i.d. RVs and Y_α denote an α-stable RV we will use ideal metrics to describe the rate of convergence,

$$\frac{X_1 + \cdots + X_n}{n^{1/\alpha}} \xrightarrow{\text{w}} Y_\alpha, \tag{15.3.3}$$

with respect to the following uniform metrics on $\mathfrak{X}$ ($\xrightarrow{\text{w}}$ stands for the weak convergence).

Total variation metrics[7]

$$\begin{aligned} \sigma(X_1, X_2) := &\sup_{A \in \mathcal{B}(U)} |\Pr\{X_1 \in A\} - \Pr\{X_2 \in A\}|, \\ := &\sup\{|Ef(X_1) - Ef(X_2)| : f : U \to \mathbb{R} \text{ is measurable and} \\ &\text{for any } x, y \in B, |f(x) - f(y)| \leq \mathbb{I}(x, y) \text{ where } \mathbb{I}(x, y) = 1 \\ &\text{if } x \neq y \text{ and } 0 \text{ otherwise}\}, \quad X_1, X_2 \in \mathfrak{X}(U), \end{aligned} \tag{15.3.4}$$

and

$$\begin{aligned} \mathbf{Var}(X_1, X_2) := &\sup\{|Ef(X_1) - Ef(X_2)| : f : U \to \mathbb{R} \text{ is measurable and} \\ &\|f\|_\infty \leq 1\} \\ = &\ 2\sigma(X_1, X_2), \quad X_1, X_2 \in \mathfrak{X}(U). \end{aligned} \tag{15.3.5}$$

---

[7] See Lemma 3.3.1, (3.4.18), and (3.3.13) in Chap. 3.

In $\mathfrak{X}(\mathbb{R}^n)$, we have $\mathbf{Var}(X_1, X_2) := \int |\mathrm{d}(F_{X_1} - F_{X_2})|$.

Uniform metric between densities: [p_X denotes the density for $X \in \mathfrak{X}(\mathbb{R}^k)$]

$$\ell(X_1, X_2) := \operatorname*{ess\,sup}_x |p_{X_1}(x) - p_{X_2}(x)|. \tag{15.3.6}$$

Uniform metric between characteristic functions:

$$\chi(X_1, X_2) := \sup_{t \in \mathbb{R}} |\phi_{X_1}(t) - \phi_{X_2}(t)| \quad X_1, X_2 \in \mathfrak{X}(\mathbb{R}), \tag{15.3.7}$$

where ϕ_X denotes the characteristic function of X. The metric χ is topologically weaker than **Var**, which is itself topologically weaker than ℓ by Schené's theorem.[8]

We will use the following simple metrics on $\mathfrak{X}(\mathbb{R})$.

Kolmogorov metric:

$$\rho(X_1, X_2) := \sup_{x \in \mathbb{R}} |F_{X_1}(x) - F_{X_2}(x)|. \tag{15.3.8}$$

Weighted χ-metric:

$$\chi_r(X_1, X_2) := \sup_{t \in \mathbb{R}} |t|^{-r} |\phi_{X_1}(t) - \phi_{X_2}(t)|. \tag{15.3.9}$$

L^p-version of ζ_m:

$$\zeta_{m,p}(X_1, X_2) := \sup\{Ef(X_1) - f(X_2)| : \|f^{(m+1)}\|_q \leq 1\},$$
$$1/p + 1/q = 1, \quad m = 0, 1, 2, \ldots \tag{15.3.10}$$

If $\zeta_{m,p}(X_1, X_2) < \infty$, then[9]

$$\zeta_{m,p}(X_1, X_2) = \left\| \int_{-\infty}^{x} \frac{(x-t)^m}{m!} \mathrm{d}(F_{X_1}(t) - F_{X_2}(t)) \right\|_p.$$

Kantorovich ℓ_p-metric:

$$\ell_p^p(X_1, X_2) := \sup\left\{ \int f \mathrm{d}F_{X_1} + \int g \mathrm{d}F_{X_2} : \|f\|_\infty + \|f\|_L \leq \infty, \right.$$
$$\left. \|g\|_\infty + \|g\|_L < \infty, \ f(x) + g(y) \leq \|x - y\|^p, \ \forall x, y \in \mathbb{R} \right\}, \ p \geq 1 \tag{15.3.11}$$

[see (3.3.11) and (3.4.18)].

---

[8] See Billingsley (1999).

[9] See Kalashnikov and Rachev (1988, Chap. 3), Sect. 8.3, and further Lemma 18.2.1.

Now we define the ideal metrics of order $r-1$ and r, respectively. Let $\theta \in \mathfrak{X}(\mathbb{R}^k)$ and $\theta \stackrel{d}{=} -\theta$, and define for every $r > 0$ the *convolution* (probability) metric

$$\boldsymbol{\mu}_{\theta,r}(X_1, X_2) := \sup_{h\in\mathbb{R}} |h|^r \ell(X_1 + h\theta, X_2 + h\theta), \quad X_1, X_2, \in \mathfrak{X}(\mathbb{R}^k). \tag{15.3.12}$$

Thus, each RV θ generates a metric $\boldsymbol{\mu}_{\theta,r}$, $r > 0$. When $\theta \in \mathfrak{X}(U)$, we will also consider convolution metrics of the form

$$\boldsymbol{\nu}_{\theta,r}(X_1, X_2) := \sup_{h\in\mathbb{R}} |h|^r \,\mathbf{Var}(X_1 + h\theta, X_2 + h\theta), \quad X_1, X_2 \in \mathfrak{X}(U). \tag{15.3.13}$$

Lemmas 15.3.1 and 15.3.2 below show that $\boldsymbol{\mu}_{\theta,r}$ and $\boldsymbol{\nu}_{\theta,r}$ are ideal of order $r-1$ and r, respectively. In general, $\boldsymbol{\mu}_{\theta,r}$ and $\boldsymbol{\nu}_{\theta,r}$ are actually only semimetrics, but this distinction is not important in what follows and so we omit it (see Sects. 2.4 and 2.5 in Chap. 2). When θ is a symmetric α-stable RV, in place of $\boldsymbol{\mu}_{\theta,r}$ and $\boldsymbol{\nu}_{\theta,r}$ we will write $\boldsymbol{\mu}_{\alpha,r}$ and $\boldsymbol{\nu}_{\alpha,r}$, or simply $\boldsymbol{\mu}_r$ when it is understood.

The remainder of this section describes the special properties of the ideal convolution (or smoothing) metrics $\boldsymbol{\mu}_{\theta,r}$ and $\boldsymbol{\nu}_{\theta,r}$. We first verify ideality.

Lemma 15.3.1. *For all $\theta \in \mathfrak{X}$ and $r > 0$, $\boldsymbol{\mu}_{\theta,r}$ is an ideal metric of order $r-1$.*

Proof. If Z does not depend upon X_1 and X_2, then $\ell(X_1+Z, X_2+Z) \le \ell(X_1, X_2)$, and hence $\boldsymbol{\mu}_{\theta,r}(X_1 + Z, X_2 + Z) \le \boldsymbol{\mu}_{\theta,r}(X_1, X_2)$. Additionally, for any $c \ne 0$

$$\begin{aligned}\boldsymbol{\mu}_{\theta,r}(cX_1, cX_2) &= \sup_{h\in\mathbb{R}} |h|^r \ell(cX_1 + h\theta, cX_2 + h\theta)\\ &= \sup_{h\in\mathbb{R}} |ch|^r \ell(cX_1 + ch\theta, cX_2 + ch\theta) = |c|^{r-1}\boldsymbol{\mu}_{\theta,r}(X_1, X_2).\end{aligned}$$

□

The proof of the next lemma is analogous to the previous one.

Lemma 15.3.2. *For all $\theta \in \mathfrak{X}$ and $r > 0$, $\boldsymbol{\nu}_{\theta,r}$ is an ideal metric of order r.*

We now show that both $\boldsymbol{\mu}_{\theta,r}$ and $\boldsymbol{\nu}_{\theta,r}$ are bounded from above by the difference pseudomoment whenever θ has a density that is smooth enough.

Lemma 15.3.3. *Let $k \in \mathbb{N}^+ := \{0, 1, 2, \ldots\}$, and suppose that $X, Y \in \mathfrak{X}(\mathbb{R})$ satisfy $EX^j = EY^j$, $j = 1, \ldots, k-2$. Then for every $\theta \in \mathfrak{X}(\mathbb{R})$ with a density g that is $k-1$ times differentiable*

$$\boldsymbol{\mu}_{\theta,k}(X_1, X_2) \le \frac{\|g^{(k-1)}\|_\infty}{(k-1)!}\kappa_{k-1}(X_1, X_2). \tag{15.3.14}$$

Proof. In view of the inequality[10]

[10] See Zolotarev (1986, Chap. 3) and Kalashnikov and Rachev (1988, Theorem 10.1.1).

$$\zeta_{k-1}(X_1, X_2) \leq \frac{1}{(k-1)!}\kappa_{k-1}(X_1, X_2), \tag{15.3.15}$$

it suffices to show that

$$\mu_{\theta,k}(X_1, X_2) \leq \zeta_{k-1}(X_1, X_2). \tag{15.3.16}$$

However, with $H(t) = F_{X_1}(t) - F_{X_2}(t)$ we have

$$\begin{aligned}
\mu_{\theta,k}(X_1, X_2) &= \sup_{h\in\mathbb{R}} |h|^k \sup_{x\in\mathbb{R}} \frac{1}{|h|} \left| \int g\left(\frac{x-y}{h}\right) \mathrm{d}H(y) \right| \\
&= \sup_{h\in\mathbb{R}} |h|^{k-1} \sup_{x\in\mathbb{R}} \left| \int H(y) g^{(1)}\left(\frac{x-y}{h}\right) \frac{1}{h} \mathrm{d}y \right| \\
&= \sup_{h\in\mathbb{R}} |h|^{k-2} \sup_{x\in\mathbb{R}} \left| \int g^{(1)}\left(\frac{x-y}{h}\right) \frac{1}{h} \mathrm{d}H^{(-1)}(y) \right| \\
&\vdots \\
&= \sup_{h\in\mathbb{R}} |h| \sup_{x\in\mathbb{R}} \left| \int g^{(k-1)}\left(\frac{x-y}{h}\right) \frac{1}{h} H^{(-k+2)}(y) \mathrm{d}y \right|,
\end{aligned} \tag{15.3.17}$$

where

$$F_X^{-k}(x) := \int_{-\infty}^{x} \frac{(x-t)^k}{k!} \mathrm{d}F_X(t). \tag{15.3.18}$$

Therefore, by (15.3.10) and $\zeta_{k-1} = \zeta_{k-2,1}$, we have

$$\mu_{\theta,k}(X_1, X_2) \leq \|g^{(k-1)}\|_\infty \int |H^{(2-k)}(y)| \mathrm{d}y = \|g^{(k-1)}\|_\infty \zeta_{k-1}(X_1, X_2). \quad \square$$

Similarly to Lemma 15.3.3, one can prove a slightly better estimate.

Lemma 15.3.4. *For every $\theta \in \mathfrak{X}(\mathbb{R})$ with a density g that is m times differentiable and for all $X_1, X_2 \in \mathfrak{X}(\mathbb{R})$*

$$\mu_{\theta,r}(X_1, X_2) \leq C(m, p, g)\zeta_{m-1,p}(X_1, X_2), \tag{15.3.19}$$

where $r = m + 1/p$, $m \in \mathbb{N}^+$, and

$$C(m, p, g) := \|g^{(m)}\|_q, \qquad 1/p + 1/q = 1. \tag{15.3.20}$$

Proof. For any $r > 0$ and X_1, X_2, $H(t) = F_{X_1}(t) - F_{X_2}(t)$ we have, using integration by parts [see (15.3.17)] and Hölder's inequality

$$\mu_{\theta,r}(X_1, X_2) = \sup_{h>0} h^r \sup_{x\in\mathbb{R}} |p_{X_1+h\theta}(x) - p_{X_2+h\theta}(x)|$$

$$= \sup_{h>0} h^{r-m-1} \sup_{x\in\mathbb{R}} \left| \int g^{(m)} \left(\frac{x-y}{h} \right) H^{(1-m)}(y) \mathrm{d}y \right|$$

$$\leq \sup_{h>0} h^{r-m-1} \sup_{x\in\mathbb{R}} \left[\int \left| g^{(m)} \left(\frac{x-y}{h} \right) \right|^q \mathrm{d}y \right]^{1/q} \|H^{(1-m)}\|_p$$

$$= C(m, p, g) \|H^{(1-m)}\|_p.$$

By Kalashnikov and Rachev (1988, Theorem 10.2.1), $\zeta_{m-1,p}(X_1, X_2) < \infty$ implies $\zeta_{m-1,p}(X_1, X_2) = \|H^{(1-m)}\|_p$, completing the proof of the lemma. □

Lemma 15.3.5. *Under the hypotheses of Lemma 15.3.4, we have*

$$\nu_{\theta,r}(X_1, X_2) \leq C(r, g) \zeta_r(X_1, X_2), \tag{15.3.21}$$

where $C(r, g)$ is a finite constant, $r \in \mathbb{N}^+$.

The proof is similar to the proof of Lemma 15.3.4 and left to the reader.

Lemma 15.3.6. [11] *Let $m \in \mathbb{N}^+$, and suppose $E(X_1^j - X_2^j) = 0$, $j = 0, 1, \ldots, m$. Then, for $p \in [1, \infty)$,*

$$\zeta_{m,p}(X_1, X_2) \leq \begin{cases} \kappa_1^{1/p}(X_1, X_2), & \text{if } m = 0, \\ \dfrac{\Gamma(1 + 1/p)}{\Gamma(r)} \kappa_r(X_1, X_2), & \text{if } m = 1, 2, \ldots, r = m + 1/p. \end{cases} \tag{15.3.22}$$

Also, for $r = m + 1/p$,

$$\zeta_{m,p}(X_1, X_2) \leq \zeta_r(X_1, X_2).$$

Lemmas 15.3.4–15.3.6 describe the conditions under which $\zeta_{\theta,r}$ (resp. $\nu_{\theta,r}$) is finite. Thus, by (15.3.19) and (15.3.22), we have that for $r > 1$

$$\begin{cases} E(X_1^j - X_2^j) = 0, \ j = 0, 1, \ldots, m-1, \\ r := m + 1/p, \\ \kappa_{r-1}(X_1, X_2) < \infty, \end{cases} \Rightarrow \mu_{\theta,r}(X_1, X_2) < \infty, \tag{15.3.23}$$

for any θ with density g such that $\|g^{(m-1)}\|_q \leq \infty$, $1/p + 1/q = 1$. In particular, if θ is α-stable, then

$$\begin{cases} \displaystyle\int x^j \mathrm{d}(F_{X_1} - F_{X_2})(x) = 0, \ j = 0, 1, \ldots, m-1, \\ r := m + 1/p, \\ \kappa_{r-1}(X_1, X_2) < \infty, \end{cases} \Rightarrow \mu_{\alpha,r}(X_1, X_2) < \infty. \tag{15.3.24}$$

[11] See Kalashnikov and Rachev (1988, Sect. 3, Theorem 10.1).

Similarly,

$$\begin{cases} \int x^j \mathrm{d}(F_{X_1} - F_{X_2})(x) = 0, \ j = 0, 1, \ldots, r-1, \\ r \in \mathbb{N}^+, \\ \boldsymbol{\kappa}_r(X_1, X_2) < \infty, \end{cases} \Rightarrow \boldsymbol{\nu}_{\alpha,r}(X_1, X_2) < \infty. \tag{15.3.25}$$

We conclude our discussion of the ideal metrics $\boldsymbol{\mu}_{\alpha,r}$ and $\boldsymbol{\nu}_{\alpha,r}$ by showing that they satisfy the same weak convergence properties as do the Kantorovich distance ℓ_p and the pseudomoments $\boldsymbol{\kappa}_r$.

Theorem 15.3.1. *Let $k \in \mathbb{N}^+$, $0 < \alpha \le 2$, and X_n, $U \in \mathfrak{X}(\mathbb{R})$ with $EX_n^j = EU^j$, $j = 1, \ldots, k-2$, and $E|X_n|^{k-1} + E|U|^{k-1} < \infty$. If k is odd, then the following expressions are equivalent as $n \to \infty$:*

(i) $\boldsymbol{\mu}_{\alpha,k}(X_n, U) \to 0$.
(ii) *(a)* $X_n \xrightarrow{\text{w}} U$ *and (b)* $E|X_n|^{k-1} \to E|U|^{k-1}$.
(iii) $\ell_{k-1}(X_n, U) \to 0$.
(iv) $\boldsymbol{\kappa}_{k-1}(X_n, U) \to 0$.
(v) $\boldsymbol{\nu}_{\alpha,k-1}(X_n, U) \to 0$.

Proof. We note that (ii) $\Longleftrightarrow$ (iii) follows immediately from Theorem 8.3.1 with $c(x,y) = |x-y|^{k-1}$ or from (8.3.21) to (8.3.24) and $\ell_{k-1} = \widehat{\mathcal{L}}_{k-1}$. Also, (ii) $\Longleftrightarrow$ (iv) follows from the three relations[12]

$$\ell_1(X,Y) = \boldsymbol{\kappa}_1(X,Y) = \int_{\mathbb{R}} |F_X(x) - F_Y(x)| \mathrm{d}x$$

$$\boldsymbol{\kappa}_r(X,Y) = \boldsymbol{\kappa}_1(X^{\uparrow r}, Y^{\uparrow r})$$

for any $r > 0$ and $X^{\uparrow r} = |X|^r \operatorname{sgn} X$, and[13]

$$\ell_1(X_n^{\uparrow r}, U^{\uparrow r}) \to 0 \iff x_n^{\uparrow r} \xrightarrow{\text{w}} U^{\uparrow r} \text{ and } E|X_n^{\uparrow r}| \to E|U^{\uparrow r}|.$$

Finally, (iv) $\Rightarrow$ (i) by (15.3.24) and (iv) $\Rightarrow$ (v) by (15.3.25).

Thus the only new results here are the implications (i) $\Rightarrow$ (ii) and (v) $\Rightarrow$ (ii).

Now (i) $\Rightarrow$ (ii) (a) follows easily from Fourier transform arguments since the Fourier transform of g never vanishes. Similarly, if (v) holds, then $X_n + Y_\alpha \xrightarrow{\text{w}} U + Y_\alpha$, and thus (ii) (a) follows. To prove (i) $\Rightarrow$ (ii) (b), we need the following estimate for $\mu_{\alpha,k}(X,U)$.

---

[12]See Corollary 5.5.1 and Theorem 6.2.1.

[13]See Theorem 6.4.1 or Theorem 8.3.1 with $c(x,y) = |x-y|$.

Claim 3. Let $0 < \alpha \leq 2$, and consider the associated metric $\boldsymbol{\mu}_r := \boldsymbol{\mu}_{r,\alpha}$. For all k there is a constant $\beta := \beta(\alpha, k) < \infty$ such that for all $X, U \in \mathfrak{X}(\mathbb{R})$

$$\boldsymbol{\mu}_k(X,U) \geq \beta \left| \int_{\mathbb{R}} F_X^{(2-k)}(z) - F_U^{(2-k)}(z)\mathrm{d}z \right|. \tag{15.3.26}$$

Here $F^{(2-k)}$ is as in (15.3.18).

Proof of claim. Integration by parts yields

$$\begin{aligned}
\boldsymbol{\mu}_k(X,U) &= \sup_{h\in\mathbb{R}} |h|^k \sup_{x\in\mathbb{R}} |p_{X+hY}(x) - p_{U+hY}(x)|, \quad (Y := Y_\alpha) \\
&= \sup_{h\in\mathbb{R}} |h|^k \sup_{x\in\mathbb{R}} \left| \int p_{hY}(z)\mathrm{d}H(x-z) \right|, \quad (H := F_X - F_U) \\
&= \sup_{h\in\mathbb{R}} |h|^k \sup_{x\in\mathbb{R}} \left| \int H^{(2-k)}(x-z) p_{hY}^{(k-1)}(z)\mathrm{d}z \right|.
\end{aligned} \tag{15.3.27}$$

Now, $2\pi p_{hY}(z) = \int \exp(-itz)\exp(-|ht|^\alpha)\mathrm{d}t$, and differentiating $p_{hY}(z)$ $k-1$ times gives (setting $\widetilde{t} = th$)

$$\begin{aligned}
2\pi |h^k p_{hY}^{(k-1)}(z)| &= \left| h^k \int (\mathrm{i}t)^{k-1} \exp(-\mathrm{i}tz - |ht|^\alpha)\mathrm{d}t \right| \\
&= \left| h^k \int \left(\mathrm{i}\frac{\widetilde{t}}{h}\right)^{k-1} \exp(-\mathrm{i}\widetilde{t}z/h - |\widetilde{t}|^\alpha)\mathrm{d}\left(\frac{\widetilde{t}}{h}\right) \right| \\
&= \left| \int (\mathrm{i}\widetilde{t})^{k-1} \exp(-\mathrm{i}\widetilde{t}z/h - |\widetilde{t}|^\alpha)\mathrm{d}\widetilde{t} \right|.
\end{aligned}$$

Since

$$\beta := \beta(\alpha,k) := \frac{1}{2\pi} \int |t|^{k-1} \exp(-|t|^\alpha)\mathrm{d}t < \infty,$$

we obtain

$$\lim_{h\to\infty} |h^k p_{hY}^{(k-1)}(z)| = \frac{1}{2\pi} \left| \int \lim_{h\to\infty} (\mathrm{i}t)^{k-1} \exp(\mathrm{i}tz/h - |t|^\alpha)\mathrm{d}t \right| = \beta.$$

Now we multiply both sides of (15.3.27) by β^{-1}. Since β and $\boldsymbol{\zeta}_{k-1}(X,U)$ are both finite,

$$\beta^{-1}\boldsymbol{\mu}_k(X,U) \geq \beta^{-1} \sup_{x\in\mathbb{R}} \left| \int H^{(2-k)}(x-z) \lim_{h\to\infty} h^k p_{hY}^{(k-1)}(z)\mathrm{d}z \right|$$

$$
\begin{aligned}
&= \sup_{x\in\mathbb{R}} \left| \int H^{(2-k)}(x-z)\mathrm{d}z \right| \\
&= \left| H^{(2-k)}(z)\mathrm{d}z \right|,
\end{aligned}
$$

which proves the claim.

Now, using equality of the first $k-2$ moments and applying (15.3.26) to X_n and U yields

$$
\begin{aligned}
\beta^{-1}\mu_k(X_n,U) &\geq \left| \int_{-\infty}^{\infty} \frac{(z-t)^{k-2}}{(k-2)!} H_n(\mathrm{d}t)\mathrm{d}z \right|, \quad (H_n := F_{X_n} - F_U) \\
&= \left| \int_{-\infty}^{0} (\bullet)\mathrm{d}t + \int_{0}^{\infty} (\bullet)\mathrm{d}t \right| := |I_1 + I_2|. \qquad (15.3.28)
\end{aligned}
$$

To estimate I_1 and I_2, we first note that, since

$$
\int_{\mathbb{R}} (z-t)^{k-2} H_n(\mathrm{d}t) = E(z-X_n)^{k-2} - E(z-U)^{k-2} = 0,
$$

we obtain

$$
\begin{aligned}
\int_{-\infty}^{z} \frac{(z-t)^{k-2}}{(k-2)!} H_n(\mathrm{d}t) &= -\int_{z}^{\infty} \frac{(z-t)^{k-2}}{(k-2)!} H_n(\mathrm{d}t) \\
&= (-1)^{k-1} \int_{z}^{\infty} \frac{(t-z)^{k-2}}{(k-2)!} H_n(\mathrm{d}t). \qquad (15.3.29)
\end{aligned}
$$

Thus by (15.3.29) and Fubini's theorem, we obtain

$$
\begin{aligned}
I_2 &= (-1)^{k-1} \int_{0}^{\infty} \int_{z}^{\infty} \frac{(t-z)^{k-2}}{(k-2)!} H_n(\mathrm{d}t)\mathrm{d}z \\
&= (-1)^{k-1} \int_{0}^{\infty} \int_{0}^{t} \frac{(t-z)^{k-2}}{(k-2)!} \mathrm{d}z H_n(\mathrm{d}t) = \int_{0}^{\infty} \frac{(-t)^{k-1}}{(k-1)!} H_n(\mathrm{d}t). \qquad (15.3.30)
\end{aligned}
$$

Another application of Fubini's theorem gives

$$
I_1 = \int_{-\infty}^{0} \int_{t}^{0} \frac{(z-t)^{k-2}}{(k-2)!} \mathrm{d}z H_n(\mathrm{d}t) = \int_{-\infty}^{0} \frac{(-1)^{k-1}}{(k-1)!} H_n(\mathrm{d}t). \qquad (15.3.31)
$$

Combining (15.3.29)–(15.3.31) gives

$$
\beta^{-1}\boldsymbol{\mu}_k(X_n,U) \geq \left| \int \frac{(-t)^{k-1}}{(k-1)!} H_n(\mathrm{d}t) \right| = \frac{1}{(k-1)!} |E(X_n^{k-1} - U^{k-1})|,
$$

which gives the desired implication (i) $\Rightarrow$ (ii) (b).

To prove (v) $\Rightarrow$ (ii) (b), we integrate by parts to obtain

$$\begin{aligned}
\boldsymbol{\nu}_k(X_n, U) &\geq \int_{\mathbb{R}} |p_{X_n+Y}(x) - p_{U+Y}(x)|\mathrm{d}x \\
&= \int_{\mathbb{R}} \left| \int p_Y^{(k)}(x-z) \int_{-\infty}^{z} \frac{(z-t)^{k-1}}{(k-1)!} \mathrm{d}H_n(t)\mathrm{d}z \right| \mathrm{d}x \\
&\geq \left| \iint p_Y^{(k)}(x-z)\mathrm{d}x \int_{-\infty}^{z} \frac{(z-t)^{k-1}}{(k-1)!} \mathrm{d}H_n(t)\mathrm{d}z \right| \\
&= \left| \iint_{-\infty}^{z} \frac{(z-t)^{k-1}}{(k-1)!} \mathrm{d}H_n(t)\mathrm{d}z \right| \left| \int p_Y^{(k)}(x)\mathrm{d}x \right|.
\end{aligned}$$

By (15.3.28) to (15.3.31), we obtain

$$\boldsymbol{\nu}_k(X_n, U) \geq \left| \int p_Y^{(k)}(x)\mathrm{d}x \right| |E(X_n^k - U^k)|,$$

showing (v) $\Rightarrow$ (ii) (b) and completing Theorem 15.3.1. □

15.4 Rates of Convergence in the CLT in Terms of Metrics with Uniform Structure

First, we develop rates of convergence with respect to the **Var**-metric defined in (15.3.5). We suppose that $X, X_1, X_2, \dots$ denotes a sequence of i.i.d. RVs in $\mathfrak{X}(U)$, where U is a separable Banach space. $Y \in \mathfrak{X}(U)$ denotes a symmetric α-stable RV. The ideal convolution metric $\boldsymbol{\nu}_r := \boldsymbol{\nu}_{\alpha,r}$ [see (15.3.13) with $\theta = Y$] will play a central role. Our main theorem is as follows.

Theorem 15.4.1. *Let Y be an α-stable RV. Let $r = s + 1/p > \alpha$ for some integer s and $p \in [1, \infty)$, $a = 1/2^{r/\alpha} A$, and $A := 2(2^{(r/\alpha)-1} + 3^{r/\alpha})$. If $X \in \mathfrak{X}(U)$ satisfies*

$$\tau_0 := \tau_0(X, Y) := \max(\mathbf{Var}(X, Y), \boldsymbol{\nu}_{\alpha,r}(X, Y)) \leq a, \tag{15.4.1}$$

then for any $n \geq 1$

$$\mathbf{Var}\left(\frac{X_1 + \cdots + X_n}{n^{1/\alpha}}, Y\right) \leq A(a)\tau_0 n^{1-r/\alpha} \leq 2^{-r/\alpha} n^{1-r/\alpha}. \tag{15.4.2}$$

Remark 15.4.1. A result of this type was proved by Senatov (1980) for the case $U = \mathbb{R}^k$, $s = 3$, and $\alpha = 2$ via the $\boldsymbol{\zeta}_r$ metric (15.3.1). We will follow Senatov's method with some refinements.

Before proving Theorem 15.4.1, we need a few auxiliary results.

Lemma 15.4.1. *For any* $X_1, X_2 \in \mathfrak{X}(U)$ *and* $\sigma > 0$

$$\mathbf{Var}(X_1 + \sigma Y, X_2 + \sigma Y) \le \sigma^{-r} \boldsymbol{\nu}_r(X_1, X_2). \tag{15.4.3}$$

Proof. Since Y and $(-Y)$ have the same distribution,

$$\boldsymbol{\nu}_r(X_1, X_2) = \sup_{h>0} h^r \, \mathbf{Var}(X_1 + hY, X_2 + hY),$$

and thus

$$\begin{aligned}\mathbf{Var}(X_1 + hY, X_2 + hY) &\le h^{-r} \sup_{h>0} h^r \, \mathbf{Var}(X_1 + hY, X_2 + hY)\\ &= h^{-r} \boldsymbol{\nu}_r(X_1, X_2).\end{aligned}$$

□

Lemma 15.4.2. *For any* $X_1, X_2, U, V \in \mathfrak{X}(U)$ *the following inequality holds:*

$$\mathbf{Var}(X_1 + U, X_2 + U) \le \mathbf{Var}(X_1, X_2)\, \mathbf{Var}(U, V) + \mathbf{Var}(X_1 + V, X_2 + V).$$

Proof. By the definition in (15.3.5) and the triangle inequality,

$$\begin{aligned}\mathbf{Var}(X_1 + U, X_2 + U) &= \sup\{|Ef(X_1 + U) - Ef(X_2 + U)| : \|f\|_\infty \le 1\}\\ &= \sup\left\{\left|\int f(u)(\Pr_{X_1+U} - \Pr_{X_2+U})(du)\right| : \|f\|_\infty \le 1\right\}\\ &\le \sup\left\{\left|\int \overline{f}(x)(\Pr_{X_1} - \Pr_{X_2})(dx)\right| : \|f\|_\infty \le 1\right\}\\ &\quad + \mathbf{Var}(X_1 + V, X_2 + V),\end{aligned}$$

where

$$\overline{f}(x) := \int f(u)(\Pr_U - \Pr_V)(du - x) = \int f(u + x)(\Pr_U - \Pr_V)(du),$$

in which $\Pr_X$ denotes the law of the U-valued RV X. Since $\|f\|_\infty \le 1$,

$$\begin{aligned}\|\overline{f}\|_\infty &= \sup_{x \in U}\left|\int f(u + x)(\Pr_U - \Pr_V)(du)\right|\\ &\le \mathbf{Var}(U, V), \quad \text{by (15.3.5)}\end{aligned}$$

and thus

$$\sup\left\{\left|\int_U \overline{f}(x)(\Pr_{X_1} - \Pr_{X_2})(dx)\right| : \|f\|_\infty \le 1\right\}$$

is bounded by

$$\leq \sup\left\{\left|\int_U g(x)(\Pr_{X_1} - \Pr_{X_2}(dx)\right| : \|g\|_\infty \leq \mathbf{Var}(U,V)\right\}$$
$$= \mathbf{Var}(X_1, X_2)\,\mathbf{Var}(U,V). \qquad \square$$

We now proceed to the proof of Theorem 15.4.1. Throughout the proof, $Y_1, Y_2, \ldots$ denote i.i.d. copies of Y.

Proof. We proceed by induction; for $n = 1$ the assertion of the theorem is trivial. For $n = 2$ the assertion follows from the inequality

$$\mathbf{Var}\left(\frac{X_1 + X_2}{2^{1/\alpha}}, Y\right) = \mathbf{Var}\left(\frac{X_1 + X_2}{2^{1/\alpha}}, \frac{Y_1 + Y_2}{2^{1/\alpha}}\right) = \mathbf{Var}(X_1 + X_2, Y_1 + Y_2)$$
$$\leq 2\,\mathbf{Var}(X_1, Y_2) \leq A(a)\tau_0 2^{1-r/\alpha}$$

since $A(a) \geq 2^{r/\alpha}$. A similar calculation holds for $n = 3$. Suppose now that the estimate

$$\mathbf{Var}\left(\frac{X_1 + \cdots + X_j}{j^{1/\alpha}}, Y\right) \leq A(a)\tau_0 j^{1-r/\alpha} \tag{15.4.4}$$

holds for all $j < n$. To complete the induction, we only need to show that (15.4.4) holds for $j = n$.

Thus assuming (15.4.4), we have by (15.4.1)

$$\mathbf{Var}\left(\frac{X_1 + \cdots + X_j}{j^{1/\alpha}}, Y\right) \leq A(a)a = 2^{-r/\alpha}. \tag{15.4.5}$$

For any integer $n \geq 4$ and $m = [n/2]$, where $[\cdot]$ denotes integer part, the triangle inequality gives

$$V := \mathbf{Var}\left(\frac{X_1 + \cdots + X_n}{n^{1/\alpha}}, Y\right) = \mathbf{Var}\left(\frac{X_1 + \cdots + X_n}{n^{1/\alpha}}, \frac{Y_1 + \cdots + Y_n}{n^{1/\alpha}}\right)$$
$$\leq \mathbf{Var}\left(\frac{X_1 + \cdots + X_m}{n^{1/\alpha}} + \frac{X_{m+1} + \cdots + X_n}{n^{1/\alpha}},\right.$$
$$\left.\frac{Y_1 + \cdots + Y_m}{n^{1/\alpha}} + \frac{X_{m+1} + \cdots + X_n}{n^{1/\alpha}}\right)$$
$$+ \mathbf{Var}\left(\frac{Y_1 + \cdots + Y_m}{n^{1/\alpha}} + \frac{X_{m+1} + \cdots + X_n}{n^{1/\alpha}},\right.$$
$$\left.\frac{Y_1 + \cdots + Y_m}{n^{1/\alpha}} + \frac{Y_{m+1} + \cdots + Y_n}{n^{1/\alpha}}\right).$$

Hence, by Lemma 15.4.2,

$$V \leq I_1 + I_2 + I_3, \tag{15.4.6}$$

where

$$I_1 := \mathbf{Var}\left(\frac{X_1 + \cdots + X_m}{n^{1/\alpha}}, \frac{Y_1 + \cdots + Y_m}{n^{1/\alpha}}\right)$$
$$\mathbf{Var}\left(\frac{X_{m+1} + \cdots + X_n}{n^{1/\alpha}}, \frac{Y_{m+1} + \cdots + Y_n}{n^{1/\alpha}}\right),$$

$$I_2 := \mathbf{Var}\left(\frac{X_1 + \cdots + X_m}{n^{1/\alpha}} + \frac{Y_{m+1} + \cdots + Y_n}{n^{1/\alpha}},\right.$$
$$\left.\frac{Y_1 + \cdots + Y_m}{n^{1/\alpha}} + \frac{Y_{m+1} + \cdots + Y_n}{n^{1/\alpha}}\right),$$

and

$$I_3 := \mathbf{Var}\left(\frac{Y_1 + \cdots + Y_m}{n^{1/\alpha}} + \frac{X_{m+1} + \cdots + X_n}{n^{1/\alpha}},\right.$$
$$\left.\frac{Y_1 + \cdots + Y_m}{n^{1/\alpha}} + \frac{Y_{m+1} + \cdots + Y_n}{n^{1/\alpha}}\right).$$

We first estimate I_1. By (15.4.5),

$$I_1 \leq 2^{-r/\alpha} A(a)\tau_0 (n-m)^{1-r/\alpha} \leq \tfrac{1}{2} A(a)\tau_0 n^{1-r/\alpha}. \tag{15.4.7}$$

To estimate I_2 and I_3, we will use Lemma 15.4.1 and the relation

$$\frac{Y_1 + \cdots + Y_n}{n^{1/\alpha}} \stackrel{\mathrm{d}}{=} Y_1. \tag{15.4.8}$$

Thus, by (15.4.8), Lemma 15.4.1, and the fact that $\boldsymbol{v}_r$ is ideal of order r, we deduce

$$\begin{aligned}
I_2 &= \mathbf{Var}\left(\frac{X_1 + \cdots + X_n}{n^{1/\alpha}} + \left(\frac{n-m}{n}\right)^{1/\alpha} Y, \frac{Y_1 + \cdots + Y_m}{n^{1/\alpha}} + \left(\frac{n-m}{n}\right)^{1/\alpha} Y\right) \\
&\leq \left(\frac{n-m}{n}\right)^{-r/\alpha} \boldsymbol{v}_r \left(\frac{X_1 + \cdots + X_m}{n^{1/\alpha}}, \frac{Y_1 + \cdots + Y_m}{n^{1/\alpha}}\right) \\
&\leq 2^{r/\alpha} m \boldsymbol{v}_r \left(\frac{X_1}{n^{1/\alpha}}, \frac{Y_1}{n^{1/\alpha}}\right) \leq 2^{(r/\alpha)-1} n^{1-r/\alpha} \boldsymbol{v}_r(X_1, Y_1).
\end{aligned} \tag{15.4.9}$$

Analogously, we estimate I_3 by

$$I_3 = \mathbf{Var}\left(\frac{X_1 + \cdots + X_{n-m}}{n^{1/\alpha}} + \left(\frac{m}{n}\right)^{1/\alpha} Y, \frac{Y_1 + \cdots + Y_{n-m}}{n^{1/\alpha}} + \left(\frac{m}{n}\right)^{1/\alpha} Y\right)$$
$$\leq \left(\frac{m}{n}\right)^{-r/\alpha} \boldsymbol{\nu}_r \left(\frac{X_1 + \cdots + X_{n-m}}{n^{1/\alpha}}, \frac{Y_1 + \cdots + Y_{n-m}}{n^{1/\alpha}}\right)$$
$$\leq 3^{r/\alpha} n^{1-r/\alpha} \boldsymbol{\nu}_r(X_1, Y_1). \tag{15.4.10}$$

Taking (15.4.6), (15.4.7), (15.4.9), and (15.4.10) into account, we obtain

$$V \leq \left(\tfrac{1}{2}A(a) + 2^{r/\alpha - 1} + 3^{r/\alpha}\right) \tau_0 n^{1-r/\alpha} \leq A(a)\tau_0 n^{1-r/\alpha}$$

since $A(a)/2 = 2^{(r/\alpha)-1} + 3^{r/\alpha}$. □

Further, we develop rates of convergence in (15.3.3) with respect to the $\boldsymbol{\chi}$ metric (15.3.7). Our purpose here is to show that the methods of proof for Theorem 15.4.1 can be easily extended to deduce analogous results with respect to $\boldsymbol{\chi}$. The metric $\boldsymbol{\chi}_r$ (15.3.9) will play a role analogous to that played by $\boldsymbol{\nu}_r$ in Theorem 15.4.1.

Theorem 15.4.2. *Let Y be an α-stable RV in $\mathfrak{X}(\mathbb{R})$. Let $r > \alpha$, $b := 1/2^{r/\alpha}B$, and $B := \max(3^{r/\alpha}, 2C_r(2^{(r/\alpha)-1} + 3^{r/\alpha}))$, where $C_r := (r/\alpha e)^{r/\alpha}$. If $X \in \mathfrak{X}(\mathbb{R})$ satisfies*

$$\tau_r := \tau_r(X, Y) := \max\{\boldsymbol{\chi}(X, Y), \boldsymbol{\chi}_r(X, Y)\} \leq b, \tag{15.4.11}$$

then for all $n \geq 1$

$$\boldsymbol{\chi}\left(\frac{X_1 + \cdots + X_n}{n^{1/\alpha}}, Y\right) \leq B\tau_r n^{1-r/\alpha} \leq 2^{-r/\alpha} n^{1-r/\alpha}. \tag{15.4.12}$$

Remark 15.4.2. When comparing conditions (15.4.1) and (15.4.11), it is useful to note that the metric $\boldsymbol{\chi}$ is topologically weaker than **Var**, i.e., $\mathbf{Var}(X_n, Y) \to 0$ implies $\boldsymbol{\chi}(X_n, Y) \to 0$ but the converse is not true. Also, if $r = m + \beta$, $m = 0, 1, \ldots, \beta \in (0, 1]$, then [see (15.3.1) and (15.3.9)],

$$\boldsymbol{\chi}_r \leq C_\beta \boldsymbol{\zeta}_r, \tag{15.4.13}$$

where $C_\beta = \sup_t |t|^{-\beta}|1 - e^{it}|$.

Proof of inequality (15.4.13). By the definitions of $\boldsymbol{\chi}_r$ and $\boldsymbol{\zeta}_r$, we have

$$\boldsymbol{\chi}_r(X, Y) := \sup_{t \in \mathbb{R}} |E(f_t(X) - f_t(Y))|,$$

where $f_t(x) := t^{-r}\exp(itx)$ and

$$\zeta_r(X,Y) := \sup\{|E(f(X) - f(Y)| : f : \mathbb{R} \to \mathbb{C}, \\ \text{and} |f^{(m)}(x) - f^{(m)}(y)| \le |x-y|^{\beta}\},$$

where $r = m + \beta$, $m = 0, 1, \ldots,$ and $\beta \in (0,1]$. For any $t \in \mathbb{R}$

$$f_t^{(m)}(x) = t^{-\beta} \mathrm{i}^m \exp(itx),$$

and thus

$$\frac{|f_t^{(m)}(x) - f_t^{(m)}(y)|}{|s|^{\beta}} = \frac{|t|^{-\beta}|\exp(itx) - \exp(ity)|}{|s|^{\beta}} = \frac{|t|^{-\beta}|1 - \exp(its)|}{|s|^{\beta}},$$

where $s := x - y$. We observe that for any $D_r > 0$

$$D_r\zeta_r(X,Y) = \sup\{|E(f(X) - f(Y))| : |f^{(m)}(x) - f^{(m)}(y)| \le D_r|x-y|^{\beta}\}$$

and

$$\sup_{x,y\in\mathbb{R}} \frac{|f_t^{(m)}(x) - f_t^{(m)}(y)|}{|x-y|^{\beta}} \le \sup_{s\in\mathbb{R}} |st|^{-\beta}|1 - \exp(its)| := C_{\beta}.$$

A simple calculation shows that $C_{\beta} < \infty$, and this completes the proof of inequality (15.4.13). □

Finally, we note that since $\zeta_m(X,Y) := \sup\{|E(f(X) - f(Y))| : |f^{(m+1)}(x)| \le 1 \text{ a.e.}\}$ and since $|f_t^{(m+1)}(x)| = |\mathrm{i}^{m+1}\exp(itx)| = 1$, we obtain $\chi_m \le \zeta_m$.

Remark 15.4.3. One may show that for $r \in \mathbb{N}^+$ the metric χ_r has a convolution-type structure. In fact, with a slight abuse of notation,

$$\chi_r(F_{X_1}, F_{X_2}) = \chi(F_{X_1} * p_r, F_{X_2} * p_r),$$

where $p_r(t) = (t^r/r!)I_{(t>0)}$ is the density of an unbounded positive measure on the half-line $[0,\infty)$.

The proof of Theorem 15.4.2 is very similar to that of Theorem 15.4.1 and uses the following auxiliary results, which are completely analogous to Lemmas 15.4.1 and 15.4.2. We leave the details to the reader to complete the proof of Theorem 15.4.2.

Lemma 15.4.3. *For any $X_1, X_2 \in \mathfrak{X}(\mathbb{R})$, $\sigma > 0$, and $r > \alpha$*

$$\chi(X_1 + \sigma Y, X_2 + \sigma Y) \le C_r\sigma^{-r}\chi_r(X_1, X_2),$$

where $C_r := (r/\alpha e)^{r/\alpha}$.

Proof. We have

$$\begin{aligned}\chi(X_1+\sigma Y, X_2+\sigma Y) &:= \sup_{t\in\mathbb{R}} |\phi_{X_1}(t)-\phi_{X_2}(t)|\phi_{\sigma Y}(t)\\ &= \sup_{t\in\mathbb{R}} |\phi_{X_1}(t)-\phi_{X_2}(t)|\exp\{-|\sigma t|^\alpha\}\\ &\le \sup_{t\in\mathbb{R}} |\sigma t|^{-r}|\phi_{X_1}(t)-\phi_{X_2}(t)| \sup_{u>0} u^r\exp(-u^\alpha)\\ &= C_r\sigma^{-r}\chi_r(X,Y)\end{aligned}$$

since $C_r = \sup_{u>0} u^r \exp(-u^\alpha)$ by a simple computation. □

Lemma 15.4.4. *For any $X_1, X_2, Z, W \in \mathfrak{X}(\mathbb{R})$ the following inequality holds:*

$$\chi(X_1+Z, X_2+Z) \le \chi(X_1,X_2)\chi(Z,W) + \chi(X_1+W, X_2+W).$$

Proof. From the inequality

$$\begin{aligned}|\phi_{X_1+Z}(t)-\phi_{X_2+Z}(t)| &\le |\phi_{X_1}(t)-\phi_{X_2}(t)|\,|\phi_Z(t)-\phi_W(t)|\\ &\quad + |\phi_{X_1}(t)-\phi_{X_2}(t)|\,|\phi_W(t)|\end{aligned}$$

we obtain the desired result. □

Finally, we develop convergence rates with respect to the ℓ-metric defined in (15.3.6), and thus we naturally restrict our attention to the subset $\mathfrak{X}^*$ of $\mathfrak{X}(\mathbb{R}^k)$ of RVs with densities. Let $X, X_1, X_2, \dots$ denote a sequence of i.i.d. RVs in $\mathfrak{X}^*$ and $Y = Y_\alpha$ denote a symmetric α-stable RV. The ideal convolution metrics $\mu_r := \mu_{\alpha,r}$ and $\nu_r := \nu_{\alpha,r}$ (i.e., $\theta = Y$) will play a central role.

Theorem 15.4.3. *Let Y be a symmetric α-stable RV in $\mathfrak{X}(\mathbb{R}^k)$. Let $r = m + 1/p > \alpha$ for some integer m and $p \in [1,\infty)$, $a := 1/2^{r/\alpha}A$, $A := 2(2^{r/\alpha-1} + 3^{(r+1)/\alpha})$, and $D := 3^{1/\alpha}2^{r/\alpha}$. If $X \in \mathfrak{X}^*$ satisfies*

(i)

$$\tau(X,Y) := \max(\ell(X,Y), \boldsymbol{\mu}_{\alpha,r}(X,Y)) \le a, \tag{15.4.14}$$

(ii)

$$\tau_0(X,Y) := \max(\mathbf{Var}(X,Y), \boldsymbol{\nu}_{\alpha,r}(X,Y)) \le \frac{1}{A(a)D},$$

then

$$\ell\left(\frac{X_1+\cdots+X_n}{n^{1/\alpha}}, Y\right) \le A(a)\tau(X,Y)n^{1-r/\alpha}. \tag{15.4.15}$$

Remark 15.4.4. (a) Conditions (i) and (ii) guarantee ℓ-closeness (of order $n^{1-r/\alpha}$) between Y and the normalized sums $n^{-1/\alpha}(X_1+\cdots+X_n)$.

(b) From Lemmas 15.3.3, 15.3.5, and 15.3.6 we know that $\boldsymbol{\mu}_{\alpha,r+1}(X,Y)$ and $\boldsymbol{\nu}_{\alpha,r}(X,Y)$, $r = m-1+1/p$, $m = 1,2,\dots$ can be approximated from above by the rth difference pseudomoment κ_r whenever X and Y share the same first $(m-1)$ moments [see (15.3.23)–(15.3.25)]. Thus conditions (i) and (ii) could be expressed in terms of difference pseudomoments, which of course amounts to conditions on the tails of X.

To prove Theorem 15.4.3, we need a few auxiliary results similar in spirit to Lemmas 15.4.1 and 15.4.2.

Lemma 15.4.5. *Let* $X_1, X_2 \in \mathfrak{X}(\mathbb{R}^k)$. *Then*

$$\ell(X_1 + \sigma Y, X_2 + \sigma Y) \le \sigma^{-r} \mu_r(X_1, X_2).$$

Proof. $\ell(X_1 + \sigma Y, X_2 + \sigma Y) \le \sigma^{-r}\sigma^r \ell(X_1 + \sigma Y, X_2 + \sigma Y) \le \sigma^{-r}\mu_r(X_1, X_2)$. □

Lemma 15.4.6. *For any (independent)* $X, Y, U, V \in \mathfrak{X}^*(\mathbb{R}^k)$ *the following inequality holds:*

$$\ell(X + U, Y + U) \le \ell(X, Y)\,\mathbf{Var}(U, V) + \ell(X + V, Y + V).$$

Proof. Using the triangle inequality we obtain

$$\begin{aligned}
&\ell(X + U, Y + U)\\
&\quad = \sup_{x\in\mathbb{R}^k} \left| \int (p_X(x-y) - p_Y(x-y)) \Pr\{U \in \mathrm{d}y\} \right|\\
&\quad \le \sup_{x\in\mathbb{R}^k} \left| \int (p_X(x-y) - p_Y(x-y))(Pr\{U \in \mathrm{d}y\} - \Pr\{V \in \mathrm{d}y\}) \right|\\
&\qquad + \sup_{x\in\mathbb{R}^k} \left| \int (p_X(x-y) - p_Y(x-y)) \Pr\{V \in \mathrm{d}y\} \right|\\
&\quad \le \ell(X, Y)\,\mathbf{Var}(U, V) + \ell(X + V, Y + V).
\end{aligned}$$

□

To prove Theorem 15.4.3, one only needs to use the method of proof for Theorem 15.4.1 combined with the preceding two auxiliary results. The complete details are left to the reader. A more general theorem will be proved in the next section (Theorem 16.3.2).

The foregoing results show that the "ideal" structure of the convolution metrics μ_r and ν_r may be used to determine the optimal rates of convergence in the general CLT. The rates are expressed in terms of the uniform metrics **Var**, χ, and ℓ and hold uniformly in n under the sufficient conditions (15.4.1), (15.4.11), and (15.4.14), respectively. We have not explored the possible weakening of these conditions or even their possible necessity.

The ideal convolution metrics μ_r and ν_r are not limited to the context of Theorems 15.4.1–15.4.3; they can also be successfully employed to study other questions of interest. For example, we only mention here that ν_r can be used to prove a Berry–Esseen type of estimate for the Kolmogorov metric ρ given in (15.3.8).

More precisely, if $X, X_1, X_2, \ldots$ denotes a sequence of i.i.d. RVs in $\mathfrak{X}(\mathbb{R})$ and $Y \in \mathfrak{X}(\mathbb{R})$ a symmetric α-stable RV, then for all $r > \alpha$ and $n \geq 1$

$$\rho\left(\frac{X_1 + \cdots + X_n}{n^{1/\alpha}}, Y\right) \leq C\boldsymbol{\nu}_{\alpha,r}(X,Y)n^{1-r/\alpha} + C\max\{\rho(X,Y), \boldsymbol{\nu}_{\alpha,1}(X,Y), \boldsymbol{\nu}_{\alpha,r}^{1/(r-\alpha)}(X,Y)\}n^{-1/\alpha}, \tag{15.4.16}$$

where C is an absolute constant. Whenever $\boldsymbol{\nu}_{\alpha,1}(X,Y) < \infty$ and $\boldsymbol{\nu}_{\alpha,r}(X,Y) < \infty$, we obtain the right order estimate in the Berry–Esseen theorem in terms of the metric $\boldsymbol{\nu}_{\alpha,r}$.

Thus, metrics of the convolution type, especially those with the *ideal* structure, are appropriate when investigating sums of independent RVs converging to a stable limit law. We can only conjecture that there are other ideal convolution metrics, other than those explored in this section, that might furnish additional results in related limit theorem problems.[14]

References

Barlow RE, Proschan F (1975) Statistical theory of reliability and life testing: probability models. Holt, Rinehart, and Winston, New York

Billingsley P (1999) Convergence of probability measures, 2nd edn. Wiley, New York

Daley DJ (1988) Tight bounds on the exponential approximation of some aging distributions. Ann Prob 16:414–423

Dunford N, Schwartz J (1988) Linear operators, vol 1. Wiley, New York

Erdös P, Spencer J (1974) Probabilistic methods in combinatorics. Academic, New York

Hein M, Lal TN, Bousquet O (2004) Hilbertian metrics on probability measures and their application in SVMs'. In: Proceedings of the 26th DAGM symposium, pp 270–277

Kalashnikov VV, Rachev ST (1988) Mathematical methods for construction of stochastic queueing models. Nauka, Moscow (in Russian) [English transl., (1990) Wadsworth, Brooks–Cole, Pacific Grove, CA]

Klebanov L, Rachev S, Szekely G (1999) Pre-limit theorems and their applications. Acta Appl Math 58:159–174

Maejima M, Rachev S (1996) Rates of convergence in the operator-stable limit theorem. J Theor Prob 9:37–85

Rachev S, Rüschendorf L (1992) A new ideal metric with applications to multivariate stable limit theorems. ProbTheor Relat Fields 94:163–187

Rachev S, Rüschendorf L (1994) On the rate of covergence in the clt with respect to the Kantorovich metric. In: Hoffman-Jorgensen J, Kuelbs J, Markus MB (eds) Probability in Banach spaces. Birkhäuser, Boston, pp 193–207

---

[14]See, for example, Rachev and Rüschendorf (1992) for an application of ideal metrics in the multivariate CLT, Rachev and Rüschendorf (1994) for an application of the Kantorovich metric, Maejima and Rachev (1996) for rates of convergence in operator-stable limit theorems, and Klebanov et al. (1999) for rates of convergence in prelimit theorems.

Senatov VV (1980) Uniform estimates of the rate of convergence in the multi-dimensional central limit theorem. Theor Prob Appl 25:745–759
Sençimen C, Pehlivan S (2009) Strong ideal convergence in probabilistic metric spaces. Proc Ind Acad Sci 119(3):401–410
Zolotarev VM (1976a) Approximation of distributions of sums of independent random variables with values in infinite-dimensional spaces. Theor Prob Appl 21:721–737
Zolotarev VM (1976b) Metric distances in spaces of random variables and their distributions. Math USSR sb 30:373–401
Zolotarev VM (1986) Contemporary theory of summation of independent random variables. Nauka, Moscow (in Russian)

Chapter 16
Ideal Metrics and Rate of Convergence in the CLT for Random Motions

The goals of this chapter are to:

- Define ideal probability metrics in the space of random motions,
- Provide examples of ideal probability metrics and describe their basic properties,
- Derive the rate of convergence in the general central limit theorem in terms of the corresponding metrics with uniform structure.

Notation introduced in this chapter:

Notation	Description
$\mathbb{M}(d)$	Group of rigid motions on $\mathbb{R}^d$
$\mathrm{SO}(d)$	Special orthogonal group in $\mathbb{R}^d$
$g = (y, u)$	Element of $\mathbb{M}(d)$
$g_1 \circ g_2$	Convolution of two motions
$H_\alpha = (Y_\alpha, U_\alpha)$	α-stable random motion

16.1 Introduction

The ideas developed in Chap. 15 are discussed in this chapter in the context of random motions defined on $\mathbb{R}^d$. We begin by defining the corresponding ideal probability metrics and discuss their basic properties, which are similar to their counterparts in Chap. 15. Finally, we provide results for the rate of convergence in the general central limit theorem (CLT) for random motions in terms of the following metrics with uniform structure: $\boldsymbol{\rho}$, **Var**, and $\boldsymbol{\ell}$.

S.T. Rachev et al., *The Methods of Distances in the Theory of Probability and Statistics*,
DOI 10.1007/978-1-4614-4869-3_16,

16.2 Ideal Metrics in the Space of Random Motions

Let $\mathbb{M}(d)$ be the group of rigid motions on $\mathbb{R}^d$, i.e., the group of one-to-one transformations of $\mathbb{R}^d$ to $\mathbb{R}^d$ that preserves the orientation of the space and the inner product. $\mathbb{M}(d)$ is known as the Euclidean group of motions of d-dimensional Euclidean space. Letting $\mathrm{SO}(d)$ denote the special orthogonal group in $\mathbb{R}^d$, any element $g \in \mathbb{M}(d)$ can be written in the form $g = (y, u)$, where $y \in \mathbb{R}^d$ represents the translation parameter and $u \in \mathrm{SO}(d)$ is a rotation about the origin. Note that for all $x \in \mathbb{R}^d$, $g(x) = y + ux$. If $g_i = (y_i, u_i)$, $1 \le i \le n$, then the product $g(n) = g_1 \circ g_2 \circ \cdots \circ g_n$ has the form $g(n) = (y(n), u(n))$, where $u(n) = u_1, \ldots, u_n$ and $y(n) = y_1 + u_1 y_2 + \cdots + u_1 \cdots u_{n-1} y_n$. For any $c \in \mathbb{R}$ and $g = (y, u) \in \mathbb{M}(d)$, define $cg = (cy, u)$.

Next, let $(\Omega, \mathcal{F}, \Pr)$ be a probability space on which is defined a sequence of i.i.d. random variables (RVs) G_i, $i \ge 1$, with values in $\mathbb{M}(d)$. A. natural problem involves finding the limiting distribution (i.e., CLT) of the product $G_1 \circ \cdots \circ G_n$, which leads to the notion of α-stable random motion. The definition of an α-stable random motion resembles that for a spherically symmetric α-stable random vector, that is, H_α is an α-stable random motion if for any sequence of i.i.d. random motions G_i, with $G_1 \stackrel{\mathrm{d}}{=} H_\alpha$,

$$H_\alpha \stackrel{\mathrm{d}}{=} n^{-1/\alpha}(G_1 \circ \cdots \circ G_n) \text{ for any } n \ge 1, \text{ and}$$
$$H_\alpha \stackrel{\mathrm{d}}{=} uH_a, \text{ for any } u \in \mathrm{SO}(d). \tag{16.2.1}$$

Baldi (1979) proved that $H_\alpha = (Y_\alpha, U_\alpha)$ is an α-stable random motion if and only if Y_α has a spherically symmetric α-stable distribution on $\mathbb{R}^d$ and U_α is uniformly distributed on $\mathrm{SO}(d)$. Henceforth, we write $H_\alpha = (Y_\alpha, U_\alpha)$ to denote an α-stable random motion. In this section, we will be interested in the rate of convergence of i.i.d. random motions to a stable random motion.[1] First we shall define and examine the properties of ideal metrics related to this particular approximation problem.

Let $\mathfrak{X}(\mathbb{M}(d))$ be the space of all random motions $G = (Y, U)$ on $(\Omega, \mathcal{F}, \Pr)$, $Y \in \mathfrak{X}(\mathbb{R}^d)$ the space of all d-dimensional random vectors, and $U \in \mathfrak{X}(\mathrm{SO}(d))$ the space of all random "rotations" in $\mathbb{R}^d$. $\mathfrak{X}^*(\mathbb{R}^d)$ denotes the subspace of $\mathfrak{X}(\mathbb{R}^d)$ of all RVs with densities: $\mathfrak{X}^*(\mathbb{M}(d))$ is defined by $\mathfrak{X}^*(\mathbb{R}^d) \times \mathfrak{X}(\mathrm{SO}(d))$.

Define the total variation distance between elements G and G^* of $\mathfrak{X}(M(d))$ by

$$\mathbf{Var}(G, G^*) := \sup_{x \in \mathbb{R}^d} \mathbf{Var}(G(x), G^*(x)), \tag{16.2.2}$$

where for X and Y in $\mathfrak{X}(\mathbb{R}^d)$

$$\mathbf{Var}(X, Y) := 2 \sup\{|\Pr\{X \in A\} - \Pr\{Y \in A\}|, \quad A \in \mathcal{B}(\mathbb{R}^d)\},$$

in which $\mathcal{B}(\mathbb{R}^d)$ denotes the Borel sets in $\mathbb{R}^d$ [see (15.3.5)].

---

[1] See Rachev and Yukich (1991).

Let $\theta \in \mathfrak{X}(\mathbb{R}^d)$ have a spherically symmetric α-stable distribution on $\mathbb{R}^d$. As in Sect. 15.3, define smoothing metrics associated with the **Var** and ℓ distances

$$\boldsymbol{\nu}_r(X,Y) := \sup_{h\in\mathbb{R}} |h|^r \, \mathbf{Var}(X + h\theta, Y + h\theta), \qquad X,Y \in \mathfrak{X}(\mathbb{R}^d) \tag{16.2.3}$$

and

$$\boldsymbol{\mu}_r(X,Y) := \sup_{h\in\mathbb{R}} |h|^r \ell(X + h\theta, Y + h\theta), \qquad X,Y \in \mathfrak{X}^*(\mathbb{R}^d), \tag{16.2.4}$$

where $\ell(X,Y)$, $X,Y \in \mathfrak{X}^*(\mathbb{R}^d)$, is the ess sup norm distance between the densities p_X and p_Y of X and Y, respectively, that is,

$$\ell(X,Y) := \operatorname*{ess\,sup}_{y\in\mathbb{R}^d} |p_X(y) - p_Y(y)| \tag{16.2.5}$$

[see (15.3.6), (15.3.12), and (15.3.13)].

Next, extend the definitions of $\boldsymbol{\nu}_r$ and $\boldsymbol{\mu}_r$ to $\mathfrak{X}(\mathbb{M}(d))$ and $\mathfrak{X}^*(\mathbb{M}(d))$, respectively,

$$\boldsymbol{\nu}_r(G_1,G_2) := \sup_{x\in\mathbb{R}^d} \boldsymbol{\nu}_r(G_1(x), G_2(x)), \quad G_1,G_2 \in \mathfrak{X}(\mathbb{M}(d)), \tag{16.2.6}$$

and

$$\boldsymbol{\mu}_r(G_1,G_2) := \sup_{x\in\mathbb{R}^d} \boldsymbol{\mu}_r(G_1(x), G_2(x)), \qquad G_1, G, \in \mathfrak{X}^*(\mathbb{M}(d)). \tag{16.2.7}$$

As in Chap. 15, $\boldsymbol{\nu}_r$ and $\boldsymbol{\mu}_r$ will play important roles in establishing rates of convergence in the integral and local CLT theorems. Zolotarev's $\boldsymbol{\zeta}_r$ metric defined by (15.3.1) on $\mathfrak{X}(\mathbb{R}^d)$ is similarly extended in $\mathfrak{X}(\mathbb{M}(d))$

$$\boldsymbol{\zeta}_r(G_1,G_2) := \sup_{x\in\mathbb{R}^d} \boldsymbol{\zeta}_r(G_1(x), G_2(x)). \tag{16.2.8}$$

The following two theorems record some special properties of $\boldsymbol{\nu}_r$ and $\boldsymbol{\mu}_r$, which are proved by exploiting their ideality on $\mathfrak{X}(\mathbb{R}^d)$ (Lemmas 15.3.1 and 15.3.2).

Theorem 16.2.1. *$\boldsymbol{\mu}_r$ is an ideal metric on $\mathfrak{X}^*(\mathbb{M}(d))$ of order $r-1$, i.e., $\boldsymbol{\mu}_r$ satisfies the following two conditions:*

(i) Regularity: $\boldsymbol{\mu}_r(G_1 \circ G, G_2 \circ G) \le \boldsymbol{\mu}_r(G_1, G_2)$ and

$$\boldsymbol{\mu}_r(G \circ G_1, G \circ G_2) \le \boldsymbol{\mu}_r(G_1, G_2)$$

for any G_1 and G_2 that are independent of G;

(ii) Homogeneity: $\boldsymbol{\mu}_r(cG_1, cG_2) \le |c|^{r-1}\boldsymbol{\mu}_r(G_1, G_2)$ for any $c \in \mathbb{R}$.

Proof. The proof rests upon two auxiliary lemmas. □

Lemma 16.2.1. *For any independent $G \in \mathfrak{X}^*(\mathbb{M}(d))$, $Y_1, Y_2 \in \mathfrak{X}^*(\mathbb{R}^d)$*

$$\boldsymbol{\mu}_r(G(Y_1), G(Y_2)) \le \boldsymbol{\mu}_r(Y_1, Y_2). \tag{16.2.9}$$

Proof of Lemma 16.2.1. By definition of $\boldsymbol{\mu}_r$ and the regularity of ℓ, we have for $G := (Y, U)$

$$\begin{aligned}\boldsymbol{\mu}_r(G(Y_1), G(Y_2)) &= \sup_{x\in\mathbb{R}} |h|^r \ell(G(Y_1) + h\theta, G(Y_2) + h\theta)\\ &= \sup_{x\in\mathbb{R}} |h|^r \ell(Y + UY_1 + h\theta, Y + UY_2 + h\theta)\\ &\leq \sup_{x\in\mathbb{R}} |h|^r \ell(UY_1 + h\theta, UY_2 + h\theta). \end{aligned} \tag{16.2.10}$$

Next, we show $\ell(UY_1, UY_2) \leq \ell(Y_1, Y_2)$ for any independent $U \in \mathfrak{X}(\mathrm{SO}(d))$. To see this, notice that

$$\begin{aligned}\ell(UY_1, UY_2) &\leq \sup_{x\in\mathbb{R}^d} \sup_{u\in\mathrm{SO}(d)} |p_{uY_1}(x) - p_{uY_2}(x)|\\ &= \sup_{u\in\mathrm{SO}(d)} \sup_{z=u\circ x\in\mathbb{R}^d} |(p_{Y_1} - p_{Y_2})(x_1(z_1,\dots,z_d),\dots,x_d(z_1,\dots,z_d))|\\ &\quad\times \left|\frac{\partial x_1 \cdots \partial x_d}{\partial z_1 \cdots \partial z_d}\right|.\end{aligned}$$

Since the determinant of the Jacobian equals 1,

$$\ell(UY_1, UY_2) \leq \ell(Y_1, Y_2). \tag{16.2.11}$$

Combining (16.2.10) and (16.2.11) and using $U^{-1}\theta \overset{\mathrm{d}}{=} \theta$, where $UU^{-1} = I$, we have

$$\begin{aligned}\boldsymbol{\mu}_r(G(Y_1), G(Y_2)) &\leq \sup_{h\in\mathbb{R}} |h|^r \ell(U(Y_1 + hU^{-1}\theta), U(Y_2 + hU^{-1}\theta))\\ &\leq \sup_{h\in\mathbb{R}} |h|^r \ell(Y_1 + hU^{-1}\theta, Y_2 + hU^{-1}\theta) = \boldsymbol{\mu}_r(Y_1, Y_2). \end{aligned}$$ □

Lemma 16.2.2. *If G_1, G_2, and Y are independent, then*

$$\boldsymbol{\mu}_r(G_1(Y), G_2(Y)) \leq \sup_{x\in\mathbb{R}^d} \boldsymbol{\mu}_r(G_1(x), G_2(x)). \tag{16.2.12}$$

Proof of Lemma 16.2.2. We have for $G_i = (Y_i, U_i)$

$$\begin{aligned}\boldsymbol{\mu}_r(G_1(Y), G_2(Y)) &= \sup_{h\in\mathbb{R}} |h|^r \ell(Y_1 + U_1Y + h\theta, Y_2 + U_2Y + h\theta)\\ &= \sup_{h\in\mathbb{R}} |h|^r \sup_{x\in\mathbb{R}^d} |p_{Y_1+U_1Y+h\theta}(x) - p_{Y_2+U_2Y+h\theta}(x)|\\ &= \sup_{h\in\mathbb{R}} |h|^r \sup_{x\in\mathbb{R}^d} \left| \int_{\mathbb{R}^d} \left\{ \int p_{Y_1+h\theta}(x - u_1y)\Pr(U_1 \in \mathrm{d}u_1) \right.\right.\end{aligned}$$

$$
\begin{aligned}
&\left. - \int p_{Y_2+h\theta}(x - u_2 y) \Pr(U_2 \in \mathrm{d}u_2) \right\} \Pr(Y \in \mathrm{d}y) \Bigg| \\
&\le \sup_{h\in\mathbb{R}} |h|^r \sup_{y\in\mathbb{R}^d} \sup_{x\in\mathbb{R}^d} \Bigg| \int p_{Y_1+h\theta}(x - u_1 y) \Pr(U_1 \in \mathrm{d}u_1) \\
&\quad - \int p_{Y_2+h\theta}(x - u_2 y) \Pr(U_2 \in \mathrm{d}u_2) \Bigg| \\
&= \sup_{y\in\mathbb{R}^d} \sup_{h\in\mathbb{R}} |h|^r \sup_{x\in\mathbb{R}^d} |p_{Y_1+h\theta+U_1 y}(x) - p_{Y_2+h\theta+U_2 y}(x)| \\
&= \sup_{y\in\mathbb{R}^d} \boldsymbol{\mu}_r(G_1(y), G_2(y)).
\end{aligned}
$$

□

Now we can prove property (i) of the theorem. By (16.2.9),

$$
\begin{aligned}
\boldsymbol{\mu}_r(G \circ G_1, G \circ G_2) &= \sup_{x\in\mathbb{R}^d} \boldsymbol{\mu}_r(G \circ G_1(x), G \circ G_2(x)) \\
&\le \sup_{x\in\mathbb{R}^d} \boldsymbol{\mu}_r(G_1(x), G_2(x)) = \boldsymbol{\mu}_r(G_1, G_2).
\end{aligned}
$$

Similarly, by (16.2.12),

$$
\begin{aligned}
\boldsymbol{\mu}_r(G_1 \circ G, G_2 \circ G) &= \sup_{x\in\mathbb{R}^d} \boldsymbol{\mu}_r(G_1 \circ G(x), G_2 \circ G(x)) \\
&\le \sup_{x\in\mathbb{R}^d} \boldsymbol{\mu}_r(G_1(x), G_2(x)) = \boldsymbol{\mu}_r(G_1, G_2),
\end{aligned}
$$

which completes the proof of the *regularity* property. To prove the *homogeneity*, observe that by the ideality of $\boldsymbol{\mu}_r$ on $\mathfrak{X}(\mathbb{R}^d)$,

$$
\begin{aligned}
\boldsymbol{\mu}_r(cG_1, cG_2) &= \sup_{x\in\mathbb{R}^d} \boldsymbol{\mu}_r(cY_1 + U_1 x, cY_2 + U_2 x) \\
&= \sup_{x\in\mathbb{R}^d} \boldsymbol{\mu}_r\left(c\left(Y_1 + U_1\left(\frac{1}{c}x\right)\right), c\left(Y_2 + U_2\left(\frac{1}{c}x\right)\right)\right) \\
&= |c|^{r-1} \boldsymbol{\mu}_r(G_1, G_2).
\end{aligned}
$$

□

Theorem 16.2.2. $\boldsymbol{\nu}_r$ *is an ideal metric on* $\mathfrak{X}(\mathbb{M}(d))$ *of order* r.

The proof is similar to that of the previous theorem.

The usefulness of ideality may be illustrated in the following way. If μ is ideal of order r on $\mathfrak{X}^*(\mathbb{M}(d))$, then for any sequence of i.i.d. random motions $G_1, G_2, \cdots$ it easily follows that

$$
\mu(n^{-1/\alpha}(G_1 \circ \cdots \circ G_n), H_\alpha) \le n^{1-(r/\alpha)} \mu(G_1, H_\alpha) \tag{16.2.13}
$$

is a "right order" estimate for the rate of convergence in the CLT. Estimates such as these will play a crucial role in all that follows.

The next result clarifies the relation between the ideal metrics μ_r, ν_r, and ζ_r. It shows that upper bounds for the rate of the convergence problem, when expressed in terms of ζ_r, are necessarily weaker than bounds expressed in terms of either μ_r or ν_r (as in Theorems 16.3.1 and 16.3.2 below).

Theorem 16.2.3. *For any G_1 and $G_2 \in \mathfrak{X}(\mathbb{M}(d))$*

$$\mu_r(G_1, G_2) \leq C_1(r)\zeta_{r-1}(G_1, G_2), \quad r \geq 1, \tag{16.2.14}$$

and

$$\nu_r(G_1, G_2) \leq C_2(r)\zeta_r(G_1, G_2), \quad r \geq 0, r - \textit{integer}, \tag{16.2.15}$$

where $C_i(r)$ is a constant depending only on r.

The proof follows from the similar inequalities between μ_r, ν_r, and ζ_r in the space $\mathfrak{X}(\mathbb{R}^d)$ (Sect. 15.3 and Lemmas 15.3.4–15.3.6). As far as the finiteness of $\zeta_r(G_1, G_2)$ is concerned, we have that the condition

$$\left| \sum_{\substack{0 \leq i_1, \ldots, i_d \leq d \\ i_1 + \cdots + i_d = j}} \int_{\mathbb{R}_d} y_1^{i_1} \circ \cdots \circ y_d^{i_d} (\Pr(G_1(x) \in \mathrm{d}y) - \Pr(G_2(x) \in \mathrm{d}y)) \right| = 0 \tag{16.2.16}$$

for all $x \in \mathbb{R}^d$, $j = 0, 1, \ldots, m$, $m + \beta = r$, $\beta \in (0, 1]$, m-integer, implies

$$\zeta_r(G_1, G_2) \leq \frac{1}{\Gamma(1 + r)} \mathbf{Var}_r(G_1, G_2), \tag{16.2.17}$$

where the metric $\mathbf{Var}_r$ is the rth absolute pseudomoment in $\mathfrak{X}(\mathbb{M}(d))$, that is,

$$\mathbf{Var}_r(G_1, G_2) := \sup_{x \in \mathbb{R}^d} \int \|y\|^r |\Pr_{G_1(x)} - \Pr_{G_2(x)}|(\mathrm{d}y). \tag{16.2.18}$$

16.3 Rates of Convergence in the Integral and Local CLTs for Random Motions

Let $G_1, G_2, \ldots$ be a sequence of i.i.d random motions and H_α an α-stable random motion. We seek precise order estimates for the rate of convergence

$$n^{-1/\alpha}(G_1 \circ \cdots \circ G_n) \to H_\alpha \tag{16.3.1}$$

in terms of Kolmogorov's metric $\boldsymbol{\rho}$, **Var**, and ℓ distances on $\mathfrak{X}(\mathbb{M}(d))$. Here, the uniform (Kolmogorov's) metric between random motions G and G^* is defined by

$$\boldsymbol{\rho}(G,G^*) := \sup_{x\in\mathbb{R}^d} \boldsymbol{\rho}(G(x),G^*(x)), \tag{16.3.2}$$

where $\boldsymbol{\rho}(X,Y)$ is the usual Kolmogorov distance between the d-dimensional random vectors X and Y in $\mathfrak{X}(\mathbb{R}^d)$, that is,

$$\boldsymbol{\rho}(X,Y) := \sup_{A\in\mathbb{C}} |\Pr\{X\in A\} - \Pr\{Y\in A\}|, \tag{16.3.3}$$

in which $\mathbb{C}$ denotes the convex Borel sets in $\mathbb{R}^d$. Recall that the total variation metric **Var** in $\mathfrak{X}(\mathbb{M}(d))$ is defined by (16.2.2) and ℓ in $\mathfrak{X}^*(\mathbb{M}(d))$ is given by

$$\ell(G,G^*) := \sup_{x\in\mathbb{R}^d} \ell(G(x),G^*(x)) \tag{16.3.4}$$

[see (16.2.5)].

The first result obtains rates with respect to $\boldsymbol{\rho}$. Here and henceforth C denotes an absolute constant whose value may change from line to line.

The next theorem establishes the estimates of the uniform rate of convergence in the *integral* CLT for random motions.

Theorem 16.3.1. *Let $r>\alpha$, and set $\rho := \boldsymbol{\rho}(G_1,H_\alpha)$ and $\tau_r := \tau_r(G_1,H_\alpha) := \max\{\rho,\nu_r,\nu_\alpha^{1/(r-\alpha)}\}$. Then,*

$$\boldsymbol{\rho}(n^{-1/\alpha}(G_1\circ\cdots\circ G_n) \leq C(\nu_r n^{1-r/\alpha} + \tau_r n^{-1/\alpha}). \tag{16.3.5}$$

Proof. As in Sect. 15.4, it is helpful to first establish three smoothing inequalities for $\boldsymbol{\rho}$ and **Var**. Throughout, recall that H_α has components Y_α ($\stackrel{d}{=}\theta$) and U_α, and let $\overline{H}_\alpha$ denote the projection of H_α on $\mathbb{R}^d$. The purpose of the next lemma is to transfer the problem of estimating the $\boldsymbol{\rho}$-distance between two random motions to the same problem involving smoothed random motions. Here and in what follows, $G\circ\widetilde{G}$ means that $G\circ\widetilde{G}$ is a random motion whose distribution is a convolution of the distributions of G and $\widetilde{G}$. □

Lemma 16.3.1. *For any G and G^* in $\mathfrak{X}(\mathbb{M}(d))$ and $\delta>0$*

$$\boldsymbol{\rho}(G,G^*) \leq C\boldsymbol{\rho}(\delta\overline{H}_\alpha\circ G, \delta\overline{H}_\alpha\circ G^*) + C\delta, \tag{16.3.6}$$

where C is an absolute constant.

Proof of Lemma 16.3.1. The required inequality is a slight extension of the *smoothing inequality* in $\mathfrak{X}(\mathbb{R}^d)$:[2]

$$\boldsymbol{\rho}(X,Y) \leq C\boldsymbol{\rho}(X+\delta\theta, Y+\delta\theta) + C\delta \quad X,Y\in\mathfrak{X}(\mathbb{R}^d), \tag{16.3.7}$$

[2]See Paulauskas (1974, 1976), Zolotarev (1986, Lemma 5.4.2), and Bhattacharya and Ranga Rao (1976, Lemma 12.1).

where θ is a spherically symmetric α-stable random vector independent of X and Y and C is a constant depending upon α and d only. By (16.3.7), we have

$$\begin{aligned}\boldsymbol{\rho}(G, G^*) &= \sup_{x\in\mathbb{R}^d} \boldsymbol{\rho}(Y + Ux, Y^* + U^*x)\\ &\leq C \sup_{x\in\mathbb{R}^d} \boldsymbol{\rho}(\delta\theta + Y + Ux, \delta\theta + Y^* + U^*x) + C\delta\\ &= C\boldsymbol{\rho}(\delta\overline{H}_\alpha \circ G, \delta\overline{H}_\alpha \circ G^*) + C\delta. \qquad \square\end{aligned}$$

The next estimate is the analog of Lemma 15.4.1 and will be used several times in the proof.

Lemma 16.3.2. *Let* $G, \widetilde{G} \in \mathfrak{X}(\mathbb{M}(d))$, $\lambda_i \geq 0$, $i = 1, 2$; $\lambda^\alpha := \lambda_1^\alpha + \lambda_2^\alpha$; $\widetilde{H}_\alpha \stackrel{\mathrm{d}}{=} H_\alpha$. *For any* $r > 0$

$$\mathbf{Var}(\lambda_1 H_\alpha \circ G \circ \lambda_2\widetilde{H}_\alpha, \lambda_1 H_\alpha \circ \widetilde{G} \circ \lambda_2\widetilde{H}_\alpha) \leq \lambda^{-r}\boldsymbol{\nu}_r(G, \widetilde{G}). \tag{16.3.8}$$

Proof of Lemma 16.3.2. Let $\widetilde{H}_\alpha := (\widetilde{Y}_\alpha, \widetilde{U}_\alpha)$, $G := (Y, U)$, and $\widetilde{G} := (\widetilde{Y}, \widetilde{U})$. Then, by the definition of the **Var** metric,

$$\begin{aligned}&\mathbf{Var}(\lambda_1 H_\alpha \circ G \circ \lambda_2\widetilde{H}_\alpha, \lambda_1 H_\alpha \circ \widetilde{G} \circ \lambda_2\widetilde{H}_\alpha)\\ &\quad= \sup_x \mathbf{Var}(\lambda_1 H_\alpha \circ G \circ (\lambda_2\widetilde{Y}_\alpha + \widetilde{U}_\alpha x), \lambda_1 H_\alpha \circ \widetilde{G} \circ (\lambda_2\widetilde{Y}_\alpha + \widetilde{U}_\alpha x))\\ &\quad= \sup_x \mathbf{Var}(\lambda_1 H_\alpha(Y + U\lambda_2, \widetilde{Y}_\alpha + U\widetilde{U}_\alpha x), \lambda_1 H_\alpha(\widetilde{Y} + \widetilde{U}\lambda_2\widetilde{Y}_\alpha + U\widetilde{U}_\alpha x))\\ &\quad= \sup_x \mathbf{Var}(\lambda_1 Y_\alpha + U_\alpha(Y + U\lambda_2\widetilde{Y}_\alpha) + U_\alpha U\widetilde{U}_\alpha x, \lambda_1 Y_\alpha\\ &\qquad\quad + U_\alpha(\widetilde{Y} + \widetilde{U}\lambda_2\widetilde{Y}_\alpha) + U_\alpha\widetilde{U}\widetilde{U}_\alpha x)\\ &\quad= \sup_x \mathbf{Var}(\lambda_1 Y_\alpha + U_\alpha Y + \lambda_2\widetilde{Y}_\alpha + U_\alpha U\widetilde{U}_\alpha x, \lambda_1 Y_\alpha\\ &\qquad\quad + U_\alpha\widetilde{Y} + \lambda_2\widetilde{Y} + U_\alpha U\widetilde{U}_\alpha x).\end{aligned}$$

Using $\lambda_1 Y_\alpha + \lambda_2\widetilde{Y}_\alpha \stackrel{\mathrm{d}}{=} \lambda Y_\alpha$, the right-hand side equals

$$\begin{aligned}&\sup_x \mathbf{Var}(\lambda Y_\alpha + U_\alpha(Y + U\widetilde{U}_\alpha x), \lambda Y_\alpha + U_\alpha(\widetilde{Y} + \widetilde{U}\widetilde{U}_\alpha x))\\ &\quad\leq \lambda^{-r} \sup_x \sup_{h\in\mathbb{R}} |\lambda h|^r\, \mathbf{Var}(h\lambda Y_\alpha + U_\alpha(Y + U\widetilde{U}_\alpha x), h\lambda Y_\alpha + U_\alpha(\widetilde{Y} + \widetilde{U}\widetilde{U}_\alpha x))\\ &\quad= \lambda^{-r} \sup_x \boldsymbol{\nu}_r(U_\alpha(Y + U\widetilde{U}_\alpha x), U_\alpha(\widetilde{Y} + \widetilde{U}\widetilde{U}_\alpha x))\\ &\quad= \lambda^{-r} \sup_x \boldsymbol{\nu}_r(Y + U\widetilde{U}_\alpha x, \widetilde{Y} + \widetilde{U}U_\alpha x))\\ &\quad= \lambda^{-r}\boldsymbol{\nu}_r(G, \widetilde{G}),\end{aligned}$$

by definition of $\boldsymbol{\nu}_r$, and since **Var** (and hence $\boldsymbol{\nu}_r$) is invariant with respect to rotations. $\square$

The third and final lemma may be considered as the analog of Lemma 15.4.2.

Lemma 16.3.3. *For any G_1^*, G_2^*, $\widetilde{G}_1$, $\widetilde{G}_2$ in $\mathfrak{X}(\mathbb{M}(d))$ and $\lambda \geq 0$*

$$\begin{aligned}\rho(\lambda\overline{H}_\alpha \circ G_1^* \circ \widetilde{G}_1, \lambda\overline{H}_\alpha \circ G_1^* \circ \widetilde{G}_2) &\leq \rho(G_1^*, G_2^*)\,\mathbf{Var}(\lambda\overline{H}_\alpha \circ \widetilde{G}_1, \lambda\overline{H}_\alpha \circ \widetilde{G}_2)\\ &\quad + \rho(\lambda\overline{H}_\alpha \circ G_2^* \circ \widetilde{G}_1, \lambda\overline{H}_\alpha \circ G_2^* \circ \widetilde{G}_2).\end{aligned} \tag{16.3.9}$$

Also,

$$\begin{aligned}\rho(\lambda\overline{H}_\alpha \circ G_1^* \circ \widetilde{G}_1, \lambda\overline{H}_\alpha \circ G_2^* \circ \widetilde{G}_1) &\leq \rho(G_1^*, G_2^*)\,\mathbf{Var}(\lambda\overline{H}_\alpha \circ \widetilde{G}_1, \lambda\overline{H}_\alpha \circ \widetilde{G}_2)\\ &\quad + \rho(\lambda\overline{H}_\alpha \circ G_1^* \circ \widetilde{G}_2, \lambda\overline{H}_\alpha \circ G_2^* \circ \widetilde{G}_2)\end{aligned}$$

and

$$\begin{aligned}\mathbf{Var}(\lambda\overline{H}_\alpha \circ G_1^* \circ \widetilde{G}_1, \lambda\widetilde{H}_\alpha \circ G_2^* \circ \widetilde{G}_1) &\leq \mathbf{Var}(G_1^*, G_2^*)\,\mathbf{Var}(\lambda\overline{H}_\alpha \circ \widetilde{G}_1, \lambda\widetilde{H}_\alpha \circ \widetilde{G}_2)\\ &\quad + \mathbf{Var}(\lambda\overline{H}_\alpha \circ G_1^* \circ \widetilde{G}_2, \lambda\overline{H}_\alpha \circ G_2^* \circ \widetilde{G}_2).\end{aligned} \tag{16.3.10}$$

Proof. We will prove only (16.3.9). The proof of the other two inequalities is similar. We have

$$\begin{aligned}&\rho(\lambda\overline{H}_\alpha \circ G_1^* \circ \widetilde{G}_1, \lambda\overline{H}_\alpha \circ G_1^* \circ \widetilde{G}_2)\\ &= \rho(G_1^* \circ \lambda\overline{H}_\alpha \circ G_1, G_1^* \circ \lambda\overline{H}_\alpha \circ G_2)\\ &= \sup_{x\in\mathbb{R}^d}\sup_{A\in\mathbb{C}}\left|\int_{\mathbb{M}(d)} \Pr\{G_1^* \circ g(x) \in A\}(\lambda\overline{H}_\alpha \circ \widetilde{G}_1 - \lambda\overline{H}_\alpha \circ \widetilde{G}_2)\mathrm{d}g\right|\\ &\leq \sup_{x\in\mathbb{R}^d}\sup_{A\in\mathbb{C}}\left|\int \Pr\{G_1^* \circ g(x) \in A\}\right.\\ &\qquad \left. - \Pr\{G_2^* \circ g(x) \in A\})(\lambda\overline{H}_\alpha \circ \widetilde{G}_1 - \lambda\overline{H}_\alpha \circ \widetilde{G}_2)\mathrm{d}g\right|\\ &\qquad + \sup_{x\in\mathbb{R}^d}\sup_{A\in\mathbb{C}}\left|\int \Pr\{G_2^* \circ g(x) \in A\}\{\lambda\overline{H}_\alpha \circ \widetilde{G}_1 - \lambda\overline{H}_\alpha \circ \tilde{G}_2)\mathrm{d}g\right|\\ &\leq \rho(G_1^*, G_2^*)\,\mathbf{Var}(\lambda\overline{H}_\alpha \circ \widetilde{G}_1, \lambda\overline{H}_\alpha \circ \widetilde{G}_2) + \rho(G_2^* \circ \lambda\overline{H}_\alpha \circ \widetilde{G}_1, G_2^* \circ \lambda\overline{H}_\alpha \circ \widetilde{G}_2)\\ &= \rho(G_1^*, G_2^*)\,\mathbf{Var}(\lambda\overline{H}_\alpha \circ \widetilde{G}_1, \lambda\overline{H}_\alpha \circ \widetilde{G}_2) + \rho(\lambda\overline{H}_\alpha \circ G_2^* \circ \widetilde{G}_1, \lambda\overline{H}_\alpha \circ G_2^* \circ \widetilde{G}_2).\end{aligned}$$

□

On the basis of these three lemmas, Theorem 16.3.1 may now be proved. The proof uses induction on n. First, note that for a fixed n_0 and $n \leq n_0$, the estimate (16.3.5) is an obvious consequence of the hypotheses. Thus, let $n \geq n_0$ and assume that for any $j < n$

$$\rho(j^{-1/\alpha}(G_1 \circ \cdots \circ G_j), H_\alpha) \leq B(\nu_r j^{1-r/\alpha} + \tau_r j^{-1/\alpha}), \tag{16.3.11}$$

where B is an absolute constant.

Remark 16.3.1. We will use the main idea behind Senatov (1980, Theorem 2), where the case $\alpha = 2$ is considered and rates of convergence for CLT of random vectors in terms of ζ_r are investigated.

Set $m = [n/2]$ and

$$\delta := A \max(\boldsymbol{\nu}_1(G_1, H_\alpha), \boldsymbol{\nu}_r^{1/(r-\alpha)}(G_1, H_\alpha))n^{-1/\alpha}, \tag{16.3.12}$$

where A is a constant to be determined later. Note that $\delta \leq A\tau_r n^{-1/\alpha}$, which will be used in the sequel.

Let $G_1', G_2', \cdots$ be a sequence of i.i.d. random motions with $G_i' \stackrel{d}{=} H_\alpha$. By the definition of symmetric α-stable random motion and Lemma 16.3.1, it follows that

$$\begin{aligned}
&\boldsymbol{\rho}(n^{-1/\alpha}(G_1 \circ \cdots \circ G_n), H_\alpha)\\
&\quad = \boldsymbol{\rho}(n^{-1/\alpha}(G_1 \circ \cdots \circ G_n), n^{-1/\alpha}(G_1' \circ \cdots \circ G_n')),\\
&\quad \leq C\boldsymbol{\rho}(\delta \overline{H}_\alpha \circ n^{-1/\alpha}(G_1 \circ \cdots \circ G_n), \delta \overline{H}_\alpha \circ n^{-1/\alpha}(G_1' \circ \cdots \circ G_n')) + C\delta.
\end{aligned} \tag{16.3.13}$$

If the triangle inequality is applied m times, then the first term in (16.3.13) is bounded by

$$\begin{aligned}
&\boldsymbol{\rho}(\delta\overline{H}_\alpha \circ n^{-1/\alpha}G_1 \circ \cdots \circ n^{-1/\alpha}G_n, \delta\overline{H}_\alpha \circ n^{-1/\alpha}G_1 \circ \cdots \circ n^{-1/\alpha}G_{n-1} \circ n^{-1/\alpha}G_n')\\
&\quad + \sum_{j=1}^{m} \boldsymbol{\rho}(\delta\overline{H}_\alpha \circ n^{-1/\alpha}G_1 \circ \cdots \circ n^{-1/\alpha}G_{n-j} \circ n^{-1/\alpha}G_{n-j+1}' \circ \cdots \circ n^{-1/\alpha}G_n',\\
&\qquad \delta\overline{H}_\alpha \circ n^{-1/\alpha}G_1 \circ \cdots \circ n^{-1/\alpha}G_{n-j-1} \circ n^{-1/\alpha}G_{n-j}' \circ \cdots \circ n^{-1/\alpha}G_n')\\
&\quad + \boldsymbol{\rho}(\delta\overline{H}_\alpha \circ n^{-1/\alpha}G_1 \circ \cdots \circ n^{-1/\alpha}G_{n-m-1} \circ n^{-1/\alpha}G_{n-m}' \circ \cdots \circ n^{-1/\alpha}G_n',\\
&\qquad \delta\overline{H}_\alpha \circ n^{-1/\alpha}G_1' \circ \cdots \circ n^{-1/\alpha}G_n')\\
&\quad := A_1 + A_2 + A_3.
\end{aligned} \tag{16.3.14}$$

Next, using Lemma 16.3.3, A_1 and A_2 may be bounded as follows:

$$\begin{aligned}
A_1 &\leq \boldsymbol{\rho}(n^{-1/\alpha}G_1 \circ \cdots \circ n^{-1/\alpha}G_{n-1}, n^{-1/\alpha}G_1' \circ \cdots \circ n^{-1/\alpha}G_{n-1}')\\
&\quad \times \mathbf{Var}(\delta\overline{H}_\alpha \circ n^{-1/\alpha}G_n, \delta\overline{H}_\alpha \circ n^{-1/\alpha}G_n')\\
&\quad + \boldsymbol{\rho}(\delta\overline{H}_\alpha \circ n^{-1/\alpha}G_1' \circ \cdots \circ n^{-1/\alpha}G_{n-1}' \circ n^{-1/\alpha}G_n,\\
&\qquad \delta\overline{H}_\alpha \circ n^{-1/\alpha}G_1' \circ \cdots \circ n^{-1/\alpha}G_n')\\
&=: I_1 + I_3'.
\end{aligned} \tag{16.3.15}$$

Similarly,

$$
\begin{aligned}
A_2 \le & \sum_{j=1}^{m} \rho(n^{-1/\alpha}(G_1 \circ \cdots \circ G_{n-j-1}), n^{-1/\alpha}(G_1' \circ \cdots \circ G_{n-j-1}')) \\
& \times \mathbf{Var}(\delta \overline{H}_\alpha \circ n^{-1/\alpha} G_{n-j} \circ n^{-1/\alpha} G_{n-j+1}' \circ \cdots \circ n^{-1/\alpha} G_n', \\
& \delta \overline{H}_\alpha \circ n^{-1/\alpha} G_{n-j}' \circ n^{-1/\alpha} G_{n-j+1}' \circ \cdots \circ n^{-1/\alpha} G_n') \\
& + \sum_{j=1}^{m} \rho(\delta \overline{H}_\alpha \circ n^{1/\alpha} G_1' \circ \cdots \circ n^{-1/\alpha} G_{n-j-1}' \circ n^{-1/\alpha} G_{n-j} \circ \cdots \circ n^{-1/\alpha} G_n', \\
& \delta \overline{H}_\alpha \circ n^{-1/\alpha}(G_1' \circ \cdots \circ G_n')) \\
:= & I_2 + I_3''. \qquad (16.3.16)
\end{aligned}
$$

Combining (16.3.13)–(16.3.16) and letting $I_3 = I_3' + I_3''$, $I_4 := A_3$ yields

$$
\rho(n^{-1/\alpha}(G_1 \circ \cdots \circ G_n), H_\alpha) \le C(I_1 + I_2 + I_3 + I_4) + C\delta. \qquad (16.3.17)
$$

Next, Lemma 16.3.2 will be used to successively estimate each of the quantities I_1, I_2, I_3, and I_4.

By the induction hypothesis, Lemma 16.3.2 (with $\lambda_1 = \lambda = \delta$ and $\lambda_2 = 0$ there), and the ideality of $\boldsymbol{\nu}_r$, it follows that

$$
\begin{aligned}
I_1 & \le B(\nu_r(n-1)^{1-r/\alpha} + \tau_r(n-1)^{1-r/\alpha})\nu_1(n^{-1/\alpha} G_1, n^{-1/\alpha} \widetilde{G}_1)/\delta \\
& \le C(B/A)(\nu_r n^{1-r/\alpha} + \tau_r n^{1-r/\alpha}) \qquad (16.3.18)
\end{aligned}
$$

by definition of δ.

To estimate I_2, apply the induction hypothesis again, Lemma 16.3.2 [with $\lambda_1 = \delta$, $\lambda_2 = (j/n)^{1/\alpha}$], the ideality of $\boldsymbol{\nu}_r$, and the definition of δ to obtain

$$
\begin{aligned}
I_2 = & \sum_{j=1}^{m} \rho(n^{-1/\alpha}(G_1 \circ \cdots \circ G_{n-j-1}), n^{-1/\alpha}(G_1' \circ \cdots \circ G_{n-j-1}')) \\
& \mathbf{Var}(\delta \overline{H}_\alpha \circ n^{-1/\alpha} G_{n-j} \circ (j/n)^{1/\alpha} H_\alpha, \delta \overline{H}_\alpha \circ n^{-1/\alpha} G_{n-j}' \circ (j/n)^{1/\alpha} H_\alpha) \\
\le & B(\nu_r(n-m)^{1-r/\alpha} + \tau_r(n-m)^{-1/\alpha}) \\
& \sum_{j=1}^{m} \frac{1}{(\delta^\alpha + j/n)^{r/\alpha}} \boldsymbol{\nu}_r(n^{-1/\alpha} G_1, n^{-1/\alpha} G_1') \\
\le & B(\nu_r n^{1-r/\alpha} + \tau_r n^{-1/\alpha}) \sum_{j=1}^{\infty} \nu_r/(A^\alpha \nu_r^{\alpha/(r-\alpha)} + j)^{r/\alpha} \\
\le & B(\nu_n n^{1-r/\alpha} + \tau_r n^{-1/\alpha}) C(A^\alpha \nu_r^{\alpha/(r-\alpha)})^{1-(r/\alpha)} \nu_r \\
\le & CB(\nu_r n^{1-r/\alpha} + \tau_r n^{-1/\alpha})/A^{\alpha-r}, \qquad (16.3.19)
\end{aligned}
$$

where C again denotes some absolute constant.

To estimate

$$I_3 = \sum_{j=0}^{m} \boldsymbol{\rho}(\delta\overline{H}_\alpha \circ n^{-1/\alpha}(G'_1 \circ \cdots \circ G'_{n-j-1} \circ G_{n-j} \circ G'_{n-j+1} \circ \cdots \circ G'_n)$$
$$\delta\overline{H}_\alpha \circ n^{-1/\alpha}(G'_1 \circ \cdots \circ G'_n)),$$

use $2\boldsymbol{\rho} < \mathbf{Var}$, Lemma 16.3.2 [with $\lambda_1 = ((n-j-1)/(n-1))^{1/\alpha}$, $\lambda_2 = (j/(n-1))^{1/\alpha}$, and $\lambda = 1$], and the ideality of $\boldsymbol{\nu}_r$ to obtain

$$\begin{aligned}
I_3 &\le \sum_{j=0}^{m} \boldsymbol{\nu}_r \left(\left(\frac{n-j-1}{n-1}\right)^{1/\alpha} H_\alpha \circ (n-1)^{-1/\alpha} G_{n-j} \circ \left(\frac{j}{n-1}\right)^{1/\alpha} H_\alpha\,, \right.\\
&\qquad \left. \left(\frac{n-j-1}{n-1}\right)^{1/\alpha} H_\alpha \circ (n-1)^{1/\alpha} G'_{n-j} \circ \left(\frac{j}{n-1}\right)^{1/\alpha} H_\alpha \right)\\
&\le \sum_{j=0}^{m} \boldsymbol{\nu}_r((n-1)^{-1/\alpha} G_1, (n-1)^{-1/\alpha} G'_1)\\
&\le n^{1-r/\alpha} \nu_r,
\end{aligned} \tag{16.3.20}$$

where it is assumed that n_0 is chosen such that $(n/(n-1))^{r/\alpha} \le 2$.

Similarly, using Lemma 16.3.2 with $\lambda_1 = 0$ and $\lambda_2 = 1$, we may bound I_4

$$\begin{aligned}
I_4 &\le \boldsymbol{\rho}(m^{-1/\alpha}(G_1 \circ \cdots \circ G_{n-m-1}) \circ m^{-1/\alpha}(G'_{n-m} \circ \cdots \circ G'_n),\\
&\qquad m^{-1/\alpha}(G'_1 \circ \cdots \circ G'_{n-m-1} \circ m^{-1/\alpha}(G'_{n-m} \circ \cdots \circ G'_n))\\
&= \boldsymbol{\rho}(\delta\overline{H}_\alpha \circ m^{-1/\alpha}(G_1 \circ \cdots \circ G_{n-m-1}) \circ H_\alpha,\\
&\qquad \delta\overline{H}_\alpha \circ m^{-1/\alpha}(G'_1 \circ \cdots \circ G'_{n-m-1}) \circ H_\alpha)\\
&\le \mathbf{Var}(m^{-1/\alpha}(G_1 \circ \cdots \circ G_{n-m-1}) \circ H_\alpha, m^{-1/\alpha}(G'_1 \circ \cdots \circ G'_{n-m-1}) \circ H_\alpha)\\
&\le \boldsymbol{\nu}_r(m^{-1/\alpha}(G_1 \circ \cdots \circ G_{n-m-1}), m^{-1/\alpha}(G'_1 \circ \cdots \circ G'_{n-m-1}))\\
&\le m^{-r/\alpha}(n-m-1)\boldsymbol{\nu}_r(G_1, G'_1) \le 2^{r/\alpha} n^{1-r/\alpha} \nu_r,
\end{aligned} \tag{16.3.21}$$

since we may assume that $((n-m-1)/n)(n/m)^{r/\alpha}$ is bounded by $2^{r/\alpha}$ for $n \ge n_0$.

Finally, combining estimates (16.3.17)–(16.3.21) and the definition of δ yields

$$\begin{aligned}
\boldsymbol{\rho}(n^{-1/\alpha}(G_1 \circ \cdots \circ G_n), H_\alpha) &\le C(I_1 + I_2 + I_3 + I_4) + C\delta\\
&\le C(A^{-1} + A^{\alpha-r})B(\nu_r n^{1-r/\alpha} + \tau_r n^{-1/\alpha})\\
&\quad + C\nu_r n^{1-r/\alpha} + CA\tau_r n^{-1/\alpha}.
\end{aligned}$$

Choosing the absolute constant A such that $C(A^{-1} + A^{\alpha-r}) \leq \frac{1}{2}$ shows, for sufficiently large B,

$$\rho(n^{-1/\alpha}(G_1 \circ \cdots \circ G_n), H_\alpha) \leq B(\nu_r n^{1-r/\alpha} + \tau_r n^{-1/\alpha}),$$

completing the proof of Theorem 16.3.1. □

The main theorem in the second part of this section deals with uniform rates of convergence in the *local* limit theorem on $\mathbb{M}(d)$ (see further Theorem 16.3.2). Again, ideal smoothing metrics play a considerable role. More precisely, if $\{G_i\} = \{Y_i, U_i\}_{i\geq 1}$ are i.i.d. random motions and $G_1(x)$ has a density $p_{G_1(x)}$ for any $X \in \mathbb{R}^d$, then ideal metrics are used to determine the rate of convergence in the limit relationship

$$\ell(n^{-1/\alpha}(G_1 \circ \cdots \circ G_n), H_\alpha) \to 0, \tag{16.3.22}$$

where ℓ is determined by (16.2.5) and (16.3.4).

The result considers rates in (16.3.22) under hypotheses on $\nu_r := \boldsymbol{\nu}_r\ (G_1, H_\alpha)$, $\ell := \ell(G_1, H_\alpha)$, and $\mu_r := \boldsymbol{\mu}_r(G_1, H_\alpha)$, where

$$\begin{aligned}\boldsymbol{\mu}_r(G_1, H_\alpha) &:= \sup_{x\in\mathbb{R}^d} \sup_{h\in\mathbb{R}} |h|^r \ell(G_1(x) + h\theta, H_\alpha + h\theta) \\ &= \sup_{x\in\mathbb{R}} |h|^r \ell((h\overline{H}_\alpha) \circ G_1, (h\overline{H}_\alpha) \circ H_\alpha)\end{aligned} \tag{16.3.23}$$

and $\overline{H}_\alpha := (Y_\alpha, I)$ denotes, as before, the projection of H_α on $\mathbb{R}^d$.

The proof of the next theorem depends heavily upon the ideality of $\boldsymbol{\nu}_r$ and $\boldsymbol{\mu}_r$. As in the proof of Theorem 16.3.1, ideality is first used to establish some critical smoothing inequalities. The first smoothing inequality provides a rate of convergence in (16.3.1) with respect to the **Var**-metric and could actually be considered a companion lemma to the main result. The proof of the next lemma is similar to that of Theorem 15.4.1 and is thus omitted.

Lemma 16.3.4. *Let $r > \alpha$ and*

$$K_r := K_r(G_1, H_\alpha) := \max\{\mathbf{Var}(G_1, H_\alpha), \boldsymbol{\nu}_r(G_1, H_\alpha)\} \leq a,$$

where $a^{-1} := 2^{1+r/\alpha}(2^{(r/\alpha)-1} + 3^{r/\alpha})$. If $A := 2(2^{(r/\alpha)-1} + 3^{r/\alpha})$, then

$$\mathbf{Var}(n^{-1/\alpha}(G_1 \circ \cdots \circ G_n), H_\alpha) \leq AK_r n^{1-r/\alpha}.$$

The next estimate, the companion to Lemma 16.3.2, is the analog of Lemma 15.4.5. The proof is similar to that of Lemma 16.3.2 and will be omitted.

Lemma 16.3.5. *Let $G_1, G_2 \in \mathfrak{X}(\mathbb{M}(d))$, $\lambda_i > 0$, $i = 1, 2$; $\lambda^\alpha = \lambda_1^\alpha + \lambda_2^\alpha$, $\widetilde{H}_\alpha \stackrel{d}{=} H_\alpha$. For all $r > 0$*

$$\ell(\lambda_1 H_\alpha \circ G_1 \circ \lambda_2 \widetilde{H}_\alpha, \lambda_1 H_\alpha \circ G_2 \circ \lambda_2 \widetilde{H}_\alpha) \leq \lambda^{r-1} \boldsymbol{\mu}_r(G_1, G_2) \tag{16.3.24}$$

and

$$\ell(\lambda_1 \widetilde{H}_\alpha \circ G_1 \lambda_2 \widetilde{H}_\alpha, \lambda_1 \overline{H}_\alpha \circ G_2 \circ \lambda_2 \widetilde{H}_\alpha) \leq \lambda^{r-1} \boldsymbol{\mu}_r(G_1, G_2). \tag{16.3.25}$$

The following smoothing inequality may be considered the analog of Lemma 15.4.6. Only (16.3.27) is used in the sequel.

Lemma 16.3.6. *Let $G_1^*, G_2^*, \widetilde{G}_1, \widetilde{G}_2 \in \mathfrak{X}(\mathbb{M}(d))$ and $\lambda \geq 0$. Then*

$$\begin{aligned}\ell(\lambda \overline{H}_\alpha \circ G_1^* \circ \widetilde{G}_1, \lambda \overline{H}_\alpha \circ G_1^* \circ \widetilde{G}_2) \leq \ell(G_1^*, G_2^*) \, \mathbf{Var}(\lambda \overline{H}_\alpha \circ \widetilde{G}_1, \lambda \overline{H}_\alpha \circ \widetilde{G}_2) \\ + \ell(\lambda \overline{H}_\alpha \circ G_2^* \circ \widetilde{G}_1, \lambda \overline{H}_\alpha \circ G_2^* \circ \widetilde{G}_2)\end{aligned} \tag{16.3.26}$$

and

$$\begin{aligned}\ell(\lambda \overline{H}_\alpha \circ G_1^* \circ \widetilde{G}_1, \lambda \overline{H}_\alpha \circ G_2^* \circ \widetilde{G}_1) \leq \ell(G_1^*, G_2^*) \, \mathbf{Var}(\lambda \overline{H}_\alpha \circ \widetilde{G}_1, \lambda \overline{H}_\alpha \circ \widetilde{G}_2) \\ + \ell(\lambda \overline{H}_\alpha \circ G_1^* \circ \widetilde{G}_2, \lambda \overline{H}_\alpha \circ G_2^* \circ \widetilde{G}_2).\end{aligned} \tag{16.3.27}$$

Proof. Since $\overline{H}_\alpha \circ G = G \circ \overline{H}_\alpha$, we see that $\ell(\lambda \overline{H}_\alpha \circ G_1^* \circ \widetilde{G}_1, \lambda \overline{H}_\alpha \circ G_2^* \circ \widetilde{G}_2)$ equals

$$\begin{aligned}&\ell(G_1^* \circ \lambda \overline{H}_\alpha \circ \widetilde{G}_1, G_1^* \circ \lambda \overline{H}_\alpha \circ \widetilde{G}_2) \\ &= \sup_{x \in \mathbb{R}^d} \operatorname*{ess\,sup}_{z \in \mathbb{R}^d} |p_{G_1^* \circ \lambda \overline{H}_\alpha \circ \widetilde{G}_1(x)}(z) - p_{G_1^* \circ \lambda \overline{H}_\alpha \circ \widetilde{G}_2(x)}(z)| \\ &= \sup_x \operatorname*{ess\,sup}_z \left| \int_{\mathbb{M}(d)} p_{G_1^* \circ g(x)}(z) [\Pr(\lambda \widetilde{H}_\alpha \circ \widetilde{G}_1 \in dg) - \Pr(\lambda \overline{H}_\alpha \circ G_2 \in d)] \right| \\ &= \sup_x \operatorname*{ess\,sup}_z |p_{G_1^* \circ g(x)}(z) - p_{G_2^* \circ g(x)}(z)| \int_{\mathbb{M}(d)} |\Pr(\lambda \overline{H}_\alpha \circ \widetilde{G}_1 \in dg) \\ &\quad - \Pr(\lambda \overline{H}_\alpha \circ \widetilde{G}_2 \circ dg)| + \ell(\lambda \overline{H}_\alpha \circ G_2^* \circ \widetilde{G}_1, \lambda \overline{H}_\alpha \circ G_2^* \circ \widetilde{G}_2) \\ &\leq \ell(G_1^*, G_2^*) \, \mathbf{Var}(\lambda \overline{H}_\alpha \circ \widetilde{G}_1, \lambda \overline{H}_\alpha \circ \widetilde{G}_2) \\ &\quad + \ell(\lambda \overline{H}_\alpha \circ G_2^* \circ \widetilde{G}_1, \lambda \overline{H}_\alpha \circ G_2^* \circ \widetilde{G}_2).\end{aligned}$$

This proves (16.3.26); (16.3.27) is proved similarly. □

With these three smoothing inequalities, the main result may now be proved.

Theorem 16.3.2. *Let the following two conditions hold:*

$$\lambda_r(G_1, H_\alpha) := \max\{\ell(G_1, H_\alpha), \boldsymbol{\mu}_{r+1}(G_1, H_\alpha)\} < \infty \tag{16.3.28}$$

and

$$K_r := K_r(G_1, H_\alpha) := \max\{\mathbf{Var}(G_1, H_\alpha), \boldsymbol{\nu}_r(G_1, H_\alpha)\} \leq 1/DA, \tag{16.3.29}$$

where $r > \alpha$, $A := 2(2^{(r/\alpha)-1} + 3^{r/\alpha})$ *and* $D := 2(3^{-1+(r+1)/\alpha})$. *Then*

$$\ell(n^{-1/\alpha}(G_1 \circ \cdots \circ G_n), H_\alpha) \leq A\lambda_r(G_1, H_\alpha)n^{1-r/\alpha}. \tag{16.3.30}$$

Proof. Let $G_1', G_2', \ldots$ be a sequence of i.i.d. random motions with $G_i' \stackrel{d}{=} H_\alpha$. Now (16.3.30) holds for $n = 1, 2$ and 3. Let $n > 3$.

Suppose that for all $j < n$

$$\ell(j^{-1/\alpha}(G_1 \circ \cdots \circ G_j), H_\alpha) \leq A\lambda_r j^{1-r/\alpha}. \tag{16.3.31}$$

To complete the induction proof, it only remains to show that (16.3.31) holds for $j = n$. By (16.3.27) with $\lambda = 0$ and $m = [n/2]$,

$$\ell(n^{-1/\alpha}(G_1 \circ \cdots \circ G_n), n^{-1/\alpha}(G_1' \circ \cdots \circ G_n'))$$

is bounded by

$$\begin{aligned}
&\leq \ell((n^{-1/\alpha}(G_1 \circ \cdots \circ G_m) \circ n^{-1/\alpha}(G_{m+1} \circ \cdots \circ G_n),\\
&\qquad n^{-1/\alpha}(G_1' \circ \cdots \circ G_m') \circ n^{-1/\alpha}(G_{m+1} \circ \cdots \circ G_n))\\
&\quad + \ell(n^{-1/\alpha}(G_1' \circ \cdots \circ G_m') \circ n^{-1/\alpha}(G_{m+1} \circ \cdots \circ G_n),\\
&\qquad n^{-1/\alpha}(G_1' \circ \cdots \circ G_m') \circ n^{-1/\alpha}(G_{m+1}' \circ \cdots \circ G_n'))\\
&\leq \ell(n^{-1/\alpha}(G_1 \circ \cdots \circ G_m), n^{-1/\alpha}(G_1' \circ \cdots \circ G_m'),\\
&\qquad \mathbf{Var}(n^{-1/\alpha}(G_{m+1} \circ \cdots \circ G_n), n^{-1/\alpha}(G_{m+1}' \circ \cdots \circ G_n'))\\
&\quad + \ell((n^{-1/\alpha}(G_1 \circ \cdots \circ G_m) \circ n^{-1/\alpha}(G_{m+1}' \circ \cdots \circ G_n'),\\
&\qquad n^{-1/\alpha}(G_1' \circ \cdots \circ G_m') \circ n^{-1/\alpha}(G_{m+1}' \circ \cdots \circ G_n'))\\
&\quad + \ell((n^{-1/\alpha}(G_1' \circ \cdots \circ G_m') \circ n^{-1/\alpha}(G_{m+1} \circ \cdots \circ G_n),\\
&\qquad n^{-1/\alpha}(G_1' \circ \cdots \circ G_m') \circ n^{-1/\alpha}(G_{m+1}' \circ \cdots \circ G_n'))\\
&:= I_1 + I_2 + I_3.
\end{aligned}$$

As in the proof of Lemma 16.3.4, it may be shown via Lemma 16.3.5 that

$$I_2 + I_3 \leq (2^{(r/\alpha)-1} + 3^{r/\alpha})\mu_{r+1}n^{1-r/\alpha} \leq \tfrac{1}{2}A\lambda_r n^{1-r/\alpha}.$$

It remains to estimate I_1. By the homogeneity property $\ell(X, Y) = c\ell(cX, cY)$ and the induction hypothesis, the first factor in I_1 is bounded by

$$\begin{aligned}
(n/m)^{1/\alpha}\ell(m^{-1/\alpha}(G_1 \circ \cdots \circ G_m), H_\alpha) &\leq (n/m)^{1/\alpha}A\lambda_r m^{1-r/\alpha}\\
&\leq 3^{(r+1)/\alpha-1}A\lambda_r n^{1-r/\alpha}.
\end{aligned}$$

By Lemma 16.3.4, the second factor in I_1 is bounded by

$$\begin{aligned}&\mathbf{Var}(n^{-1/\alpha}(G_{m+1}\circ\cdots\circ G_n), n^{-1/\alpha}(G'_{m+1}\circ\cdots\circ G'_n))\\ &\le AK_r(n-m)^{1-r/\alpha}\\ &\le AK_r \le D^{-1}.\end{aligned}$$

Hence, $I_1 \le \frac{1}{2}A\lambda_r n^{1-r/\alpha}$. Combining this with the displayed bound on $I_1 + I_2$ shows that

$$\ell(n^{-1/\alpha}(G_1\circ\cdots\circ G_n), H_\alpha) \le A\lambda_r n^{1-r/\alpha}$$

as desired. □

Conditions (16.3.28) and (16.3.29) in Theorem 16.3.2 and the conditions $\nu_r = \boldsymbol{\nu}_r(G_1, H_\alpha) < \infty$ and $\tau_r = \boldsymbol{\tau}_r(G_1, H_\alpha) < \infty$ in Theorem 16.3.1 can be examined via Theorem 16.2.3 and estimate (16.2.16).

References

Baldi P (1979) Lois stables sur les deplacements de R^n. In: Probability measures on groups. Lecture notes in mathematics, vol 706. Springer, Berlin, pp 1–9

Bhattacharya RM, Ranga Rao R (1976) Normal approximation and asymptotic expansions. Wiley, New York

Paulauskas VI (1974) Estimates of the remainder term in limit theorems in the case of stable limit law. Lith Math Trans XIV:127–147

Paulauskas VI (1976) On the rate of convergence in the central limit theorem in certain Banach spaces. Theor Prob Appl 21:754–769

Rachev ST, Yukich JE (1991) Rates of convergence of α-stable random motions. J Theor Prob 4(2):333–352

Senatov VV (1980) Uniform estimates of the rate of convergence in the multi- dimensional central limit theorem. Theor Prob Appl 25:745–759

Zolotarev VM (1986) Contemporary theory of summation of independent random variables. Nauka, Moscow (in Russian)

Chapter 17
Applications of Ideal Metrics for Sums of i.i.d. Random Variables to the Problems of Stability and Approximation in Risk Theory

The goals of this chapter are to:

- Formulate and analyze the mathematical problem behind insurance risk theory,
- Consider the problem of continuity and provide a solution based on ideal probability metrics,
- Consider the problem of stability and provide a solution based on ideal probability metrics.

Notation introduced in this chapter:

Notation	Description
$N(t)$	Number of claims up to time t
$X(t)$	Total amount of claims

17.1 Introduction

In this chapter, we present applications of ideal probability metrics to insurance risk theory. First, we describe and analyze the mathematical framework. When building a model, we must consider approximations that lead to two main issues: the problem of continuity and the problem of stability. We solve the two problems using the techniques of ideal probability metrics.

17.2 Problem of Stability in Risk Theory

When using a stochastic model in insurance risk theory, one must consider the model as an approximation of real insurance activities. The stochastic elements derived from these models represent an idealization of real insurance phenomena

S.T. Rachev et al., *The Methods of Distances in the Theory of Probability and Statistics*, DOI 10.1007/978-1-4614-4869-3_17,

under consideration. Hence the problem arises of establishing the limits in which one can use our *ideal* model. The practitioner must know the accuracy of our recommendations, which have resulted from our investigations based on the ideal model.

Mostly one deals with real insurance phenomena including the following main elements: input data (epochs of claims, size of claims, ...) and, resulting from these, output data (number of claims up to time t, total claim amount ...).

In this section we apply the method of metric distances to investigate the "horizon" of widely used stochastic models in insurance mathematics. The main stochastic elements of these models are as follows.

(a) *Model input elements:* the *epochs of the claims*, denoted by $T_0 = 0,\ T_1, T_2, \dots$, where $\{W_i = T_i - T_{i-1},\ i = 1, 2, \dots\}$ is a sequence of positive RVs, and the sequences of *claim sizes* $X_0 = 0,\ X_1, X_2, \dots$, where X_n is the claim occurring at time T_n.

(b) *Model output elements:* the *number of claims* up to time t

$$N(t) = \sup\{n : T_n \leq t\}, \tag{17.2.1}$$

and the *total claim* amount at time t

$$X(t) = \sum_{i=0}^{N(t)} X_i. \tag{17.2.2}$$

In particular, let us consider the problem of calculating the distribution of $X(t)$. Teugels (1985) writes that it is generally extremely complicated to evaluate the compound distribution $G_t(x)$ of $X(t)$

$$G_t(x) = \sum_{n=1}^{\infty} \Pr\Big\{\sum_{i=1}^{n} X_i \leq x | N(t) = n\Big\} \Pr\{N(t) = n\} + \Pr\{N(t) = 0\},\ x \geq 0. \tag{17.2.3}$$

This forces one to rely on approximations, even in the case when the sequences $\{X_i\}$ and $\{W_i\}$ are independent and consist of i.i.d. RVs.

Here, using approximations means that we investigate *ideal* models that are rather simple but nevertheless close in some sense to the real (disturbed) model. For example, as an ideal model we can consider $\widetilde{W}_i = \widetilde{T}_i - \widetilde{T}_{i-1}, i = 1, 2, \dots$, to be independent with a common simple distribution (e.g., an exponential). Moreover, one often supposes that the claim sizes $\widetilde{X}_i$ in the ideal model are i.i.d. and independent of $\widetilde{W}_i$.

We consider $\widetilde{W}_i$ and $\widetilde{X}_i$ as input elements for our ideal model. Correspondingly, we define

$$\widetilde{N}(t) = \sup\{n : \widetilde{T}_n \leq t\}, \tag{17.2.4}$$

$$\widetilde{X}(t) = \sum_{i=0}^{\widetilde{N}(t)} \widetilde{X}_i \tag{17.2.5}$$

as the output elements of our ideal model, related to the output elements $N(t)$ and $X(t)$ of the real model. More concretely, our approximation problem can be stated in the following way: if the input elements of the ideal and real models are *close* to each other, then can we estimate the deviation between the corresponding outputs? Translating the concept of closeness in a mathematical way one uses some measures of comparisons between the characteristics of the random elements involved.

In this section, we confine ourselves to investigating the sketched problems when the sequences $\{X_i\}$ and $\{W_i\}$ have i.i.d. components and are mutually independent. Then we can state our mathematical problem in the following way.

PR I. Let μ, ν, τ be simple probability metrics on $\mathfrak{X}(\mathbb{R})$, i.e., metrics in the distribution function space.[1] Find a function $\psi : [0,\infty) \times [0,\infty) \to [0,\infty)$, nondecreasing in both arguments, vanishing, and continuous at the origin such that for every $\varepsilon, \delta > 0$

$$\left.\begin{array}{l}\mu(W_1, \widetilde{W}_1) < \varepsilon \\ \nu(X_1, \widetilde{X}_1) < \delta\end{array}\right\} \Rightarrow \tau(X(t), \widetilde{X}(t)) \leq \psi(\varepsilon, \delta). \tag{17.2.6}$$

The choice of τ is dictated by the user, who also wants to be able to check the left-hand side of (17.2.6). For this reason, the next stability problem is relevant.

PR II. Find a qualitative description of the ε-neighborhood (resp. δ-neighborhood) of the set of ideal model distributions $F_{\widetilde{W}_1}$ (resp. $F_{\widetilde{X}_1}$).

17.3 Problem of Continuity

In this section we consider *PR I* as described in (17.2.6). Usually, in practice, the metric τ is chosen to be the Kolmogorov (uniform) metric,

$$\boldsymbol{\rho}(X,Y) = \sup_{x\in\mathbb{R}} |F_X(x) - F_Y(x)|. \tag{17.3.1}$$

Moreover, we will choose $\mu = \nu = \boldsymbol{\kappa}_r$, where

$$\boldsymbol{\kappa}_r(X,Y) = r\int_{\mathbb{R}} |x|^{r-1}|F_X(x) - F_Y(x)|\mathrm{d}x, \quad r > 0, \tag{17.3.2}$$

is the difference pseudomoment.[2] The usefulness of $\boldsymbol{\kappa}_r$ will follow from the considerations in the next section, where *PR II* is treated. The metric $\boldsymbol{\kappa}_r$ metrizes the

---

[1]As before, we will write $\mu(X,Y)$, $\nu(X,Y)$, $\tau(X,Y)$ instead of $\mu(F_X,F_Y)$, $\nu(F_X,F_Y)$, $\tau(F_X,F_Y)$.

[2]See Case D in Sect. 4.4 of Chap. 4.

weak convergence, plus the convergence of the rth absolute moments in the space of RVs X with $E|X|^r < \infty$, i.e.,[3]

$$\kappa_r(X_n, X) \to 0 \iff \begin{cases} X_n \xrightarrow{w} X \\ E|X_n|^r \to E|X|^r \end{cases} \quad \text{as } n \to \infty.$$

Also, note that

$$\kappa_r(X, Y) = \kappa_1(X|X|^{r-1}, Y|Y|^{r-1}). \tag{17.3.3}$$

First, let us simplify the right-hand side of (17.2.6). Using the triangle inequality we get

$$\begin{aligned}\rho(X(t), \widetilde{X}(t)) &= \rho\left(\sum_{i=0}^{N(t)} X_i, \sum_{i=0}^{\widetilde{N}(t)} \widetilde{X}_i\right) \\ &\leq \rho\left(\sum_{i=0}^{N(t)} X_i, \sum_{i=0}^{N(t)} \widetilde{X}_i\right) + \rho\left(\sum_{i=0}^{N(t)} \widetilde{X}_i, \sum_{i=0}^{\widetilde{N}(t)} \widetilde{X}_i\right) \\ &:= I_1 + I_2. \end{aligned} \tag{17.3.4}$$

Assuming $H(t) = EN(t)$ to be finite, we have

$$I_1 = \rho\left(\sum_{i=0}^{N(t)} \widetilde{X}_i, \sum_{i=0}^{N(t)} X_i\right) = \rho\left(\frac{1}{H(t)}\sum_{i=0}^{N(t)} X_i, \frac{1}{H(t)}\sum_{i=0}^{N(t)} \widetilde{X}_i\right). \tag{17.3.5}$$

From this expression we are going to estimate I_1 from above, by $\kappa_r(X_1, \widetilde{X}_1)$. This will be achieved in two steps:

1. Estimation of the closeness between the RVs

$$Z(t) = \frac{1}{H(t)}\sum_{i=0}^{N(t)} X_i, \quad \widetilde{Z}(t) = \frac{1}{H(t)}\sum_{i=0}^{N(t)} \widetilde{X}_i, \tag{17.3.6}$$

   in terms of an appropriate (*ideal* for this purpose) metric.
2. Passing from the ideal metric to ρ and κ_r, respectively, via inequalities of the type

$$\phi_1(\rho) \leq \text{ideal metric} \leq \phi_2(\kappa_r) \tag{17.3.7}$$

   for some nonnegative, continuous functions $\phi_i : [0,\infty) \to [0,\infty)$ with $\phi_i(0) = 0$, $\phi_i(t) > 0$ if $t > 0$, $i = 1, 2$.

[3] See Theorems 5.5.1 and 6.4.1 in Chaps. 5 and 6, respectively.

Considering the first step, we choose $\zeta_{m,p}$ ($m = 0, 1, \ldots, p \geq 1$) as our ideal metric, where $\zeta_{m,p}(X, Y)$ is given by (15.3.10). The metric $\zeta_{m,p}$ is ideal of order $r = m + 1/p$, i.e., for each X, Y, Z, with Z independent of X and Y and every $c \in \mathbb{R}$,[4]

$$\zeta_{m,p}(cX + Z, cY + Z) \leq |c|^r \zeta_{m,p}(X, Y). \tag{17.3.8}$$

These and other properties of $\zeta_{m,p}$ will be considered in the next chapter.[5]

Lemma 17.3.1. *Let $\{X_i\}$ and $\{\widetilde{X}_i\}$ be two sequences of i.i.d. RVs, and let $N(t)$ be independent of the sequences $\{X_i\}$, $\{\widetilde{X}_i\}$ and have finite moment $H(t) = EN(t) < \infty$. Then,*

$$\zeta_{m,p}(Z(t), \widetilde{Z}(t)) \leq H(t)^{1-r} \zeta_{m,p}(X_1, \widetilde{X}_1\}, \tag{17.3.9}$$

where $r = m + 1/p$.

Proof. The following chain of inequalities proves the required estimate.

$$\zeta_{m,p}(Z(t), \widetilde{Z}(t))$$

(a)

$$\leq H(t)^{-r} \zeta_{m,p}\left(\sum_{i=0}^{N(t)} X_i, \sum_{i=0}^{N(t)} \widetilde{X}_i\right)$$

(b)

$$\leq H(t)^{-r} \sum_{k=1}^{\infty} \Pr(N(t) = k) \zeta_{m,p}\left(\sum_{i=1}^{k} X_i, \sum_{i=1}^{k} \widetilde{X}_i\right)$$

(c)

$$\begin{aligned}
&\leq H(t)^{-r} \sum_{k=1}^{\infty} \Pr(N(t) = k) \sum_{i=1}^{k} \zeta_{m,p}(X_i, \widetilde{X}_i) \\
&= H(t)^{-r} \sum_{k=1}^{\infty} k \Pr(N(t) = k) \zeta_{m,p}(X_1, \widetilde{X}_1) \\
&= H(t)^{-r} \zeta_{m,p}(X_1, \widetilde{X}_1).
\end{aligned}$$

---

[4] See Definition 15.3.1 in Chap. 15.

[5] More specifically, see Lemma 18.2.2.

Here (a) follows from (17.3.8) with $Z = 0$ and $c = H(t)^{-1}$. Inequality (b) results from the independence of $N(t)$ with respect to $\{X_i\}, \{\widetilde{X}_i\}$. Finally, (c) can be proved by induction using the triangle inequality and (17.3.8) with $c = 1$. □

The obtained estimate (17.3.9) is meaningful if $\boldsymbol{\zeta}_{m,p}(X_1, \widetilde{X}_1) \leq \infty$. This implies, however, that[6]

$$\int_{\mathbb{R}} x^j \mathrm{d}(F_{X_1}(x) - F_{\widetilde{X}_1}(x)) = 0, \quad \text{for } j = 0, 1, \ldots, m. \tag{17.3.10}$$

Let us now find a lower bound for $\boldsymbol{\zeta}_{m,p}(Z(t), \widetilde{Z}(t))$ in terms of $\boldsymbol{\rho}$.[7]

Lemma 17.3.2. *If Y has a bounded density p_Y, then*

$$\boldsymbol{\rho}(X, Y) \leq \left(1 + \sup_{x \in \mathbb{R}} p_Y(x)\right) (c_{m,p} \boldsymbol{\zeta}_{m,p}(X, Y))^{1/(r+1)}, \tag{17.3.11}$$

where

$$c_{m,p} = \frac{(2m+2)!(2m+3)^{1/2}}{(m+1)!(3 - 2/p)^{1/2}}.$$

Proof. To prove (17.3.11), we use similar estimates between the Lévy metric $\mathbf{L} = \mathbf{L}_1$ [see (4.2.3)] and $\boldsymbol{\zeta}_{m,p}$. For any RVs X and Y[8]

$$\mathbf{L}(X, Y)^{r+1} \leq c_{m,p} \boldsymbol{\zeta}_{m,p}(X, Y). \tag{17.3.12}$$

Next, since the density of Y exists and is bounded, we have

$$\boldsymbol{\rho}(X, Y) \leq \left(1 + \sup_{x \in \mathbb{R}} p_Y(x)\right) \mathbf{L}(X, Y), \tag{17.3.13}$$

which implies (17.3.11). □

In addition, let us remark that $\boldsymbol{\zeta}_{0,\infty} = \boldsymbol{\rho}$ and $\boldsymbol{\zeta}_{0,1} = \boldsymbol{\kappa}_1$. So, combining Lemmas 15.3.6, 17.3.1, and 17.3.2, we prove immediately the following lemma.

Lemma 17.3.3. *Let $\{X_i\}$, $\{\widetilde{X}_i\}$ be two sequences of i.i.d. RVs and let $N(t)$ be independent of $\{X_i\}$, $\{\widetilde{X}_i\}$ with $H(t) = EN(t) < \infty$. Suppose that*

$$\boldsymbol{\kappa}_r(X_1, \widetilde{X}_1) < \infty$$

---

[6]Indeed, if (17.3.10) fails for some $j = 0, 1, \ldots, m$, then $\boldsymbol{\zeta}_{m,p}(X_1, \widetilde{X}_1) \geq \sup_{c>0} |E(cX_1^j - x\widetilde{X}_1^j)| = +\infty$.

[7]An upper bound for $\boldsymbol{\zeta}_{m,p}(X_1, \widetilde{X}_1)$ in terms of $\boldsymbol{\kappa}_r$ $(r = m + 1/p)$ is given by Lemma 15.3.6.

[8]See Kalashnikov and Rachev (1988, Theorem 3.10.2).

and

$$\int x^j \mathrm{d}(F_{X_1}(x) - F_{\widetilde{X}_1}(x)) = 0, \qquad j = 0, 1, \ldots, m, \tag{17.3.14}$$

for some $r = m + 1/p \geq 1$ $(m = 1, 2, \ldots; 1 \leq p < \infty)$. *Moreover, let* $\widetilde{Z}(t)$ *[see (17.3.6)] have a bounded density* $p_{\widetilde{Z}(t)}$. *Then*

$$\begin{aligned} I_1 &= \boldsymbol{\rho}\left(\sum_{i=0}^{N(t)} X_i, \sum_{i=0}^{N(t)} \widetilde{X}_i\right) \leq \psi_1(\kappa_r(X_1, \widetilde{X}_1)) \\ &:= (1 + \sup p_{\widetilde{Z}(t)}(x))(c_{m,p}\phi_2(\kappa_r(X_1, \widetilde{X}_1)))^{1/(1+r)} H(t)^{(1-r)/(1+r)}, \end{aligned} \tag{17.3.15}$$

where

$$\phi_2(\kappa_r) = \begin{cases} \kappa_1^{1/p}, & m = 0, 1 \leq p < \infty, \\ \dfrac{\Gamma(1 + p^{-1})}{\Gamma(r)} \kappa_r, & m > 0, 1 \leq p < \infty. \end{cases} \tag{17.3.16}$$

Now, going back to (17.3.4), *we need also to estimate*

$$I_2 = \boldsymbol{\rho}\left(\sum_{i=0}^{N(t)} \widetilde{X}_i, \sum_{i=0}^{\widetilde{N}(t)} \widetilde{X}_i\right)$$

from above by some function, ψ_2, *say, of* $\boldsymbol{\kappa}_r(W_1, \widetilde{W}_1)$.

Lemma 17.3.4. *Let* $\{W_i\}$, $\{\widetilde{W}_i\}$ *be two sequences of i.i.d. positive RVs, both independent of* $\{\widetilde{X}_i\}$. *Suppose that* $H(t) = EN(t) < \infty$, $\widetilde{H}(t) = E\widetilde{N}(t) < \infty$,

$$\theta(\widetilde{W}_1) = \sup_k \sup_x p_{k^{-1/2}\sum_{i=1}^k \widetilde{W}_i}(x) < \infty, \quad \kappa_r(W_1, \widetilde{W}_1) < \infty, \tag{17.3.17}$$

and

$$\int_0^\infty x^j \mathrm{d}(F_{W_1}(x) - F_{\widetilde{W}_1}(x)) = 0, \quad j = 0, 1, \ldots, m, \tag{17.3.18}$$

for some $r = m + 1/p (\geq 2)$ $(m = 1, 2, \ldots; 1 \leq p < \infty)$. *Finally, let* $F_{\widetilde{X}_1}(a) < 1$ $\forall a > 0$, *and* $E\widetilde{X}_1 < \infty$. *Then*

$$\begin{aligned} I_2 &= \boldsymbol{\rho}\left(\sum_{i=0}^{N(t)} \widetilde{X}_i, \sum_{i=0}^{\widetilde{N}(t)} \widetilde{X}_i\right) \leq \psi_2(\boldsymbol{\kappa}_r(W_1, \widetilde{W}_1)) \\ &:= (1 + \theta(\widetilde{W}_1))\boldsymbol{\kappa}_r(W_1, \widetilde{W}_1)^{1/(r+1)} \end{aligned}$$

$$+ \inf_{a>0}\{2(c_{m,p}(1+\theta(\widetilde{W}_1))\phi_2(\kappa_r(W_1,\widetilde{W}_1)))^{1/(1+r)}\chi_{\widetilde{X}_{1,r}}(a)$$
$$+a^{-1}E\widetilde{X}_1\max(H(t),\widetilde{H}(t))\}, \qquad (17.3.19)$$

where ϕ_2 *is given by* (17.3.16) *and*

$$\chi_{\widetilde{X}_{1,r}}(a) := \sum_{k=1}^{\infty} k^{(1-r/2)(1+r)} F^k_{\widetilde{X}_1}(a).$$

Remark 17.3.1. The normalization $k^{-1/2}$ of the sum $\sum_{i=1}^k \widetilde{W}_1$ in (17.3.17) comes from the quite natural assumption that the $\widetilde{W}_i$s – the claim's interarrival times for the ideal model – are in the domain of attraction of the normal law. Actually, this case will be considered in the next section. However, for example, if we need to approximate W_is with $\widetilde{W}_i$s, where $\widetilde{W}_i$ are in the normal domain of attraction of symmetric α-stable distribution with $\alpha < 2$, then we should use the normalization $k^{-1/2}$ in (17.3.17).

Remark 17.3.2. Note that if $\kappa_r(W_1,\widetilde{W}_1)$ tends to zero, then the right-hand side of (17.3.19) also tends to zero since, for each $a > 0$, $\chi_{\widetilde{X}_{1,r}}(a) < \infty$.

Proof of Lemma 17.3.4. By the independence of $N(t)$ and $\widetilde{N}(t)$ with respect to $\{\widetilde{X}_i\}$, we find that, for every $a > 0$,

$$I_2 = \sup_{0\le x\le a}\left|\sum_{k=1}^{\infty}[\Pr(N(t)=k)-\Pr(\widetilde{N}(t)=k)]\Pr\left(\sum_{i=1}^k \widetilde{X}_i \le x\right)\right|$$
$$+\sup_{x>a}\left|\Pr\left(\sum_{i=1}^{N(t)}\widetilde{X}_i > x\right)-\Pr\left(\sum_{i=1}^{\widetilde{N}(t)}\widetilde{X}_i > x\right)\right|$$
$$+|\Pr(N(t)=0)-\Pr(\widetilde{N}(t)=0)| =: J_{1,a}+J_{2,a}+J_3.$$

Estimating $J_{1,a}$ we get

(a)

$$J_{1,a} \le \sum_{k=1}^{\infty}\left(\left|\Pr\left(\sum_{i=1}^k W_i \le T\right)-\Pr\left(\sum_{i=1}^k \widetilde{W}_i \le t\right)\right|\right.$$
$$\left.+\left|\Pr\left(\sum_{i=1}^{k+1} W_i \le t\right)-\Pr\left(\sum_{i=1}^{k+1}\widetilde{W}_i \le t\right)\right|\right)\Pr\left(\max_{i=1,\dots,k}\widetilde{X}_i \le x\right)$$
$$\le \sum_{k=1}^{\infty}\left\{\rho\left(\sum_{i=1}^k W_i,\sum_{i=1}^k \widetilde{W}_i\right)+\rho\left(\sum_{i=1}^{k+1} W_i,\sum_{i=1}^{k+1}\widetilde{W}_i\right)\right\} F^k_{\widetilde{X}_1}(a);$$

(b)

$$\leq \sum_{k=1}^{\infty} (c_{m,p}\phi_2(\kappa_r(W_1, \widetilde{W}_1)))^{1/(1+r)}$$

$$\times\{k^{(1-r/2)(1+r)} + (k+1)^{(1-r/2)(1+r)}\}(1+\theta(\widetilde{W}_1))F^{k}_{\widetilde{X}_1}(a)$$

$$\leq 2(1+\theta(\widetilde{W}_1))(c_{m,p}\phi_2(\kappa_r(W_1, \widetilde{W}_1)))^{1/(1+r)}\chi_{\widetilde{X}_{1,r}}(a).$$

Inequality (a) follows from

$$\Pr(N(t) = k) = \Pr\left(\sum_{i=1}^{k} W_i \leq t\right) - \Pr\left(\sum_{i=1}^{k+1} W_i \leq t\right).$$

We derived (b) from Lemmas 17.3.2 and 15.3.6 [see also (17.3.16) and (17.3.17)]. Furthermore, one finds with Chebyshev's inequality that

$$J_{2,a} \leq \max\left(\Pr\left(\sum_{i=1}^{N(t)} \widetilde{X}_i > a\right), \Pr\left(\sum_{i=1}^{\widetilde{N}(t)} \widetilde{X}_i > a\right)\right)$$

$$\leq a^{-1}(EX_1)\max(H(t), \widetilde{H}(t)).$$

Inequality (15.3.22) can be extended in the case $m = 1$, $p = \infty$ (so $\zeta_{0,\infty} = \rho$) to

$$\rho(W_1, \widetilde{W}_1) \leq \left(1 + \sup_x P_{\widetilde{W}_1}(x)\right)\kappa_r(W_1, \widetilde{W}_1)^{1/(r+1)}. \tag{17.3.20}$$

By virtue of (17.3.13), we see that to prove (17.3.20), it is enough to show the following estimate.

Claim 3. For any nonnegative RVs X and Y

$$\mathbf{L}(X, Y) \leq \kappa_r(X, Y)^{1/(1+r)}. \tag{17.3.21}$$

Indeed, if the Lévy metric $\mathbf{L}(X, Y)$ is greater than $\varepsilon \in (0, 1)$, then there is $x_0 \geq 0$ such that $|F_X(x) - F_Y(x)| \geq \varepsilon\ \forall x \in [x_0, x_0 + \varepsilon]$. Thus

$$\kappa_r(X, Y) \geq r\int_{x_0}^{x_0+\varepsilon} x^{r-1}|F_X(x) - F_Y(x)|\mathrm{d}x \geq \varepsilon^{r+1}.$$

Letting $\varepsilon \to \mathbf{L}(X, Y)$ proves the claim. Finally, since $J_3 \leq \rho(W_1, \widetilde{W}_1)$, the lemma follows. □

We can conclude with the following theorem, which follows immediately by combining (17.3.4) and Lemmas 17.3.3 and 17.3.4.

Theorem 17.3.1. *Under the conditions of Lemmas 17.3.3 and 17.3.4,*

$$\rho(X(t), \widetilde{X}(t)) \leq \psi(\kappa_r(W_1, \widetilde{W}_1), \kappa_r(X_1, \widetilde{X}_1))$$
$$=: \psi_1(\kappa_r(X_1, \widetilde{X}_1)) + \psi_2(\kappa_r(W_1, \widetilde{W}_1)),$$

where ψ_1 (resp. ψ_2) is given by (17.3.15) *(resp.* (17.3.19)*).*

The preceding theorem gives us a solution to *PR I* [see (17.2.6)] with $\mu = \nu = \kappa_r$, and $\tau = \rho$ under some moment conditions (see Lemmas 17.3.3 and 17.3.4).

17.4 Stability of Input Characteristics

To solve *PR II* (Sect. 17.2), we will investigate the conditions on the real input characteristics that imply $\mu(W_1, \widetilde{W}_1) < \varepsilon$ and $\nu(X_1, \widetilde{X}_1) < \delta$ for $\mu = \nu = \kappa_r$ [see (17.2.6)]. We consider only $r = 2$ and qualitative conditions on the distribution of W_1, implying $\kappa_2(W_1, \widetilde{W}_1) < \varepsilon$. One can follow the same idea to check $\kappa_r(W_1, \widetilde{W}_1) < \varepsilon$, $r \neq 2$, and $\kappa_r(X_1, \widetilde{X}_1) < \delta$. We will characterize the input *ideal* distribution F_{W_1} supposing that the *real* F_{W_1} belongs to one of the so-called *aging* classes of distributions[9]

$$\text{IFR} \subset \text{IFRA} \subset \text{NBU} \subset \text{NBUE} \subset \text{HNBUE}. \tag{17.4.1}$$

These kinds of quantitative conditions on F_{W_1} are quite natural in an insurance risk setting.[10] For example, $F_{W_1} \in$ IFR if and only if the residual lifelength distribution $\Pr(W_1 \leq x + t | W_1 > t)$ is nondecreasing in t for all positive x.

The preceding assumption leads in a natural way to the choice of an exponential ideal distribution in view of the characterizations of the exponential law given in the next lemma, Lemma 17.4.1. Moreover, we emphasize here the use of the NBUE and HNBUE classes as we want to impose the weakest possible conditions on the *real* (unknown) F_{W_1}. Let us recall the definitions of these classes.

Definition 17.4.1. Let W be a positive RV with $EW < \infty$, and denote $\overline{F} = 1 - F$. Then $F_W \in$ NBUE if

$$\int_t^\infty \overline{F}_W(u)\mathrm{d}u \leq (EW)\overline{F}_W(t), \qquad \forall t > 0, \tag{17.4.2}$$

and $F_W \in$ HNBUE if

$$\int_t^\infty \overline{F}_W(u)\mathrm{d}u \leq (EW)\exp(-t/EW), \qquad \forall t > 0. \tag{17.4.3}$$

[9] See Sect. 15.2 in Chap. 15.

[10] See Barlow and Proschan (1975) and Kalashnikov and Rachev (1988).

Lemma 17.4.1. *(i) If $F_W \in$ NBUE and $m_i = EW^i < \infty$, $i = 1,2,3$, then*

$$\overline{F}_W(t) = \exp(-t/m_1) \quad \textit{iff} \quad \alpha := m_1^2 + \frac{m_2}{2} - \frac{m_3}{3m_1} = 0. \tag{17.4.4}$$

(ii) If $F_W \in$ HNBUE and $m_i = EW^i < \infty$, $i = 1,2$, then

$$\overline{F}_W(t) = \exp(-t/m_1) \quad \textit{iff} \quad \beta := 2 - \frac{m_2}{m_1^2} = 0. \tag{17.4.5}$$

The *only if* parts of Lemma 17.4.1 are obvious. The *iff* parts result from the following estimates of the stability of exponential law characterizations (i) and (ii) in Lemma 17.4.1. Further, denote $E(\lambda)$, an exponentially distributed RV, by parameter $\lambda > 0$.

Lemma 17.4.2. *(i) If $F_W \in$ NBUE and $m_i = EW^i < \infty$, $i = 1,2,3$, then*

$$\kappa_2(W, E(\lambda)) \leq 2\alpha + 2|\lambda^{-2} - m_1^2|. \tag{17.4.6}$$

(ii) If $F_W \in$ HNBUE and $m_i = EW^i < \infty$, $i = 1,2$, then

$$\kappa_2(W, E(\lambda)) \leq C(m_1, m_2)\beta^{1/8} + 2|\lambda^{-2} - m_1^2|, \tag{17.4.7}$$

where

$$C(m_1, m_2) = 8\sqrt{6}m_1(\sqrt{m_2} + m_1\sqrt{2}). \tag{17.4.8}$$

Proof. (i) The proof of the first part relies on the following claim concerning the stability of the exponential law characterizations in the class NBU. Let us recall that if F_W has a density, then $F_W \in$ NBU if the *hazard rate function* $h_W(t) = F_W'(t)/\overline{F}_W(t)$ satisfies

$$h_W(t) \geq h = h_W(0), \qquad \forall t \geq 0. \tag{17.4.9}$$

Claim. Let $F_W \in$ NBU and $\mu_i = \mu_i(W) = EW^i < \infty$, $i = 1,2$. Then

$$\int_0^\infty t|F_W'(t) - h\exp(ht)|\mathrm{d}t \leq \mu_1 - h\mu_2 + h^{-1}. \tag{17.4.10}$$

Proof of the claim. By (17.4.9), it follows that $H(t) = h\overline{F}_W(t) - F_W'(t)$ is a nonpositive function on $[0,\infty)$. Clearly,

$$\overline{F}_W(t) = \exp(-ht)\left(1 + \int_0^t H(u)\exp(hu)\mathrm{d}u\right).$$

Hence

$$\int_0^\infty t|F'_W(t) - h\exp(-ht)|\mathrm{d}t = \int_0^\infty t\left|h\exp(-ht)\int_0^t H(u)\exp(hu)\mathrm{d}u - H(t)\right|\mathrm{d}t$$
$$\leq \int_0^\infty ht\exp(-ht)\int_0^t |H(u)|\exp(hu)\mathrm{d}u\mathrm{d}t + \int_0^\infty t|h(t)|\mathrm{d}t$$
$$= -\int_0^\infty \left(\int_0^\infty ht\exp(ht)\mathrm{d}t\right)H(u)\exp(hu)\mathrm{d}u - \int_0^\infty tH(t)\mathrm{d}t.$$

Integrating by parts in the first integral and replacing $H(t)$ by $h\overline{F}_W(t) - F'_W(t)$ we obtain the required inequality (17.4.10).

Now, continuing the proof of Lemma 17.4.2 (i), note that $F_W \in$ NBUE implies $F_{W^*} \in$ NBU, where $F_{W^*}(t) = m_1^{-1}\int_0^t \overline{F}_W(u)\mathrm{d}u$, $t \geq 0$. Also

$$\kappa_2(W, E(m_1^{-1})) = 2m_1\int_0^\infty t|F'_{W^*}(t) - h_{W^*}(0)\exp(-t/h_{W^*}(0))|\mathrm{d}t, \tag{17.4.11}$$

where

$$h_{W^*}(0) = m_1^{-1}, \qquad EW^* = m_2/2m_1, \tag{17.4.12}$$

and

$$E(W^*)^2 = m_3/3m_1. \tag{17.4.13}$$

Using claim (17.4.11)–(17.4.13), we get

$$\frac{1}{2}\kappa_2(W, E(m_1^{-1}) \leq \frac{m_2}{2} - \frac{m_3}{3m_1} + m_1^2. \tag{17.4.14}$$

On the other hand, for each $\lambda > 0$ one easily shows that

$$\kappa_2(E(\lambda), E(m_1^{-1})) = 2|m_1^2 - \lambda^{-2}|. \tag{17.4.15}$$

From (17.4.14) and (17.4.15), using the triangle inequality, (17.4.6) follows.

(ii) To derive (17.4.7), we use the representation of κ_2 as a minimal metric: for any two nonnegative RVs X and Y with finite second moment[11]

$$\kappa_2(X,Y) = \inf\{E|\widetilde{X}^2 - \widetilde{Y}^2| : \widetilde{X} \stackrel{\mathrm{d}}{=} X, \widetilde{Y} \stackrel{\mathrm{d}}{=} Y\}. \tag{17.4.16}$$

Similarly,[12]

[11] Apply Theorem 8.2.2 of Chap. 8 with $c(x, y) = |x - y|$ and the representation (17.3.3). See also Remark 7.2.3.

[12] Apply Theorem 8.2.2 of Chap. 8 with $c(x, y) = |x - y|^2$.

$$\ell_2(X,Y) = \left(\int_0^1 |F_X^{-1}(t) - F_Y^{-1}(t)|^2 \mathrm{d}t\right)^{1/2}$$
$$= \inf\{(E(\widetilde{X} - \widetilde{Y})^2)^{1/2} : \widetilde{X} \stackrel{\mathrm{d}}{=} X, \widetilde{Y} \stackrel{\mathrm{d}}{=} Y\}. \quad (17.4.17)$$

By Holder's inequality, we obtain that

$$E|\widetilde{X}^2 - \widetilde{Y}^2| \le (E(\widetilde{X} - \widetilde{Y})^2)^{1/2}((E\widetilde{X}^2)^{1/2} + (E\widetilde{Y}^2)^{1/2}.$$

Hence, by (17.4.16) and (17.4.17),

$$\kappa_r(X,Y) \le \ell_2(X,Y)((EX^2)^{1/2} + (EY^2)^{1/2}). \quad (17.4.18)$$

In Kalashnikov and Rachev (1988, Lemma 4.2.1), it is shown that for $W \in$ NBUE

$$\ell_2(W, E(m_1^{-1})) \le 8\sqrt{6}m_1\beta^{1/8}. \quad (17.4.19)$$

By (17.4.18) and (17.4.19), we now get that

$$\kappa_2(W, E(m_1^{-1})) \le C(m_1, m_2)\beta^{1/8}. \quad (17.4.20)$$

The result in (ii) is a consequence of (17.4.15) and (17.4.20). □

Remark 17.4.1. Note that the term $|\lambda^{-2} - m_1^2|$ in (17.4.6) and (17.4.7) is zero if we choose the parameter λ in our *ideal* exponential distribution F_W to be m_1^{-1}, and hence the *if* parts of Lemma 17.4.1 follow.

Reformulating Lemma 17.4.2 toward our original problem *PR II*, we can state the following theorem.

Theorem 17.4.1. *Let* $\widetilde{W} \stackrel{\mathrm{d}}{=} E(\lambda)$. *Then*

$$\kappa_2(W, \widetilde{W}) \le \varepsilon, \quad (17.4.21)$$

where $\varepsilon = 2\alpha + 2|\lambda^{-2} - m_1^2|$ *if* $F_W \in$ NBUE, *and*

$$\varepsilon = C(m_1, m_2)\beta^{1/8} + 2|\lambda^{-2} - m_1^2|$$

if $F_W \in$ HNBUE.

Remark 17.4.2. In the case where F_W belongs to IFR, IFRA, or NBU, the preceding estimate (17.4.21) can be improved using more refined estimates than (17.4.19).[13]

[13]See Kalashnikov and Rachev (1988, Lemma 4.2.1).

The preceding results concerning *PR I* and *PR II* lead to the following recommendations:

(i) One checks if F_{W_1} belongs to some of the classes in (17.4.1). There are statistical procedures for checking that $F_{W_1} \in \text{HNBUE}$.[14]
(ii) If, for example, $F_{W_1} \in \text{HNBUE}$, then one computes $m_1 = EW_1$, $m_2 = EW_2$, and $\beta = 2 - m_2/m_1^2$. If β is close to zero, then we can choose the *ideal* distribution $F_{\widetilde{W}}(x) = 1 - \exp(x/m_1)$. Then the possible deviation between F_{W_1} and $F_{\widetilde{W}_1}$ in $\boldsymbol{\kappa}_2$-metric is given by Theorem 17.4.1:

$$\boldsymbol{\kappa}_2(W_1, \widetilde{W}_1) \leq C(m_1, m_2)\beta^{1/8} = \varepsilon. \tag{17.4.22}$$

(iii) In a similar way, choose $F_{\widetilde{X}_1}$ and estimate the deviation

$$\boldsymbol{\kappa}_2(X_1, \widetilde{X}_1) \leq \delta. \tag{17.4.23}$$

(iv) Compute the approximating compound Poisson distribution $F_{\sum_{i=1}^{\widetilde{N}(t)} \widetilde{X}_1}$.[15] Then the possible deviation between the *real* compound distribution $F_{\sum_{i=1}^{N(t)} X_1}$ the *ideal* $F_{\sum_{i=1}^{\widetilde{N}(t)} X_1}$ in terms of the uniform metric is[16]

$$\boldsymbol{\rho}\left(\sum_{i=1}^{N(t)} X_i, \sum_{i=1}^{\widetilde{N}(t)} \widetilde{X}_i\right) \leq \psi(\varepsilon, \delta). \tag{17.4.24}$$

If F_W does not belong to any of the classes in (17.4.1), then one can compute the empirical distribution function $\widehat{F}_{W_1}^{(N)}(\cdot, \omega)$ based on N observations $W_1, W_2, \ldots, W_N$. Choosing $\lambda > 0$ [or $F_{\widetilde{W}_1}(x) = 1 - \exp(-\lambda x)$] such that $E\boldsymbol{\kappa}_2(\widehat{F}_{W_1}^{(N)}, F_{\widetilde{W}_1}) < \varepsilon$, we get that

$$\boldsymbol{\kappa}_2(F_{W_1}, F_{\widetilde{W}_1}) < \varepsilon + E\boldsymbol{\kappa}_2(\widehat{F}_{W_1}^{(N)}, F_{\widetilde{W}_1}). \tag{17.4.25}$$

Dudley's theorem[17] implies that the second term on the right-hand side of (17.4.25) can be estimated by some function $\phi(N)$, tending to zero as $N \to \infty$.

---

[14] See Basu and Ebrahimi (1985) and the references therein for testing whether F_{W_1} belongs to the aging classes.

[15] See Teugels (1985).

[16] See Theorem 17.3.1.

[17] See Kalashnikov and Rachev (1988, Theorems 4.9.7 and 4.9.8).

References

Barlow RE, Proschan F (1975) Statistical theory of reliability and life testing: probability models. Holt, Rinehart, and Winston, New York

Basu AP, Ebrahimi N (1985) Testing whether survival function is harmonic new better than used in expectation. Ann Inst Stat Math 37:347–359

Kalashnikov VV, Rachev ST (1988) Mathematical methods for construction of stochastic queueing models. Nauka, Moscow (in Russian) [English transl., (1990) Wadsworth, Brooks–Cole, Pacific Grove, CA]

Teugels JL (1985) Selected topics in insurance mathematics. Katholieke Universiteit Leuven, Leuven

Chapter 18
How Close Are the Individual and Collective Models in Risk Theory?

The goals of this chapter are to:

- Describe individual and collective models in insurance risk theory,
- Define stop-loss probability metrics and discuss their properties,
- Provide estimates of the distance between individual and collective models in terms of stop-loss metrics.

Notation introduced in this chapter:

Notation	Description
$F_1 * F_2$	Convolution of distribution functions
S^{ind}	Aggregate claim in individual model
S^{coll}	Aggregate claim in collective model
$\mathbf{d}_m$	Stop-loss metric of order d
$\mathbf{d}_{m,p}$	L_p-version of $\mathbf{d}_m$

18.1 Introduction

The subject of this chapter is individual and collective models in insurance risk theory and how ideal probability metrics can be employed to calculate the distance between them. We begin by describing stop-loss distances and their properties. We provide a Berry–Esseen-type theorem for the convergence rate in the central limit theorem (CLT) in terms of stop-loss distances using the general method of ideal probability metrics. Finally, we consider approximations in risk theory by means of compound Poisson distributions and estimate the distance between the individual and the collective models using stop-loss metrics of different orders.

DOI 10.1007/978-1-4614-4869-3_18,

18.2 Stop-Loss Distances as Measures of Closeness Between Individual and Collective Models

In Chaps. 16 and 17, we defined and used an ideal metric of order $r = m + 1/p > 0$,

$$\zeta_{m,p}(X,Y) = \sup\{|Ef(X) - Ef(Y)| : \|f^{(m+1)}\|_q \le 1\}, \tag{18.2.1}$$

$m = 0, 1, 2, \ldots$, $p \in [1, \infty]$, $1/p + 1/q = 1$. The *dual* representation of $\zeta_{1,\infty}(X,Y)$ gives for any X and Y with equal means

$$\zeta_{1,\infty}(X,Y) = \sup_{x\in\mathbb{R}} \left| \int_x^{\infty} (x-t)\mathrm{d}(F_X(t) - F_Y(t)) \right|, \tag{18.2.2}$$

where F_X stands for the distribution functions (DF) of X.

The distance $\zeta_{1,\infty}(X,Y)$ in (18.2.2) is well known in risk theory as the *stop-loss metric*[1] and is used to measure the distance between the so-called *individual and collective models*. More precisely, let $X_1, \ldots, X_n$ be independent real-valued variables with DFs F_i, $1 \le i \le n$, of the form

$$F_i = (1 - p_i)E_0 + p_i V_i, \quad 0 \le p_i \le 1. \tag{18.2.3}$$

Here E_0 is the one-point mass DF concentrated at zero and V_i is any DF on $\mathbb{R}$. We can, therefore, write $X_i = C_i D_i$, where C_i has a DF V_i, D_i is Bernoulli distributed with success probability p_i, and C_i and D_i are independent. Then

$$S^{\mathrm{ind}} := \sum_{i=1}^{n} X_i = \sum_{i=1}^{n} C_i D_i \tag{18.2.4}$$

has a DF $F = F_1 * \cdots * F_n$, where $*$ denotes the convolution of DFs.

The notation S^{ind} comes from risk theory,[2] where S^{ind} is the so-called *aggregate claim in the individual model*. Each of n policies leads with (small) probability p_i to a *claim amount* C_i with DF V_i.

Consider approximations of S^{ind} by compound Poisson distributed random variables (RVs)

$$S^{\mathrm{coll}} := \sum_{i=1}^{N} Z_i, \tag{18.2.5}$$

where $\{Z_i\}$ are i.i.d., $Z_i \stackrel{\mathrm{d}}{=} V$ (i.e., Z_i has DF V), N is Poisson distributed with parameter μ and $\{Z_i\}$, and N are independent. The empty sum in (18.2.5) is defined

---

[1] See Gerber (1981, p. 97).

[2] See Gerber (1981, Chap. 4).

to be zero. In risk theory, S^{coll} is referred to as a *collective model*. The usual choice of V and μ in a collective model is[3]

$$\mu = \widetilde{\mu} := \sum_{i=1}^{n} p_i, \quad V = \widetilde{V} := \sum_{i=1}^{n} \frac{p_i}{\mu} V_i = \sum_{i=1}^{n} \frac{p_i}{\mu} F_{C_i}. \tag{18.2.6}$$

This choice leads to the following representation of S^{coll}:

$$S^{\text{coll}} = \sum_{i=1}^{n} S_i^{\text{coll}}. \tag{18.2.7}$$

Here, $S_i^{\text{col}} = \sum_{j=1}^{N} Z_{ij}$, $N_i \stackrel{\text{d}}{=} \mathcal{P}(p_i)$ (i.e., Poisson distribution with parameter p_i), $Z_{ij} \stackrel{\text{d}}{=} V_i$, and N_i, Z_{ij} are independent (i.e., one approximates each summand X_i by a compound Poisson distributed RV S_i^{coll}).

Our further objective is to replace the usual choice (18.2.6) in the compound Poisson model by a *scaled model*, i.e., we choose $Z_{ij} \stackrel{\text{d}}{=} u_i C_i$, $\mu = \sum_{i=1}^{n} \mu_i$, with scale factors u_i and with μ_i such that the first two moments of S^{ind} and S^{coll} coincide.

Remark 18.2.1. In the usual collective model (18.2.6),

$$ES^{\text{ind}} = \sum_{i=1}^{n} p_i EC_i = ES^{\text{coll}}, \tag{18.2.8}$$

and if $q_i = 1 - p_i$, then[4]

$$\begin{aligned} \text{Var}(S^{\text{ind}}) &= \sum_{i=1}^{n} p_i \,\text{Var}(C_i) + \sum_{i=1}^{n} p_i q_i (EC_i)^2 \\ &< \text{Var}(S^{\text{coll}}) = \sum_{i=1}^{n} p_i (\text{Var}(C_i)) + \sum_{i=1}^{n} p_i (EC_i)^2. \end{aligned} \tag{18.2.9}$$

To compare the scaled and individual models, we will use several distances well known in risk theory. Among them is the *stop-loss metric of order m*

$$\begin{aligned} \mathbf{d}_m(X,Y) &:= \sup_t \left| \int_t^{\infty} \frac{(x-t)^m}{m!} \mathrm{d}(F_X(x) - F_Y(x)) \right| \\ &= \sup_t (1/m!) |E(X-t)_+^m - E(Y-t)_+^m|, m \in \mathbb{N} := \{1,2,\ldots\}, (\cdot)_+ \\ &:= \max(\cdot, 0). \end{aligned} \tag{18.2.10}$$

---

[3]See Gerber (1981, Sect. 1, Chap. 4).

[4]See Gerber (1981, p. 50).

This choice is motivated by risk theory and allows us to estimate the difference of two stop-loss premiums.[5]

We will also consider the L_p-version of $\mathbf{d}_s$, namely,

$$\mathbf{d}_{m,p}(X,Y) := \left(\int |D_m(t)|^p \mathrm{d}t\right)^{1/p}, \quad 1 \le p < \infty,$$
$$\mathbf{d}_{m,\infty}(X,Y) := \mathbf{d}_m(X,Y), \tag{18.2.11}$$

where

$$D_m(t) := D_{m,X,Y}(t) := (1/m!)(E(X-t)_+^m - E(Y-t)_+^m). \tag{18.2.12}$$

The rest of this section is devoted to the study of the stop-loss metrics $\mathbf{d}_m$ and $\mathbf{d}_{m,p}$.

Lemma 18.2.1. *If $E(X^j - Y^j) = 0$, $1 \le j \le m$, then*

$$\mathbf{d}_m(X,Y) = \zeta_{m,\infty}(X,Y),$$
$$|E(X-Y)| < \mathbf{d}_1(X,Y) \le \int |F_X(x) - F_Y(x)|\mathrm{d}x, \tag{18.2.13}$$

and

$$\mathbf{d}_{m,p}(X,Y) = \zeta_{m,p}(X,Y). \tag{18.2.14}$$

Proof. We will prove (18.2.13) only. The proof of (18.2.14) is similar.

Here and in what follows, we use the notation

$$H_0(t) := H(t) := F_X(t) - F_Y(t) \tag{18.2.15}$$

and

$$H_1(t) := \int_t^\infty H(u)\mathrm{d}u \; H_k(t) := \int_t^\infty H_{k-1}(u)\mathrm{d}u \text{ for } k \ge 2. \tag{18.2.16}$$

Claim 1. (a) If $xH(x) \to 0$ for $x \to \infty$, then for $k = 1, \ldots, m$

$$\begin{aligned} D_m(t) &= -\frac{1}{(m-1)!}\int_t^\infty (x-t)^{m-1} H(x)\mathrm{d}x \\ &= -\frac{1}{(m-k)!}\int_t^\infty (x-t)^{m-k} H_{k-1}(x)\mathrm{d}x \\ &= -H_m(t). \end{aligned} \tag{18.2.17}$$

(b) $|EX - EY| \le \mathbf{d}_1(X,Y) \le \int_{-\infty}^\infty |H(x)|\mathrm{d}x$.

---

[5]See Gerber (1981, p. 97) for $s = 1$.

The proof of (a) follows from repeated partial integration, and (b) follows from (a).

Claim 2. If f is $(m+1)$-times differentiable, $E(X^j - Y^j)$ exists, $1 \le j \le m$, and $f(X)$ and $f(Y)$ are integrable, then

$$E(f(X) - f(Y)) = \sum_{j=0}^{m} \frac{f^{(j)}(0)}{j!} E(X^j - Y^j) + (-1)^{m+1} \int_{-\infty}^{0} \overline{D}_m(t) f^{(m+1)}(t)\mathrm{d}t + \int_{0}^{\infty} D_m(t) f^{(m+1)}(t)\mathrm{d}t \qquad (18.2.18)$$

and

$$E(f(X) - f(Y)) = \int_{\mathbb{R}} D_m(t) f^{(m+1)}(t)\mathrm{d}t = (-1)^{m+1} \int_{\mathbb{R}} \overline{D}_m(t) f^{(m+1)}(t)\mathrm{d}t, \qquad (18.2.19)$$

where

$$\overline{D}_m(t) := \overline{D}_{m,X,Y}(t) := (1/m!)(E(t-X)_+^m - E(t-Y)_+^m) \; s \ge 1. \qquad (18.2.20)$$

The proof of (18.2.18) follows from the Taylor series expansion,

$$\begin{aligned} E(f(X) - f(Y)) &= \int_{\mathbb{R}} f(x)\mathrm{d}H(x) \\ &= \int_{\mathbb{R}} \left[f(0) + \cdots + \frac{x^m}{m!} f^{(m)}(0) + \int_0^x \frac{(x-t)^m}{m!} f^{(m+1)}(t)\mathrm{d}t \right] \mathrm{d}H(x) \\ &= \sum_{j=0}^{m} \frac{f^{(j)}(0)}{j!} E(X^j - Y^j) + \int_{-\infty}^{0} (-1)^{m+1} \overline{D}_m(t) f^{(m+1)}(t)\mathrm{d}t \\ &\quad + \int_0^{\infty} D_m(t) f^{(m+1)}(t)\mathrm{d}t. \end{aligned}$$

To prove (18.2.19), observe that if $E(X^j - Y^j)$ is finite, $1 \le j \le m$, then

$$D_m(t) = (1/m!) \sum_{j=0}^{m} \binom{m}{j} E(X^j - Y^j)(-t)^{m-j} + (-1)^{m+1} \overline{D}_m(t). \qquad (18.2.21)$$

Now (18.2.19) follows from (18.2.18) and (18.2.21), and thus the proof of Claim 2 is completed.

It is known that for a function h on $\mathbb{R}$ with

$$\|h\|_\infty = \operatorname*{ess\,sup}_{x \in \mathbb{R}} |h(x)|$$

the following dual representation holds:[6]

$$\|h\|_\infty = \sup\left\{\int h(t)g(t)\mathrm{d}t : \|g\|_1 \le 1\right\}. \tag{18.2.22}$$

Recall that

$$\zeta_{m,\infty}(X,Y) := \sup\{|E(f(X) - f(Y))| : f \in \mathcal{F}_m\}, \tag{18.2.23}$$

where $\mathcal{F}_m := \{f : \mathbb{R}^1 \to \mathbb{R}^1, f^{(m+1)}$ exists and $\|f^{(m+1)}\|_1 < 1\}$.

Thus (18.2.19), (18.2.22), and (18.2.23) imply

$$\begin{aligned}\zeta_{m,\infty}(X,Y) &= \sup_{f\in\mathcal{F}_m}\left|\int D_m(t) f^{(m+1)}(t)\mathrm{d}t\right| \\ &= \|D_m\|_\infty = \|\overline{D}_m\|_\infty = \mathbf{d}_m(X,Y).\end{aligned}$$

□

The next lemma shows that the moment condition in Lemma 18.2.1 is necessary for the finiteness of $\zeta_{m,\infty}$.[7]

Lemma 18.2.2. *(a)* $\zeta_{m,\infty}(X,Y) < \infty$ *implies that*

$$E(X^j - Y^j) = 0 \quad 1 \le j \le m. \tag{18.2.24}$$

(b) $\zeta_{m,\infty}$ *is an ideal metric of order m, i.e.,* $\zeta_{m,\infty}$ *is a simple probability metric such that*

$$\zeta_{m,\infty}(X+Z, Y+Z) \le \zeta_{m,\infty}(X,Y)$$

for Z independent of X, Y and[8]

$$\zeta_{m,\infty}(cX, cY) = |c|^m \zeta_{m,\infty}(X,Y), \quad \textit{for } c \in \mathbb{R}. \tag{18.2.25}$$

(c) For independent $X_1,\dots,X_n$ *and* $Y_1,\dots,Y_n$ *and for* $c_i \in \mathbb{R}$ *the following inequality holds:*

$$\zeta_{m,\infty}\left(\sum_{i=1}^n c_i X_i, \sum_{i=1}^n c_i Y_i\right) \le \sum_{i=1}^n |c_i|^m \zeta_{m,\infty}(X_i, Y_i). \tag{18.2.26}$$

---

[6] See, for example, Dunford and Schwartz (1988, Sect. IV.8) and Neveu (1965).

[7] See condition (17.3.10) for $\zeta_{m,p}$ in Chap. 17.

[8] See Definition 15.3.1 in Chap. 15.

Proof. (a) For any $a > 0$ and $1 \le j \le m$, $f_a(x) := ax^j \in \mathcal{F}_m$, and therefore

$$\zeta_{m,\infty}(X,Y) \ge \sup_{a>0} a|E(X^j - Y^j)|,$$

i.e., $E(X^j - Y^j) = 0$.

(b) Since for $z \in \mathbb{R}$ and $f \in \mathcal{F}_m$, $f_z(x) := f(x+z) \in \mathcal{F}_m$, the first part follows from conditioning on $Z = z$. For the second part note that for $c \in \mathbb{R}^1 : f \in \mathcal{F}_m$ if and only if $|c|^{-m} f_c \in \mathcal{F}_m$ with $f_c(x) = f(cx)$.

Finally, (c) follows from (b) and the triangle inequality for $\zeta_{m,\infty}$. □

The proof of the next lemma is similar.

Lemma 18.2.3. *(a) $\mathbf{d}_m$ is an ideal metric of order m.*
(b) For $X_1, \dots, X_n$ independent, $Y_1, \dots, Y_n$ independent, and $c_i > 0$

$$\mathbf{d}_m\left(\sum_{i=1}^n c_i X_i, \sum_{i=1}^n c_i Y_i\right) \le \sum_{i=1}^n c_i^m \mathbf{d}_m(X_i, Y_i). \tag{18.2.27}$$

(c) $\mathbf{d}_m(X+a, Y+a) = \mathbf{d}_m(X,Y)$ for all $a \in \mathbb{R}$.
(d) If $EX = EY = \mu$, $\sigma^2 = \mathrm{Var}(X) = \mathrm{Var}(Y)$, then with $\widetilde{X} = (X-\mu)/\sigma$, $\widetilde{Y} = (Y-\mu)\sigma$

$$\mathbf{d}_m(\widetilde{X}, \widetilde{Y}) = \sigma^{-m}\mathbf{d}_m(X,Y). \tag{18.2.28}$$

Recall the definition of the difference pseudomoment of order m:

$$\boldsymbol{\kappa}_m(X,Y) := m\int_{-\infty}^{\infty} |x|^{m-1}|H(x)|\mathrm{d}x. \tag{18.2.29}$$

In the next lemma, we prove that the finiteness of $\mathbf{d}_{m+1}$ implies the moment condition (18.2.24).

Lemma 18.2.4. *(a) If $X, Y \ge 0$ a.s., $E(X^j - Y^j)$ exists and is finite, $1 \le j \le m$, and $\mathbf{d}_m(X,Y) < \infty$, then $E(X^j - Y^j) = 0$, $1 \le j \le m-1$.*
(b) If $\mathbf{d}_m(X,Y) < \infty$ and $\boldsymbol{\kappa}_m(X,Y) < \infty$, then $E(X^j - Y^j) = 0$, $1 \le j \le m-1$.

Proof. (a) From (18.2.16) we obtain for $t \ge 0$

$$\begin{aligned}
&(m-1)!D_m(t) \\
&= \int_t^{\infty} (x-t)^{m-1} H(x)\mathrm{d}x \\
&= \int_0^{\infty} (x-t)^{m-1} H(x)\mathrm{d}x
\end{aligned}$$

$$= \sum_{j=0}^{m-1} (-t)^{m-1-j} \left(\int_0^\infty x^j H(x) dx \right) \binom{m-1}{j}$$

$$= \sum_{j=0}^{m-1} \binom{m-1}{j} (-t)^{m-j-1} \frac{E(Y^{j+1} - X^{j+1})}{j+1}.$$

Since $\mathbf{d}_m(X,Y) = \sup_t D_m(t) < \infty$, all coefficients of the foregoing polynomial for $j = 0, \ldots, m-2$ must be zero.

(b) By $\mathbf{d}_m(X,Y) < \infty$

$$m!\mathbf{d}_m(X,Y) = \sup_{x \in \mathbb{R}} \left| \int_x^\infty \frac{(t-x)^{m-1}}{(m-1)!} H(t) dt \right| < \infty,$$

and thus

$$\limsup_{x \to -\infty} \left| \sum_{j=0}^{m-1} \binom{m-1}{j} (-x)^{m-1-j} \int_x^\infty t^j H(t) dt \right| < \infty. \tag{18.2.30}$$

Further, by $\boldsymbol{\kappa}_m(X,Y) < \infty$ [see (18.2.29)],

$$\limsup_{x \to -\infty} \left| \int_x^\infty t^{m-1} H(t) dt \right| \le (1/m) \boldsymbol{\kappa}_m(X,Y) < \infty.$$

Thus,

$$\limsup_{x \to -\infty} \left| \sum_{j=0}^{m-2} \binom{m-j}{j} (-x)^{m-2-j} \int_x^\infty t^j H(t) dt \right| = 0. \tag{18.2.31}$$

Since

$$\limsup_{x \to -\infty} \left| \int_x^\infty t^{m-2} H(t) dt \right| \le \frac{1}{m-1} \boldsymbol{\kappa}_{m-1}(X,Y) \le 2 + (1/m) \boldsymbol{\kappa}_m(X,Y) < \infty,$$

by (18.2.31), we have

$$\limsup_{x \to -\infty} \left| \sum_{j=0}^{m-3} \binom{m-1}{j} (-x)^{m-3-j} \int_x^\infty t^j H(t) dt \right| = 0. \tag{18.2.32}$$

Similarly to (18.2.31) and (18.2.32), we obtain

$$\limsup_{x \to -\infty} \left| \sum_{j=0}^{m-k} \binom{m-1}{j} (-x)^{m-k-j} \int_x^\infty t^j H(t) dt \right| = 0 \tag{18.2.33}$$

for all $k = 2, \ldots, m$. In the case where $k = m$, we have

$$0 = \limsup_{x \to -\infty} \left| \int_x^\infty H(t) \mathrm{d}t \right| = \limsup_{x \to -\infty} \left| \int_x^\infty t \mathrm{d}H(t) \right|, \qquad (18.2.34)$$

and thus $E(X - Y) = 0$. Set $k = m - 1$ in (18.2.33); then

$$0 = \limsup_{x \to -\infty} \left| (-x) \int_x^\infty H(t) \mathrm{d}t + (m-1) \int_x^\infty t H(t) \mathrm{d}t \right|. \qquad (18.2.35)$$

By (18.2.34) and $\boldsymbol{\kappa}_m(X, Y) < \infty$,

$$\begin{aligned} \limsup_{x \to -\infty} \left| x \int_x^\infty H(t) \mathrm{d}t \right| &= \limsup_{x \to -\infty} \left| x \int_{-\infty}^x H(t) \mathrm{d}t \right| \\ &\leq \limsup_{x \to -\infty} \int_{-\infty}^x |t||H(t)| \mathrm{d}t = 0. \end{aligned} \qquad (18.2.36)$$

Combining (18.2.35) and (18.2.36) implies

$$\limsup_{x \to -\infty} \left| \int_x^\infty t H(t) \mathrm{d}t \right| = 0,$$

and hence $E(X^2 - Y^2) = 0$. In the same way we get $E(X^j - Y^j) = 0$ for all $j = 1, \ldots, m-1$. □

We next establish some relations between the different metrics considered so far. We use the notation $\zeta_m := \zeta_{m,1}$ for the Zolotarev metric.[9]

Lemma 18.2.5. *(a) If $X, Y \geq 0$ a.s, $E(X^j - Y^j)$ is finite, $1 \leq j \leq m$, and $\mathbf{d}_m(X, Y) < \infty$, then*

$$\mathbf{d}_m(X, Y) \leq (1/m!) \max\{|E(X^m - Y^m)|, \boldsymbol{\kappa}_m(X, Y)\}. \qquad (18.2.37)$$

(b) $$\mathbf{d}_m(X, Y) \leq \mathbf{d}_{m-1,1}(X, Y) \text{ if } x^s H(x) \underset{x \to \infty}{\longrightarrow} 0,$$

$$\zeta_{m,\infty}(X, Y) \leq \zeta_m(X, Y) \text{ if } \zeta_{m,\infty}(X, Y) < \infty, \qquad (18.2.38)$$

$$\mathbf{d}_{m,p}(X, Y) = \zeta_{m,p}(X, Y) \leq \zeta_{m+1/p}(X, Y) \text{ if } 1 \leq p < \infty, \text{ and } \zeta_{m,p}(X, Y) < \infty.$$

[9] See (15.3.1) in Chap. 15.

(c) If $E(X^j - Y^j) = 0$, $1 < j \leq m$, *then*

$$\mathbf{d}_m(X,Y) = \zeta_{m,\infty}(X,Y) \leq (1/m!)\kappa_m(X,Y). \tag{18.2.39}$$

(d)
$$\kappa_m(X,Y) \leq E|X|X|^{m-1} - Y|Y|^{m-1}| \leq E|X|^m + E|Y|^m.$$

Proof. (a) For $t \geq 0$ it follows from (18.2.16) that

$$\begin{aligned}(m-1)!|D_m(t)| &= \left|\int_t^\infty (x-t)^{m-1}H(x)\mathrm{d}x\right| \leq \int_t^\infty (x-t)^{m-1}|H(x)|\mathrm{d}x \\ &\leq \int_0^\infty x^{m-1}|H(x)|\mathrm{d}x = (1/m)\kappa_m(X,Y).\end{aligned}$$

For $t \leq 0$ it follows from Lemma 18.2.4 (a) that

$$\begin{aligned}(m-1)!D_m(t) &= \int_t^\infty (x-t)^{m-1}H(x)\mathrm{d}x = \int_0^x (x-t)^{m-1}H(x)\mathrm{d}x \\ &= (1/m)E(Y^m - X^m).\end{aligned}$$

(b) From (18.2.16) it follows that if $x^m H(x) \to 0$, then

$$\begin{aligned}\mathbf{d}_m(X,Y) &= \sup_t |D_m(t)| = \sup_t |H_m(t)| = \sup_t \left|\int_t^\infty H_{m-1}(u)\mathrm{d}u\right| \\ &\leq \sup_t \int_t^\infty |D_{m-1}(u)|\mathrm{d}u = \mathbf{d}_{m-1,1}(X,Y).\end{aligned}$$

If $E(X^j - Y^j) = 0$, $1 \leq j \leq m$, then $\zeta_{m,\infty}(X,Y) = \mathbf{d}_m(X,Y) \leq \mathbf{d}_{m-1,1}(X,Y) = \zeta_m(X,Y)$. The relation $\zeta_{m,p}(X,Y) \leq \zeta_{m+1/p}(X,Y)$ follows from the inequality

$$|f^m(x) - f^m(y)| \leq \|f^{(m1)}\|_q |x-y|^{1/p} \leq |x-y|^{1/p}$$

for any function f with $\|f^{(m+1)}\|_q \leq 1$ and $1/p + 1/q = 1$.

(c) By (b) and Lemma 18.2.1,

$$\mathbf{d}_m(X,Y) = \zeta_{m,\infty}(X,Y) \leq \zeta_m(X,Y). \tag{18.2.40}$$

Further, by (18.2.14) with $p = 1$,

$$\zeta_m(X,Y) = \int_{-\infty}^\infty \left|\int_x^\infty \frac{(t-x)^m}{m!}\mathrm{d}H(t)\right| \mathrm{d}x. \tag{18.2.41}$$

By the assumption $E(X^j - Y^j) = 0$, $j = 1, \ldots, m$,

$$\begin{aligned}
\zeta_m(X,Y) &= \int_{-\infty}^{\infty} \left| \int_x^{\infty} \frac{(t-x)^{m-1}}{(m-1)!} H(t)\mathrm{d}t \right| \mathrm{d}x \\
&= \int_0^{\infty} \left| \int_x^{\infty} \frac{(t-x)^{m-1}}{(m-1)!} H(t)\mathrm{d}t \right| \mathrm{d}x + \int_{-\infty}^{0} \left| \int_{-\infty}^{x} \frac{(x-t)^{m-1}}{(m-1)!} H(t)\mathrm{d}t \right| \mathrm{d}x \\
&\leq \int_0^{\infty} \int_x^{\infty} \frac{(x-t)^{m-1}}{(m-1)!} |H(t)|\mathrm{d}t\mathrm{d}x + \int_{-\infty}^{0} \int_{-\infty}^{x} \frac{(x-t)^{m-1}}{(m-1)!} |H(t)|\mathrm{d}t\mathrm{d}x \\
&= (1/m!)\kappa_m(X,Y).
\end{aligned}$$

(d) Clearly, for any X and Y[10]

$$\kappa_1(X,Y) = \int_{-\infty}^{\infty} |F_X(x) - F_Y(x)|\mathrm{d}x \leq E|X-Y|.$$

Now use the representation

$$\kappa_m(X,Y) = \kappa_1(X|X|^{m-1}, Y|Y|^{m-1})$$

to complete the proof of (d). □

The next relations concern the *uniform metric*

$$\rho(X,Y) := \sup_{x\in\mathbb{R}} |F_X(x) - F_Y(x)|, \tag{18.2.42}$$

the stop-loss distance $\mathbf{d}_m$ defined in (18.2.10), and the pseudomoment $\kappa_m(X,Y)$ defined in (18.2.29).

Lemma 18.2.6. *(a) If X has a bounded Lebesgue density f_X, $|f_X(t)| \leq M$, then*

$$\mathbf{d}_m(X,Y) \geq K(m)(1+M)^{-m-1}\rho(X,Y)^{m+1}, \tag{18.2.43}$$

where $K(m) = \dfrac{(m+1)\sqrt{3}}{(2m+2)!\sqrt{2m+3}}$.

(b) If for some $\delta > 0$, $\widetilde{m}_\delta := E(|X|^{m+\delta} + |Y|^{m+\delta}) < \infty$, then

$$\kappa_m(X,Y) \leq 2\left(\frac{\delta\widetilde{m}_\delta}{2m}\right)(\rho(X,Y))^{d/(m+d)}\frac{m+\delta}{\delta}. \tag{18.2.44}$$

Proof. (a) We first apply Lemma 18.2.1, $\mathbf{d}_m = \zeta_{m,\infty}$. Then Lemma 17.3.2 completes the proof of (18.2.43).

[10]See, for example, (6.5.11) in Chap. 6.

(b) For $\alpha > 0$ and $\rho = \boldsymbol{\rho}(X, Y)$

$$\begin{aligned}\kappa_m(X,Y) &= m\int_{-\infty}^{\infty}|x|^{m-1}|H(x)|\mathrm{d}x\\ &\leq m\int_{-\alpha}^{\alpha}|x|^{m-1}|H(x)|\mathrm{d}x + E|X|^m\{|X|>\alpha\} + E|Y|^m\{|Y|>\alpha\}\\ &\leq 2\rho\alpha^m + \frac{\widetilde{m}_\delta}{\alpha^\delta} =: f(\alpha).\end{aligned}$$

Minimizing $f(\alpha)$ we obtain (18.2.44). □

Remark 18.2.2. Estimate (18.2.44), combined with (18.2.39), shows that

$$\mathbf{d}_m(X,Y) \leq (2/m!)\left(\frac{\delta\widetilde{m}_\delta}{2m}\right)^{m/(m+\delta)}(\boldsymbol{\rho}(X,Y))^{\delta/(m+\delta)}\frac{m+\delta}{m},$$
$$\text{if } \boldsymbol{\zeta}_{m,\infty}(X,Y) < \infty. \tag{18.2.45}$$

Under the assumption of a finite-moment-generating function, this can be improved to $\boldsymbol{\rho}(X,Y)\{\log(\boldsymbol{\zeta}(X,Y))\}^\alpha$ for some $\alpha > 0$.

An important step in the proof of the precise rate of convergence in the CLT, the so-called *Berry–Esseen-type theorems*, is the smoothing inequality.[11] For the stop-loss metrics there are some similar inequalities that also lead to Berry–Esseen-type theorems.

Lemma 18.2.7 (Smoothing inequality).

(a) Let Z be independent of X and Y, $\zeta_{m,\infty}(X,Y) < \infty$; then for any $\varepsilon > 0$ the following inequality holds:

$$\mathbf{d}_m(X,Y) \leq \mathbf{d}(X+\varepsilon Z, Y+\varepsilon Z) + 2\frac{\varepsilon^m}{m!}E|Z|^m. \tag{18.2.46}$$

(b) If X, Y, Z, W are independent, $x^m H(x) \to 0$, $x \to \infty$, then

$$\mathbf{d}_m(X+Z, Y+Z) \leq 2\mathbf{d}_m(Z,W)\boldsymbol{\sigma}(X,Y) + \mathbf{d}_m(X+W, Y+W) \tag{18.2.47}$$

and

$$\mathbf{d}_m(X+Z, Y+Z) \leq 2\mathbf{d}_m(X,Z)\boldsymbol{\sigma}(W,Z) + \mathbf{d}_m(X+W, Z+W), \tag{18.2.48}$$

where $\boldsymbol{\sigma}$ is the total variation metric [see (15.3.4)*].*

[11] See Lemmas 16.3.1 and 16.3.3 and (16.3.7).

Proof. (a) From Lemmas 18.2.3 and 18.2.5

$$\begin{aligned}\mathbf{d}_m(X,Y) &\le \mathbf{d}_m(X,X+\varepsilon Z)+\mathbf{d}_m(X+\varepsilon Z,Y+\varepsilon Z)+\mathbf{d}_m(Y+\varepsilon Z,Y)\\ &\le \mathbf{d}_m(X+\varepsilon Z,Y+\varepsilon Z)+2\mathbf{d}_m(0,\varepsilon Z)\\ &\le \mathbf{d}_m(X+\varepsilon Z,Y+\varepsilon Z)+2\frac{\varepsilon^m}{m!}\kappa_m(0,Z)\\ &= \mathbf{d}_m(X+\varepsilon Z,Y+\varepsilon Z)+2\frac{\varepsilon^m}{m!}E|Z|^m.\end{aligned}$$

(b) $\mathbf{d}_m(X+Z,Y+Z)$

$$\begin{aligned}&=\frac{1}{(m-1)!}\sup_x\left|\int_x^\infty (t-x)^{m-1}(F_{X+Z}(t)-F_{Y+Z}(t))\mathrm{d}t\right|\\ &=\frac{1}{(m-1)!}\sup_x\left|\int_x^\infty\left[\int_{-\infty}^\infty (t-x)^{m-1}F_Z(t-u)\mathrm{d}(F_X(u)-F_Y(u))\right]\mathrm{d}t\right|\\ &\le\frac{1}{(m-1)!}\sup_x\left|\int_x^\infty\left[\int_{-\infty}^\infty (t-x)^{m-1}\{F_Z(t-u)-F_W(t-u)\}\mathrm{d}H(u)\right]\mathrm{d}t\right|\\ &\quad+\frac{1}{(m-1)!}\sup_x\left|\int_x^\infty\left[\int_{-\infty}^\infty (t-x)^{m-1}F_W(t-u)\mathrm{d}H(u)\right]\mathrm{d}t\right|\\ &\le\int_{-\infty}^\infty \mathbf{d}_m(Z,W)|\mathrm{d}H(u)|+\mathbf{d}_m(X+W,Y+W)\\ &=2\mathbf{d}_m(Z,W)\sigma(X,Y)+\mathbf{d}_m(X+W,Y+W).\end{aligned}$$

Inequality (18.2.48) is derived similarly. □

From the smoothing inequality we obtain the following relation between $\mathbf{d}_1$ and $\mathbf{d}_m$.

Lemma 18.2.8. *If* $E(X^j-Y^j)=0$, $1\le j\le m$, *then*

$$\mathbf{d}_1(X,Y)\le\lambda_m(\mathbf{d}_m(X,Y))^{1/m}, \tag{18.2.49}$$

where

$$\lambda_m:=\mathbb{K}_m^{1/m}\left(\frac{2\mathbb{K}_2}{m-1}\right)^{(m-1)/m}m,\quad \mathbb{K}_m:=\int|\mathbb{H}_{m-1}(x)|\frac{1}{\sqrt{2\pi}}\exp(-x^2/2)\mathrm{d}x,$$

in which $\mathbb{H}_m$ *is the Hermite polynomial of order* m, $\mathbb{K}_1=1$, $\mathbb{K}_2=(2/\pi)^{1/2}$. *In particular,*

$$\mathbf{d}_1(X,Y)\le(4/\sqrt{\pi})(\mathbf{d}_2(X,Y))^{1/2}. \tag{18.2.50}$$

Proof. Let Z be a $N(0,1)$-distributed RV independent of X, Y. Then for $\varepsilon > 0$ from (18.2.46)

$$\mathbf{d}_1(X,Y) \le \mathbf{d}_1(X+\varepsilon Z, Y+\varepsilon Z) + 2\varepsilon(2/\pi)^{1/2}. \tag{18.2.51}$$

With $W := \varepsilon Z$ it follows from Lemma 18.2.1 that[12]

$$\begin{aligned}\mathbf{d}_1(X+W,Y+W) &= \sup\{|E(f(X+W)-f(Y+W))|;\ \|f^{(2)}\|_1 \le 1\}\\ &= \sup\{|E(g_f(X)-g_f(Y)|;\ \|f^{(2)}\|_1 \le 1\},\end{aligned}$$

where

$$g_f(t) := \int_{-\infty}^{\infty} f(x) f_W(x-t)\mathrm{d}x = f * f_W(t), \quad f_W := F_W'.$$

The derivatives of g_f have the following representation:

$$\begin{aligned}g_f^{(m)}(t) &= (-1)^m \int_{-\infty}^{\infty} f(x) F_W^{(m)}(x-t)\mathrm{d}x = (-1)^m \int_{-\infty}^{\infty} f(x+t) f_W^{(m)}(x)\mathrm{d}x\\ &= (-1)^{m-1}\int_{-\infty}^{\infty} f^{(1)}(x+t) f_W^{(m-1)}(x)\mathrm{d}x\end{aligned}$$

and

$$g_f^{(m+1)}(t) = (-1)^{m-1}\int_{-\infty}^{\infty} f^{(2)}(x+t) f_W^{(m-1)}(x)\mathrm{d}x = (-1)^{m-1} f^{(2)} * f_W^{(m-1)}(t).$$

For the L^1-norm we therefore obtain

$$\begin{aligned}\|g_f^{(m-1)}\|_1 &= \int |g_f^{(m+1)}(t)|\mathrm{d}t = \|f^{(2)} * f_W^{(m-1)}\|_1\\ &\le \|f^{(2)}\|_1 \|f_W^{(m-1)}\|_1 \le \frac{1}{\varepsilon^{m-1}}\|f_W^{(m-1)}\|_1 = \frac{1}{\varepsilon^{m-1}}\mathbb{K}_m.\end{aligned}$$

Therefore, from Lemma 18.2.1,

$$\mathbf{d}_1(X+\varepsilon Z, Y+\varepsilon Z) = \zeta_{1,\infty}(X+\varepsilon Z, Y+\varepsilon Z) \le \frac{1}{\varepsilon^{m-1}}\mathbb{K}_m \mathbf{d}_m(X,Y). \tag{18.2.52}$$

From the smoothing inequality (18.2.46) we obtain

$$\mathbf{d}_1(X,Y) \le \frac{1}{\varepsilon^{m-1}}\mathbb{K}_m \mathbf{d}_m(X,Y) + 2\mathbb{K}_2\varepsilon.$$

Minimizing the right-hand side with respect to ε, we obtain (18.2.49). □

---

[12]See (18.2.13).

Lemma 18.2.9. *Let Z be independent of X and Y with Lebesgue density f_Z.*

(a) If $C_{3,Z} := \|f_Z^{(3)}\|_1 < \infty$, then

$$\sigma(X+Z, V+Z) \le \frac{1}{2} C_{3,Z} \mathbf{d}_{2,1}(X,Y). \tag{18.2.53}$$

(b) If $C_{s,Z} := \|f_Z^{(s)}\|_1 < \infty$, and if $\zeta_{m,\infty}(X,Y) < \infty$, then for $m \ge 1$

$$\mathbf{d}_m(X+Z, Y+Z) \le C_{s,Z} \zeta_{m+s}(X,Y). \tag{18.2.54}$$

Proof. (a) With $H(t) = F_X(t) - F_Y(t)$,

$$\begin{aligned}
&2\sigma(X+Z, Y+Z) \\
&= \int |f_{X+Z}(x) - f_{Y+Z}(x)| \mathrm{d}x = \int_{\mathbb{R}} \left| \int_{\mathbb{R}} f_Z(x-t) \mathrm{d}H(t) \right| \mathrm{d}x \\
&= \int_{\mathbb{R}} \left| \int_{\mathbb{R}} H(t) f_Z'(x-t) \mathrm{d}t \right| \mathrm{d}x = \int_{\mathbb{R}} \left| \int_{\mathbb{R}} f_Z'(x-t) \mathrm{d} \left(\int_x^\infty H(u) \mathrm{d}u \right) \right| \mathrm{d}x \\
&= \int_{\mathbb{R}} \left| \int_{\mathbb{R}} \left(\int_x^\infty H(u) \mathrm{d}u \right) f_Z''(x-t) \mathrm{d}t \right| \mathrm{d}x \\
&\le \int_{\mathbb{R}} \int_{\mathbb{R}} \left| \int_x^\infty \frac{(u-x)}{1!} H(u) \mathrm{d}u \right| |f_Z^{(3)}(x-t)| \mathrm{d}t \mathrm{d}x \\
&= \frac{1}{2} C_{3,Z} \int_{\mathbb{R}} |E(X-t)_+^2 - E(Y-t)_+^2| \mathrm{d}t = C_{3,Z} \mathbf{d}_{2,1}(X,Y).
\end{aligned}$$

(b) If $C_{s,Z} = \|f_Z^{(s)}\|_1 < \infty$ and $\zeta_{m,\infty}(X,Y) < \infty$, then by (18.2.38), similarly to (a), we get[13]

$$\mathbf{d}_m(X+Z, Y+Z) \le \zeta_m(X+Z, Y+Z) \le C_{s,Z} \zeta_{m+s}(X,Y). \qquad \square$$

Theorem 18.2.1. *Let $\{X_n\}$ be i.i.d., set $EX_1 = 0$, $EX_1^2 = 1$, and $S_n = n^{-1/2} \sum_{i=1}^{n} X_i$, and let Y be a standard normal RV. Then, for $m = 1, 2$,*

$$\mathbf{d}_m(S_n, Y) \le C(m) \max\{\mathbf{d}_{m,1}(X_1, Y), \mathbf{d}_{3,1}(X,Y)\} n^{-1/2}, \tag{18.2.55}$$

[13] See Zolotarev (1986, Theorem 1.4.5) and Kalashnikov and Rachev (1988, Chap. 3, p. 10, Theorem 10).

where $C(m)$ is an absolute constant, and

$$\mathbf{d}_3(S_n, Y) \leq \mathbf{d}_3(X_1, Y)n^{-1/2}. \tag{18.2.56}$$

Proof. Inequality (18.2.56) is a direct consequence of Lemma 18.2.3(b). The proof of (18.2.55) is based on Lemmas 18.2.3, 18.2.5, 18.2.7, and 18.2.9 and follows step by step the proof of Theorem 16.3.1. □

Remark 18.2.3. In terms of ζ_s metrics, a similar inequality is given by Zolotarev (1986, Theorem 5.4.7):

$$\zeta_1(S_n, Y) < 11.5 \max(\zeta_1(X_1, Y)\zeta_3(X_1, Y))n^{-1/2}. \tag{18.2.57}$$

Open Problem 18.2.1. Regarding the right-hand side of (18.2.55), one could expect that the better bound should be $C(m)\max\{\mathbf{d}_m(X_1, Y), \mathbf{d}_{3,1}(X_1, Y)\}$ $n^{-1/2}$. Is it true that for $m = 1, 2$, $p \in (1, \infty]$,

$$\mathbf{d}_{m,p}(S_n, Y) \leq C(m, p)\max\{\mathbf{d}_{m,p}(X_1, Y), \mathbf{d}_{3,1}(X_1, Y)\}n^{-1/2}? \tag{18.2.58}$$

What is a good bound for $C(m, p)$ in (18.2.58)?

18.3 Approximation by Compound Poisson Distributions

We now consider the problem of approximation of the individual model $S^{\text{ind}} = \sum_{i=1}^n X_i = \sum_{i=1}^n C_i D_i$ by a compound model, i.e., by a compound Poisson distributed RV[14]

$$S^{\text{coll}} = \sum_{i=1}^{N} Z_i \stackrel{\text{d}}{=} \sum_{i=1}^{n} S_i^{\text{coll}}, \qquad S_i^{\text{coll}} = \sum_{j=1}^{N_i} Z_{ij}.$$

Choose $Z_{ij} \stackrel{\text{d}}{=} u_i C_i$ and N_i to be Poisson (μ_i)-distributed. Then N is Poisson $(\mu)(N \stackrel{\text{d}}{=} \mathcal{P}(\mu))$, $\mu = \sum_{i=1}^n \mu_i$, and

$$F_{Z_j} = \sum_{i=1}^{n} \frac{\mu_i}{\mu} F_{u_i C_i}. \tag{18.3.1}$$

We choose μ_i, u_i in such a way that the first two moments of S_i^{coll} coincide with the corresponding moments of X_i.

[14]See (18.2.3), (18.2.4), and (18.2.5).

Lemma 18.3.1. *Let* $a_i := EC_i$, $b_i := EC_i^2$, *and define*

$$\mu_i := \frac{p_i b_i}{b_i - p_i a_i^2}, \qquad u_i := \frac{p_i}{\mu_i} = \frac{b_i - p_i a_i^2}{b_i}. \tag{18.3.2}$$

Then

$$ES_i^{\text{coll}} = EX_i = p_i a_i \text{ and } \operatorname{Var}(S_i^{\text{coll}}) = \operatorname{Var}(X_i) = p_i b_i - (p_i a_i)^2. \tag{18.3.3}$$

Proof. Since $N_i \stackrel{\text{d}}{=} \mathcal{P}(\mu_i)$ and $Z_{ij} \stackrel{\text{d}}{=} u_i C_i$, we obtain $EZ_{ij} = u_i a_i$, $EZ_{ij}^2 = u_i^2 b_i$,

$$ES_i^{\text{coll}} = E\sum_{j=1}^{N_i} Z_{ij} = \mu_i u_i a_i = p_i a_i = EX_i,$$

and

$$\operatorname{Var}(X_i) = p_i b_i - (p_i a_i)^2 = p_i(b_i - p_i a_i^2) = \frac{p_i^2 b_i}{\mu_i} = \mu_i u_i^2 b_i = (EN_i)EZ_{ij}^2$$
$$= \operatorname{Var}(S_i^{\text{coll}}). \qquad \square$$

So in contrast to the "usual" choice (18.2.6) of $\mu = \widetilde{\mu}$ and $\nu = \widetilde{\nu}$, we use a scaling factor u_i and μ_i such that the first two moments agree. We see that

$$\mu_i > p_i \text{ for } p_i > 0 \tag{18.3.4}$$

and $u_i < 1$.

Theorem 18.3.1. *Let* μ_i, u_i *be as defined in* (18.3.2); *then*

$$\mathbf{d}_1(S^{\text{ind}}, S^{\text{coll}}) \le \frac{4}{\sqrt{\pi}}\left(\sum_{i=1}^n p_i b_i\right)^{1/2},$$
$$\mathbf{d}_2(S^{\text{ind}}, S^{\text{coll}}) \le \sum_{i=1}^n p_i b_i. \tag{18.3.5}$$

Proof. By (18.2.50), we have $\mathbf{d}_1(S^{\text{ind}}, S^{\text{coll}}) \le (4/\sqrt{\pi})\,(\mathbf{d}_2(S^{\text{ind}}, S^{\text{coll}})^{1/2}$. Now

$$\mathbf{d}_2(S^{\text{ind}}, S^{\text{coll}}) = \mathbf{d}_2\left(\sum_{i=1}^n X_i, \sum_{i=1}^n S_i^{\text{coll}}\right) \le \sum_{i=1}^n \mathbf{d}_2(X_i, S_i^{\text{coll}})$$

by Lemma 18.2.3. By Lemma 18.2.5 (b) and (d), it follows that

$$\mathbf{d}_2(X_i, S_i^{\text{coll}}) \le \tfrac{1}{2}(EX_i^2 + E(S_i^{\text{coll}})^2) = EX_i^2 = p_i b_i. \qquad \square$$

Remark 18.3.1. Note that in our model, $\mathbf{d}_2(S^{\text{ind}}, S^{\text{coll}}) = \frac{1}{2}\sup_t |E(S^{\text{ind}} - t)_+^2 - E(S^{\text{coll}} - t)_+^2|$ is finite. In view of Lemma 18.2.4, this is not necessarily true for the usual model. By Lemma 18.2.2, the $\zeta_{2,\infty}$-distance between S^{ind} and S^{coll} is infinite in the usual model, while $\zeta_{2,\infty}(S^{\text{ind}}, S^{\text{coll}}) = \mathbf{d}_2(S^{\text{ind}}, S^{\text{coll}})$ is finite in our *scaled* model determined by (18.3.1)–(18.3.3). Moreover, the $\mathbf{d}_3$-metric for the usual model is infinite, as follows from Lemma 18.2.4. This indicates more stability in our new scaled approximation.

We will next consider the special case where $\{X_i\}_{i\ge 1}$ are i.i.d. For this purpose we will use a Berry–Esseen-type estimate for $\mathbf{d}_1$.[15] In the next theorem, we use the following moment characteristic:

$$\tau_3(X,Y) := \max((E|\widetilde{X}| + E|Y|), \tfrac{1}{3}(E|\widetilde{X}|^3 + E|Y|^3)), \quad \widetilde{X} := \frac{X - EX}{\text{Var}(X)}.$$

Theorem 18.3.2. *If $\{X_i\}$ are i.i.d. with $a = EC_1$, $\sigma^2 = \text{Var}(C_1)$, $p = \Pr(D_i = 1)$, then*

$$\mathbf{d}_1(S^{\text{ind}}, S^{\text{coll}}) \le 11.5[p\sigma^2 + p(1-p)a^2]^{1/2}\left(\tau_3(X_1, Y) + \tau_3\left(\sum_{i=1}^{N_1} C_i, Y\right)\right), \tag{18.3.6}$$

where Y has a standard normal distribution and $N_1 \overset{\text{d}}{=} \mathcal{P}(\mu_1)$.

Proof. By the ideality of $\mathbf{d}_1$ (Lemma 18.2.3),

$$\mathbf{d}_1(S^{\text{ind}}, S^{\text{coll}}) = \mathbf{d}_1\left(\sum_{i=1}^n X_i, \sum_{i=1}^n S_i^{\text{coll}}\right)$$
$$= (n\,\text{Var}(X_1))^{1/2}\mathbf{d}_1\left(\frac{1}{\sqrt{n}}\sum_{i=1}^n \widetilde{X}_i, \frac{1}{\sqrt{n}}\sum_{i=1}^n Y_i\right),$$

where $Y_i := (\widetilde{S_i^{\text{coll}}})$. By the triangle inequality, (18.2.57), and Lemma 18.2.5 (c),

$$\mathbf{d}_1\left(\frac{1}{\sqrt{n}}\sum_{i=1}^n \widetilde{X}_i, \frac{1}{\sqrt{n}}\sum_{i=1}^n Y_i\right) \le \kappa_1\left(\frac{1}{\sqrt{n}}\sum_{i=1}^n \widetilde{X}_i, Y\right) + \kappa_1\left(\frac{1}{\sqrt{n}}\sum_{i=1}^n Y_i, Y\right)$$

[15] See Theorem 18.2.1 and Remark 18.2.2.

$$\leq 11.5\{\max(\boldsymbol{\kappa}_1(\widetilde{X}_1, Y), \mathbf{d}_{2,1}(\widetilde{X}_1, Y)) + \max(\boldsymbol{\kappa}_1(Y_1, Y), \mathbf{d}_{2,1}(Y_1, Y))\} n^{-1/2},$$

where $\boldsymbol{\kappa}_1 = \boldsymbol{\zeta}_1 = \boldsymbol{\zeta}_{1,1}$ is the first difference pseudomoment [see (18.2.29)]. With

$$\boldsymbol{\kappa}_1(\widetilde{X}_1, Y) \leq E|\widetilde{X}_1| + E|Y|, \quad d_{2,1}(\widetilde{X}_1, Y) \leq \tfrac{1}{2}(E|\widetilde{X}_1|^3 + |E|Y|^3),$$

and similarly for the second term, we get

$$\max(\boldsymbol{\kappa}_1(Y_1, Y), \mathbf{d}_{2,1}(Y_1, Y)) \leq \tau_3(Y_1, Y) \leq \tau_3\left(\sum_{i=1}^{N_1} C_i, Y\right). \qquad \square$$

The next theorem gives a better estimate for $\mathbf{d}_1(S^{\text{ind}}, S^{\text{coll}})$ when p_i are relatively small.

Theorem 18.3.3. *Let μ_i, u_i be as in (18.3.2), and let $C_i \geq 0$ a.s.; then for any $\Delta_i > 1$*

$$\mathbf{d}_1(S^{\text{coll}}, S^{\text{ind}}) \leq \sum_{i=1}^{n} p_i^2 \tau_i, \tag{18.3.7}$$

where $\tau_i := a_i + \Delta_i v_i + \max(\Delta_i a_i v_i, 2a_i \widetilde{v}_i + (1 + \Delta_i a_i v_i p_i) u_i)$, $v_i := a_i^2 / b_i \leq 1/p_i$, $p_i v_i \leq 1 - \Delta_i^{-1}$, and $\widetilde{v}_i := a_i^2/(b_i - p_i a_i^2)$.

Proof. Since $\mathbf{d}_1$ is an ideal metric, for the proof it is enough to establish[16]

$$\mathbf{d}_1(S_i^{\text{coll}}, X_i) \leq p_i^2 \tau_i. \tag{18.3.8}$$

We will omit the index i in what follows. Since the first moments of S^{ind} and S^{coll} are the same, $|D_{1,S^{\text{coll}},X}(t)|$ determined by (18.2.12) admits the form

$$|D_{1,S^{\text{coll}},X}(t)| = \left| \int_{-\infty}^{t} (t - x)(\mathrm{d}F_{S^{\text{coll}}}(x) - \mathrm{d}F_X(x)) \right|.$$

Further, we will consider only the case where $t > 0$ since the case where $t < 0$ can be handled in the same manner using the preceding equality. For $t > 0$

$$\begin{aligned}
|D_{1,S^{\text{coll}},X}(t)| &:= \left| \int_t^{\infty} (x - t)(\mathrm{d}F_{S^{\text{coll}}}(x) - \mathrm{d}F_X(x)) \right| \\
&= \left| \sum_{k=1}^{\infty} \frac{\mu^k}{k!} \exp(-\mu) \int_t^{\infty} (x - t) \mathrm{d}F_{uC}^{*k}(x) - p \int_t^{\infty} (x - t) \mathrm{d}F_C(x) \right| \\
&\leq I_1 + I_2,
\end{aligned}$$

---

[16] See (18.2.4), (18.2.7), and (18.2.26).

where

$$I_1 := \left| \mu \exp(-\mu) \int_t^\infty (x-t)\mathrm{d}F_{uC}(x) - p \int_t^\infty (x-t)\mathrm{d}F_C(x) \right| \tag{18.3.9}$$

and

$$I_2 := \sum_{n=2}^{\infty} \frac{\mu^k}{k!} \exp(-\mu) \int_t^\infty (x-t)\mathrm{d}F_{uC}^{*k}(x).$$

Since $u = p/\mu$, it follows that

$$I_2 \leq \sum_{k=2}^{\infty} \frac{\mu_k}{k!} \exp(-\mu) k u a = \mu \exp(-\mu) u a (\exp(\mu) - 1) \leq u a \mu^2 = p a \mu. \tag{18.3.10}$$

Using $\mu = pb/(b - pa^2) = p/(1 - pa^2/b)$ we obtain

$$p \leq \mu \leq p\left(1 + \Delta \frac{a^2 p}{b}\right); \tag{18.3.11}$$

therefore,

$$I_2 \leq p a \mu \leq p a p \left(1 + \Delta \frac{a^2 p}{b}\right) = p^2 a + \Delta \frac{a^2 p^3}{b}. \tag{18.3.12}$$

For the estimate of I_1 we use $\overline{F}_C := 1 - F_C$ to obtain

$$I_1 = \left| \mu \exp(-\mu) \int_t^\infty \overline{F}_{uC}(x)\mathrm{d}x - p \int_t^\infty \overline{F}_C(x)\mathrm{d}x \right|.$$

Since, by (18.3.11), $u = p/\mu \leq 1$ and, therefore, $\overline{F}_{uC}(x) \leq \overline{F}_C(x)$, we obtain

$$\begin{aligned} &\mu \exp(-\mu) \int_t^\infty \overline{F}_{uC}(x)\mathrm{d}x - p \int_t^\infty \overline{F}_C(x)\mathrm{d}x \\ &\leq p\left(1 + \Delta \frac{a^2 p}{b}\right) \exp(-p) \int_t^\infty \overline{F}_C(x)\mathrm{d}x - p \int_t^\infty \overline{F}_C(x)\mathrm{d}x \\ &\leq \Delta \left(\frac{a^2}{b}\right) p^2 (EC) = \Delta \frac{a^3 p^2}{b}. \end{aligned} \tag{18.3.13}$$

On the other hand, by (18.3.11), $\exp(-\mu) \geq 1 - \mu \geq 1 - p(1 + \Delta a^2 p/b)$, implying

$$A := p \int_t^\infty \overline{F}_C(x)\mathrm{d}x - \mu \exp(-\mu) \int_t^\infty \overline{F}_{uC}(x)\mathrm{d}x$$

$$
\begin{aligned}
&\le p\int_t^\infty \overline{F}_C(x)\mathrm{d}x - p\left(1-p-\Delta\frac{a^2p^2}{b}\right)\int_t^\infty \overline{F}_{uC}(x)\mathrm{d}x\\
&\le p\left(\int_t^\infty \overline{F}_C(x)\mathrm{d}x - \int_t^\infty \overline{F}_{uC}(x)\mathrm{d}x\right) + p^2\left(1+\Delta\frac{a^2p}{b}\right)ua.
\end{aligned}
\tag{18.3.14}
$$

Now, since

$$
u = p/\mu = \frac{(b-pa^2)}{b} = 1-\frac{pa^2}{b},
$$

then

$$
\begin{aligned}
&p\left(\int_t^\infty \overline{F}_C(x)\mathrm{d}x - \int_{t/u}^\infty \overline{F}_C(y)\mathrm{d}y\right) + p^2\left(\frac{a^2}{b}\right)\int_{t/u}^\infty \overline{F}_C(x)\mathrm{d}x\\
&\quad\le p\int_t^{t/u} \overline{F}_C(x)\mathrm{d}x + \frac{p^2a^3}{b} \le p\overline{F}_C(t)t\left(\frac{1}{u}-1\right)+\frac{p^2a^3}{b}\\
&\quad\le pa\left(\frac{1}{u}-1\right)+\frac{p^2a^3}{b} \le 2p^2\frac{a^3}{b-pa^2}.
\end{aligned}
\tag{18.3.15}
$$

Thus, by (18.3.14) and (18.3.15),

$$
A \le 2p^2\frac{a^3}{b-pa^2} + p^2\left(1+\Delta\frac{a^2p}{b}\right)au.
\tag{18.3.16}
$$

Estimates (18.3.16) and (18.3.14) imply

$$
\begin{aligned}
I_1 &\le \max\left(\Delta\frac{a^3p^2}{b}, 2p^2\frac{a^3}{b-pa^2} + ap^2\left(1+\Delta\frac{a^2p}{b}\right)u\right)\\
&= ap^2\max\left(\Delta\frac{a^2}{b}, 2\frac{a^2}{b-pa^2} + u + \Delta\frac{a^2u}{b}p\right).
\end{aligned}
\tag{18.3.17}
$$

Thus the required bound (18.3.7) follows from (18.3.12) and (18.3.17). □

Remark 18.3.2. From the regularity of $\mathbf{d}_1$ it follows that

$$
\begin{aligned}
\mathbf{d}_1(S^{\text{coll}}, S^{\text{ind}}) &\le \sum_{i=1}^n \mathbf{d}_1(S_i^{\text{coll}}, X_i)\\
&\le \sum_{i=1}^n (ES_i^{\text{coll}} + EX_i) = 2\sum_{i=1}^n a_i p_i.
\end{aligned}
\tag{18.3.18}
$$

Clearly, for small p_i estimate (18.3.7) is a refinement of the preceding bound.

We next give a direct estimate for $\mathbf{d}_2$ and use the relation between $\mathbf{d}_2$ and $\mathbf{d}_1$ to obtain an improved estimate for $\mathbf{d}_1$ for p_i not too small.

Theorem 18.3.4. *Let $C_i \geq 0$ a.s., and let μ_i and u_i be as in (18.3.2). Then*

$$\mathbf{d}_2(S^{\mathrm{ind}}, S^{\mathrm{coll}}) \leq \frac{1}{2}\sum_{i=1}^{n} p_i^2 \tau_i^*, \tag{18.3.19}$$

where

$$\tau_i^* := b_i + 3a_i^2 + \Delta_i a_i^2 + 2\widetilde{v}_i b_i^2 + b_i u_i^2 + \Delta_i a_i p_i, \tag{18.3.20}$$

and $\Delta_i, \widetilde{v}_i$ are defined as in Theorem 18.3.3. Moreover,

$$\mathbf{d}_1(S^{\mathrm{ind}}, S^{\mathrm{coll}}) \leq (4/\sqrt{\pi})\left(\sum_{i=1}^{n} p_i^2 \tau_i^*\right)^{1/2}. \tag{18.3.21}$$

Proof. Again, it is enough to consider $\mathbf{d}_1(S_i^{\mathrm{coll}}, X_i)$, and we will omit the index i. Then, for $t > 0$,

$$\begin{aligned}
&\left|\int_t^\infty (x-t)^2 \mathrm{d}(F_{S^{\mathrm{coll}}}(x) - F_X(x))\right| \\
&= \left|\sum_{k=1}^{\infty} \frac{\mu^k}{k!}\exp(-\mu)\int_t^\infty (x-t)^2 \mathrm{d}F_{uC}^{*k}(x) - p\int_t^\infty (x-t)^2 \mathrm{d}F_C(x)\right| \\
&\leq I_1 + I_2,
\end{aligned} \tag{18.3.22}$$

where

$$I_1 := \left|\mu\exp(-\mu)\int_t^\infty (x-t)^2 \mathrm{d}F_{uC}(x) - p\int_t^\infty (x-t)^2 \mathrm{d}F_C(x)\right|$$

and

$$I_2 := \sum_{k=2}^{\infty} \frac{\mu^k}{k1}\exp(-\mu)\int_t^\infty (x-t)^2 \mathrm{d}F_{uC}^{*k}(x).$$

Since $u = p/\mu$, we obtain

$$\begin{aligned}
I_2 &\leq \sum_{k=2}^{\infty} \frac{\mu^k}{k!}\exp(-\mu) E\left(\sum_{i=1}^{k} uC_i\right)^2 = u^2 \sum_{k=2}^{\infty} \frac{\mu^k}{k!}\exp(-\mu)(kb + k(k-1)a^2) \\
&= u^2 b\mu\exp(-\mu)(\exp(\mu) - 1) + u^2 a^2 \mu^2 \exp(-\mu)\exp(\mu) \leq u^2(b + a^2)\mu^2 \\
&= p^2(b + a^2).
\end{aligned} \tag{18.3.23}$$

Furthermore, from the fact that $uC \le C$ and (18.3.11),

$$\frac{1}{2}I_1 = \left|\mu\exp(-\mu)\int_t^{\infty}(x-t)\overline{F}_{uC}(x)\mathrm{d}x - p\int_t^{\infty}(x-t)\overline{F}_C(x)\mathrm{d}x\right| =: |A| \tag{18.3.24}$$

and

$$\begin{aligned} A &\le p\left(1+\Delta\frac{a^2p}{b}\right)\int_t^{\infty}(x-t)\overline{F}_C(x)\mathrm{d}x - p\int_t^{\infty}(x-t)\overline{F}_C(x)\mathrm{d}x \\ &\le \frac{1}{2}\Delta a^2p^2. \end{aligned} \tag{18.3.25}$$

On the other hand, by $\mu \ge p$, $\exp(-\mu) \ge 1-\mu \ge 1-p(1+\Delta(a^2/b)p)$, it follows that

$$\begin{aligned} -A &\le p\int_t^{\infty}(x-t)\overline{F}_C(x)\mathrm{d}x - p\left(1-p-\Delta\frac{a^2p^2}{b}\right)\int_t^{\infty}(x-t)\overline{F}_{uC}(x)\mathrm{d}x \\ &\le p\left(\int_t^{\infty}(x-t)\overline{F}_C(x)\mathrm{d}x - \int_t^{\infty}(x-t)\overline{F}_{uC}(x)\mathrm{d}x\right) \\ &\quad +p^2\left(1+\Delta\frac{a^2p}{b}\right)u^2b/2 \end{aligned}$$

and

$$\begin{aligned} &\int_t^{\infty}(x-t)\overline{F}_C(x)\mathrm{d}x - \int_t^{\infty}(x-t)\overline{F}_{uC}(x)\mathrm{d}x \\ &\quad = \int_t^{\infty}x\overline{F}_C(x)\mathrm{d}x - \int_t^{\infty}x\overline{F}_{uC}(x)\mathrm{d}x + \int_t^{\infty}t(\overline{F}_{uC}(x)-\overline{F}_C(x))\mathrm{d}x \\ &\quad \le \int_t^{\infty}x\overline{F}_C(x)\mathrm{d}x - \int_{t/u}^{\infty}y\overline{F}_C(y)u^2\mathrm{d}y \\ &\quad \le \int_t^{\infty}x\overline{F}_C(x)\mathrm{d}x - \int_{t/u}^{\infty}y\overline{F}_C(y)\left(1-\frac{pa^2}{b}\right)^2\mathrm{d}y \\ &\quad = \int_t^{t/u}x\overline{F}_C(x)\mathrm{d}x + \left(\frac{2pa^2}{b}\right)\int_t^{\infty}x\overline{F}_C(x)\mathrm{d}x \\ &\quad \le \frac{t}{u}\overline{F}_C(t)\left(\frac{t}{u}-t\right) + \left(\frac{2pa^2}{b}\right)\frac{b}{2} \\ &\quad \le \frac{b(1-u)}{u^2} + (pa^2) = \frac{b^2pa^2}{(b-pa^2)^2} + pa^2. \end{aligned}$$

Thus,

$$-A \leq \frac{b^2 p^2 a^2}{(b-pa^2)^2} + p^2 a^2 + p^2 \left(1 + \frac{\Delta a^2 p}{b}\right) \frac{bu^2}{2}.$$

So we obtain

$$I_1 \leq \max\left(\Delta a^2 p^2, \frac{2b^2 p^2 a^2}{(b-pa^2)^2} + 2p^2 a^2 + p^2 \left(1 + \frac{\Delta a^2 p}{b}\right) bu^2\right),$$

which, together with (18.3.23), implies (18.3.19).

Equation (18.3.21) is a consequence of (18.3.19) and (18.2.50). □

As a corollary, we obtain an estimate for

$$\mathbf{V}_2(S^{\text{ind}}, S^{\text{coll}}) := \sup_t |\,\mathbf{Var}((S^{\text{ind}} - t)_+) - \mathbf{Var}((S^{\text{coll}} - t)_+)|.$$

Corollary 18.3.1.

$$\mathbf{V}^2(S^{\text{ind}}, S^{\text{coll}}) \leq 2\sum_{i=1}^n p_i^2 \tau_i^* + \left(\sum_{i=1}^n p_i^2 \tau_i\right) 2\sum_{i=1}^n p_i a_i,$$

where τ_i^* *is defined by* (18.3.20) *and* τ_i *is the same as in* (18.3.7).

Proof.

$$\begin{aligned}
&\mathbf{V}_2(S^{\text{ind}}, S^{\text{coll}}) \\
&\leq \sup_t |E(S^{\text{coll}}-t)_+^2 - E(S^{\text{ind}} - t)_+^2| + \sup_t |(E(S^{\text{coll}} - t)_+)^2 - (E(S^{\text{ind}} - t)_+)^2| \\
&\leq 2\mathbf{d}_2(S^{\text{coll}}, S^{\text{ind}}) + \mathbf{d}_1(S^{\text{ind}}, S^{\text{coll}}) \sup_t (E(S^{\text{coll}} - t)_+ + E(S^{\text{ind}} - t)_+) \\
&\leq 2\mathbf{d}_2(S^{\text{coll}}, S^{\text{ind}}) + \mathbf{d}_1(S^{\text{ind}}, S^{\text{coll}})(ES^{\text{coll}} + ES^{\text{ind}}) \\
&\leq 2\sum_{i=1}^n p_i^2 \tau_i^* + \left(\sum_{i=1}^n p_i^2 \tau_i\right) 2\sum_{i=1}^n p_i a_i.
\end{aligned}$$

The last inequality follows from (18.3.19) and (18.3.7). □

Remark 18.3.3. One could try to find the RV $\{Z_{ij}\}_{j\geq 1}$ (not necessarily scaled versions of X_i) such that the first k moments of S_i^{coll} coincide with those of X_i. For this purpose (omitting the index i) let $\phi_X(s) = Es^X$ denote the generating function of X. Then, for $Y = \sum_{j=1}^N Z_j$, a compound Poisson distributed RV with N, Poisson $\mathcal{P}(\mu)$, we have $\phi_Y(s) = \phi_N(\phi_Z(s))$, where $\phi_Z := \phi_{Z_1}$.

Now, for the Poisson RV N we obtain the factorial moments $\phi_N^{(k)}(1) = EN(N-1)\cdots(N-k+1) = \mu^k$. Denote the factorial moments by $b_k := EX(X-1)\cdots(X-$

$k+1)$, $a_k := EZ(Z-1)\cdots(Z-k+1)$. This implies $EY = \phi_Y^{(1)}(1) = \mu a_1$, $EY(Y-1) = \phi_Y^{(2)}(1) = \mu^2 a_1^2 + \mu a_2$, $EY(Y-1)(Y-2) = \phi_Y^{(3)}(1) = \mu^3 a_1 a_2 + \mu a_3$.

Thus, we obtain the equations

$$\phi_Y^{(1)}(1) = \mu a_1 = b_1,$$
$$\phi_Y^{(2)}(1) = \mu^2 a_1^2 + \mu a_2 = b_2,$$
$$\phi_Y^{(3)}(1) = \mu^3 a_1^3 + 3\mu^2 a_1 a_2 + \mu a_3 = b_3,$$

and so on; that is,

$$\mu a_1 = b_1, \quad \mu a_2 = b_2 - b_1^2, \quad \mu a_3 = b_3 - b_1^3 - 3b_1(b_2 - b_1^2) = b_3 - 3b_1 b_2 + 2b_1^3,$$

and so on. In contrast to the scaled model, where we have two free parameters μ and u, here we have more *nearly* free parameters. These equations can easily be solved, but one must find solutions $\mu > 0$ such that $\{a_i\}$ are factorial moments of a distribution. In our case where $X = CD$, this is seen to be possible for p small. With $\lambda = p/\mu$ we obtain for the first three moments A_i of $Z : A_1 = \lambda c_1$, $A_2 = \lambda(c_2 + 2c_1 - pc_1)$, and $A_3 = \lambda(c_3 - O(p))$, where c_i are the corresponding moments of C. For p small A_1, A_2, A_3 is a moment sequence. For an example concerning the approximation of a binomial RV by compound Poisson distributed RVs with three coinciding moments and further three moments close to each other, see Arak and Zaitsev (1988, p. 80). They used the closeness in this case to derive the optimal bounds for the variation distance.

By Lemmas 18.2.5 (c) and (d) and 18.2.8, it follows that if one can match the first s moments of X_i and S_i^{coll}, then

$$\mathbf{d}_1(S^{\text{ind}}, S^{\text{coll}}) \le \lambda_s(\mathbf{d}_s(S^{\text{ind}}, S^{\text{coll}}))^{1/s} \le \lambda_s \left[\frac{1}{s!}\sum_{i=1}^{n}(E|X_i|^s + E|S_i^{\text{coll}}|^s)\right]^{1/s}. \tag{18.3.26}$$

This implies that in the case of $E|X_i|^s + E|S_i^{\text{coll}}|^s \le C$, we have the order $n^{1/s}$ as $n \to \infty$ and, in particular, the finiteness of the $\mathbf{d}_s$ distance.

References

Arak TV, Zaitsev AYu (1988) Uniform limit theorems for sums of independent random variables. In: Proceedings of the Steklov Institute of Mathematics, vol 174, AMS

Dunford N, Schwartz J (1988) Linear operators, vol 1. Wiley, New York

Gerber H (1981) An introduction to mathematical risk theory. Huebner Foundation Monograph, Philadelphia

Kalashnikov VV, Rachev ST (1988) Mathematical methods for construction of stochastic queueing models. Nauka, Moscow (in Russian) [English transl., (1990) Wadsworth, Brooks–Cole, Pacific Grove, CA]
Neveu J (1965) Mathematical foundations of the calculus of probability. Holden-Day, San Francisco
Zolotarev VM (1986) Contemporary theory of summation of independent random variables. Nauka, Moscow (in Russian)

Chapter 19
Ideal Metric with Respect to Maxima Scheme of i.i.d. Random Elements

The goals of this chapter are to:

- Introduce max-ideal and max-smoothing metrics and derive rates of convergence in the max-stable limit theory of random vectors in terms of the Kolmogorov metric,
- Provide infinite-dimensional analogs to the convergence rate theorems for random vectors,
- Discuss probability metrics that are ideal with respect to both summation and maxima.

Notation introduced in this chapter:

Notation	Description
$\vee, \wedge$	"Max" and "min" operators
$\boldsymbol{\rho}_r$	Weighted Kolmogorov metric
$M(x)$	Equals $\min_{1 \le i \le m} \lvert x^{(i)} \rvert$ for $x := (x^{(1)}, \ldots, x^{(m)})$
$\boldsymbol{\rho}_\psi$	Kolmogorov metric with weight function $M(x)$
$\boldsymbol{\mu}_\psi$	Weighted metric between the logarithms of distribution functions
$\mathbf{B}$	Separable Banach space
$\mathbf{L}_r[T]$	$\mathbf{L}_r$-space of measurable functions on T
$\mathbf{X}$	Sequence of random processes S_n, $s \geq 1$
$\mathbf{C}$	Sequence of constants satisfying normalizing conditions
$\mathbf{Y}$	Sequence of i.i.d. max-stable processes
$\mathcal{L}_{p,r}$	$\mathcal{L}_p$ on $\mathfrak{X}(\mathbf{L}_r[T])$
$\boldsymbol{\ell}_{p,r}$	Minimal metric with respect to $\mathcal{L}_{p,r}$
$\boldsymbol{\chi}_{p,r}$	Weighted version of Ky Fan metric
$\boldsymbol{\xi}_{p,r}$	Minimal metric with respect to $\chi_{p,r}$
$\boldsymbol{\eta}_{p,r}$	Weighted version of Prokhorov metric
$\boldsymbol{\Delta}_{p,r}$	Compound max-ideal metric

DOI 10.1007/978-1-4614-4869-3_19,

19.1 Introduction

In this chapter, we discuss the problem of estimating the rate of convergence in limit theorems arising from the maxima scheme of independent and identically distributed (i.i.d.) random elements. The chapter is divided into three parts.

We begin with an extreme-value theory of random vectors. We introduce max-ideal and max-smoothing metrics, specifically designed for the maxima scheme, which play a role in the theory similar to the role of the corresponding counterparts in the scheme of summation discussed in Chap. 15. Using the universal methods of the theory of probability metrics, we derive convergence rates in the max-stable limit theorem in terms of the Kolmogorov metric.

Next, we consider the rate of convergence to max-stable processes. We provide infinite-dimensional analogs of the convergence rate theorems for random vectors (RVs).

Finally, we consider probability metrics that are ideal with respect to both summation and maxima, the so-called *doubly ideal metrics*. This question is interesting for the theory of probability metrics itself. It turns out that such metrics exist; the order of ideality, however, is bounded by 1.

19.2 Rate of Convergence of Maxima of Random Vectors Via Ideal Metrics

Suppose $X_1, X_2, \ldots, X_n$ are i.i.d. RVs in $\mathbb{R}^m$ with a distribution function (DF) F. Define the sample maxima as $M_n = (M_n^{(1)}, \ldots, M_n^{(m)})$, where $M_n^{(i)} = \max_{1\leq j\leq n} X_j^{(i)}$. For many DFs F there exist normalizing constants $a_n^{(i)} > 0$, $b_n^{(i)} \in \mathbb{R}$ $(n \geq 1, 1 < i \leq m)$ such that

$$\left(\frac{M_n^{(1)} - b_n^{(1)}}{a_n^{(1)}}, \ldots, \frac{M_n^{(m)} - b_n^{(m)}}{a_n^{(m)}}\right) \stackrel{\mathrm{d}}{\Longrightarrow} Y, \tag{19.2.1}$$

where Y is an RV with nondegenerate marginals. The DF H of Y is said to be a *max-extreme value DF*. The marginals H_i of H must be one of the three extreme value types $\phi_\alpha(x) = \exp(-x^{-\alpha}), (x \geq 0, \alpha > 0), \psi_\alpha(x) = \phi_\alpha(-x^{-1})$, or $\Lambda(x) = \phi_1(\mathrm{e}^x)$. Moreover, necessary and sufficient conditions on F for convergence in (19.2.1) are known.[1]

Throughout we will assume that the limit DF H of Y in (19.2.1) is *simple max-stable*, i.e., each marginal $Y^{(i)}$ has DF $H_i(x) = \phi_1(x) = \exp(-x^{-1})$ $(x \geq 0)$. Note that if $Y_1, Y_2, \ldots$ are i.i.d. copies of Y, then

[1]See, for example, Resnick (1987a) and references therein.

$$\frac{1}{n}\left(\max_{1\le j\le n} Y_j^{(i)},\ldots,\max_{1\le j\le n} Y_j^{(m)}\right) \stackrel{\mathrm{d}}{\Longrightarrow} Y.$$

In this section, we are interested in the rate of convergence in (19.2.1) with respect to different "max-ideal" metrics.[2] In the next section, we will investigate similar rate-of-convergence problems but with respect to compound metrics and their corresponding minimal metrics.

Definition 19.2.1. A probability metric μ on the space $\mathfrak{X} := \mathfrak{X}(\mathbb{R}^n)$ of RVs is called a *max-ideal metric of order* $r > 0$ if, for any RVs $X, Y, Z \in \mathfrak{X}$ and positive constant c, the following two properties are satisfied:

(i) *Max-regularity*: $\mu(X_1 \vee Z, X_2 \vee Z) \le \mu(X_1, X_2)$, where $x \vee y := (x^{(1)} \vee y^{(1)}, \ldots, x^{(m)} \vee y^{(m)})$ for $x, y \in \mathbb{R}^m$, $\vee := \max$.
(ii) *Homogeneity of order* r: $\mu(cX_1, cX_2) = c^r \mu(X_1, Y_2)$.

If μ is a simple p. metric, i.e., $\mu(X_1, X_2) = \mu(\Pr_{X_1}, \Pr_{X_2})$, it is assumed that Z is independent of X and Y in (i).

The preceding definition is similar to Definition 15.3.1 in Chap. 15 of an ideal metric of order r w.r.t. the summation scheme. Taking into account the metric structure of the convolution metrics $\boldsymbol{\mu}_{\theta,r}$ and $\boldsymbol{\nu}_{\theta,r}$,[3] we can construct in a similar way a *max-smoothing metric* $(\widetilde{\boldsymbol{\nu}}_r)$ *of order* r as follows: for any RVs X' and X'' in $\mathfrak{X}$, and Y being a simple max-stable RV, define

$$\begin{aligned}\widetilde{\boldsymbol{\nu}}_r(X', X'') &= \sup_{h>0} h^r \boldsymbol{\rho}(X' \vee hY, X'' \vee hY)\\ &= \sup_{h>0} h^r \sup_{x\in\mathbb{R}^m} |F_{X'}(x) - F_{X''}(x)| F_Y(x/h),\end{aligned} \qquad (19.2.2)$$

where $\boldsymbol{\rho}$ is the Kolmogorov metric in $\mathfrak{X}$,

$$\boldsymbol{\rho}(X', X'') = \sup_{x\in\mathbb{R}^m} |F_{X'}(x) - F_{X''}(x)|. \qquad (19.2.3)$$

Here and in what follows in this section, $X' \vee X''$ means an RV with DF $F_{X'}(x)F_{X''}(x)$.

Lemma 19.2.1. *The max-smoothing metric* $\widetilde{\boldsymbol{\nu}}_r$ *is max-ideal of order* $r > 0$.

The proof is similar to that of Lemma 15.3.1 and is thus omitted.

Another example of a max-ideal metric is given by the weighted *Kolmogorov metric*

[2]See Maejima and Rachev (1997) for a discussion of the convergence rates in the multivariate max-stable limit theorem.

[3]See (15.3.12) and (15.3.13).

$$\boldsymbol{\rho}_r(X', X'') := \sup_{x \in \mathbb{R}^m} M^r(x)|F_{X'}(x) - F_{X''}(x)|, \tag{19.2.4}$$

where $M(x) := \min_{1 \le i \le m} |x^{(i)}|$ for $x := (x^{(1)}, \ldots, x^{(m)})$.

Lemma 19.2.2. *$\boldsymbol{\rho}_r$ is a max-ideal metric of order $r > 0$.*

Proof. The max-regularity property follows easily from $|F_{X' \vee Z}(x) - F_{X'' \vee Z}(x)| \le |F_{X'}(x) - F_{X''}(x)|$ for any Z independent of X' and X''. The homogeneity property is also obvious. □

Next we consider the rate of convergence in (19.2.1) with $a_n^{(i)} = 1/n$ and $b_n^{(i)} = 0$ by means of a max-ideal metric μ. In the sequel, for any X we write $\widetilde{X} := n^{-1}X$.

Lemma 19.2.3. *Suppose $X_1, X_2, \ldots$ are i.i.d. RVs, $M_n := \bigvee_{i=1}^n X_i$, Y is simple max-stable, and μ_r is a max-ideal simple p. metric of order $r > 1$. Then*

$$\mu_r(\widetilde{M}_n, Y) \le n^{1-r}\mu_r(X_1, Y). \tag{19.2.5}$$

Proof. Take $Y_1, Y_2, \ldots$ to be i.i.d. copies of Y, $N_n := Y_1 \vee \cdots \vee Y_n$. Then

$$\begin{aligned}
\mu_r(\widetilde{M}_n, Y) &= \mu_r(\widetilde{M}_n, \widetilde{N}_n) \text{ (by the max-stability of } Y) \\
&= n^{-r}\mu_r(M_n, N_n) \text{ (by the homogeneity property)} \\
&\le n^{-r}\sum_{i=1}^n \mu_r(X_i, Y_i) \\
&= n^{1-r}\mu_r(X_1, Y).
\end{aligned}$$

The inequality follows from the triangle inequality and max-regularity of μ_r. □

By virtue of Lemmas 19.2.1–19.2.3, we have that for $r > 1$ and $n \to \infty$

$$\widetilde{\boldsymbol{\nu}}_r(X_1, Y) < \infty \Rightarrow \widetilde{\boldsymbol{\nu}}_r(\widetilde{M}_n, Y) \le n^{1-r}\widetilde{\boldsymbol{\nu}}_r(X_1, Y) \to 0 \tag{19.2.6}$$

and

$$\boldsymbol{\rho}_r(X_1, Y) < \infty \Rightarrow \boldsymbol{\rho}_r(\widetilde{M}_n, Y) \le n^{1-r}\boldsymbol{\rho}_r(X_1, Y) \to 0. \tag{19.2.7}$$

The last two implications indicate that the right order of the rate of the uniform convergence $\boldsymbol{\rho}(\widetilde{M}_n, Y) \to 0$ should be $O(n^{1-r})$ provided that $\boldsymbol{\nu}_r(X_1, Y) < \infty$ or $\boldsymbol{\rho}_r(X_1, Y) < \infty$. The next theorem gives the proof of this statement for $1 < r \le 2$.

Theorem 19.2.1. *Let $r > 1$.*

(a) If

$$\widetilde{\boldsymbol{\nu}}_r(X_1, Y) < \infty, \tag{19.2.8}$$

then

$$\boldsymbol{\rho}(\widetilde{M}_n, Y) \leq A(r)[\widetilde{\nu}_r n^{1-r} + \kappa n^{-1}] \to 0 \quad as \quad n \to \infty. \tag{19.2.9}$$

In (19.2.9)*, the absolute constant* $A(r)$ *is given by*

$$A(r) := 2[c_1(4^r + 2^r) \vee c_1 c_2 4(3/2)^{\widetilde{r}} \vee c_2(4c_1 4^{\widetilde{r}}/(r-1))^{1/(r-1)}], \tag{19.2.10}$$

where $c_1 := 1+4e^{-2}m$, $c_2 := mc_1, \widetilde{r} := 1 \vee (r-1)$, *and* $\widetilde{\nu}_r$, k *are the following measures of deviation of* $F_{X'}$ *from* F_Y,

$$\kappa := \max(\rho, \widetilde{\nu}_1, \widetilde{\nu}_r^{r/(r-1)}), \quad \rho := \boldsymbol{\rho}(X_1, Y),$$
$$\widetilde{\nu}_1 := \widetilde{\boldsymbol{\nu}}_1(X_1, Y), \quad \widetilde{\nu}_r := \widetilde{\boldsymbol{\nu}}_r(X_1, Y). \tag{19.2.11}$$

(b) If $\boldsymbol{\rho}_r(X_1, Y) < \infty$, *then*

$$\boldsymbol{\rho}(\widetilde{M}_n, Y) \leq B(r)[\rho_r n^{1-r} + \tau n^{-1}] \to 0 \quad as \quad n \to \infty, \tag{19.2.12}$$

where

$$B(r) := (1 \vee K_1 \vee K_r \vee K_r^{1/(r-1)})A(r), \quad K_r := (r/e)^r, \tag{19.2.13}$$

and

$$\tau := \max(\rho, \rho_1, \rho_r^{1/(r-1)}), \quad \rho := \boldsymbol{\rho}(X_1, Y),$$
$$\rho_1 := \boldsymbol{\rho}_1(X_1, Y), \quad \rho_r := \boldsymbol{\rho}_r(X_1, Y). \tag{19.2.14}$$

Remark 19.2.1. Since the simple max-stable RV Y is concentrated on $\mathbb{R}^m_+$, then

$$\boldsymbol{\rho}(\widetilde{M}_n, Y) = \boldsymbol{\rho}\left(\left(\bigvee_{i=1}^n \widetilde{X}_i\right)_+, Y\right) = \boldsymbol{\rho}\left(n^{-1}\bigvee_{i=1}^n (X_i)_+, Y\right), \tag{19.2.15}$$

where $(x)_+ := ((x^{(1)})_+, \ldots, (x^{(m)})_+)$, $(x^{(i)})_+ := 0 \vee x^{(i)}$. Therefore, without loss of generality we may consider X_i as being nonnegative RVs. Thus, subsequently we assume that all RVs X under consideration are nonnegative.

Similar to the proof of Theorem 16.3.1, the proof of the preceding theorem is based on relationships between the max-ideal metrics $\boldsymbol{\nu}_r$ and $\boldsymbol{\rho}_r$ and the uniform metric $\boldsymbol{\rho}$. These relationships have the form of max-smoothing-type inequalities; see further Lemmas 19.2.4–19.2.7. Recall that in our notations $X' \vee X''$ means maximum of independent copies of X' and X''. The first lemma is an analog of Lemma 16.3.1 concerning the smoothing property of stable random motion.

Lemma 19.2.4 (Max-smoothing inequality). *For any $\delta > 0$*

$$\rho(X,Y) < c_1\rho(X \vee \delta Y, Y \vee \delta Y) + c_2\delta, \tag{19.2.16}$$

where

$$c_1 = 1 + 4\mathrm{e}^{-2}m, \quad c_2 = mc_1. \tag{19.2.17}$$

Proof. Let $\mathbf{L}(X'X'')$ be the Lévy metric,

$$\mathbf{L}(X',X'') = \inf\{\varepsilon > 0 : F_{X'}(x-\varepsilon\mathbf{e})-\varepsilon \leq F_{X''}(x) \leq F_{X'}(x+\varepsilon\mathbf{e})+\varepsilon\}, \tag{19.2.18}$$

in $\mathfrak{X}^m = \mathfrak{X}(\mathbb{R}^m_+)$, $\mathbf{e} = (1,1,\ldots,1) \in \mathbb{R}^m$.[4] □

Claim 1. If c_1 is given by (19.2.17), then

$$\rho(X,Y) \leq (1+c_1)\mathbf{L}(X,Y). \tag{19.2.19}$$

Since $F_{Y^{(i)}}(t) = \exp(-1/t)$, $t > 0$, it is easy to see that

$$\begin{aligned}\rho(X,Y) &\leq \left(1 + \sum_{j=1}^{m} \sup_{t>0}\left(\frac{\mathrm{d}}{\mathrm{d}t}F_{Y^{(j)}}(t)\right)\right)\mathbf{L}(X,Y)\\ &= (1+4\mathrm{e}^{-2}m)\mathbf{L}(X,Y),\end{aligned}$$

which proves (19.2.19).

Claim 2. For any $X \in \mathfrak{X}^m$ and a simple max-stable RV Y

$$\mathbf{L}(X,Y) \leq \rho(X \vee \delta Y, Y \vee \delta Y) + \delta m, \quad \delta > 0. \tag{19.2.20}$$

Proof of Claim 2. Let $\mathbf{L}(X,Y) > \gamma$. Then there exists $x_0 \in \mathbb{R}^m_+$ [i.e., $x_0 \geq \overline{0}$, i.e., $x_0^{(i)} \geq 0, i = 1,\ldots,m$] such that

$$|F_X(x) - F_Y(x)| > \gamma, \quad \text{for any} \quad x_0 \leq x \leq x_0 + \gamma\mathbf{e}. \tag{19.2.21}$$

By (19.2.21) and the Hoeffding–Fréchet inequality

$$F_Y(x) \geq \max\left(0, \textstyle\sum_{j=1}^m F_{Y^{(j)}}(x) - m + 1\right),$$

we have that

[4] See Example 4.2.3 in Chap. 4.

$$|F_X(x_0+\gamma\mathbf{e}) - F_Y(x_0+\gamma\mathbf{e})|F_{\delta Y}(x_0+\gamma\mathbf{e})$$

$$\geq \gamma F_{\delta Y}(\gamma\mathbf{e}) = \gamma F_Y\left(\frac{\gamma}{\delta}\mathbf{e}\right) \geq \gamma\left(\sum_{j=1}^{m} F_{Y^{(j)}}(\gamma/\delta) - m + 1\right)$$

$$= \gamma\left(\sum_{j=1}^{m}\exp(-\delta/\gamma) - m + 1\right) \geq \gamma(m(1-\delta/\gamma) - m + 1) = \gamma - m\delta.$$

Therefore, $\boldsymbol{\rho}(X \vee \delta Y, Y \vee \delta Y) \geq \gamma - \delta m$. Letting $\gamma \to \mathbf{L}(X,Y)$ we obtain (19.2.20).

Now, the inequality in (19.2.16) is a consequence of Claims 1 and 2. □

The next lemma is an analog of Lemmas 16.3.2 and 15.4.1.

Lemma 19.2.5. *For any $X', X'' \in \mathfrak{X}^m$*

$$\boldsymbol{\rho}(X' \vee \delta Y, X'' \vee \delta Y) \leq \delta^{-r}\widetilde{\boldsymbol{\nu}}_r(X', X'') \tag{19.2.22}$$

and

$$\boldsymbol{\rho}(X' \vee \delta Y, X'' \vee \delta Y) \leq K_r\delta^{-r}\boldsymbol{\rho}_r(X', X''), \tag{19.2.23}$$

where

$$K_r := (r/e)^r. \tag{19.2.24}$$

Proof of Lemma 19.2.5. Inequality (19.2.22) follows immediately from the definition of $\widetilde{\boldsymbol{\nu}}_r$ [see (19.2.2)]. Using the Hoeffding–Fréchet inequality

$$F_Y(x) \leq \min_{1\leq i\leq m} F_{Y^{(i)}}(x^{(i)}) = \min_{1\leq i\leq m}\exp\{-1/x^{(i)}\} \tag{19.2.25}$$

we have

$$\begin{aligned}
\boldsymbol{\rho}(X' \vee \delta Y, X'' \vee \delta Y) &= \sup_{x\in\mathbb{R}^n} F_{\delta Y}(x)|F_{X'}(x) - F_{X''}(x)| \\
&\leq \sup_{x\in\mathbb{R}^m}\min_{1\leq i\leq m}\exp(-\delta/x^{(i)})|F_{X'}(x) - F_{X''}(x)| \\
&= \sup_{x\in\mathbb{R}^m}\min_{1\leq i\leq m}\left[\left(\frac{\delta}{x^{(i)}}\right)^r\exp\left(-\frac{\delta}{x^{(i)}}\right)\right) \\
&\quad\times\left(\frac{\delta}{x^{(i)}}\right)^{-r}|F_{X'}(x) - F_{X''}(x)| \\
&\leq K_r\sup_{x\in\mathbb{R}^m}\min_{1\leq i\leq m}\left(\frac{\delta}{x^{(i)}}\right)^{-r}|F_{X'}(x) - F_{X''}(x)| \\
&= K_r\delta^{-r}\boldsymbol{\rho}_r(X', X''),
\end{aligned}$$

which proves (19.2.23). □

Lemma 19.2.6. *For any X' and X''*

$$\widetilde{\nu}_r(X', X'') \leq K_r \rho_r(X', X''). \tag{19.2.26}$$

Proof. Apply (19.2.23) and (19.2.2) to get the preceding inequality. □

Lemma 19.2.7. *For any independent RVs X', X'', Z, $W \in \mathfrak{X}^m$*

$$\rho(X' \vee Z, X'' \vee Z) \leq \rho(Z, W)\rho(X', X'') + \rho(X' \vee W, X'' \vee W). \tag{19.2.27}$$

Proof. For any $x \in \mathbb{R}^m$

$$\begin{aligned} &F_Z(x)|F_{X'}(x) - F_{X''}(x)| \\ &\quad \leq |F_Z(x) - F_W(x)||F_{X'}(x) - F_{X''}(x)| + F_W(x)|F_{X'}(x) - F_{X''}(x)|, \end{aligned}$$

which proves (19.2.27). □

The last lemma resembles Lemmas 15.4.2 and 15.4.4 dealing with *smoothing for sums of i.i.d.* Now we are ready for the proof of the theorem.

Proof of Theorem 19.2.1.

(a) Let $Y_1, Y_2, \ldots$ be a sequence of i.i.d. copies of Y, $N_n := \bigvee_{i=1}^{N} Y_i$. Hence

$$\rho(\widetilde{M}_n, Y) = \rho(\widetilde{M}_n, \widetilde{N}_n). \tag{19.2.28}$$

By the smoothing inequality (19.2.16),

$$\rho(\widetilde{M}_n, \widetilde{N}_n) \leq c_1 \rho(\widetilde{M}_n \vee \delta Y, \widetilde{N}_n \vee \delta Y) + c_2 \delta. \tag{19.2.29}$$

Consider the right-hand side of (19.2.29), and obtain for $n \geq 2$

$$\begin{aligned} &\rho(\widetilde{M}_n \vee \delta Y, \widetilde{N}_n \vee \delta Y) \\ &\quad \leq \sum_{j=1}^{m+1} \rho\left(\bigvee_{i=1}^{j-1} \widetilde{Y}_i \vee \bigvee_{i=j}^{n} \widetilde{X}_i \vee \delta Y, \bigvee_{i=1}^{j} \widetilde{Y}_i \vee \bigvee_{i=j+1}^{n} \widetilde{X}_i \vee \delta Y\right) \\ &\qquad + \rho\left(\bigvee_{j=1}^{m+1} \widetilde{Y}_j \vee \bigvee_{j=m+2}^{n} \widetilde{X}_j \vee \delta Y, \bigvee_{j=1}^{m+1} \widetilde{Y}_j \vee \bigvee_{j=m+2}^{n} \widetilde{Y}_j \vee \delta Y\right), \end{aligned} \tag{19.2.30}$$

where m is the integer part of $n/2$ and $\bigvee_{j=1}^{0} := 0$. By Lemma 19.2.7, we can estimate each term on the right-hand side of (19.2.30) as follows:

$$\rho\left(\bigvee_{i=1}^{j-1}\widetilde{Y}_i \vee \bigvee_{i=j}^{n}\widetilde{X}_i \vee \delta Y, \bigvee_{i=1}^{j}\widetilde{Y}_i \vee \bigvee_{i=j+1}^{n}\widetilde{X}_i \vee \delta Y\right)$$

$$\leq \rho\left(\bigvee_{i=j+1}^{n}\widetilde{X}_i, \bigvee_{i=j+1}^{n}\widetilde{Y}_i\right)\rho\left(\bigvee_{i=1}^{j-1}\widetilde{Y}_i \vee \widetilde{X}_j \vee \delta Y, \bigvee_{i=1}^{j}\widetilde{Y}_i \vee \delta Y\right)$$

$$+\rho\left(\bigvee_{i=1}^{j-1}\widetilde{Y}_i \vee \widetilde{X}_j \vee \delta Y \vee \bigvee_{i=j+1}^{n}\widetilde{Y}_i, \bigvee_{i=1}^{j}\widetilde{Y}_i \vee \delta Y \vee \bigvee_{i=j+1}^{n}\widetilde{Y}_i\right). \tag{19.2.31}$$

Combining (19.2.28)–(19.2.31) and using Lemma 19.2.7, again we have

$$\rho\left(\bigvee_{j=1}^{n}\widetilde{X}_j, Y\right) \leq c_1(I_1 + I_2 + I_3 + I_4) + c_2\delta, \tag{19.2.32}$$

where

$$I_1 := \rho\left(\bigvee_{i=2}^{n}\widetilde{X}_i, \bigvee_{i=2}^{n}\widetilde{Y}_i\right)\rho(\widetilde{X}_1 \vee \delta Y, \widetilde{Y}_1 \vee \delta Y),$$

$$I_2 := \sum_{j=2}^{m+1}\rho\left(\bigvee_{i=j+1}^{n}\widetilde{X}_i, \bigvee_{i=j+1}^{n}\widetilde{Y}_i\right)\rho\left(\bigvee_{i=1}^{j-1}\widetilde{Y}_i \vee \widetilde{X}_j \vee \delta Y, \bigvee_{i=1}^{j}\widetilde{Y}_i \vee \delta Y\right),$$

$$I_3 := \sum_{j=1}^{m+1}\rho\left(\widetilde{X}_j \vee \bigvee_{i=m+2}^{n}\widetilde{Y}_i, \widetilde{Y}_j \vee \bigvee_{i=m+2}^{n}\widetilde{Y}_i\right),$$

and

$$I_4 := \rho\left(\bigvee_{j=1}^{m+1}\widetilde{Y}_j \vee \bigvee_{j=m+2}^{n}\widetilde{X}_j, \bigvee_{j=1}^{m+1}\widetilde{Y}_j \vee \bigvee_{j=m+2}^{n}\widetilde{Y}_j\right).$$

Take $n \geq 3$. We estimate I_3 by making use of Lemmas 19.2.1 and 19.2.5,

$$I_3 \leq \sum_{j=1}^{m+1}\widetilde{\boldsymbol{\nu}}_r(\widetilde{X}_j, \widetilde{Y}_j)\left(\frac{n-m-1}{n}\right)^{-r} \leq \sum_{j=1}^{m+1}\widetilde{\boldsymbol{\nu}}_r(\widetilde{X}_j, \widetilde{Y}_j)4^r$$

$$\leq (m+1)n^{-r}\widetilde{\boldsymbol{\nu}}_r(X_1, Y_1)4^r \leq 4^r n^{1-r}\nu_r. \tag{19.2.33}$$

In the same way, we estimate I_4,

$$I_4 \le \widetilde{\boldsymbol{\nu}}_r\left(\bigvee_{j=m+2}^{n} \widetilde{X}_j, \bigvee_{j=m+2}^{n} \widetilde{Y}_j\right)\left(\frac{m+1}{n}\right)^{-r}$$

$$\le 2^r(n-m)n^{-r}\widetilde{\boldsymbol{\nu}}_r(X_1, Y_1) \le 2^r n^{1-r}\widetilde{\boldsymbol{\nu}}_r. \tag{19.2.34}$$

Set

$$\delta := A\max(\widetilde{\upsilon}_r, \widetilde{\upsilon}_r^{1/(r-1)})n^{-1}, \tag{19.2.35}$$

where $A > 0$ will be chosen later. Suppose that for all $k < n$

$$\boldsymbol{\rho}\left(k^{-1}\bigvee_{j=1}^{k} X_j, k^{-1}\bigvee_{j=1}^{k} Y_j\right), \le A(r)(\widetilde{\upsilon}_r k^{1-r} + \kappa k^{-1}), \tag{19.2.36}$$

where $\widetilde{\upsilon}_r = \widetilde{\boldsymbol{\nu}}_r(X_1, Y)$, $\kappa = \kappa(X_1, Y) = \max(\rho, \widetilde{\upsilon}_1, \widetilde{\upsilon}_r^{1/(r-1)})$ [see (19.2.11)]. Here $A(r)$ is an absolute constant to be determined later. For $k = 1$ the inequality in (19.2.36) holds with $A(r) \ge 1$. For $k = 2$

$$\boldsymbol{\rho}\left(2^{-1}\bigvee_{j=1}^{2} X_j, 2^{-1}\bigvee_{j=1}^{2} Y_j\right) \le 2\boldsymbol{\rho}(X_1, Y_2),$$

which means (19.2.36) is valid with $A(r) \ge 4 \vee 2^r$.

Let us estimate I_1 in (19.2.32). By (19.2.22), (19.2.35), and (19.2.36),

$$I_1 \le A(r)(\widetilde{\upsilon}_r(n-1)^{1-r} + \kappa(n-1)^{-1})\widetilde{\boldsymbol{\nu}}_1(n^{-1}X_1, n^{-1}Y_1)\frac{1}{A\widetilde{\upsilon}_1 n^{-1}}$$

$$\le \left(\frac{3}{2}\right)^{(r-1)\vee 1}\frac{A(r)}{A}(\widetilde{\upsilon}_r n^{1-r} + \kappa n^{-1}).$$

Similarly, we estimate I_2:

$$I_2 = \sum_{j=2}^{m+1}\boldsymbol{\rho}\left((n-j)^{-1}\bigvee_{i=1}^{n-j} X_i, (n-j)^{-1}\bigvee_{i=1}^{n-j} Y_i\right)\boldsymbol{\rho}\left(\left(\frac{j-1}{n}+\delta\right)Y \vee \widetilde{X}_j,\right.$$
$$\left.\left(\frac{j-1}{n}+\delta\right)Y \vee \widetilde{Y}_j\right)$$

$$\le \sum_{j=2}^{m+1} A(r)(\widetilde{\upsilon}_r(n-j)^{1-r} + \kappa(n-j)^{-1})\frac{\widetilde{\boldsymbol{\nu}}_r(\widetilde{X}_j, \widetilde{Y}_j)}{\left(\dfrac{j-1}{n}+\delta\right)^r}$$

$$\leq \sum_{j=2}^{m+1} A(r)(\widetilde{\nu}_r(n-m-1)^{1-r} + \kappa(n-m-1)^{-1})\frac{n^{-r}\widetilde{\nu}_r(X_1, Y)}{n^{-r}(j-1+\delta n)^r}$$

$$\leq A(r)(4^{r-1}\widetilde{\nu}_r n^{1-r} + 4\kappa n^{-1})\sum_{j=2}^{\infty}\frac{\nu_r}{(j-1+A\nu_r^{1/(r-1)})^r}$$

$$\leq 4^{(r-1)\vee 1}\frac{1}{r-1}\frac{A(r)}{A^{r-1}}(\widetilde{\nu}_r n^{1-r} + \kappa n^{-1}).$$

Now we can use the preceding estimates for I_1 and I_2 and combine them with (19.2.33)–(19.2.35) and (19.2.32) to get

$$\boldsymbol{\rho}\left(\bigvee_{j=1}^{n}\widetilde{X}_j, Y\right) \leq \left(c_1\left(\frac{3}{2}\right)^{\widetilde{r}}(1/A) + c_1 4^{\widetilde{r}}\frac{1}{r-1}\frac{1}{A^{r-1}}\right) A(r)\widetilde{\nu}_r n^{1-r} + \kappa n^{-2})$$

$$+c_2(4^r + 2^r)\widetilde{\nu}_r n^{1-r} + c_2 A\kappa n^{-1}, \quad \widetilde{r} := \max(1, r-1).$$

Now choose $A = \max\left(4c_1(3/2)^{\widetilde{r}}, \left(4c_1 4^{\widetilde{r}}\frac{1}{r-1}\right)^{1/(r-1)}\right)$. Then

$$\boldsymbol{\rho}\left(\bigvee_{j=1}^{n}\widetilde{X}_j, Y\right) \leq \frac{1}{2}A(r)(\widetilde{\nu}_r n^{1-r} + \kappa n^{-1})$$

$$+(c_1(4^r + 2^r) \vee c_2 A)(\widetilde{\nu}_r n^{1-r} + \kappa n^{-2}).$$

Finally, letting $\frac{1}{2}A(r) := c_1(4^r + 2^r) \vee c_2 A$ completes the proof. □

(b) By (a) and Lemma 19.2.6,

$$\boldsymbol{\rho}\left(\bigvee_{j=1}^{n}\widetilde{X}_j, Y\right) \leq A(r)[K_r\rho_r n^{1-r} + \max(\rho, K_1\rho_r, K_r^{1/(r-1)}\rho_r)n^{-1}]$$

$$\leq (1 \vee K_1 \vee K_r \vee K_r^{1/(r-1)})A(r)[\rho_r n^{1-r} + \tau n^{-1}]. \quad \square$$

Further, we will prove that the order $O(n^{1-r})$ of the rate of convergence in (19.2.9) and (19.2.12) is precise for any $r > 1$ under the conditions $\widetilde{\nu}_r < \infty$ or $\boldsymbol{\rho}_r < \infty$. Moreover, we will investigate more general *tail* conditions than $\rho_r = \boldsymbol{\rho}_r(X_1, Y) < \infty$.

Let $\psi : [0,\infty) \to [0,\infty)$ denote a continuous and increasing function. Let us consider the metrics $\boldsymbol{\rho}_\psi$ and $\boldsymbol{\mu}_\psi$ defined by

$$\boldsymbol{\rho}_{\psi}(X', X'') := \sup_{x \in \mathbb{R}_+^m} \psi(M(x)|F_{X'}(x) - F_{X''}(x)|, \qquad X', X'' \in \mathfrak{X}^m,$$

and[5]

$$\boldsymbol{\mu}_{\psi}(X', X'') := \sup_{x \in \mathbb{R}_+^m} \psi(M(x))| \log F_{X'}(x) - \log F_{X''}(x)|,$$

and recall that $M(x) := \min\{x^{(i)} : i = 1, \ldots, m\}$, $x \in \mathbb{R}_+^m$.

We will investigate the rate of convergence in $\widetilde{M}_n \stackrel{\mathrm{d}}{\Longrightarrow} Y$, assuming that either $\boldsymbol{\rho}_{\psi}(X_1, Y) < \infty$ or $\boldsymbol{\mu}_{\psi}(X_1, Y) < \infty$. Obviously, $\boldsymbol{\rho}_{\psi}(X_1, Y) < \infty$ implies that for each i, $\boldsymbol{\rho}_{\psi}(X_1^{(i)}, Y^{(i)}) := \sup\{\psi(x)|F_i(x) - \phi_1(x)| : x \in \mathbb{R}_+\} \le \boldsymbol{\rho}_{\psi}(X_1, Y) < \infty$, where F_i is the DF of $X_1^{(i)}$. We also define

$$\widetilde{\rho}_{\psi} := \max\{\boldsymbol{\rho}_{\psi}(X_1^{(i)}, Y^{(i)}) : i = 1, \ldots, m\},$$

and whenever $\boldsymbol{\mu}_{\psi}(X_1^{(i)}, Y^{(i)}) := \sup\{\psi(x)| \log F_i(x) - \log \phi_1(x)| : x \in \mathbb{R}_+\} < \infty$, we also define

$$\widetilde{\mu}_{\psi} := \max\{\boldsymbol{\mu}_{\psi}(X_1^{(i)}, Y^{(i)}) : i = 1, \ldots, k\}.$$

In the proofs of the results below, we will often use the following inequalities. Since $H(x) := \Pr(Y \le x) \le H_i(x^{(i)}) := \Pr(Y^{(i)} \le x^{(i)}) = \phi_1(x^{(i)})$ for each i, we have

$$H(x) \le \phi_1(M(x)). \tag{19.2.37}$$

For $a, b > 0$ we have

$$n|a - b| \min(a^{n-1}, b^{n-1}) \le |a^n - b^n| \le n|a - b| \max(a^{n-1}, b^{n-1}) \tag{19.2.38}$$

and

$$\min(a, b) \left|\log \frac{a}{b}\right| \le |a - b| \le \max(a, b) \left|\log \frac{a}{b}\right|. \tag{19.2.39}$$

Theorem 19.2.2. *Assume that*

$$g(a) := \sup_{x \ge 0} \frac{\phi_1(xa)}{\psi(x)}$$

is finite for all $a \ge 0$. For $n \ge 2$ define $R(n) := ng(1/(n-1))$.

(i) If $\rho_{\psi} := \boldsymbol{\rho}_{\psi}(X_1, Y) < \infty$ and $\widetilde{\mu}_{\psi} < \infty$, then for all $n \ge 2$, $\boldsymbol{\rho}(\widetilde{M}_n, Y) \le R(n)\tau$, where $\tau := \max(\rho_{\psi} \exp \widetilde{\mu}_{\psi}, 1 + \exp \widetilde{\mu}_{\psi})$.

---

[5] See the definition of $\boldsymbol{\rho}_r$ provided in (19.2.4).

(ii) *If* $\rho_\psi < \infty$, *then*

$$\limsup_{n\to\infty} \frac{1}{R(n)} \rho(\widetilde{M}_n, Y) \le \overline{\tau},$$

where $\overline{\tau} := \max(\rho_\psi \exp \widetilde{\rho}_\psi, 1 + \exp \widetilde{\rho}_\psi)$.

(iii) *If* $\rho_\psi < \infty$ *and if there exists a sequence* δ_n *of positive numbers such that*

$$\lim_{n\to\infty} \frac{n}{\psi(\delta_n)} = \lim_{n\to\infty} \frac{1}{R(n)} \phi_1\left(\frac{\delta_n}{n-1}\right) = 0,$$

then in (ii) $\overline{\tau}$ *may be replaced by* ρ_ψ.

Remark 19.2.2. (a) In Theorem 19.2.2, we normalize the partial maxima M_n by n. In Theorem 19.2.4 below, we prove a result in which other normalizations are allowed.

(b) If Y is not simple max-stable but has marginals $H_i = \phi_{\alpha_i}(x)$ $(\alpha_i > 0)$, then, by means of simple monotone transformations, Theorem 19.2.2 can be used to estimate

$$\rho((M_n^{(1)} n^{-1/\alpha_1}, \ldots, M_n^{(k)} n^{-1/\alpha_k}), Y), \quad \text{where } M_n^{(i)} := \bigvee_{j=1}^{n} X_j^{(i)}.$$

Proof of Theorem 19.2.2. Using (19.2.38) and $H(x) = H^n(nx)$ we have

$$I := |F^n(nx) - H(x)| \le n|F(nx) - H(nx)| \max(F^{n-1}(nx), H^{n-1}(nx)),$$

where F is the DF of X_i. Let us consider $I_1 := n|F(nx) - H(nx)|H^{n-1}(nx))$. Using (19.2.37) we have

$$H^{n-1}(nx) \le \phi_1\left(\frac{nM(x)}{n-1}\right).$$

Hence, $I_1 \le ng(1/(n-1))\psi(nM(x))|F(nx) - H(nx)|$, and we obtain

$$I_1 \le R(n)\rho_\psi. \tag{19.2.40}$$

Next, consider $I_2 := n|F(nx) - H(nx)|F^{n-1}(nx)$, and let δ_n denote a sequence of positive numbers to be determined later. Observe that for each i and $u \ge \delta_n$ we have

$$|\log F_i(u) - \log \phi_1(u)| \le \frac{1}{\psi(\delta_n)} \sup_{u \ge \delta_n} \psi(u) \left|\log \frac{F_i(u)}{\phi_1(u)}\right| =: \frac{1}{\psi(\delta_n)} \mu_n^{(i)},$$

so that

$$F_i(u) \le \phi_1(u) \exp \frac{1}{\psi(\delta_n)} \mu_n^{(i)}.$$

If $nx_i \geq \delta_n$, then for each i we obtain

$$F^{n-1}(nx) \leq F_i^{n-1}(nx_i) \leq \phi_1^{n-1}(nx_i) \exp \frac{n-1}{\psi(\delta_n)} \mu_n^{(i)}. \tag{19.2.41}$$

This implies that

$$I_2 \leq R(n)\psi(nx_i)|F(nx) - H(nx)| \exp \frac{n-1}{\psi(\delta_n)} \mu_n^{(i)}.$$

Choosing i such that $x_i = M(x)$, it follows that

$$I_2 \leq R(n)\rho_\psi \exp \frac{n-1}{\psi(\delta_n)} \mu_n^{(i)}. \tag{19.2.42}$$

On the other hand, if $nx_i \leq \delta_n$ for some index i, we have $I \leq F_i^n(\delta_n) + \phi_1^n(\delta_n)$. Using (19.2.41) with $nx_i = \delta_n$, it follows that

$$I \leq \phi_1^{n-1}(\delta_n) \left(1 + \exp \frac{n-1}{\psi(\delta_n)} \mu_n^{(i)} \right). \tag{19.2.43}$$

Using $\phi_1^{n-1}(\delta_n) = \phi_1 \left(\frac{\delta_n}{n-1} \right) \leq \phi(\delta_n) g(1/(n-1))$ we obtain

$$I \leq \psi(\delta_n) g(1/(n-1)) \left(1 + \exp \frac{n-1}{\psi(\delta_n)} \mu_n^{(i)} \right). \tag{19.2.44}$$

Proof of (i). Choose δ_n such that $n-1 \leq \psi(\delta_n) \leq n$; since $\mu_n^{(i)} \leq \widetilde{\mu}_\psi$, inequalities (19.2.40), (19.2.42), and (19.2.44) yield

$$I \leq \begin{cases} R(n)\rho_\psi \exp \widetilde{\mu}_\psi & \text{if } nM(x) \geq \delta_n, \\ R(n)(1 + \exp \widetilde{\mu}_\psi) & \text{if } nM(x) < \delta_n. \end{cases}$$

This proves (i).

Proof of (ii). Again choose δ_n such that $n-1 \leq \psi(\delta_n) \leq n$. Using (19.2.39) we obtain

$$\limsup_{n\to\infty} \mu_n^{(i)} \leq \rho_\psi(X_1^{(i)}, Y^{(i)}) \leq \widetilde{\rho}_\psi. \tag{19.2.45}$$

Combining (19.2.40), (19.2.42), (19.2.44), and (19.2.45) we obtain the proof of (ii).

Proof of (iii). Using the sequence δ_n satisfying the assumption of the theorem, it follows from (19.2.40), (19.2.42), (19.2.43), and (19.2.45) that

$$\limsup_{n\to\infty} \frac{1}{R(n)} \rho(\widetilde{M}_n, Y) \le \rho_\psi,$$

which completes the proof. □

Suppose now that ψ is *regularly varying* with index $r \ge 1$,[6] i.e., $\phi(x) \sim x^r L(x)$ as $x \to \infty$ and $L(x)$ varying slowly, $\psi \in RV_r$. We may assume that ψ' is positive and $\psi' \in RV_{r-1}$. In this case, $g(a) = \phi_1(\overline{x}a)/\psi(\overline{x})$, where $\overline{x}$ is a solution of the equation $x^2\psi'(x)/\psi(x) = 1/a$. It follows that $\overline{x}a \to 1/r$ as $a \to 0$ and, hence, that $g(a) \sim K(r)1/\psi(1/a)$ $(a \to 0)$, where $K(r) = (r/e)^r$. In particular, if $\psi(t) = t^r$, then $\boldsymbol{\rho}_\psi = \boldsymbol{\rho}_r$ [see (19.2.4)] and both Theorem 19.2.1 (for $1 < r \le 2$) and Theorem 19.2.2 (for any $r > 1$) state that $\boldsymbol{\rho}_r(X_1, Y) < \infty$ implies $\boldsymbol{\rho}(\widetilde{M}_n, Y) = O(n^{1-r})$. Moreover, in Theorem 19.2.1 we obtain an estimate for $\boldsymbol{\rho}(\widetilde{M}_n, Y)$ [see (19.2.12)], which is *uniform on* $n = 1, 2, \ldots$. The next theorem shows that the condition $\boldsymbol{\rho}_r(X_1, Y) < \infty$ is necessary for having rate $O(n^{1-r})$ in the uniform convergence $\boldsymbol{\rho}(\widetilde{M}_n, Y) \to 0$ as $n \to \infty$.

Theorem 19.2.3. *Assume that $\psi \in RV_r$, $r \ge 1$, and that $\lim_{x\to\infty} \psi(x)/x = \infty$. Let Y denote an RV with a simple max-stable DF H, and let $X_1, X_2, \ldots$ be i.i.d. with common DF F. Then*

(i) $\boldsymbol{\rho}_\psi(X_1, Y) < \infty$ holds if and only if $\limsup_{n\to\infty}(\psi(n)/n)\boldsymbol{\rho}(\widetilde{M}_n, Y) < \infty$

and

(ii) If $r > 1$, then $\limsup_{M(x)\to\infty} \psi(M(x))|F(x) - H(x)| = 0$ if and only if

$$\lim_{n\to\infty} \frac{\psi(n)}{n} \boldsymbol{\rho}(\widetilde{M}_n, Y) = 0.$$

Proof. (i) If $\boldsymbol{\rho}_\psi < \infty$, then the result is a consequence of Theorem 19.2.2. To prove the "if" part, use inequality (19.2.38) to obtain

$$n|F(x) - H(x)| \min(F^{n-1}(x), H^{n-1}(x)) \le \boldsymbol{\rho}(\widetilde{M}_n, Y).$$

Now, if $M(x) \to \infty$, then choose n such that $n \le M(x) < n + 1$; then $F^{n-1}(x) \ge F^{n-1}(n\mathbf{e})$ and $H^{n-1}(x) \ge H^{n-1}(n\mathbf{e})$, and it follows that

$$\psi(M(x))|F(x) - H(x)| \le \frac{\psi(n+1)}{\psi(n)} \frac{(\psi(n)/n)\boldsymbol{\rho}(\widetilde{M}_n, Y)}{\min(F^{n-1}(n\mathbf{e}, H^{n-1}(n\mathbf{e}))}.$$

Since $\psi(n+1) \sim \psi(n)$ $(n \to \infty)$, it follows that

$$\limsup_{M(x)\to\infty} \psi(M(x))|F(x) - H(x)| < \infty$$

and, consequently, that $\boldsymbol{\rho}_\psi(X_1, Y) < \infty$.

[6]See Resnick (1987a).

(ii) If

$$\lim_{n\to\infty} \frac{\psi(n)}{n} \boldsymbol{\rho}(\widetilde{M}_n, Y) = 0,$$

then it follows as in the proof of (i) that $\limsup_{M(x)\to\infty} \psi(M(x))|F(x) - H(x)| = 0$. To prove the "only if" part, choose A such that $\psi(M(x))|F(x) - H(x)| \le \varepsilon$, $M(x) \ge A$.

Now we proceed as in the proof of Theorem 19.2.2: if $M(nx) \ge \delta_n > A$, then we have

$$I_1 \le \varepsilon \boldsymbol{\rho}_\psi R(n) \quad I_2 \le \varepsilon \boldsymbol{\rho}_\psi R(n) \exp \frac{n-1}{\psi(\delta_n)} \mu_n^{(i)}.$$

If $M(nx) \le \delta_n$, (19.2.43) remains valid. If we choose δ_n such that $\psi(\delta_n) = n^s$ with $1 < s < r$, then it follows that

$$\lim_{n\to\infty} \frac{\psi(n)}{n} \phi_1 \left(\frac{\delta_n}{n-1} \right) = \lim_{n\to\infty} \frac{n}{\psi(\delta_n)} = 0$$

and, hence, that

$$\limsup_{n\to\infty} \frac{1}{R(n)} \boldsymbol{\rho}(\widetilde{M}_n, Y) \le \varepsilon \boldsymbol{\rho}_\psi.$$

Now let $\varepsilon \downarrow 0$ to obtain the proof of (ii). □

Remark 19.2.3. (a) In a similar way one can prove that $\rho_\psi < \infty$ holds if and only if for each marginal

$$\limsup_{n\to\infty} \frac{\psi(n)}{n} \boldsymbol{\rho} \left(\frac{M_n^{(i)}}{n}, Y^{(i)} \right) < \infty.$$

(b) If $\psi(0) = 0$, if ψ is 0-*regularly varying* ($\psi \in ORV$, i.e., for any $x > 0$, $\limsup_{t\to\infty} \psi(xt)/\psi(t) < \infty$), and if $\limsup_{x\to\infty}(\psi(x)/x) = \infty$. Theorem 19.2.3 (i) remains valid. To prove this assertion, we only prove that $\limsup_{a\to\infty} \psi(a)g(1/a) < \infty$. Indeed, since ψ is increasing, we have $\psi(a) \le \psi(x)$ if $a \le x$, and since $\psi \in ORV$, we have $\psi(a) < A(a/x)^\alpha \psi(x)$ if $x \ge a \ge x_0$ for some positive numbers x_0, A, and α. Using $\sup_{p\ge0} p^\alpha \phi_1(1/p) < \infty$ we obtain

$$\limsup_{a\to\infty} \psi(a)g(1/a) = \limsup_{a\to\infty} \sup_{x\ge0} \frac{\psi(a)\phi_1(x/a)}{\psi(x)} < \infty.$$

Remark 19.2.4. Up to now, we have normalized all partial maxima by n^{-1} and have always assumed that the limit DF H of Y in (19.2.1) is simple max-stable. We can remove these restrictions as follows. For simplicity we analyze the situation in $\mathbb{R}$. Assume that $H(x)$ is a simple max-stable and that there exists an increasing and continuous function $r : [0,\infty) \to [0,\infty)$ with an inverse s such that for the DF F of X we have

$$F(r(x)) = H(x) \tag{19.2.46}$$

or, equivalently,

$$F(x) = H(s(x)). \tag{19.2.47}$$

For a sequence a_n of positive numbers to be determined later, it follows from (19.2.47) that

$$\Pr(M_n \le a_n x) = F^n(a_n x) = H^n(s(a,x)) = H\left(\frac{s(a_n x)}{n}\right).$$

For $a > 0$ we obtain

$$|F^n(a_n x) - H(x^\alpha)| = \left|\phi_1\left(\frac{s(a_n x)}{n}\right) - \phi_1(x^\alpha)\right|. \tag{19.2.48}$$

If $s \in RV_\alpha$ (or, equivalently, $r \in RV_{1/\alpha}$) and if we choose $a_n = r(n)$, then it follows that (19.2.1) holds, i.e.,

$$\lim_{n\to\infty} F^n(a_n x) = H(x^\alpha). \tag{19.2.49}$$

If s is regularly varying, then we expect to obtain a rate of convergence that results in (19.2.49). We quote the following result from the theory of regular variation functions.

Lemma 19.2.8 (Omey and Rachev 1991). *Suppose $h \in RV_\eta$ ($\eta > 0$) and that h is bounded on bounded intervals of $[0,\infty)$. Suppose $0 \le p \in ORV$ and such that*

$$A_1(x/y)^\zeta \le \frac{p(x)}{p(y)} \le A_2(x/y)^\xi, \quad \textit{for each } x \ge y \ge x_0$$

for some constants $A_i > 0$, $x_0 \in \mathbb{R}$, $\xi < \eta$ and $\zeta \in \mathbb{R}$. If for each $x > 0$

$$\limsup_{t\to\infty} \frac{h(t)}{t^\eta p(t)}\left|\frac{h(tx)}{h(t)} - x^\eta\right| < \infty, \tag{19.2.50}$$

then

$$\limsup_{t\to\infty} \frac{h(t)}{t^\eta p(t)} \sup_{x\ge 0}\left|\phi_1\left(\frac{h(tx)}{h(t)}\right) - \phi_1(x^\eta)\right| < \infty. \tag{19.2.51}$$

If s satisfies the hypothesis of Lemma 19.2.8 (with an auxiliary function p and $\eta = \alpha$), then take $h(t) = s(t) = n$, $t = a_n$, in (19.2.51) to obtain

$$\limsup_{n\to\infty} \frac{n}{a_n^\alpha p(a_n)} \sup_{x\ge\infty}\left|\phi_1\left(\frac{s(a_n x)}{n}\right) - \phi_1(x^\alpha)\right| < \infty.$$

Combining these results with (19.2.48), we obtain the following theorem.

Theorem 19.2.4. *Suppose* $H(x) = \phi_1(x)$, *and assume there exists an increasing and continuous function* $r : [0,\infty) \to [0,\infty)$ *with an inverse* s *such that* $F(r(x)) = H(x)$.

(a) *If* $s \in RV_\alpha$ $(\alpha > 0)$, *then* $\lim_{n\to\infty} \Pr\{M_n \le a_n x\} = H(x^\alpha)$, *where* $a_n = r(n)$.
(b) *If* $s \in RV_\alpha$ *with a remainder term as in* (19.2.50), *then*

$$\limsup_{n\to\infty} \frac{n}{a_n^\alpha p(a_n)} \rho(M_n/a_n, Y_\alpha) < \infty,$$

where Y_α *has DF* $H(x^\alpha)$.

19.3 Ideal Metrics for the Problem of Rate of Convergence to Max-Stable Processes

In this section, we extend the results on the rate of convergence for maxima of random vectors developed in Sect. 19.2 by investigating maxima of random processes. In the new setup, we need another class of ideal metrics simply because the weighted Kolmogorov metrics $\boldsymbol{\rho}_r$ and $\boldsymbol{\rho}_\psi$[7] cannot be extended to measure the distance between processes (see Open problem 4.4.1 in Chap. 4).

Let $\mathbf{B} = (\mathbf{L}_r[T], \|\cdot\|_r)$, $1 \le r \le \infty$, be the separable Banach space of all measurable functions $x : T \to \mathbb{R}$ (T is a Borel subset of $\mathbb{R}$) with finite norm $\|x\|_r$, where

$$\|x\|_r = \left\{\int_T |x(t)|^r \mathrm{d}t\right\}^{1/r}, \qquad 1 \le r < \infty, \tag{19.3.1}$$

and if $r = \infty$, $\mathbf{L}_\infty(T]$ is assumed to be the space of all continuous functions on a compact subset T with the norm

$$\|x\|_\infty = \sup_{t\in\mathbb{T}} |x(t)|. \tag{19.3.2}$$

Suppose $\mathbf{X} = \{X_n, n \ge 1\}$ is a sequence of (dependent) random variables taking values in $\mathbf{B}$. Let $\mathcal{C}$ be the class of all sequences $\mathbf{C} = \{c_j(n); j,n = 1,2,\ldots\}$ satisfying the conditions

$$c_1(n) > 0, \quad c_j(n) > 0, \quad j = 1,2,\ldots, \quad \sum_{j=1}^{\infty} c_j(n) = 1. \tag{19.3.3}$$

For any $\mathbf{X}$ and $\mathbf{C}$ define the normalized maxima $\widetilde{X}_n := \bigvee_{j=1}^\infty c_j(n) X_j$, where $\bigvee :=$ max and $\overline{X}'_n(t) := \bigvee_{j=1}^\infty c_j(n) X_j(t)$, $t \in T$.

[7]See the definition in (19.2.4) and Theorems 19.2.1 and 19.2.2.

In the previous section we considered a special case of the sequence $c_j(n)$, namely, $c_j(n) = 1/n$ for $j \le n$ and $c_j(n) = 0$ for $j > n$, and that X_n were i.i.d. random vectors. Here we are interested in the limit behavior of $\overline{X}_n$ in the general setting determined previously. To this end, we explore an approximation (denoted by $\overline{Y}_n$) of $\overline{X}_n$ with a known limit behavior. More precisely, let $\mathbf{Y} = \{Y_n, n \ge 1\}$ be a sequence of i.i.d. RVs, and define $\overline{Y}_n = \bigvee_{j=1}^{\infty} c_j(n) Y_j$. Assuming that

$$\overline{Y}_n \stackrel{\mathrm{d}}{=} Y_1, \quad \text{for any } \mathbf{C} \in \mathcal{C}, \tag{19.3.4}$$

we are interested in estimates of the deviation between $\overline{X}_n$ and $\overline{Y}_n$. The RV Y_1 satisfying (19.3.4) is called a *simple max-stable process.*

Example 19.3.1 (de Haan 1984). Consider a Poisson point process on $\mathbb{R}_+ \times [0,1]$ with intensity measure $(\mathrm{d}x/x^2)\mathrm{d}y$. With probability 1 there are denumerably many points in the point process. Let $\{\xi_k, \eta_k\}$, $k = 1, 2, \ldots$, be an enumeration of the points in the process. Consider a family of nonnegative functions $\{f_t(\cdot), t \in T\}$ defined on $[0,1]$. Suppose for fixed $t \in T$ the function $f_t(\cdot)$ is measurable and $\int_0^1 f_t(v)\mathrm{d}v < \infty$. We claim that the family of RVs $Y(t) := \sup_{k \ge 1} f_t(\eta_k)\xi_k$ form a simple max-stable process. Clearly, it is sufficient to show that for any $\mathbf{C} \in \mathcal{C}$ and any $0 < t_1 < \cdots < t_k \in T$ the joint distribution of $(Y(t_1), \ldots, Y(t_k))$ satisfies the equality

$$\prod_{j=1}^{\infty} \Pr\{c_j Y(t_1) \le y_1, \ldots, c_j Y(t_k) \le y_k\}$$
$$= \Pr\{Y(t_1) \le y_1, \ldots, Y(t_k) < y_k\}, \quad \text{where } c_j = c_j(n).$$

Now

$$\prod_{j=1}^{\infty} \Pr\{c_j Y(t_1) \le y_1, \ldots, c_j Y(t_k) \le y_k\}$$
$$= \prod_{j=1}^{\infty} \Pr\{f_{t_i}(\eta_m)\xi_m \le y_i/c_j,\ i = 1, \ldots, k;\ m = 1, 2, \ldots\}$$
$$= \prod_{j=1}^{\infty} \Pr\{\text{there are no points of the point process above the graph of}$$
$$\text{the function } g(v) = (1/c_j) \min_{i<k} y_i/f_{t_i}(v), v \in [0,1]\}.$$

As a consequence, we can write

$$\prod_{j=1}^{\infty} \Pr\{c_j Y(t_1) \le y_1, \ldots, c_j Y(t_k) \le y_k\}$$
$$= \prod_{j=1}^{\infty} \exp\left(-\int_0^1 \left[\int_{\{x > g(v)\}} x^{-2}\mathrm{d}x\right]\mathrm{d}v\right)$$

$$= \prod_{j=1}^{\infty} \exp\left(-\int_0^1 \left(c_j \max_{i\le k} f_{t_i}(v)/y_i\right) \mathrm{d}v\right)$$

$$= \exp\left(\sum_{j=1}^{\infty} c_j \left(-\int_0^1 \max_{i\le k} f_{t_i}(v)/y_i \mathrm{d}v\right)\right)$$

$$= \exp\left(-\int_0^1 \left(\max_{i\le k} f_{t_i}(v)/y_i\right) \mathrm{d}v\right)$$

$$= \Pr\{Y(t_1) \le y_1, \ldots, Y(t_k) \le y_k\}.$$

In this section, we seek the weakest conditions providing an estimate of the deviation $\mu(\overline{X}_n, \overline{Y}_n)$ with respect to a given compound or simple p. metric μ. Such a metric will be defined on the space $\mathfrak{X}(\mathbf{B})$ of all RVs $X : (\Omega, \mathcal{A}, \Pr) \to (\mathbf{B}, \mathfrak{B}(\mathbf{B}))$, where the probability space $(\Omega, \mathcal{A}, \Pr)$ is assumed to be nonatomic.[8]

Our method is based on exploring *compound max-ideal metrics of order* $r > 0$, i.e., compound p. metrics μ_r satisfying[9]

$$\mu_r(c(X_1 \vee Y), c(X_2 \vee Y)) \le c^r \mu_r(X_1, X_2), \quad X_1, X_2, Y \in \mathfrak{X}(\mathbf{B}), \quad c > 0. \tag{19.3.5}$$

In particular, if the sequence $\mathbf{X}$ consists of i.i.d. RVs, then we will derive estimates of the rate of convergence of $\overline{X}_n$ to Y_i in terms of the *minimal metric* $\widehat{\mu}_r$ defined by[10]

$$\begin{aligned}\widehat{\mu}_r(X, Y) &:= \widehat{\mu}_r(\Pr_X, \Pr_Y)\\ &:= \inf\{\mu_r(X', Y') : X', Y' \in \mathfrak{X}(\mathbf{B}), X' \stackrel{\mathrm{d}}{=} X, Y' \stackrel{\mathrm{d}}{=} Y\}.\end{aligned} \tag{19.3.6}$$

By virtue of μ_r-ideality, $\widehat{\mu}_r$ is a *simple max-ideal metric of order* $r > 0$, i.e., (19.3.5) holds for Y independent of X_i.

We start with estimates of the deviation between $\overline{X}_n$ and $\overline{Y}_n$ in terms of the $\mathcal{L}_p$-*probability compound metric*.[11] For any $r \in [1, \infty]$ define

$$\mathcal{L}_{p,r}(X, Y) := [E\|X - Y\|_r^p]^{1/p}, \quad p \ge 1, \tag{19.3.7}$$

$$\mathcal{L}_{\infty,r}(X, Y) := \operatorname{ess\,sup} \|X - Y\|_r. \tag{19.3.8}$$

Let

$$\ell_{p,r} := \widehat{\mathcal{L}}_{p,r}. \tag{19.3.9}$$

---

[8]See Sect. 2.7 and Remark 2.7.2 in Chap. 2.

[9]See Definition 19.2.1.

[10]See Definition 3.3.2 in Chap. 3.

[11]See Example 3.4.1 in Chap. 3 with $d(x, y) = \|x - y\|_r$.

Let us recall some of the metric and topological properties of $\ell_{p,r}$ and $(\mathcal{P}(\mathbf{B}), \ell_{p,r})$. The duality theorem for the minimal metric w.r.t. $\mathcal{L}_{p,r}$ implies[12]

$$\begin{aligned}\ell_{p,r}^{p}(X,Y) &= \sup\{Ef(X)+Eg(Y): f:\mathbf{B}\to\mathbb{R}, g:\mathbf{B}\to\mathbb{R},\\ &\qquad \|f\|_\infty := \sup\{|f(x)| : x\in\mathbf{B}\} < \infty, \|g\|_\infty < \infty\\ \mathrm{Lip}(f) := \sup_{x\neq y}\frac{|f(x)-f(y)|}{\|x-y\|_r} &< \infty, \mathrm{Lip}(g)<\infty,\ f(x)+g(x)\le \|x-y\|_r\\ &\qquad \text{for any } x,y\in\mathbf{B}\},\ \text{for any } p\in[1,\infty). \qquad (19.3.10)\end{aligned}$$

Moreover, by Corollary 6.2.1 in Chap. 6, representation (19.3.10) can be refined in the special case of $p = 1$:

$$\ell_{1,r}(X,V) = \sup\{|Ef(X)-Ef(Y)| : f:\mathbf{B}\to\mathbb{R}, \|f\|_\infty \le \infty, \mathrm{Lip}(f)\le 1\}.$$

Corollary 7.5.2 and (7.5.15) in Chap. 7 give the dual form for $\ell_{\infty,r}$,

$$\ell_{\infty,r}(X,Y) = \inf\{\varepsilon>0 : \Pi_\varepsilon(X,Y)=0\}, \qquad (19.3.11)$$

where $\Pi_e(X,Y) := \sup\{\Pr\{X\in A\} - \Pr\{Y\in A^\varepsilon\} : A\in\mathfrak{B}(\mathbf{B})\}$ and A^ε is the ε-neighborhood of A w.r.t. the norm $\|\cdot\|_r$.

If $\mathbf{B}=\mathbb{R}$, then $\ell_p = \ell_{p,r}$ has the explicit representation

$$\ell_p(X,Y) = \left[\int_0^1 |F_X^{-1}(x)-F_Y^{-1}(x)|^p \mathrm{d}x\right]^{1/p}, \quad 1\le p<\infty, \qquad (19.3.12)$$

$$\ell_\infty(X,Y) = \sup\{|F_X^{-1}(x)-F_Y^{-1}(x)| : x\in[0,1]\}, \qquad (19.3.13)$$

where F_X^{-1} is the generalized inverse of the DF F_X of X.[13]

As far as the ℓ_p-convergence in $\mathcal{P}(\mathbf{B})$ is concerned, if π is the Prokhorov metric

$$\pi(X,Y) := \inf\{\varepsilon>0 : \Pi_\varepsilon(X,Y)<\varepsilon\} \qquad (19.3.14)$$

and $\omega_X(N) := \{E\|X\|_r^p I\{\|X\|_r > N\}\}^{1/p}$, $N>0$, $X\in\mathfrak{X}(\mathbf{B})$, then for any $N>0$, $X,Y\in\mathfrak{X}(\mathbf{B})$

$$\ell_{p,r}(X,Y) \le \pi(X,Y) + 2N\pi^{1/p}(X,Y) + \omega_X(N) + \omega_Y(N), \qquad (19.3.15)$$

$$\ell_{p,r}(X,Y) \ge \pi(X,Y)^{(p+1)/p}, \quad \ell_{\infty,r}(X,Y) \ge \pi(X,Y), \qquad (19.3.16)$$

---

[12] See Corollary 5.3.2 in Chap. 5 and (3.3.12) in Chap. 3.

[13] See Corollary 7.4.2 and (7.5.15) in Chap. 7.

and

$$\omega_X(N) \le 3(\ell_{p,r}(X,Y) + \omega_Y(N)). \tag{19.3.17}$$

In particular, if $E\|X_n\|^p + E\|X\|^p < \infty, n = 1, 2, \ldots$, then[14]

$$\ell_{p,r}(X_n, X) \to 0 \iff \underline{\pi}(X_n, X) \to 0 \text{ and } \lim_{N\to\infty} \sup_n \omega_{X_n}(N) = 0. \tag{19.3.18}$$

Define the sample maxima with normalizing constants $c_j(n)$ by

$$\overline{X}_n = \bigvee_{j=1}^{\infty} c_j(n)X_j, \quad \overline{Y}_n = \bigvee_{j=1}^{\infty} c_j(n)Y_j. \tag{19.3.19}$$

In the next theorem, we obtain estimates of the closeness between $\overline{X}_n$ and $\overline{Y}_n$ in terms of the metric $\mathcal{L}_{p,r}$. In particular, if $\mathbf{X}$ and $\mathbf{Y}$ have i.i.d. components and Y_1 is a simple max-stable process [see (19.3.4)], then we obtain the rate of convergence of $\overline{X}_n$ to Y_1 in terms of the minimal metric $\ell_{p,r}$. With this aim in mind, we need some conditions on the sequences $\mathbf{X}$, $\mathbf{Y}$, and $\mathbf{C}$ [see (19.3.3)].

Condition 1. Let

$$a_p(n) := \left[\sum_{j=1}^{\infty} c_j^p(n)\right]^{\overline{p}}, \text{ for } p \in (0,\infty),\ \overline{p} := \min(1, 1/p), \tag{19.3.20}$$

and

$$a_\infty(n) := \sup_{j\ge 1} c_j(n). \tag{19.3.21}$$

Assume that

$$a_\alpha(n) < \infty \text{ for some fixed } \alpha \in (0,1) \text{ and all } n \ge 1,$$

$$a_1(n) = 1,\ \forall n > 1,\ a_p(n) \to 0 \text{ as } n \to \infty,\ \forall p > 1.$$

The main examples of $\mathbf{C}$ satisfying Condition 1 are the Cesàro and Abel summation schemes.

Cesàro sum:

$$c_j(n) = \begin{cases} 1/n, & j = 1, 2, \ldots, n, \\ 0, & j = n+1, n+2, \ldots, \end{cases} \tag{19.3.22}$$

$$a_p(n) = \begin{cases} n^{1-p} & \text{for } p \in (0,1], \\ n^{-1+1/p} & \text{for } p \in [1,\infty]. \end{cases} \tag{19.3.23}$$

[14]See Lemma 8.3.1 and Corollary 8.3.1 in Chap. 8.

Abel sum:

$$c_j(n) = (\exp(1/n) - 1)\exp(-j/n), \ j = 1, 2, \ldots, \ n = 1, 2, \ldots, \qquad (19.3.24)$$

$$a_p(n) = (1 - \exp(-1/n))^p/(1 - exp(-p/n)) \sim (1/p)n^{1-p}$$
$$\text{as } n \to \infty \text{ for any } p \in (0, 1),$$
$$a_p(n) = (1 - \exp(-1/n)(1 - \exp(-p/n))^{-1/p} \sim p^{-1/p}n^{-1+1/p}$$
$$\text{as } n \to \infty \text{ for any } p \in [1, \infty),$$
$$a_p(n) = 1 - \exp(-1/n) \sim 1/n \text{ as } n \to \infty \text{ for } p = \infty. \qquad (19.3.25)$$

The following condition concerns the sequences **X** and **Y**.

Condition 2. Let $\alpha \in (0, 1)$ be such that $a_\alpha(n) < \infty$ [see (19.3.22)], and assume that

$$\sup_{j \geq 1} E|X_j(t)|^\alpha < \infty \text{ for any } t \in T, \qquad (19.3.26)$$

$$\sup_{j \geq 1} E|Y_j(t)|^\alpha < \infty \text{ for any } t \in T. \qquad (19.3.27)$$

Condition 2 is quite natural. For example, if Y_j, $j \geq 1$, are independent copies of a max-stable process,[15] then all one-dimensional marginal DFs are of the form $\exp(-\beta(t)/x)$, $x > 0$ (for some $\beta(t) \geq 0$), and hence (19.3.27) holds. In the simplest m-dimensional case, $T = \{t_k\}_{k=1}^m$ and $X_j = \{X_j(t_k)\}_{k=1}^m$ $j > 1$ are i.i.d. RVs and as $Y_j = \{Y_j(t_k)\}_{k=1}^m$, $j \geq 1$, are i.i.d. RVs with a simple max-stable distribution (Sect. 19.2). One can check that condition (19.3.26) is necessary to have a rate $O(n^{1-r})$ $(r > 1)$ of the uniform convergence of the DF of $(1/n)\bigvee_{j=1}^n X_j$ to the simple max-stable distribution F_{Y_1}, see Theorem 19.2.3.

Theorem 19.3.1. *(a) Let* **X**, **Y**, *and* **C** *satisfy Conditions 1 and 2. Let* $1 < p \leq r \leq \infty$ *and*

$$\mathcal{L}_{p,r}(X_j, Y_j) \leq \mathcal{L}_{p,r}(X_1, Y_1) < \infty, \quad \forall j = 1, 2, \ldots \qquad (19.3.28)$$

Then

$$\mathcal{L}_{p,r}(\overline{X}_n, \overline{Y}_n) \leq a_p(n)\mathcal{L}_{p,r}(X_1, Y_1) \to 0 \ \text{as } n \to \infty. \qquad (19.3.29)$$

(b) If **X** *and* **Y** *have i.i.d. components,* $1 < p \leq t \leq \infty$, *and* $\ell_{p,r}(X_1, Y_1) < \infty$, *then*

$$\ell_{p,r}(\overline{X}_n, \overline{Y}_n) \leq a_p(n)\ell_{p,r}(X_1, Y_1) \to 0 \ \text{as } n \to \infty, \qquad (19.3.30)$$

where $\ell_{p,r}$ *is determined by* (19.3.10).

---

[15]See, for example, Resnick (1987a).

In particular, if $\mathbf{Y}$ *satisfies the max-stable property*

$$\overline{Y}_n \stackrel{\mathrm{d}}{=} Y_1, \tag{19.3.31}$$

then

$$\ell_{p,r}(\overline{X}_n, Y_1) \le a_p(n)\ell_{p,r}(X_1, Y_1) \to 0 \ \ as \ \ n \to \infty. \tag{19.3.32}$$

Proof. (a) Let $1 < p < r < \infty$. By Conditions 1 and 2 and Chebyshev's inequality, we have

$$\Pr\{\overline{X}_n(t) > \lambda\} \le \lambda^{-\alpha} a_\alpha(n) \sup_{j\ge 1} EX_j(t)^\alpha \to 0 \text{ as } \lambda \to \infty,$$

and hence

$$\Pr\{\overline{X}_n(t) + \overline{Y}_n(t) < \infty\} = 1, \quad \text{for any } \ t \in T.$$

For any $\omega \in \Omega$ such that $\overline{X}_n(t)(\omega) + \overline{Y}_n(t)(\omega) < \infty$ we have

$$\overline{X}_n(t)(\omega) = \bigvee_{j=1}^{m} c_j(n) X_j(t)(\omega) + \varepsilon_\omega(m), \qquad \lim_{m\to\infty} \varepsilon_\omega(m) = 0,$$

$$\overline{Y}_n(t) * (\omega) = \bigvee_{j=1}^{m} c_j(n) Y_j(t)(\omega) + \delta_\omega(m), \qquad \lim_{m\to\infty} \delta_\omega(m) = 0,$$

and hence

$$|\overline{X}_n(t)(\omega) - \overline{Y}_n(t)(\omega)|$$
$$\le \bigvee_{j=1}^{m} |c_j(n) X_j(t)(\omega) - c_j(n) Y_j(t)(\omega)| + |\varepsilon_\omega(m)| + |\delta_\omega(m)|.$$

So, with probability 1,

$$|\overline{X}_n(t) - \overline{Y}_n(t)| \le \bigvee_{j=1}^{\infty} c_j(n) |X_j(t) - Y_j(t)|. \tag{19.3.33}$$

Using the Minkowski inequality and the fact that $p/r \le 1$ we obtain

$$\mathcal{L}_{p,r}(\overline{X}_n, \overline{Y}_n) = \left\{ E \left| \int_T |\overline{X}_n(t) - \overline{Y}_n(t)|^r \mathrm{d}t \right]^{p/r} \right\}^{1/p}$$

$$\leq \left\{ E \left| \int_T \left[\bigvee_{j=1}^{\infty} c_j(n)|X_j(t) - Y_j(t)| \right]^r \mathrm{d}t \right]^{p/r} \right\}^{1/p}$$

$$\leq \left\{ E \left| \int_T | \sum_{j=1}^{\infty} c_j^r(n)|X_j(t) - Y_j(t)|^r \mathrm{d}t \right]^{p/r} \right\}^{1/p}$$

$$\leq \left\{ E \sum_{j=1}^{\infty} c_j^p(n) \left[\int_T |X_j(t) - Y_j(t)|^r \mathrm{d}t \right]^{p/r} \right\}^{1/p}$$

$$\leq a_p(n) \mathcal{L}_{p,r}(X_1, Y_1).$$

If $p < r = \infty$, then

$$\mathcal{L}_{p,\infty}(\overline{X}_n, \overline{Y}_n) \leq \left\{ E \left[\sup_{t \in T} \bigvee_{j=1}^{\infty} c_j(n)|X_j(t) - Y_j(t)| \right]^p \right\}^{1/p}$$

$$\leq \left\{ E \sum_{i=1}^{\infty} c_j^p(n) \sup_{t \in T} |X_j(t) - Y_j(t)| \right\}^{1/p}$$

$$\leq \sum_{i=1}^{\infty} c_j(n) \mathcal{L}_{P,\infty}(X_1, Y_1).$$

The statement for $p = r = \infty$ can be proved in an analogous way.

(b) By the definition of the minimal metric,[16] we have

$$\widehat{\mathcal{L}}_{p,r}(\overline{X}_n, \overline{Y}_n) = \inf\{\mathcal{L}_{p,r}(\widetilde{X}, \widetilde{Y}) : \widetilde{X} \stackrel{\mathrm{d}}{=} \overline{X}_n, \widetilde{Y} \stackrel{\mathrm{d}}{=} \overline{Y}_n\}$$

$$\leq \inf \left\{ \left[\sum_{j=1}^{\infty} c_j^p(n) \mathcal{L}_{p,r}^p(\widetilde{X}_j, \widetilde{Y}_j) \right]^{1/p} : \{\widetilde{X}_j, j \geq 1\} \text{ are i.i.d.,} \right.$$

$$\left. \{\widetilde{Y}_j, j \geq 1\} \text{ are i.i.d.}, (\widetilde{X}_j, \widetilde{Y}_j) \stackrel{\mathrm{d}}{=} (\widetilde{X}_1, \widetilde{Y}_1), \widetilde{X}_1 \stackrel{\mathrm{d}}{=} X_1, \widetilde{Y}_1 \stackrel{\mathrm{d}}{=} Y_1 \right\}$$

$$\leq \inf \left\{ \left[\sum_{j=1}^{\infty} c_j^p(n) \mathcal{L}_{p,r}^p(\widetilde{X}_1, \widetilde{Y}_1) \right]^{1/p} : \widetilde{X}_1 \stackrel{\mathrm{d}}{=} X_1, \widetilde{Y}_1 \stackrel{\mathrm{d}}{=} Y_1 \right\}$$

$$= a_p(n) \widehat{\mathcal{L}}_{p,r}(X_1, Y_1).$$

[16] See (19.3.6) and Sect. 7.2 in Chap. 7.

By (19.3.9) and (19.3.10), we obtain (19.3.30).

Finally, (19.3.32) follows immediately from (19.3.30) and (19.3.31). □

Corollary 19.3.1. *Let* $\{X_j, j \geq 1\}$ *and* $\{Y_j, j \geq 1\}$ *be random sequences with i.i.d. real-valued components and* $F_{Y_1}(x) = \exp\{-1/x\}$, $x \geq 0$. *Then*

$$\ell_p\left(\bigvee_{j=1}^{\infty} c_j(n)X_j, Y_1\right) \leq a_p(n)\ell_p(X_1, Y_1), \qquad p \in [1, \infty], \tag{19.3.34}$$

where the metric ℓ_p *is given by* (19.3.12) *and* (19.3.13)*. In particular, if, for some* $1 < p \leq \infty$, $\ell_p(X_1, Y_1) < \infty$, *then* $\ell_p(\bigvee_{j=1}^{\infty} c_j(n)X_j, Y_1) \to 0$ *as* $n \to \infty$.

Note that $\ell_p(X_1, Y_1) < \infty$ *for* $1 < p < \infty$ *may be viewed as a tail condition similar to the condition* $\boldsymbol{\rho}_r(X_1, Y_1) < \infty$ $(r > 1)$ *in Theorem 19.2.1 (b).*

Open Problem 19.3.1. It is not difficult to check that if $E|X_n|^p + E|X|^p < \infty$, then, as $n \to \infty$,

$$\ell_p(X_n, X) = \left\{\int_0^1 |F_{X_n}^{-1}(t) - F_X^{-1}(t)|^p \mathrm{d}t\right\}^{1/p} \to 0, \tag{19.3.35}$$

provided that for some $r > p$

$$\boldsymbol{\rho}_r(X_n, X) := \sup_{x \in \mathbb{R}} |x|^r |F_{X_n}(x) - F_X(x)| \to 0. \tag{19.3.36}$$

Since on the right-hand side of (19.3.34) the conditions $\ell_p(X_1, Y_1) < \infty$ and $E|Y_1|^p = \infty$ imply $E|X_1|^p = \infty$, it is a matter of interest to find necessary and sufficient conditions for $\boldsymbol{\rho}_r(X_n, X) \to 0$ $(r > p)$, resp. $\ell_p(X_n, X) \to 0$, in the case of X_n and X having infinite pth absolute moments, for example, under the assumption $\ell_p(X_n, Y) + \ell_p(X, Y) < \infty$, $p > 1$, where Y is a simple max-stable RV.

Let $\boldsymbol{\pi}$ be the Prokhorov metric (19.3.14) in the space $\mathfrak{X}(\mathbf{B}, \|\cdot\|_r)$. Using the relationship between $\boldsymbol{\pi}$ and $\ell_{p,r}$ [see (19.3.16)], we get the following rate of convergence of $\overline{X}(n)$ to Y, under the assumptions of Theorem 19.3.1 (b).

Corollary 19.3.2. *Suppose the assumptions of Theorem 19.3.1 (b) are valid and that* (19.3.31) *holds. Then,*

$$\boldsymbol{\pi}(\overline{X}(n), Y_1) \leq a_p(n)^{p/(1+p)} \ell_{p,r}(X_1, Y_1)^{p/(1+p)}. \tag{19.3.37}$$

The next theorem is devoted to a similar estimate of the closeness between $\overline{X}_n$ and Y_n, but now in terms of the compound Q-difference pseudomoment

$$\boldsymbol{\tau}_{p,r}(X, Y) = E\|Q_p X - Q_p Y\|, \quad p > 0, \tag{19.3.38}$$

where the homeomorphism Q_p on $\mathbf{B}$ is defined by[17]

$$(Q_p x)(t) = |x(t)|^p \operatorname{sgn} x(t). \tag{19.3.39}$$

Recall that the minimal metric $\boldsymbol{\kappa}_{p,r} = \widetilde{\boldsymbol{\tau}}_{p,r}$ admits the following form of Q_p-*difference pseudomoment*:[18]

$$\begin{aligned}\boldsymbol{\kappa}_{p,r}(X,Y) = \sup\{&|Ef(X) - Ef(Y)| : f : \mathbf{B} \to \mathbb{R}, \|f\|_\infty < \infty,\\ &|f(x) - f(y)| \le \|Q_p x - Q_p y\|_r, \quad \forall x, y \in \mathbf{B}\},\end{aligned} \tag{19.3.40}$$

and if $\mathbf{B} = \mathbb{R}$, then $\boldsymbol{\kappa}_{p,r} =: \boldsymbol{\kappa}_p$ is the pth *difference pseudomoment*

$$\begin{aligned}\boldsymbol{\kappa}_p(X,Y) &= p\int_{-\infty}^{\infty} |x|^{p-1}|F_X(x) - F_Y(x)|\mathrm{d}x\\ &= \int_{-\infty}^{\infty} |F_{Q_pY}(x) - F_{Q_pY}(x)|\mathrm{d}x.\end{aligned} \tag{19.3.41}$$

Recall also that[19]

$$\boldsymbol{\kappa}_{p,r}(X,Y) = \ell_{1,p}(Q_p X, Q_p Y) = \widehat{\boldsymbol{\tau}}_{p,r}(X,Y), \quad \forall X, Y \in \mathfrak{X}(\mathbf{B}),$$

and thus, by (19.3.18), if $E\|X_n\|_r^p + E\|X\|_r^p < \infty, n = 1, 2, \ldots$, then

$$\boldsymbol{\kappa}_{p,r}(X_n, X) \to 0 \iff \boldsymbol{\pi}(X_n, X) \to 0 \text{ and } E\|X_n\|_r^p \to E\|X\|_r^p.$$

In the next theorem we relax the restriction $1 < p \le r \le \infty$ imposed in Theorem 19.3.1.

Theorem 19.3.2. *(a) Let Conditions 1 and 2 hold, $p > 0$, and $1/p < r \le \infty$. Assume that*

$$\boldsymbol{\tau}_{p,r}(X_j, Y_j) \le \boldsymbol{\tau}_{p,r}(X_1, Y_1) < \infty, \quad j = 1, 2, \ldots. \tag{19.3.42}$$

Then

$$\boldsymbol{\tau}_{p,r}(\overline{X}_n, \overline{Y}_n) \le \alpha_{\overline{p}}(n)\boldsymbol{\tau}_{p,r}(X_1, Y_1) \to 0 \quad \text{as } n \to \infty, \tag{19.3.43}$$

where $\alpha_{\overline{p}}(n) = \sum_{j=1}^{\infty} c_j^{\overline{p}}(n)$, $\overline{p} := p\min(1, r)$.

---

[17]See Example 4.4.3 and (4.4.41) in Chap. 4.

[18]See (4.4.42) and (4.4.43) in Chap. 4 and Remark 7.2.3 in Chap. 7.

[19]See Remark 7.2.3 in Chap. 7.

(b) *If* $\mathbf{X}$ *and* $\mathbf{Y}$ *consist of i.i.d. RVs, then* $\boldsymbol{\kappa}_{p,r}(X_1, Y_1) < \infty$ *implies*

$$\boldsymbol{\kappa}_{p,r}(\overline{X}_n, \overline{Y}_n) \leq \alpha_{\overline{p}}(n)\boldsymbol{\kappa}_{p,r}(X_1, Y_1) \to 0 \quad \text{as } n \to \infty. \tag{19.3.44}$$

Moreover, assuming that (19.3.31) *holds, we have*

$$\boldsymbol{\kappa}_{p,r}(\overline{X}_n, Y_1) \leq \alpha_{\overline{p}}(n)\boldsymbol{\kappa}_{p,r}(X_1, Y_1) \to 0 \quad \text{as } n \to \infty. \tag{19.3.45}$$

Proof. (a) By Conditions 1 and 2,

$$\Pr\left(\bigvee_{j=1}^{\infty} c_j^p(n)(Q_p X_j)(t) + \bigvee_{j=1}^{\infty} c_j^p(Q_p Y_j)(t) < \infty\right) = 1.$$

Hence, as in Theorem 19.3.1, we have

$$\begin{aligned}
&\left| Q_p\left(\bigvee_{j=1}^{\infty} c_j(n)X_j\right)(t) - Q_p\left(\bigvee_{j=1}^{\infty} c_j(n)Y_j\right)(t)\right| \\
&\quad = \left|\bigvee_{j=1}^{\infty} c_j^p(n)(Q_p X_j)(t) - \bigvee_{j=1}^{\infty} c_j^p(n)(Q_p Y_j)(t)\right| \\
&\quad \leq \bigvee_{j=1}^{\infty} c_j^p(n)|(Q_p X_j)(t) - (Q_p Y_j)(t)|.
\end{aligned}$$

Next, denote $\widetilde{r} = \min(1, 1/r)$ and then

$$\begin{aligned}
\boldsymbol{\tau}_{p,r}(\overline{X}_n, \overline{Y}_n) &= E\left[\int_T \left| Q_p\left(\bigvee_{j=1}^{\infty} c_j(n)X_j\right)(t) - Q_p\left(\bigvee_{j=1}^{\infty} c_j(n)Y_j\right)(t)\right|^r \mathrm{d}t\right]^{\widetilde{r}} \\
&\leq E\left[\sum_{j=1}^{\infty} \int_T c_j^{pr}(n)|(Q_p X_j)(t) - (Q_p Y_j)(t)|^r \mathrm{d}t\right]^{\widetilde{r}} \\
&\leq \sum_{j=1}^{\infty} c_j^{pr\widetilde{r}}(n)\boldsymbol{\tau}_{p,r}(X_j, Y_j) \leq \alpha_{\overline{p}}(n)\boldsymbol{\tau}_{p,r}(X_1, Y_1).
\end{aligned}$$

(b) Passing to the minimal metrics, as in Theorem 19.3.1 (b), we obtain (19.3.44) and (19.3.45). □

The next corollary can also be proved directly using Lemma 19.2.3, noting that $\boldsymbol{\kappa}_p$ is a max-ideal metric of order p.

Corollary 19.3.3. *Let* $\mathbf{X}$ *and* $\mathbf{Y}$ *consist of i.i.d. real-valued RVs and* $F_{Y_1}(x) = \exp\{-1/x\}$, $x \geq 0$. *Then*

$$\kappa_p(X_n, Y_1) \leq \alpha_p(n)\kappa_p(X_1, Y_1), \quad p > 1, \tag{19.3.46}$$

where $\alpha_p(n) = \sum_{j=1}^{\infty} c_j^p(n)$ *and* κ_p *is given by* (19.3.41).

The main assumption in Corollary 19.3.3 is $\ell_{p,r}(X_1, Y_1) < \infty$. To relax it, we will consider a more refined estimate than (19.3.37). For this purpose we introduce the following metric:

$$\chi_{p,r}(X, Y) := \left[\sup_{t>0} t^p \Pr\{\|X - Y\|_r > t\}\right]^{1/(1+p)}, \quad p > 0, \ r \in [1, \infty]. \tag{19.3.47}$$

Lemma 19.3.1. *For any* $p > 0$, $\chi_{p,r}$ *is a compound probability metric in* $\mathfrak{X}(\mathbf{B})$.

Proof. Let us check the triangle inequality. For any $\alpha \in [0, 1]$ and any $f > 0$

$$\Pr\{\|X - Y\|_r > t\} \leq \Pr\{\|X - Z\|_r > \alpha t\} + \Pr\{\|Z - Y\|_r > (1 - \alpha)t\},$$

and hence $\chi_{p,r}^{p+1}(X, Y) \leq \alpha^{-p}\chi_{p,r}^{p+1}(X, Z) + (1 - \alpha)^{-p}\chi_{p,r}^{p+1}(Z, Y)$. Minimizing the right-hand side of the last inequality over all $\alpha \in (0, 1)$, we obtain $\chi_{p,r}(X, Y) \leq \chi_{p,r}(X, Z) + \chi_{p,r}(Z, Y)$. □

We will also use the *minimal metric* w.r.t. $\chi_{p,r}$

$$\xi_{p,r}(X, Y) := \widehat{\chi}_{p,r}(X, Y), \quad p > 0. \tag{19.3.48}$$

The fact that $\xi_{p,r}$ is a metric follows from Theorem 3.3.1 in Chap. 3.

Lemma 19.3.2. *(a) Let*

$$\widetilde{\omega}_X(N) := \left[\sup_{t>N} t^p \Pr\{\|X\|_r > t\}\right]^{1/(1+p)}, \quad N > 0, \ X \in \mathfrak{X}(\mathbf{B}), \tag{19.3.49}$$

and

$$\eta_{p,r}(X, Y) := \left[\sup_{t>0} t^p \Pi_t(X, Y)\right]^{1/(1+p)}, \tag{19.3.50}$$

where Π_t *is defined as in* (19.3.11). *Then for any* $N > 0$ *and* $p > 0$

$$\pi \leq \eta_{p,r} \leq \xi_{p,r} \leq \begin{cases} \ell_{p,r}^{p/(1+p)} & \text{if } p \geq 1, \\ \ell_{p,r}^{1/(1+p)} & \text{if } p \leq 1, \end{cases} \tag{19.3.51}$$

where $\ell_{p,r}$, $p < 1$, *is determined by* (3.3.12) *and* (3.4.18) *with* $d(x, y) = \|x - y\|_r^p$:

$$\widehat{\mathcal{L}}_p(X,Y) = \ell_{p,r}(X,Y) := \sup\{|Ef(X) - Ef(Y)| : f : \mathbb{B} \to \mathbb{R} \text{ bounded}, \\ |f(x) - f(y)| \leq \|x-y\|_r^p, \quad \forall x, y \in \mathbb{B}\}.$$

Moreover,

$$\widetilde{\omega}_X(N) \leq 2^{p/(1+p)}[\boldsymbol{\eta}_{p,r}(X,Y) + \widetilde{\omega}_Y(N/2)] \tag{19.3.52}$$

and

$$\boldsymbol{\xi}_{p,r}^{p+1}(X,Y) \leq \max[\boldsymbol{\pi}^p(X,Y), (2N)^p \boldsymbol{\pi}(X,Y), 2^p(\widetilde{\omega}_X^p(N) + \widetilde{\omega}_Y^p(N))]. \tag{19.3.53}$$

(b) *In particular, if* $\lim_{N\to\infty}(\widetilde{\omega}_{X_n}(N) + \widetilde{\omega}_X(N)) = 0$, $n \geq 1$, *then the following statements are equivalent:*

$$\boldsymbol{\xi}_{p,r}(X_n, X) \to 0, \tag{19.3.54}$$

$$\boldsymbol{\eta}_{p,r}(X_n, X) \to 0, \tag{19.3.55}$$

$$\boldsymbol{\pi}(X_n, X) \to 0 \text{ and } \lim_{N\to\infty} \sup_{n\geq 1} \widetilde{\omega}_{X_n}(N) = 0. \tag{19.3.56}$$

Proof. Suppose $\boldsymbol{\pi}(X,Y) > \varepsilon > 0$. Then $\Pi_\varepsilon(X,Y) > \varepsilon$ [see (19.3.14)], and thus $\boldsymbol{\eta}_{p,r}(X,Y) \geq \varepsilon$, which gives $\boldsymbol{\eta}_{p,r} \geq \boldsymbol{\pi}$. Using $\boldsymbol{\eta}_{p,r} \leq \boldsymbol{\chi}_{p,r}$ and passing to the minimal metric $\boldsymbol{\xi}_{p,r} = \widehat{\boldsymbol{\chi}}_{p,r}$ we get $\boldsymbol{\eta}_{p,r} \leq \boldsymbol{\xi}_{p,r}$. For $p \geq 1$, by Chebyshev's inequality, $\boldsymbol{\chi}_{p,r} \leq \mathcal{L}_{p,r}^{p/(1+p)}$, which implies $\boldsymbol{\xi}_{p,r} \leq \ell_{p,r}^{p/(1+p)}$. The case of $p \in (0,1)$ is handled in the same way, which completes the proof of (19.3.51).

The proof of (19.3.53) and (b) is similar to that of Lemma 8.3.1 and Theorem 8.3.1.[20] □

Open Problem 19.3.2. The equality $\boldsymbol{\eta}_{p,r} = \boldsymbol{\xi}_{p,r}$ may fail in general. The problem of getting dual representation for $\boldsymbol{\xi}_{p,r}$ similar to that of $\widehat{\mathcal{L}}_{p,r}$ [see (19.3.9) and (19.3.10)] is open.

The main purpose of the next theorem is to refine the estimate (19.3.37) in the case of $r = \infty$. By Lemma 19.3.2 (b) and (19.3.18), we know that $\ell_{p,\infty}$ is topologically stronger than $\boldsymbol{\xi}_{p,\infty} = \widehat{\boldsymbol{\chi}}_{p,\infty}$. Thus, in the next theorem we will show that it is possible to replace $\ell_{p,\infty}$ with $\boldsymbol{\xi}_{p,\infty}$ on the right-hand side of inequality (19.3.37) with $r = \infty$.

Theorem 19.3.3. *(a) Let Conditions 1 and 2 hold and* **X** *and* **Y** *be sequences of RVs taking values in* $\mathfrak{X}(\mathbf{L}_\infty)$ *such that*

$$\boldsymbol{\chi}_{p,\infty}(X_j, Y_j) \leq \boldsymbol{\chi}_{p,\infty}(X_1, Y_1) < \infty, \quad \forall j \geq 1. \tag{19.3.57}$$

[20] For additional details, see Kakosyan et al. (1988, Lemmas 2.4.1 and 2.4.2 and Theorem 2.4.1).

Then,

$$\chi_{p,\infty}(X_n,\overline{Y}_n) \leq \alpha_p^{1/(1+p)}\chi_{p,\infty}(X_1,Y_1) \to 0 \quad \text{as } n \to \infty, \tag{19.3.58}$$

where $\alpha_p := a_p^p$, $p > 1$.

(b) If $\mathbf{X}$ *and* $\mathbf{Y}$ *have i.i.d. components and* $\overline{Y}_n \stackrel{\mathrm{d}}{=} Y_1$, *then*

$$\xi_{p,\infty}(\overline{X}_n, Y_1) \leq \alpha_p^{1/(1+p)}(n)\xi_{p,\infty}(X_1,Y_1). \tag{19.3.59}$$

In particular,

$$\begin{aligned}\pi(\overline{X}_n, Y_1) &\leq \alpha_p^{1/(1+p)}\xi_{p,\infty}(X_1,Y_1)\\ &\leq \alpha_p^{1/(1+p)}\ell_{p,\infty}(X_1,Y_1)^{p/(1+p)}.\end{aligned} \tag{19.3.60}$$

Proof. (a) By (19.3.2) and (19.3.33),

$$\begin{aligned}\chi_{p,\infty}^{1+p}(\overline{X}_n,\overline{Y}_n) &\leq \sup_{u>0} u^p \Pr\left\{\sup_{t\in T}\bigvee_{j=1}^{\infty}|c_j(n)X_j(t)-c_j(n)Y_j(t)|>u\right\}\\ &\leq \sum_{j=1}^{\infty}\sup_{u>0}u^p\Pr\left\{\sup_{t\in T}|X_j(t)-Y_j(t)|>u/c_j(n)\right\}\\ &= \sum_{j=1}^{\infty}c_j^p(n)\chi_{p,\infty}^{1+p}(X_j,Y_j) \leq \alpha_p(n)\chi_{p,\infty}1+p(X_1,Y_1).\end{aligned}$$

(b) Passing to the minimal metrics in (19.3.58), similar to part (b) of Theorem 19.3.1, we get (19.3.59). Finally, using inequality (18.2.52) we obtain (19.3.60). □

Further, we will investigate the uniform rate of convergence of the distributions of maxima of random sequences. Here we assume that $\mathbf{X} := \{X, X_j,\ j \geq 1\}$, $\mathbf{Y} := \{Y, Y_j,\ j \geq 1\}$ are sequences of i.i.d. RVs taking on values in $\mathbb{R}_+^\infty$ and

$$\overline{X}_n := \bigvee_{j=1}^{\infty} c_j(n)X_j, \quad \overline{Y}_n := \bigvee_{j=1}^{\infty} c_j(n)Y_j, \tag{19.3.61}$$

where the components $Y^{(i)}$, $i \geq 1$, of $\mathbf{Y}$ follow an extreme-value distribution $F_{Y^{(i)}}(x) = \phi_1(x)\exp(-1/x)$, $x \geq 0$.

In addition, we will consider $\mathbf{C} \in \mathcal{C}$ [see (19.3.3)] subject to the condition

$$\alpha_p(n) := \sum_{j=1}^{\infty} c_j^p(n) \to 0 \text{ as } n \to \infty \quad \text{for any } p > 1. \tag{19.3.62}$$

Denote $a \circ x := (a^{(1)}x^{(1)}, a^{(2)}x^{(2)}, \ldots)$, $bx := (bx^{(1)}, bx^{(2)}, \ldots)$ for any $a = (a^{(1)}, a^{(2)}, \ldots) \in \mathbb{R}^\infty$, $x = (x^{(1)}, x^{(2)}, \ldots) \in \mathbb{R}^\infty$, $b \in \mathbb{R}$.

We will examine the uniform rate of convergence $\boldsymbol{\rho}(\overline{X}_n, Y) \to 0$ (as $n \to \infty$) where $\boldsymbol{\rho}$ is the *Kolmogorov (uniform) metric*

$$\boldsymbol{\rho}(X, Y) := \sup\{|F_X(x) - F_Y(x)| : x \in \mathbb{R}^\infty\}. \tag{19.3.63}$$

Here, $F_X(x) := \Pr\{\bigcap_{i=1}^{\infty}[X^{(i)} \leq x^{(i)}]\}$, and $x = (x^{(1)}, x^{(2)}, \ldots)$ is the DF of X. Our aim is to prove an infinite-dimensional analog of Theorem 19.2.1 concerning the uniform rate of convergence for maxima of m-dimensional random vectors [see (19.2.12)].

First, note that the assumption that the components $X_j^{(k)}$ of X are nonnegative is not a restriction since $\boldsymbol{\rho}(\overline{X}_n, Y) = \boldsymbol{\rho}(\bigvee_{j=1}^{\infty} c_j(n)\widetilde{X}_j, Y)$, where $\widetilde{X}_j^{(k)} = \max(X_j^{(k)}, 0)$, $k \geq 1$.[21] As in (19.2.4), we define the *weighted Kolmogorov probability metric*

$$\boldsymbol{\rho}_p(X, Y) := \sup\{M^p(x)|F_X(x) - F_Y(x)| : x \in \mathbb{R}^\infty\}, \quad p > 0, \tag{19.3.64}$$

where $M(x) := \inf_{i \geq 1} |x^{(i)}|$, $x \in \mathbb{R}^\infty$.

First, we will obtain an estimate of the rate of convergence of $\overline{X}_n$ to Y in terms $\boldsymbol{\rho}_p$, $p > 1$.[22]

Lemma 19.3.3. *Let $p > 1$. Then*

$$\boldsymbol{\rho}_p(\overline{X}_n, Y) \leq \alpha_p(n)\boldsymbol{\rho}_p(X, Y). \tag{19.3.65}$$

Proof. For any $x \in \mathbb{R}^\infty$

$$\begin{aligned} M^p(x)|F_{\overline{X}_n}(x) - F_Y(x)| &= M^p(x)|F_{\overline{X}_n}(x) - F_{\overline{Y}_n}(x)| \\ &\leq \sum_{j=1}^{\infty} M^p(x)|F_{X_j}(x/c_j(n)) - F_{Y_j}(x/c_j(n))| \leq \alpha_p(n)\boldsymbol{\rho}_p(X, Y). \end{aligned}$$

□

The problem now is how to pass from estimate (19.3.62) to a similar estimate for $\boldsymbol{\rho}(\overline{X}_n, Y)$. We were able to solve this problem for the case of finite-dimensional random vectors (Theorem 19.2.1). A close look at the proof of Theorem 19.2.1 shows that in the infinite-dimensional case, the max-smoothing inequality (Lemma 19.2.4) is not valid.[23] Further, we will use relationships between $\boldsymbol{\rho}$, $\boldsymbol{\rho}_p$, and other metric

---

[21] See Remark 19.2.1.

[22] See Lemmas 19.2.2 and 19.2.3 for similar results.

[23] The same is true for the summation scheme; see (16.3.7) in Chap. 16.

structures that will provide estimates for $\boldsymbol{\rho}(\overline{X}_n, Y)$ "close" to that on the right-hand side of (19.3.62).

The next lemma deals with inequalities between $\boldsymbol{\rho}$, $\boldsymbol{\rho}_p$, and the Lévy metric in the space $\mathfrak{X}^\infty = \mathfrak{X}(\mathbb{R}^\infty)$ of random sequences. We define the *Lévy metric* as follows:

$$\mathbf{L}(X,Y) := \inf\{\varepsilon > 0 : F_X(x - \varepsilon\mathbf{e}) - \varepsilon \le F_Y(x) \le F_X(x + \varepsilon\mathbf{e}) + \varepsilon\} \quad (19.3.66)$$

for all $x \in \mathbb{R}^\infty$, where $\mathbf{e} := (1, 1, \ldots)$.

Open Problem 19.3.3. What are the convergence criteria for $\mathbf{L}$, $\boldsymbol{\rho}$, and $\boldsymbol{\rho}_p$ in $\mathfrak{X}^\infty$? Since $\mathbf{L}$, $\boldsymbol{\rho}$, and $\boldsymbol{\rho}_p$ are simple metrics, the answer to this question depends on the choice of the norm

$$\|x\|_p = \left[\sum_{i=1}^{\infty} |x^{(i)}|^p\right]^{1/p}, \quad \|x\|_\infty = \sup_{1 \le i < \infty} |x^{(i)}|$$

in the space of probability laws $\mathcal{P}(\mathbb{R}^\infty, \|\cdot\|_p)$.

Lemma 19.3.4. *(a) For any $\beta > 0$, $X, Y \in \mathfrak{X}^\infty$*

$$\mathbf{L}^{\beta+1}(X,Y) \le E\|X - Y\|_\infty^\beta, \quad (19.3.67)$$

where $\|x\|_\infty := \sup_{i \ge 1} |x^{(i)}|$,

$$\mathbf{L}(X,Y) \le \boldsymbol{\rho}(X,Y), \quad (19.3.68)$$

and

$$\mathbf{L}^{p+1}(X,Y) \le 2^p \boldsymbol{\rho}_p(X,Y). \quad (19.3.69)$$

(b) If $Y = (Y^{(1)}, Y^{(2)}, \ldots)$ has bounded marginal densities $p_{Y^{(i)}}$, $i = 1, 2, \ldots$, with $A_i := \sup_{x \in \mathbb{R}} p_{Y^{(i)}}(x) < \infty$ and $A := \sum_{i=1}^\infty A_i$, then

$$\boldsymbol{\rho}(X,Y) \le (1 + A)\mathbf{L}(X,Y). \quad (19.3.70)$$

Moreover, if $X, Y \in \mathfrak{X}_+^\infty = \mathfrak{X}(\mathbb{R}_+^\infty)$ (i.e., X, Y have nonnegative components), then

$$\mathbf{L}^{p+1}(X,Y) \le \boldsymbol{\rho}_p(X,Y) \quad (19.3.71)$$

and

$$\boldsymbol{\rho}(X,Y) \le \Lambda(p) A^{p/(1+p)} \boldsymbol{\rho}_p^{1/(1+p)}(X,Y), \quad p > 0, \quad (19.3.72)$$

where

$$\Lambda(p) := (1 + p) p^{-p/(1+p)}. \quad (19.3.73)$$

Proof. (a) Inequalities (19.3.67) and (19.3.68) are obvious. The first follows from Chebyshev's inequality, the second from the definitions of $\mathbf{L}$ and $\boldsymbol{\rho}$. One can obtain (19.3.69) in the same manner as (19.3.70), which we will prove completely.

(b) Let $\mathbf{L}(X,Y) < \varepsilon$. Further, for each $x \in \mathbb{R}^\infty$ and $n = 1,2,\ldots$, let $x_n := (x^{(1)},\ldots,x^{(n)},\infty,\infty,\ldots)$. Then $F_X(x_n) - F_Y(x_n) < \varepsilon + F_Y(x_n + \varepsilon\mathbf{e}) - F_Y(x_n) \le \varepsilon + [A_1 + \cdots + A_n]\varepsilon$. Analogously,

$$F_Y(x_n) - F_X(x_n) \le F_Y(x_n) - F_Y(x_n - \varepsilon\mathbf{e}) + \varepsilon \le \varepsilon + [A_1 + \cdots + A_n]\varepsilon.$$

Letting $n \to \infty$, we obtain $\boldsymbol{\rho}(X,Y) < (1+A)\varepsilon$, which proves (19.3.70).

Further, let $\mathbf{L}(Y,Y) > \varepsilon > 0$. Then there exists $x_0 \in \mathbb{R}^\infty_+$ such that $|F_X(x) - F_Y(x)| > \varepsilon$ for all $x \in [x_0, x_0 + \varepsilon\mathbf{e}]$ [i.e., $x^{(i)} \in [x_0^{(i)}, x_0^{(i)} + \varepsilon]$ for all $i \ge 1$]. Hence

$$\begin{aligned}\boldsymbol{\rho}_p(X,Y) &\ge \sup\{M^p(x)\varepsilon : x \in [x_0, x_0 + \varepsilon\mathbf{e}]\}\\ &\ge \varepsilon \inf_{z \in \mathbb{R}^\infty_+} \sup_{x \in [z, z+\varepsilon\mathbf{e}]} M^p(x) = \varepsilon^{1+p}.\end{aligned}$$

Letting $\varepsilon \to \mathbf{L}(X,Y)$ we obtain (19.3.71).

By (19.3.70) and (19.3.71), we obtain

$$\boldsymbol{\rho}(X,Y) < (1+A)\boldsymbol{\rho}_p^{1/(1+p)}(X,Y). \tag{19.3.74}$$

Next we will use the homogeneity of $\boldsymbol{\rho}$ and $\boldsymbol{\rho}_p$ to improve (19.3.74). That is, using the equality

$$\boldsymbol{\rho}(cX,cY) = \boldsymbol{\rho}(X,Y), \quad \boldsymbol{\rho}_p(cX,cY) = c^p\boldsymbol{\rho}_p(X,Y), \quad c > 0, \tag{19.3.75}$$

we have, by (19.3.74),

$$\begin{aligned}\boldsymbol{\rho}(X,Y) &\le \left(1 + \frac{1}{c}A\right)\boldsymbol{\rho}_p^{1/(1+p)}(cX,cY)\\ &= (c^{p/(1+p)} + c^{1/(1+p)}A)\boldsymbol{\rho}_p^{1/(1+p)}(X,Y).\end{aligned} \tag{19.3.76}$$

Minimizing the right-hand side of (19.3.76) w.r.t. $c > 0$ we obtain (19.3.72).

□

Theorem 19.3.4. *Let $\gamma > 0$ and $a = (a^{(1)}, a^{(2)}, \ldots) \in \mathbb{R}^\infty_+$ be such that $A(a,\gamma) := \sum_{k=1}^\infty (a^{(k)})^{1/\gamma} < \infty$. Then for any $p > 1$ there exists a constant $c = c(a,p,\gamma)$ such that*

$$\boldsymbol{\rho}(\overline{X}_n, Y) \le c\alpha_p(n)^{1/(1+p\gamma)}\boldsymbol{\rho}_p(a \circ X, a \circ Y)^{1/(1+p\gamma)}. \tag{19.3.77}$$

Remark 19.3.1. In estimate (19.3.77) the *convergence index* $\alpha_p(n)^{1/(1+p\gamma)}$ tends to the correct one $\alpha_p(n)$ as $\gamma \to 0$ (Lemma 19.3.3). The constant c has the form

$$c := (1+\widetilde{p})\widetilde{p}^{\widetilde{p}/(1+\widetilde{p})}[A(a,\gamma)\lambda(\gamma)]^{\widetilde{p}/(1+\widetilde{p})}, \tag{19.3.78}$$

where $\widetilde{p} := p\gamma$ and

$$\lambda(\gamma) := \gamma \exp[(1+1/\gamma)(\ln(1+1/\gamma)-1)]. \tag{19.3.79}$$

Choosing $a = a(\gamma) \in \mathbb{R}^\infty$ such that $(a^{(k)})^{-1/\gamma}\lambda(\gamma) = k^{-\theta}$ for any $k \geq 1$ and some $\theta > 1$, one can obtain that $c = c(a(\gamma), p, \gamma) \to 1$ as $\gamma \to 0$. However, in this case, $a^{(k)} = a^{(k)}(\gamma) \to \infty$ as $\gamma \to 0$ for any $k \geq 1$, and hence $\boldsymbol{\rho}_p(a \circ X, a \circ Y) \to \infty$ as $\gamma \to 0$.

Proof of Theorem 19.3.4. Denote

$$\widetilde{X}_j := a \circ X_j, \quad \widetilde{Y}_j := a \circ Y_j, \quad p_k(\gamma) := \sup_{x \geq 0} p_{(\widetilde{Y}^{(i)})^{1/\gamma}}(x), \tag{19.3.80}$$

where $p_X(\cdot)$ means the density of a real-valued RV X. Using inequality (19.3.72), we have that for any $\widetilde{p} > \gamma$, i.e., $p > 1$,

$$\begin{aligned}\boldsymbol{\rho}(\overline{X}_n, Y) &= \boldsymbol{\rho}\left(\bigvee_{j=1}^{\infty} c_j(n)^{1/\gamma}\widetilde{X}_j^{1/\gamma}, \widetilde{Y}_j^{1/\gamma}\right)\\ &\leq \Lambda(\widetilde{p})\left(\sum_{k=1}^{\infty} p_k(\gamma)\right)\boldsymbol{\rho}_{\widetilde{p}}^{1/(1+\widetilde{p})}\left(\bigvee_{j=1}^{\infty} c_j(n)^{1/\gamma}\widetilde{X}_j^{1/\gamma}, \widetilde{Y}^{1/\gamma}\right),\end{aligned} \tag{19.3.81}$$

where $\Lambda(\widetilde{p})$ is given by (19.3.73). Next we exploit Lemma 19.3.3 and obtain

$$\boldsymbol{\rho}_{\widetilde{p}}\left(\bigvee_{j=1}^{\infty} c_j(n)^{1/\gamma}\widetilde{X}^{1/\gamma}, \widetilde{Y}^{1/\gamma}\right) \leq \alpha_{\widetilde{p}/\gamma}(n)\boldsymbol{\rho}_{\widetilde{p}/\gamma}(\widetilde{X}_j, \widetilde{Y}_j). \tag{19.3.82}$$

Now we can choose $\widetilde{p} := p\gamma$. Then, by (19.3.81) and (19.3.82),

$$\boldsymbol{\rho}(X_n, Y) \leq \Lambda(\widetilde{p})\left(\sum_{k=1}^{\infty} p_k(\gamma)\right)^{\widetilde{p}/(1+\widetilde{p})} \alpha_p(n)^{1/(1+\widetilde{p})}\boldsymbol{\rho}_p(\widetilde{X}_j, \widetilde{Y}_j)^{1/(1+\widetilde{p})}. \tag{19.3.83}$$

Finally, note that since the components of Y have common DF ϕ_1, then $p_k(\gamma) = (a^{(k)})^{-1/\gamma}\lambda(\gamma)$, where $\lambda(\gamma)$ is given by (19.3.79). □

In Theorem 19.3.4, we have no restrictions on the sequence of **C** of normalizing constants $c_j(n)$ [see (19.3.3) and (19.3.62)]. However, the rate of convergence $\alpha_p(n)^{1/(1+p\gamma)}$ is close but not equal to the exact rate of convergence, namely, $\alpha_p(n)$.

In the next theorem, we impose the following conditions on **C**, which will allow us to reach the exact rate of convergence.

(A.1) There exist absolute constants $K_1 > 0$ and a sequence of integers $m(n)$, $n = 2, 3, \ldots$, such that

$$\sum_{j=1}^{m(n)} c_j(n) \geq K_1 \leq \sum_{j=m(n)+1}^{\infty} c_j(n) \tag{19.3.84}$$

and $m(n) < n$.

(A.2) There exist constants $\beta \in (0, 1)$, $\theta \geq 0$, $\varepsilon_m(n)$, and $\delta_{im}(n)$, $i = 1, 2, \ldots$, $n = 2, 3, \ldots$, such that

$$c_{i+m}(n) = \varepsilon_m(n) c_i(n-m) + \delta_{im}(n) \tag{19.3.85}$$

and

$$\left\{\sum_{i=1}^{\infty} |\delta_{im}(n)|^{\beta}\right\}^{1/(1+\beta)} \leq \theta \alpha_p(n) \tag{19.3.86}$$

for all $i = 1, 2, \ldots, n = 2, 3, \ldots$, and $m = m(n)$ defined by (A.1).

(A.3) There exists a constant K_2 such that

$$\alpha_p(n - m(n)) \leq K_2 \alpha_p(n). \tag{19.3.87}$$

We will check now that the Cesàro sum (for any $p > 1$) satisfies (A.1) to (A.3).

Example 19.3.2. Cesàro sum [see (19.3.22)]. For any $p \geq 1$ we have $\alpha_p(n) = n^{1-p}$.

(A.1) Take $m(n) = [n/2]$, where $[a]$ means the integer part of $a \geq 0$. Then (19.3.84) holds with $K_1 \leq \frac{1}{2}$ and, obviously, $m(n) < n$.

(A.2) Equality (19.3.85) is valid with $\varepsilon_m(n) = (n-m)/n$ and $\delta_{im} = 0$. Hence, $\theta = 0$ in (19.3.86).

(A.3) $K_2 := 2^{p-1}$.

Theorem 19.3.5. *Let Y be max-stable sequences [see* (19.3.4)*] and* **C** *satisfy (A.1) to (A.3). Let $a \in \mathbb{R}_+^{\infty}$ be such that*

$$\mathcal{A}(a) := \sum_{i=1}^{\infty} 1/a^{(i)} < \infty. \tag{19.3.88}$$

Let $p > 1$, $\widetilde{X} = a \circ X$, $\widetilde{Y} = a \circ Y$, and

$$\lambda_p := \lambda_p(\widetilde{X}, \widetilde{Y}) := \max(\boldsymbol{\rho}_p^{1/(p+1)}(\widetilde{X}, \widetilde{Y}), \boldsymbol{\rho}_p(\widetilde{X}, \widetilde{Y}), \Gamma_{\beta}),$$

where

$$\Gamma_\beta := \theta\{[E\|\widetilde{X}\|_\infty^\beta]^{1/(1+\beta)} + [E\|\widetilde{Y}\|_\infty^\beta]^{1/(1+\beta)}\},$$

and β, θ are given by (A.2). Then there exist absolute constants A and B such that

$$\lambda_p \le A \Rightarrow \boldsymbol{\rho}(\overline{X}_n, Y) \le B\lambda_p \alpha_p(n). \tag{19.3.89}$$

Remark 19.3.2. As appropriate pairs (A, B) satisfying (19.3.89) one can take any A and B such that $A \le C_8(p,a)$, $B \ge C_9(p,a)$, where the constants C_8 and C_9 are defined in the following way. Denote

$$C_1(a) := 1 + (2/e)^2 \mathcal{A}(a)/K_1, \quad C_2(a) := C_1(a)(1+K_2), \tag{19.3.90}$$

$$C_3(a) := (2/e)^2 \mathcal{A}(a), \quad C_4(p,a) := (p/e)^p \mathcal{B}(a)^{-p}, \tag{19.3.91}$$

where $\mathcal{B}(a) := \min_{i \ge 1} a^{(i)} > 0$ [see (19.3.88)],

$$C_5(p,a) := 4C_4(p,a)K_1^{-p}, \quad C_6(p,a) := \Lambda(p)\left(\frac{C_3(a)}{K_1}\right)^{p/(1+p)}, \tag{19.3.92}$$

where $\Lambda(p)$ is given by (19.3.73),

$$C_7(p,a) = \Lambda(p) C_3(a)^{p/(1+p)},$$

$$C_8(p,a) := (2C_6(p,a)C_2(a))^{-1-p}, \tag{19.3.93}$$

and

$$C_9(p,d) := \max\{1, C_5(p,a), C_7(p,a)(1 \vee \alpha_p(2))^{-p/(1+p)}\}.$$

The proof of Theorem 19.3.5 is essentially based on the next lemma. In what follows, $X' \vee X''$, for $X', X'' \in \mathfrak{X}(\mathbb{R}_+^\infty)$, always means a random sequence with DF $F_{X'}(x)F_{X''}(x)$, $x \in \mathbb{R}_+^\infty$, and $\widetilde{X}$ means $a \circ X$, where $a \in \mathbb{R}_+^\infty$ satisfies (19.3.88).

Lemma 19.3.5. *(a) ($\boldsymbol{\rho}_p$ is a max-ideal metric of order $p > 0$.) For any X', X'', $Z \in \mathfrak{X}(\mathbb{R}_+^\infty)$ and $c > 0$, $\boldsymbol{\rho}_p(cX', cX'') = c^p \boldsymbol{\rho}_p(X', X'')$, $p > 0$, and*

$$\boldsymbol{\rho}_p(X' \vee Z, X'' \vee Z) \le \boldsymbol{\rho}_p(X', X'').$$

(b) (Max-smoothing inequality.) If Y is a simple max-stable sequence, then for any X', $X'' \in \mathfrak{X}(\mathbb{R}_+^\infty)$ and $\delta > 0$

$$\boldsymbol{\rho}(X' \vee \delta\widetilde{Y}, X'' \vee \delta\widetilde{Y}) \le C_4(p,a)\delta^{-p}\boldsymbol{\rho}_p(X', X''), \tag{19.3.94}$$

$$\boldsymbol{\rho}(X', \widetilde{Y}) \le C_7(p,a)\boldsymbol{\rho}_p^{1/(1+p)}(X', \widetilde{Y}), \tag{19.3.95}$$

where C_4 and C_7 are given in Remark 19.3.2.

(c) For any X', X'', U, $V \in \mathfrak{X}(\mathbb{R}_+^\infty)$

$$\rho(X' \vee U, X'' \vee U) \le \rho(X', X'')\rho(U, V) + \rho(X' \vee V, X'' \vee V). \quad (19.3.96)$$

Remark 19.3.3. Lemma 19.3.5 is the analog of Lemmas 15.3.2, 15.4.1, and 15.4.2 concerning the summation scheme of i.i.d. RVs.

Proof. (a) and (c) are obvious; see Lemmas 19.2.2 and 19.2.7.

(b). Let $G(x) := \exp(-1/x)$, $x \ge 0$, and

$$C(p) := (p/e)^p = \sup_{x>0} x^{-p} G(x). \quad (19.3.97)$$

Then

$$F_{\widetilde{Y}}(x/\delta) \le \min_{i\ge1} F_{a^{(i)}Y^{(i)}}(x^{(i)}/\delta) = \min_{i\ge1} G(x^{(i)}/a^{(i)}\delta) \le C(p)\mathcal{B}(a)^{-p} M(x)^p \delta^{-p}.$$

Hence, by (19.3.91) and (19.3.97), $\rho(X' \vee \delta\widetilde{Y}, X'' \vee \delta\widetilde{Y}) \le C_4(p,a)\delta^{-p}\rho_p(X', X'')$, which proves (19.3.94). Further, by Lemma 19.3.4 [see (19.3.72)], we have

$$\rho(X', \widetilde{Y}) \le \Lambda(p)\left(C(2)\sum_{i=1}^{\infty} 1/a^{(i)}\right)^{p/(1+p)} \rho_p(X', \widetilde{Y})^{1/(1+p)}$$
$$= C_7(p,a)\rho_p(X', \widetilde{Y})^{1/(1+p)}. \qquad \square$$

Proof of Theorem 19.3.5. The main idea of the proof is to follow the *max-Bergstrom* method as in Theorem 19.2.1 but avoiding the use of max-smoothing inequality (19.2.16). If $n = 1, 2$, then by (19.3.95) and Lemma 19.3.4 we have

$$\rho(\overline{X}_n, Y) \le C_7(p,a)\rho_p^{1/(1+p)}\left(\bigvee_{i=1}^{\infty} c_i(n)\widetilde{X}_i, \widetilde{Y}_i\right)$$
$$\le C_7(p,a)\alpha_p(n)^{1/(1+p)}\rho_p^{1/(1+p)}(\widetilde{X}, \widetilde{Y}).$$

Since $\lambda_p \ge \rho_p^{1/(1+p)}(\widetilde{X}, \widetilde{Y})$ and $C_7(p,a)\alpha_p(n)^{1/(1+p)} \le \mathcal{B}\alpha_p(n)$ for $n = 1, 2$, we have proved (19.3.89) for any A and $n = 1, 2$.

We now proceed by induction. Suppose that

$$\rho\left(\bigvee_{j=1}^{\infty} c_j(k)\widetilde{X}_j, Y\right) \le \mathcal{B}\lambda_p\alpha_p(k), \quad \forall k = 1, \ldots, n-1. \quad (19.3.98)$$

Let $m = m(n)$, $n \ge 3$, be given by (A.1). Then using the triangle inequality we obtain

$$\rho\left(\bigvee_{j=1}^{\infty} c_j(n)\widetilde{X}_j, \widetilde{Y}\right) \leq J_1 + J_2, \tag{19.3.99}$$

where

$$J_1 := \rho\left(\bigvee_{j=1}^{m} c_j(n)\widetilde{X}_j \vee \bigvee_{j=m+1}^{\infty} c_j(n)\widetilde{X}_j, \bigvee_{j=1}^{m} c_j(n)\widetilde{Y}_j \vee \bigvee_{j=m+1}^{\infty} c_j(n)\widetilde{X}_j\right)$$

and

$$J_2 := \rho\left(\bigvee_{j=1}^{m} c_j(n)\widetilde{Y}_j \vee \bigvee_{j=m+1}^{\infty} c_j(n)\widetilde{X}_j, \widetilde{Y}\right).$$

Now we will use inequality (19.3.96) to estimate J_1

$$J_1 \leq J_1' + J_1'', \tag{19.3.100}$$

where

$$J_1' := \rho\left(\bigvee_{j=1}^{m} c_j(n)\widetilde{X}_j, \bigvee_{j=1}^{m} c_j(n)\widetilde{Y}_j\right) \rho\left(\bigvee_{j=m+1}^{\infty} c_j(n)\widetilde{X}_j, \bigvee_{j=m+1}^{\infty} c_j(n)\widetilde{Y}_j\right)$$

and

$$J_1'' := \rho\left(\bigvee_{j=1}^{m} c_j(n)\widetilde{X}_j \vee \bigvee_{j=m+1}^{\infty} c_j(n)\widetilde{Y}_j, \bigvee_{j=1}^{m} c_j(n)\widetilde{Y}_j \vee \bigvee_{j=m+1}^{\infty} c_j(n)\widetilde{Y}_j\right).$$

Let us estimate J_1'. Since Y is a simple max-stable sequence,[24]

$$\bigvee_{j=m+1}^{\infty} c_j(n)\widetilde{Y}_j \stackrel{\mathrm{d}}{=} a \circ \left(\sum_{j=m+1}^{\infty} c_j(n)\right) Y. \tag{19.3.101}$$

Hence, by (19.3.101), (19.3.70), (A.1), and (A.2), we have

$$\rho\left(\bigvee_{j=m+1}^{\infty} c_j(n)\widetilde{X}_j, \bigvee_{j=m+1}^{\infty} c_j(n)\widetilde{Y}_j\right)$$

[24]See (19.3.3) and (19.3.4).

$$
\begin{aligned}
&\leq \left(1+\left(\frac{2}{e}\right)^2 \sum_{i=1}^{\infty}\left(a^{(i)} \sum_{j=m+1}^{\infty} c_j(n)\right)^{-1}\right) \mathbf{L}\left(\bigvee_{j=1}^{\infty} c_{j+m}(n)\widetilde{X}_j, \bigvee_{j=1}^{\infty} c_{j+m}(n)\widetilde{Y}_j\right)\\
&\leq C_1(a)\mathbf{L}\left(\bigvee_{j=1}^{\infty}(\varepsilon_m(n)_j(n-m)+\delta_{jm}(n))\widetilde{X}_j, \bigvee_{j=1}^{\infty}(\varepsilon_m(n)c_j(n-m)+\delta_{jm}(n))\widetilde{Y}_j\right)\\
&\leq C_1(a)\left[\mathbf{L}\left(\bigvee_{j=1}^{\infty}(\varepsilon_m(n)c_j(n-m)+\delta_{jm}(n))\widetilde{X}_j, \bigvee_{j=1}^{\infty}\varepsilon_m(n)c_j(n-m)\widetilde{X}_j\right)\right.\\
&\quad +\mathbf{L}\left(\bigvee_{j=1}^{\infty}\varepsilon_m(n)c_j(n-m)\widetilde{X}_j, \bigvee_{j=1}^{\infty}\varepsilon_m(n)c_j(n-m)\widetilde{Y}_j\right)\\
&\quad \left.+\mathbf{L}\left(\bigvee_{j=1}^{\infty}\varepsilon_m(n)c_j(n-m)\widetilde{Y}_j, \bigvee_{j=1}^{\infty}(\varepsilon_m(n)c_j(n-m)+\delta_{jm}(n))\widetilde{Y}_j\right)\right]\\
&=: C_1(a)(I_1+I_2+I_3), \qquad (19.3.102)
\end{aligned}
$$

where $C_1(a)$ is given by (19.3.90). Let us estimate I_1 using (A.2) and inequality (19.3.67):

$$
\begin{aligned}
I_1 &\leq \left\{E\bigvee_{j=1}^{\infty}\|(\varepsilon_m(n)c_j(n-m)+\delta_{jm}(n))\widetilde{X}_j-\varepsilon_m(n)c_j(n-m)\widetilde{X}_j\|_\infty^\beta\right\}^{1/(1+\beta)}\\
&\leq \left\{E\sum_{j=1}^{\infty}|\delta_{jm}|^\beta\|\widetilde{X}_j\|_\infty^\beta\right\}^{1/(1+\beta)} \leq \theta\alpha_p(n)\{E\|\widetilde{X}\|_\infty^\beta\}^{1/(1+\beta)}. \qquad (19.3.103)
\end{aligned}
$$

Analogously,

$$
I_3 \leq \theta\alpha_p(n)\{E\|\widetilde{Y}\|_\infty^\beta\}^{1/(1+p)}. \qquad (19.3.104)
$$

To estimate I_2, we use the inductive assumption (19.3.98), condition (A.3), and (19.3.68):

$$
\begin{aligned}
I_2 &\leq \boldsymbol{\rho}\left(\bigvee_{j=1}^{\infty}\varepsilon_m(n)c_j(n-m)\widetilde{X}_j, \bigvee_{j=1}^{\infty}\varepsilon_m(n)c_j(n-m)\widetilde{Y}_j\right)\\
&\leq \mathcal{B}\lambda_p\alpha_p(n-m) \leq K_2\mathcal{B}\lambda_p\alpha_p(n). \qquad (19.3.105)
\end{aligned}
$$

Hence, by (19.3.102)–(19.3.105) and (19.3.90), we have

$$\rho\left(\bigvee_{j=m+1}^{\infty} c_j(n)\widetilde{X}_j, \bigvee_{j=m+1}^{\infty} c_j(n)\widetilde{Y}_j\right) \le C_1(a)[\Gamma_\beta + K_2\mathcal{B}\lambda_p]\alpha_a(n)$$
$$\le C_2(a)\mathcal{B}\lambda_p\alpha_p(n). \qquad (19.3.106)$$

Next, let us estimate $\rho(\bigvee_{j=1}^m c_j(n)\widetilde{X}_j, \bigvee_{j=1}^m c_j(n)\widetilde{Y}_j)$ in J_1'. Since Y is a simple max-stable sequence,[25] we have

$$\bigvee_{j=1}^{m} c_j(n)\widetilde{Y}_j \stackrel{\mathrm{d}}{=} \sum_{j=1}^{m} c_j(n)\widetilde{Y}. \qquad (19.3.107)$$

Thus, by (19.3.72), (19.3.92), and (A.1),

$$\rho\left(\bigvee_{j=1}^{m} c_j(n)\widetilde{X}_j, \bigvee_{j=1}^{m} c_j(n)\widetilde{Y}_j\right)$$
$$\le \Lambda(p)\left[(2/e)^2\sum_{i=1}^{\infty}\left(a^{(i)}\sum_{j=1}^{m} c_j(n)\right)^{-1}\right]^{p/(1+p)} \rho_p(\widetilde{X}, Y)^{1/(1+p)}$$
$$\le C_6(p,a)\lambda_p^{1/(1+p)} \le C_6(p,a)A^{1/(1+p)}. \qquad (19.3.108)$$

Using the estimates in (19.3.106) and (19.3.108) we obtain the following bound for J_1':

$$J_1' \le C_6(p,a)A^{1/(1+p)}C_2(a)\mathcal{B}\lambda_p\alpha_p(n) \le \frac{1}{2}\mathcal{B}\lambda_p\alpha_p(n). \qquad (19.3.109)$$

Now let us estimate J_1''. By (19.3.94), (19.3.101), (A.1), and (19.3.65), we have

$$J_1'' \le C_4(p,a)\rho_p\left(\bigvee_{j=1}^{m} c_j(n)\widetilde{X}_j, \bigvee_{j=1}^{m} c_j(n)\widetilde{Y}_j\right)\left(\sum_{j=m+1}^{\infty} c_j(n)\right)^{-p}$$
$$\le C_4(p,a)K_1^{-p}\lambda_p\alpha_p(n). \qquad (19.3.110)$$

Analogously, we estimate J_2 [see (19.3.99)]

[25]See (19.3.3) and (19.3.4).

$$J_2 \le C_4(p,a)\boldsymbol{\rho}_p\left(\bigvee_{j=m+1}^{\infty} c_j(n)\widetilde{X}_j, \bigvee_{j=m+1}^{\infty} c_j(n)\widetilde{Y}_j\right)\left(\sum_{j=1}^{m} c_j(n)\right)^{-p}$$
$$\le C_4(p,a)K_1^{-p}\lambda_p\alpha_p(n). \tag{19.3.111}$$

Since $2C_4(p,a)K_1^{-p} \le \mathcal{B}/2$ (Remark 19.3.2),

$$J_1'' + J_2 \le \frac{1}{2}\mathcal{B}\lambda_p\alpha_p(n) \tag{19.3.112}$$

by (19.3.110) and (19.3.111). Finally, using (19.3.99), (19.3.100), (19.3.109), and (19.3.112) we obtain (19.3.98) for $k = n$. □

In the case of the Cesàro sum (19.3.22), one can refine Theorem 19.3.5 following the proof of the theorem and using some simplifications (Example 19.3.1). That is, the following assertion holds.

Corollary 19.3.4. *Let $X, X_1, X_2, \ldots$ be a sequence of i.i.d. RVs taking values in $\mathbb{R}_+^\infty$. Let $Y = (Y^{(1)}, Y^{(2)}, \ldots)$ be a max-stable sequence*[26] *with $F_{Y^{(i)}}(x) = \exp(-1/x)$, $x > 0$. Let $a \in \mathbb{R}_+^\infty$ satisfy (19.3.88). Denote $\overline{\lambda}_p := \overline{\lambda}_p(\widetilde{X}, \widetilde{Y}) := \max\{\boldsymbol{\rho}(\widetilde{X}, \widetilde{Y}), \boldsymbol{\rho}_p(\widetilde{X}, \widetilde{Y})\}$, $\widetilde{X} := a \circ X$, $\widetilde{Y} := a \circ Y$. Then there exist constants C and D such that*

$$\overline{\lambda}_p \le C \Rightarrow \boldsymbol{\rho}\left((1/n)\bigvee_{k=1}^{n} X_k, Y\right) \le D\overline{\lambda}_p n^{1-p}. \tag{19.3.113}$$

Remark 19.3.4. As an example of a pair (C, D) that fulfills (19.3.113) one can choose any (C, D) satisfying the inequalities

$$CD(\tfrac{2}{3})^{p-1} \le \tfrac{1}{2}, \quad D \ge \max(2^p, 4C_4(p,a)(2^{p-1} + 6^p)),$$

where $C_4(p,a)$ is defined by (19.3.91).

Remark 19.3.5. Let $Z_1, Z_2, \ldots$ be a sequence of i.i.d. RVs taking values in the Hilbert space $H = (\mathbb{R}^\infty, \|\cdot\|_2)$ with $EZ_1 = 0$ and covariance operator $\mathbf{V}$. The CLT in H states that the distribution of the normalized sums $\widetilde{Z}, = n^{-1/2}\sum_{i=1}^n Z_i$ weakly tends to the normal distribution of an RV $Z \in \mathfrak{X}(H)$ with mean 0 and covariance operator $\mathbf{V}$. However, the uniform convergence

$$\boldsymbol{\rho}(F_{\widetilde{Z}_n}, F_Z) := \sup_{x \in \mathbb{R}^\infty} |F_{\widetilde{Z}_n}(x) - F_Z(x)| \to \infty \quad n \to \infty$$

[26]See (19.3.4).

may fail.[27] In contrast to the summation scheme, Theorem 19.3.5 shows that under some tail conditions the distribution function of the normalized maxima $\overline{X}_n$ of i.i.d. RVs $X_i \in \mathfrak{X}(\mathbb{R}^\infty)$ converges uniformly to the DF of a simple max-stable sequence Y. Moreover, the rate of uniform convergence is nearly the same as in the finite-dimensional case (Theorems 19.2.1 and 19.2.3). Furthermore, in our investigations we did not assume that $\mathbb{R}^\infty$ had the structure of a Hilbert or even normed space.

Open Problem 19.3.4. Smith (1982), Cohen (1982), Resnick (1987b), and Balkema and de Haan (1990) consider the univariate case $(X, X_1, X_2, \dots \in \mathfrak{X}(\mathbb{R}))$ of general normalized maxima[28]

$$\boldsymbol{\rho}\left(a_n \bigvee_{i=1}^{n} X_i - b_n, Y\right) \le c(X_1, Y)\phi_{X_1}(n), \quad n = 1, 2, \dots.$$

To extend results of this type to the multivariate case $(X, X_1, X_2, \dots \in \mathfrak{X}(\mathbf{B}))$ using the method developed here, one needs to generalize the notions of compound and simple max-stable metrics[29] by determining a metric μ_ϕ in $\mathfrak{X}(\mathbf{B})$ such that for any $X_1, X_2, Y \in \mathfrak{X}(\mathbf{B})$ and $c > 0$

$$\mu_\phi(c(X_1 \vee Y), c(X_2 \vee Y)) \le \phi(c)\mu_\phi(X_1, X_2),$$

where $\phi : [0, \infty) \to [0, \infty)$ is a suitably chosen regular-varying with nonnegative index, strictly increasing continuous function, $\phi(0) = 0$.

19.4 Double Ideal Metrics

The minimal $\widehat{\mathcal{L}}_p$-metrics are ideal w.r.t. summation and maxima of order $r_p = \min(p, 1)$. Indeed, by Definition 15.3.1 in Chap. 15 and Definition 19.2.1, the *p-average probability metrics*[30]

$$\mathcal{L}_p(X, Y) = (E\|X - Y\|^p)^{\min(1, 1/p)}, \quad 0 < p < \infty,$$

$$\mathcal{L}_\infty(X, Y) = \operatorname{ess\,sup} \|X - Y\|, \quad X, Y \in \mathfrak{X}^d := \mathfrak{X}(\mathbb{R}^d), \tag{19.4.1}$$

are compound ideal metrics w.r.t. the sum and maxima of random vectors, i.e., for any $X, Y, Z \in \mathfrak{X}^d$

$$\mathcal{L}_p(cX + Z, cY + Z) \le |c|^{r_p}\mathcal{L}_p(X, Y), \quad c \in \mathbb{R}, \tag{19.4.2}$$

---

[27] See, for example, Sazonov (1981, pp. 69–70).

[28] See also Theorem 19.3.4.

[29] See (19.3.5) and Lemma 19.3.5(a).

[30] See also Example 3.4.1 in Chap. 3.

and

$$\mathcal{L}_p(cX \vee Z, cY \vee Z) \leq c^{r_p} \mathcal{L}_p(X, Y), \quad c \geq 0, \tag{19.4.3}$$

where $x \vee y := (x^{(1)} \vee y^{(1)}, \ldots, x^{(d)} \vee y^{(d)})$. Denote as before by $\widehat{\mathcal{L}}_p$ the corresponding minimal metric, i.e,

$$\widehat{\mathcal{L}}_p(X, Y) = \inf\{\mathcal{L}_p(\widetilde{X}, \widetilde{Y}); \widetilde{X} \stackrel{d}{=} X, \widetilde{Y} \stackrel{d}{=} Y\}, \quad 0 < p < \infty. \tag{19.4.4}$$

Then, by (19.4.2) and (19.4.4), the *ideality* properties hold:

$$\widehat{\mathcal{L}}(cX + Z, cY + Z) \leq |c|^{r_p} \widehat{\mathcal{L}}(X, Y), \quad c \in \mathbb{R}, \tag{19.4.5}$$

and

$$\widehat{\mathcal{L}}_p(cX \vee Z, cY \vee Z) \leq c^{r_p} \widehat{\mathcal{L}}_p(X, Y), \quad c > 0, \tag{19.4.6}$$

for any $X, Y \in \mathfrak{X}^d$ and Z independent of X and Y.[31] In particular, if $X_1, X_2, \ldots$ are i.i.d. RVs and $Y_{(\alpha)}$ has a symmetric stable distribution with parameter $\alpha \in (0, 1)$, and $p \in (\alpha, 1]$, then one gets from (19.4.2)

$$\widehat{\mathcal{L}}_p\left(n^{-1/\alpha} \sum_{i=1}^{n} X_i, Y_{(\alpha)}\right) \leq n^{1-p/\alpha} \widehat{\mathcal{L}}_p(X_1, Y_{(\alpha)}), \tag{19.4.7}$$

which gives a precise estimate in the CLT under the only assumption that $\widehat{\mathcal{L}}_p(X_1, Y_{(\alpha)}) < \infty$. Note that $\widehat{\mathcal{L}}_p(X, Y) < \infty$ $(0 < p \leq 1)$ does not imply the finiteness of pth moments of $\|X\|$ and $\|Y\|$. For example, in the one-dimensional case, $d = 1$,[32]

$$\widehat{\mathcal{L}}_1(X, Y) = \int_{\mathbb{R}} |F_X(x) - F_Y(x)| \mathrm{d}x, \quad X, Y \in \mathfrak{X}^1, \tag{19.4.8}$$

and therefore, $\widehat{\mathcal{L}}_1(X_1, Y_{(\alpha)}) < \infty$ is a *tail* condition on the DF F_X implying $E|X_1| = +\infty$. Similarly, by (19.4.6), if $Z_{(\alpha)}$ is α-max-stable distributed RV on $\mathbb{R}^1$ (i.e., $F_{Z_{(\alpha)}} := \exp(-x^{-\alpha})$, $x \geq 0$), then for $0 < \alpha < p \leq 1$

$$\widehat{\mathcal{L}}_p\left(n^{-1/\alpha} \bigvee_{i=1}^{n} X_i, Z_{(\alpha)}\right) \leq n^{1-p/\alpha} \widehat{\mathcal{L}}_p(X_1, Z_{(\alpha)}), \tag{19.4.9}$$

for any i.i.d. RVs X_i.

---

[31] See, for example, Theorem 7.2.2 in Chap. 7.

[32] See Corollary 7.4.2 in Chap. 7.

In this section we will investigate the following problems posed by Zolotarev (1983, p. 300):

> "It is known that there are ideal metrics or order $s \leq 1$ both in relation to the operation of ordinary addition of random variables and in the relation to the operation $\max(X, Y)$. Such a metric of first order is the Kantorovich metric. 'Doubly ideal metrics' may be useful in analyzing schemes in which both operations are present (schemes of this kind are actually encountered in certain queueing systems). However, not a single 'doubly ideal' metric of order $s > 1$ is known. The study of the properties of these doubly ideal metrics and those of general type is an important and interesting problem."[33]

We will prove that the problem of the existence of doubly ideal metrics of order $r > 1$ has an essential negative answer. In spite of this, the minimal $\widehat{\mathcal{L}}_p$-metrics behave like ideal metrics of order $r > 1$ with respect to maxima and sums.[34]

First, we will show that $\widehat{\mathcal{L}}_p$, in spite of being only a *simple* $(r_p, +)$*-ideal metric*, i.e., ideal metric of order r_p w.r.t. a summation scheme,[35] $r_p = \min(1, p)$, it acts as an ideal $(r, +)$ metric of order $r = 1 + \alpha - \alpha/p$ for $0 < \alpha \leq p \leq 2$. We formulate this result for Banach spaces U of type p. Let $\{Y_i\}_{i \geq 1}$ be a sequence of independent random *signs*,

$$P(Y_i = 1) = P(Y_i = -1) = 1/2.$$

Definition 19.4.1 (See Hoffman-Jorgensen 1977). For any $p \in [1, 2]$ a separable *Banach space* $(U, \|\cdot\|)$ *is said to be of type* p if there exists a constant C such that for all $n \in \mathbb{N}$ and $x_1, \ldots, x_n \in U$

$$E\left\|\sum_{i=1}^{n} Y_i x_i\right\|^p \leq C \sum_{i=1}^{n} \|x_i\|^p. \tag{19.4.10}$$

The preceding definition implies the following condition:[36] there exists $A > 0$ such that for all $n \in \mathbb{N} := \{1, 2, \ldots\}$ and independent $X_1, \ldots, X_n \in \mathfrak{X}(U)$ with $EX_i = 0$ and finite $E\|X_i\|^p$ the following relation holds:

$$E\left\|\sum_{i=1}^{n} X_i\right\|^p \leq A \sum_{i=1}^{n} E\|X_i\|^p. \tag{19.4.11}$$

Remark 19.4.1. (a) Every separable Banach space is of type 1.
(b) Every finite-dimensional Banach space and every separable Hilbert space is of type 2.
(c) $\mathcal{L}^q := \{X \in \mathfrak{X}^1 : E|X|^q < \infty\}$ is of the type $p = \min(2, q)$ $\forall q \geq 1$.

---

[33]The Kantorovich metric referred to by Zolotarev in the quote is (19.4.8) in this chapter.

[34]See Rachev and Rüschendorf (1992) for applications of double ideal metrics in estimating convergence rates.

[35]See Definition 15.3.1.

[36]See Hoffman-Jorgensen and Pisier (1976).

(d) $\ell_q := \left\{ X \in \mathbb{R}^\infty, \|x\|_q^q := \sum_{j=1}^{\infty} |x^{(j)}|^q < \infty \right\}$ is of type $p = \min(2, q), q \geq 1$.

Theorem 19.4.1. *If U is of type p, $1 \leq p \leq 2$, and $0 < \alpha < p \leq 2$, then for any i.i.d. RVs $X_1, \ldots, X_n \in \mathfrak{X}(U)$ with $EX_i = 0$ and for a symmetric stable RV $Y_{(\alpha)}$ the following bound holds:*

$$\widehat{\mathcal{L}}_p \left(n^{-1/\alpha} \sum_{i=1}^{n} X_i, Y_{(\alpha)} \right) \leq B_p n^{1/p - 1/\alpha} \widehat{\mathcal{L}}_p(X_1, Y_{(\alpha)}), \tag{19.4.12}$$

where B_p is an absolute constant.

Proof. We use the following result of Woyczynski (1980): if U is of type p, then for some constant B_p and any independent $Z_1, \ldots, Z_n \in \mathfrak{X}(U)$ with $EZ_i = 0$

$$E \left\| \sum_{i=1}^{n} Z_i \right\|^q \leq B_p^p E \left(\sum_{i=1}^{n} \|Z_i\|^p \right)^{q/p}, \quad q \geq 1. \tag{19.4.13}$$

Let $Y_1, \ldots, Y_n \in \mathfrak{X}(U)$ be independent, $Y_i \stackrel{\mathrm{d}}{=} Y_{(\alpha)}$; then $Z_i = X_i - Y_i$, $1 \leq i \leq n$ are also independent. Take $Y_{(\alpha)} = n^{-1/\alpha} \sum_{i=1}^{n} Y_i$. Then, from (19.4.13) with $q = p$ it follows that

$$\mathcal{L}_p^p \left(n^{-1/\alpha} \sum_{i=1}^{n} X_i, Y_{(\alpha)} \right) \leq B_p^p n^{1-p/\alpha} \mathcal{L}_p^p(X_1, Y_1). \tag{19.4.14}$$

Passing to the minimal metrics in the last inequality we establish (19.4.12). □

From the well-known inequality between the Prokhorov metric $\boldsymbol{\pi}$ and $\widehat{\mathcal{L}}_p$,[37]

$$\boldsymbol{\pi}^{p+1} \leq (\widehat{\mathcal{L}}_p)^p, \qquad p \geq 1, \tag{19.4.15}$$

we immediately obtain the following corollary.

Corollary 19.4.1. *Under the assumptions of Theorem 19.4.1,*

$$\boldsymbol{\pi} \left(n^{-1/\alpha} \sum_{i=1}^{n} X_i, Y_{(\alpha)} \right) \leq B_p^{p/(p+1)} n^{(1-p/\alpha)/(p+1)} \widehat{\mathcal{L}}_p(X_1, Y_{(\alpha)})^{p/(p+1)} \tag{19.4.16}$$

for any $p \in [1, 2]$ and $p > \alpha$.

Remark 19.4.2. For $r = p \in [1, 2]$ the rates in (19.4.16) and in Zolotarev's estimate

[37] See (8.3.7) and (8.3.21) in Chap. 8.

$$\pi\left(n^{-1/\alpha}\left\|\sum_{i=1}^{n} X+i\right\|, \|Y_{(\alpha)}\|\right) \leq Cn^{(1-r/\alpha)/(r+1)}\zeta_r^{1/(r+1)}(X_1, Y_{(\alpha)}), \quad (19.4.17)$$

where $0 \leq \alpha \leq r < \infty$ and C is an absolute constant, are the same. On the right-hand side, ζ_r is Zolotarev's metric.[38] A problem with the application of ζ_r for $r > 1$ in the infinite-dimensional case was pointed out by Bentkus and Rackauskas (1985). In Banach spaces, the convergence w.r.t. ζ_r, $r > 1$, does *not* imply weak convergence. Gine and Leon (1980) showed that in Hilbert spaces ζ_r does imply the weak convergence, while by results of Senatov (1981) there is no inequality of the type $\zeta_r \geq c\pi^a$, $a > 0$, where c is an absolute constant. Under some smoothness conditions on the Banach space, Zolotarev (1976) obtained the estimate[39]

$$\pi^{1+r}(\|X\|, \|Y\|) \leq C\zeta_r(X, Y), \quad (19.4.18)$$

where $C = C(r)$. Therefore, under these conditions, (19.4.17) follows from the ideality of $\zeta_r : \zeta_r(n^{-1/\alpha}\sum_{i=1}^{n} X_i, Y_{(\alpha)})$. It was proved by Senatov (1981) that the order in (19.4.17) is the right one for $r = 3$, $\alpha = 2$, namely, $n^{-1/8}$. The only known upper estimate for ζ_r applicable in the stable case is[40]

$$\zeta_r \leq \frac{\Gamma(1+\alpha)}{\Gamma(1+r)}\nu_r, \quad r = m + \alpha, \quad 0 < \alpha \leq 1, \quad m \in \mathbb{N}, \quad (19.4.19)$$

where

$$\nu_r(X, Y) = \int \|x\|^r |\Pr_X - \Pr_Y|(dx) \quad (19.4.20)$$

is the rth *absolute pseudomoment*. So $\nu_r(X_1, Y_{(\alpha)}) < \infty$ ensures the validity of (19.4.17).

In contrast to the bound (19.4.17), which concerns only the distance between the norms of X and Y, estimate (19.4.16) concerns the Prokhorov distance $\pi(X, Y)$ itself, which is topologically strictly stronger than $\pi(\|X\|, \|Y\|)$ in the Banach space setting and is more informative. Furthermore, it follows that[41]

$$\widehat{\mathcal{L}}_p^p(X, Y) \leq 2^p\kappa_p(X, Y) \leq 2^p\nu_p(X, Y), \quad (19.4.21)$$

where κ_r, $r > 0$, is the rth *difference pseudomoment*,

$$\kappa_r(X, Y) = \inf\{Ed_r(\widetilde{X}, \widetilde{Y}); \widetilde{X} \stackrel{d}{=} X, \widetilde{Y} \stackrel{d}{=} Y\}$$

---

[38]See (15.3.1) in Chap. 15.

[39]See Zolotarev (1976, Theorem 5).

[40]See Zolotarev (1978, Theorem 4).

[41]See Zolotarev (1978, p. 272).

$$= \sup\{|Ef(X) - Ef(Y)| : f : U \to \mathbb{R} \text{ bounded}$$
$$|f(x) - f(y)| \le d_r(x, y),\ x, y \in U\}, \qquad (19.4.22)$$

and $d_r(x, y) = \| x \| x \|^{r-1} - y \| y \|^{r-1} \|$.[42] Since *the problem of whether* $\kappa_r(X, Y) < \infty$, $E(X - Y) = 0$ *implies* $\zeta_r(X, Y) < \infty$ *is still open for* $1 < r < 2$, the right-hand side of (19.4.16) seems to contain weaker conditions than the right-hand side of (19.4.17).

Remark 19.4.3. If $U = \mathcal{L}^p$ [see Remark 19.4.1 (c)], then with an appeal to the Burkholder inequality one can choose the constants B_p in (19.4.16) as follows:[43]

$$B_1 = 1, \quad B_p = 18p^{3/2}/(p-1)^{1/2}, \quad \text{for } 1 < p \le 2. \qquad (19.4.23)$$

Remark 19.4.4. Let $1 \le p \le 2$, let $(E, \mathcal{E}, \mu)$ be a measurable space, and define

$$\ell_{p,\mu} := \{X : (E, \mathcal{E} \times (\omega, \mathcal{A} \to (\mathbb{R}^1, \mathcal{B}^1) : \|X\|_{p,\mu} < \infty\}, \qquad (19.4.24)$$

where $\|X\|_{p,\mu} := E(\int |X(t)|^p \mathrm{d}\mu(t))^{1/p}$; $(\ell_{p,\mu}, \| \cdot \|_{p,\mu})$ is a Banach space of stochastic processes.[44] Let $X_1, \dots, X_n \in \mathfrak{X}(\ell_{p,\mu})$ with $EX_i = 0$. Recall the *Marcinkiewicz–Zygmund inequality*: if $\{\xi_n, n > 1\}$ are independent integrable RVs with $E\xi_n = 0$, then for every $p > 1$ there exist positive constants A_p and B_p such that[45]

$$A_p \left\| \left(\sum_{j=1}^n \xi_j^2 \right)^{1/2} \right\|_p \le \left\| \sum_{j=1}^n \xi_j \right\|_p \le B_p \left\| \left(\sum_{j=1}^n \xi_j^2 \right)^{1/2} \right\|_p . \qquad (19.4.25)$$

By the Marcinkiewicz–Zygmund inequality,

$$E \left\| \sum_{i=1}^n X_i \right\|_{p,\mu}^p = E \int \left| \sum_{i=1}^n X_i(t) \right|^p \mathrm{d}\mu(t) \le \int B_p^p E \left(\sum_{i=1}^n X_i^2(t) \right)^{p/2} \mathrm{d}\mu(t).$$

Since $p \le 2$, we obtain from the Minkowski inequality

$$E \left\| \sum_{i=1}^n X_i \right\|_{p,\mu}^p \le B_p \sum_{i=1}^n E \int |X_i(t)|^p \mathrm{d}\mu(t) = B_p \sum_{i=1}^n \|X_i\|_{p,\mu}^p,$$

---

[42]See Remark 7.2.3 in Chap. 7.

[43]See Chow and Teicher (1978, p. 396).

[44]It is identical to $\mathcal{L}^p$ for one-point measures μ.

[45]See Shiryayev (1984, p. 469) and Chow and Teicher (1978, p. 367).

i.e., $\ell_{p,\mu}$ is of type p, and therefore one can apply Theorem 19.4.1 and Corollary 19.4.1 to stochastic processes in $\ell_{p,\mu}$.

For $0 < \alpha < 2p \leq 1$ we have the following analog of Theorem 19.4.1 using the same metric as in Sect. 19.3 (Lemma 19.3.2 and Theorem 19.3.3). Again, let $(U, \|\cdot\|)$ be of type p and let ξ_p be the minimal metric w.r.t. the compound metric

$$\chi_p(X, Y) := \left[\sup_{t>0} t^p \Pr\{\|X - Y\| > t\}\right]^{1/(1+p)}, \quad p > 0.$$

Then the following bound for the $\mathcal{L}_p$-distance between the normalized sums of i.i.d. random elements in $\mathfrak{X} = \mathfrak{X}(U)$ holds.

Theorem 19.4.2. *Let $X_1, \ldots, X_n \in \mathfrak{X}$ be i.i.d, let $Y_1, \ldots, Y_n \in \mathfrak{X}$ be i.i.d., and let $0 < \alpha < 2p < 1$. Then*

$$\widehat{\mathcal{L}}_p\left(n^{-1/\alpha}\sum_{i=1}^n X_i, n^{-1/\alpha}\sum_{i=1}^n Y_i\right) \leq B_p n^{1/2p-1/\alpha}(\xi_{2p}(X_1, Y_1))^{p+1/2}, \quad (19.4.26)$$

where B_p is an absolute constant.

Proof. We have

$$\begin{aligned} E\left\|n^{-1/\alpha}\sum_{i=1}^n X_i - n^{1/\alpha}\sum_{i=1}^n Y_i\right\|^p &= n^{-p/\alpha} E\left\|\sum_{i=1}^n (X_i - Y_i)\right\|^p \\ &\leq n^{-p/\alpha} E\left(\sum_{i=1}^n \|X_i - Y_i\|\right)^p \\ &\leq B_p n^{-p/\alpha}\sqrt{n}\left(\sup_{c>0} c^2 \Pr(\|X_1 - Y_1\|^p > c)\right)^{1/2} \\ &= B_p n^{-p/\alpha}\sqrt{n}(\chi_{2p}(X, Y))^{p+1/2}; \end{aligned}$$

the last inequality follows from Pisier and Zinn (1977, Lemma 5.3). Passing to the minimal metrics, (19.4.26) follows. □

Remark 19.4.5. From the ideality of order p of $\widehat{\mathcal{L}}_p$ [see (19.4.5)] for $0 < p \leq 1$ one obtains for $0 < \alpha < 2p \leq 1$ the bound

$$\widehat{\mathcal{L}}_p\left(n^{-1/\alpha}\sum_{i=1}^n X_i, n^{-1/\alpha}\sum_{i=1}^n Y_i\right) \leq n^{1-p/\alpha}\widehat{\mathcal{L}}_p(X_1, Y_1), \quad (19.4.27)$$

and by the Holder inequality,

$$\widehat{\mathcal{L}}_p\left(n^{-1/\alpha}\sum_{i=1}^n X_i, n^{-1/\alpha}\sum_{i=1}^n Y_i\right) \le \widehat{\mathcal{L}}_{2p}\left(n^{-1/\alpha}\sum_{i=1}^n X_i, n^{-1/\alpha}\sum_{i=1}^n Y_i\right)$$
$$\le n^{1-2p/\alpha}\widehat{\mathcal{L}}_{2p}(X_1, Y_1). \tag{19.4.28}$$

Since $(\xi_{2p}(X_1, Y_1))^{1+2p} \le \widehat{\mathcal{L}}_{2p}(X_1, Y_1)$ for $p < 1/2$ [see (19.3.51)], the condition $\xi_{2p}(X_1, Y_1) < \infty$ is weaker than the condition $\widehat{\mathcal{L}}_{2p}(X_1, Y_1) < \infty$. Comparing the estimates (19.4.27) and (19.4.26), it is clear that (19.4.26) has the better order, $(1 - p/\alpha > 1 - 2p/\alpha > (1/2p) - (1/\alpha))$. However,

$$\widehat{\mathcal{L}}_p(X_1, Y_1) \le 2\xi_{2p}(X_1, Y_1)^{(p+1)/2}, \tag{19.4.29}$$

and thus the *tail condition* in (19.4.27) is weaker than that in (19.4.26). To prove (19.4.29), it is enough to show that

$$\widehat{\mathcal{L}}_p(X_1, Y_1) \le 2\chi_{2p}(X_1, Y_1)^{(p+1)/2}. \tag{19.4.30}$$

The last inequality follows from the bound

$$Ed^p(X_1, Y_1) \le T^p + \int_T^\infty \Pr(\mathrm{d}(X_1, Y_1) > t)pt^{p-1}\mathrm{d}t$$
$$\le T^p + (\chi_{2p}(X_1, Y_1))^{p+1}T^{-p}, \quad T > 0,$$

after a minimization with respect to T.

Up to now we have investigated the ideal properties of $\widehat{\mathcal{L}}_p$ w.r.t. the sums of i.i.d. RVs. Next we will look at the max-ideality of $\widehat{\mathcal{L}}_p$, and this will lead us to the *doubly ideal* properties of $\widehat{\mathcal{L}}_p$.

First, let us point out that *there is no compound ideal metric of order $r > 1$ for the summation scheme while compound max-ideal metrics of order $r > 1$ exist.*

Remark 19.4.6. It is easy to see that there is no nontrivial compound ideal metric μ w.r.t. the summation scheme when $r > 1$ since the ideality (Definition 15.3.1) would imply

$$\mu(X, Y) = \mu\left(\frac{X + \cdots + X}{n}, \frac{Y + \cdots + Y}{n}\right) \le n^{1-r}\mu(X, Y), \quad \forall n \in \mathbb{N},$$

i.e., $\mu(X, Y) \in \{0, \infty\}$, $\forall X, Y \in \mathfrak{X}(U)$.

On the other hand, the following metrics are examples of compound max-ideal metrics of any order. For $U = \mathbb{R}^1$ and any $0 < p \le \infty$ define for $X, Y \in \mathfrak{X}(\mathbb{R}^1)$

$$\boldsymbol{\Delta}_{r,p}(X, Y) = \left(\int_{-\infty}^\infty \phi_{X,Y}^p(x)|x|^{rp-1}\mathrm{d}x\right)^q \tag{19.4.31}$$

and

$$\mathbf{\Delta}_{r,\infty}(X,Y) = \sup_{x\in\mathbb{R}^1} |x|^r \phi_{X,Y}(x),$$

where $q = \min(1, 1/p)$ and $\phi_{X,Y}(x) = \Pr(X \leq x < Y) + \Pr(Y \leq x < X)$. It is easy to see that $\mathbf{\Delta}_{r,p}$ is a compound probability metric. Obviously, for any $c > 0$ the following relation holds:

$$\mathbf{\Delta}_{r,p}(cX, cY) = \left(\int_{-\infty}^{\infty} \phi_{X,Y}^p(x/c)|x|^{rp-1}\mathrm{d}x\right)^q = c^{rpq}\mathbf{\Delta}_{r,p}(X,Y),$$

and $\mathbf{\Delta}_{r,\infty}(cX, cY) = c^r \mathbf{\Delta}_{r,\infty}(X,Y)$. Furthermore, from $\{X \vee Z \leq x < Y \vee Z\} \subset \{X \leq x < Y\}$, which can be established for any RVs X, Y, Z by considering the different possible order relations between X, Y, Z, it follows that $\mathbf{\Delta}_{r,p}$ is a *compound max-ideal metric of order $r(1 \wedge p)$ for $0 < p \leq \infty$ and $0 < r < \infty$.*

Note that $\mathbf{\Delta}_{r,p}$ is an extension of the metric $\mathbf{\Theta}_p$ defined in Example 3.4.3 in Chap. 3; in fact, $\mathbf{\Theta}_p = \mathbf{\Delta}_{1,p}$. Following step by step the proof of Theorem 7.4.4 one can see that the minimal metric $\widehat{\mathbf{\Delta}}_{r,p}$ has the form of the difference pseudomoment

$$\widehat{\mathbf{\Delta}}_{r,p}(X,Y) = \left(\int_{-\infty}^{\infty} |F_X(x) - F_Y(x)|^p |x|^{rp-1}\mathrm{d}x\right)^q \tag{19.4.32}$$

for $p \in (0,\infty)$, and $\widehat{\mathbf{\Delta}}_{r,\infty}(X,Y) = \sup_{x\in\mathbb{R}^1} |x|^r |F_X(x) - F_Y(x)|$ is the weighted Kolmogorov metric $\boldsymbol{\rho}_r$ [see (19.2.4)]. Thus, if $Z_{(\alpha)}$ is an α-max-stable distributed RV, then as in (19.2.5) and (19.4.9) we obtain

$$\widehat{\mathbf{\Delta}}_{r,p}\left(n^{-1/\alpha}\bigvee X_i, Z_{(\alpha)}\right) \leq n^{1-r^*/\alpha}\widehat{\mathbf{\Delta}}_{r,p}(X_1, Z_{(\alpha)}),$$

where $r^* := r(1 \wedge p)$.

Next we want to investigate the properties of the $\mathcal{L}_p$-metrics w.r.t. maxima.[46] Following the notations in Remark 19.4.4 we consider for $0 < \lambda \leq \infty$ the Banach space $U = \ell_{\lambda,\mu} = \{X : (E,\mathcal{E}) \times (\Omega,\mathcal{A}) \to (\mathbb{R}^1,\mathcal{B}^1); \|X\|_{\lambda,\mu} < \infty\}$, where

$$\|X\|_{\lambda,\mu} := E\left(\int |X(t)|^\lambda \mathrm{d}\mu(t)\right)^{1/\lambda^*} \quad \text{for } 0 < \lambda < \infty,\ \lambda^* = 1 \vee \lambda,$$

and define, for $X, Y \in U$, $X \vee Y$ as the pointwise maximum, $(X \vee Y)(t) = X(t) \vee Y(t)$, $t \in E$. Following the definition of a simple max-stable process [see (19.3.40)] we call $Z_{(\alpha)}$ an *α-max-stable process* if

$$Z_{(\alpha)} \stackrel{\mathrm{d}}{=} n^{-1/\alpha}\bigvee_{i=1}^{n} Y_i \tag{19.4.33}$$

[46] See (19.4.1) and Example 3.4.1 in Chap. 3.

for any $n \in \mathbb{N}$ and the Y_i are i.i.d. copies of $Z_{(\alpha)}$.

The proof of the next lemma and theorem are similar to that in Theorem 19.4.1 and thus left to the reader.

Lemma 19.4.1. *(a) For $0 < \lambda \leq \infty$ and $0 < p \leq \infty$, $\mathcal{L}_p$ is a compound ideal metric of order $r = 1 \wedge p$, with respect to a maxima scheme, i.e.,* (19.4.6) *holds.*

(b) If $X_1, \dots, X_n \in \mathfrak{X}(\ell_{\lambda,\mu})$ are i.i.d. and if $Z_{(\alpha)}$ is an α-max-stable process, then for $r = 1 \wedge p$

$$\widehat{\mathcal{L}}_p\left(n^{-1/\alpha}\bigvee_{i=1}^{n} X_i, Z_{(\alpha)}\right) \leq n^{1-r/\alpha}\widehat{\mathcal{L}}_p(X_1, Z_{(\alpha)}). \tag{19.4.34}$$

Estimate (19.4.34) is interesting for $r \leq \alpha$ only; for $1 < p \leq \lambda < \infty$ one can improve it as follows (Theorem 19.3.1).

Theorem 19.4.3. *Let $1 \leq p \leq \lambda < \infty$; then for $X_1, \dots, X_n \in \mathfrak{X}(\ell_{\lambda,\mu})$ i.i.d. the following relation holds:*

$$\widehat{\mathcal{L}}_p\left(n^{-1/\alpha}\bigvee_{i=1}^{n} X_i, Z_{(\alpha)}\right) \leq n^{1/p-1/\alpha}\widehat{\mathcal{L}}_p(X_1, Z_{(\alpha)}). \tag{19.4.35}$$

Remark 19.4.7. (a) Comparing (19.4.35) with (19.4.34) we see that actually $\widehat{\mathcal{L}}_p$ "acts" in this important case as a simple max-ideal metric of order $\alpha + 1 - \alpha/p$. For $1 < p$ it holds that $1/p - 1/\alpha < 1 - 1/\alpha$, i.e., (19.4.35) is an improvement over (19.4.34).

(b) An analog of Theorem 19.4.3 holds also for the sequence space $\ell_\lambda \subset \mathbb{R}^\infty$ [Remark 19.4.1 (d)].

Now we are ready to investigate the question of the existence and construction of doubly ideal metrics. Let U be a Banach space with maximum operation $\vee$.

Definition 19.4.2 (Double ideal metrics). A probability metric μ on $\mathfrak{X}(U)$ is called

(a) (r, I)-ideal if μ is *compound* $(r, +)$*-ideal and compound* $(r, \vee)$*-ideal*, i.e., for any X_1, X_2, Y, and $Z \in \mathfrak{X}(U)$ and $c > 0$

$$\mu(X_1 + Y, X_2 + Y) \leq \mu(X_1, X_2), \tag{19.4.36}$$

$$\mu(X_1 \vee Z, X_2 \vee Z) \leq \mu(X_1, X_2), \tag{19.4.37}$$

and

$$\mu(cX_1, cX_2) = c^r \mu(X_1, X_2); \tag{19.4.38}$$

(b) (r, II)-ideal if μ is compound $(r, \vee)$-ideal and simple $(r, +)$-ideal, i.e., (19.4.36)–(19.4.38) hold with Y independent of X_i;

(c) (r, III)-ideal if μ is simple $(r, \vee)$-ideal and simple $(r, +)$-ideal, i.e., (19.4.36)–(19.4.38) hold with Y and Z independent of X_i.

Remark 19.4.8. In the preceding definition (c) the metric μ can be compound or simple. An example of a compound $(1/p, \mathrm{III})$-ideal metric is the $\boldsymbol{\Theta}_p$-metric $(p \geq 1)$[47]

$$\boldsymbol{\Theta}_p(X,Y) := \left\{\int_{-\infty}^{\infty} (\Pr(X_1 \leq t < X_2) + \Pr(X_2 \leq t < X_1))^p \mathrm{d}t\right\}^{1/p}, \quad 1 \leq p < \infty,$$

$$\boldsymbol{\Theta}_\infty(X,Y) := \sup_{t \in \mathbb{R}^1} (\Pr(X_1 \leq t < X_2) + \Pr(X_2 \leq t < X_1)).$$

Remark 19.4.9. Note that if μ is an (r, II)-ideal metric, then one obtains for $\{X_i\}$ i.i.d., $\{X_i^*\}$ i.i.d.

$$S_k; = \sum_{i=1}^{k} X_i,\ S_k^* := \sum_{i=1}^{k} X_i^*,\ Z_n := n^{1/\alpha} \bigvee_{k=1}^{n} S_k,\ Z_n^* := n^{-1/\alpha} \bigvee_{k=1}^{n} S_k^* \tag{19.4.39}$$

the estimate

$$\begin{aligned}\mu(Z_n, Z_n^*) &\leq n^{-r/\alpha} \mu\left(\bigvee_{k=1}^{n} S_k, \bigvee_{k=1}^{n} S_k^*\right)\\ &\leq n^{-r/\alpha} \sum_{k=1}^{n} \mu(S_k, S_k^*) \leq n^{-r/\alpha} \sum_{k=1}^{n}\sum_{j=1}^{k} \mu(X_j, X_j^*),\end{aligned} \tag{19.4.40}$$

and, hence, for the minimal metric $\widehat{\mu}$ we get

$$\widehat{\mu}(Z_n, Z_n^*) \leq \frac{n(n+1)}{2} n^{-r/\alpha} \widehat{\mu}(X_1, X_1^*) < n^{2-r/\alpha} \widehat{\mu}(X_1, X_1^*), \tag{19.4.41}$$

which gives us a rate of convergence if $0 < \alpha < r/2$. Therefore, from the known ideal metrics of order $r \leq 1$ one gets a rate of convergence for $\alpha \in (0, \frac{1}{2})$. It is therefore of interest to study Zolotarev's question for the construction of doubly ideal metrics of order $r > 1$.

Remark 19.4.10. $\mathcal{L}_p$, $0 < p < \infty$, is an example of a $(1 \wedge p, \mathrm{I})$-ideal metric. We saw in Remark 19.4.6 that there is no (r, I)-ideal metric for $r > 1$. $\widehat{\mathcal{L}}_p$ is (r, III)-ideal metric of order $r = \min(1, p)$.

We now show that Zolotarev's question on the existence of an (r, II)- or an (r, III)-ideal metric has essentially a negative answer.

[47]See (3.4.12) and (19.4.31).

Theorem 19.4.4. *Let $r > 1$, let the simple probability metric μ be (r, III)-ideal on $\mathfrak{X}(\mathbb{R})$, and assume that it satisfies the following regularity conditions.*

Condition 1. If X_n (resp. Y_n) converges weakly to a constant a (resp. b), then

$$\overline{\lim_{n\to\infty}}\, \mu(X_n, Y_n) \geq \mu(a, b). \tag{19.4.42}$$

Condition 2. $\mu(a, b) = 0 \iff a = b$.
Then for any integrable $X, Y \in \mathfrak{X}(\mathbb{R})$ the following holds: $\mu(X, Y) \in \{0, \infty\}$.

Proof. If μ is a simple $(r, +)$-ideal metric, then for integrable $X, Y \in \mathfrak{X}(\mathbb{R}^1)$ the following holds: $\mu((1/n)\sum_{i=1}^n X_i, (1/n)\sum_{i=1}^n Y_i) \leq n^{1-r}\mu(X, Y)$, where (X_i, Y_i) are i.i.d. copies of (X, Y). By the weak law of large numbers and Condition 1, we have

$$\mu(EX, EY) \leq \overline{\lim}\, \mu\left(\frac{1}{n}\sum_{i=1}^n X_i, \frac{1}{n}\sum_{i=1}^n Y_i\right).$$

Assuming that $\mu(X, Y) < \infty$, we have $\mu(EX, EY) = 0$, i.e., $EX = EY$ by Condition 2. Therefore, $\mu(X, Y) < \infty$ implies that $EX = EY$. Therefore, by $\mu(X \vee a, Y \vee a) \leq \mu(X, Y)$, we have that $E(X \vee a) = E(Y \vee a)$ for all $a \in \mathbb{R}^1$, i.e., $\int_{-\infty}^a \Pr(X < x) - \Pr(Y < x)\mathrm{d}x = 0$ for all $a \in \mathbb{R}^1$. Thus $X \stackrel{\mathrm{d}}{=} Y$, and therefore $\mu(X, Y) = 0$. □

Remark 19.4.11. Condition 1 seems to be quite natural. For example, let $\mathcal{F}$ be a class of nonnegative lower semicontinuous (LSC) functions on $\mathbb{R}^2$ and $\phi : [0, \infty) \to [0, \infty)$ continuous, nondecreasing. Suppose μ has the form of a *minimal* functional,

$$\mu(X, Y) = \inf\left\{\phi\left(\sup_{f\in\mathcal{F}} Ef(\widetilde{X}, \widetilde{Y})\right) : \widetilde{X} \stackrel{\mathrm{d}}{=} X, \widetilde{Y} \stackrel{\mathrm{d}}{=} Y\right\}, \tag{19.4.43}$$

with respect to a compound metric $Ef(\widetilde{X}, \widetilde{Y})$ with a $\overline{\xi}$-structure.[48] Then μ is LSC on $\mathfrak{X}(\mathbb{R}^2)$, i.e., $(X_n, Y_n) \xrightarrow{\mathrm{w}} (X, Y)$ implies

$$\liminf_{n\to\infty} \mu(X_n, Y_n) \geq \mu(X, Y), \tag{19.4.44}$$

so Condition 1 is fulfilled. Actually, suppose $\liminf_{n\to\infty} \mu(X_n, Y_n) < \mu(X, Y)$. Then for some subsequence $\{m\} \subset \mathbb{N}$, $\mu(X_n, Y_n)$ converges for some $a < \mu(X, Y)$. For $f \in \mathcal{F}$ the mapping $h_f : \mathfrak{X}(\mathbb{R}^2) \to \mathbb{R}$, $h_f(X, Y) := Ef(X, Y)$ is LSC. Therefore, also $\phi(\sup_{f\in\mathcal{F}} h_f)$ is LSC and there exists a sequence $(\widetilde{X}_m, \widetilde{Y}_m)$ with

[48]See (4.4.64) in Chap. 4.

$\widetilde{X}_m \stackrel{\mathrm{d}}{=} X_m, \widetilde{Y}_m \stackrel{\mathrm{d}}{=} Y_m$ such that $\mu(X_m, Y_m) = \phi(\sup_{f \in \mathcal{F}} h_f(\widetilde{X}_m, \widetilde{Y}_m))$. The sequence $\{\lambda_m := \Pr_{\widetilde{X}_m, \widetilde{Y}_m}\}_{m \geq 1}$ is tight. For any weakly convergent subsequence λ_{m_k} with limit λ, obviously λ has marginals $\Pr_X$ and $\Pr_Y$. Then for $(\widetilde{X}, \widetilde{Y})$ with distribution λ

$$a = \liminf_k \mu(X_{m_k}, Y_{m_k}) = \liminf_k E\phi\left(\sup_{f \in \mathcal{F}} h_f(\widetilde{X}_{m_k}, \widetilde{Y}_{m_k})\right)$$

$$\geq E\phi\left(\sup_{f \in \mathcal{F}} h_f(\widetilde{X}, \widetilde{Y})\right) \geq \mu(X, Y),$$

which contradicts our assumption. Therefore, (19.4.44) holds.

Despite the fact that (r, III)-ideal and, thus, (r, II)-ideal metrics do not exist, we will show next that for $0 < \alpha \leq 2$ the metrics $\mathcal{L}_p$ for $1 < p \leq 2$ "act" as (r, II)-ideal metrics in terms of the rate of convergence problem $\mathcal{L}_p(Z_n, Z_n^*) \to 0$ $(n \to \infty)$, where Z_n and Z_n^* are given by (19.4.39). The order of (r, II)-ideality is $r = 2\alpha + 1 - \alpha/p > 2\alpha$, and therefore we obtain a rate of convergence of $n^{2-r/\alpha}$ [see below (19.4.48)].

We consider first the case where $\{X_i\}$, $\{X_i^*\}$ in (19.4.39) are i.i.d. RVs in $(U, \|\cdot\|) = (\ell_p, \|\cdot\|_p)$, where for $x = \{x^{(j)}\} \in \ell_p$, $\|x\|_p := \left(\sum_{j=1}^{\infty} |x^{(j)}|^p\right)^{1/p}$ [Remark 19.4.1 (d)]. For $x, y \in \ell_p$ we define $x \vee y = \{x^{(j)} \vee y^{(j)}\}$.

Theorem 19.4.5. *Let $0 \leq \alpha < p \leq 2$, $1 \leq p \leq 2$, and $E(X_1 - X_1^*) = 0$; then for Z_n and Z_n^* given by* (19.4.39)

$$\widehat{\mathcal{L}}_p(Z_n, Z_n^*) \leq (p/(p-1)) B_p n^{1/p - 1/\alpha} \widehat{\mathcal{L}}_p(X_1, X_1^*), \tag{19.4.45}$$

where the constant B_p is the same as in the Marcinkiewicz–Zygmund inequality (19.4.25)*. In the Hilbert space $(\ell_2, \|\cdot\|_2)$ the following relation holds:*

$$\widehat{\mathcal{L}}_2(Z_n, Z_n^*) \leq \sqrt{2} n^{1/2 - 1/\alpha} \widehat{\mathcal{L}}_2(X_1, X_1^*). \tag{19.4.46}$$

In particular, for the Prokhorov metric π we have

$$\pi(Z_n, Z_n^*) \leq (p/(p-1))^{p/(p+1)} B_p^{p/(p+1)} n^{(1-p/\alpha)/(p+1)} \widehat{\mathcal{L}}_p^{p/(p+1)}(X_1, X_1^*). \tag{19.4.47}$$

Proof. Let $(\widetilde{X}_i, \widetilde{X}_i^*)$ be independent pairs of random variables in $\mathfrak{X}(\ell_p)$. Then for $\widetilde{S}_k = \sum_{i=1}^k \widetilde{X}_i$, $\widetilde{S}_k^* = \sum_{i=1}^k \widetilde{X}_i^*$ we have

$$\mathcal{L}_p^p\left(n^{-1/\alpha} \bigvee_{k=1}^n \widetilde{S}_k, n^{-1/\alpha} \bigvee_{i=1}^n \widetilde{S}_k^*\right) = n^{-p/\alpha} \mathcal{L}_p^p\left(\bigvee_{k=1}^n \widetilde{S}_k, \bigvee_{k=1}^n \widetilde{S}_k^*\right)$$

$$= n^{-p/\alpha} E \left[\sum_{j=1}^{\infty} \left| \bigvee_{k=1}^{n} \widetilde{S}_k^{(j)} - \bigvee_{i=1}^{n} \widetilde{S}_k^{*(j)} \right|^p \right]$$

$$\leq n^{-p/\alpha} E \sum_{j=1}^{\infty} \bigvee_{k=1}^{n} |\widetilde{S}_k^{(j)} - \widetilde{S}_k^{*(j)}|^p$$

$$= n^{-p/\alpha} \sum_{j=1}^{\infty} E \bigvee_{k=1}^{n} |\widetilde{S}_k^{(j)} - \widetilde{S}_k^{*(j)}|^p$$

$$\leq n^{-p/\alpha} \sum_{j=1}^{\infty} (p/(p-1))^p E|\widetilde{S}_n^{(j)} - \widetilde{S}_n^{*(j)}|^p.$$

The last inequality follows from Doob's inequality.[49] Therefore, we can continue applying the Marcinkiewicz–Zygmund inequality (19.4.25) with

$$\leq n^{-p/\alpha} \sum_{j=1}^{\infty} (p/(p-1))^p B_p^p E \left[\sum_{i=1}^{n} (\widetilde{X}_i^{(j)} - \widetilde{X}_i^{*(j)})^2 \right]^{p/2}$$

$$\leq (p/(p-1))^p B_p^p n^{-p/\alpha} \sum_{j=1}^{\infty} \sum_{i=1}^{n} E|\widetilde{X}_i^{(j)} - \widetilde{X}_i^{*(j)}|^p$$

$$= (p/(p-1))^p B_p^p n^{1-p/\alpha} \mathcal{L}_p^p(\widetilde{X}_1, \widetilde{X}_1^*); \tag{19.4.48}$$

the last inequality follows from the assumption that $p/2 \leq 1$. Passing to the minimal metrics we obtain (19.4.45) and (19.4.46). Finally, by means of $\pi^{p+1} \leq \widehat{\mathcal{L}}_p^p$, we obtain (19.4.47). □

The same proof also applies to the Banach space $\ell_{p,\mu}$ [(19.4.24) and Theorem 19.4.3].

Theorem 19.4.6. *If* $0 \leq \alpha < p \leq 2$, $1 \leq p \leq 2$, *and* $X_1, \ldots, X_n \in \mathfrak{X}(\ell_{p,\mu})$ *are i.i.d. and* $X_1^*, \ldots, X_n^* \in \mathfrak{X}(\ell_{p,\mu})$ *are i.i.d. such that* $E(X_1 - X_1^*) = 0$, *then*

$$\widehat{\mathcal{L}}_p \left(n^{-1/\alpha} \bigvee_{k=1}^{n} S_k, n^{-1/\alpha} \bigvee_{k=1}^{n} S_k^* \right) \leq (p/(p-1)) B_p n^{1/p-1/\alpha} \widehat{\mathcal{L}}_p(X_1, X_1^*) \tag{19.4.49}$$

and

$$\pi(Z_1, Z_n^*) \leq (p(p-1))^{p/(1+p)} B_p^{p/(1+p)} n^{(1-p/\alpha)/(p+1)} \widehat{\mathcal{L}}_p^{p/(p+1)}(X_1, X_1^*). \tag{19.4.50}$$

[49]See Chow and Teicher (1978, p. 247).

For an application of Theorem 19.4.6 to the problem of stability for queueing models, refer to Sect. 13.3 of Chap. 13.

References

Balkema AA, de Haan L (1990) A convergence rate in extreme-value theory. J. Applied Probability 27, 577–585

Bentkus VY, Rackauskas A (1985) Estimates of the distance between sums of independent random elements in banach spaces. Theor Prob Appl 29:50–65

Chow YS, Teicher H (1978) Probability theory: independence, interchangeability, martingales. Springer, New York

Cohen JP (1982) Convergence rate for the ultimate and penultimate approximation in extreme-value theory. Adv Appl Prob 14:833–854

de Haan L (1984) A spectral representation for max-stable processes. Ann Prob 12:1197–1204

Gine E, Leon JR (1980) On the central limit theorem in Hilbert space. Stochastica 4:43–71

Hoffman-Jorgensen J (1977) Probability in banach spaces. Ecole d 'Ete de Probabilités de Saint-Flow VI. In: Lecture Notes in Mathematics, vol 598, pp 1–186. Springer, Berlin

Hoffman-Jorgensen J, Pisier G (1976) The law of large numbers and the central limit theorem in Banach spaces. Ann Prob 4:587–599

Kakosyan AB, Klebanov L, Rachev ST (1988) Quantitative criteria for convergence of probability measures. Ayastan Press, Erevan (in Russian)

Maejima M, Rachev S (1997) Rates-of-convergence in the multivariate max-stable limit theorem. Stat Probab Lett 32:115–123

Omey E, Rachev S (1991) Rates of convergence in multivariate extreme value theory. J Multivar Anal 37:36–50

Pisier G, Zinn J (1977) On the limit theorems for random variables with values in the spaces L^p. Z Wahrsch Verw Geb 41:289–305

Rachev S, Rüschendorf L (1992) Rate of convergence for sums and maxima and doubly idcal metrics. Theor Probab Appl 37(2):276–289

Resnick SI (1987a) Extreme values: regular variation and point processes. Springer, New York

Resnick SI (1987b) Uniform rates of convergence to extreme-value distributions. In: Srivastava J (ed) Probability and statistics: essays in honor of Franklin A. Graybill. North-Holland, Amsterdam

Sazonov VV (1981) Normal approximation – some recent advances. In: Lecture Notes in Mathematics, vol 879. Springer, New York

Senatov VV (1981) Some lower estimates for the rate of convergence in the central limit theorem in Hilbert space. Sov Math Dokl 23:188–192

Shiryayev AN (1984) Probability. Springer, New York

Smith RL (1982) Uniform rates of convergence in extreme-value theory. Adv Appl Prob 14:600–622

Woyczynski WA (1980) On Marcinkiewicz-Zygmund laws of large numbers in Banach spaces and related rates of convergence. Prob Math Stat 1:117–131

Zolotarev VM (1976) Metric distances in spaces of random variables and their distributions. Math. USSR sb. 30:373–401

Zolotarev VM (1978) On pseudomoments. Theor Prob Appl 23:269–278

Zolotarev VM (1983) Probability metrics. Theory Prob Appl 28:278–302

Chapter 20
Ideal Metrics and Stability of Characterizations of Probability Distributions

The goals of this chapter are to:

- Describe the general problem of stability of probability distributions when a set of assumptions characterizing them has been perturbed,
- Characterize and study the stability of the class of exponential distributions through ideal probability metrics,
- Characterize the stability in de Finetti's theorem,
- Provide as an example a characterization of stability of environmental processes.

Notation introduced in this chapter:

Notation	Description
$B(\alpha, \beta)$	Beta distribution with parameters α and β
$\Gamma(p)$	Gamma function
$\Gamma(\alpha, \nu)$	Gamma density with parameters α and ν
$S_{p,n,s}$, $S_{p,n} := S_{p,n,n}$	p-spheres on $\mathbb{R}^n$
$\\|P - Q\\|$	**Var**(P, Q)
$\ll$	Absolute continuity

20.1 Introduction

No probability distribution is a true representation of the probabilistic law of a given random phenomenon: assumptions such as normality, exponentiality, and the like are seldom if ever satisfied in practice. This is not necessarily a cause for concern because many stochastic models characterizing certain probability distributions are relatively insensitive to "small" violations of the assumptions. On the other hand, there are models where even a slight perturbation of the assumptions that determine the choice of a distribution will cause a substantial change in the properties of

S.T. Rachev et al., *The Methods of Distances in the Theory of Probability and Statistics*,
DOI 10.1007/978-1-4614-4869-3_20,

the model. It is therefore of interest to investigate the invariance or *stability* of the set of assumptions characterizing certain distributions by examining the effects of perturbations of the assumptions.

There are several approaches to this problem. One is based on the concept of statistical robustness,[1] another makes use of information measures,[2] and a third one utilizes different measures of distance.[3] It is this third approach that we adopt in this book; it allows us to derive not merely qualitative results but also bounds on the distance between a particular attribute of the *ideal distribution*, the theoretical representation of the law of the physical random phenomenon under consideration, and a *perturbed* distribution obtained from the ideal distribution by an appropriate weakening of the assumptions.

This *stability analysis* is formalized as follows: given a specific ideal model, we denote by $\mathcal{U}$ the class of all possible *input* distributions and by $\mathcal{V}$ the class of all possible *output* distributions of interest. Let $\mathcal{F} : \mathcal{U} \to \mathcal{V}$ be a transformation that maps $\mathcal{U}$ on $\mathcal{V}$. For example, in the next section, $\mathcal{U}$ is the class of all distribution functions (DFs) F on $(0, \infty)$ satisfying the moment-normalizing condition $\int x^p \mathrm{d}F(x) = 1$ for some positive p. For a given $F \in \mathcal{U}$, the output $\mathcal{F}(F) \in \mathcal{V}$ is the set of distributions of random variables (RVs)

$$X_{k,n,p} := \sum_{j=1}^{k} \zeta_j^p \Big/ \sum_{j=1}^{n} \zeta_j^p, \quad 1 \le k \le n, \quad n \in \mathbb{N} := \{1, 2, \dots\},$$

where $\zeta_1, \zeta_2, \dots$ is a sequence of independent and identically distributed (i.i.d.) RVs with DF F.

The characterization problem we are interested in is as follows: *Does there exist a (unique) DF $F = F_p$ such that $X_{k,n,p}$ has a beta $B(k/p, (n-k)/p)$-distribution for any $k \le n$, $n \in \mathbb{N}$?*

It is well known that F_1 is the standard exponential distribution and F_2 is the absolute value of a standard normal RV.[4] Having a positive answer to the problem, our next task is to investigate the *stability of the characterization of the input distribution F_p*. The stability analysis may be described as follows: given $\varepsilon > 0$, we seek conditions under which there exist strictly increasing functions f_1 and f_2, both continuous strictly increasing and vanishing at the origin, such that the following two implications hold:

---

[1] See, for example, Hampel (1971), Huber (1977), Papantoni-Kazakos (1977), and Roussas (1972).

[2] See, for example, Akaike (1981), Csiszar (1967), Kullback (1959), Ljung (1978), and Wasserstein (1969).

[3] See, for example, Zolotarev (1977a,b, 1983), Kalashnikov and Rachev (1985, 1986a,b, 1988), Hernandez-Lerma and Marcus (1984), and Rachev (1989).

[4] See, for example, Cramer (1946, Sect. 18) and Diaconis and Freedman (1987).

(a) Given a simple probability metric μ_1 on $\mathfrak{X}(\mathbb{R})$, (i) $\mu_1(\widetilde{F}_p, F_p) = \mu_1(\widetilde{\zeta}_1, \zeta_1) < \varepsilon$ implies (ii) $\sup_{k,n} \mu_1(\widetilde{X}_{k,n,p}, X_{k,n,p}) < f_1(\varepsilon)$.

In (ii), $X_{k,n,p}$ is determined as previously where the ζ_i are F_p-distributed [and thus $X_{k,n,p}$ has a $B(k/p,(n-k)/p)$-distribution]. Further, in (ii) the RV $\widetilde{X}_{k,n,p} := \sum_{i=1}^{k} \widetilde{\zeta}_i^p / \sum_{i=1}^{n} \widetilde{\zeta}_i^p$ is determined by a "disturbed" sequence $\widetilde{\zeta}_1, \widetilde{\zeta}_2, \ldots$ of i.i.d. nonnegative RVs with common DF $\widetilde{F}_p$ close to F_p in the sense that (i) holds for some "small" $\varepsilon > 0$.

Along with (a), we will prove the continuity of the inverse mapping $\mathcal{F}^{-1}$:

(b) Given a simple p. metric μ_2 on $\mathfrak{X}(\mathbb{R})$, the following implication holds:

$$\sup_{k,n} \mu_2(\widetilde{X}_{k,n,p}, X_{k,n,p}) < \varepsilon \quad \Rightarrow \quad \mu_2(\widetilde{F}_p, F_p) < f_2(\varepsilon).$$

If a small value of $\varepsilon > 0$ yields a small value of $f_2(\varepsilon) > 0$, $i = 1, 2, \ldots$, then the *characterization of the input distribution* $U \in \mathcal{U}$ (in our case $U = F_p$) can be regarded as being relatively insensitive to small perturbations of the assumptions, or *stable*. In practice, the principal difficulty in performing such a stability analysis is in determining the appropriate metrics μ_i such that (a) and (b) hold. The procedure we use is first to determine the *ideal metrics* μ_1 and μ_2. These are the metrics most appropriate for the characterization problem under consideration. What is meant by *most appropriate* will vary from characterization to characterization, but ideal metrics have so far been identified for a large class of problems (Chaps. 15–18). The detailed discussion of the preceding problem of stability will be given in Sects. 20.2 and 20.3.

In Sect. 20.4, we will consider the *stability of the input distributions*. Here the characterization problem arises from the soil erosion model developed by Todorovic and Gani (1987) and its generalization.[5] The outline of the generalized erosion model is as follows. Let $Y, Y_1, Y_2, \ldots$ be an i.i.d. sequence of random variables; Y_i represents the yield of a given crop in the ith year. Let $Z, Z_1, Z_2, \ldots$ be, independent of Y, a sequence of i.i.d. RVs; Z_i represents the proportion of crop yield maintained in the year i, $Z_1 < 1$ corresponds to a "bad" year due to erosion, and $Z_i > 1$ corresponds to a "good" year in which rain comes at the right time. Further, let τ be a geometric RV independent of Y and Z representing a disastrous event such as a drought. The total crop yield until the disastrous year is

$$G = \sum_{k=1}^{\tau} Y_k \prod_{i=1}^{k} Z_i.$$

Now, the input distributions are $U := (F_Y, F_Z)$ and the output distribution $V = F(U)$ is the law of G. In general, the description of the class of compound

[5] See Rachev and Todorovic (1990) and Rachev and Samorodnitsky (1990).

distributions $V(x) = \Pr(G \leq x)$ is a complicated problem.[6] Consider the simple example of V being $E(\lambda)$, i.e., exponential with parameter $\lambda > 0$.[7] Here, the *input* $U = (F_Y, F_Z)$ consists of a constant $Z = z \in (0,1)$ and the mixture $F_Y(x) = F_{\overline{Y}}(x) := zE(\lambda/p) + (1-z)(E(\lambda/p) * E(\lambda z))$, where $p := (1 + E\tau)^{-1}$ and $*$ stands for the convolution operator. Again, we can pose the problem of stability of the exponential distribution $E(\lambda)$ as an output of the characterization problem

$$U = (F_{\overline{Y}}, F_{\overline{Z}}) \xrightarrow{\mathcal{F}} V = E(\lambda).$$

As in the previous example, the problem is to choose an *ideal metric* providing the implication

$$\left.\begin{array}{l} \nu(Y^*, \overline{Y}) \leq \varepsilon \\ \nu(Z^*, z) \leq \delta \end{array}\right\} \Rightarrow \nu(F_{V^*}, E(\lambda)) \leq \phi(\varepsilon, \delta),$$

where $V^* = \mathcal{F}(F_{Y^*}, F_{Z^*})$ and ϕ is a continuous strictly increasing function in both arguments on $\mathbb{R}^2_+$ and vanishing at the origin.

20.2 Characterization of an Exponential Class of Distributions $\{F_p,\ 0 < p \leq \infty\}$ and Its Stability

Let $\zeta_1, \zeta_2, \ldots$ be a sequence of i.i.d. RVs with DF F satisfying the normalization $E\zeta_1^p = 1, 0 < p < \infty$, and define

$$X_{k,n,p} := \sum_{j=1}^{k} \zeta_j^p \Big/ \sum_{j=1}^{n} \zeta_j^p, \quad 1 \leq k \leq n, \quad n \in \mathbb{N} := \{1, 2, \ldots, \}. \tag{20.2.1}$$

Theorem 20.2.1. *For any $0 < p < \infty$ there exists exactly one distribution $F = F_p$ such that for all $k \leq n$, $n \in \mathbb{N}$, $X_{k,n,p}$ has a beta distribution $B(k/p, (n-k)/p)$*[8]. *F_p has the density*

$$f_p(x) = \frac{p^{1-1/p}}{\Gamma(1/p)} \exp\left(-\frac{x^p}{p}\right), \quad x \geq 0. \tag{20.2.2}$$

Proof. Let the RVs $\{\zeta_i\}_{i \in \mathbb{N}}$ have a common density f_p. Then

[6] See the problem of stability in risk theory in Sect. 17.2 of Chap. 17.

[7] For the general case, see Sect. 20.4.

[8] The density of a beta distribution with parameters α and β is given by

$$\frac{\Gamma(\alpha+\beta)}{\Gamma(\alpha)\Gamma(\beta)} x^{\alpha-1}(1-x)^{\beta-1},\ 0 < x < 1.$$

$$f_{\zeta_1^p}(x) = \frac{1}{p^{1/p}\Gamma(1/p)} x^{-1+1/p} \exp(-x/p), \quad x \geq 0,$$

is the $\Gamma(1/p, 1/p)$-density. Recall that $\Gamma(\alpha, \nu)$-density is given by

$$\frac{1}{\Gamma(\nu)} \alpha^{\nu} x^{\nu-1} \exp(-\alpha x), \quad x \geq 0, \quad \nu > 0, \quad \alpha > 0.$$

The family of gamma densities is closed under convolutions, $\Gamma(\alpha, \mu) * \Gamma(\alpha, \nu) = \Gamma(\alpha, \mu + \nu)$,[9] and hence $\sum_{i=1}^{k} \zeta_i^p$ is $\Gamma(1/p, k/p)$-distributed.

The usual calculations show that

$$f_\kappa(x) = B\left(\frac{k}{p}, \frac{n-k}{p}\right) \frac{x^{-1+k/p}}{(x+1)^{n/p}}, \quad x > 0,$$

where

$$\kappa := \frac{\sum_{i=1}^{n} \zeta_i^p}{\sum_{i=k+1}^{n} \zeta_i^p}$$

and

$$B\left(\frac{k}{p}, \frac{n-k}{p}\right) := \frac{\Gamma(n/p)}{\Gamma(k/p)\Gamma((n-k)/p)}.$$

This leads to the $B(k/p, (n-k)/p)$-distribution of $X_{k,n,p}$.[10]

On the other hand, assuming that $X_{1,n,p}$ has a $B(1/p, (n-1)/p)$-distribution for all $n \in \mathbb{N}$, by the strong law of large numbers (SLLN), $nX_{1,n,p} \to \zeta_1^p$ a.s. Further, the density of $(nX_{1,n,p})^{1/p}$, given by

$$\frac{\Gamma\left(\frac{n}{p}\right)}{\Gamma\left(\frac{1}{p}\right)\Gamma\left(\frac{n-1}{p}\right)} \times \left(\frac{x^p}{n}\right)^{-1+1/p} \left(1 - \frac{x^p}{n}\right)^{(n-1)/p-1} \frac{p}{n} x^{p-1},$$

converges pointwise to $f_p(x)$ since

$$\left(\frac{p}{n}\right)^{1/p} \frac{\Gamma\left(\frac{n}{p}\right)}{\Gamma\left(\frac{n-1}{p}\right)} \to 1$$

as $n \to \infty$.[11] Thus, $f_{\zeta_1} = f_p$, as required. □

---

[9]See, for example, Feller (1971).

[10]See Cramer (1946, Sect. 18) for the case $p = 2$.

[11]See, for example, Abramowitz and Stegun (1970, p. 257).

Remark 20.2.1. One simple extension is to the case where $\widetilde{\zeta}_1, \widetilde{\zeta}_2, \ldots$ are i.i.d. RVs on the whole real line satisfying the conditions $E\widetilde{\zeta}_1 = 0$, $E|\widetilde{\zeta}_1|^p = 1$. Then $\overline{X}_{k,n,p} = \sum_{j=1}^{k} |\widetilde{\zeta}_j|^p \Big/ \sum_{j=1}^{n} |\widetilde{\zeta}_j|^p$ is $B(k/p,(n-k)/p)$-distributed if and only if the density $\widetilde{f}_p$ of ζ_1 satisfies $\widetilde{f}_p(x) + \widetilde{f}_p(-x) = f_p(|x|)$. In this way, for $p = 2$ one gets the normal distribution[12] and for $p = 1$ the Laplace distribution. Uniqueness can be obtained by the additional assumption of symmetry of F.

Remark 20.2.2. To obtain a meaningful result for $p = \infty$, we must normalize $X_{k,n,p}$ in (20.2.1) by looking at the limit distribution of

$$X_{k,n,p}^{1/p} = \frac{\left(\sum\limits_{j=1}^{k} \zeta_j^p\right)^{1/p}}{\left(\sum\limits_{j=1}^{n} \zeta_j^p\right)^{1/p}}$$

as $p \to \infty$.

Let β be a $B(k/p,(n-k)/p)$-distributed RV, and define $\gamma_{k,n,p} = \beta^{1/p}$; then $\gamma_{k,n,p}$ has a density given by

$$f_{\gamma_{k,n,p}}(x) = B\left(\frac{k}{p}, \frac{n-k}{p}\right) p x^{k-1}(1-x^p)^{(n-1)/p}, \quad 0 \le x \le 1.$$

By Theorem 20.2.1, ζ_j are F_p-distributed if and only if $X_{k,n,p}^{1/p} \stackrel{\mathrm{d}}{=} \gamma_{k,n,p}$.

Let $\gamma_{k,n,\infty}$ be the weak limit of $\gamma_{k,n,p}$ as $p \to \infty$, i.e.,

$$\Pr(\gamma_{k,n,\infty} \le x) = \begin{cases} \dfrac{n-k}{n} x^k, & \text{if } 0 \le x < 1 \\ 1, & \text{if } x \ge 1. \end{cases} \tag{20.2.3}$$

Thus the preceding DF plays the role of a normalized $B(k/p,(n-k)/p)$-distribution as $p \to \infty$. Clearly, $X_{k,n,p}^{1/p}$ converges to

$$X_{k,n,\infty} := \bigvee_{i=1}^{k} \zeta_i \Big/ \bigvee_{i=1}^{n} \zeta_i, \quad \left(\bigvee \zeta_i := \max \zeta_i\right), \tag{20.2.4}$$

as $p \to \infty$. Now, similarly to the case where $p \in (0,\infty)$, we pose the following question: does there exist a (unique) DF F_∞ of ζ_1 such that $X_{k,n,\infty} \stackrel{\mathrm{d}}{=} \gamma_{k,n,\infty}$ for any $k \le n$, $n \in \mathbb{N}$?

---

[12] See Cramer (1946, Sect. 18).

Theorem 20.2.2. *Let $\zeta_1, \zeta_2, \ldots$ be a sequences of positive i.i.d. RVs, and let F_∞ stand for a uniform distribution on $[0, 1]$. Then $X_{k,n,\infty}$ and $\gamma_{k,n,\infty}$ are equally distributed for any $k \le n$, $n \in \mathbb{N}$, if and only if ζ_1 is E_∞-distributed.*

Proof. Assuming that ζ_1 is F_∞-distributed, the DF of $X_{k,n,\infty}$ has the form $\Pr(X \le x(X \vee Y))$, where X and Y are independent with DFs $F_X(t) = t^k$ and $F_Y(t) = t^{n-k}$, $0 \le t \le 1$. Therefore, for $0 \le x \le 1$

$$\begin{aligned} F_{X_{k,n,\infty}}(x) &= \int_0^x \Pr(t \le x(t \vee Y))\mathrm{d}t^k \\ &= \int_0^x \Pr(t \le xY, Y > t)\mathrm{d}t^k + \int_0^x \Pr(t \le xt, T \le t)\mathrm{d}t^k \\ &=: I_1(x) + I_2(x). \end{aligned}$$

Now $I_1(x) = [(n-k)/n]x^k$ for $x \in [0, 1]$ and $I_2(x) = 0$ for $0 < x < 1$, $I_2(1) = k/n$. This implies that $X_{k,n,\infty}$ has a distribution given by (20.2.3).

On the other hand, if $X_{1,n,\infty} := \zeta_1 / \bigvee_{i=1}^n \zeta_i$ has the same distribution as $\gamma_{1,n,\infty}$, then if we let $n \to \infty$, the distribution of $\bigvee_{i=1}^n \zeta_i$ converges weakly to 1, and therefore the limit of $F_{X_{1,n,\infty}} = F_{\gamma_{1,n,\infty}}$ is $F_{\zeta_1} = F_\infty$. □

Theorems 20.2.1 and 20.2.2 show that the basic probability distributions – exponential, normal, and uniform – correspond respectively to F_1, F_2, and F_∞ in our characterization problem. Next, we will examine the stability of the exponential class F_p, $0 < p \le \infty$.

We now consider a *disturbed* sequence $\widetilde{\zeta}_1, \widetilde{\zeta}_2, \ldots$ of i.i.d. nonnegative RVs with common DF $\widetilde{F}_p$ close to F_p in the sense that the uniform metric

$$\rho := \rho(\widetilde{\zeta}_1, \zeta_1) = \rho(\widetilde{F}_p, F_p) \tag{20.2.5}$$

is close to zero.[13] The next theorem says that the distribution of $\widetilde{X}_{k,n,p} = \sum_{i=1}^k \widetilde{\zeta}_i^p / \sum_{i=1}^n \widetilde{\zeta}_i^p$ is close to the beta $B(k/p, (n-k)/p)$-distribution w.r.t. the uniform metric. In what follows, c denotes absolute constants that may be different in different places and $c(\ldots)$ denotes quantities depending only on the arguments in parentheses.

Remark 20.2.3. In view of the comments at the beginning of the section, the choice of the metric ρ as a *suitable* metric for the problem of stability is dictated by the following observation. In the stability analysis of the characterization of the *input* distribution F_p, we require the existence of simple metrics μ_1 and μ_2 such that[14]

$$\mu_1(\widetilde{F}_p, F_p) \le \varepsilon \quad \Rightarrow \quad \sup_{k,n} \mu_1(\widetilde{X}_{k,n,p}, X_{k,n,p}) \le f_1(\varepsilon) \tag{20.2.6}$$

[13] Here, as before, $\rho(X, Y) := \sup_x |F_X(x) - F_Y(x)|$.

[14] See (i) and (ii) in implications (a) and (b) in Sect. 20.1 of this chapter.

and

$$\sup_{k,n} \mu_2(\widetilde{X}_{k,n,p}, X_{k,n,p}) \leq \varepsilon \quad \Rightarrow \quad \mu_2(\widetilde{F}_p, F_p) \leq f_2(\varepsilon). \tag{20.2.7}$$

Clearly, we would like to select metrics μ_1 and μ_2 in such a way that as $n \to \infty$,

$$\mu_1(X_n, Y_n) \to 0 \quad \Longleftrightarrow \quad \mu_2(X_n, Y_n) \to 0,$$

i.e., the μ_i generate the exact same uniformities[15] and, in particular, μ_i metrize the exact same topology in the space of laws. The *ideal* choice will be to find a metric such that both (20.2.6) and (20.2.7) are valid with $\mu = \mu_1 = \mu_2$. The next two theorems show that this choice is possible with $\mu = \boldsymbol{\rho}$.

Theorem 20.2.3. *For any* $0 < p < \infty$ *and* $\{\widetilde{\zeta}_i\}$ *i.i.d. with* $E\widetilde{\zeta}_1^p = 1$ *and* $\widetilde{m}_\delta := E\widetilde{\zeta}_1^{(2+\delta)p} < \infty$ $(\delta > 0)$ *we have*

$$\Delta := \sup_{k,n} \boldsymbol{\rho}(X_{k,n,p}, \widetilde{X}_{k,n,p}) \leq c(\delta, \widetilde{m}_\delta, p)\rho^{\delta/(3(2+\delta))}. \tag{20.2.8}$$

Proof. The proof follows the two-stage approach of the method of metric distances (Fig. 1.1 in Chap. 1).

(a) First stage: solution of problem in terms of *ideal* metric (Claim 2).
(b) Transition from the *ideal* metric to the *traditional* metric (Claims 1, 2, and 4).

We start with the first claim.

Claim 1. *The* traditional *metric* $\boldsymbol{\rho}$ *is a regular metric.*[16] In particular,

$$\boldsymbol{\rho}(X_{k,n,p}, \widetilde{X}_{k,n,p})$$
$$\leq \boldsymbol{\rho}\left(\sum_{i=1}^{k} \zeta_i^p, \sum_{i=1}^{k} \widetilde{\zeta}_i p\right) + \boldsymbol{\rho}\left(\sum_{i=k+1}^{n} \zeta_i^p, \sum_{k+1}^{n} \widetilde{\zeta}_i^p\right) \leq n\boldsymbol{\rho}(\zeta_1, \widetilde{\zeta}_1). \tag{20.2.9}$$

To prove (20.2.9), observe that

$$X_{k,n,p} = \frac{X_1}{X_1 + X_2}, \quad \widetilde{X}_{k,n,p} = \frac{\widetilde{X}_1}{\widetilde{X}_1 + \widetilde{X}_2},$$

where $X_1 = \sum_{i=1}^k \zeta_i^p$, $X_2 = \sum_{i=k+1}^n \zeta_i^p$, $\widetilde{X}_1 = \sum_{i=1}^k \widetilde{\zeta}_i^p$, and $\widetilde{X}_2 = \sum_{i=k+1}^n \widetilde{\zeta}_i^p$. Since $\phi(t) = t/(1+t)$ is strictly monotone and $X_{k,n,p} = \phi(X_1/X_2)$, we have that

$$\boldsymbol{\rho}(X_{k,n,p}, \widetilde{X}_{k,n,p}) = \boldsymbol{\rho}\left(\frac{X_1}{X_2}, \frac{\widetilde{X}_1}{\widetilde{X}_2}\right).$$

[15] See Dudley (2002, Sect. 11.7).

[16] See Definition 15.3.1(i) in Chap. 15.

Choosing $X_1^* \stackrel{d}{=} X_1$, X_1^* independent of $\widetilde{X}_2$, we obtain

$$\begin{aligned}\rho\left(\frac{X_1}{X_2}, \frac{\widetilde{X}_1}{\widetilde{X}_2}\right) &\leq \rho\left(\frac{X_1}{X_2}, \frac{X_1^*}{\widetilde{X}_2}\right) + \rho\left(\frac{X_1^*}{\widetilde{X}_2}, \frac{\widetilde{X}_1}{\widetilde{X}_2}\right) \\ &= \sup_{x\geq 0}\left|\int_0^\infty \left[\Pr\left(\frac{y}{X_2} \leq x\right) - \Pr\left(\frac{y}{\widetilde{X}_2} \leq x\right)\right] \mathrm{d}F_{X_1}(y)\right| \\ &\quad + \sup_{x\geq 0}\left|\int_0^\infty \left[\Pr\left(\frac{X_1}{y} \leq x\right) - \Pr\left(\frac{\widetilde{X}_1}{y} \leq x\right)\right] \mathrm{d}F_{\widetilde{X}_2}(y)\right| \\ &\leq \int_0^\infty \sup_{x\geq 0}\left|P\left(X_2 \geq \frac{y}{x}\right) - P\left(\widetilde{X}_2 \geq \frac{y}{x}\right)\right| \mathrm{d}F_{X_1}(y) \\ &\quad + \int_0^\infty \sup_{x\geq 0}|P(X_1 \leq xy) - P(\widetilde{X}_1 \leq xy)| \mathrm{d}F_{\widetilde{X}_2}(y) \\ &= \rho(X_1, \widetilde{X}_1) + \rho(X_2, \widetilde{X}_2).\end{aligned}$$

The second part of (20.2.9) follows from the regularity of $\boldsymbol{\rho}$, i.e.,

$$\boldsymbol{\rho}(X + Z, Y + Z) \leq \boldsymbol{\rho}(X, Y)$$

for Z independent of X, Y.

Claim 2. (*Bound from above of the* traditional *metric* $\boldsymbol{\rho}$ *by the* ideal *metric* $\boldsymbol{\zeta}_2$). Let $n > p$, $E\zeta_1^p = E\widetilde{\zeta}_1^p = 1$, $\sigma_p^2 := \mathbf{Var}(\zeta_1^p)$, $\widetilde{\sigma}_p^2 := \mathbf{Var}(\widetilde{\zeta}_1^p) < \infty$. Then

$$\boldsymbol{\rho}\left(\sum_{i=1}^n \zeta_i^p, \sum_{i=1}^n \widetilde{\zeta}_i^p\right) \leq 3\sigma_p^{2/3}\left(2\pi\left(1 - \frac{p}{n}\right)\right)^{-1/3} \boldsymbol{\zeta}_2^{1/3}\left(\frac{1}{\sqrt{n}}\sum_{i=1}^n Z_i, \frac{1}{\sqrt{n}}\sum_{i=1}^n \widetilde{Z}_i\right), \tag{20.2.10}$$

where

$$Z_i := \frac{\zeta_i^p - 1}{\sigma^p}, \quad \widetilde{Z}_i := \frac{\widetilde{\zeta}_i^p - 1}{\sigma_p},$$

and

$$\boldsymbol{\zeta}_2(X, Y) := \int_{-\infty}^\infty \left|\int_{-\infty}^x (F_X(t) - F_Y(t))\mathrm{d}t\right| \mathrm{d}x$$

is the Zolotarev $\boldsymbol{\zeta}_2$-metric.[17]

---

[17]See (15.2.1) and (15.2.2) in Chap. 15.

Proof. For any $n = 1, 2, \ldots$ the following relation holds:

$$\rho\left(\sum_{i=1}^{n} \zeta_i^p, \sum_{i=1}^{n} \widetilde{\zeta}_i^p\right) = \rho\left(\frac{1}{\sqrt{n}} \sum_{i=1}^{n} Z_i, \frac{1}{\sqrt{n}} \sum_{i=1}^{n} \widetilde{Z}_i\right).$$

From (15.2.16) we have

$$\rho(X, Y) \leq 3M^{2/3}(\zeta_2(X, Y))^{1/3}, \tag{20.2.11}$$

where $M = \sup_{x\in\mathbb{R}} f_X(x)$ and the density of X is assumed to exist. We have

$$f_{1/\sqrt{n}\sum_{i=1}^{n} Z_i}(x) = \sigma_p \sqrt{n} f_{\sum_{i=1}^{n} \zeta_i^p}(\sqrt{n}\sigma_p x + 1)$$

and

$$f'_{\sum_{i=1}^{n} \zeta_i^p}(x)$$
$$= \frac{1}{p^{n/p}\Gamma\left(\frac{n}{p}\right)} \left[\left(\frac{n}{p} - 1\right) x^{-2+n/p} \exp(-x/p) - \frac{1}{p} x^{-1+n/p} \exp(-x/p)\right] = 0$$

if and only if $(n/p) - 1 = (1/p)x$.

The sum $\sum_{i=1}^{n} \zeta_i^p$ is $\Gamma(1/p, n/p)$-distributed and, hence, for $n > p$ the following inequality holds:

$$f_{\sum_{i=1}^{n} \zeta_i^p}(x) \leq \frac{p^{(n-p)/p}(-1 + n/p)^{(n-p)/p} \exp(1 - n/p)}{p^{n/p}\left(\frac{n}{p} - 1\right)\left[\left(\frac{n}{p} - 1\right)^{n/p-3/2} \exp\left(-\frac{n}{p} + 1\right)(2\pi)^{1/2}\right]}$$

using $\Gamma(z) \geq z^{z-1/2}\mathrm{e}^{-z}(2\pi)^{1/2}$.

This implies that

$$\sigma_p \sqrt{n} f_{\sum_{i=1}^{n} \zeta_i^p}(x) \leq \sigma_p \frac{\sqrt{n}\, p^{n/p-1}}{p^{n/p}\left(\frac{n}{p} - 1\right)^{1/2} (2\pi)^{1/2}}$$
$$= \sigma_p \left(2\pi\left(1 - \frac{p}{n}\right)\right)^{-1/2}, \tag{20.2.12}$$

and thus (20.2.11) and (20.2.12) together imply (20.2.10).

Since the metric ζ_2 is an ideal metric of order 2, we obtain the following claim.[18]

Claim 3. (*Solution of estimation problem in terms of ideal metric* ζ_2).

$$\zeta_2\left(\frac{1}{\sqrt{n}}\sum_{i=1}^{n} Z_i, \frac{1}{\sqrt{n}}\sum_{i=1}^{n} \widetilde{Z}_i,\right) \leq \zeta_2(Z_1, \widetilde{Z}_1). \tag{20.2.13}$$

Claim 4. (*Bound from above of ideal metric* ζ_2 *by* traditional *metric* $\boldsymbol{\rho}$). If $m_\delta < \infty$, then

$$\zeta_2(Z_1, \widetilde{Z}_1) \leq c(\delta, \widetilde{m}_\delta, p)\rho^{\delta/(2+\delta)}. \tag{20.2.14}$$

Proof. For RVs X, Y with $E(X - Y) = 0$ the following inequality holds:

$$\begin{aligned}\zeta_2(X, Y) &\leq \int_{-\infty}^{\infty} |x||F_X(x) - F_Y(x)|\mathrm{d}x \\ &\leq N^2\boldsymbol{\rho}(X, Y) + \frac{1}{2}EX^2 I\{|X| > N\} + \frac{1}{2}EY^2 I\{|Y| > N\} \\ &\leq N^2\boldsymbol{\rho}(X, Y) + \frac{1}{2}N^{-\delta}(E|X|^2 + \delta + E|Y|^2 + \delta).\end{aligned}$$

Minimizing the right-hand side over $N > 0$, we get (20.2.14).

Combining Claims 2–4 we get $\boldsymbol{\rho}\left(\sum_{i=1}^{n}\zeta_i^p, \sum_{i=1}^{n}\widetilde{\zeta}_i^p\right) \leq c(\delta, \widetilde{m}_\delta, p)^{\delta/3(2+\delta)}$ if $p/n<1$. From Claim 1 we then obtain

$$\boldsymbol{\rho}(X_{k,n,p}, \widetilde{X}_{k,n,p}) \leq \begin{cases} 2p\rho & \text{if } p \geq \frac{n}{2}, \\ p\rho + c\rho^{\delta/3(2+\delta)} & \text{if } p \geq k,\ p < \frac{n}{2}, \\ c\rho^{\delta/3(2+\delta)} & \text{if } p < k, \end{cases} \tag{20.2.15}$$

which proves (20.2.8). □

Remark 20.2.4. Claim 1 of the proof of Theorem 20.2.3 also remains true for the total variation metric $\boldsymbol{\sigma}$.[19] But $\boldsymbol{\rho}$ seems to be the appropriate metric for this problem since $\boldsymbol{\rho}$ is related to the ideal metric ζ_2 of order 2 [see (20.2.11)], while the total variation metric is too "strong" to be estimated from above by ζ_2 or any other ideal metric of order 2.

Open Problem 20.2.1. (*Topological structure of metric space* $(\mathcal{F}(\mathbb{R}), \mu)$ *of DFs where* μ *is an ideal metric of order* $r > 1$). Consider the space $\mathfrak{X}_r(X_0)$, $r > 1$, of all RVs X such that $EX^j = EX_0^j$, $j = 0, 1, \ldots, [r]$, and $E|X|^r < \infty$. Let μ be an

---

[18]See (15.2.18) in Chap. 15.

[19]See (3.3.13) in Chap. 3.

ideal metric of order $r > 1$ in $\mathfrak{X}(X_0)$, i.e., μ is a simple metric, and for any X, Y, and $Z \in \mathfrak{X}_r(X_0)$ (Z is independent of X and Y) and any $c \in \mathbb{R}$[20]

$$\mu(cX + Z, cY + Z) \leq |c|^r \mu(X, Y).$$

What is the topological structure of the space of laws of $X \in \mathfrak{X}_r(X_0)$ endowed with the metric μ?

Theorem 20.2.3 implies the following result on qualitative stability:[21]

$$\zeta_1 \xrightarrow{\text{w}} \zeta_1,\ m_\delta < \infty \quad \Rightarrow \quad \widetilde{X}_{k,n} \xrightarrow{\text{w}} X_{k,n}.$$

For the stability in the opposite direction we prove the following result.[22]

Theorem 20.2.4. *For any $0 < p < \infty$ and any i.i.d. sequences $\{\zeta_i\}$, $\{\widetilde{\zeta}_i\}$ with $E\zeta_1^p = E\widetilde{\zeta}_1^p = 1$ and ζ_1, $\widetilde{\zeta}_1$ having continuous distribution functions, the following relation holds:*

$$\boldsymbol{\rho}(\zeta_1, \widetilde{\zeta}_1) \leq \sup_{k,n} \boldsymbol{\rho}(X_{k,n,p}, \widetilde{X}_{k,n,p}). \tag{20.2.16}$$

Proof. Denote $X_i = \zeta_i^p$ and $\widetilde{X}_i = \widetilde{X}_i^p$. Then

$$\sup_{k,n} \boldsymbol{\rho}\left(\sum_{i=1}^k \zeta_i^p \Big/ \sum_{i=1}^n \zeta_i^p, \sum_{i=1}^k \widetilde{\zeta}_i^p \Big/ \sum_{i=1}^n \widetilde{\zeta}_i^p\right)$$

$$\geq \sup_n \boldsymbol{\rho}\left(\frac{X_1}{\sum_{i=1}^n X_i}, \frac{\widetilde{X}_1}{\sum_{i=1}^n \widetilde{X}_i}\right) = \sup_n \boldsymbol{\rho}\left(\frac{X_1}{\frac{1}{n}\sum_{i=2}^{n+1} X_i}, \frac{\widetilde{X}_1}{\frac{1}{n}\sum_{i=2}^{n+1} \widetilde{X}_i}\right)$$

$$\geq \boldsymbol{\rho}(X_1, \widetilde{X}_1) - \overline{\lim_{n\to\infty}}\, \boldsymbol{\rho}\left(\frac{X_1}{\frac{1}{n}\sum_{i=2}^{n+1} X_i}, X_1\right) - \overline{\lim_{n\to\infty}}\, \boldsymbol{\rho}\left(\frac{\widetilde{X}_1}{\frac{1}{n}\sum_{i=2}^{n+1} \widetilde{X}_i}, \widetilde{X}_1\right).$$

By the strong law of large numbers and the assumption $EX_1 = E\widetilde{X}_1 = 1$,

$$\frac{X_1}{\frac{1}{n}\sum_{i=2}^{n+1} X_i} \to X_1 \text{ a.s.} \quad \text{and} \quad \frac{\widetilde{X}_1}{\frac{1}{n}\sum_{i=2}^{n+1} \widetilde{X}_i} \to \widetilde{X}_1 \text{ a.s.} \tag{20.2.17}$$

[20]See Remark 19.4.6 in Chap. 19.

[21]See (20.2.6) with $\mu_1 = \boldsymbol{\rho}$.

[22]See (20.2.7) with $\mu_2 = \boldsymbol{\rho}$.

Since X_1 and $\widetilde{X}_1$ have continuous DFs, the convergence in (20.2.17) is valid w.r.t. the uniform metric $\boldsymbol{\rho}$. Hence $\sup_{k,n} \boldsymbol{\rho}(X_{k,n,p}, \widetilde{X}_{k,n,p}) \geq \boldsymbol{\rho}(X_1, \widetilde{X}_1) = \boldsymbol{\rho}(\zeta_1, \widetilde{\zeta}_1)$, as required. □

Next we would like to prove similar results for the case $p = \infty$ and $\widetilde{X}_{k,n,\infty} = \bigvee_{i=1}^{k} \widetilde{\zeta}_i \big/ \bigvee_{i=1}^{n} \widetilde{\zeta}_i$. In this case, the structure of the ideal metric is totally different. Instead of $\boldsymbol{\zeta}_2$, which is an ideal metric for the summation scheme, we will explore the weighted Kolmogorov metrics $\boldsymbol{\rho}_r$, $r > 0$,[23] which are ideal for the maxima scheme.

We will use the following condition.

Condition 1. There exists a nondecreasing continuous function $\phi(t) = \phi_{\widetilde{\zeta}_1}(t) : [0, 1] \to [0, \infty)$, $\phi(0) = 0$ and such that

$$\phi(t) \geq \sup_{1-t \leq x \leq 1} (-\log x)^{-1} |F_{\widetilde{\zeta}_1}(x) - x|.$$

Obviously Condition 1 is satisfied for $\widetilde{\zeta}_1 \stackrel{d}{=} \zeta_1$, uniformly distributed on $[0, 1]$. Let $\psi(t) = -\log(1-t)\phi(t)$, and let ψ^{-1} be the inverse of ψ.

Theorem 20.2.5. *(i) If Condition 1 holds and if $F_{\widetilde{\zeta}_1}(1) = 1$, then*

$$\Delta := \sup_{k,n} \boldsymbol{\rho}(X_{k,n,\infty}, \widetilde{X}_{k,n,\infty}) \leq c(\phi \circ \psi^{-1}(\rho))^{1/2} \quad \textit{where} \quad \rho := \boldsymbol{\rho}(\zeta_1, \widetilde{\zeta}_1).$$

(ii) If $\widetilde{\zeta}_1$ has a continuous DF, then $\Delta \geq \rho$.

Proof. (i) **Claim 1.** For any $1 \leq k \leq n$ the following inequality holds:

$$\boldsymbol{\rho}(X_{k,n,\infty}, \widetilde{X}_{k,n,\infty}) \leq \boldsymbol{\rho}\left(\bigvee_{i=1}^{k} \zeta_i, \bigvee_{i=1}^{k} \widetilde{\zeta}_i\right) + \boldsymbol{\rho}\left(\bigvee_{i=k+1}^{n} \zeta_i, \bigvee_{i=k+1}^{n} \widetilde{\zeta}_i\right). \tag{20.2.18}$$

Proof. We use the representation

$$X_{k,n,\infty} = \frac{X_1}{X_1 \vee X_2}, \quad \widetilde{X}_{k,n,\infty} = \frac{\widetilde{X}_1}{\widetilde{X}_1 \vee \widetilde{X}_2},$$

where

$$X_1 = \bigvee_{i=1}^{k} \zeta_i, \quad X_2 = \bigvee_{i=k+1}^{n} \zeta_i, \quad \widetilde{X}_1 = \bigvee_{i=1}^{k} \widetilde{\zeta}_i, \quad \widetilde{X}_2 = \bigvee_{i=k+1}^{n} \widetilde{\zeta}_i.$$

---

[23]See (19.2.4) in Chap. 19.

Following the proof of (20.2.9), since $\boldsymbol{\rho}$ is a simple metric, we may assume (X_1, X_2) is independent of $(\widetilde{X}_1, \widetilde{X}_2)$. Thus, by the regularity of the uniform metric and its invariance w.r.t. monotone transformations, we get

$$\begin{aligned}\boldsymbol{\rho}(X_{k,n,\infty}, \widetilde{X}_{k,n,\infty}) &= \boldsymbol{\rho}\left(\frac{X_1}{X_1 \vee X_2}, \frac{\widetilde{X}_1}{\widetilde{X}_1 \vee \widetilde{X}_2}\right) = \boldsymbol{\rho}\left(1 \vee \frac{X_2}{X_1}, 1 \vee \frac{\widetilde{X}_2}{\widetilde{X}_1}\right)\\ &\le \boldsymbol{\rho}\left(\frac{X_2}{X_1}, \frac{\widetilde{X}_2}{\widetilde{X}_1}\right) \le \boldsymbol{\rho}\left(\frac{X_2}{X_1}, \frac{\widetilde{X}_2}{X_1}\right) + \boldsymbol{\rho}\left(\frac{\widetilde{X}_2}{X_1}, \frac{\widetilde{X}_2}{\widetilde{X}_1}\right)\\ &\le \boldsymbol{\rho}(X_2, \widetilde{X}_2) + \boldsymbol{\rho}(X_1, \widetilde{X}_1),\end{aligned}$$

with the last inequality obtained by taking conditional expectations.

Claim 2. Let[24]

$$\rho_* = \boldsymbol{\rho}_*(\zeta_1, \widetilde{\zeta}_1) := \sup_{0 \le x \le 1} (-\log x)^{-1} |F_{\zeta_1}(x) - F_{\widetilde{\zeta}_1}(x)|.$$

Then

$$\boldsymbol{\rho}\left(\bigvee_{i=1}^{n} \zeta_i, \bigvee_{i=1}^{n} \widetilde{\zeta}_i\right) \le c\sqrt{\rho_*}. \tag{20.2.19}$$

Proof. Consider the transformation $f(t) = (-\log t)^{-1/\alpha}$ $(0 < t < 1)$. Then

$$\boldsymbol{\rho}\left(\bigvee_{i=1}^{n} \zeta_i, \bigvee_{i=1}^{n} \widetilde{\zeta}_i\right) = \boldsymbol{\rho}\left(f\left(\bigvee_{i=1}^{n} \zeta_i\right), f\left(\bigvee_{i=1}^{n} \widetilde{\zeta}_i\right)\right) = \boldsymbol{\rho}\left(\bigvee_{i=1}^{n} X_i, \bigvee_{i=1}^{n} \widetilde{X}_i\right), \tag{20.2.20}$$

where $X_i = f(\zeta_i)$, $\widetilde{X}_i = f(\widetilde{\zeta}_i)$. Since X_1 has extreme-value distribution with parameter α, so does $Z_n := n^{-1/\alpha} \bigvee_{i=1}^{n} X_i$. The density of Z_n is given by

$$F_{Z_n}(x) = \frac{\mathrm{d}}{\mathrm{d}x} \exp(-x^{-\alpha}) = \alpha x^{-\alpha-1} \exp(-x^{-\alpha}),$$

and thus

$$C_n := \sup_{x>0} f_{Z_n}(x) = \alpha \left(\frac{\alpha+1}{\alpha}\right)^{\alpha+1/\alpha} \exp\left(-\frac{\alpha+1}{\alpha}\right). \tag{20.2.21}$$

Let $\boldsymbol{\rho}_\alpha$ be the *weighted Kolmogorov metric*

$$\boldsymbol{\rho}_\alpha(X, Y) = \sup_{x>0} x^\alpha |F_X(x) - F_Y(x)| \tag{20.2.22}$$

[24]In fact, $\boldsymbol{\rho}_*$ plays the role of *ideal metric* for our problem.

(Lemma 19.2.2 in Chap. 19). Then by (19.3.72) and Lemma 19.3.4,

$$\boldsymbol{\rho}(X,Y) \le \Lambda_\alpha A^{\alpha/(1+\alpha)} \boldsymbol{\rho}_\alpha^{1/(1+\alpha)}(X,Y), \tag{20.2.23}$$

where $\Lambda_\alpha := (1+\alpha)\alpha^{-\alpha(1+\alpha)}$ and $A := \sup_{x>0} F_Y'(x)$ (the existence of density being assumed). Hence, by (20.2.20)–(20.2.23),

$$\boldsymbol{\rho}\left(\bigvee_{i=1}^{n}\zeta_i, \bigvee_{i=1}^{n}\widetilde{\zeta}_i\right) = \boldsymbol{\rho}(Z_n,\widetilde{Z}_n) \le \Lambda_\alpha C_n^{\alpha/(1+\alpha)} \boldsymbol{\rho}_\alpha^{1/(1+\alpha)}(Z_n,\widetilde{Z}_n), \tag{20.2.24}$$

where $\widetilde{Z}_n = n^{-1/\alpha}\bigvee_{i=1}^{n}\widetilde{X}_i$. The metric $\boldsymbol{\rho}_\alpha$ is an ideal metric of order α w.r.t. the maxima scheme for i.i.d. RVs (Lemma 19.2.2) and, in particular,

$$\boldsymbol{\rho}_\alpha(Z_n,\widetilde{Z}_n) \le \boldsymbol{\rho}(X_1,\widetilde{X}_1) = \boldsymbol{\rho}_*(\zeta_1,\widetilde{\zeta}_1). \tag{20.2.25}$$

From Condition 1 we now obtain the following claim.

Claim 3. $\rho_* \le \phi \circ \psi^{-1}(\rho)$.

Proof. For any $0 \le t \le 1$ the following relation holds:

$$\begin{aligned}\rho_* &= \max\left\{\sup_{0\le x\le 1-\varepsilon}(-\log x)^{-1}|F_{\zeta_1}(x)-x|,\ \sup_{1-\varepsilon\le x\le 1}(-\log x)^{-1}|F_{\widetilde{\zeta}_1}(x)-x|\right\}\\ &\le \max((-\log(1-\varepsilon))^{-1}\rho, \phi(\varepsilon)).\end{aligned} \tag{20.2.26}$$

Choosing ε by $\phi(\varepsilon) = (-\log(1-\varepsilon))^{-1}\rho$, i.e., $\rho = \psi(\varepsilon)$, one proves the claim.

From Claims 1–3 we obtain

$$\boldsymbol{\rho}(X_{k,n,p},\widetilde{X}_{k,n,p}) \le \min(n\rho, c(\phi\circ\psi^{-1}(\rho)^{1/2}), \tag{20.2.27}$$

which proves (i).

(ii) For the proof of (ii) observe that $F_{\bigvee_{i=1}^{n}\widetilde{\zeta}_i}(x) = F_{\widetilde{\zeta}_i}^n \to 1$ for any x with $F_{\widetilde{\zeta}_i}(x) > 0$. As in the proof of Theorem 20.2.4, we then obtain

$$\begin{aligned}\sup_{k,n}\boldsymbol{\rho}\left(\frac{\bigvee_{i=1}^{k}\zeta_i}{\bigvee_{i=1}^{n}\zeta_i}, \frac{\bigvee_{i=1}^{k}\widetilde{\zeta}_i}{\bigvee_{i=1}^{n}\widetilde{\zeta}_i}\right) &\ge \limsup_n \boldsymbol{\rho}\left(\frac{\zeta_1}{\bigvee_{i=1}^{n}\zeta_i}, \frac{\widetilde{\zeta}_1}{\bigvee_{i=1}^{n}\widetilde{\zeta}_i}\right)\\ &\ge \boldsymbol{\rho}(\zeta_1,\widetilde{\zeta}_1) - \overline{\lim_n}\,\boldsymbol{\rho}\left(\frac{\zeta_1}{\bigvee_{i=1}^{n}\zeta_i},\zeta_1\right) - \overline{\lim_n}\,\boldsymbol{\rho}\left(\frac{\widetilde{\zeta}_1}{\bigvee_{i=1}^{n}\widetilde{\zeta}_i},\widetilde{\zeta}_1\right)\\ &= \boldsymbol{\rho}(\zeta_1,\widetilde{\zeta}_1)\end{aligned}$$

since ζ_1 and $\widetilde{\zeta}_1$ have continuous DFs. □

Remark 20.2.5. In Theorem 20.2.5 (i) the constant c depends on $\alpha > 0$ [see (20.2.23)]. Thus, one can optimize c by choosing α appropriately in (20.2.22).

20.3 Stability in de Finetti's Theorem

In this section, we apply the characterization of distributions F_p $(0 < p \leq \infty)$[25] to show that the uniform distribution on the *positive p-sphere* $S_{p,n}$,

$$S_{p,n} := \left\{ x = (x_1, \ldots, x_n) \in \mathbb{R}_+^n : \sum_{i=1}^n x_i^p = n \right\},$$

$$S_{\infty,n} := \left\{ x \in \mathbb{R}_+^n : \bigvee_{i=1}^n x_i = n \right\}, \tag{20.3.1}$$

has approximately independent F_p-distributed components.[26] This will lead us to the stability of the following de Finetti-type theorem. Let $\zeta = (\zeta_1, \ldots, \zeta_n)$ be nonnegative RVs and $C_{n,p}$ the class of ζ-laws with the property that given $\sum_{i=1}^n \zeta_i^p = s$ (for $p = \infty$ given $\bigvee_{i=1}^n \zeta_i = s$), the conditional distribution of ζ is uniform on $S_{p,n}$. Then, the joint distribution of i.i.d. ζ_i with common F_p-distribution is in the class $C_{n,p}$. Moreover, if $P \in \mathcal{P}(\mathbb{R}_+^\infty)$ and for any $n \geq 1$ the projection $T_{1,2,\ldots,n}P$ on the first n-coordinates belongs to C_n, then p is a mixture of i.i.d F_p-distributed RVs (de Finetti's theorem).

The de Finetti theorem will follow from the following stability theorem: if n nonnegative RVs ζ_i are conditionally uniform on $S_{p,n}$ given $\sum_{i=1}^n \zeta_i^p = s$ (resp. $\bigvee_{i=1}^n \zeta_i = s$ for $p = \infty$), then the total variation metric $\boldsymbol{\sigma}$ between the law of $(\zeta_1, \ldots, \zeta_k)$ (k fixed, n large enough) and a mixture of i.i.d. F_p-distributed RV $(\zeta_1, \ldots, \zeta_k)$ is less than const $\times\, k/n$.

Remark 20.3.1. An excellent survey on de Finetti's theorem is given by Diaconis and Freedman (1987), where the cases $p = 1$ and $p = 2$ are considered in detail.

We start with another characterization of the exponential class of distributions F_p (Theorem 20.2.1). Let

$$S_{p,s,n} := \left\{ x \in \mathbb{R}_+^n : \sum_{i=1}^n x_i^p = s \right\}$$

denote the p-sphere of radius s in $\mathbb{R}_+^n$, $0 < p < \infty$. The next two lemmas are simple applications of the well-known formulae for conditional distributions.

[25] See (20.2.2) and Theorems 20.2.1 and 20.2.2.

[26] Rachev and Rüschendorf (1991) discuss the approximate independence of distributions on spheres and their stability properties.

Lemma 20.3.1. *Let $\zeta_1, \ldots, \zeta_n$ be i.i.d. RVs with common DF F_p, where $0 < p < \infty$. Then the conditional distribution of $(\zeta_1, \ldots, \zeta_n)$ given $\sum_{i=1}^n \zeta_i^p = s$, denoted by*

$$P_{s,p} := P_{(\zeta_1,\ldots,\zeta_n)|\sum_{i=1}^n \zeta_i^p = s},$$

is uniform on $S_{p,s,n}$.

Similarly, we examine the case $p = \infty$; let $\zeta_1, \ldots, \zeta_n$ be i.i.d. F_∞-distributed [recall that F_∞ is the $(0, 1)$-uniform distribution]. Denote the conditional distribution of $(\zeta_1, \ldots, \zeta_n)$ given $\bigvee_{i=1}^n \zeta_i = s$ by $P_{s,\infty} := \Pr_{(\zeta_1,\ldots,\zeta_n)|\bigvee_{i=1}^n \zeta_i = s}$.

Lemma 20.3.2. *$P_{s,\infty}$ is uniform on $S_{\infty,s,n} := \left\{x \in \mathbb{R}_+^n : \bigvee_{i=1}^n x_i = s\right\}$ for almost all $s \in [0, 1]$.*

Now, using the preceding lemma, we can prove a stability theorem related to de Finetti's theorem for $p = \infty$.

Let $P_\sigma^{n,\infty}$ for $\sigma > 0$ be the law of $(\sigma\zeta_1, \ldots, \sigma\zeta_n)$, and let $Q_{n,s,k}^{(\infty)}$ be the law of $(\eta_1, \ldots, \eta_k)$, where $\eta = (\eta_1, \ldots, \eta_n)$ $(n > k)$ is uniform on $S_{\infty,s,n}$. In the next theorem, we evaluate the deviation between $Q_{n,s,k}^{(\infty)}$ and $P_s^{k,\infty}$ in terms of the *total variation metric*

$$\sigma(Q_{n,s,k}^{(\infty)}, P_s^{k,\infty}) := \sup_{A \in \mathcal{B}^k} |Q_{n,s,k}^{(\infty)}(A) - P_s^{k,\infty}(A)|,$$

where $\mathcal{B}^k$ is the σ-algebra of Borel sets in $\mathbb{R}^k$.

Theorem 20.3.1. *For any $s > 0$ and $0 < k \leq n$*

$$\sigma(Q_{n,s,k}^{(\infty)}, P_s^{k,\infty}) = k/n. \tag{20.3.2}$$

Proof. We need the following invariant property of the total variation metric $\boldsymbol{\sigma}$.

Claim 1 (Sufficiency theorem). If $T : \mathbb{R}^n \to \mathbb{R}$ is a sufficient statistic for P, $Q \in \mathcal{P}(\mathbb{R}^n)$, then

$$\boldsymbol{\sigma}(P, Q) = \boldsymbol{\sigma}(P \circ T^{-1}, Q \circ T^{-1}). \tag{20.3.3}$$

Proof. Take $\mu = \frac{1}{2}(P + Q)$ and let $f := \mathrm{d}P/\mathrm{d}\mu$, $g := \mathrm{d}Q/\mathrm{d}\mu$. Since T is sufficient, then $f = h_1 \circ T$, $g = h_2 \circ T$, and

$$h_1 = \frac{\mathrm{d}P \circ T^{-1}}{\mathrm{d}\mu \circ T^{-1}}, \quad h_2 = \frac{\mathrm{d}Q \circ T^{-1}}{\mathrm{d}\mu \circ T^{-1}}.$$

Clearly, $\boldsymbol{\sigma}(P \circ T^{-1}, Q \circ T^{-1}) \leq \boldsymbol{\sigma}(P, Q)$. On the other hand,

$$\begin{aligned} \boldsymbol{\sigma}(P, Q) &= \sup_{A \in \mathcal{B}^k} \left| \int_A (h_1 \circ T - h_2 \circ T) \mathrm{d}\mu \right| \\ &\leq \sup_{A \in \mathcal{B}^k} \left| \int_{T \circ A} (h_1 - h_2) \mathrm{d}\mu \circ T^{-1} \right| \end{aligned}$$

$$= \sup_{A \in \mathcal{B}^k} \left| \int_{T \circ A} \left(\frac{\mathrm{d}P \circ T^{-1}}{\mathrm{d}\mu \circ T^{-1}} - \frac{\mathrm{d}Q \circ T^{-1}}{\mathrm{d}\mu \circ T^{-1}} \right) \mathrm{d}\mu \circ T^{-1} \right|$$

$$\leq \sigma(P \circ T^{-1}, Q \circ T^{-1}),$$

which proves the claim.

Further, without loss of generality, we may assume $s = 1$ since

$$\sigma(Q^{(\infty)}_{n,s,k}, P^{k,\infty}_s) = \sigma(\Pr_{(\eta_1,\ldots,\eta_k)/\bigvee_{i=1}^n \eta_i = s}, \Pr_{(s\zeta_1,\ldots,s\zeta_k)}) = \sigma(Q^{(\infty)}_{n,1,k}, P^{k,\infty}_1)$$

by the zero-order ideality of σ.[27] Let $\widetilde{Q}$ be the law of $\eta_1 \vee \cdots \vee \eta_k$ determined by $Q^{(\infty)}_{n,1,k}$, the distribution of $(\eta_1, \ldots, \eta_k)$, where the vector $\eta = (\eta_1, \ldots, \eta_k, \eta_{k+1}, \ldots, \eta_n)$ is uniformly distributed on the simplex $S_{\infty,1,n} = \{x \in \mathbb{R}^n_+ : \bigvee_{i=1}^{\infty} x_i = 1\}$. Let $\widetilde{P}$ be the law of $\zeta_1 \vee \cdots \vee \zeta_n$, where ζ_is are i.i.d. uniforms. Then with $\gamma_{k,n,\infty} = \bigvee_{i=1}^k \zeta_i / \bigvee_{i=1}^n \zeta_i$, $\widetilde{Q} = \Pr_{\gamma_{k,n,\infty}}$ and $\widetilde{Q}$ has a DF given by (20.2.3). On the other hand, $\widetilde{P}((-\infty, x]) = x^k, 0 \leq x \leq 1$. Hence,

$$\widetilde{Q} = \frac{n-k}{n}\widetilde{P} + \frac{k}{n}\delta_1$$

is the mixture of $\widetilde{P}$ and δ_1, the point measure at 1. Consider the total variation distance

$$\sigma(Q^{(\infty)}_{n,1,k}, P^{k,\infty}_1) = \sup_{A \in \mathcal{B}^k} \left| \Pr\left((\eta_1, \ldots, \eta_k) \in A \,\middle|\, \bigvee_{i=1}^n \eta_i = 1 \right) - \Pr((\zeta_1, \ldots, \zeta_k) \in A) \right|.$$

We realize $Q_{n,1,k}$ is the law of $\zeta_1/M, \ldots, \zeta_k/M$, where $M = \bigvee_{i=1}^n \zeta_i$, so $\widetilde{Q}$ is the law of $\max(\zeta_1/M, \ldots, \zeta_k/M)$. By Claim 1,

$$\sigma(Q^{(\infty)}_{n,1,k}, P^{k,\infty}_1) = \sigma(\widetilde{Q}, \widetilde{P}) = \sup_{A \in \mathcal{B}^k} \left| \frac{n-k}{n}\widetilde{P}(A) + \frac{k}{n}\delta_1(A) - \widetilde{P}(A) \right|$$

$$= \frac{k}{n} \sup_{A \in \mathcal{B}^k} |\delta_1(A) - \widetilde{P}(A)| = \frac{k}{n},$$

as required. □

Let C_n be the class of distributions of $X = (X_1, \ldots, X_n)$ on R^n_+, which share with the i.i.d. uniforms[28] the property that, given $M := \bigvee_{i=1}^n X_i = s$, the conditional joint distribution of X is uniform on $S_{\infty,s,n}$. Clearly, $P^{n,\infty}_s \in C_n$. As a consequence of Theorem 20.3.1, we get the following *stability form of de Finetti's theorem.*

[27]See Definition 15.3.1 in Chap. 15.

[28]See Lemma 20.3.1.

Corollary 20.3.1. *If $P \in C_n$, then there is a μ such that for all $k < n$*

$$\|P_k - P_{\mu k}\| \le k/n, \tag{20.3.4}$$

where P_k is the P-law of the first k-coordinates $(X_1, \dots, X_k)$ and $P_{\mu k} = \int P_\sigma^{k,\infty} \mu(ds)$.

Proof. Define $\mu = \Pr_{\bigvee_{i=1}^n X_i}$; then $P_k = \int Q_{n,s,k}^{(\infty)} \mu(ds)$, $P_{\mu k} = \int P_s^{k,\infty} \mu(ds)$, and therefore $\sigma(P_k, P_{\mu k}) \le \int \sigma(Q_{n,s,k}^{(\infty)}, P_s^{k,\infty}) \mu(ds) = k/n$. □

In particular, one gets the de Finetti-type characterization of scale mixtures of i.i.d. uniform variables.

Corollary 20.3.2. *Let P be a probability on $\mathbb{R}_+^\infty$ with P_n being the P-law of the first n coordinates. Then P is a uniform scale mixture of i.i.d. uniform distributed RVs if and only if $P_n \in C_n$ for every n.*

Following the same method we will consider the case $p \in (0, \infty)$. Let $\zeta_1, \zeta_2, \dots$ be i.i.d. RVs with DF F_p given by Theorem 20.2.1. Then, by Lemma 20.3.1, the conditional distribution of $(\zeta_1, \dots, \zeta_n)$ given $\sum_{i=1}^n \zeta_i^p = s$ is $Q_{n,s,k}^{(p)}$, where $Q_{n,s,k}^{(p)}$ is the distribution of the first k coordinates of a random vector $(\eta_1, \dots, \eta_n)$ uniformly distributed on the p-sphere of radius s, denoted by $S_{p,s,n}$. Let $P_\sigma^{n,p}$ be the law of the vector $(\sigma\zeta_1, \dots, \sigma\zeta_n)$. The next result shows that $Q_{n,s,k}^{(p)}$ is close to $P_{(s/n)^{1/p}}^{k,p}$ w.r.t. the total variation metric.

Theorem 20.3.2. *Let $0 < p < \infty$; then for $k < n - p$ and k, n big enough,*

$$\sigma(Q_{n,s,k}^{(p)}, P_{(s/n)^{1/p}}^{k,p}) \le \text{const} \times k/n. \tag{20.3.5}$$

Proof. By the zero-order ideality of σ,

$$\begin{aligned}
\sigma(Q_{n,s,k}^{(p)}, P_{(s/n)^{1/p}}^{k,p}) &= \sup_{A \in \mathcal{B}^k} |\Pr(\eta_1, \dots, \eta_k) \in (A/\eta_1^p + \cdots + \eta_n^p = s) \\
&\quad - \Pr(((s/n)^{1/p}\zeta_1, \dots, (s/n)^{1/p}\zeta_k) \in A)| \\
&= \sup_{A \in \mathcal{B}^k} \Bigg| \Pr(((n/s)^{1/p}\eta_1, \dots, (n/s)^{1/p}\eta_k) \in A) \\
&\quad \Big/ \sum_{i=1}^n ((n/s)^{1/p}\eta_i)^p = n) - \Pr((\zeta_1, \dots, \zeta_k) \in A) \Bigg| \\
&= \sigma(Q_{n,n,k}^{(p)}, P_1^{k,p}).
\end{aligned}$$

Thus, it suffices to take $s = n$. Let $\widetilde{Q}_k$ be the $Q_{n,n,k}^{(p)}$-law of $\eta_1^p + \cdots + \eta_k^p$ and $\widetilde{P}_k$ be the $P_1^{k,p}$-law of $\zeta_1^p + \cdots + \zeta_k^p$. Then $\sigma(Q_{n,n,k}^{(p)}, P_1^{k,p}) = \sigma(\widetilde{Q}_k, \widetilde{P}_k)$, as in the proof of Theorem 20.2.1. By Lemma 20.3.1, we may consider $Q_{n,n,k}$

as the law of $\zeta_1/R, \ldots, \zeta_k/R)$, where $R^p := (1/n)\sum_{i=1}^n \zeta_i^p$. Thus, $\widetilde{Q}_k$ is the law of $\sum_{i=1}^k (\zeta_i/R)^p = n\sum_{i=1}^k \zeta_i^p \Big/ \left(\sum_{i=1}^n \zeta_i^p\right)$. Hence, as in the proof of Theorem 20.2.1, $\widetilde{Q}_k$ has a density

$$f(x) = \frac{1}{n} B\left(\frac{k}{p}, \frac{n-k}{p}\right) \left(\frac{x}{n}\right)^{(k/p)-1} \left(1 - \frac{x}{n}\right)^{-1+(n-k)/p},$$

$$B\left(\frac{k}{p}, \frac{n-k}{p}\right) := \frac{\Gamma\left(\frac{n}{p}\right)}{\Gamma\left(\frac{k}{p}\right)\Gamma\left(\frac{n-k}{p}\right)} \tag{20.3.6}$$

for $0 \le x \le n$ and $f(x) = 0$ for $x > n$. On the other hand, $\widetilde{P}_k$ has a gamma $(1/p, k/p)$-density

$$g(x) := \frac{1}{p^{k/p}\Gamma(k/p)} \exp(-x/p) x^{-1+k/p}, \quad \text{for } 0 \le x \le \infty. \tag{20.3.7}$$

By Scheffe's theorem [see Billingsley (1999)],

$$\begin{aligned} \sigma(\widetilde{Q}_k, \widetilde{P}_k) &= \int_0^\infty |f(x) - g(x)| \mathrm{d}x \\ &= 2\int_0^\infty \max(0, f(x) - g(x)) \mathrm{d}x \\ &= \int_0^\infty \max\left(0, \frac{f(x)}{g(x)} - 1\right) g(x) \mathrm{d}x. \end{aligned} \tag{20.3.8}$$

By (20.3.6) and (20.3.7), $f/g = Ah$, where

$$A = \left(\frac{p}{n}\right)^{k/p} \Gamma\left(\frac{n}{p}\right) \Big/ \Gamma\left(\frac{n-k}{p}\right)$$

and

$$h(x) = \exp\left(\frac{x}{p}\right)\left(1 - \frac{x}{n}\right)^{-1+(n-k)/p}$$

for $x \in [0, n]$ and $h(x) = 0$ for $x > n$. We have

$$\log h(x) = \frac{x}{p} + (-1 + (n-k)/p) \log\left(1 - \frac{x}{n}\right)$$

and

$$\frac{\partial}{\partial x} \log h(x) \ge 0$$

if and only if $x \le k + p$. Hence, if $k + p \le n$, then

$$\log h(x) \le \frac{k+p}{p} + \left(\frac{n-k}{p} - 1\right) \log\left(1 - \frac{k+p}{n}\right). \tag{20.3.9}$$

We use the following consequence of the Stirling expansion of the gamma function:[29]

$$\Gamma(x) = \exp(-x)x^{x\ 1/2}(2\pi)^{1/2}\exp(\theta/12x), \quad 0 \le \theta < 1. \tag{20.3.10}$$

This implies that

$$A = \left(\frac{n}{n-k}\right)^{(n-k)/p+1/2} \exp\left(-\frac{k}{p}\right)\widetilde{\theta}$$

with

$$\widetilde{\theta} = \exp\left[\frac{p}{12}\left(\frac{\theta_1}{n} - \frac{\theta_2}{n-k}\right)\right] \le \exp\left(\frac{p}{12n}\right)$$

and $0 \le \theta_i < 1$. Hence,

$$\begin{aligned} Ah &\le e\left(\frac{n}{n-k}\right)^{(n-k)/p+1/2}\left(\frac{n-k-p}{n}\right)^{(n-k)/p-1}\widetilde{\theta} \\ &= e\left(\frac{n-k-p}{n-k}\right)^{(n-k)/p}\frac{n}{n-k-p}\left(\frac{n}{n-k}\right)^{1/2}\widetilde{\theta} \\ &= e\left(1 - \frac{p}{n-k}\right)^{(n-k)/p}\frac{n}{n-k-p}\left(\frac{1}{1-k/n}\right)^{1/2}\widetilde{\theta}. \end{aligned}$$

We use the following estimate:

$$\sup_{0 \le x < a}\left|\exp(-x) - \left(1 - \frac{x}{a}\right)^a\right| \le c/a \tag{20.3.11}$$

with $c := \sup_{0 \le x < a} x\exp(-x) = 1/e, a > 1$, implying that

$$\left|e\left(1 - \frac{p}{n-k}\right)^{(n-k)/p} - 1\right| \le \frac{p}{n-k}.$$

Furthermore, we use the estimates

$$\left(1 - \frac{k}{n}\right)^{-1/2} \le 1 + \frac{k}{2n} \quad \text{and} \quad \widetilde{\theta} \le \exp\left(\frac{p}{12n}\right) \le 1 + \frac{p}{12n}\exp(1/12)$$

[29]See Abramowitz and Stegun (1970, p. 257).

to obtain

$$Ah \leq \left(1 + \frac{p}{n-k}\right) \frac{n}{n-k-p} \left(1 + \frac{k}{n}\right) \left(1 + \frac{p \exp(1/12)}{12n}\right),$$

implying that $Ah - 1$ is bounded by the right-hand side of (20.3.5). □

Analogously to Corollary 20.3.1 and 20.3.2, we can state de Finetti's theorem (and its stable version) for the class $C_{n,p}$ of distributions of $X_1, \ldots, X_n$, which share with i.i.d. F_p-distributed RVs $(\zeta_1, \ldots, \zeta_n)$ the property that given $\sum_{i=1}^{n} X_i^p = s$, the conditional joint distribution of X is uniform on the positive pth sphere $S_{p,s,n}$.

20.4 Characterization and Stability of Environmental Processes

The objective of this section is the study of four stochastic models that take into account the effect of erosion on annual crop production. More precisely, we are concerned with the limit behavior of four recursive equations modeling environmental processes:

$$S_0 = 0, \quad S_n \stackrel{\mathrm{d}}{=} (Y + S_{n-1})Z, \tag{20.4.1}$$

$$M_0 = 0, \quad M_n \stackrel{\mathrm{d}}{=} (Y \vee M_{n-1})Z, \tag{20.4.2}$$

$$G \stackrel{\mathrm{d}}{=} (Y + \delta G)Z, \tag{20.4.3}$$

and

$$H \stackrel{\mathrm{d}}{=} (Y \vee \delta H)Z, \tag{20.4.4}$$

where the RVs on the right-hand sides of (20.4.1)–(20.4.4) are assumed to be independent. Y, Z, $S_{(\cdot)}$, $M_{(\cdot)}$, G, and H are RVs taking on values in the Banach space $\mathbb{B} = C(T)$ of continuous functions x on the compact set T with the usual supremum norm $\|x\|$. For any $x, y \in \mathbb{B}$ define the pointwise maximum and multiplication: $(x \vee y)(t) = x(t) \vee y(t)$ and $(x \cdot y)(t) = x(t) \cdot y(t)$. Z in (20.4.2) and (20.4.4) is assumed to be nonnegative, i.e., $Z(t) \geq 0$ for all $t \in T$. Finally, $\delta = \delta(d)$ is a Bernoulli RV independent of Y, G, H, Z with success probability d.

Equation (20.4.1) arises in modeling the total crop yield over n years. That is, consider a set of crop-producing areas A_t $(t \in T)$, and denote by $\{Y_n(t)\}_{n \geq 1}$ the sequence of annual yields. For fixed n, the real-valued RVs $Y_n(t)$, $t \in T$, are dependent. Let $Z_n(t)$ be the proportion of crop yield maintained in year n after the environmental effect from the previous year: $Z_n(t) < 1$ corresponds to a "bad" year, probably due to erosion, while $Z_n(t) \geq 1$ corresponds to a "good" year. The RVs Z_n are assumed to be i.i.d. and independent of $\{Y_n\}$. Assuming that the crop-growing area A_t is subject to environmental effects, the resulting sequence of annual yields is

$$X_n(t) = Y_n(t) \prod_{i=1}^{n} Z_i(t). \tag{20.4.5}$$

Let us denote by

$$S_n(t) = \sum_{k=1}^{n} X_k(t), \quad n \in \mathbb{N}, \tag{20.4.6}$$

the total crop yield over n years. Then, clearly, the process S_n satisfies the recursive Eq. (20.4.1), where here and in what follows Y and Z are generic independent RVs with $Y \stackrel{\mathrm{d}}{=} Y_1$ and $Z \stackrel{\mathrm{d}}{=} Z_1$, and independent of the Y_i and Z_i.

Analogously, the maximal crop yield over n years

$$M_n = \bigvee_{k=1}^{n} X_k \tag{20.4.7}$$

has a distribution determined by (20.4.2).

Next we consider the situation where each year a disastrous event may occur with probability $1 - d \in (0, 1)$. The year of the disaster is a geometric RV $\tau = \tau(d)$, $\Pr(\tau(d) = k) = (1 - d)d^{k-1}$, $k \in \mathbb{N}$. Thus, the total crop yield until the disastrous year can be modeled by

$$\begin{aligned} G := S_\tau = \sum_{k=1}^{\tau} X_k \stackrel{\mathrm{d}}{=} \sum_{k=1}^{1+\delta\tau} X_k \stackrel{\mathrm{d}}{=} X_1 + \delta \sum_{k=2}^{1+\tau} X_k \stackrel{\mathrm{d}}{=} Y_1 Z_1 + \delta \sum_{k=2}^{1+\tau} Y_k \prod_{i=1}^{k} Z_i \\ \stackrel{\mathrm{d}}{=} YZ + \delta Z \sum_{k=2}^{1+\tau} Y_{k-1} \prod_{i=2}^{k} Z_i \stackrel{\mathrm{d}}{=} (Y + \delta G)Z, \end{aligned} \tag{20.4.8}$$

i.e., G satisfies the recurrence (20.4.3). Analogously, the maximal crop yield until the year of the disaster

$$H := M_\tau = \bigvee_{k=1}^{\tau} X_k \tag{20.4.9}$$

satisfies (20.4.4).

Further, our goal is to prove that S_n has a limit S (a.s.) and S satisfies

$$S \stackrel{\mathrm{d}}{=} (Y + S)Z. \tag{20.4.10}$$

Similarly, the limit M of M_n in (20.4.2) satisfies

$$M \stackrel{\mathrm{d}}{=} (Y \vee M)Z. \tag{20.4.11}$$

The problem is characterizing the set of solutions of (20.4.10), (20.4.11), (20.4.3), and (20.4.4). Since the general solution seems to be difficult to obtain, we will use appropriate approximations and evaluate the error involved in these approximations.

Following the main idea of this book that each approximation problem has *natural* (*suitable*, *ideal*) metrics in terms of which a problem can be solved easily and completely, we choose $\mathcal{L}_p$-metric and its minimal ℓ_p for our approximation problem. Recall that $\mathfrak{X}(B)$ is the set of all random elements on a nonatomic probability space $\{\Omega, \mathcal{A}, \Pr\}$ with values in $\mathbb{B}$ and

$$\mathcal{L}_p(X,Y) := \begin{cases} (E\|X-Y\|^p)^{p'}, & \text{if } 0<\infty,\ p' = \min(1,p^{-1}) \\ \Pr\{X \neq Y\}, & \text{if } p = 0 \\ \operatorname{ess\,sup}\|X-Y\|, & \text{if } p=\infty,\ X,Y \in \mathfrak{X}(\mathbb{B}). \end{cases} \tag{20.4.12}$$

The corresponding minimal (simple) metric $\ell_p(X,Y) = \ell_p(\Pr_X, \Pr_Y)$ is given by[30]

$$\ell_p(X,Y) = \inf\{\mathcal{L}_p(\widetilde{X},\widetilde{Y}); \widetilde{X},\widetilde{Y} \in \mathfrak{X}(B), \widetilde{X} \stackrel{\mathrm{d}}{=} X, \widetilde{Y} \stackrel{\mathrm{d}}{=} Y\}. \tag{20.4.13}$$

In what follows we will need some analogs to the $\mathcal{L}_p$-metric in the space $\mathfrak{X}(B^\infty)$. The space B^∞ is a Banach space with the usual supremum norm defined by $\|\overline{X}\| = \sup\{\|X_i\| : i \geq 1\}$, where $\overline{X} = (X_1, X_2, \ldots)$. Now, on $\mathfrak{X}(B^\infty)$ we consider the following metrics:

$$\mathbf{K}(\overline{X},\overline{Y}) = \inf\{\varepsilon > 0 : \Pr(\|\overline{X}-\overline{Y}\| > \varepsilon) < \varepsilon\} \quad \text{(Ky Fan)}, \tag{20.4.14}$$

$$\mathcal{L}_p(\overline{X},\overline{Y}) = (E\|\overline{X}-\overline{Y}\|^p)^{1\wedge p^{-1}} \quad \text{for } 0<p<\infty, \tag{20.4.15}$$

$$\mathcal{L}_0(\overline{X},\overline{Y}) = \Pr\{\overline{X} \neq \overline{Y}\}, \quad \text{and } \mathcal{L}_\infty(\overline{X},\overline{Y}) = \operatorname{ess\,sup}\|\overline{X}-\overline{Y}\|.$$

Clearly, if X_n and X are random elements in $\mathfrak{X}(B)$, then $X_n \to X$ (Pr-a.s.) if and only if $\mathbf{K}(X_n^*, X^*) \to 0$, where $X_n^* := (X_n, X_{n+1}, \ldots)$ and $X^* = (X, X, \ldots)$. Similarly to the proof of Lemma 8.3.1 in Chap. 8, we have that if

$$E\{\sup_{n\geq 1}\|X_n\|^p\} + E\|X\|^p < \infty$$

for some $p \in [1,\infty)$, then as $n \to \infty$,

$$\begin{aligned} &\mathcal{L}_p(X_n^*, X^*) \to 0 \text{ if and only if } X_n \to X \text{ (Pr-a.s.) and} \\ &E\sup_{m\geq n}\|X_m\|^p \to E\|X\|^p. \end{aligned} \tag{20.4.16}$$

In both *limit* cases $p=0$, $p=\infty$,

$$\mathcal{L}_p(X_n^*, X^*) \to 0 \Rightarrow X_n \to X(\text{Pr-a.s.}). \tag{20.4.17}$$

[30] The basic properties of ℓ_p-metrics were summarized in Chap. 19; see (19.3.9)–(19.3.18).

Theorem 20.4.1. *(a) (Existence of limit S). Suppose that $\{Y_n\}_{n\in\mathbb{N}} \subset \mathfrak{X}(B)$ is an i.i.d. sequence with $N_p(Y) < \infty$, where $0 \le p \le \infty$, and*

$$N_p(Y) := \mathcal{L}_p(Y,0) = \ell_p(Y,0) = \begin{cases} [E\|Y\|]^{\min(1,1/p)}, & 0 < p < \infty, \\ \text{ess sup } \|Y\|, & p = \infty, \\ \Pr(Y \neq 0), & p = 0. \end{cases} \tag{20.4.18}$$

Assume also that $\{Z_n\}_{n\in\mathbb{N}} \subset \mathfrak{X}(B)$ is an i.i.d. sequence independent of $\{Y_n\}_{n\in\mathbb{N}}$ such that $N_p(Z) < 1$. Given S_n by (20.4.6)*, there exists S such that $S_n \to S$ (Pr-a.s.). Moreover, S satisfies* (20.4.10) *with Y, Z, and S mutually independent.*

(b) (Rate of convergence of S_n to S). Let $p \in [0,\infty]$, $N_p(Y) < \infty$, and $N_p(Z) < 1$. Assume that the laws of S_n and S are specified by (20.4.1) *and* (20.4.10)*, respectively. Then,*

$$\ell_p(S_n, S) \le N_p^n(Z) \frac{N_p(Y)}{1 - N_p(Z)}. \tag{20.4.19}$$

Proof. (a) For any $k, n \in \mathbb{N}$, $1 \le p < \infty$,

$$\begin{aligned} \mathcal{L}_p(S_n^*, S_{n+1}^*) &= \mathcal{L}_p((S_n, S_{n+1}, \ldots), (S_{n+k}, S_{n+k+1}, \ldots)) \\ &= \left(E \max_{m \ge n} \left\| \sum_{i=m}^{m+k} Y_i \prod_{j=1}^{i} Z_j \right\|^p \right)^{1/p} \\ &\le \sum_{i \ge n} \left(E\|Y_i\|^p \prod_{j=1}^{i} \|Z_j\|^p \right)^{1/p} \\ &= N_p(Y)(N_p(Z))^n / (1 - N_p(Z)). \end{aligned}$$

On the other hand, the space of all sequences $\overline{X}$ with $E\|X\|^p < \infty$ is complete with respect to $\mathcal{L}_p$ and, thus, $S^* = (S, S, \ldots)$ exists. Finally, notice that $\mathcal{L}_p(S_n^*, S^*) \le (N_p(Z))^n \, N_p(Y)/(1 - N_p(Z))$ holds. This proves the assertion for $1 \le p < \infty$. The cases $0 \le p < 1$ and $p = \infty$ are treated analogously. Equation (20.4.10) follows from $S_n \to S$ (Pr-a.s.) and (20.4.1).

(b) From (20.4.1) and (20.4.10) we have for $0 < p < \infty$

$$\begin{aligned} \ell_p(S_n, S) &\le \mathcal{L}_p((Y + S_{n-1})Z, (Y + S)Z) \le \mathcal{L}(S_{n-1}Z, SZ) \\ &\le \{E\|S_{n-1} - S\|^p \|Z\|^p\}^{1 \wedge p^{-1}} = \mathcal{L}_p(S_{n-1}, S) N_p(Z), \end{aligned} \tag{20.4.20}$$

where the last inequality follows from the independence of (S_{n-1}, S) and Z. Taking the minimum of the right-hand side of (20.4.20) over all the joint distributions of S_{n-1} and S we obtain

$$\ell_p(S_n, S) \leq \ell(S_{n-1}, S)N_p(Z). \tag{20.4.21}$$

Hence,

$$\ell_p(S_n, S) \leq \ell_p(0, S)N_p^n(Z) = N_p(S)N_p^n(Z). \tag{20.4.22}$$

From the Minkowski inequality

$$N_p(S) \leq N_p(Z)N_p(Y + S) \leq N_p(Z)\{N_p(Y) + N_p(S)\},$$

which implies that $N_p(S) \leq N_p(Y)\,\{1 - N_p(Z)\}^{-1}$. This and (20.4.22) prove (20.4.10). The cases $p = 0$ and $p = \infty$ can be handled similarly. □

The problem of characterizing the distribution of S as a solution of $S \stackrel{\mathrm{d}}{=} (S + Y)Z$ is still open. Here we consider two examples.

Example 20.4.1. Let the distribution of $S \in \mathfrak{X}(B)$ be symmetric α-stable. In other words, the characteristic function of $S_{\bar{t}} = (S(t_1), \ldots, S(t_n)), \bar{t} = (t_1, \ldots, t_n), 0 < t_1 < \cdots < t_n \leq 1$, is

$$E\exp\{i(\theta, S_{\bar{t}})\} = \exp\left\{-\int_{\mathbb{R}^n} |(\theta, \bar{s})|^\alpha \Gamma_{S_{\bar{t}}}(\mathrm{d}\bar{s})\right\},$$

where $\Gamma_{S_{\bar{t}}}(\cdot)$ is the spectral, finite symmetric measure of a symmetric α-stable random vector $S_{\bar{t}}$.[31] For any $z \in (0, 1)$ let us choose an α-stable $Y_{\bar{t}} = (Y(t_1), \ldots, Y(t_n))$ with spectral measure

$$\Gamma_{Y_{\bar{t}}}(\mathrm{d}\bar{s}) = \frac{1 - z^\alpha}{z^\alpha}\Gamma_{S_{\bar{t}}}(\mathrm{d}\bar{s}).$$

Then S satisfies (20.4.8), with $Z = z$ and Y having marginals $Y_{\bar{t}}$.[32]

Example 20.4.2 (Rachev and Samorodnitsky 1990). Let $\mathbb{B} = \mathbb{R}$, Z be a uniformly $(0, 1)$-distributed RV. Consider (20.4.10) with nonnegative Y and S. If ϕ_X stands for the Laplace transform of a nonnegative RV X, then, by (20.4.10), $\theta\phi_S(\theta) = \int_0^\theta \phi_S(x)\theta_Y(x)\mathrm{d}x$ for all $\theta > 0$. Differentiating we obtain

$$\theta_Y(\theta) = 1 + \theta\phi_S'(\theta)/\phi_S(\theta),$$

and thus

[31] See Kuelbs (1973) and Samorodnitski and Taqqu (1994).

[32] For other similar examples, see class L, Feller (1971, Sect. 8, Chap. XVII).

$$\phi_S(\theta) = \exp\left(-\int_0^\theta \frac{1-\theta_Y(x)}{x}\mathrm{d}x\right). \tag{20.4.23}$$

It follows from (20.4.23) that

$$\begin{aligned}\infty > \int_0^\theta \frac{(1-\phi_Y(x))}{x}\mathrm{d}x &= \int_0^\theta \left(\int_0^\infty \exp(-xy)(1-F_Y(y))\mathrm{d}y\right)\mathrm{d}x \\ &= \int_0^\infty (1-F_Y(y))\frac{(1-\exp(-y\theta)}{y}\mathrm{d}y.\end{aligned}$$

Thus,

$$\int_1^\infty \frac{1-F_Y(y))}{y}\mathrm{d}y < \infty \quad \text{or} \quad \int_1^\infty (\ln y)F_Y(\mathrm{d}y) < \infty.$$

Thus, in the equation

$$S \stackrel{\mathrm{d}}{=} (Y+S)Z, \tag{20.4.24}$$

where Z is uniformly distributed, Y must satisfy

$$E\ln(l+Y) < \infty. \tag{20.4.25}$$

With an appeal to Feller (1971, Theorem XIII 4.2), we draw the following cnclusions.

(a) *Any RV Y satisfying* (20.4.25) *gives a unique solution F_S of* (20.4.24) *for which the Laplace transform is given by $\phi_S(\theta)$ in* (20.4.23). More detailed analysis of (20.4.24) shows:
(b) *Any distribution F_S determined by* (20.4.24) *is infinitely divisible. More precisely, let Y correspond to S in* (20.4.24), *and let $0 < \beta < 1$. Then there is a distribution F_{S_β} with Laplace transform $\phi_S(\cdot)^\beta$; F_S solves* (20.4.24), *and the corresponding F_{Y_β} is the mixture $F_{Y_\beta}(x) = (1-\beta)F_0(x) + \beta F_y(x)$.*
(c) *The class $\mathbb{S}$ of RVs S solving* (20.4.24) *consists of infinitely divisible RVs whose Lèvy measure λ is of the following form:*[33]

$$\lambda \ll \text{Leb} \quad \text{and} \quad \lambda(\mathrm{d}x) = H(x)\mathrm{d}x, \tag{20.4.26}$$

where $H(0) \in [0,1]$, H is nonincreasing, and $H(x) \downarrow 0$ as $x \to \infty$. The corresponding Y has $1-H$ as its distribution function.

Finally, note that if S is the solution of (20.4.24) with given Y and uniformly distributed Z, then for any $\alpha > 0$

$$S \stackrel{\mathrm{d}}{=} (S+Y_\alpha)Z_\alpha, \tag{20.4.27}$$

[33]See Shiryayev (1984, p. 337).

where S, Y_α, and Z_α are independent, F_{Y_α} is the mixture

$$\frac{\alpha}{1+\alpha}F_0 + \frac{1}{1+\alpha}F_Y,$$

and Z_α has density $f_{Z_\alpha}(z) = (1+\alpha)z^\alpha, 0 \leq z \leq 1$.

As we have seen, in general the problem of evaluating the distribution of S is a difficult one, and in most cases we must resort to approximations. Here we start with the analysis of the stability of the set of solutions $\Pr_S$ of

$$S \stackrel{\mathrm{d}}{=} (Y+S)Z, \tag{20.4.28}$$

where Y, S, and Z are independent RVs in $\mathfrak{X}(\mathbb{B})$ with some Y and Z for which we only know that they are *close* to some given Y^* and Z^*.

Suppose we want to approximate the distribution of S in (20.4.28) by the distribution of $S*$ defined by

$$S^* \stackrel{\mathrm{d}}{=} (Y^*+S^*)Z^*, \tag{20.4.29}$$

where Y^*, S^*, and Z^* are independent and such that we can evaluate the law of S^* given the laws of Z^* and Y^*, respectively. Assume also that the distributions of Z and Z^* (resp. Y and Y^*) are close w.r.t. the minimal metric ℓ_p, i.e., for some small $\varepsilon > 0$ and $\delta > 0$

$$\ell_p(Z, Z^*) < \varepsilon \quad \text{and} \quad \ell_p(Y, Y^*) < \delta. \tag{20.4.30}$$

Theorem 20.4.2. *Assume that* (20.4.30) *holds,*

$$N_p(Z^*) < 1 - \varepsilon,$$

and

$$N_p(Y^*) + N_p(S^*) < \infty.$$

Then

$$\ell_p(S, S^*) \leq \frac{(\varepsilon + N_p(Z^*))\delta + \{N_p(Y^*) + N_p(S^*)\}\varepsilon}{1 - \varepsilon - N_p(Z^*)}. \tag{20.4.31}$$

Proof. From the definition of S and S^*,

$$\begin{aligned}
\ell_p(S, S^*) &= \ell_p(Z(Y+S), Z^*(Y^*+S^*)) \\
&\leq \ell_p(Z(Y+S), Z(Y^*+S^*)) + \ell_p(Z(Y^*+S^*), Z^*(Y^*+S^*)) \\
&\leq N_p(Z)\ell_p(Y+S, Y^*+S^*) + N_p(Y^*+S^*)\ell_p(Z, Z^*) \\
&\leq N_p(Z)[\ell_p(Y, Y^*) + \ell_p(S, S^*)] + \ell_p(Z, Z^*)[N_p(Y^*) + N_p(S^*)],
\end{aligned}$$

and thus

$$\ell_p(S,S^*) \le \frac{N_p(Z)\delta + \{N_p(Y^*) + N_p(S^*)\}\varepsilon}{1 - N_p(Z)}.$$

Finally, by the triangle inequality and (20.4.18), $N_p(Z) = \ell_p(Z,0) \le \varepsilon + N_p Z^*)$, which proves the assertion. □

In a similar fashion, one may evaluate the rate of convergence of $M_n \to M$, where $M_n = \sup_{1\le k\le n} X_k$, $M_n \stackrel{\mathrm{d}}{=} (Y \vee M_{n-1})Z$ (here $Z \ge 0$ and the product and maximum are pointwise). Similarly to Theorem 20.4.1, letting $n \to \infty$, one obtains (20.4.11). Further, since Z and Y are independent,

$$\begin{aligned}\ell_p(M_n, M) &\le \mathcal{L}_p((Y \vee M_{n-1})Z, (Y \vee M)Z)\\ &\le \mathcal{L}_p(M_{n-1}Z, M \vee Z) \le \mathcal{L}_p(M_{n-1}, M)N_p(Z).\end{aligned}$$

From this, as in Theorem 20.4.1(b), we get

$$\ell_p(M_n, M) \le N_p^n(Z)\frac{N_p(Y)}{1 - N_p(Z)}.$$

Suppose now that the assumption of Theorem 20.4.1 (b) holds; then $M_n \to M$ (a.s.), and

$$\mathcal{L}_p(M_n^*, M^*) \le N_p^n(Z)\frac{N_p(Z)}{1 - N_p(Z)},$$

where $M_n^* = (M_n, M_{n+1}, \dots)$, $M^* = (M, M, \dots)$.

Example 20.4.3. All simple max-stable processes are solutions of $M \stackrel{\mathrm{d}}{=} (Y \vee M)Z$. Given an α-max-stable process M, i.e., one whose marginal distributions are specified by

$$\begin{aligned}&\Pr\{M(t_1) \le x_1, \dots, M(t_n) \le x_n\}\\ &\quad = \exp\left\{-\int_\Omega \left(\max_{1\le i\le n} (\lambda_i x_i^{-\alpha}) U_{\bar t}(\mathrm{d}\lambda_1 \dots, \mathrm{d}\lambda_n)\right)\right\},\end{aligned}$$

where $\alpha > 0$, $\bar t = (t_1, \dots, t_n)$ and $U_{(\bullet)}$ is a finite measure on[34]

$$\Omega = \left\{(\lambda_1, \dots, \lambda_n); \lambda_i > 0, i = 1, \dots, n, \sum_1^n \lambda_i = 1\right\}.$$

For any $z \in (0, 1)$, if we define the max-stable

$$Y_{\bar t} = (Y(t_1), \dots, Y(t_n))$$

[34]See Resnick (1987).

with max-stable measure

$$U_{Y_{\bar{t}}}(\mathrm{d}\lambda) = \frac{1 - z^{\alpha}}{z^{\alpha}} U_{M_{\bar{t}}}(\mathrm{d}\lambda),$$

then M satisfies (20.4.11), with $Z = z$ and Y being α-max-stable with marginals having spectral measure $M_{Y_{\bar{t}}}$. A more general example of M as a solution of the preceding equation is given by Balkema et al. (1993), where the class L for the maxima scheme is studied.

Example 20.4.4. Suppose $\mathbb{B} = \mathbb{R}$ and Z is $(0, 1)$-uniformly distributed. Then $M \stackrel{\mathrm{d}}{=} (Y \vee M)Z$ implies that

$$F_M(x) = \exp\left(-\int_x^{\infty} \frac{1}{t}\overline{F}_Y(t)\mathrm{d}t\right), \quad \overline{F} := 1 - F_Y. \tag{20.4.32}$$

For example, if Y has a *Pareto distribution*, $\overline{F}_Y(t) = \min(1, t^{-\beta}, \beta > 1$, then M has a truncated extreme-value distribution

$$F_M(x) = \begin{cases} \exp\left(-1 + x - \dfrac{x^{1-\beta}}{\beta - 1}\right), & \text{for } 0 \le x \le 1 \\[2ex] \exp\left(-\dfrac{x^{1-\beta}}{\beta - 1}\right), & \text{for } x \ge 1. \end{cases} \tag{20.4.33}$$

From (20.4.11) it also follows that

$$F_M(x) > 0,\ \forall x > 0 \quad \Longleftrightarrow \quad E\ln(1 + Y) < \infty. \tag{20.4.34}$$

Note that if M has an atom at its origin, then

$$0 < \Pr(M \le 0) = \Pr(M \le 0)\Pr(Y \le 0),$$

i.e., $Y \equiv 0$, the degenerate case. Moreover, *the condition* $E\ln(1 + Y) < \infty$ *is necessary and sufficient for the existence of the nondegenerate solution of* $M \stackrel{\mathrm{d}}{=} (Y \vee M)Z$ *given by* (20.4.32).[35] Clearly, the latter assertion can be extended for any Z such that Z^{α} is $(0, 1)$-uniformly distributed for some $\alpha > 0$ since $M \stackrel{\mathrm{d}}{=} (M \vee Y)Z \Rightarrow M^{\alpha} \stackrel{\mathrm{d}}{=} (M^{\alpha} \vee Y^{\alpha})Z^{\alpha}$.

As far as the approximation of M is concerned, we have the following theorem.

Theorem 20.4.3. *Suppose the distribution of M is determined by* (20.4.11) *and*

$$M^* \stackrel{\mathrm{d}}{=} Z^*(Y^* \vee M^*), \quad \ell_p(Z, Z^*) < \varepsilon, \quad \ell_p(Y, Y^*) < \delta. \tag{20.4.35}$$

---

[35] See Rachev and Samorodnitsky (1990).

Assume also that $N_p(Z^*) \leq 1 - \varepsilon$ and $N_p(Y) + N_p(M^*) < \infty$. Then, as in Theorem 20.4.2, we arrive at

$$\ell_p(M, M^*) \leq \frac{(\varepsilon + N_p(Z^*))\delta + [N_p(Y^*) + N_p(M^*)]\varepsilon}{1 - N_p(Z^*) - \varepsilon}. \tag{20.4.36}$$

Proof. Recall that ℓ_p-metric is regular with respect to the sum and maxima of independent RVs, i.e., $\ell_p(X + Z, Y + Z) \leq \ell_p(X, Y)$ and $\ell_p(X \vee Z, Y \vee Z) \leq \ell_p(X, Y)$ for any $X, Y, Z \in \mathfrak{X}(\mathbb{B})$, and Z-independent of X and Y [see (19.4.2)–(19.4.6)]. Thus, one can repeat step by step the proof of Theorem 20.4.2 by replacing the equation $S \stackrel{\mathrm{d}}{=} (S + Y)Z$ with $M \stackrel{\mathrm{d}}{=} (M \vee Y)Z$. □

Note that in both Theorems 20.4.2 and 20.4.3, the ℓ_p-metric was chosen as a suitable metric for the stability problems under consideration. The reason is the *double ideality* of ℓ_p, i.e., ℓ_p plays the role of ideal metric for both summation and maxima schemes.[36]

Next, we consider the relation

$$G \stackrel{\mathrm{d}}{=} Z(Y + \delta G), \tag{20.4.37}$$

where, as before, Z, Y, and G are independent elements of $B = C(T)$, and δ is a Bernoulli RV, independent of them, with $\Pr\{\delta = 1\} = d$. If $Z \equiv 1$, then G could be chosen to have a *geometric infinitely divisible distribution*, i.e., the law of G admits the representation

$$G \stackrel{\mathrm{d}}{=} \sum_{i=1}^{\tau(d)} Y_i, \tag{20.4.38}$$

where the variables Y_i are i.i.d. and $\tau(d)$ is independent of the Y_i geometric RVs with mean $1/(1 - d)$ [see (20.4.8)].

Lemma 20.4.1. *In the finite-dimensional case $B = \mathbb{R}^m$, a necessary and sufficient condition for G to be geometric infinitely divisible is that its characteristic function is of the form*

$$f_G(t) = (1 - \log \phi(t))^{-1}, \tag{20.4.39}$$

where $\phi(\bullet)$ is an infinitely divisible characteristic function.

The proof is similar (but slightly more complicated) to that of Lemma 20.4.2, and we will write it in detail.

Example 20.4.5. Suppose Z has a density

$$p_Z(z) = (1 + z)z^{\alpha}, \quad \text{for } z \in (0, 1). \tag{20.4.40}$$

---

[36]See Sect. 19.4 of Chap. 19.

Then, from (20.4.37) we have

$$\theta^{\alpha+1} f_G(\theta) = (\alpha+1) \int_0^\theta u^\alpha f_Y(u)\{(1-d) + \mathrm{d} f_G(u)\} du,$$

where $f_{(\bullet)}$ stands for the characteristic function of the RV $(\bullet)$. Differentiating we obtain the equation

$$f'_G(\theta) + \frac{\alpha+1}{\theta}[1 - \mathrm{d} f_Y(\theta)] f_G(\theta) = \frac{\alpha+1}{\theta}(1-d) f_Y(\theta)$$

whose solution clearly describes the distribution of G for given Z and Y.

Next, let us consider the approximation problem assuming that $Z = z$ is a constant *close to 1*. Suppose further that the distribution of Y belongs to the class of "aging" distributions HNBUE.[37] Then our problem is to approximate the distribution of G defined by

$$G \overset{\mathrm{d}}{=} Z(Y + \delta G), \quad Z = z(\text{const}), \quad F_Y \in \text{HNBUE}, \qquad (20.4.41)$$

where Y and G are independent, by means of G^* specified by

$$G^* \overset{\mathrm{d}}{=} Y^* + \delta G, \quad F_Y^*(t) = 1 - \exp(-t/\mu)\ t \geq 0, \qquad (20.4.42)$$

where Y^* and G^* are independent. Given that $EY = \mu$ and $\mathbf{Var}\, Y = \nu^2$, we obtain the following estimate of the deviation between the distributions of Y and Y^* in terms of the metric ℓ_p,[38]

$$\ell_1(Y, Y^*) \leq 2(\mu^2 - \nu^2)^{1/2}, \qquad (20.4.43)$$

and for $p > 1$

$$\ell_p(Y, Y^*) \leq (\mu^2 - \nu^2)^{1/4p} 8\mu\Gamma(2p)^{1/p}. \qquad (20.4.44)$$

The following proposition gives an estimate of the distance between G and G^*.

Theorem 20.4.4. *Suppose that* G *satisfies* (20.4.37) *where* Z, Y, *and* G *are independent elements of* $B = C(T)$ *and* δ *is a Bernoulli RV independent of them, and consider*

$$G* \overset{\mathrm{d}}{=} (Y^* + \delta G^*) Z^*, \quad G^*, Z^*, Y^* \in \mathfrak{X}(B),$$

[37] See Definition 17.4.1 in Chap. 17.

[38] See Kalashnikov and Rachev (1988, Chap. 4, Sect. 2, Lemma 10).

where G^*, Z^*, Y^*, and δ are again independent. Assume also that

$$\ell_p(Z,Z^*) \le \varepsilon, \quad \ell_p(Y,Y^*) \le \delta, \quad N_p(Z^*)d < 1-\varepsilon d.$$

Then

$$\ell_p(G,G^*) \le \frac{(\varepsilon + N_p(Z^*))\delta + [N_p(Y^*) + \mathrm{d}N_p(G^*)]\varepsilon}{1 - \mathrm{d}N_p(Z^*) - \mathrm{d}\varepsilon}.$$

Proof. Analogously to Theorem 20.4.2,

$$\begin{aligned}\ell_p(G,G^*) &\le \ell_p(Z(Y+\delta G), Z(Y^*+\delta G^*)) + \ell_p(Z(Y^*+\delta G^*), Z^*(Y^*+\delta G^*))\\ &\le N_p(Z)\ell_p(Y+\delta G, Y^*+\delta G^*) + N_p(Y^*+\delta G^*)\ell_p(Z,Z^*)\\ &\le N_p(Z)[\ell_p(Y,Y^*) + d\ell_p(G,G^*)] + [N_p(Y^*) + dN_p(G^*)]\ell_p(Z,Z^*).\end{aligned} \tag{20.4.45}$$

The assertion follows from this and $N_p(Z) \le \varepsilon + N_p(Z^*)$. □

In the special case given in (20.4.41) and (20.4.42), the inequality

$$\ell_p(G,G^*) \le \frac{N_p(Z)\delta + [N_p(Y^*) + \mathrm{d}N_p(G^*)]\varepsilon}{1 - \mathrm{d}N_p(Z)} \tag{20.4.46}$$

holds and, moreover, $N_p(Y^*) + \mathrm{d}N_p(G^*) \le (1+d)/(1-d)\mu(\Gamma(p+1))^{1/p}$. Finally, since $\varepsilon = \ell_p(Z,Z^*) = 1-z$, and δ can be defined as the right-hand side of (20.4.43) or (20.4.44), we have the following theorem.

Theorem 20.4.5. *If G and G^* are given by* (20.4.41) *and* (20.4.42)*, respectively, then*

$$\ell_p(G,G^*) \le \frac{1-z}{1-zd}\mu(\Gamma(p+1))^{1/p}\frac{1+d}{1-d} + \frac{z\delta_p}{1-zd}, \tag{20.4.47}$$

where

$$\delta_p := \begin{cases} 2(\mu^2-\nu^2)^{1/2}, & \text{if } p=1\\ \Gamma(2p)^{1/p}(\mu^2-\nu^2)^{1/4p}8\mu, & \text{if } p>1\end{cases}$$

and $N_p(Y) = (EY^p)^{1/p}$.

For $p=1$ we obtain from (20.4.47)

$$\int_{-\infty}^{\infty} |F_G(x) - F_{G^*}(x)|\mathrm{d}x \le \frac{2}{1-zd}\left[\left(\frac{1-z}{1-d}\right)\mu + z(\mu^2-\nu^2)^{1/2}\right].$$

Finally, consider the geometric maxima H defined by

$$H \stackrel{d}{=} Z(Y \vee \delta H), \quad \text{or equivalently,} \quad H \stackrel{d}{=} \bigvee_{k=1}^{\tau(d)} Y_k \prod_{j=1}^{k} Z_j, \tag{20.4.48}$$

where Z, Y, δ, H, $\tau(d)$, Y_k, and Z_j are assumed to be independent, $Y_k \stackrel{d}{=} Y$, $Z_k \stackrel{d}{=} Z$, $Z > 0$, $Y > 0$, and $H > 0$.

Example 20.4.6. If $Z = 1$, then H has a *geometric maxima infinitely divisible (GMID) distribution*, i.e., for any $d \in (0, 1)$

$$H \stackrel{d}{=} \bigvee_{1}^{\tau(d)} Y_k, \tag{20.4.49}$$

where $Y_k = Y_k^{(d)}$, $k \in \mathbb{N}$, are i.i.d. nonnegative RVs and $\tau(d)$ is independent of Y_k geometric RVs

$$\Pr(\tau(d) = k) = (1 - d)d^{k-1}, \qquad k \geq 1. \tag{20.4.50}$$

Let $\mathbb{B} = \mathbb{R}^m$. Let $\Pr(H \leq x) = G(x)$, $x \in \mathbb{R}_+^m$, and $\Pr(Y_1^{(d)} \leq x) = G_d(x)$. Then (20.4.49) is the same as

$$G(x) = \sum_{j=1}^{\infty}(1 - d)d^{k-1}G_d(x) = \frac{(1 - d)G_d(x)}{1 - dG_d(x)}. \tag{20.4.51}$$

If we solve for G_d in (20.4.51), then we get

$$G_d(x) = G(x)/(1 - d + \mathrm{d}G(x)), \tag{20.4.52}$$

which is clearly equivalent to

$$H \stackrel{d}{=} Y \vee \delta H, \tag{20.4.53}$$

where $Y \stackrel{d}{=} Y_1^{(d)}$ [see (20.4.49)].

We now characterize the class GMID. We will consider the slightly more general case where H and Y_k are not necessarily nonnegative. The characterizations are in terms of max-infinitely divisible (MID) distributions, exponent measures, and multivariate extremal processes.[39] A MID distribution F with exponent measure μ has the property that the support $[x : F(x) > 0]$ is a rectangle. Let $\ell \in \mathbb{R}^m$ be the *bottom* of this rectangle [see Resnick (1987, p. 260)]. Clearly, in the one-dimensional case $m = 1$, any DF F is a MID distribution.

---

[39] For background on these concepts, see Resnick (1987, Chap. 5).

Lemma 20.4.2. *For a distribution G on $\mathbb{R}^m$ the following are equivalent:*

(i) $G \in GMID$.
(ii) $\exp(1 - 1/G)$ *is a MID distribution.*
(iii) There exists $\ell \in [-\infty, \infty)^m$ *and an exponent measure* μ *concentrated on the rectangle* $\{x \in \mathbb{R}^m,\ x \in \ell\}$*, such that for any* $x \geq \ell$

$$G(x) = \frac{1}{1 + \mu(\mathbb{R}^m \setminus [\ell, x])}.$$

(iv) There exists an extremal process $\{Y(t),\ t > 0\}$ *with values in* $\mathbb{R}^m$ *and an independent exponential RV* E *with mean* 1 *such that* $G(x) = \Pr(Y(E) \leq x)$.

Proof. (i) $\Rightarrow$ (ii). We have the following identity:

$$G = \lim_{\alpha \downarrow 0} 1 \Bigg/ \left[1 + \frac{1}{\alpha}\left(1 - \frac{G}{\alpha + (1-\alpha)G}\right)\right]. \tag{20.4.54}$$

Therefore,

$$\exp(1 - 1/G) = \lim_{\alpha \downarrow 0} \exp\left[-\frac{1}{\alpha}\left(1 - \frac{G}{\alpha + (1-\alpha)G}\right)\right].$$

If $G \in GMID$, then $G/(\alpha + (1-\alpha)G)$ is a DF for any $\alpha \in (0, 1)$, which implies that[40]

$$\exp\left[-\frac{1}{\alpha}\left(1 - \frac{G}{\alpha + (1-\alpha)G}\right)\right]$$

is a MID distribution. Since the class of MID distributions is closed under weak convergence, it follows that $\exp(1 - 1/G)$ is a MID distribution.
(ii) $\Rightarrow$ (iii). If $\exp(1 - 1/G)$ is a MID distribution, then, by the characterization of MID distribution, there exists $\ell \in [-\infty, \infty)^m$ and an exponent measure μ concentrating on $\{x : x \geq \ell\}$ such that for

$$x \geq \ell, \quad \exp\left\{1 - \frac{1}{G(x)}\right\} = \exp\{-\mu(\mathbb{R}^m \setminus [\ell, x])\}$$

and equating exponents yields (iii).
(iii) $\Rightarrow$ (iv). Suppose μ is the exponent measure assumed to exist by (iii), and let $\{Y(t),\ t > 0\}$ be an extremal process with

$$\Pr(Y(t) \leq x) = \exp\{-t\mu(\mathbb{R}^m \setminus [\ell, x])\}. \tag{20.4.55}$$

[40] See Resnick (1987, pp. 257–258).

Then

$$\Pr(Y(E) \le x) = \int_0^\infty \mathrm{e}^{-t} \Pr(Y(t) \le x)\mathrm{d}t = \int_0^\infty \mathrm{e}^{-t} \exp\{-t\mu(\mathbb{R}^m \setminus [\ell, x])\}\mathrm{d}t$$
$$= 1/(1 + \mu(\mathbb{R}^m \setminus [\ell, x])),$$

as required.
(iv) $\Rightarrow$ (i). Suppose $G(x) = \Pr(Y(E) \le x)$. If (20.4.55) holds, then

$$G(x) = 1/(1 + \mu(\mathbb{R}^m \setminus [\ell, x])).$$

To show $G \in G$MID, we need to show that

$$\frac{G(x)}{1 - d + \mathrm{d}G(x)} = \frac{1}{1 + (1 - d)\mu(\mathbb{R}^m \setminus [\ell, x])}$$

is a distribution, and this follows readily by observing

$$\Pr(Y((1 - d)E) \le x) = \frac{1}{1 + (1 - d)\mu(\mathbb{R}^m \setminus [\ell, x])}.$$

□

In particular, Lemma 20.4.2 implies that the real-valued RV H has a GMID distribution if and only if its DF F_H can be represented as $F_H(t) = (1 - \log \Phi(t))^{-1}$, where $\Phi(t)$ is an arbitrary DF. For instance, if

$$\Phi(x) = \exp(-x^{-\alpha}), \quad x > 0,$$

then

$$F_H(x) = \frac{x^\alpha}{1 + x^\alpha}, \quad x \ge 0,$$

is the *log logistic* distribution with parameter $\alpha > 0$. If Φ is the Gumbel distribution, i.e., $\Phi(x) = \exp(-\mathrm{e}^{-x})$, $x \in \mathbb{R}$, then clearly F_H is the exponential distribution with parameter 1.

Example 20.4.7. Consider (20.4.53) for real-valued RVs Z, Y, and H.
Assume Z has the density (20.4.40). Then

$$F_H(x) = \int_0^1 F_Y\left(\frac{x}{z}\right)\left[1 + \mathrm{d}F_H\left(\frac{x}{z}\right)\right](\alpha + 1)z^\alpha \mathrm{d}z$$

or

$$x^{-\alpha-1} F_H(x) = (\alpha + 1)\int_x^\infty F_Y(y)[1 + \mathrm{d}F_H(y)]y^{-\alpha-2}\mathrm{d}y.$$

This equation is easily solved, and we obtain

$$F_H(x) = \left(\exp -(\alpha+1)\int_x^\infty \frac{1}{u}[1-\mathrm{d}F_Y(u)]\mathrm{d}u\right)$$
$$\times(\alpha+1)\int_x^\infty \left(\exp(\alpha+1)\int_y^\infty \frac{1}{u}[1-\mathrm{d}F_Y(u)]\mathrm{d}u\right)\frac{1}{y}F_Y(y)\mathrm{d}y. \tag{20.4.56}$$

The stability analysis is handled in a similar way to Theorem 20.4.4. Consider the equations

$$H \stackrel{\mathrm{d}}{=} (Y \vee H)Z \quad \text{and} \quad H^* \stackrel{\mathrm{d}}{=} (Y^* \vee H^*)Z^*, \tag{20.4.57}$$

where Y, H, Z (resp. Y^*, H^*, Z^*) are independent nonnegative elements of $\mathfrak{X}(\mathbb{B})$. Following the model from the beginning of Chap. 20, suppose that the *input* distributions $(\Pr_Y, \Pr_Z)$ and $(\Pr_{Y^*}, \Pr_{Z^*})$ are close in the sense that

$$\ell_p(Z, Z^*) \le \varepsilon \qquad \ell_p(Y, Y^*) \le \delta. \tag{20.4.58}$$

Then the *output* distributions $\Pr_H$, $\Pr_{H^*}$ are also close, as the following theorem asserts.

Theorem 20.4.6. *Suppose H and H^* satisfy* (20.4.57) *and* (20.4.58) *holds. Suppose also that $N_p(Z^*) < 1 - \varepsilon d$. Then*

$$\ell_p(H, H^*) \le \frac{(\varepsilon + N_p(Z^*))\delta + [N^p(Y^*) + \mathrm{d}N_p(H^*)]\varepsilon}{1 - \mathrm{d}N_p(Z^*) - \mathrm{d}\varepsilon}.$$

The proof is similar to that of Theorem 20.4.4.

References

Abramowitz M, Stegun IA (1970) Handbook of mathematical functions, 9th Printing. Dover, New York

Akaike H (1981) Modern development of statistical methods. In: Eykhoff P (ed) Trends and progress in system identification. Pergamon, Oxford, pp 169–184

Balkema AA, De Haan L, Karandikar RL (1993) The maximum of n independent stochastic processes. J. Applied Probability 30, 66–81

Billingsley, P (1999) Convergence of probability measures, 2nd edn. Wiley, New York

Cramer H (1946) Mathematical methods of statistics. Princeton University Press, New York

Csiszar I (1967) Information-type measures of difference of probability distributions and indirect observations. Studio Sci Math Hungar 2:299–318

Diaconis P, Freedman D (1987) A dozen de Finetti-style results in search of a theory. Ann Inst H Poincare 23:397–623

Dudley RM (2002) Real analysis and probability, 2nd edn. Cambridge University Press, New York

Feller W (1971) An introduction to probability theory and its applications, vol II, 2nd edn. Wiley, New York

Hampel FR (1971) A general qualitative definition of robustness. Ann Math Stat 42:1887–1896

Hernandez-Lerma O, Marcus SI (1984) Identification and approximation of Queueing systems. IEEE Trans Automat Contr AC29:472–474

Huber PJ (1977) Robust statistical procedures. CBMS regional conference series in applied mathematics, 27, Society for Industrial and Applied Mathematics, Philadelphia

Kalashnikov VV, Rachev ST (1985) Characterization problems in queueing and their stability. Adv Appl Prob 17:868–886

Kalashnikov VV, Rachev ST (1986a) Characterization of inverse problems in queueing and their stability. J Appl Prob 23:459–473

Kalashnikov VV, Rachev ST (1986b) Characterization of queueing models and their stability. In: Prohorov YK, et al. (eds) Probability theory and mathematical statistics, vol 2. VNU Science Press, Utrecht, pp 37–53

Kalashnikov VV, Rachev ST (1988) Mathematical methods for construction of stochastic queueing models. Nauka, Moscow (in Russian) [English transl., (1990) Wadsworth, Brooks–Cole, Pacific Grove, CA]

Kuelbs T (1973) A representation theorem for symmetric stable processes and stable measures. Z Wahrsch Verw Geb 26:259–271

Kullback S (1959) Information theory and statistics. Wiley, New York

Ljung L (1978) Convergence analysis of parametric identification models. IEEE Trans Automat Contr AC-23:770–783

Papantoni-Kazakos P (1977) Robustness in parameter estimation. IEEE Trans Inf Theor IT-23:223–231

Rachev S, Rüschendorf L (1991) Approximate independence of distributions on spheres and their stability properties. Ann Prob 19:1311–1337

Rachev ST (1989) The stability of stochastic models. Appl Prob Newslett ORSA 12:3–4

Rachev ST, Samorodnitsky G (1990) Distributions arising in modelling environmental processes. Operations research and industrial engineering. Cornell University (Preprint)

Rachev ST, Todorovic P (1990) On the rate of convergence of some functionals of a stochastic process. J Appl Prob 28:805–814

Resnick SI (1987) Extreme values: regular variation and point processes. Springer, New York

Roussas GG (1972) Contiguity of probability measures: some applications in statistics. Cambridge University Press, Cambridge

Samorodnitski G, Taqqu MS (1994) Stable non-gaussian random processes: stochastic models with infinite variance. Chapman & Hall, Boca Raton, FL

Shiryayev AN (1984) Probability. Springer, New York

Todorovic P, Gani J (1987) Modeling of the effect of erosion on crop production. J Appl Prob 24:787–797

Wasserstein LN (1969) Markov processes over denumerable products of spaces describing large systems of automata. Prob Inf Transmiss 5:47–52

Zolotarev VM (1977a) General problems of mathematical models. In: Proceedings of 41st session of ISI, New Delhi, pp 382–401

Zolotarev VM (1977b) Ideal metrics in the problem of approximating distributions of sums of independent random variables. Theor Prob Appl 22:433–449

Zolotarev VM (1983) Foreword. In: Lecture Notes in Mathematics, vol 982. Springer, New York, pp VII-XVII

Part V
Euclidean-Like Distances and Their Applications

Chapter 21
Positive and Negative Definite Kernels and Their Properties

The goals of this chapter are to:

- Formally introduce positive and negative definite kernels,
- Describe the properties of positive and negative definite kernels,
- Provide examples of positive and negative definite kernels and to characterize coarse embeddings in a Hilbert space,
- Introduce strictly and strongly positive and negative definite kernels.

Notation introduced in this chapter:

Notation	Description
$\mathcal{K}$	Positive definite kernel
$\mathrm{Re}\,\mathcal{K}$	Real part of a positive definite kernel
$\bar{c}$	Complex conjugate of complex number $c \in \mathbb{C}$
$\mathcal{H}(\mathcal{K})$	Hilbert space with reproducing kernel $\mathcal{K}$
$(\varphi, \psi)_{\mathcal{H}}$	Inner product between two elements of Hilbert space $\mathcal{H}$
$\mathcal{L}$	Negative definite kernel
$(\mathfrak{X}, \mathfrak{A}, \mu)$	Space $\mathfrak{X}$ with measure μ defined on algebra $\mathfrak{A}$ of Bair subsets of $\mathfrak{X}$

21.1 Introduction

In this chapter, we introduce positive and negative definite kernels, describe their properties, and provide theoretical results that characterize coarse embeddings in a Hilbert space. This material prepares the reader for the subsequent chapters in which we describe an important class of probability metrics with a Euclidean-like structure.

We begin with positive definite kernels and then continue with negative definite kernels. Finally, we discuss necessary and sufficient conditions under which a metric space admits a coarse embedding in a Hilbert space.

S.T. Rachev et al., *The Methods of Distances in the Theory of Probability and Statistics*, DOI 10.1007/978-1-4614-4869-3_21,

21.2 Definitions of Positive Definite Kernels

One of the main notions of Part V of this book is that of the positive definite kernel. Some definitions and results can be found, for example, in Vakhaniya et al. (1985).

Let $\mathfrak{X}$ be a nonempty set. A map $\mathcal{K} : \mathfrak{X}^2 \to \mathbb{C}$ is called a *positive definite kernel* if for any $n \in \mathbb{N}$ an arbitrary set $c_1, \dots, c_n$ of complex numbers and an arbitrary set $x_1, \dots, x_n$ of points of $\mathfrak{X}$ the following inequality holds:

$$\sum_{i=1}^{n} \sum_{j=1}^{n} \mathcal{K}(x_i, x_j) c_i \bar{c}_j \geq 0. \tag{21.2.1}$$

Here and subsequently the notation $\bar{c}$ denotes the complex conjugate of c.

The 12 main properties of positive definite kernels are explained below.

Property 21.2.1. Let $\mathcal{K}$ be a positive definite kernel. Then for all $x, y \in \mathfrak{X}$

$$\mathcal{K}(x,x) \geq 0, \quad \mathcal{K}(x,y) = \bar{\mathcal{K}}(y,x).$$

It follows from here that if a positive definite kernel $\mathcal{K}$ is real-valued, then it is symmetric.

Property 21.2.2. If $\mathcal{K}$ is a real positive definite kernel, then (21.2.1) holds if and only if it holds for real $c_1, \dots, c_n$.

Property 21.2.3. If $\mathcal{K}$ is a positive definite kernel, then $\bar{\mathcal{K}}$ and $\mathrm{Re}\, \mathcal{K}$ are positive definite kernels.

Property 21.2.4. If $\mathcal{K}_1$ and $\mathcal{K}_2$ are positive definite kernels and α_1, α_2 are nonnegative numbers, then $\alpha_1 \mathcal{K}_1 + \alpha_2 \mathcal{K}_2$ is a positive definite kernel.

Property 21.2.5. Suppose that $\mathcal{K}$ is a positive definite kernel. Then

$$\left|\mathcal{K}(x,y)\right|^2 \leq \mathcal{K}(x,x)\mathcal{K}(y,y)$$

holds for all $x, y \in \mathfrak{X}$.

Property 21.2.6. If $\mathcal{K}$ is a positive definite kernel, then

$$\left|\mathcal{K}(x,x_1) - \mathcal{K}(x,x_2)\right|^2 \leq \mathcal{K}(x,x)\big(\mathcal{K}(x_1,x_1) + \mathcal{K}(x_2,x_2) - 2\mathrm{Re}\, \mathcal{K}(x_1,x_2)\big)$$

holds for all $x, x_1, x_2 \in \mathfrak{X}$.

One can easily prove Properties 21.2.1–21.2.6 on the basis of (21.2.1) for specially chosen $n \geq 1$ and $c_1, \dots, c_n$.

Property 21.2.7. Let $\mathcal{K}_\alpha$ be a generalized sequence of positive definite kernels such that the limit

$$\lim_{\alpha} \mathcal{K}_\alpha(x, y) = \mathcal{K}(x, y)$$

exists for all $x, y \in \mathfrak{X}$. Then $\mathcal{K}$ is a positive definite kernel.
Property 21.2.7 follows immediately from the definition of positive definite kernel.

For further study of positive definite kernels we will need the following two theorems.

Theorem 21.2.1 (Aronszajn 1950). *Let $\mathfrak{X}$ be a set, and let $\mathcal{K} : \mathfrak{X}^2 \to \mathbb{R}^1$ be a positive definite kernel. Then there exists a unique Hilbert space $\mathcal{H}(\mathcal{K})$ for which the following statements hold:*

(a) Elements of $\mathcal{H}(\mathcal{K})$ are real functions given on $\mathfrak{X}$.
(b) Denoting $\mathcal{K}_x(y) = \mathcal{K}(x, y)$ we have

$$\{\mathcal{K}_x(y) : x \in \mathfrak{X}\} \subset \mathcal{H}(\mathcal{K});$$

(c) For all $x \in \mathfrak{X}$ and $\varphi \in \mathcal{H}(\mathcal{K})$ we have

$$(\varphi, \mathcal{K}_x) = \varphi(x).$$

Note that the space $\mathcal{H}(\mathcal{K})$ is called a *Hilbert space* with *reproducing kernel* and the statement in (c) is also called a *reproducing property*.

Proof. Let $\mathcal{H}_o$ be a linear span of the family

$$\{\mathcal{K}_x : x \in \mathfrak{X}\}.$$

Define on $\mathcal{H}_o$ a bilinear form in the following way: if $\varphi = \sum_{i=1}^n \alpha_i \mathcal{K}_{x_i}$ and $\psi = \sum_{j=1}^m \beta_j \mathcal{K}_{y_j}$, then set

$$s(\varphi, \psi) = \sum_{i=1}^n \sum_{j=1}^m \alpha_i \beta_j \mathcal{K}(x_i, y_j),$$

where $\alpha_i, \beta_j \in \mathbb{R}^1$ and $x_i, y_j \in \mathfrak{X}$. It is easy to see that the value $s(\varphi, \psi)$ does not depend on the concrete representations of elements φ and ψ. It is obvious that s is a symmetric positive form satisfying the condition

$$s(\varphi, \mathcal{K}_x) = \varphi(x), \quad \varphi \in \mathcal{H}_o, \quad x \in \mathfrak{X}.$$

The last relation and Cauchy–Bunyakovsky inequality imply that $\varphi = 0$ if $s(\varphi, \varphi) = 0$. Therefore,

$$(\varphi, \psi) = s(\varphi, \psi)$$

is the inner product in $\mathcal{H}_o$.

Denote by $\mathcal{H}$ the completion of $\mathcal{H}_o$, and let $\mathcal{H}(\mathcal{K})$ be a set of real-valued functions given on $\mathfrak{X}$ of the form

$$\varphi(x) = (h, \mathcal{K}_x)_{\mathcal{H}},$$

where $h \in \mathcal{H}$ and $(.,.)_{\mathcal{H}}$ is the inner product in $\mathcal{H}$. Define the following inner product in $\mathcal{H}(\mathcal{K})$:

$$(\varphi_1, \varphi_2)_{\mathcal{H}(\mathcal{K})} = (h_1, h_2)_{\mathcal{H}}.$$

The definition is correct because the linear span of elements $\mathcal{K}_x$ is dense everywhere in $\mathcal{H}$. The space $\mathcal{H}(\mathcal{K})$ is complete because it is isometric to $\mathcal{H}$. We have

$$\mathcal{K}_x(y) = (\mathcal{K}_x, \mathcal{K}_y)_{\mathcal{H}},$$

and therefore $\mathcal{K}_x \in \mathcal{H}(\mathcal{K})$, that is, $\mathcal{H}_o \subset \mathcal{H}(\mathcal{K})$.

The reproducing property follows now from the equalities

$$(\varphi, \mathcal{K}_x)_{\mathcal{H}(\mathcal{K})} = (h, \mathcal{K}_x)_{\mathcal{H}} = \varphi(x).$$

The uniqueness of a Hilbert space satisfying (a)–(c) follows from the fact that the linear span of the family $\{\mathcal{K}_x : x \in \mathfrak{X}\}$ must be dense (according to the reproducing property) in that space. □

Remark 21.2.1. Repeating the arguments of Theorem 21.2.1, it is easy to see that if $\mathcal{K}$ is a complex-valued positive definite kernel, then there exists a unique complex Hilbert space satisfying (b) and (c) whose elements are complex-valued functions.

Theorem 21.2.2 (Aronszajn 1950; Kolmogorov 1941). *A function $\mathcal{K} : \mathfrak{X}^2 \to \mathbb{R}^1$ is a positive definite kernel if and only if there exist a real Hilbert space $\mathcal{H}$ and a family $\{a_x : x \in \mathfrak{X}\} \subset \mathcal{H}$ such that*

$$\mathcal{K}(x, y) = (a_x, a_y) \tag{21.2.2}$$

for all $x, y \in \mathfrak{X}$.

Proof. Suppose that $\mathcal{K}(x, y)$ has the form (21.2.2). Then $\mathcal{K}(x, y) = \mathcal{K}(y, x)$. Let $n \in \mathbb{N}$, $x_1, \dots, x_n \in \mathfrak{X}$, $c_1, \dots, c_n \in \mathbb{R}^1$. We have

$$\sum_{i=1}^{n}\sum_{j=1}^{n} \mathcal{K}(x_i, x_j) c_i c_j = \sum_{i=1}^{n}\sum_{j=1}^{n} (a_{x_i}, a_{x_j}) c_i c_j = \left\| \sum_{i=1}^{n} c_i a_{x_i} \right\|_{\mathcal{H}}^2 \geq 0.$$

According to Property 21.2.1 of positive definite kernels, $\mathcal{K}$ is positive definite. Conversely, if $\mathcal{K}$ is a positive definite kernel, then we can choose $\mathcal{H}$ as the Hilbert space with reproducing kernel $\mathcal{K}$ and set $a_x = \mathcal{K}_x$. □

Remark 21.2.2. Theorem 21.2.2 remains true for complex-valued functions $\mathcal{K}$, but the Hilbert space $\mathcal{H}$ must be complex in this case.

Let us continue studying the properties of positive definite kernels. We need to use the notion of summable family.

Suppose that I is a nonempty set and $\tilde{I}$ is a set of all finite nonempty subsets of I. $\tilde{I}$ is a directed set with respect to inclusion. A family $\{u_i\}_{i\in I}$ of elements of a Banach space U is called summable if the generalized sequence $\sum_{i\in\alpha} u_i$, $\alpha \in \tilde{I}$, converges in U. If u is the limit of this generalized sequence, then we write

$$\sum_{i\in I} u_i = u.$$

Property 21.2.8. A function $\mathcal{K}: \mathfrak{X}^2 \to \mathbb{C}$ is a positive definite kernel if and only if there exists a family $\{f_i\}_{i\in I}$ of complex-valued functions such that $\sum_{i\in I} |f_i(x)|^2 < \infty$ for any $x \in \mathfrak{X}$ and

$$\mathcal{K}(x,y) = \sum_{i\in I} f_i(x)\bar{f}_i(y), \quad x, y \in \mathfrak{X}. \tag{21.2.3}$$

If $\mathcal{K}$ is real-valued, then the functions f_i, $i \in I$, may be chosen as real-valued.

To prove the positive definiteness of kernel (21.2.3), it is sufficient to note that each summand is a positive definite kernel and apply Properties 21.2.5 and 21.2.7.

Theorem 21.2.2 implies the existence of a Hilbert space $\mathcal{H}$ and of the family $\{a_x,\ x \in \mathfrak{X}\} \subset \mathcal{H}$ such that $\mathcal{K}(x,y) = (a_x, a_y)$ for all $x, y \in \mathfrak{X}$. Let us take the orthonormal basis $\{u_i\}_{i\in I}$ in $\mathcal{H}$, and set

$$f_i(x) = (a_x, u_i), \quad x \in \mathfrak{X}, \quad i \in I.$$

It is clear that

$$\sum_{i\in I} |f_i(x)|^2 = \|a_x\|^2 < \infty$$

and

$$\mathcal{K}(x,y) = (a_x, a_y) = \sum_{i\in I} (a_x, u_i)\overline{(a_y, u_i)} = \sum_{i\in I} f_i(x)\bar{f}_i(y).$$

Property 21.2.9. Suppose that $\mathcal{K}_1$ and $\mathcal{K}_2$ are positive definite kernels. Then $\mathcal{K}_1 \cdot \mathcal{K}_2$ is a positive definite kernel. In particular, $|\mathcal{K}_1|^2$ is a positive definite kernel, and for any integer $n \geq 1$ $\mathcal{K}_1^n$ is a positive definite kernel.
The proof follows from the fact that the product of two functions of the form (21.2.3) has the same form.

Property 21.2.10. If $\mathcal{K}$ is a positive definite kernel, then $\exp(\mathcal{K})$ is a positive definite kernel, too.

The proof follows from the expansion of $\exp(\mathcal{K})$ in power series and Properties 21.2.7 and 21.2.9.

Property 21.2.11. Let $\mathfrak{X} = \mathfrak{X}_1 \times \mathfrak{X}_2$, where $\mathfrak{X}_1$ and $\mathfrak{X}_2$ are nonempty sets. Suppose that $\mathcal{K}_j : \mathfrak{X}_j^2 \to \mathbb{C}$ is a positive definite kernel ($j = 1, 2$). Then $\mathcal{K} : \mathfrak{X}^2 \to \mathbb{C}$ defined as

$$\mathcal{K}(x, y) = \mathcal{K}_1(x_1, y_1) \cdot \mathcal{K}_2(x_2, y_2)$$

for all $x = (x_1, x_2) \in \mathfrak{X}$, $y = (y_1, y_2) \in \mathfrak{X}$ is a positive definite kernel.

The proof follows immediately from (21.2.3).

Property 21.2.12. Let $(\mathfrak{X}, \mathfrak{A})$ be a measurable space, and let μ be a σ-finite measure on it. Suppose that $\mathcal{K} : \mathfrak{X}^2 \to \mathbb{C}$ is a positive definite kernel on $\mathfrak{X}^2$, which is measurable and integrable with respect to $\mu \times \mu$. Then

$$\int_{\mathfrak{X}} \int_{\mathfrak{X}} \mathcal{K}(x, y) \mathrm{d}\mu(x) \mathrm{d}\mu(y) \geq 0.$$

Proof. If $\mathcal{K}$ is a measurable (with respect to the product of σ-fields) function of two variables, then the function $\mathcal{K}(t, t)$ of one variable is measurable, too. Therefore, there exists a set $\mathfrak{X}_o \in \mathfrak{A}$ such that $\mu(\mathfrak{X}_o) < \infty$, and the function $\mathcal{K}(t, t)$ is bounded on $\mathfrak{X}_o$. Because $\mathcal{K}$ is positive definite, we have

$$\sum_{i=1}^{n} \mathcal{K}(t_i, t_i) + \sum_{i \neq j} \mathcal{K}(t_i, t_j) \geq 0$$

for all $n \geq 2, t_1, \ldots, t_n \in \mathfrak{X}$. Integrating both sides of the last inequality over the set $\mathfrak{X}_o$ with respect to the n-times product $\mu \times \cdots \times \mu$ we obtain

$$n(\mu(\mathfrak{X}_o))^{n-1} \int_{\mathfrak{X}_o} \mathcal{K}(t, t) \mathrm{d}\mu(t) + n(n-1)(\mu(\mathfrak{X}_o))^{n-2} \int_{\mathfrak{X}_o} \int_{\mathfrak{X}_o} \mathcal{K}(s, t) \mathrm{d}\mu(s) \mathrm{d}\mu(t) \geq 0,$$

and, in view of the arbitrariness of n,

$$\int_{\mathfrak{X}_o} \int_{\mathfrak{X}_o} \mathcal{K}(s, t) \mathrm{d}\mu(s) \mathrm{d}\mu(t) \geq 0.$$

□

21.3 Examples of Positive Definite Kernels

Let us give some important examples of positive definite kernels.

Example 21.3.1. Let F be a nondecreasing bounded function on the real line. Define

$$\mathcal{K}(x, y) = \int_{-\infty}^{\infty} \mathrm{e}^{i(x-y)u} \mathrm{d}F(u),$$

where i is the imaginary unit. Then $\mathcal{K}$ is a positive definite kernel because of

$$\sum_{s=1}^{n}\sum_{t=1}^{n}\mathcal{K}(x_s,x_t)c_s\bar{c}_t = \int_{-\infty}^{\infty}\sum_{s=1}^{n}\mathrm{e}^{ix_su}c_s\overline{\sum_{t=1}^{n}\mathrm{e}^{ix_tu}c_t}\mathrm{d}F(u)$$
$$= \int_{-\infty}^{\infty}\left|\sum_{s=1}^{n}\mathrm{e}^{ix_su}c_s\right|^2\mathrm{d}F(u) \geq 0.$$

The kernel

$$\mathcal{K}_1(x,y) = \mathrm{Re}\,\mathcal{K}(x,y) = \int_{-\infty}^{\infty}\cos((x-y)u)\mathrm{d}F(u)$$

is also positive definite.

Example 21.3.2. Let F be a nondecreasing bounded function on $\mathbb{R}^1$ such that

$$\int_{-\infty}^{\infty}\mathrm{e}^{xu}\mathrm{d}F(u) < \infty$$

for all $x \in \mathbb{R}^1$. Define

$$\mathcal{K}(x,y) = \int_{-\infty}^{\infty}\mathrm{e}^{(x+y)u}\mathrm{d}F(u).$$

It is easy to see that $\mathcal{K}$ is a positive definite kernel.

Let, as usual, $\mathbb{N}_o$ be the set of all nonnegative integers.

Example 21.3.3. Suppose that F is a nondecreasing bounded function on $\mathbb{R}^1$ such that

$$\int_{-\infty}^{\infty}u^n\mathrm{d}F(u)$$

converges for all $u \in \mathbb{N}_o$. Define $\mathcal{K}: \mathbb{N}^2 \to \mathbb{R}^1$ as

$$\mathcal{K}(m,n) = \int_{-\infty}^{\infty}u^{m+n}\mathrm{d}F(u).$$

It is easy to see that $\mathcal{K}$ is a positive definite kernel.

Example 21.3.4. The inner product (x,y) in the Hilbert space $\mathcal{H}$ as a function of two variables is a positive definite kernel on $\mathcal{H}^2$. From here it follows that $\exp\{(x,y)\}$ and $\exp\{\mathrm{Re}\,(x,y)\}$ are positive definite kernels.

Example 21.3.5. The kernel

$$\mathcal{K}(x,y) = \exp\{-\|x-y\|^2\},$$

where x, y are elements of the Hilbert space $\mathcal{H}$, is positive definite. Indeed, for all $x_1, \dots, x_n \in \mathcal{H}$ and $c_1, \dots, c_n \in \mathbb{C}$ we have

$$\sum_{i=1}^{n}\sum_{j=1}^{n} c_i \bar{c}_j \exp\{-\|x_i - x_j\|^2\}$$
$$= \sum_{i=1}^{n}\sum_{j=1}^{n} c_i \bar{c}_j \exp\{-\|x_i\|^2\} \cdot \exp\{-\|x_j\|^2\} \cdot \exp\{2\mathrm{Re}\,(x_i, x_j)\}$$
$$= \sum_{i=1}^{n}\sum_{j=1}^{n} c_i' \bar{c}_j' \exp\{2\mathrm{Re}\,(x_i, x_j)\} \geq 0,$$

where $c_i' = c_i \exp\{-\|x_i\|^2\}$, and we used the positive definiteness of the kernel $\exp\{2\mathrm{Re}\,(x, y)\}$.

Example 21.3.6. Suppose that x, y are real numbers. Denote by $x \vee y$ the maximum of x and y. For any fixed $a \in \mathbb{R}^1$ set

$$U_a(x) = \begin{cases} 1 & \text{for } x < a, \\ 0 & \text{for } x \geq a. \end{cases}$$

Suppose that F is a nondecreasing bounded function on $\mathbb{R}^1$, and introduce the kernel

$$\mathcal{K}(x, y) = \int_{-\infty}^{\infty} U_a(x \vee y) \mathrm{d}F(a).$$

For all sets $x_1, \dots, x_n$ and $c_1, \dots, c_n$ of real numbers we have

$$\sum_{i=1}^{n}\sum_{j=1}^{n} \mathcal{K}(x_i, x_j) c_i c_j = \int_{-\infty}^{\infty} \left(\sum_{i=1}^{n} U_a(x_i) c_i \right)^2 \mathrm{d}F(a) \geq 0,$$

that is, $\mathcal{K}$ represents a positive definite kernel.

The following example provides a generalization of Example 21.3.6.

Example 21.3.7. Let $\mathfrak{X}$ be an arbitrary set and $\mathcal{A}$ a subset of $\mathfrak{X}$. Define

$$\mathcal{K}(x, y) = \begin{cases} 1 & \text{for } x \in \mathcal{A} \text{ and } y \in \mathcal{A}, \\ 0 & \text{otherwise.} \end{cases}$$

Then $\mathcal{K}$ is a positive definite kernel on $\mathfrak{X}^2$,

$$\sum_{i=1}^{n}\sum_{j=1}^{n}\mathcal{K}(x_i,x_j)c_ic_j = \sum_{j:x_j\in\mathcal{A}}\Big(\sum_{i:x_i\in\mathcal{A}} c_i\Big)c_j = \Big(\sum_{i:x_i\in\mathcal{A}} c_i\Big)^2 \geq 0.$$

21.4 Positive Definite Functions

Suppose that $\mathfrak{X} = \mathbb{R}^d$ is a d-dimensional Euclidean space. Let f be a complex-valued function on $\mathbb{R}^d$. We will say that f is a *positive definite function* if $\mathcal{K}(s,t) = f(s-t)$ is a positive definite kernel on $\mathbb{R}^d \times \mathbb{R}^d$.

Theorem 21.4.1 (Bochner 1933). *Let f be a complex-valued function on $\mathbb{R}^d$. f is a positive definite continuous function under condition $f(0) = 1$ if and only if f is a characteristic function of a probability measure on $\mathbb{R}^d$.*

Proof. For simplicity, we consider the case of $d = 1$.

(a) Let $f(t) = E\mathrm{e}^{itX}$, where X is a random variable. Then for all $t_1,\dots,t_n \in \mathbb{R}^1$ and $c_1,\dots,c_n \in \mathbb{C}$ we have

$$\sum_{j=1,k=1}^{n} f(t_j-t_k)c_j\bar{c}_k = E\sum_{j=1}^{n}\sum_{k=1}^{n}\big(c_j\mathrm{e}^{it_jX}\big)\big(\bar{c}_k\mathrm{e}^{-it_kX}\big)$$

$$= E\Big|\sum_{j=1}^{n} c_j\mathrm{e}^{it_jX}\Big|^2 \geq 0.$$

Therefore, the characteristic function of an arbitrary random variable is positive definite.

(b) Suppose now that f is a continuous positive definite function such that $f(0) = 1$. It is easy to calculate that for any $\sigma > 0$ the function

$$\varphi_\sigma(t) = \begin{cases} 1-\frac{|t|}{\sigma} & \text{for } |t| \leq \sigma; \\ 0 & \text{for } |t| \geq \sigma \end{cases}$$

is a characteristic function of the density

$$p(x) = \frac{1}{\pi\sigma}\frac{\sin^2(\sigma x/2)}{x^2}.$$

Let us consider the expression

$$p_\sigma(x) = \frac{1}{2\pi\sigma}\int_0^\sigma \mathrm{d}u \int_0^\sigma f(u-v)\mathrm{e}^{-iux}\mathrm{e}^{ivx}\mathrm{d}v. \qquad (21.4.1)$$

According to Property 21.2.12, we have $p_\sigma(x) \geq 0$. But, changing the variables in (21.4.1), we easily find

$$p_\sigma(x) = \frac{1}{2\pi}\int_{-\sigma}^{\sigma} \mathrm{e}^{-itx}\left(1 - \frac{|t|}{\sigma}\right) f(t)\mathrm{d}t \geq 0. \tag{21.4.2}$$

From the general properties of characteristic functions, we see from (21.4.2) that $p_\sigma \in L^1(\mathbb{R}^1)$ is the probability density function with characteristic function

$$\left(1 - \frac{|t|}{\sigma}\right) f(t).$$

But

$$f(t) = \lim_{\sigma\to\infty}\left(1 - \frac{|t|}{\sigma}\right) f(t), \quad f(0) = 1,$$

and f is a characteristic function in view of its continuity. □

Let us now consider a complex-valued function f given on the interval $(-a, a)$ $(a > 0)$ on the real line. We will say that f is a positive definite function on $(-a, a)$ if $f(x - y)$ is a positive definite kernel on $(-a, a) \times (-a, a)$. The following result was obtained by Krein (1940).

Theorem 21.4.2. *Let f be given on $(-a, a)$ and continuous at the origin. Then f is positive definite on $(-a, a)$ if and only if*

$$f(x) = \int_{-\infty}^{\infty} \mathrm{e}^{ixt}\mathrm{d}\sigma(t),$$

where $\sigma(t)$ $(-\infty < t < \infty)$ is a nondecreasing function of bounded variation.

We omit the proof of this theorem.

21.5 Negative Definite Kernels

Let $\mathfrak{X}$ be a nonempty set, and $\mathcal{L} : \mathfrak{X}^2 \to \mathbb{C}$. We will say that $\mathcal{L}$ is a *negative definite kernel* if for any $n \in \mathbb{N}$, arbitrary points $x_1, \ldots, x_n \in \mathfrak{X}$, and any complex numbers $c_1, \ldots, c_n$, under the condition $\sum_{j=1}^{n} c_j = 0$, the following inequality holds:

$$\sum_{i=1}^{n}\sum_{j=1}^{n} \mathcal{L}(x_i, x_j)c_i\bar{c}_j \leq 0. \tag{21.5.1}$$

The nine properties of negative definite kernels are as follows:

Property 21.5.1. If $\mathcal{L}$ is a real symmetric function on $\mathfrak{X}^2$, then $\mathcal{L}$ is a negative definite kernel if and only if (21.5.1) holds for arbitrary real numbers $c_1, \ldots, c_n$ under the condition $\sum_{j=1}^n c_j = 0$.

Property 21.5.1 follows from the definition of a negative definite kernel.

Property 21.5.2. If $\mathcal{L}$ is a negative definite kernel satisfying the condition

$$\mathcal{L}(x, y) = \overline{\mathcal{L}(y, x)}$$

for all $x, y \in \mathfrak{X}$, then the function $\operatorname{Re} \mathcal{L}$ is a negative definite kernel.
This property is an obvious consequence of Property 21.5.1.

Property 21.5.3. If the negative definite kernel $\mathcal{L}$ satisfies the conditions

$$\mathcal{L}(x, x) = 0, \quad \mathcal{L}(x, y) = \overline{\mathcal{L}(y, x)}$$

for all $x, y \in \mathfrak{X}$, then $\operatorname{Re} \mathcal{L} \geq 0$.
For the proof it is sufficient to put in (21.5.1) $n = 2$, $x_1 = x$, $x_2 = y$, $c_1 = 1$, and $c_2 = -1$.

Property 21.5.4. If $\mathcal{K} : \mathfrak{X}^2 \to \mathbb{C}$ is a positive definite kernel, then the function $\mathcal{L}$ defined by

$$\mathcal{L}(x, y) = \mathcal{K}(x, x) + \mathcal{K}(y, y) - 2\mathcal{K}(x, y), \quad x, y \in \mathfrak{X},$$

represents a negative definite kernel such that

$$\mathcal{L}(x, x) = 0, \ \mathcal{L}(x, y) = \overline{\mathcal{L}(y, x)}, \quad x, y \in \mathfrak{X}.$$

The proof follows from the definitions of positive and negative definite kernels.

Property 21.5.5. Suppose that $\mathcal{L}$ is a negative definite kernel such that $\mathcal{L}(x_o, x_o) = 0$ for some $x_o \in \mathfrak{X}$. Then the function

$$\mathcal{K}(x, y) = \mathcal{L}(x, x_o) + \mathcal{L}(x_o, y) - \mathcal{L}(x, y), \quad x, y \in \mathfrak{X},$$

is a positive definite kernel.

Proof. Take $n \in \mathbb{N}$, $x_1, \ldots, x_n \in \mathfrak{X}$, $c_1, \ldots, c_n \in \mathbb{C}$, and $c_o = -\sum_{j=1}^n c_j$. Then

$$\sum_{i=1}^{n}\sum_{j=1}^{n} \mathcal{K}(x_i, x_j) c_i \bar{c}_j = \sum_{i=0}^{n}\sum_{j=0}^{n} \mathcal{K}(x_i, x_j) c_i \bar{c}_j = \sum_{i=0}^{n}\sum_{j=0}^{n} \mathcal{L}(x_i, x_j) c_i \bar{c}_j \geq 0.$$

□

Property 21.5.6. Suppose that a real-valued negative definite kernel $\mathcal{L}$ satisfies the conditions

$$\mathcal{L}(x,x) = 0, \quad \mathcal{L}(x,y) = \mathcal{L}(y,x), \quad x, y \in \mathfrak{X}.$$

Then $\mathcal{L}$ can be represented in the form

$$\mathcal{L}(x,y) = \mathcal{K}(x,x) + \mathcal{K}(y,y) - 2\mathcal{K}(x,y), \quad x, y \in \mathfrak{X}, \tag{21.5.2}$$

where $\mathcal{K}$ is a real-valued positive definite kernel.
Let us fix an arbitrary $x_o \in \mathfrak{X}$. Set

$$\mathcal{K}(x,y) = \frac{1}{2}\Big(\mathcal{L}(x,x_o) + \mathcal{L}(x_o,y) - \mathcal{L}(x,y)\Big), \quad x, y \in \mathfrak{X}.$$

According to Property 21.5.5, $\mathcal{K}$ represents a positive definite kernel. It is easy to verify that $\mathcal{K}$ satisfies (21.5.2).

Property 21.5.7. Let $\mathcal{H}$ be a Hilbert space and $(a_x)_{x\in\mathfrak{X}}$ a family of elements of $\mathcal{H}$. Then the kernel

$$\mathcal{L}(x,y) = \|a_x - a_y\|^2$$

is negative definite. Conversely, if a negative definite kernel $\mathcal{L} : \mathfrak{X}^2 \to \mathbb{R}^1$ satisfies the conditions $\mathcal{L}(x,x) = 0$, $\mathcal{L}(x,y) = \mathcal{L}(y,x)$, then there exists a real Hilbert space $\mathcal{H}$ and a family $(a_x)_{x\in\mathfrak{X}}$ of its elements such that

$$\mathcal{L}(x,y) = \|a_x - a_y\|^2, \quad x, y \in \mathfrak{X}.$$

The first part of this statement follows from Property 21.5.1. The second part follows from Property 21.5.6 and the Aronszajn–Kolmogorov theorem.

Property 21.5.8. Let $\mathcal{L} : \mathfrak{X}^2 \to \mathbb{C}$ satisfy the condition $\mathcal{L}(x,y) = \overline{\mathcal{L}(y,x)}$ for all $x, y \in \mathfrak{X}$. Then the following statements are equivalent:

(a) $\exp\{-\alpha\mathcal{L}\}$is a positive definite kernel for all $\alpha > 0$.
(b) $\mathcal{L}$ is a negative definite kernel.

Proof. Suppose that statement (a) is true. Then it is easy to see that for any $\alpha > 0$ the kernel $\mathcal{L}_\alpha = (1 - \exp\{-\alpha\mathcal{L}\})/\alpha$ is negative definite. It is clear that the limit function $\mathcal{L} = \lim_{\alpha\to 0}\mathcal{L}_\alpha$ is negative definite, too.

Now let us suppose that statement (b) holds. Passing from the kernel $\mathcal{L}$ to the function $\mathcal{L}_o = \mathcal{L} - \mathcal{L}(x_o,x_o)$, we may suppose that $\mathcal{L}(x_o,x_o) = 0$ for some $x_o \in \mathfrak{X}$. According to Property 21.5.5, we have

$$\mathcal{L}(x,y) = \mathcal{L}(x,x_o) + \overline{\mathcal{L}(y,x_o)} - \mathcal{K}(x,y),$$

where $\mathcal{K}$ is a positive definite kernel. Let $\alpha > 0$. For any $n \in \mathbb{N}$, $x_1, \ldots, x_n \in \mathfrak{X}$, $c_1, \ldots, c_n \in \mathbb{C}$ we have

$$\sum_{i=1}^{n}\sum_{j=1}^{n} \exp\{-\alpha\mathcal{L}(x_i,x_j)\}c_i\bar{c}_j$$
$$= \sum_{i=1}^{n}\sum_{j=1}^{n} \exp\{-\alpha\mathcal{L}(x_i,x_o)\} \times \exp\{-\alpha\overline{\mathcal{L}(x_j,x_o)}\} \times \exp\{\alpha\mathcal{K}(x_i,x_j)\}c_i\bar{c}_j$$
$$= \sum_{i=1}^{n}\sum_{j=1}^{n} \exp\{\alpha\mathcal{K}(x_i,x_j)\}c_i'\bar{c}_j' \geq 0,$$

where $c_i' = \exp\{-\alpha\mathcal{L}(x_i,x_o)\}c_i$. □

Property 21.5.9. Suppose that a negative definite kernel $\mathcal{L} : \mathfrak{X}^2 \to \mathbb{R}^1_+$ satisfies the conditions

$$\mathcal{L}(x,x) = 0,\ \mathcal{L}(x,y) = \mathcal{L}(y,x),\quad x,y \in \mathfrak{X}.$$

Let ν be a measure on $\mathbb{R}^1_+$ such that

$$\int_{\mathbb{R}^1_+} \min(1,t)\mathrm{d}\nu(t) < \infty.$$

Then the kernel

$$\mathcal{L}_\nu(x,y) = \int_{\mathbb{R}^1_+} (1 - \exp\{-t\mathcal{L}(x,y)\})\mathrm{d}\nu(t),\quad x,y \in \mathfrak{X},$$

is negative definite. In particular, if $\alpha \in [0,1]$, then $\mathcal{L}^\alpha$ is a negative definite kernel. According to Property 21.5.8, the function $\exp\{-t\mathcal{L}(x,y\}$ is a positive definite kernel; therefore, $1 - \exp\{-t\mathcal{L}(x,y\}$ is negative definite for all $t \geq 0$. Hence, $\mathcal{L}_\nu(x,y)$ is a negative definite kernel. To complete the proof, it is sufficient to note that $\mathcal{L}^\alpha = C_\alpha\mathcal{L}_{\nu_\alpha}$, where $\nu_\alpha(B) = \int_B x^{-(\alpha+1)}\mathrm{d}x$ for any Borel set $B \subset \mathbb{R}^1_+$ and C_α is a positive constant.

Theorem 21.5.1 (Schoenberg 1938). *Let $(\mathfrak{X},d)$ be a metric space. $(\mathfrak{X},d)$ is isometric to a subset of a Hilbert space if and only if d^2 is a negative definite kernel on $\mathfrak{X}^2$.*

Proof. Let us suppose that d^2 is a negative definite kernel. According to Property 21.5.7, there exists a Hilbert space $\mathcal{H}$ and a family $(a_x)_{x\in\mathfrak{X}}$ such that

$$d^2(x,y) = \|a_x - a_y\|^2,$$

that is, $d(x,y) = \|a_x - a_y\|$. Therefore, the map $x \to a_x$ is an isometry from $\mathfrak{X}$ to $\mathcal{Y} = \{a_x : x \in \mathfrak{X}\} \subset \mathcal{H}$.

Let us now suppose that f is an isometry from $(\mathfrak{X}, d)$ to a subset $\mathcal{Y}$ of a Hilbert space $\mathcal{H}$. Set $a_x = f(x)$. We have

$$d(x, y) = \|a_x - a_y\|,$$

that is,

$$d^2(x, y) = \|a_x - a_y\|^2,$$

which is a negative definite kernel by Property 21.5.7. □

Let us now give one important example of negative definite kernels.

Example 21.5.1. Let $(\mathcal{X}, \mathfrak{A}, \mu)$ be a space with a measure (μ is not necessarily a finite measure). Define the function $\psi_p : L^p(\mathfrak{X}, \mathfrak{A}, \mu) \to \mathbb{R}^1_+$ by setting

$$\psi_p(x) = \|x\|_p^p = \int_{\mathcal{X}} |x(t)|^p \mathrm{d}\mu(t), \quad x \in \mathfrak{X} = L^p.$$

Then

$$\mathcal{L}(x, y) = \psi_p(x - y), \quad x, y \in \mathfrak{X}$$

is a negative definite kernel for any $p \in (0, 2]$.

Proof. Indeed, the kernel $(u, v) \to |u - v|^p$ is negative definite on $\mathbb{R}^1$, and therefore

$$\sum_{i=1}^{n}\sum_{j=1}^{n} \|x_i - x_j\|^p c_i c_j = \int_{\mathcal{X}} \left(\sum_{i,j} c_i c_j |x_i(t) - x_j(t)|^p \right) \mathrm{d}\mu(t) \le 0$$

for all $x_1, \ldots, x_n \in L^p$, $c_1, \ldots, c_n \in \mathbb{R}^1$. □

From Property 21.5.9 it follows that $\mathcal{L}_p^\alpha$ is a negative definite kernel for $\alpha \in [0, 1]$.

Corollary 21.5.1. *For any measure μ, the space $L^p(\mu)$ with $1 \le p \le 2$ is isometric to some subspace of a Hilbert space.*

Proof. The proof follows immediately from Example 21.5.1 and Schoenberg's theorem 21.5.1. □

21.6 Coarse Embeddings of Metric Spaces into Hilbert Space

Definition 21.6.1. Let $(\mathfrak{X}, d_1)$ and $(\mathfrak{Y}, d_2)$ be metric spaces. A function f from $\mathfrak{X}$ to $\mathfrak{Y}$ is called a coarse embedding if there exist two nondecreasing functions ρ_1 and ρ_2 from $\mathbb{R}^1_+$ into itself and such that

$$\rho_1(d_1(x, y)) \le d_2(f(x), f(y)) \le \rho_2(d_1(x, y)) \text{ for all } x, y \in \mathfrak{X}, \tag{21.6.1}$$

$$\lim_{z\to\infty} \rho_1(z) = \infty. \tag{21.6.2}$$

Our goal here is to prove the following theorem.

Theorem 21.6.1. *A metric space $(\mathfrak{X}, d)$ admits a coarse embedding into a Hilbert space if and only if there exist a negative definite symmetric kernel $\mathcal{L}$ on $\mathfrak{X}^2$ and nondecreasing functions ρ_1, ρ_2 such that*

$$\mathcal{L}(x,x) = 0, \quad \forall x \in \mathfrak{X}; \tag{21.6.3}$$

$$\rho_1(d(x,y)) \le \mathcal{L}(x,y) \le \rho_2(d(x,y)); \tag{21.6.4}$$

$$\lim_{z\to\infty} \rho_1(z)) = \infty. \tag{21.6.5}$$

Proof. Suppose that there exists a negative definite kernel $\mathcal{L}$ satisfying (21.6.3)–(21.6.5). According to Theorem 21.5.1, there exists a Hilbert space $\mathcal{H}$ and a map $f : \mathfrak{X} \to \mathcal{H}$ such that

$$\mathcal{L}(x,y) = \|f(x) - f(y)\|^2 \text{ for all } x, y \in \mathfrak{X}.$$

Therefore,

$$\sqrt{\rho_1(d(x,y))} \le \|f(x) - f(y)\| \le \sqrt{\rho_2(d(x,y))},$$

which means that f is a coarse embedding.

Suppose now that there exists a coarse embedding f from $\mathfrak{X}$ into a Hilbert space $\mathcal{H}$. Set

$$\mathcal{L}(x,y) = \|f(x) - f(y)\|^2.$$

According to Property 21.5.7 of Sect. 21.5, $\mathcal{L}$ is a negative definite kernel satisfying (21.6.3). This kernel satisfies (21.6.4) and (21.6.5) by the definition of a coarse embedding. □

21.7 Strictly and Strongly Positive and Negative Definite Kernels

Let $\mathfrak{X}$ be a nonempty set, and let $\mathcal{L} : \mathfrak{X}^2 \to \mathbb{C}$ be a negative definite kernel. As we know, this means that for arbitrary $n \in \mathbb{N}$, any $x_1, \dots, x_n \in \mathfrak{X}$, and any complex numbers $c_1, \dots, c_n$, under the condition $\sum_{j=1}^n c_j = 0$, the following inequality holds:

$$\sum_{i=1}^n \sum_{j=1}^n \mathcal{L}(x_i, x_j) c_i \bar{c}_j \le 0. \tag{21.7.1}$$

We will say that a negative definite kernel $\mathcal{L}$ is strictly negative definite if the equality in (21.7.1) is true for $c_1 = \dots = c_n = 0$ only.

If $\mathcal{L}$ is a real-valued symmetric function given on $\mathfrak{X}^2$, then bearing in mind Property 21.5.1 of negative definite kernels shows us that $\mathcal{L}$ is a negative definite kernel if and only if (21.7.1) is true for all real numbers $c_1, \dots, c_n$ ($\sum_{j=1}^{n} c_j = 0$), and the equality holds for $c_1 = \dots = c_n = 0$ only.

Let $\mathcal{K}$ be a positive definite kernel. We will say that $\mathcal{K}$ is a strictly positive definite kernel if the function

$$\mathcal{L}(x, y) = \mathcal{K}(x, x) + \mathcal{K}(y, y) - 2\mathcal{K}(x, y), \quad x, y \in \mathfrak{X} \tag{21.7.2}$$

is a strictly negative definite kernel.

Let $\mathcal{K}$ be real-valued symmetric function given on $\mathfrak{X}^2$. Suppose that $\mathcal{K}$ is a strictly negative definite kernel and $\mathcal{L}$ is defined by (21.7.2). Then $\mathcal{L}(x, x) = 0$ for any $x \in \mathfrak{X}$. Choosing in (21.7.1) $n = 2$, $c_1 = 1 = -c_2$, we obtain $\mathcal{L}(x, y) \geq 0$ for all $x, y \in \mathfrak{X}$, and $\mathcal{L}(x, y) = 0$ if and only if $x = y$. Let us now fix arbitrary $x, y, z \in \mathfrak{X}$ and set in (21.7.1) $n = 3$, $x_1 = x$, $x_2 = y$, $x_3 = z$, $c_1 = \lambda/(\mathcal{L}(x, z))^{1/2}$, $c_2 = \lambda/(\mathcal{L}(y, z))^{1/2}$, $c_3 = -(c_1 + c_2)$, $\lambda = \big((\mathcal{L}(x, z))^{1/2} + (\mathcal{L}(y, z))^{1/2}\big)/(\mathcal{L}(x, y))^{1/2}$. Then (21.7.1) implies that

$$(\mathcal{L}(x, y))^{1/2} \leq (\mathcal{L}(x, z))^{1/2} + (\mathcal{L}(z, y))^{1/2}.$$

As $\mathcal{K}(x, y) = \mathcal{K}(y, x)$, then $\mathcal{L}(x, y) = \mathcal{L}(y, x)$. Therefore, bearing in mind Schoenberg's theorem 21.5.1, we obtain the following theorem.

Theorem 21.7.1. *Let $\mathfrak{X}$ be a nonempty set and $\mathcal{K}$ a real-valued symmetric function on $\mathfrak{X}^2$. Suppose that $\mathcal{K}$ is a strictly positive definite kernel and $\mathcal{L}$ is defined by (21.7.2). Then*

$$d(x, y) = (\mathcal{L}(x, y))^{1/2} \tag{21.7.3}$$

is a metric on $\mathfrak{X}$. The metric space $(\mathfrak{X}, d)$ is isometric to a subset of a Hilbert space.

Later in this section we will suppose that $\mathfrak{X}$ is a metric space. We will denote by $\mathfrak{A}$ the algebra of its Bair subsets. When discussing negative definite kernels, we will suppose they are continuous, symmetric, and real-valued. Denote by $\mathcal{B}$ the set of all probability measures on $(\mathfrak{X}, \mathfrak{A})$.

Suppose that $\mathcal{L}$ is a real continuous function, and denote by $\mathcal{B}_{\mathcal{L}}$ the set of all measures $\mu \in \mathcal{B}$ for which the integral

$$\int_{\mathfrak{X}} \int_{\mathfrak{X}} \mathcal{L}(x, y) \mathrm{d}\mu(x) \mathrm{d}\mu(y)$$

exists.

Theorem 21.7.2. *Let $\mathcal{L}$ be a real continuous function on $\mathfrak{X}^2$ under the condition*

$$\mathcal{L}(x, y) = \mathcal{L}(y, x), \quad x, y \in \mathfrak{X}. \tag{21.7.4}$$

The inequality

$$2\int_{\mathfrak{X}}\int_{\mathfrak{X}}\mathcal{L}(x,y)\mathrm{d}\mu(x)\mathrm{d}\nu(y)-\int_{\mathfrak{X}}\int_{\mathfrak{X}}\mathcal{L}(x,y)\mathrm{d}\mu(x)\mathrm{d}\mu(y)$$
$$-\int_{\mathfrak{X}}\int_{\mathfrak{X}}\mathcal{L}(x,y)\mathrm{d}\nu(x)\mathrm{d}\nu(y)\geq 0 \tag{21.7.5}$$

holds for all $\mu,\nu\in\mathcal{B}_{\mathcal{L}}$ *if and only if* $\mathcal{L}$ *is a negative definite kernel.*

Proof. It is obvious that the definition of a negative definite kernel is equivalent to the condition that

$$\int_{\mathfrak{X}}\int_{\mathfrak{X}}\mathcal{L}(x,y)h(x)h(y)\mathrm{d}Q(x)\mathrm{d}Q(y)\leq 0 \tag{21.7.6}$$

for any probability measure Q given on $(\mathfrak{X},\mathfrak{A})$ and arbitrary integrable function h satisfying the condition $\int_{\mathfrak{X}}h(x)\mathrm{d}Q(x)=0$. Let Q_1 be an arbitrary measure from $\mathcal{B}$ dominating both μ and ν. Denote

$$h_1=\frac{\mathrm{d}\mu}{\mathrm{d}Q_1},\quad h_2=\frac{\mathrm{d}\nu}{\mathrm{d}Q_1},\quad h=h_1-h_2.$$

Then inequality (21.7.5) may be written in the form (21.7.6) for $Q=Q_1$, $h=h_1-h_2$. The measure Q_1 and the function h with zero mean are arbitrary in view of the arbitrariness of μ and ν. Therefore, (21.7.5) and (21.7.6) are equivalent. □

Definition 21.7.1. Let Q be a measure on $(\mathfrak{X},\mathfrak{A})$, and let h be a function integrable with respect to Q and such that

$$\int_{\mathfrak{X}}h(x)\mathrm{d}Q(x)=0.$$

We will say that $\mathcal{L}$ is a strongly negative definite kernel if $\mathcal{L}$ is negative definite and the equality

$$\int_{\mathfrak{X}}\int_{\mathfrak{X}}\mathcal{L}(x,y)h(x)h(y)\mathrm{d}Q(x)\mathrm{d}Q(y)=0$$

implies that $h(x)=0$ Q-almost everywhere (a.e.) for any measure Q.

Theorem 21.7.3. *Let* $\mathcal{L}$ *be a real continuous function satisfying* (21.7.4). *Inequality* (21.7.5) *holds for all measures* $\mu,\nu\in\mathcal{B}$, *with equality in the case* $\mu=\nu$ *only, if and only if* $\mathcal{L}$ *is a strongly negative definite kernel.*

Proof. The proof is obvious in view of the equivalency of (21.7.5) and (21.7.6). □

Of course, a strongly negative definite kernel is at the same time a strictly negative definite kernel.

Here are some examples of strongly negative definite kernels.

Example 21.7.1. Let $\mathfrak{X} = \mathbb{R}^1$. Set

$$U(z) = \int_0^\infty \left(1 - \cos(zx)\right)\frac{1+x^2}{x^2}\mathrm{d}\theta(x),$$

where $\theta(x)$ is a real nondecreasing function, $\theta(-0) = 0$. It is easy to verify that the kernel

$$\mathcal{L}(x, y) = U(x - y)$$

is negative definite. $\mathcal{L}$ is strongly negative definite if and only if $\operatorname{supp}\theta = [0, \infty)$.

Because

$$|x|^r = c_r \int_0^\infty \left(1 - \cos(xt)\right)\frac{\mathrm{d}t}{t^{r+1}}$$

for $0 < r < 2$, where

$$c_r = 1/\int_0^\infty \left(1 - \cos t\right)\frac{\mathrm{d}t}{t^{r+1}} = -1/\left(\Gamma(-r)\cos\frac{\pi r}{2}\right),$$

then $|x - y|^r$ is a *strongly negative definite kernel* for $0 < r < 2$. It is a negative definite kernel (but not strongly) for $r = 0$ and $r = 2$.

Example 21.7.2. Let $\mathfrak{X}$ be a separable Hilbert space. Assume that $f(t)$ is a real characteristic functional of an infinitely divisible measure on $\mathfrak{X}$. Then $\mathcal{L}(t) = -\log f(t)$ is a negative definite function on $\mathfrak{X}$ (i.e., $\mathcal{L}(x - y), x, y \in \mathfrak{X}$, is a negative definite kernel). We know that

$$\mathcal{L}(t) = \frac{1}{2}(Bt, t) - \int_{\mathfrak{X}} \left(\mathrm{e}^{i\langle t,x\rangle} - 1 - \frac{i\langle t, x\rangle}{1 + \|x\|^2}\right)\frac{1 + \|x\|^2}{\|x\|^2}\mathrm{d}\theta(x),$$

where B is the kernel operator and θ is a finite measure for which $\theta(\{0\}) = 0$. Clearly, if $\operatorname{supp}\theta = \mathfrak{X}$, then $\mathcal{L}$ is a strongly negative definite function on $\mathfrak{X}$.

Example 21.7.3. Let $\mathcal{L}(z)$ be a survival function on $\mathbb{R}^1$ [i.e., $1 - \mathcal{L}(x)$ is a distribution function]. Then the function $\mathcal{L}(x \bigwedge y)$ is a negative definite kernel (here $x \bigwedge y$ is the minimum of x and y). Suppose that

$$g_a(z) = \begin{cases} 0 & \text{for } z \le a, \\ 1 & \text{for } z > a, \end{cases}$$

and for all $x_1 \le x_2 \le \cdots \le x_n$ we have

$$\sum_{i=1}^n \sum_{j=1}^n g_a(x_i \wedge x_j) h_i h_j = \sum_{i=k}^n \sum_{j=k}^n h_i h_j = \Big(\sum_{i=k}^n h_i\Big)^2 \ge 0,$$

where k is determined by the conditions $x_k > a$, $x_{k-1} \le a$. The foregoing conclusion now follows from the obvious equality

$$\mathcal{L}(z) = \int_{-\infty}^{\infty} (1 - g_a(x))d\sigma(a),$$

where σ is a suitable distribution function. Clearly, $\mathcal{L}(x \wedge y)$ is a strongly negative definite kernel if and only if σ is decreasing and strictly monotone.

References

Aronszajn N (1950) Theory of reproducing kernels. Trans Am Math Soc 68:337–404

Bochner S (1933) Integration von funktionen, deren werte die elemente eines vektorraumes sind. Fund Math 20:262–276

Kolmogorov AN (1941) Stationary sequences in Hilbert space. Bull MGU (in Russian) 2:1–40

Krein MG (1940) On the problem of prolongation of Hermitian positive functions. Dokl Akad Nauk (in Russian) 26:17–22

Schoenberg IJ (1938) Metric spaces and positive definite functions. Trans Am Math Soc 44(3):552–563

Vakhaniya NN, Tarieladze VI, Chobanyan SA (1985) Probability distributions in Banach spaces. Nauka, Moscow (in Russian)

Chapter 22
Negative Definite Kernels and Metrics: Recovering Measures from Potentials

The goals of this chapter are to:

- Introduce probability metrics through strongly negative definite kernel functions and provide examples,
- Introduce probability metrics through m-negative definite kernels and provide examples,
- Introduce the notion of potential corresponding to a probability measure,
- Present the problem of recovering a probability measure from its potential,
- Consider the relation between the problems of convergence of measures and the convergence of their potentials,
- Characterize probability distributions using the theory of recovering probability measures from potentials.

Notation introduced in this chapter:

Notation	Description
$\mathcal{B}_{\mathcal{K}}$	Set of all measures defined on measurable space $(\mathfrak{X}, \mathfrak{A})$ such that $\int_{\mathfrak{X}} \mathcal{K}(x,x) d\mu(x) < \infty$
$\mathfrak{K}(\mu, \nu)$	Positive definite kernel defined on $\mathcal{B}_{\mathcal{K}}^2$
$\mathcal{B}(\mathcal{L})$	Set $\mathcal{B}_{\mathcal{K}}$ when kernel $\mathcal{K}$ arises from a strongly negative kernel $\mathcal{L}$
$(\mathcal{B}(\mathcal{L}), \mathfrak{N})$	Metric space in which the distance $\mathfrak{N}$ is defined through a strongly negative kernel $\mathcal{L}$
$\mathfrak{R}$	Set of all signed measures on $(\mathfrak{X}, \mathfrak{A})$
$\|\|R\|\|$	Norm of a signed measure $R \in \mathfrak{R}$
$\mathcal{N}_m^{1/m}$	Probability metric arising from m-negative definite kernel
$\varphi(x;\mu),\ x \in \mathfrak{X}$	Potential of a measure μ
$f(u;\mu)$	Characteristic function of μ

S.T. Rachev et al., *The Methods of Distances in the Theory of Probability and Statistics*, DOI 10.1007/978-1-4614-4869-3_22, © Springer Science+Business Media, LLC 2013

22.1 Introduction

In this chapter, we introduce special classes of probability metrics through negative definite kernel functions discussed in the previous chapter. Apart from generating distance functions with interesting mathematical properties, kernel functions are central to the notion of potential of probability measures. It turns out that for strongly negative definite kernels, a probability measure can be uniquely determined by its potential. The distance between probability measures can be bounded by the distance between their potentials, meaning that, under some technical conditions, a sequence of probability measures converges to a limit if and only if the sequence of their potentials converges to the potential of the limiting probability measure. Finally, the problem of characterizing classes of probability distributions can be reduced to the problem of recovering a measure from potential. Examples are provided for the normal distribution, for symmetric distributions, and for distributions symmetric to a group of transformations.

22.2 $\mathfrak{N}$-Metrics in the Set of Probability Measures

In this section, we introduce distances generated by negative definite kernels in the set of probability measures. The corresponding metric space is isometric to a convex subset of a Hilbert space.[1]

22.2.1 A Class of Positive Definite Kernels in the Set of Probabilities and $\mathfrak{N}$-Distances

Let $(\mathfrak{X}, \mathfrak{A})$ be a measurable space. Denote by $\mathcal{B}$ the set of all probability measures on $(\mathfrak{X}, \mathfrak{A})$. Suppose that $\mathcal{K}$ is a positive definite symmetric kernel on $\mathfrak{X}$, and let us define the following function on $\mathfrak{X}^2$:

$$\mathfrak{K}(\mu, \nu) = \int_{\mathfrak{X}} \int_{\mathfrak{X}} \mathcal{K}(x, y) \mathrm{d}\mu(x) \mathrm{d}\nu(y). \tag{22.2.1}$$

Denote by $\mathcal{B}_{\mathcal{K}}$ the set of all measures $\mu \in \mathcal{B}$ for which

$$\int_{\mathfrak{X}} \mathcal{K}(x, x) \mathrm{d}\mu(x) < \infty.$$

Proposition 22.2.1. *The function $\mathfrak{K}$ given by* (22.2.1) *is a positive definite kernel on $\mathcal{B}_{\mathcal{K}}^2$.*

[1]Sriperumbudur et al. (2010) discuss metrics similar to the $\mathfrak{N}$-distances that we cover in this chapter. However, the results they present were already reported in the literature.

Proof. If $\mu, \nu \in \mathcal{B}_{\mathcal{K}}$, then, according to Property 21.2.5 of positive definite kernels provided in Sect. 21.2 of Chap. 21, the integral on the right-hand side of (22.2.1) exists. In view of the symmetry of $\mathfrak{K}$, we must prove that for arbitrary $\mu_1, \dots, \mu_n \in \mathcal{B}_{\mathcal{K}}$ and arbitrary $c_1, \dots, c_n \in \mathbb{R}^1$ we have

$$\sum_{i=1}^{n}\sum_{j=1}^{n} \mathfrak{K}(\mu_i, \mu_j)c_i c_j \geq 0.$$

Approximating measures μ_i, μ_j by discrete measures we can write

$$\mathfrak{K}(\mu_i, \mu_j) = \int_{\mathfrak{X}}\int_{\mathfrak{X}} \mathcal{K}(x, y)\mathrm{d}\mu_i(x)\mathrm{d}\mu_j(y) = \lim_{m\to\infty} \sum_{s=1}^{m}\sum_{t=1}^{m} \mathcal{K}(x_{s,i}, x_{t,j})a_{s,i}a_{t,j}.$$

Therefore,

$$\sum_{i=1}^{n}\sum_{j=1}^{n} \mathfrak{K}(\mu_i, \mu_j)c_i c_j = \lim_{m\to\infty} \sum_{s=1}^{m}\sum_{t=1}^{m} \left[\sum_{i=1}^{n}\sum_{j=1}^{n} \mathcal{K}(x_{s,i}, x_{t,j})(a_{s,i}c_i)(a_{t,j}c_j)\right].$$

The double summation in the square brackets on the right-hand side of the preceding equality is nonnegative in view of the positive definiteness of $\mathcal{K}$. Therefore, the limit is nonnegative. □

Consider now a negative definite kernel $\mathcal{L}(x, y)$ on $\mathfrak{X}^2$ such that $\mathcal{L}(x, y) = \mathcal{L}(y, x)$ and $\mathcal{L}(x, x) = 0$ for all $x, y \in \mathfrak{X}$. Then for any fixed $x_o \in \mathfrak{X}$ the kernel

$$\mathcal{K}(x, y) = \mathcal{L}(x, x_o) + \mathcal{L}(x_o, y) - \mathcal{L}(x, y)$$

is positive definite (see Property 21.5.5 of negative definite kernels explained in Chap. 21). According to Proposition 22.2.1, the function

$$\begin{aligned} \mathfrak{K}(\mu, \nu) &= \int_{\mathfrak{X}}\int_{\mathfrak{X}} \mathcal{K}(x, y)\mathrm{d}\mu(x)\mathrm{d}\nu(y) \\ &= \int_{\mathfrak{X}} \mathcal{L}(x, x_o)\mathrm{d}\mu(x) + \int_{\mathfrak{X}} \mathcal{L}(x_o, y)\mathrm{d}\nu(y) \\ &\quad - \int_{\mathfrak{X}}\int_{\mathfrak{X}} \mathcal{L}(x, y)\mathrm{d}\mu(x)\mathrm{d}\nu(y) \end{aligned} \tag{22.2.2}$$

is a positive definite kernel on $\mathcal{B}_{\mathcal{K}}^2$. Property 21.5.4 for negative definite kernels explained in Chap. 21 shows us that

$$\mathcal{N}(\mu, \nu) = \mathfrak{K}(\mu, \mu) + \mathfrak{K}(\nu, \nu) - 2\mathfrak{K}(\mu, \nu)$$

is a negative definite kernel on $\mathcal{B}_{\mathcal{K}}^2$. Bearing in mind (22.2.2), we can write $\mathcal{N}$ in the form

$$\mathcal{N}(\mu,\nu) = 2\int_{\mathfrak{X}}\int_{\mathfrak{X}} \mathcal{L}(x,y)\mathrm{d}\mu(x)\mathrm{d}\nu(y)$$
$$-\int_{\mathfrak{X}}\int_{\mathfrak{X}} \mathcal{L}(x,y)\mathrm{d}\mu(x)\mathrm{d}\mu(y) - \int_{\mathfrak{X}}\int_{\mathfrak{X}} \mathcal{L}(x,y)\mathrm{d}\nu(x)\mathrm{d}\nu(y), \tag{22.2.3}$$

which is independent of the choice of x_o.

In the case where $\mathcal{L}$ is a strongly negative definite kernel, Theorem 21.7.3 in Chap. 21 shows that $\mathcal{N}(\mu,\nu) = 0$ if and only if $\mu = \nu$. For any given $\mathcal{L}$ set

$$\mathcal{K}(x,y) = \mathcal{L}(x,x_o) + \mathcal{L}(x_o,y) - \mathcal{L}(x,y)$$

and denote by $\mathcal{B}(\mathcal{L})$ the set $\mathcal{B}_{\mathcal{K}}$. Therefore, we have the following result.

Theorem 22.2.1. *Let $\mathcal{L}$ be a strongly negative definite kernel on $\mathfrak{X}^2$ satisfying*

$$\mathcal{L}(x,y) = \mathcal{L}(y,x), \quad \textit{and} \quad \mathcal{L}(x,x) = 0 \quad \textit{for all} \quad x,y \in \mathfrak{X}. \tag{22.2.4}$$

Let $\mathcal{N}$ be defined by (22.2.3)*. Then $\mathfrak{N} = \mathcal{N}^{1/2}(\mu,\nu)$ is a distance on $\mathcal{B}(\mathcal{L})$.*

In the remaining part of this chapter, we suppose that $\mathcal{L}$ satisfies (22.2.4).

Suppose now that $(\mathfrak{X},d)$ is a metric space. Assume that $d^2(x,y) = \mathcal{L}(x,y)$, where $\mathcal{L}$ is a strongly negative definite kernel on $\mathfrak{X}^2$. As we already noted, in this case $\mathcal{N}(\mu,\nu)$ is a strictly negative definite kernel on $\mathcal{B}(\mathcal{L}) \times \mathcal{B}(\mathcal{L})$ and, according to Schonenberg's theorem, the metric space $(\mathcal{B}(\mathcal{L}),\mathfrak{N})$, where $\mathfrak{N} = \mathcal{N}^{1/2}$ is isometric to a subset of the Hilbert space $\mathcal{H}$. We can identify $\mathfrak{X}$ with some subset of $\mathcal{B}(\mathcal{L})$ by letting a point from $\mathfrak{X}$ correspond to the measure concentrated at that point.

Remark 22.2.1. It is easy to see that under such isometry, the image $\tilde{\mathcal{B}}(\mathcal{L})$ of the set $\mathcal{B}(\mathcal{L})$ is a convex subset of $\mathcal{H}$. Every point of this image is a barycenter of a set of points from image $\tilde{\mathfrak{X}}$ of the space $\mathfrak{X}$. Thus, the distance (the metric) $\mathfrak{N}$ between two measures can be described as the distance between the corresponding barycenters in the Hilbert space $\mathcal{H}$.

The converse is also true. That is, if there exists an isometry of space $\mathcal{B}(\mathcal{L})$ (with the distance on $\mathfrak{X}$ preserved) onto some subset $\tilde{\mathcal{B}}(\mathcal{L})$ of the Hilbert space $\mathcal{H}$ such that $\tilde{\mathcal{B}}(\mathcal{L})$ is a convex set and the distance between measures is the distance between the corresponding barycenters in $\mathcal{H}$, then $\mathcal{L}(x,y) = d^2(x,y)$ is a strongly negative definite kernel on $\mathfrak{X}^2$ and $\mathcal{N}(\mu,\nu)$ is calculated from (22.2.3).

Let X, Y be two independent random variables (RVs) with cumulative distribution functions μ, ν, respectively. Denote by X', Y' independent copies of X, Y, i.e., X and X' are identically distributed (notation $X \stackrel{d}{=} X'$), $Y \stackrel{d}{=} Y'$, and all RVs X, X', Y, Y' are mutually independent. Now we can write $\mathcal{N}(\mu,\nu)$ in the form

$$\mathcal{N}(\mu,\nu) = 2E\mathcal{L}(X,Y) - E\mathcal{L}(X,X') - E\mathcal{L}(Y,Y').$$

Sometimes we will write $\mathcal{N}(X,Y)$ instead of $\mathcal{N}(\mu,\nu)$ and $\mathfrak{N}(X,Y)$ instead of $\mathfrak{N}(\mu,\nu)$.

Let us give some examples of $\mathfrak{N}$ distances.

Example 22.2.1. Consider random vectors taking values in $\mathbb{R}^d$. As was shown in Sect. 21.2 of Chap. 21, the function $\mathcal{L}(x,y) = \|x-y\|^r$ $(0 < r < 2)$ is a strongly negative definite kernel on $\mathbb{R}^d$. Therefore,

$$\mathcal{N}(X,Y) = 2E\mathcal{L}(X,Y) - E\mathcal{L}(X,X') - E\mathcal{L}(Y,Y') \tag{22.2.5}$$

is a negative definite kernel on the space of probability distributions with a finite rth absolute moment, and $\mathfrak{N}(X,Y) = \mathcal{N}^{1/2}(X,Y)$ is the distance, generated by $\mathcal{N}$.

Let us calculate the distance (22.2.5) for the one-dimensional case. Denote by $f_1(t)$ and $f_2(t)$ the characteristic functions of X and Y, respectively. Further, let

$$u_j(t) = \operatorname{Re} f_j(t), \quad j = 1,2,$$
$$v_j(t) = \operatorname{Im} f_j(t), \quad j = 1,2.$$

Using the well-known formula

$$E|X|^r = c_r \int_0^\infty (1-u(t))t^{-1-r}\,dt,$$

where

$$c_r = 1\Big/\int_0^\infty (1-\cos t)\frac{dt}{t^{r+1}} = -1\Big/\left(\Gamma(-r)\cos\frac{\pi r}{2}\right)$$

depends only on r, we can transform the left-hand side of (22.2.5) as follows:

$$\begin{aligned}\mathcal{N}(X,Y) &= 2E|X-Y|^r - E|X-X'|^r - E|Y-Y'|^r\\ &= c_r \int_0^\infty [2-(1-u_1(t)u_2(t)-v_1(t)v_2(t))\\ &\quad -(1-u_1^2(t)-v_1^2(t))-(1-u_2^2(t)-v_2^2(t))]t^{-1-r}\,dt\\ &= c_r\int_0^\infty |f_1(t)-f_2(t)|^2 t^{-1-r}\,dt \geq 0.\end{aligned}$$

Clearly, the equality is attained if and only if $f_1(t) = f_2(t)$ for all $t \in \mathbb{R}^1$, so that $X \overset{d}{=} Y$.

Example 22.2.2. Let $\mathcal{L}(z)$ be a survival function on $\mathbb{R}^1$ [i.e., $1 - \mathcal{L}(x)$ is a distribution function]. Then the function $\mathcal{L}(x \bigwedge y)$ is a negative definite kernel (here $x \bigwedge y$ is the minimum of x and y). Suppose that

$$g_a(z) = \begin{cases} 0 & \text{for } z \leq a, \\ 1 & \text{for } z > a, \end{cases}$$

and for all $x_1 \leq x_2 \leq \cdots \leq x_n$ we have

$$\sum_{i=1}^{n}\sum_{j=1}^{n} g_a(x_i \wedge x_j)h_i h_j = \sum_{i=k}^{n}\sum_{j=k}^{n} h_i h_j = \left(\sum_{i=k}^{n} h_i\right)^2 \geq 0,$$

where k is determined by the conditions $x_k > a$, $x_{k-1} \leq a$. The preceding conclusion now follows from the obvious equality

$$\mathcal{L}(z) = \int_{-\infty}^{\infty} (1 - g_a(x))\mathrm{d}\sigma(a),$$

where σ is a suitable distribution function. Clearly, $\mathcal{L}(x \wedge y)$ is a strongly negative definite kernel if and only if σ is decreasing and strictly monotone. In this case,

$$\mathcal{N}(\mu, \nu) = \int_{-\infty}^{\infty} (F_\mu(a) - F_\nu(a))^2 \mathrm{d}\theta(a),$$

where F_μ, F_ν are distribution functions corresponding to the measures μ and ν.

22.3 *m*-Negative Definite Kernels and Metrics

In this section, we first introduce the notion of m-negative definite kernels and then proceed with a class of probability metrics generated by them.

22.3.1 *m*-Negative Definite Kernels and Metrics

We now turn to the generalization of the concept of a negative definite kernel. Let m be an even integer and $(\mathfrak{X}, d)$ a metric space. Assume that $\mathcal{L}(x_1, \ldots, x_m)$ is a real continuous function on $\mathfrak{X}^m$ satisfying the condition $\mathcal{L}(x_1, x_2, \ldots, x_{m-1}, x_m) = \mathcal{L}(x_2, x_1, \ldots, x_m, x_{m-1})$. We say that function $\mathcal{L}$ is an m-negative definite kernel if for any integer $n \geq 1$, any collection of points $x_1, \ldots, x_n \in \mathfrak{X}$, and any collection of complex numbers $h_1, \ldots, h_n$ satisfying the condition $\sum_{j=1}^n h_j = 0$ the following inequality holds:

$$(-1)^{m/2} \sum_{i_1=1}^{n} \ldots \sum_{i_m=1}^{n} \mathcal{L}(x_{i_1}, \ldots, x_{i_m}) h_{i_1} \ldots h_{i_m} \geq 0. \tag{22.3.1}$$

If the equality in (22.3.1) implies that $h_1 = \cdots = h_n = 0$, then we call $\mathcal{L}$ a strictly m-negative definite kernel. By passing to the limit, we can prove that $\mathcal{L}$ is an m-negative definite kernel if and only if

$$(-1)^{m/2} \int_{\mathfrak{X}} \ldots \int_{\mathfrak{X}} \mathcal{L}(x_1, \ldots, x_m) h(x_1) \ldots h(x_m) \mathrm{d}Q(x_1) \ldots \mathrm{d}Q(x_m) \geq 0 \tag{22.3.2}$$

for any measure $Q \in \mathcal{B}$ and any integrable function $h(x)$ such that

$$\int_{\mathfrak{X}} h(x)\mathrm{d}Q(x) = 0. \tag{22.3.3}$$

We say that $\mathcal{L}$ is a strongly m-negative definite kernel if the equality in (22.3.2) is attained only for $h = 0$, Q-almost everywhere.

We will denote by $\mathcal{B}(\mathcal{L})$ the set of all measures $\mu \in \mathcal{B}$ for which

$$\int_{\mathfrak{X}} \cdots \int_{\mathfrak{X}} \mathcal{L}(x_1, \ldots, x_m)\mathrm{d}\mu(x_1)\ldots\mathrm{d}\mu(x_m) < \infty.$$

Let μ, ν belong to $\mathcal{B}(\mathcal{L})$. Assume that Q is some measure from $\mathcal{B}(\mathcal{L})$ that dominates μ and ν, and denote

$$h_1(x) = \frac{\mathrm{d}\mu}{\mathrm{d}Q}, \quad h_2(x) = \frac{\mathrm{d}\nu}{\mathrm{d}Q}, \quad h(x) = h_1(x) - h_2(x).$$

Let

$$\mathcal{N}_m(\mu, \nu) = (-1)^{m/2} \int_{\mathfrak{X}} \cdots \int_{\mathfrak{X}} \mathcal{L}(x_1, \ldots, x_m)h(x_1)\ldots h(x_m)\mathrm{d}Q(x_1)\ldots\mathrm{d}Q(x_m). \tag{22.3.4}$$

It is easy to see that if $\mathcal{L}$ is a strongly m-negative definite kernel, then $\mathcal{N}_m^{1/m}(\mu, \nu)$ is a metric on the convex set of measures $\mathcal{B}(\mathcal{L})$.

We need one additional definition. Let $\mathcal{K}(x_1, \ldots, x_m)$ be a continuous real function given on $\mathfrak{X}^m$. We say that $\mathcal{K}$ is an m-positive definite kernel if for any integer $n \geq 1$, any collection of points $x_1, \ldots, x_n \in \mathfrak{X}$, and any real constants $h_1, \ldots, h_n$ the following inequality holds:

$$\sum_{i_1=1}^{n} \cdots \sum_{i_m=1}^{n} \mathcal{K}(x_{i_1}, \ldots, x_{i_m})h_{i_1}\ldots h_{i_m} \geq 0.$$

Lemma 22.3.1. *Assume that $\mathcal{L}$ is an m-negative definite kernel and for some $x_0 \in \mathfrak{X}$ the equality $\mathcal{L}(x_0, \ldots, x_0) = 0$ is fulfilled. Then there exists an m-positive definite kernel $\mathcal{K}$ such that*

$$\begin{aligned}(-1)^{m/2} &\int_{\mathfrak{X}} \cdots \int_{\mathfrak{X}} \mathcal{L}(x_1, \ldots, x_m)h(x_1)\ldots h(x_m)\mathrm{d}Q(x_1)\ldots\mathrm{d}Q(x_m) \\ &= \int_{\mathfrak{X}} \cdots \int_{\mathfrak{X}} \mathcal{K}(x_1, \ldots, x_m)h(x_1)\ldots h(x_m)\mathrm{d}Q(x_1)\ldots\mathrm{d}Q(x_m)\end{aligned} \tag{22.3.5}$$

for any measure $Q \in \mathcal{B}(\mathcal{L})$ and any integrable function $h(x)$ satisfying condition (22.3.3).

Proof. For simplicity we will consider only the case of $m = 2$. The function $K(x_1, x_2)$ defined by

$$K(x_1, x_2) = \mathcal{L}(x_1, x_0) + \mathcal{L}(x_0, x_2) - \mathcal{L}(x_1, x_2)$$

is positive definite. If $x_1, \dots, x_n \in \mathfrak{X}$ and $c_1, \dots, c_n$ are real numbers, then letting $c_0 = -\sum_{j=1}^{n} c_j$ we have

$$\sum_{i,j=1}^{n} c_i c_j K(x_i, x_j) = \sum_{i,j=0}^{n} c_i c_j K(x_i, x_j)$$
$$= -\sum_{i,j=0}^{n} c_i c_j \mathcal{L}(x_i, x_j) \geq 0.$$

Equality (22.3.5) is fulfilled since

$$\int_{\mathfrak{X}} \int_{\mathfrak{X}} \mathcal{L}(x_1, x_0) h(x_1) h(x_2) \mathrm{d}Q(x_1) \mathrm{d}Q(x_2)$$
$$= \int_{\mathfrak{X}} \mathcal{L}(x_1, x_0) h(x_1) \mathrm{d}Q(x_1) \int_{\mathfrak{X}} h(x_2) \mathrm{d}Q(x_2) = 0$$

and, analogously,

$$\int_{\mathfrak{X}} \int_{\mathfrak{X}} \mathcal{L}(x_0, x_i) h(x_1) h(x_2) \mathrm{d}Q(x_1) \mathrm{d}Q(x_2) = 0\,.$$

□

Let us now consider the set $\mathfrak{R}$ of all signed measures R on $(\mathfrak{X}, \mathfrak{A})$ for which the measures R_+ and R_- (the positive and negative parts of R) belong to $\mathcal{B}(\mathcal{L})$, where $\mathcal{L}$ is a strongly m-negative definite kernel on $\mathfrak{X}^m$. According to Lemma 22.3.1, there exists an m-positive definite kernel K for which (22.3.5) holds. For $R \in \mathfrak{R}$ let

$$\|R\| = \left(\int_{\mathfrak{X}} \int_{\mathfrak{X}} K(x_1, \dots, x_m) \mathrm{d}R(x_1) \dots \mathrm{d}R(x_m) \right)^{1/m}. \tag{22.3.6}$$

Clearly, the set $\mathfrak{R}$ forms a linear space in which $\|R\|$ is a norm and, therefore, $\mathfrak{R}$ is a normed space. However, $\mathfrak{R}$ is not yet a Banach space because it may not be complete with respect to that norm. We obtain the corresponding Banach space $\mathfrak{R}_c$ after carrying out the procedure of completion.

Thus, if for some strongly m-negative definite kernel $\mathcal{L}$ the metric d admits the representation

$$d(x, y) = \mathcal{N}_m^{1/m}(\delta_x, \delta_y)\,, \tag{22.3.7}$$

where $\mathcal{N}_m(\mu,\nu)$ is determined by (22.3.4), then $\mathfrak{X} \in \mathcal{B}(\mathcal{L})$. The set $\mathcal{B}(\mathcal{L})$, in turn, is isometric to a subset of a Banach space [namely, the space $\mathfrak{R}_c$ with norm (22.3.6)]. It is easy to verify that the value $\mathcal{N}_m(\mu,\nu)$ is equal to the mth degree of the distance between their barycenters corresponding to μ and ν in the space $\mathfrak{R}_c$. Below are some examples of m-negative definite kernels and the corresponding metrics $\mathcal{N}_m^{1/m}$.

Example 22.3.1. Let $\mathfrak{X} = \mathbb{R}^1$ and let

$$\mathcal{L}(x_1,\ldots,x_m) = |x_1 - x_2 + x_3 - x_4 + \cdots + x_{m-1} - x_m|^r. \tag{22.3.8}$$

For $r \in [0,m]$ this function is m-negative definite, and for $r \in (0,2) \cup (2,4) \cup \ldots \cup (m-2,m)$ it is a strongly m-negative definite kernel. This is clear for $r = 0,2,\ldots,m$. Let us prove it for $r \in (0,m), r \neq 2,4,\ldots,m-2$. Let $k \in [0,m]$ be an even integer such that $k-2 < r < k$. We have

$$|x|^r = A_{r,k} \int_0^\infty \left(\sum_{j=0}^{(k-2)/2} (-1)^j \frac{(xu)^{2j}}{(2j)!} - \cos(xu) \right) \frac{du}{u^{1+r}}, \tag{22.3.9}$$

where

$$A_{r,k} = \left(\int_0^\infty \left(\sum_{j=0}^{(k-2)/2} (-1)^j \frac{(u^2 j)^{2j}}{(2j)!} - \cos u \right) \frac{du}{u^{1+r}} \right)^{-1}. \tag{22.3.10}$$

If $Q \in \mathcal{B}(\mathcal{L})$ and $h(x)$ is a real function such that $\int_{\mathbb{R}^1} h(x)\mathrm{d}Q(x) = 0$, then taking (22.3.9) into account we have

$$(-1)^{m/2} \int_{\mathbb{R}^1} \cdots \int_{\mathbb{R}^1} \mathcal{L}(x_1,\ldots,x_m) h(x_1)\ldots h(x_m) \mathrm{d}Q(x_1)\ldots \mathrm{d}Q(x_m)$$
$$= A_{r,k} \int_0^\infty \left| \int_{\mathbb{R}^1} e^{ixz} h(x) \mathrm{d}Q(x) \right|^m \frac{\mathrm{d}}{z} z^{1+r} \geq 0.$$

It is clear that equality is attained if and only if $h(x) = 0$, Q-almost everywhere. Consequently, $\mathcal{L}$ is a strongly m-negative definite kernel. For the kernel (22.3.8) and $r \in (0,2) \cup (2,4) \cup \ldots \cup (m-2,m)$ there exists a corresponding metric $\mathfrak{N}_m = \mathcal{N}_m^{1/m}(\mu,\nu)$ admitting the representation

$$\mathcal{N}_m(\mu,\nu) = A_{r,k} \int_0^\infty |f(t) - g(t)|^m \frac{dt}{t^{1+r}}, \tag{22.3.11}$$

where $f(t)$ and $g(t)$ are the characteristic functions of the measures μ and ν, respectively.

Example 22.3.2. Let $\mathfrak{X} = \mathbb{R}^1$, and let

$$\mathcal{L}(x_1,\ldots,x_m) = g(x_1 - x_2 + x_3 - x_4 + \cdots + x_{m-1} - x_m), \tag{22.3.12}$$

where g is an even, continuous function. This is an m-negative definite kernel if and only if

$$g(u) = \int_0^\infty \left(\sum_{k=0}^{(m-2)/2} (-1)^k u^{2k} x^{2k}/(2k)! - \cos(ux) \right) \frac{1+x^m}{x^m} \mathrm{d}\theta(x) + P_{m-2}(u), \tag{22.3.13}$$

where $\theta(x)$ is a nondecreasing bounded function, $\theta(-0) = 0$, and $P_{m-2}(n)$ is a polynomial of at most $m-2$ degrees in the even powers of u. Here, $\mathcal{L}$ is strongly m-negative definite if and only if $\operatorname{supp}\theta = [0, \infty)$.

The distance $\mathfrak{N}_m$ corresponding to the function $\mathcal{L}$ defined in (22.3.12) and (22.3.13) admits the representation

$$\mathfrak{N}_m(\mu, \mu) = \left(\int_0^\infty |f_\mu(t) - f_\nu(t)|^m \frac{1+t^m}{t^m} \mathrm{d}\theta(t) \right)^{1/m}, \tag{22.3.14}$$

where f_μ and f_ν are the characteristic functions of the measures μ and ν, respectively. We do not present a proof here. We only note that conceptually it is close to the proof of a Lévy–Khinchin-type formula that gives the representation of negative definite functions.[2]

Example 22.3.2 implies that if the metric $\mathfrak{N}_m$ corresponds to the kernel $\mathcal{L}$ of (22.3.12) and (22.3.13), then, by (22.3.14), the Banach space $\mathfrak{R}_c$ is isometric to the space $L^m(\mathbb{R}^1, \frac{1+t^m}{t^m}\mathrm{d}\theta(t))$. Thus, if $\mathcal{L}$ is determined by (22.3.12) and (22.3.14), then the set of measures $\mathcal{B}(\mathcal{L})$ with metric $\mathfrak{N}_m$ is isometric to some convex subset $\underline{\mathcal{B}}(\mathcal{L})$ of the space $L^m(\mathbb{R}^1, \frac{1+t^m}{t^m}\mathrm{d}\theta(t))$. Of course, $\mathfrak{N}_m(\mu, \nu)$ is equal to the distance between the barycenters corresponding to μ and ν in the space $L^m(\mathbb{R}^1, \frac{1+t^m}{t^m}\mathrm{d}\theta(t))$, and the points of the real line correspond to the extreme points of the set $\tilde{\mathcal{B}}(\mathcal{L})$.

22.4 $\mathfrak{N}$-Metrics and the Problem of Recovering Measures from Potentials

We will refer to the metrics constructed in Sects. 22.2 and 22.3 as the $\mathfrak{N}$-metrics. They enable us to provide a simple solution to the problem of the uniqueness of a measure with a given potential. The question of the uniqueness of a measure having a given potential is essentially a question of the uniqueness of the solution of an integral equation of a special form. This question arises in certain problems of mathematical physics, functional analysis (especially in connection with the extension of isometry), and the theory of random processes and in the construction of characterizations of probability distributions.

[2]See, for example, Akhiezer (1961).

22.4.1 Recovering Measures from Potentials

Suppose first that $\mathcal{L}(x, y) = \mathcal{L}(y, x)$ is a strongly negative definite kernel on $\mathfrak{X}^2$, and $\mu \in \mathcal{B}(\mathcal{L})$. The quantity

$$\varphi(x) = \int_{\mathfrak{X}} \mathcal{L}(x, y) \mathrm{d}\mu(y), \quad x \in \mathfrak{X}, \tag{22.4.1}$$

is the potential of the measure μ corresponding to the kernel $\mathcal{L}$ (in short, the potential of μ). We are interested in the question of whether different measures can have the same potential. We will provide conditions guaranteeing the coincidence of measures with equal potentials and offer certain generalizations.

Theorem 22.4.1. *If $\mathcal{L}$ is a strongly negative definite kernel, then $\mu \in \mathcal{B}(\mathcal{L})$ is uniquely determined by the potential φ given by* (22.4.1).

Proof. Assume that two measures $\mu, \nu \in \mathcal{B}(\mathcal{L})$ have the same potential. Then

$$\int_{\mathfrak{X}} \mathcal{L}(x, y) \mathrm{d}\mu(y) = \int_{\mathfrak{X}} \mathcal{L}(x, y) \mathrm{d}\nu(y), \quad x \in \mathfrak{X}. \tag{22.4.2}$$

Integrating both sides of (22.4.2) with respect to $\mathrm{d}\mu(x)$ we obtain

$$\int_{\mathfrak{X}} \int_{\mathfrak{X}} \mathcal{L}(x, y) \mathrm{d}\mu(x) \mathrm{d}\mu(y) = \int_{\mathfrak{X}} \int_{\mathfrak{X}} \mathcal{L}(x, y) \mathrm{d}\mu(x) \mathrm{d}\nu(y). \tag{22.4.3}$$

Similarly, integrating both sides of (22.4.3) with respect to $\mathrm{d}\nu$ leads to

$$\int_{\mathfrak{X}} \int_{\mathfrak{X}} \mathcal{L}(x, y) \mathrm{d}\nu(x) \mathrm{d}\mu(y) = \int_{\mathfrak{X}} \int_{\mathfrak{X}} \mathcal{L}(x, y) \mathrm{d}\nu(x) \mathrm{d}\nu(y). \tag{22.4.4}$$

Adding the corresponding sides of (22.4.3) and (22.4.4) and taking into account that $\mathcal{L}(x, y) = \mathcal{L}(y, x)$ we obtain

$$2 \int_{\mathfrak{X}} \int_{\mathfrak{X}} \mathcal{L}(x, y) \mathrm{d}\mu(x) \mathrm{d}\nu(y) = \int_{\mathfrak{X}} \int_{\mathfrak{X}} \mathcal{L}(x, y) \mathrm{d}\mu(x) \mathrm{d}\nu(y) + \int_{\mathfrak{X}} \int_{\mathfrak{X}} \mathcal{L}(x, y) \mathrm{d}\nu(x) \mathrm{d}\nu(y).$$

By the definition of the metric $\mathfrak{N}$, we see that

$$\mathcal{N}(\mu, \nu) = 0,$$

that is, $\mu = \nu$. □

Let us consider some consequences of Theorem 22.4.1. Let $\mathfrak{X} = \mathbb{R}^1$ and d be the standard distance on the real line, and let

$$\varphi(x) = \int_{-\infty}^{\infty} |y - x|^r \mathrm{d}\mu(y). \tag{22.4.5}$$

We know that for $r \in (0, 2)$ the function $\mathcal{L}(x, y) = |x - y|^r$ is a strongly negative definite kernel. Therefore, by Theorem 22.4.1, μ is uniquely determined from its potential φ in (22.4.5).

The problem of recovering a measure from its potential (22.4.5) was first considered by Plotkin (1970, 1971), who proved the uniqueness of the recovery for all $r \geq 0$, $r \neq 2k$, $k = 0, 1, 2, \ldots$. This result was rediscovered by Rudin (1976). Their results can be derived from Theorem 22.4.1 since the case $r > 2$ can be reduced to $0 < r < 2$ by differentiating (22.4.2) with respect to x for $\mathcal{L}(x, y) = |x - y|^r$. It is clear that for $r = 2k$ the recovery of μ is impossible. In this case, (22.4.2) only shows the coincidence of some moments of the measures μ and ν.

A generalization of Plotkin's and Rudin's results can be found in Linde (1982), Koldobskii (1991), and Gorin and Koldobskii (1987). Their considerations are mostly related to the study of norms in the spaces L^p, L^∞, and C. They also consider certain other Banach spaces. Our method is also useful in the study of the L^p spaces, as shown by the following lemmas.

Lemma 22.4.1. *Let $\mathcal{L}(x, y)$ be a negative definite kernel on $\mathfrak{X}^2$ taking nonnegative values and such that $\mathcal{L}(x, x) = 0$, $\mathcal{L}(x, y) = \mathcal{L}(y, x)$. Assume that ν is a measure (not necessarily finite) on $\mathbb{R}^1_+ = [0, \infty)$ satisfying the condition*

$$\int_0^\infty \min(1, x)\mathrm{d}\nu(x) < \infty.$$

Then the kernel

$$\mathcal{L}_\nu(x, y) = \int_0^\infty (1 - \exp(-u\mathcal{L}(x, y)))\mathrm{d}\nu(u) \tag{22.4.6}$$

is negative definite. In particular, if $\alpha \in [0, 1]$, then $\mathcal{L}^\alpha(x, y)$ is a negative definite kernel.

Proof. We first show that the function $\exp\{-\lambda\mathcal{L}(x, y)\}$ is positive definite for all $\lambda > 0$. For any x_0 define

$$K(x, y) = \mathcal{L}(x, x_0) + \mathcal{L}(x_0, y) - \mathcal{L}(x, y)$$

so that

$$\mathcal{L}(x, y) = \mathcal{L}(x, x_0) + \mathcal{L}(x_0, y) - K(x, y).$$

The proof of Lemma 22.3.1 implies that $K(x,y)$ is a positive definite kernel. It can be easily verified that $\exp(\lambda K(x,y))$ is a positive definite kernel as well.[3] Let $x_1,\dots,x_n \in \mathfrak{X}$, and let $c_1,\dots,c_n$ be complex constants. We have (the bar denotes the complex conjugate)

$$\begin{aligned}\sum_{i=1}^{n}\sum_{j=1}^{n} c_i\bar{c}_j \exp\{-\lambda\mathcal{L}(x_i,x_j)\}\\ &= \sum_{i=1}^{n}\sum_{j=1}^{n} c_i\bar{c}_j \exp\{-\lambda\mathcal{L}(x_i,x_0)\}\exp\{-\lambda\mathcal{L}(x_j,x_0)\}\exp\{\mathcal{L}K(x_i,x_j)\}\\ &= \sum_{i=1}^{n}\sum_{j=1}^{n} c_i'\bar{c}_j' \exp\{\lambda K(x_i,x_j)\}\\ &\geq 0,\end{aligned}$$

where $c_j' = c_i \exp\{-\lambda\mathcal{L}(x_i,x_0)\}$. Thus, $\exp\{-\lambda\mathcal{L}(x,y)\}$ is indeed positive definite. This implies that $1-\exp\{-\lambda\mathcal{L}(x,y)\}$ is a negative definite kernel, and hence so is $\mathcal{L}_\nu(x,y)$.

If $\alpha \in (0,1)$, consider the measure

$$\nu_\alpha(A) = \int_A x^{-(\alpha+1)}\mathrm{d}x, \quad A \in \mathfrak{A}(\mathbb{R}^1_+).$$

Then

$$\mathcal{L}_{\nu_\alpha}(x,y) = c_\alpha\mathcal{L}^\alpha(x,y),$$

where c_α is a positive constant. This concludes the proof. □

Lemma 22.4.2. *Let (Λ,Σ,σ) be a measure space (where the measure σ is not necessarily finite). Then for $0 < p < 2$ the function $\mathcal{L}_p(x,y)$ defined on $L^p(\Lambda,\Sigma,\sigma)\times L^p(\Lambda,\Sigma,\sigma)$ by*

$$\mathcal{L}_p(x,y) = \|x-y\|_p^p = \int_\Lambda |x(u)-y(u)|^p\mathrm{d}\sigma(u), \quad x,y \in L^p, \tag{22.4.7}$$

is a strongly negative definite kernel (it is also a negative definite kernel for $p=2$, but not in the strong case).

Proof. Note that for $p \in (0,2)$ the function $(u,v) \to |u-v|^p$ is a strongly negative definite kernel on $\mathbb{R}^1 \times \mathbb{R}^1$. Therefore, for $x_1,\dots,x_n \in L^p$ and $h_1,\dots,h_n \in \mathbb{R}^1$ such that $\sum_{j=1}^n h_j = 0$ we obtain

[3]See Vakhaniya et al. (1985).

$$\sum_{i,j} \|x_i - x_j\|_p^p h_i h_j = \int_\Lambda \sum_{i,j} |x_i(u) - x_j(u)|^p h_i h_j \mathrm{d}\sigma(u) \leq 0.$$

The lemma is proved. □

From Lemmas 22.4.1 and 22.4.2 we conclude that

$$\mathcal{L}(x, y) = \|x - y\|_p^\alpha, \quad x, y \in L^p,$$

is a strongly negative definite kernel for $p \in (0, 2)$ and $0 < \alpha < p$. Theorem 22.4.1 now implies that the measure μ on L^p is uniquely determined by its potential

$$\varphi(x) = \int_{L^p} \|x - y\|_p^\alpha \mathrm{d}\mu(y), \quad x \in L^p, \tag{22.4.8}$$

in the case of $p \in (0, 2)$ and $\alpha \in (0, p)$. It is clear that if we want to recover μ in the class of measures with fixed support supp μ, then it is sufficient to consider only the restriction of φ to supp μ, that is, we need to know $\varphi(x)$, $x \in \operatorname{supp} \mu$. Although the uniqueness of μ for a given φ in (22.4.8) was obtained by Linde (1982) and Koldobskii (1982) (using a different method), our result concerning the recovery of μ with a given support from the values of $\varphi(x)$, $x \in \operatorname{supp} \mu$, appears to be new.

Can we relax the conditions $p \in (0, 2), \alpha \in (0, p)$ when considering the potential (22.4.8), or the constraint $r \in (0, 2)$ when studying (22.4.5)? We will try to answer these questions by introducing potentials related to m-negative definite kernels. Although we already noted that for (22.4.5) the case $r > 2$ can be reduced to $r \in (0, 2)$ when the potential is determined for all $x \in \mathbb{R}^1$, it is interesting to study the uniqueness of the recovery of measures with fixed support from the values of $\varphi(x)$ on the support. In this case, using differentiation to reduce powers may prove impossible.

Let $\mathcal{L}(x_1, \ldots, x_m)$, with an even $m \geq 2$, be a strongly m-negative definite kernel on $\mathfrak{X}^m$, and let $\mu \in \mathcal{B}(\mathcal{L})$. Assume that $\mathcal{L}$ is symmetric in its arguments and real-valued, and $\mathcal{L}(x, \ldots, x) = 0$ for any $x \in \mathfrak{X}$. Consider the function

$$\varphi(x_1, \ldots, x_{m-1}) = \int_{\mathfrak{X}} \mathcal{L}(x_1, \ldots, x_{m-1}, x_m) \mathrm{d}\mu(x_m), \quad x_1, \ldots, x_{m-1} \in \mathfrak{X}. \tag{22.4.9}$$

We will refer to φ as the potential of μ corresponding to the kernel $\mathcal{L}$ (if we need to stress that φ corresponds to an m-negative definite kernel, we will refer to it as the m-potential of μ). Let us consider a natural question of whether different measures can have the same m-potential.

Theorem 22.4.2. *If $\mathcal{L}$ is a strongly m-negative definite kernel, then $\mu \in \mathcal{B}(\mathcal{L})$ is uniquely determined by the potential* (22.4.9).

Proof. Assume that the measures $\mu, \nu \in \mathcal{B}(\mathcal{L})$ have the same potential. Then

$$\int_{\mathfrak{X}} \mathcal{L}(x_1, x_2, \ldots, x_m) \mathrm{d}\mu(x_m) = \int_{\mathfrak{X}} \mathcal{L}(x_1, x_2, \ldots, x_m) \mathrm{d}\nu(x_m). \tag{22.4.10}$$

Integrate successively both sides of (22.4.10) with respect to $\mathrm{d}\mu(x_1) \ldots \mathrm{d}\mu(x_{m-1})$, then with respect to $\mathrm{d}\nu(x_1)\mathrm{d}\mu(x_2) \ldots \mathrm{d}\mu(x_{m-1})$, and so on, and finally with respect to $\mathrm{d}\nu(x_1) \ldots \mathrm{d}\nu(x_{m-1})$. This leads to

$$\begin{aligned}
&\int_{\mathfrak{X}} \cdots \int_{\mathfrak{X}} \mathcal{L}(x_1, x_2, \ldots, x_m) \mathrm{d}\mu(x_1) \ldots \mathrm{d}\mu(x_m) \\
&\quad = \int_{\mathfrak{X}} \cdots \int_{\mathfrak{X}} \mathcal{L}(x_1, \ldots, x_m) \mathrm{d}\mu(x_1) \ldots \mathrm{d}\mu(x_{m-1}) \mathrm{d}\nu(x_m) \\
&\quad \cdots \int_{\mathfrak{X}} \cdots \int_{\mathfrak{X}} \mathcal{L}(x_1, \ldots, x_m) \mathrm{d}\nu(x_1) \ldots \mathrm{d}\nu(x_{m-1}) \mathrm{d}\mu(x_m) \\
&\quad = \int_{\mathfrak{X}} \cdots \int_{\mathfrak{X}} \mathcal{L}(x_1, \ldots, x_m) \mathrm{d}\nu(x_1) \ldots \mathrm{d}\nu(x_m),
\end{aligned}$$

which implies that

$$\mathcal{N}_m(\mu, \nu) = 0,$$

that is, $\mu = \nu$. □

Consider again the potential (22.4.5),

$$\varphi(x) = \int_{-\infty}^{\infty} |y - x|^r \mathrm{d}\mu(y), \quad x \in \mathbb{R}^1,$$

where μ is a measure on the σ-algebra of Borel subsets of the real line. Without making the assumption $r \in (0, 2)$, suppose only that $r \neq 2k, k = 0, 1, \ldots$. There exists an even m such that $m - 2 < r < m$. In this case, the function $\mathcal{L}(x_1, \ldots, x_m) = |x_1 - x_2 + \cdots + x_{m-1} - x_m|^r$ is a strongly m-negative definite kernel (Example 22.3.1). If the function $\varphi(x), x \in \mathbb{R}^1$, is known, then we also know the function

$$\begin{aligned}
\varphi_m(x_1, \ldots, x_{m-1}) &= \int_{-\infty}^{\infty} \mathcal{L}(x_1, \ldots, x_m) \mathrm{d}\mu(x_m) \\
&= \int_{-\infty}^{\infty} |x_1 - x_2 + \cdots + x_{m-1} - x_m|^r \mathrm{d}\mu(x_m) \\
&= \varphi(x_1 - x_2 + \cdots + x_{m-1}).
\end{aligned}$$

Theorem 22.4.2 implies that μ can be uniquely recovered from its m-potential φ_m, and hence μ can also be uniquely recovered from its potential φ. Similar reasoning allows us to verify that the measure μ on L^p is uniquely determined by its potential

(22.4.8) for any $p > 0$ and $\alpha \in (0, p)$. However, we cannot consider φ only on the support of μ. For us it is enough to know φ on the set $\{x_1 - x_2 + \cdots + x_{m-1} : x_j \in \mathrm{supp}\mu,\ j = 1, 2, \ldots, m_1\}$.

22.4.2 *Stability in the Problem of Recovering a Measure from its Potential*

We saw in Sect. 22.4 that $\mathfrak{N}$-metrics enable us to obtain a relatively simple solution to the problem of recovering a measure from its potential. It seems plausible that if two measures have *close* potentials, then the measures themselves are close in the corresponding $\mathfrak{N}$-metric. If this is actually the case, then the convergence of the corresponding sequence of potentials can be used as a criterion for convergence of a sequence of measures. In this section, we will consider in greater detail the case where the potentials are close in a uniform sense while the closeness of the measures is stated in terms of $\mathfrak{N}$-metrics.

Suppose first that $\mathcal{L}(x, y)$ is a symmetric strongly negative definite kernel on $\mathfrak{X}^2$, and that $\mu \in \mathcal{B}(\mathcal{L})$. For the sake of convenience, the potential (22.4.1) will now be denoted by $\varphi(x; \mu)$, that is,

$$\varphi(x; \mu) = \int_{\mathfrak{X}} \mathcal{L}(x, y) \mathrm{d}\mu(y). \tag{22.4.11}$$

Theorem 22.4.3. *Suppose that $\mathcal{L}(x, y)$ is a symmetric strongly negative definite kernel on $\mathfrak{X}$. Then for any $\mu, \nu \in \mathcal{B}(\mathcal{L})$ we have*

$$\mathcal{N}^{\frac{1}{2}}(\mu, \nu) \le (2 \sup_{x \in \mathfrak{X}} |\ \varphi(x; \mu) - \varphi(x; \nu)|)^{\frac{1}{2}}. \tag{22.4.12}$$

Proof. Let

$$\varepsilon = \sup_{x \in \mathfrak{X}} |\ \varphi(x; \mu) - \varphi(x; \nu)|)^{\frac{1}{2}}.$$

Clearly,

$$\left| \int_{\mathfrak{X}} \mathcal{L}(x, y) \mathrm{d}\mu(y) - \int_{\mathfrak{X}} \mathcal{L}(x, y) \mathrm{d}\nu(y) \right| \le \varepsilon. \tag{22.4.13}$$

Integrating both sides of (22.4.13) with respect to $\mathrm{d}\mu(x)$ we obtain

$$\left| \int_{\mathfrak{X}} \int_{\mathfrak{X}} \mathcal{L}(x, y) \mathrm{d}\mu(x) \mathrm{d}\mu(y) - \int_{\mathfrak{X}} \int_{\mathfrak{X}} \mathcal{L}(x, y) \mathrm{d}\mu(x) \mathrm{d}\nu(y) \right| \le \varepsilon. \tag{22.4.14}$$

If both sides of (22.4.13) are integrated with respect to $\mathrm{d}\nu(x)$, then

$$\left|\int_{\mathfrak{X}}\int_{\mathfrak{X}} \mathcal{L}(x,y)\mathrm{d}\nu(x)\mathrm{d}\mu(y) - \int_{\mathfrak{X}}\int_{\mathfrak{X}} \mathcal{L}(x,y)\mathrm{d}\nu(x)\mathrm{d}\nu(y)\right| \le \varepsilon. \tag{22.4.15}$$

The result now follows from (22.4.14), (22.4.15), and the definition of $\mathfrak{N}$. □

An analogous result for the potential (22.4.9) and metric $\mathfrak{N}_m$ holds as well. The proof of the following result is similar to that of Theorem 22.4.3.

Theorem 22.4.4. *Suppose that $\mathcal{L}(x_1,\dots,x_m)$, where $m \ge 2$ is even, is a strongly m-negative definite kernel on $\mathfrak{X}$, and $\mu,\nu \in \mathcal{B}(\mathcal{L})$. Then*

$$\mathfrak{N}_m(\mu,\nu) \le \Big(m \sup_{(x_1,\dots,x_{m-1})\in\mathfrak{X}^{m-1}} |\varphi(x_1,\dots,x_{m-1};\mu) - \varphi(x_1,\dots,x_{m-1};\nu)|\Big)^{\frac{1}{m}}, \tag{22.4.16}$$

where

$$\varphi(x_1,\dots,x_{m-1};\theta) = \int_{\mathfrak{X}} \mathcal{L}(x_1,\dots x_{m-1};x_m)\mathrm{d}\theta(x_m),\ \ \theta \in \mathcal{B}(\mathcal{L}). \tag{22.4.17}$$

To obtain quantitative criteria for the convergence of probability measures in terms of the convergence of the corresponding potentials, we need a lower bound for $\mathfrak{N}(\mu,\nu)$. Since we cannot obtain such an estimate in general, we will consider only functions $\mathcal{L}(x,y)$ that depend on the difference $x-y$ of the arguments $x,y \in \mathbb{R}^1 = \mathfrak{X}$.

Recall that (Example 21.7.1) when $\mathfrak{X} = \mathbb{R}^1$, then an even continuous function $\mathcal{L}(z)$ with $\mathcal{L}(0) = 0$ is strongly negative definite if and only if

$$\mathcal{L}(z) = \int_0^\infty (1-\cos(zu))\frac{1+u^2}{u^2}\mathrm{d}\theta(u),\ \ \operatorname{supp}\theta = [0,\infty), \tag{22.4.18}$$

where $\theta(u)$ is a real bounded nondecreasing function with $\theta(-0) = 0$. If $\mu \in \mathcal{B}(\mathcal{L})$, then the integral

$$\int_{\mathfrak{X}}\int_{\mathfrak{X}} \mathcal{L}(x-y)\mathrm{d}\mu(x)\mathrm{d}\nu(y)$$

is finite. This integral can be written as

$$\int_{\mathfrak{X}}\int_{\mathfrak{X}} \mathcal{L}(x-y)\mathrm{d}\mu(x)\mathrm{d}\nu(y) = \int_0^\infty\int_{\mathfrak{X}}\int_{\mathfrak{X}}\Big(1-\cos(xu)\cos(yu) - \sin(xu)\sin(yu)\Big)\frac{1+u^2}{u^2}\mathrm{d}\mu(x)\mathrm{d}\mu(y)\mathrm{d}\theta(u).$$

Let $f(u;\mu)$ be the characteristic function corresponding to the measure μ. Then the preceding equality becomes

$$\int_{\mathfrak{X}}\int_{\mathfrak{X}} \mathcal{L}(x-y)\mathrm{d}\mu(x)\mathrm{d}\nu(y) = \int_0^\infty (1-|f(u;\mu)|^2)\frac{1+u^2}{u^2}\mathrm{d}\theta(u).$$

Since the left-hand-side of the preceding equality is finite by the assumption that $\mu \in \mathcal{B}(\mathcal{L})$, the right-hand side is also finite. Consequently,

$$\lim_{\delta\to+0}\int_\delta^\infty (1-|f(u;\mu)|^2)\frac{1+u^2}{u^2}\mathrm{d}\theta(u) = 0. \qquad (22.4.19)$$

This holds for every measure $\mu \in \mathcal{B}(\mathcal{L})$. It is convenient for us to consider a subset of $\mathcal{B}(\mathcal{L})$ such that convergence in (22.4.19) is uniform with respect to this subset. For this purpose we introduce the function $\omega(\delta)$, defined for $\delta \in [0,\infty)$ and satisfying the conditions

$$\omega(0) = \lim_{\delta\to+0}\omega(\delta) = 0,$$

$$\omega(\delta_1) \le \omega(\delta_2) \text{ for } 0 \le \delta_1 \le \delta_2.$$

Let $\mathcal{B}(\mathcal{L};\omega)$ be the set of all measures $\mu \in \mathcal{B}(\mathcal{L})$ for which

$$\sup_{\mu\in\mathcal{B}(\mathcal{L};\omega)}\int_0^\infty (1-|f(u;\mu)|^2)\frac{1+u^2}{u^2}\mathrm{d}\theta(u) \le \omega(\delta). \qquad (22.4.20)$$

Here, θ is the function that appears in (22.4.18).

Theorem 22.4.5. *Suppose that $\mathcal{L}$ is defined by* (22.4.18) *and that $\mu,\nu \in \mathcal{B}(\mathcal{L};\omega)$. Then*

$$\sup_x|\varphi(x;\mu)-\varphi(x;\nu)| \le \inf_{\delta>0}\left[\sqrt{2}\mathcal{N}^{\frac{1}{2}}(\mu,\nu)\left(\int_\delta^\infty \frac{1+u^2}{u^2}\mathrm{d}\theta(u)\right)^{\frac{1}{2}} + 2\sqrt{2}\omega(\delta)\right].$$

(22.4.21)

Proof. We have

$$\varphi(x;\mu) = \int_{\mathfrak{X}} \mathcal{L}(x-y)\mathrm{d}\mu(y)$$

$$= \int_0^\infty (1-\cos(ux)\mathrm{Re} f(u;\mu) - \sin(ux)\mathrm{Im} f(u;\mu))\frac{1+u^2}{u^2}\mathrm{d}\theta(u),$$

where $\operatorname{Re} f$ and $\operatorname{Im} f$ are the real and the imaginary parts of f, respectively. Therefore, the difference of the potentials of $\mu, \nu \in \mathcal{B}(\mathcal{L};\omega)$ can be represented as

$$\begin{aligned}|\varphi(x;\mu) - \varphi(x;\nu)| &= \Big|\int_0^\infty [\cos(ux)(\operatorname{Re} f(u;\mu) - \operatorname{Re} f(u;\nu)) \\ &\quad + \sin(ux)(\operatorname{Im} f(u;\mu) - \operatorname{Im} f(u;\nu))]\frac{1+u^2}{u^2}\mathrm{d}\theta(u)\Big| \\ &\le \sqrt{2}\int_0^\infty |f(u;\mu) - f(u;\nu)|\frac{1+u^2}{u^2}\mathrm{d}\theta(u). \qquad (22.4.22)\end{aligned}$$

Let us represent the integral on the right-hand side of (22.4.22) as the sum of integrals over the intervals $[0,\delta]$ and (δ,∞), where δ is for now an arbitrary positive number. Applying the Cauchy–Buniakowsky inequality to the integral over (δ,∞) we obtain

$$\begin{aligned}&\int_\delta^\infty |f(u;\mu) - f(u;\nu)|\frac{1+u^2}{u^2}\mathrm{d}\theta(u) \\ &\le \left(\int_\delta^\infty |f(u;\mu) - f(u;\nu)|^2\,\frac{1+u^2}{u^2}\mathrm{d}\theta(u)\right)^{1/2}\left(\int_\delta^\infty \frac{1+u^2}{u^2}\mathrm{d}\theta(u)\right)^{1/2} \\ &\le \mathfrak{N}(\mu,\nu)\left(\int_\delta^\infty \frac{1+u^2}{u^2}\mathrm{d}\theta(u)\right)^{1/2}. \qquad (22.4.23)\end{aligned}$$

For the integral over $[0,\delta]$ we have

$$\begin{aligned}&\int_0^\delta |f(u;\mu) - f(u;\nu)|\frac{1+u^2}{u^2}\mathrm{d}\theta(u) \\ &\le \int_0^\delta |1 - f(u;\mu)|\,\frac{1+u^2}{u^2}\mathrm{d}\theta(u) + \int_0^\delta |1 - f(u;\nu)|\,\frac{1+u^2}{u^2}\mathrm{d}\theta(u) \\ &\le \int_0^\delta (1 - |f(u;\mu)|^2)\frac{1+u^2}{u^2}\mathrm{d}\theta(u) + \int_0^\delta (1 - |f(u;\nu)|^2)\frac{1+u^2}{u^2}\mathrm{d}\theta(u) \\ &\le 2\omega(\delta). \qquad (22.4.24)\end{aligned}$$

By the arbitrariness of $\delta > 0$, inequality (22.4.21) now follows from (22.4.22)–(22.4.24). □

Corollary 22.4.1. *Suppose that in the statement of Theorem 22.4.5 the function $\theta(u)$ is such that the integral $\int_0^\infty \frac{1+u^2}{u^2} \mathrm{d}\theta(u)$ converges. Then*

$$\mathfrak{N}(\mu;\nu) \le \sup_{x\in X} |\,\varphi(x;\mu) - \varphi(x;\nu)|$$

$$\le \sqrt{2}\Big(\int_0^\infty \frac{1+u^2}{u^2} \mathrm{d}\theta(u)\Big)^{1/2} \cdot \mathfrak{N}(\mu,\nu). \qquad (22.4.25)$$

Proof. The result follows directly from Theorems 22.4.3 and 22.4.5. Note that instead of $\mathcal{B}(\mathcal{L};\omega)$, the whole space $\mathcal{B}$ can be considered here. □

We can now state the quantitative criteria for the convergence of a sequence of measures in terms of the convergence of a sequence of their potentials. The result below follows directly from Theorems 22.4.3 and 22.4.5.

Theorem 22.4.6. *Let $\mathfrak{X} = \mathbb{R}^1$ and the function $\mathcal{L}$ be defined by* (22.4.18), *and let $\mu_1, \mu_2, \dots, \mu_n$ be a sequence of measures from $\mathcal{B}(\mathcal{L};\omega)$. The sequence $\{\mu_n, n \ge 1\}$ converges in $\mathfrak{N}$ to some measure ν if and only if the sequence of potentials $\{\varphi(x;\mu_n), n \ge 1\}$ converges in the uniform metric to the potential $\varphi(x;\nu)$ of ν. Here,*

$$\mathfrak{N}(\mu_n;\nu) \le \sup_{x\in X} |\varphi(x;\mu_n) - \varphi(x;\nu)|$$

$$\le \inf_{\delta\ge 0}\left[\sqrt{2}\mathfrak{N}(\mu_n,\nu)\left(\int_\delta^\infty \frac{1+u^2}{u^2} \mathrm{d}\theta(u)\right)^{1/2} + 2\sqrt{2}\omega(\delta)\right]. \qquad (22.4.26)$$

Corollary 22.4.2. *Suppose that in the statement of Theorem 22.4.6 the integral*

$$\int_0^\infty \frac{1+u^2}{u^2} \mathrm{d}\theta(u)$$

converges and $\{\mu_n, n \ge 1\}$ is a sequence of arbitrary measures from $\mathcal{B}$. This sequence converges in $\mathfrak{N}$ to ν if and only if the sequence of potentials $\{\phi(x;\mu_n), n \ge 1\}$ converges in the uniform metric to the potential $\phi(x;\nu)$. Here,

$$\mathfrak{N}(\mu_n,\nu) \le \sup_{x\in X} |\varphi(x;\mu_n) - \phi(x;\nu)|$$

$$\le \sqrt{2}\left(\int_0^\infty \frac{1+u^2}{u^2} \mathrm{d}\theta(u)\right)^{1/2} \cdot \mathfrak{N}(\mu_n,\nu)\,. \qquad (22.4.27)$$

Note that for a bounded, continuous, and symmetric function $\mathcal{L}$ of the form (22.4.18), the convergence in $\mathfrak{N}$ is equivalent to the weak convergence of measures. Therefore, weak convergence of measures is equivalent to the uniform convergence of their potentials corresponding to the kernels $\mathcal{L}$.

Note that our main focus is the theoretical issues of the uniqueness and stability of the recovery of a measure from its potential. Of course, explicit reconstruction formulas are of interest as well. Such results can be found in Koldobskii (1982, 1991).

22.5 $\mathfrak{N}$-Metrics in the Study of Certain Problems of the Characterization of Distributions

The problem of characterizing probability distributions involves the description of all probability laws with a certain property $\mathcal{P}$. In cases where this property can be stated as a functional equation, the characterization problem reduces to the description (finding) of the probabilistic solutions of the equation. This approach can be found in many publications devoted to characterization problems, including the well-known monograph by Kagan et al. (1973).

Situations in which a certain class of distributions with a property $\mathcal{P}$ is known and it must be established that there are no other distributions possessing this property are fairly common. In such cases, one can apply results about positive solutions of functional equations. Such an approach was developed in Kakosyan et al. (1984).

Problems of recovering a distribution from the distributions of suitable statistics, or from certain functionals of distributions of these statistics, also belong to characterization problems. These particular problems are related to the problem of recovering a measure from the potential as well as to $\mathfrak{N}$-metrics.[4] Below we show that it is possible to use $\mathfrak{N}$-metrics in such problems.

22.5.1 *Characterization of Gaussian and Related Distributions*

Let us begin with the question of whether it is possible to recover a distribution of independent identically distributed (i.i.d.) RVs $X_1, \ldots, X_n$ from the function

$$U_r(a_1, \ldots, a_n) = E\left|\sum_{j=1}^{n} a_j X_j\right|^r, r \in (0, 2), \qquad (22.5.1)$$

[4]See Klebanov and Zinger (1990).

where the parameter r is fixed. Here, we assume the existence of the first absolute moment of X. We can write $U_r(a_1, \dots a_n)$ as follows:

$$\begin{aligned}U_r(a_1,\dots,a_n) &= \int_{\mathfrak{X}^n} \left|\sum_{j=1}^{n} a_j x_j\right|^r \mathrm{d}F(x_1,\dots,x_n)\\ &= \int_{\mathfrak{X}^n} \left|\sum_{j=1}^{n-1} \frac{a_j x_j}{x_n} + a_n\right|^r |x_n|^r \,\mathrm{d}F(x_1,\dots,x_n)\\ &= \int_{\mathfrak{X}^n} \left|\sum_{j=1}^{n-1} \frac{a_j x_j}{x_n} + a_n\right|^r \mathrm{d}F_1(x_1,\dots,x_n),\end{aligned}$$

where $\mathrm{d}F_1(x_1,\dots,x_n) = |x_n|^r \,\mathrm{d}F(x_1,\dots,x_n)$. Clearly, the value $E|x_n|^r$ is known because it is the value $U_r(0,\dots,0,1)$. The measure

$$\mathrm{d}F_1(x_1,\dots,x_n)/E|X_n|^r$$

is a probability measure, and therefore the problem of recovering F from the known function $U_r(a_1,\dots,a_n)$ reduces to the problem of recovering the distribution of $Y = \sum_{j=1}^{n-1} \frac{a_j X_j}{X_n}$ from the potential. As we already saw in Sect. 22.4, such a recovery is unique. Since the coefficients a_j, $(j = l,\dots,n-l)$, are arbitrary, we can recover the distribution of X_1 from the distribution of Y (to within a scale parameter).[5] However, since $E|X_1|^r$ is known, we can uniquely determine the scale parameter as well. Note that the problem of recovering the distribution of X_1 from (22.5.1) was considered in Braverman (1987).

The preceding arguments enable us to reduce this problem to one of recovering a measure from the potential. Below we demonstrate the possibilities of this approach and the connections of $\mathfrak{N}$-metrics to related characterization problems, including those in Banach spaces. Our first result is a formalization of arguments given previously.

Theorem 22.5.1. *Let* $\mathbf{B}$ *be a Banach space, and let* $X_1,\dots,X_n$ $(n \geq 3)$ *be i.i.d. random vectors with values in* $\mathbf{B}$. *Suppose that for any* $a_1,\dots,a_n$ *from the conjugate space* $\mathbf{B}^*$, *the RVs* $\langle a_j, X_j\rangle$ *have an absolute moment of order* $r \in (0,2)$, $j = 1,\dots,n$. *Then the function*

$$\varphi(a_1,\dots,a_n) = E\left|\sum_{j=1}^{n} \langle a_j, X_j\rangle\right|^r$$

on $\mathbf{B}^{*n}$ *uniquely determines the distribution of* X_1.

[5] See, for example, Kagan et al. (1973).

Proof. Proceed by following the outline given peviously. □

Remark 22.5.1. If in Theorem 22.5.1 we have $n > 3$, then the a_j for $j > 3$ can be set to zero, so that we can consider only φ on $\mathbf{B}^{*3}$. As a_3, we can choose only vectors that are collinear to a fixed vector from $\mathbf{B}^*$.

Corollary 22.5.1. *Suppose that* $\mathbf{B}$ *is a Banach space and* $X_1, \ldots, X_n$ $(n > 3)$ *are i.i.d. random vectors with values in* $\mathbf{B}$ *and such that* $E\,\|X_1\|^r$ *exists for some* $r \in (0,2)$. *Let*

$$\Psi(A_1, \ldots, A_n) = E\left\|\sum_{j=1}^{n} A_j X_j\right\|^r,$$

where $A_1, \ldots, A_n$ *are linear continuous operators acting from* $\mathbf{B}$ *into* $\mathbf{B}$. *Then the distribution of* X_1 *is uniquely determined by* Ψ.

Proof. It is enough to consider operators A_j mapping $\mathbf{B}$ into its one-dimensional subspace and then use Theorem 22.5.1. □

Corollary 22.5.2. *Let* $X_1, \ldots, X_n$ $(n \geq 3)$ *be i.i.d. RVs variables (with values in* $\mathbb{R}^1$*) with* $E|X_1|^r < \infty$ *for some fixed* $r \in (0,2)$. *If for all real* $a_1, \ldots, a_n$,

$$E\left|\sum_{j=1}^{n} a_j X_j\right| = C_r\left(\sum_{j=1}^{n} a_j^2\right)^{r/2}, \tag{22.5.2}$$

where C_r *is positive and depends only on* r, *then* X_1 *follows a normal distribution with mean 0.*

Proof. It is enough to note that (22.5.2) holds for a normal distribution with mean 0 and then use Theorem 22.5.1. □

The result of Corollary 22.5.2 in a somewhat more general setting $r \neq 2k$, $k = 0, 1, 2, \ldots$, was obtained in Braverman (1987). We now present its substantial generalization.

Theorem 22.5.2. *Suppose that a Banach space* $\mathbf{B}$ *and a real number* $r > 0$ *are such that* $\|x - y\|^r$, $x, y \in \mathbf{B}$, *is a strongly negative definite function. Let* X_1, X_2, X_3, X_4 *be i.i.d. random vectors with values in* $\mathbf{B}$ *and such that* $E\|X_1\|^r < \infty$. *Assume that for some real function* h *the relation*

$$E\left\|\sum_{j=1}^{4} a_j X_j\right\|^r = h\left(\sum_{j=1}^{4} a_j^2\right) \tag{22.5.3}$$

holds for at least the following collections of parameters $a_1, a_2, a_3, a_4 \in \mathbb{R}^1$:

$$a_1 = 1,\ a_2 = a_3 = -\frac{1}{\sqrt{2}},\ a_4 = 0; \tag{22.5.4}$$

$$a_1 = -a_2 = 1,\ a_3 = a_4 = 0; \tag{22.5.5}$$

$$a_1 = a_2 = \frac{1}{\sqrt{2}},\ a_3 = a_4 = -\frac{1}{\sqrt{2}}. \tag{22.5.6}$$

Then X_1 has a Gaussian distribution with mean 0.

Proof. By (22.5.3)–(22.5.6), we have

$$E\left\|X_1 - \frac{X_2 + X_3}{\sqrt{2}}\right\|^r = h(2),$$

$$E\,\|X_1 - X_2\|^r = h(2),$$

$$E\left\|\frac{X_1 + X_2}{\sqrt{2}} - \frac{X_3 + X_4}{\sqrt{2}}\right\|^r = h(2).$$

These three equalities imply that

$$2E\left\|X_1 - \frac{X_2 + X_3}{\sqrt{2}}\right\|^r - E\,\|X_1 - X_2\|^r - E\left\|\frac{X_1 + X_2}{\sqrt{2}} - \frac{X_3 + X_4}{\sqrt{2}}\right\|^r = 0$$

or, equivalently,

$$N\left(X_1, \frac{X_1 + X_2}{\sqrt{2}}\right) = 0,$$

where $\mathfrak{N}$ is the metric corresponding to the strongly negative definite kernel $\mathcal{L}(u, v) = \|u - v\|^r$. Therefore,

$$X_1 \stackrel{d}{=} \frac{X_2 + X_3}{\sqrt{2}}. \tag{22.5.7}$$

Let $x^* \in \mathbf{B}^*$. From (22.5.7) we find

$$\langle x^*, X_1\rangle \stackrel{d}{=} \frac{\langle x^*, X_2\rangle + \langle x^*, X_3\rangle}{\sqrt{2}}.$$

Now by the famous Pólya theorem, the RV $\langle x^*, X_1\rangle$ has a Gaussian distribution with mean 0. Since $x^* \in \mathbf{B}^*$ was chosen arbitrarily, the result follows. □

Similar arguments can be used to characterize symmetric distributions in $\mathbb{R}^n$.

Theorem 22.5.3. *Let X, Y be i.i.d. random vectors in $\mathbb{R}^n$ with $E\|X\|^r < \infty$ for some $r \in (0, 2)$. Then we have*

$$E\,\|X+Y\|^r \ge E\,\|X-Y\|^r, \tag{22.5.8}$$

with equality if and only if X has a symmetric distribution.

Proof. This result can be obtained from Theorem 22.2.1 and Example 22.2.1. However, we present an alternative proof. Consider first the scalar case, where X and Y are i.i.d. RVs taking real values and having distribution function $F(x)$. Suppose that x, y are two real numbers and $r \in (0,2)$. It is easy to verify that

$$|x+y|^r - |x-y|^r = C_r \int_0^\infty \sin\frac{xt}{2}\sin\frac{yt}{2}\frac{dt}{t^{1+r}}, \tag{22.5.9}$$

where C_r is a positive constant that depends only on r. Integrating both sides of (22.5.9) with respect to $\mathrm{d}F(x) - \mathrm{d}F(y)$ we obtain

$$E\,|X+Y|^r - E\,|X-Y|^r = C_r \int_0^\infty \varphi^2\left(\frac{t}{2}\right)\frac{dt}{t^{1+r}}, \tag{22.5.10}$$

where $\varphi(t) = \int_{-\infty}^\infty \sin(tx)\mathrm{d}F(x)$ is the sine-Fourier transform of F. Thus, in the scalar case, (22.5.8) follows from (22.5.10) since $C_r > 0$. If the right-hand side of (22.5.10) is equal to zero, then the sine-Fourier transform of $F(x)$ is identically zero. This is equivalent to the symmetry of X, which concludes the scalar case.

The vector case is easily reduced to the scalar one by noting that for $x \in \mathbb{R}^n$

$$\|x\|^r = \int_{S^{n-1}} |(x,\tau)|^r\,\mathrm{d}M(\tau), \tag{22.5.11}$$

where M is a measure on the unit sphere S^{n-1} in $\mathbb{R}^d$, and then using the result in the one-dimensional case. □

Here is a generalization of (22.5.8), which extends the range of variation of r.

Theorem 22.5.4. *Suppose that $m = 2k$ is an even positive integer and $X_1, \dots, X_m$ are i.i.d. vectors in $\mathbb{R}^n$. Let $E\,\|X_1\|^r < \infty$, where $r \in (m-2, m)$ is fixed. Then*

$$\sum_{j=0}^m (-1)^j \binom{m}{j} E\,\|X_1 + \cdots + X_{m-j} - X_{m-j+1} - \cdots - X_m\|^r \ge 0, \tag{22.5.12}$$

with equality if and only if X_1 has a symmetric distribution.

Proof. The result is derived from the following facts.

(a) For $r \in (m-2, m)$ the function

$$\mathcal{L}(x_1, \dots, x_m) = |x_1 - x_2 + \cdots + x_{m-1} - x_m|^r \tag{22.5.13}$$

is a strongly m-negative definite kernel.

(b) Suppose that μ, ν are two measures in $\mathbb{R}^n$ and $\mathcal{L}$ is a strongly m-negative definite kernel. Let

$$\mathcal{N}_m(\mu,\nu) = (-1)^{m/2} \int_{\mathbb{R}^n} \dots \int_{\mathbb{R}^n} \mathcal{L}(x_1,\dots,x_m)\mathrm{d}Q(x_1)\dots\mathrm{d}Q(x_m), \tag{22.5.14}$$

where $Q = \mu - \nu$. Then $\mathfrak{N}_m(\mu,\nu)$ is a metric on $\mathcal{B}(\mathcal{L})$.

(c) If μ is a measure generated by X_1, ν is a measure generated by $-X_1$, and $\mathcal{L}$ is determined by (22.5.13), then $\mathcal{N}_m(\mu,\nu)$ in (22.5.14) coincides with the left-hand side of (22.5.12).

The theorem is proved. □

Let us now study the case of a separable Hilbert space $\mathfrak{H}$. Let $\mathcal{L}(x-y)$, $x, y \in \mathfrak{H}$ be a real strongly negative definite function. The following result can be obtained by substituting $Y = -X'$ in Theorem 22.2.1.

Theorem 22.5.5. *If X, Y are i.i.d. random vectors in $\mathfrak{H}$ for which $E\mathcal{L}(X+Y) < \infty$, then*

$$E\,\mathcal{L}(X+Y) \ge E\,\mathcal{L}(X-Y)\,, \tag{22.5.15}$$

with equality if and only if X has a symmetric distribution.

Observe that (22.5.8) is a special case of (22.5.15) with $\mathcal{L}(x) = \|x\|^r$, $\mathfrak{H} = \mathbb{R}^n$. We note that Theorems 22.5.3–22.5.5 are set forth in Zinger and Klebanov (1991).

Theorem 22.5.3 can be used to obtain a criterion for convergence of a sequence of random vectors in $\mathbb{R}^n$ to a set S of random vectors with a symmetric distribution.

Theorem 22.5.6. *Suppose that $\{X_m,\ m \ge 1\}$ is a sequence of random vectors in $\mathbb{R}^n$, $\mathcal{L}(x,y) = \|x-y\|^r$ $(r \in (0,2),\ x, y \in \mathbb{R}^n)$, and $\mathfrak{N}$ is a metric generated by $\mathcal{L}$. The sequence $\{X_m,\ m \ge 1\}$ approaches the set S of random vectors in $\mathbb{R}^n$ with symmetric distributions if and only if*

$$\lim_{m\to\infty} [E\,\|X_m + X_m'\|^r - E\,\|X_m - X_m'\|^r] = 0\,,$$

where X_m' is an independent copy of X_m.

This result becomes almost trivial in view of the following lemma.

Lemma 22.5.1. *Let $\mathcal{L}$, $\mathfrak{N}$, and S be the same as in Theorem 22.5.6, and let X be a random vector in $\mathbb{R}^n$. Then*

$$\mathfrak{N}(X,S) = \frac{1}{2^{r/2}}[E\,\|X + X'\|]^r - E\,\|X - X'\|^r]^{1/2}, \tag{22.5.16}$$

where X' is an independent copy of X.

Proof. Similar to the proof of Theorem 22.5.3, we have

$$E\,\|X + X'\|^r - E\,\|X - X'\| = C_r \int_{S^{n-1}} \mathrm{d}M(s) \int_0^\infty \varphi^2\left(\frac{t}{2}s\right) \frac{\mathrm{d}t}{t^{1+r}}.$$

This identity, which follows from (22.5.10) and (22.5.11), can be rewritten as

$$E\,\|X + X'\|^r - E\,\|X - X'\| = 2^r C_r \int_{S^{n-1}} \mathrm{d}M(s) \int_0^\infty \varphi^2(ts) \frac{\mathrm{d}t}{t^{1+r}}, \quad (22.5.17)$$

where φ and M were defined in the proof of Theorem 22.5.3. On the other hand,

$$\begin{aligned} N(X,S) &= \inf_{Y \in S} N(X,Y) \\ &= \inf_{Y \in S} C_r \int_{S^{n-1}} \mathrm{d}M(s) \int_0^\infty |f(ts;X) - f(ts;Y)|^2 \frac{\mathrm{d}t}{t^{1+r}}, \end{aligned}$$

where $f(u;X)$ and $f(u;Y)$ are the characteristic functions of X and Y, respectively. Since Y has a symmetric distribution, $f(u;Y)$ is real, so that $\mathrm{Im} f(u;Y) = 0$ and $\mathrm{Re} f(u;Y) = f(u;Y)$. Therefore,

$$\begin{aligned} N(X,S) &= \inf_{Y \in S} C_r \int_{S^{n-1}} \mathrm{d}M(s) \\ &\quad \times \int_0^\infty [(\mathrm{Re} f(ts;X) - f(ts;Y))^2 + (\mathrm{Im} f(ts;X))^2] \frac{\mathrm{d}t}{t^{1+r}} \\ &\geq C_r \int_{S^{n-1}} \mathrm{d}M(s) \int_0^\infty (\mathrm{Im} f(ts;X))^2 \frac{\mathrm{d}t}{t^{1+r}} \\ &= C_r \int_{S^{n-1}} \mathrm{d}M(s) \int_0^\infty \varphi^2(ts) \frac{\mathrm{d}t}{t^{1+r}}. \end{aligned}$$

It is clear that if $f(u,Y) = \mathrm{Re} f(u;X)$, then we obtain an equality in the preceding inequality. Hence, taking into account (22.5.17), we obtain the result. □

Incidentally, the proof of Lemma 22.5.1 implies that the closest (in the $\mathfrak{N}$ metric) symmetric random vector to X is the vector Y with the characteristic function $f(u;Y) = \mathrm{Re} f(u;X)$. This vector can be constructed as a mixture of X and $-X$ taken with equal probabilities:

$$Y = \epsilon X - (1 - \epsilon) X,$$

where ϵ is an RV independent of X taking on values 0 or 1 with probability $1/2$.

Most of the results presented in this section are concerned with moments of sums of RVs (or vectors). However, other operations on RVs can be studied as well using a suitable choice for $\mathcal{L}$. For example, if we use $\mathcal{L}$ given in Example 22.2.2, then we obtain an analog of Theorem 22.5.2 that characterizes the exponential distribution through the mean values of order statistics.

Theorem 22.5.7. *Let $X_1, \ldots, X_n$ ($n \geq 4$) be i.i.d. nonnegative RVs with finite first moment. Assume that there exists a finite limit $\lim_{x \to +0} F(x)/x = \lambda$ not equal to zero, where $F(x)$ is the distribution function of X_1. If the expectation $E\left(\bigwedge_{j=1}^n \frac{x_j}{a_j}\right)$ depends only on the sum of real positive parameters $a_1, \ldots, a_n$ (that are chosen arbitrarily), then X_1 has an exponential distribution with parameter λ.*

Proof. Let $\bar{F}(x) = 1 - F(x)$. It is easy to see that

$$E\left(\bigwedge_{j=1}^n \frac{x_j}{a_j}\right) = \int_0^\infty \prod_{j=1}^n \bar{F}(a_j x)\mathrm{d}x.$$

The assumptions of the theorem imply that

$$\int_0^\infty \prod_{j=1}^n \bar{F}(a_j x)\mathrm{d}x = \text{const.} \tag{22.5.18}$$

whenever $\sum_{j=1}^n a_j = \text{const}$. Set the following values successively in (22.5.18):

$$a_1 = a_2 = 1,\ a_3 = \cdots = a_n = 0;$$
$$a_1 = 1,\ a_2 = a_3 = 1/2,\ a_4 = \cdots = a_n = 0;$$
$$a_1 = a_2 = a_3 = a_4 = 1/2,\ a_5 = \cdots = a_n = 0.$$

Then, from the doubled second equality obtained in this way we calculate the first and the third, finding

$$\int_0^\infty \left[\bar{F}(x) - \bar{F}^2\left(\frac{x}{2}\right)\right]^2 \mathrm{d}x = 0\,,$$

so that

$$\bar{F}(x) = \bar{F}^2\left(\frac{x}{2}\right). \tag{22.5.19}$$

Equations of the form (22.5.19) are well known.[6] When the limit specified in the hypothesis of the theorem exists, the only solution of the preceding equation is $\bar{F}(x) = \exp(-\lambda x)$. □

The foregoing proof demonstrates that if $E(\bigwedge_{j=1}^n \frac{x_j}{a_j})$ depends on the sum of $a_1, \ldots, a_n$ only for the three sets of parameters specified previously, then $F(x)$ is a function of the exponential distribution. Instead of the expectation of the minimum, we can take the expectation of any strictly increasing function of it (as long as the expectation exists).

[6]See, for example, Kakosyan et al. (1984).

22.5.2 Characterization of Distributions Symmetric to a Group of Transformations

Consider two i.i.d. random vectors X and Y in $\mathbb{R}^d$, and a real orthogonal matrix A, that is $AA^T = A^T A = I$. Here, A^T stands for the transpose of the matrix A, and I is the unit matrix.

Theorem 22.5.8. *Suppose that A is an orthogonal $d \times d$ matrix and that $\mathcal{L}$ is a negative definite kernel such that $\mathcal{L}(Ax, Ay) = \mathcal{L}(x, y)$. Then for the i.i.d. RVs X, Y we have*

$$E\mathcal{L}(X, AY) \geq E\mathcal{L}(X, Y). \tag{22.5.20}$$

In addition, if $\mathcal{L}$ is a strongly negative definite kernel, then the equality in (22.5.20) is attained if and only if the distribution of the vector X is invariant with respect to the group G generated by the matrix A.

Proof. Let us consider the corresponding $\mathcal{N}$ kernel on the space of corresponding probability distributions

$$\mathcal{N}(X, AY) = 2E\mathcal{L}(X, AY) - E\mathcal{L}(X, X') - E\mathcal{L}(AY, AY').$$

As we saw in Chap. 21, $\mathcal{N}$ is negative definite if $\mathcal{L}$ is such, and it is a square of distance if $\mathcal{L}$ is a strongly negative definite kernel. But $\mathcal{L}(Ax, Ay) = \mathcal{L}(x, y)$, and therefore

$$\mathcal{N}(X, AY) = 2\big(E\mathcal{L}(X, AY) - E\mathcal{L}(X, X')\big),$$

which implies the statement of the theorem. □

Corollary 22.5.3. *Suppose that X and Y are two i.i.d. random vectors such that the moment $E\|X\|^r$ exists for some $r \in (0, 2)$ and A is a real orthogonal matrix. Then*

$$E\|X - AY\|^r \geq E\|X - Y\|^r, \tag{22.5.21}$$

with equality if and only if the distribution of X is invariant with respect to the group G generated by the matrix A.

Proof. It is clear that

$$\mathcal{L}(x, y) = \|x - y\|^r$$

is a strongly negative definite kernel in $\mathbb{R}^d \times \mathbb{R}^d$, and $\mathcal{L}(Ax, Ay) = \mathcal{L}(x, y)$ because the Euclidean distance is invariant under orthogonal transformations. □

Remark 22.5.2. Note that for $r = 2$, the equality in (22.5.21) does not characterize any property of invariance. It imposes some restrictions on the first moments of the distribution of X.

Example 22.5.1. Let $g(t)$ be a real characteristic function of an infinitely divisible probability distribution on $\mathbb{R}^d$. Then $\mathcal{L}(x, y) = -\log g(x - y)$ is a negative definite kernel. Further, if the support of the corresponding spectral measure in the Lévy–Khinchin representation of $g(t)$ coincides with the whole $\mathbb{R}^d$, then the kernel is strongly negative definite.

Let us take $\mathcal{L}(x, y) = 1 - \exp\{-\|x - y\|^2\}$. Since $g(t) = \exp(-\|t\|^2)$ is the characteristic function of a multivariate normal distribution, then the function $\mathcal{L}(x, y)$ is a strongly negative definite kernel. Therefore,

$$E\exp\{-\|X - Y\|^2\} \geq E\exp\{-\|X - AY\|^2\} \tag{22.5.22}$$

with equality if and only if the distribution of X is invariant with respect to the group G. Note that here we do not need any moment-type restrictions.

A type of generalization arises in the following way.[7] Let $B = C^T C$ be a positive definite $d \times d$ matrix, and let $\|x\|_B = (x^T Bx)^{1/2}$ be the corresponding norm in $\mathbb{R}^d$. Suppose now that A is a $d \times d$ real matrix satisfying the condition $A^T BA = B$ (which is a generalization of orthogonality).

Theorem 22.5.9. *Let X and Y be i.i.d. random vectors in $\mathbb{R}^d$ having finite absolute rth moment* $(0 < r < 2)$. *Then*

$$E\|X - AY\|_B^r \geq E\|X - Y\|_B^r, \tag{22.5.23}$$

with equality if and only if the distribution of X is invariant with respect to the group generated by the matrix A.

Proof. Apply Theorem 22.5.3 to the random vectors CX and CY and ordinary Euclidean norm. □

We can now characterize the distributions invariant with respect to a group generated by a finite set of matrices.[8]

Theorem 22.5.10. *Suppose that $B_j = C_j^T C_j$, $j = 1, \dots, m$, are positive definite $d \times d$ matrices and $A_j^T B_j A_j = B_j$. Let X, Y be i.i.d. random vectors in $\mathbb{R}^d$ having finite absolute rth moment* $(0 < r < 2)$. *Then*

$$\sum_{j=1}^{m} \left(E\|X - A_j Y\|_{B_j}^r - E\|X - Y\|_{B_j}^r \right) \geq 0, \tag{22.5.24}$$

with equality if and only if the distribution of X is invariant with respect to the group G generated by the matrices A_j, $j = 1, \dots, m$.

---

[7] See Klebanov et al. (2001).

[8] See Klebanov et al. (2001).

References

Akhiezer NI (1961) The classical moment problem. Gosudarstv. Izdat. Fiz-Mat. Lit., Moscow (in Russian)

Braverman MSh (1987) A method for the characterization of probability distributions. Teor Veroyatnost i Primenen (in Russian) 32:552–556

Gorin EA, Koldobskii AL (1987) Measure potentials in Banach spaces. Sibirsk Mat Zh 28(1):65–80

Kagan AM, Linnik YV, Rao CR (1973) Characterization problems of mathematical statistics. Wiley, New York

Kakosyan AV, Klebanov LB, Melamed IA (1984) Characterization problems of mathematical statistics. In: Lecture notes in mathematics, vol 1088. Springer, Berlin

Klebanov LB, Zinger AA (1990) Characterization of distributions: problems, methods, applications. In: Grigelionis B, et al (eds) Probability theory and mathematical Statistics VSP/TEV, vol 1, pp 611–617

Klebanov LB, Kozubowski TJ, Rachev ST, Volkovich VE (2001) Characterization of distributions symmetric with respect to a group of transformations and testing of corresponding statistical hypothesis. Statist Prob Lett 53:241–247

Koldobskii L (1982) Isometric operators in vector-valued L^p-spaces. Zap Nauchn Sem Leningrad Otdel Mat Inst Steklov 107:198–203

Koldobskii L (1991) Convolution equations in certain Banach spaces. Proc Amer Math Soc 111:755–765

Linde W (1982) Moments and measures on Banach spaces. Math Ann 258:277–287

Plotkin AI (1970) Isometric operators on L^p-spaces. Dokl Akad Nauk 193(3):537–539 (in Russian)

Plotkin AI (1971) Extensions of L^p isometries. Zap Nauchn Sem Leningrad Otdel Mat Inst Steklov (in Russian) 22:103–129

Rudin W (1976) L^p isometries and equimeasurability. Indiana Univ Math J 25(3):215–228

Sriperumbudur BA, Fukumizu GK, Schölkopf B (2010) Hilbert space embeddings and metrics on probability measures. J Mach Learn Res 11:1517–1561

Vakhaniya NN, Tarieladze VI, Chobanyan SA (1985) Probability distributions in Banach spaces. Nauka, Moscow (in Russian)

Zinger AA, Klebanov L (1991) Characterization of distribution symmetry by moment properties. In: Stability problems of stochastic models, Moscow: VNII Sistemnykh Isledovaniii pp 70–72 (in Russian)

Chapter 23
Statistical Estimates Obtained by the Minimal Distances Method

The goals of this chapter are to:

- Consider the problem of parameter estimation by the method of minimal distances,
- Study the properties of the estimators.

Notation introduced in this chapter:

Notation	Description
w°	Brownian bridge
$F_\theta(x) = F(x, \theta)$	Distribution function with parameter θ
$p_\theta(x) = p(x, \theta)$	Density of $F_\theta(x)$

23.1 Introduction

In this chapter, we consider minimal distance estimators resulting from using the $\mathfrak{N}$-metrics and compare them with classical M-estimators. This chapter, like Chap. 22, is not directly related to quantitative convergence criteria, although it does demonstrate the importance of $\mathfrak{N}$-metrics.

23.2 Estimating a Location Parameter: First Approach

Let us begin by considering a simple case of estimating a one-dimensional location parameter. Assume that

$$\mathcal{L}(x, y) = \mathcal{L}(x - y)$$

S.T. Rachev et al., *The Methods of Distances in the Theory of Probability and Statistics*, DOI 10.1007/978-1-4614-4869-3_23,

is a strongly negative definite kernel and

$$N(F,G) = -\int_{-\infty}^{\infty}\int_{-\infty}^{\infty} \mathcal{L}(x,y)\mathrm{d}R(x)\mathrm{d}R(y), \quad R = F - G,$$

is the corresponding kernel defined on the class of distribution functions (DFs). As we noted in Chap. 22, $\mathfrak{N}(F,G) = \mathcal{N}^{1/2}(F,G)$ is a distance on the class $\mathbf{B}(\mathcal{L})$ of DFs under the condition

$$\int_{-\infty}^{\infty}\int_{-\infty}^{\infty} \mathcal{L}(x,y)\mathrm{d}F(x)\mathrm{d}F(y) < \infty.$$

Suppose that $x_1,\dots,x_n$ is a random sample from a population with DF $F_\theta(x) = F(x-\theta)$, where $\theta \in \Theta \subset \mathbb{R}^1$ is an unknown parameter (Θ is some interval, which may be infinite). Assume that there exists a density $p(x)$ of $F(x)$ (with respect to the Lebesgue measure). Let $F_n^*(x)$ be the empirical distribution based on the random sample, and let θ^* be a minimum distance estimator of θ, so that

$$N(F_n^*, F_{\theta^*}) = \min_{\theta\in\Theta} N(F_n, F_\theta) \tag{23.2.1}$$

or

$$\theta^* = \mathrm{argmin}_{\theta\in\Theta} N(F_n^*, F_\theta). \tag{23.2.2}$$

We have

$$\begin{aligned} N(F_n^*, F_\theta) = {} & \frac{2}{n}\sum_{j=1}^{n}\int_{-\infty}^{\infty} \mathcal{L}(x_j - \theta - y)p(y)\mathrm{d}y \\ & -\frac{1}{n^2}\sum_{ij}\mathcal{L}(x_i - x_j) \\ & -\int_{-\infty}^{\infty}\int_{-\infty}^{\infty} \mathcal{L}(x-y)p(x)p(y)\mathrm{d}x\mathrm{d}y. \end{aligned}$$

Suppose that $\mathcal{L}(u)$ is differentiable and $\mathcal{L}$ and p are such that

$$\begin{aligned} \int_{-\infty}^{\infty} \mathcal{L}(x)p'(x+\theta)\mathrm{d}x &= \frac{\mathrm{d}}{\mathrm{d}\theta}\int_{-\infty}^{\infty} \mathcal{L}(x-\theta)p(x)\mathrm{d}x \\ &= -\int_{-\infty}^{\infty} \mathcal{L}'(x-\theta)p(x)\mathrm{d}x. \end{aligned} \tag{23.2.3}$$

Then, (23.2.2) implies that θ^* is the root of

$$\frac{\mathrm{d}}{\mathrm{d}\theta} N(F_n^*, F_\theta)|_{\theta=\theta^*} = 0$$

or

$$\sum_{j=1}^{n} \int_{-\infty}^{\infty} \mathcal{L}'(x_j - \theta^* - v) p(v) \mathrm{d}v = 0. \tag{23.2.4}$$

Since the estimator θ^* satisfies the equation

$$\sum_{j=1}^{n} g_1(x_j - \theta) = 0, \tag{23.2.5}$$

where

$$g_1(x) = \int_{-\infty}^{\infty} \mathcal{L}'(x - v) p(v) \mathrm{d}v),$$

it is an M-estimator.[1] It is well known [see, e.g., Huber (1981)] that (23.2.4) [or (23.2.5)] determines a consistent estimator only if

$$\int_{-\infty}^{\infty} g_1(x) p(x) \mathrm{d}x = 0,$$

that is,

$$\int_{-\infty}^{\infty} \int_{-\infty}^{\infty} \mathcal{L}'(u - v) p(u) p(v) \mathrm{d}u \mathrm{d}v = 0. \tag{23.2.6}$$

We show that if (23.2.3) holds, then (23.2.6) does as well. The integral

$$\int_{-\infty}^{\infty} \int_{-\infty}^{\infty} \mathcal{L}(u - v) p(u + \theta) p(v + \theta) \mathrm{d}u \mathrm{d}v = \int_{-\infty}^{\infty} \mathcal{L}(u - v) p(u) p(v) \mathrm{d}u \mathrm{d}v$$

does not depend on θ. Therefore,

$$\frac{\mathrm{d}}{\mathrm{d}\theta} \int_{-\infty}^{\infty} \int_{-\infty}^{\infty} \mathcal{L}(u - v) p(u + \theta) p(v + \theta) \mathrm{d}u \mathrm{d}v = 0. \tag{23.2.7}$$

---

[1] See, for example, Huber (1981) for the definition and properties of M-estimators.

On the other hand,

$$\begin{aligned}
&\frac{\mathrm{d}}{\mathrm{d}\theta}\int_{-\infty}^{\infty}\int_{-\infty}^{\infty}\mathcal{L}(u-v)p(u+\theta)p(v+\theta)\mathrm{d}u\mathrm{d}v\\
&=\int_{-\infty}^{\infty}\int_{-\infty}^{\infty}\mathcal{L}(u-v)p'(u+\theta)p(v+\theta)\mathrm{d}u\mathrm{d}v\\
&\quad+\int_{-\infty}^{\infty}\int_{-\infty}^{\infty}\mathcal{L}(u-v)p(u+\theta)p'(v+\theta)\mathrm{d}u\mathrm{d}v\\
&=2\int_{-\infty}^{\infty}\int_{-\infty}^{\infty}\mathcal{L}(u-v)p'(u+\theta)p(v+\theta)\mathrm{d}u\mathrm{d}v.
\end{aligned}$$

Here, we used the equality $\mathcal{L}(u-v)=\mathcal{L}(v-u)$. Comparing this with (23.2.7), we find that for $\theta=0$

$$\int_{-\infty}^{\infty}\int_{-\infty}^{\infty}\mathcal{L}(u-v)p'(u)p(v)\mathrm{d}u\mathrm{d}v=0. \tag{23.2.8}$$

However,

$$\begin{aligned}
\int_{-\infty}^{\infty}\int_{-\infty}^{\infty}\mathcal{L}(u-v)p'(u)p(v)\mathrm{d}u\mathrm{d}v &= \int_{-\infty}^{\infty}\left(\frac{\mathrm{d}}{\mathrm{d}u}\int_{-\infty}^{\infty}\mathcal{L}(u-v)p(v)\mathrm{d}v\right)p(u)\mathrm{d}u\\
&=\int_{-\infty}^{\infty}\int_{-\infty}^{\infty}\mathcal{L}'(u-v)p(u)p(v)\mathrm{d}u\mathrm{d}v.
\end{aligned}$$

Consequently [see (23.2.8)],

$$\int_{-\infty}^{\infty}\int_{-\infty}^{\infty}\mathcal{L}(u-v)p(u)p(v)\mathrm{d}u\mathrm{d}v=0,$$

which proves (23.2.6).

We see that the minimum $\mathfrak{N}$-distance estimator is an M-estimator, and the necessary condition for its consistency is automatically fulfilled.

The standard theory of M-estimators shows that the asymptotic variance of θ^* [i.e., the variance of the limiting random variable of $\sqrt{n}(\theta^*-\theta)$ as $n\to\infty$] is

$$\sigma_{\theta^*}^2 = \frac{\int\limits_{-\infty}^{\infty}\left[\int\limits_{-\infty}^{\infty} \mathcal{L}'(u-v)p(v)\mathrm{d}v\right]^2 p(u)\mathrm{d}u}{\left[\int\limits_{-\infty}^{\infty}\int\limits_{-\infty}^{\infty} \mathcal{L}''(u-v)p(u)p(v)\mathrm{d}u\mathrm{d}v\right]^2},$$

where we assumed the existence of $\mathcal{L}''$ and that the differentiation can be carried out under the integral. Note that when the parameter space Θ is compact, it is clear from geometric considerations that $\theta^* = \mathrm{argmin}_{\theta\in\Theta} N(F_n^*, F_\theta)$ is unique for sufficiently large n.

23.3 Estimating a Location Parameter: Second Approach

We now consider another method for estimating a location parameter θ. Let

$$\theta' = \mathrm{argmin}_{\theta\in\Theta} N(F_n^*, \delta_\theta), \tag{23.3.1}$$

where δ_θ is a distribution concentrated at the point θ and F_n^* is an empirical DF. Proceeding as in Sect. 23.2, it is easy to verify that θ' is a root of

$$\sum_{j=1}^{n} \mathcal{L}'(x_j - \theta) = 0, \tag{23.3.2}$$

and so it is a classic M-estimator. A consistent solution of (23.3.2) exists only if

$$\int\limits_{-\infty}^{\infty} \mathcal{L}'(u)p(u)\mathrm{d}u = 0. \tag{23.3.3}$$

What is a geometric interpretation of (23.3.3)? More precisely, how is the measure parameter δ_θ related to the family parameter, that is, to the DF F_θ? This must be the same parameter, that is, for all θ_1 we must have

$$N(F_\theta, \delta_\theta) \le N(F_\theta, \delta_{\theta_1}).$$

Otherwise,

$$\frac{\mathrm{d}}{\mathrm{d}\theta_1} N(F_\theta, \delta_{\theta_1})|_{\theta_1=\theta} = 0.$$

It is easy to verify that the last condition is equivalent to (23.3.3). Thus, (23.3.3) has to do with the accuracy of parameterization and has the following geometric interpretation. The space of measures with metric $\mathfrak{N}$ is isometric to some simplex in a Hilbert space. In this case, δ-measures correspond to the extreme points (vertices) of the simplex. Consequently, (23.3.3) signifies that the vertex closest to the measure with DF F_θ corresponds to the same value of the parameter θ (and not to some other value θ_1).

23.4 Estimating a General Parameter

We now consider the case of an arbitrary one-dimensional parameter, which is approximately the same as the case of a location parameter. We just carry out formal computations assuming that all necessary regularity conditions are satisfied.

Let $x_1, \dots, x_n$ be a random sample from a population with DF $F(x,\theta)$, $\theta \in \Theta \subset \mathbb{R}^1$. Assume that $p(x,\theta) = p_\theta(x)$ is the density of $F(x,\theta)$. The estimator

$$\theta^* = \operatorname{argmin}_{\theta\in\Theta} N(F_n^*, F_\theta)$$

is an M-estimator defined by the equation

$$\frac{1}{n}\sum_{j=1}^{n} g(x_j,\theta) = 0, \tag{23.4.1}$$

where

$$g(x,\theta) = \int_{-\infty}^{\infty} \mathcal{L}(x,v)p'_\theta(v)\mathrm{d}v - \int_{-\infty}^{\infty}\int_{-\infty}^{\infty} \mathcal{L}(u,v)p_\theta(u)p'_\theta(v)\mathrm{d}u\mathrm{d}v.$$

Here, $\mathcal{L}(u,v)$ is a negative definite kernel, which does not necessarily depend on the difference of arguments, and the prime $'$ denotes the derivative with respect to θ. As in Sect. 23.2, the necessary condition for consistency,

$$E_\theta g(x,\theta) = 0,$$

is automatically fulfilled. The asymptotic variance of θ^* is given by

$$\sigma^2_{\theta^*} = \frac{\operatorname{Var}\left(\int_{-\infty}^{\infty} \mathcal{L}(x,v)p'_\theta(v)\mathrm{d}v\right)}{\left(\int_{-\infty}^{\infty}\int_{-\infty}^{\infty} \mathcal{L}(u,v)p'_\theta(u)p'_\theta(v)\mathrm{d}u\mathrm{d}v\right)^2}.$$

We can proceed similarly to Sect. 23.3 to obtain the corresponding results in this case. Since the calculations are quite similar, we do not state these results explicitly. Note that to obtain the existence and uniqueness of θ^* for sufficiently large n, we do not need standard regularity conditions such as the existence of variance, differentiability of the density with respect to θ, and so on. These are used only to obtain the estimating equation and to express the asymptotic variance of the estimator.

In general, from the construction of θ^* we have

$$N(F_n^*, F_{\theta^*}) \le N(F_n^*, F_\theta) \text{ a.s.},$$

and hence

$$E_\theta N(F_n^*, F_{\theta^*}) \le E_\theta N(F_n^*, F_\theta)$$
$$= \frac{1}{n} \int_{-\infty}^{\infty} \int_{-\infty}^{\infty} \mathcal{L}(x, y) \mathrm{d}F(x, \theta) \mathrm{d}F(y, \theta) \xrightarrow[n \to \infty]{} 0. \quad (23.4.2)$$

In the case of a bounded kernel $\mathcal{L}$, the convergence is uniform with respect to θ. In this case it is easy to verify that $nN(F_n^*, F_\theta)$ converges to

$$-\int_{-\infty}^{\infty} \int_{-\infty}^{\infty} \mathcal{L}(x, y) \mathrm{d}w^\circ(F(x, \theta)) \mathrm{d}w^\circ(F(y, \theta))$$

as $n \to \infty$, where w° is the Brownian bridge.

23.5 Estimating a Location Parameter: Third Approach

Let us return to the case of estimating a location parameter. We will present an example of an estimator obtained by minimizing the $\mathfrak{N}$-distance, which has good robust properties. Let

$$\mathcal{L}_r(x) = \begin{cases} |x| & \text{for } |x| < r \\ r & \text{for } |x| \ge r, \end{cases}$$

where $r > 0$ is a fixed number. The famous Pólya criterion[2] implies that the function $f(t) = 1 - \frac{1}{r}\mathcal{L}_r(t)$ is the characteristic function of some probability distribution. Consequently, $\mathcal{L}_r(t)$ is a negative definite function. This implies that for a sufficiently large sample size n there exists an estimator θ^* of minimal $\mathfrak{N}^r$ distance, where $\mathcal{N}^r$ is the kernel constructed from $\mathcal{L}_r(x - y)$. If the distribution function $F(x - \theta)$ has a symmetric unimodal density $p(x - \theta)$ that is absolutely continuous and has a finite Fisher information

$$I = \int_{-\infty}^{\infty} \left(\frac{p'(x)}{p(x)} \right)^2 p(x) \mathrm{d}x,$$

then we conclude by (23.4.2) that θ^* is consistent and is asymptotically normal. The estimator θ^* satisfies (23.2.5), where

[2] See, for example, Lukacs (1969).

$$g_1(x) = \int_{-\infty}^{\infty} \mathcal{L}'(x - v)p(v)dv$$

and

$$\mathcal{L}'(u) = \begin{cases} 0 & \text{for } |u| \geq r, \\ 1 & \text{for } 0 < u < r, \\ 0 & \text{for } u = 0, \\ -1 & \text{for } -r < u < 0. \end{cases}$$

This implies that θ^* has a bounded influence function and, hence, is B-robust.[3]

Consider now the estimator θ' obtained by the method discussed in Sect. 23.3. It is easy to verify that this estimator is consistent under the same assumptions. However, θ' satisfies the equation

$$\sum_{j=1}^{n} \mathcal{L}'(x_j - \theta) = 0,$$

so that it is a trimmed median. It is well known that a trimmed median is the most B-robust estimator in the corresponding class of M-estimators.[4]

23.6 Semiparametric Estimation

Let us now briefly discuss semiparametric estimation. This problem is similar to that considered in Sect. 23.4, except that here we do not assume that the sample comes from a parametric family. Let $x_1, \ldots, x_n$, be a random sample from a population given by DF $F(x)$, which belongs to some distribution class $\mathcal{P}$. Suppose that the metric $\mathfrak{N}$ is generated by the negative definite kernel $\mathcal{L}(x, y)$ and that $\mathbf{P} \subset \mathcal{B}(\mathcal{L})$. $\mathcal{B}(\mathcal{L})$ is isometric to some subset of the Hilbert space $\mathfrak{H}$. Moreover, Aronszajn's theorem implies that $\mathfrak{H}$ can be chosen to be minimal in some sense. In this case, the definition of $\mathfrak{N}$ is extended to the entire $\mathfrak{H}$.

We assume that the distributions under consideration lie on some "nonparametric curve." In other words, there exists a nonlinear functional φ on $\mathfrak{H}$ such that the distributions F satisfy the condition

$$\varphi(F) = c = \text{const}.$$

The functional φ is assumed to be smooth. For any $H \in \mathfrak{H}$

[3]See Hampel et al. (1986).

[4]See Hampel et al. (1986).

$$\lim_{t\to 0}\frac{N(F+tH,G)-N(F,G)}{t}=2\int_{-\infty}^{\infty}\int_{-\infty}^{\infty}\mathcal{L}(x,y)\mathrm{d}(G(x)-F(x))\mathrm{d}H(y)$$
$$=\langle\ \mathrm{grad}\ N(F,G),H\rangle,$$

where G is fixed.

Under the parametric formulation of Sect. 23.4, the equation for θ has the form

$$\frac{\mathrm{d}}{\mathrm{d}\theta}N(F_\theta,F_n^*)=0,$$

that is,

$$\left\langle \mathrm{grad}\ N(F,F_n^*)|_{F=F_\theta},\ \frac{\mathrm{d}}{\mathrm{d}\theta}F_\theta\right\rangle=0.$$

Here, the equation explicitly depends on the gradient of the functional $N(F,F_n^*)$. However, under the nonparametric formulation, we work with the conditional minimum of the functional $N(F,F_n^*)$, assuming that F lies on the surface $\varphi(F)=C$. Here, our estimator is

$$\tilde{F}^*=\underset{F\in\{F:\varphi(F)=c\}}{\mathrm{argmin}}\ N(F,F_n^*).$$

According to general rules for finding conditional critical points, we have

$$\mathrm{grad}\ N(\tilde{F}^*,\tilde{F}_n^*)=\lambda\ \mathrm{grad}\ \phi(\tilde{F}^*), \tag{23.6.1}$$

where λ is a number. Thus, in the general case, (23.6.1) is an eigenvalue problem. This is a general framework of semiparametric estimation.

References

Hampel FR, Ronchetti EM, Rousseeuw PJ, Stahel WA (1986) Robust statistics: the approach based on influence functions. Wiley, New York

Huber P (1981) Robust statistics. Wiley, New York

Lukacs E (1969) Characteristic functions. Griffin, London

Chapter 24
Some Statistical Tests Based on $\mathfrak{N}$-Distances

The goals of this chapter are to:

- Construct statistical tests based on the theory of $\mathfrak{N}$-distances,
- Study properties of multivariate statistical tests.

24.1 Introduction

In this chapter, we construct statistical tests based on the theory of $\mathfrak{N}$-distances. We consider a multivariate two-sample test, a test to determine if two distributions belong to the same additive type, and tests for multivariate normality with unknown mean and covariance matrix.

24.2 A Multivariate Two-Sample Test

Here we introduce a class of free-of-distribution multivariate statistical tests closely connected to $\mathfrak{N}$-distances.

Let $\mathcal{L}(x,y)$ be a strongly negative definite kernel on $\mathbb{R}^d \times \mathbb{R}^d$. As always, we suppose that $\mathcal{L}$ satisfies

$$\mathcal{L}(x,y) = \mathcal{L}(y,x) \text{ and } \mathcal{L}(x,x) = 0 \text{ for all } x, y \in \mathfrak{X}.$$

Suppose that X, Y are two independent random vectors in $\mathbb{R}^d$, and define one-dimensional independent random variables (RVs) U, V by the relation

$$U = \mathcal{L}(X,Y) - \mathcal{L}(X,X'), \tag{24.2.1}$$

$$V = \mathcal{L}(Y',Y'') - \mathcal{L}(X'',Y''). \tag{24.2.2}$$

S.T. Rachev et al., *The Methods of Distances in the Theory of Probability and Statistics*, DOI 10.1007/978-1-4614-4869-3_24, © Springer Science+Business Media, LLC 2013

Here, $X \stackrel{d}{=} X' \stackrel{d}{=} X''$, and all vectors X, X', X'', Y, Y', Y'' are mutually independent.

It is clear that the condition $\mathfrak{N}(X, Y) = 0$ is equivalent to $\mathcal{N}(X, Y) = 0$, which is equivalent to $EU = EV$. But

$$\mathcal{N}(X, Y) = 0 \iff X \stackrel{d}{=} Y \implies U \stackrel{d}{=} V.$$

Therefore, under the conditions

$$E\mathcal{L}(X, X') < \infty, \quad E\mathcal{L}(Y, Y') < \infty, \tag{24.2.3}$$

we have

$$X \stackrel{d}{=} Y \iff U \stackrel{d}{=} V. \tag{24.2.4}$$

Assume now that we are interested in testing the hypothesis $H_o : X \stackrel{d}{=} Y$ for multivariate random vectors X, Y. We have seen that, theoretically, this hypothesis is equivalent to $H_o' : U \stackrel{d}{=} V$, where U, V are random variables taking values in $\mathbb{R}^1$. To test H_o', we can use a arbitrary one-dimensional free-of-distribution test, say the Kolomogorov–Smirnov test. It is clear that if the distributions of X and Y are continuous, then U and V have continuous distributions, too. Therefore, the test for H_o' will appear to be free of distribution in this case.

Consider now the two independent samples

$$X_1, \ldots, X_n; \quad Y_1, \ldots, Y_n \tag{24.2.5}$$

from general populations X and Y, respectively. To apply a one-dimensional test to U and V, we must construct (or simulate) the samples from these populations based on observations (24.2.5). We can proceed using one of the following two methods:

Method 1. Split each sample into three equal parts, and consider each of the parts as a sample from X, X', X'' and from Y, Y', Y'', respectively. Of course, this methods leads to essential loss of information but is unobjectionable from a theoretical point of view.

Method 2. Simulate the samples from X' and X'' (as well as from Y' and Y'') by independent choices from observations $X_1, \ldots, X_n$ (and from $Y_1, \ldots, Y_n$, respectively). Theoretically, the drawback of this approach is that now we do not test the hypothesis $X \stackrel{d}{=} Y$, but one of the identities of the corresponding empirical distributions. Therefore, the test is, obviously, asymptotically free of distribution (as $n \to \infty$) but generally is not free of distribution for a fixed value of sample size n.

Let us start with the studies of test properties based on Method 1. We simulated 5,000 pairs of samples of volume $n = 300$ from two-dimensional Gaussian vectors, calculated values of U and V (the splitting into three equal parts had been done), and applied Kolmogorov–Smirnov statistics. The values of U and V were calculated

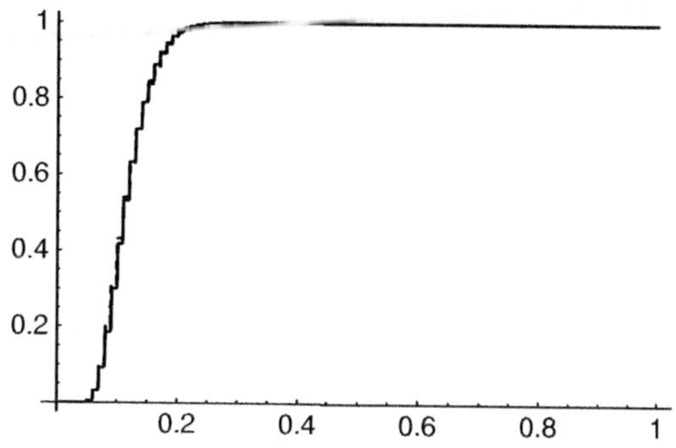

Fig. 24.1 The p-values for simulated (*dashed line*) and theoretical Kolmogorov–Smirnov (*solid line*) test

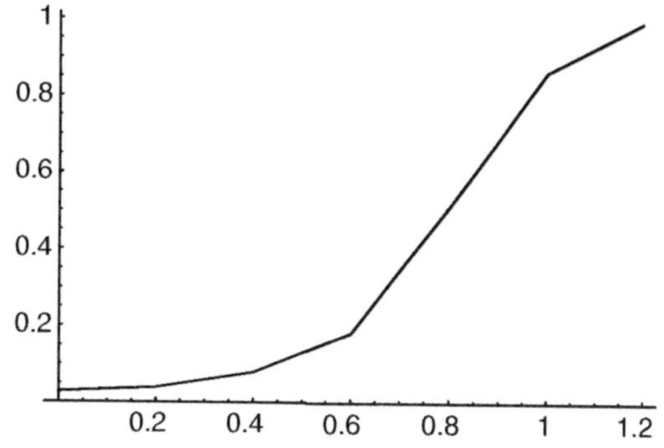

Fig. 24.2 Power of test under two-component location alternatives

for the kernel $\mathcal{L}(x, y) = \|x - y\|$ with an ordinary Euclidean norm. The results of the simulation for the p-values are shown in Fig. 24.1 by the dashed line. The solid line corresponds to theoretical p-values of the Kolmogorov–Smirnov test when the sample size equals 100.

As can be seen in Fig. 24.1, the graphs appear to be almost identical. In full agreement with theory, simulations show that the distribution of the test under zero hypothesis does not depend either on the parameters of the underlying distribution or on its dimensionality. We omit the corresponding graphs (they are identical to those of Fig. 24.1).

Let us now discuss the simulation study of the power of the proposed test using Method 2. We start with location alternatives for X and Y. In other words, we test the hypothesis $H_o : X \stackrel{d}{=} Y$ against the alternative $X \stackrel{d}{=} Y + \theta$, where θ is a known vector.

Figure 24.2 shows the plot of the power of our test for the following case. We simulated samples of volume $n = 100$ from two-dimensional Gaussian distributions. The first sample was taken from a distribution with zero mean vector and covariance matrix

$$\Lambda = \begin{pmatrix} 1 & \alpha \\ \alpha & 1 \end{pmatrix},$$

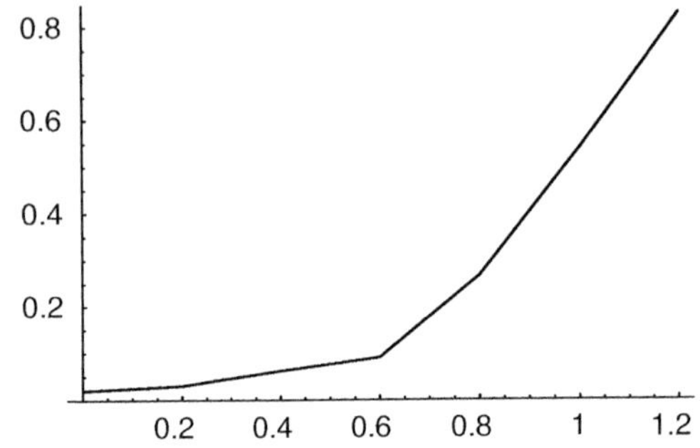

Fig. 24.3 Power of test under one-component location alternatives, correlation = 0.5

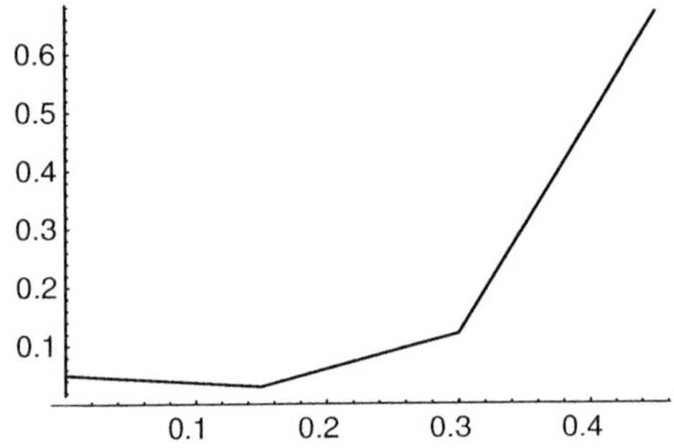

Fig. 24.4 Power of test under correlation alternatives

where $\alpha = 0.5$. For the other sample, the Gaussian distribution was with the same covariance matrix but having mean vector $(0.2m, 0.2m)$, $m = 0, 1, \ldots, 6$. The procedure was repeated 500 times for each sample. The portion of rejected hypotheses is shown in Fig. 24.2.

Figure 24.3 shows a plot of the power of our test for almost the same case as in the previous figure, but we changed only the first coordinate of the mean vector, i.e., we had mean vector $(0.2m, 0)$, $m = 0, 1, \ldots, 6$. The reduction of the power is natural in this case because the distance between simulated distributions is approximately $1/\sqrt{2}$ times smaller in the second case.

As we can see from the results of the simulations, the correlation between the components of the Gaussian vector do not essentially affect the power for the scale alternatives. The simulations for different correlation coefficients show us that the sensitivity of the statistic to such alternatives is essentially lower than that for the location alternatives.

Figure 24.4 shows a plot of the power of our test for the following case. We simulated samples of volume $n = 2{,}000$ from two-dimensional Gaussian distributions. The first sample was taken from the distribution with zero mean vector and covariance matrix

$$\Lambda = \begin{pmatrix} 1 & \alpha \\ \alpha & 1 \end{pmatrix},$$

where $\alpha = 0$, and the second one with zero mean vector and covariance matrix

$$\Xi = \begin{pmatrix} 1 & \beta \\ \beta & 1 \end{pmatrix},$$

where $\beta = 0.3m$, $m = 0, 1, 2, 3$. We see from the figure that the power is not as high as it was for the location alternatives (here we have $n = 2{,}000$ while for the location alternatives $n = 100$). This finding is expected because the distributions being compared have the same marginals.

24.3 Testing If Two Distributions Belong to the Same Additive Type

Suppose that $z_1, \dots, z_k$ $(k \geq 3)$ are independent and identically distributed random vectors in $\mathbb{R}^d$ having the DF $F(x)$. Consider the vector $Z = (z_2 - z_1, \dots, z_d)$. It is clear that the distribution of the vector Z is the same as for random vectors $z_j + \theta$, $j = 1, \dots, d$, $\theta \in \mathbb{R}^d$. In other words, the distribution of Z is the same for the additive type of F, i.e., for all DFs of the form $F(x - \theta)$. The problem of recovering the additive type of a distribution on the basis of the distribution of Z was considered by Kovalenko (1960), who proved that recovery is possible if the characteristic function has "not too many" zeros, i.e., the set of zeros is not dense in any d-dimensional ball.

Based on the result by Kovalenko, it is possible to conduct a test to determine if two distributions belong to the same additive type. Let $X_1, \dots, X_n$ and $Y_1, \dots, Y_n$ be independent samples from the populations X and Y, respectively. We want to test the hypothesis $X \stackrel{d}{=} Y + \theta$ for a constant unknown vector θ against the alternative $X \stackrel{d}{\neq} Y + \theta$ for all θ. To construct the test we can do the following steps:

1. By independent sampling or by permutations from the values $X_1, \dots, X_n$, generate two independent samples $X_1', \dots, X_n'$ and $X_1'', \dots, X_n''$.
2. By independent sampling or by permutations from the values $Y_1, \dots, Y_n$, generate two independent samples $Y_1', \dots, Y_n'$ and $Y_1'', \dots, Y_n''$.
3. Form vector samples

$$Z_X = ((X_1' - \text{Mean}(X'), X_1'' - \text{Mean}(X'')), \dots,$$
$$(X_n' - \text{Mean}(X'), X_n'' - \text{Mean}(X'')))$$

and

$$Z_Y = ((Y_1' - \text{Mean}(Y'), Y_1'' - \text{Mean}(Y'')), \dots,$$
$$(Y_n' - \text{Mean}(Y'), Y_n'' - \text{Mean}(Y''))).$$

4. Using the two methods described in the previous section, test the hypothesis that the samples Z_X and Z_Y are taken from the same population.

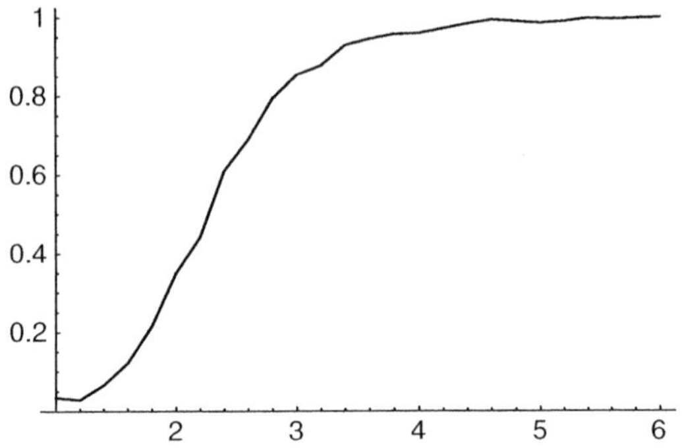

Fig. 24.5 Power of test under scale alternatives (split sample)

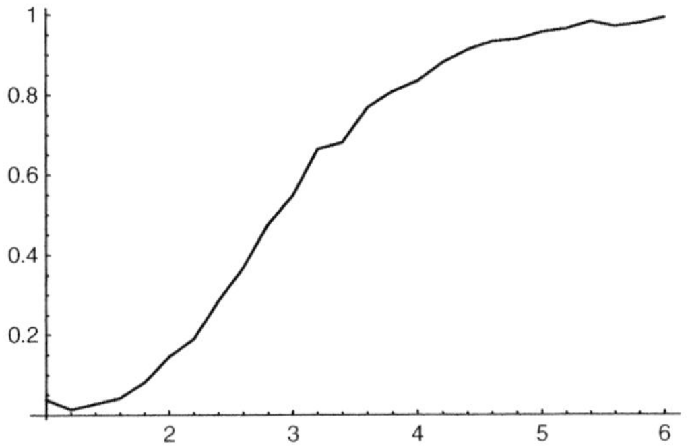

Fig. 24.6 Power of test under scale alternatives (permuted sample)

It is clear that Method 2 is theoretically good only asymptotically because of the impact associated with sampling from the observed data. To avoid this impact, we can (as we did in the previous section) split the original samples into a corresponding number of parts. But our simulations show that Method 2 of permuting original data works rather well. Consequently, usually we do not need to split the original sample.

We simulated 500 pairs of samples from Gaussian distributions $(0, 1)$ and $(3, \sigma)$ of size $n = 300$ each. Figure 24.5 shows a plot of the power of our test for the case of split samples. The parameter σ changes from 1 to 7.5 with step 0.3. On the abscissa-axis we have $m = 1 + (\sigma - 1)/0.3$. We used the kernel $\mathcal{L}(x, y) = \|x - y\|$.

We also simulated 500 pairs of samples from Gaussian distributions $(0, 1)$ and $(3, \sigma)$ of size $n = 100$ each. Figure 24.6 shows a plot of the power of our test for the case of permuted samples. The parameter σ changes from 1 to 6 with step 0.2. On the abscissa-axis we have $m = 1 + (\sigma - 1)/0.2$. We used the kernel $\mathcal{L}(x, y) = \|x - y\|$. Comparing Figs. 24.5 and 24.6, we find that there is almost the same power for both split and permuted samples.

Figure 24.7 shows a plot of the power of our test for the same case as for Fig. 24.6, but we used the kernel $\mathcal{L}(x, y) = 1 - \exp(-\|x - y\|^2)$. A comparison to Fig. 24.6 indicates that the last kernel produces a higher power. But this effect depends on the underlying distribution (recall that the Gaussian is used to generate both figures).

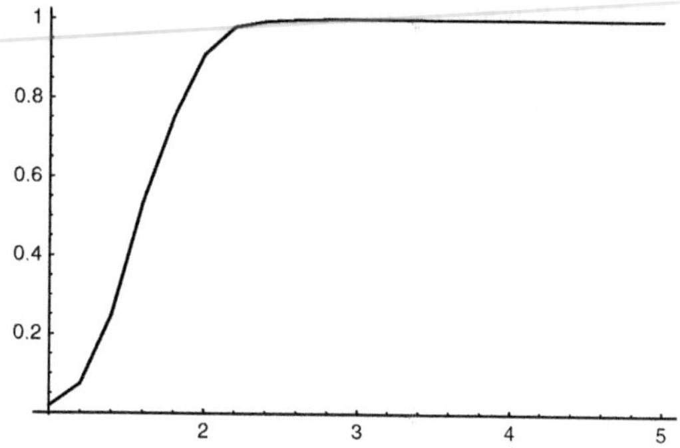

Fig. 24.7 Power of test under scale alternatives, $L = 1 - \exp \|.\|^2$

24.4 A Test for Multivariate Normality

Undoubtedly, there is interest in tests to assess whether a vector of observations is Gaussian with unknown mean and covariance matrix. Such a test may be constructed based on the following characterization of Gaussian law.

Proposition 24.4.1. *Let Z, Z', Z'', Z''' denote four independent and identically distributed random vectors in $\mathbb{R}^d$. The vector Z has a Gaussian distribution if and only if*

$$Z \stackrel{d}{=} \frac{2}{3}Z' + \frac{2}{3}Z'' - \frac{1}{3}Z'''. \tag{24.4.1}$$

Suppose now that $Z_1, \ldots, Z_n$ is a random sample from the population Z. We can construct the following test for determining if Z is Gaussian.

1. Choosing independently from the values $Z_1, \ldots, Z_n$ (or using permutations of those values), generate $Z'_1, \ldots, Z'_n$, $Z''_1, \ldots, Z''_n$, and $Z'''_1, \ldots, Z'''_n$.
2. Build two samples

$$X = (Z_1, \ldots, Z_n)$$

and

$$Y = \left(\left(\frac{2}{3}Z'_1 + \frac{2}{3}Z''_1 - \frac{1}{3}Z'''_1\right), \ldots, \left(\frac{2}{3}Z'_n + \frac{2}{3}Z''_n - \frac{1}{3}Z'''_n\right)\right).$$

3. Test the hypothesis that X and Y are taken from the same population. According to Proposition 24.4.1, this hypothesis is equivalent to one of the normality of Z.

Figure 24.8 shows the power of our test for the case where we simulated samples of volume $n = 300$ from the mixture of two Gaussian distributions, both with unit variance and mean 1 and 5, respectively. The mixture proportion p changed from 0 to 1 with step 0.1. Of course, the power is small near $p = 0$ and $p = 1$ because the mixture almost corresponds to a Gaussian distribution [with the parameters $(0, 1)$ for p close to 0, and with parameters $(5, 1)$ for p close to 1]. But the power is close to 1 for $p \in (0.3, 0.7)$.

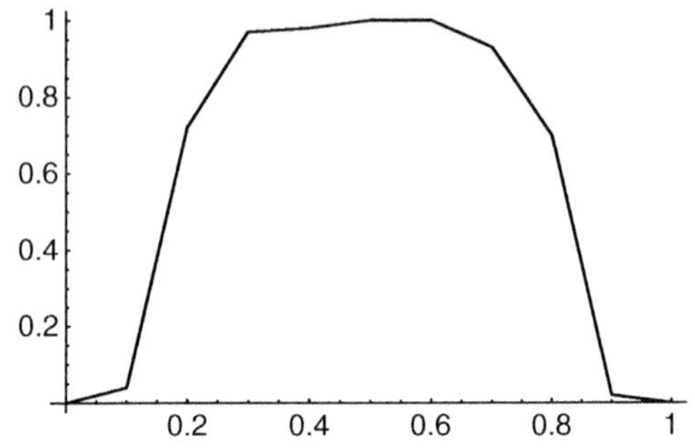

Fig. 24.8 Power of test for normality with arbitrary parameters

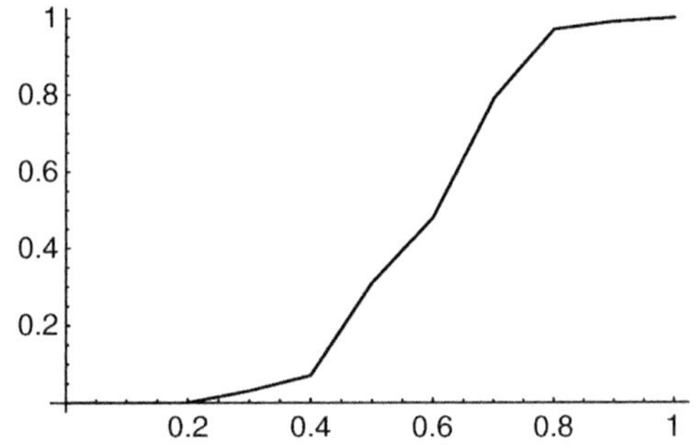

Fig. 24.9 Power of test for normality with zero mean

We can use another characterization of the normal distribution with zero mean to construct a corresponding statistical test. To do this, we can change the definition

$$Y = \left(\left(\frac{2}{3} Z_1' + \frac{2}{3} Z_1'' - \frac{1}{3} Z_1''' \right), \dots, \left(\frac{2}{3} Z_n' + \frac{2}{3} Z_n'' - \frac{1}{3} Z_n''' \right) \right)$$

by

$$Y = \left(\frac{Z_1' + Z_1''}{\sqrt{2}}, \dots, \frac{Z_n' + Z_n''}{\sqrt{2}} \right).$$

Samples X and Y are taken from the same population if and only if Z is Gaussian with zero mean and arbitrary variance.

Figure 24.9 demonstrates the power of our test for the case where we simulated samples of volume $n = 200$ from a Gaussian distribution with parameters $(a, 1)$. Parameter a (mean value of the distribution) changed from 0 to 1 with step 0.1.

References

Kovalenko IN (1960) On reconstruction of the additive type of distribution by a successive run of independent experiments. In: Proceedings of all-union meeting on probability theory and math statistics, Yerevan

Chapter 25
Distances Defined by Zonoids

The goals of this chapter are to:

- Introduce $\mathfrak{N}$-distances defined by zonoids,
- Explain the connections between $\mathfrak{N}$-distances and zonoids.

Notation introduced in this chapter:

Notation	Description
$h(K, u)$	Support function of a convex body
$K_1 \oplus K_2$	Minkowski sum of sets K_1 and K_2
S^{d-1}	Unit sphere in $\mathbb{R}^d$

25.1 Introduction

Suppose that $\mathfrak{X}$ is a metric space with the distance ρ. It is well known (Schoenberg 1938) that $\mathfrak{X}$ is isometric to a subspace of a Hilbert space if and only if ρ^2 is a negative definite kernel. The so-called $\mathfrak{N}$-distance (Klebanov 2005) is a variant of a construction of a distance on a space of measures on $\mathfrak{X}$ such that $\mathfrak{N}^2$ is a negative definite kernel. Such a construction is possible if and only if ρ^2 is a strongly negative definite kernel on $\mathfrak{X}$.

In this chapter, we show that the supporting function of any zonoid in $\mathbb{R}^d$ is a negative definite first-degree homogeneous function. The inverse is also true. If the support of a generating measure of a zonoid coincides with the unit sphere, then the supporting function is strongly negative definite, and therefore it generates a distance on the space of Borel probability measures on $\mathbb{R}^d$.

S.T. Rachev et al., *The Methods of Distances in the Theory of Probability and Statistics*,
DOI 10.1007/978-1-4614-4869-3_25,

25.2 Main Notions and Definitions

Here we review some known definitions and facts from stochastic geometry.[1]

Let $\mathfrak{C}$ (resp. $\mathfrak{C}'$) be the system of all compact convex sets (resp. nonempty compact convex sets) in $\mathbb{R}^d$. A set $K \in \mathfrak{C}'$ is called a convex body if $K \in \mathfrak{C}'$; then for each $u \in S^{d-1}$ there is exactly one number $h(K,u)$ such that the hyperplane

$$\{x \in \mathbb{R}^d : \langle x, u\rangle - h(K,u) = 0\} \tag{25.2.1}$$

intersects K, and $\langle x, u\rangle - h(K,u) \le 0$ for each $x \in K$. This hyperplane is called the *support hyperplane*, and the function $h(K,u)$, $u \in S^{d-1}$ (where S^{d-1} is the unit sphere), is the *support function* (restricted to S^{d-1}) of K. Equivalently, one can define

$$h(K,u) = \sup\{\langle x, u\rangle,\ x \in K\},\quad u \in \mathbb{R}^d. \tag{25.2.2}$$

Its geometrical meaning is the signed distance of the support hyperplane from the coordinate origin.

An important property of $h(K,u)$ is its additivity:

$$h(K_1 \oplus K_2, u) = h(K_1,u) + h(K_2,u),$$

where $K_1 \oplus K_2 = \{a+b : a \in K_1, b \in K_2\}$ is the Minkowski sum of K_1 and K_2. For $K \in \mathfrak{C}'$ let $\check{K} = \{-k, k \in K\}$. We say that K is *centrally symmetric* if $K' = \check{K}'$ for some translate K', i.e., if K has a center of symmetry.

The Minkowski sum of finitely many centered line segments is called a *zonotope*. Consider a zonotope

$$\mathcal{Z} = \bigoplus_{i=1}^{k} a_i[v_i, -v_i], \tag{25.2.3}$$

where $a_i > 0$, $v_i \in \mathbb{S}^{d-1}$. Its support function is given by

$$h(\mathcal{Z},u) = h_{\mathcal{Z}}(u) = \sum_{i=1}^{k} a_i|\langle u, v_i\rangle|. \tag{25.2.4}$$

We use the notation $\mathcal{K}'$ for the space of all compact subsets of $\mathbb{R}^d$ with the Hausdorff metric

$$d_H(K_1,K_2) = \max\{\sup_{x\in K_1} \mathrm{dist}(x,K_2),\ \sup_{y\in K_2} \mathrm{dist}(y,K_1)\}, \tag{25.2.5}$$

where $\mathrm{dist}(x,K) = \inf_{z\in K} \|x-z\|$.

[1]See, for example, Ziegler (1995) and Beneš and Rataj (2004).

A set $\mathcal{Z} \in \mathfrak{C}'$ is called a *zonoid* if it is a limit in a d_H distance of a sequence of zonotopes.

It is known that a convex body $\mathcal{Z}$ is a zonoid if and only if its support function has a representation

$$h(\mathcal{Z}, u) = \int_{\mathbb{S}^{d-1}} |\langle u, v\rangle| \mathrm{d}\mu_{\mathcal{Z}}(v) \tag{25.2.6}$$

for an even measure $\mu_{\mathcal{Z}}$ on $\mathbb{S}^{d-1}$. The measure $\mu_{\mathcal{Z}}$ is called the *generating measure* of $\mathcal{Z}$. It is known that the generating measure is unique for each zonoid $\mathcal{Z}$.

25.3 $\mathfrak{N}$-Distances

Suppose that $(\mathfrak{X}, \mathfrak{A})$ is a measurable space and $\mathcal{L}$ is a strongly negative definite kernel on $\mathfrak{X}$. Denote by $\mathcal{B}_{\mathcal{L}}$ the set of all probabilities μ on $(\mathfrak{X}, \mathfrak{A})$ for which there exists the integral

$$\int_{\mathfrak{X}} \int_{\mathfrak{X}} \mathcal{L}(x, y) \mathrm{d}\mu(x) \mathrm{d}\mu(y) < \infty. \tag{25.3.1}$$

For $\mu, \nu \in \mathcal{B}_{\mathcal{L}}$ consider

$$\begin{aligned} \mathcal{N}(\mu, \nu) = 2 &\int_{\mathfrak{X}} \int_{\mathfrak{X}} \mathcal{L}(x, y) \mathrm{d}\mu(x) \mathrm{d}\nu(y) \\ &- \int_{\mathfrak{X}} \int_{\mathfrak{X}} \mathcal{L}(x, y) \mathrm{d}\mu(x) \mathrm{d}\mu(y) \\ &- \int_{\mathfrak{X}} \int_{\mathfrak{X}} \mathcal{L}(x, y) \mathrm{d}\nu(x) \mathrm{d}\nu(y). \end{aligned} \tag{25.3.2}$$

It is known (Klebanov 2005) that

$$\mathfrak{N}(\mu, \nu) = \left(\mathcal{N}(\mu, \nu)\right)^{1/2}$$

is a distance on $\mathcal{B}_{\mathcal{L}}$.

Described below are some examples of negative definite kernels.

Example 25.3.1. Let $\mathfrak{X} = \mathbb{R}^1$. For $r \in [0, 2]$ define

$$\mathcal{L}_r(x, y) = |x - y|^r.$$

The function $\mathcal{L}_r$ is a negative definite kernel. For $r \in (0, 2)$, $\mathcal{L}_r$ is a strongly negative definite kernel.

For the proof of the statement in this example and the statement in the next example (Example 25.3.2), see Klebanov (2005).

Example 25.3.2. Let $\mathcal{L}(x, y) = f(x - y)$, where $f(t)$ is a continuous function on $\mathbb{R}^d$, $f(0) = 0$, $f(-t) = f(t)$. $\mathcal{L}$ is a negative definite kernel if and only of

$$f(t) = \int_{\mathbb{R}^d} \left(1 - \cos\langle t, u\rangle\right) \frac{1 + \|u\|^2}{\|u\|^2} \mathrm{d}\Theta(u), \tag{25.3.3}$$

where Θ is a finite measure on $\mathbb{R}^d$. Representation (25.3.3) is unique. Kernel $\mathcal{L}$ is strongly negative definite if the support of the measure Θ coincides with the whole space $\mathbb{R}^d$.

We will give an alternative proof for the fact that $|x - y|$ is a negative definite kernel. For the case $\mathfrak{X} = \mathbb{R}^1$ define

$$\mathcal{L}(x, y) = 2\max(x, y) - x - y = |x - y|. \tag{25.3.4}$$

Then $\mathcal{L}$ is a negative definite kernel.

Proof. It is sufficient to show that $\max(x, y)$ is a negative definite kernel. For arbitrary $a \in \mathbb{R}^1$ consider

$$u_a(x) = \begin{cases} 1, & x < a, \\ 0, & x \geq a. \end{cases} \tag{25.3.5}$$

It is clear that

$$u_a(\max(x, y)) = u_a(x)u_a(y).$$

Let $F(a)$ be a nondecreasing bounded function on $\mathbb{R}^1$. Define

$$\mathcal{K}(x, y) = \int_{-\infty}^{\infty} u_a(\max(x, y))\mathrm{d}F(a).$$

For any integer $n > 1$ and arbitrary $c_1, \ldots, c_n$ under condition $\sum_{j=1}^n c_j = 0$ we have

$$\sum_{i=1}^n \sum_{j=1}^n \mathcal{K}(x_i, x_j)c_i c_j = \int_{-\infty}^{\infty} \sum_{i=1}^n \sum_{j=1}^n u_a(x_i)u_a(x_j)c_i c_j \mathrm{d}F(a)$$

$$= \int_{-\infty}^{\infty} \left(\sum_{i=1}^n u_a(x_i)c_i \right)^2 \mathrm{d}F(a) \geq 0.$$

But

$$\mathcal{K}(x, y) = \int_{-\infty}^{\infty} u_a(\max(x, y))\mathrm{d}F(a)$$
$$= F(+\infty) - F(\max(x, y)).$$

Let us fix arbitrary $A > 0$ and apply the previous equality to the function

$$F(a) = F_A(a) = \begin{cases} A & \text{for } a > A, \\ a & \text{for } -A \leq a \leq A, \\ -A & \text{for } a < -A. \end{cases} \tag{25.3.6}$$

In this case, $\mathcal{K}(x, y) = A - \max(x, y)$ for $x, y \in [-A, A]$, and, as $A \to \infty$, we obtain that $\max(x, y)$ is a negative definite kernel. □

Directly from the definition of a negative definite kernel and Example 25.3.1 we obtain the next example.

Example 25.3.3. Let $x, y \in \mathbb{R}^d$, and $f : \mathbb{R}^d \to \mathbb{R}^1$. Define

$$\mathcal{L}(x, y) = |f(x) - f(y)|.$$

Then $\mathcal{L}$ is a negative definite kernel.

Of course, the mixture of negative definite kernels is again a negative definite kernel.

Example 25.3.4. Let us choose and fix a vector $\theta \in \mathbb{S}^{d-1}$ and consider the kernel

$$\mathcal{L}_\theta(x, y) = |\langle x, \theta\rangle - \langle y, \theta\rangle|.$$

From previous considerations it is clear that $\mathcal{L}_\theta$ is a negative definite kernel on $\mathbb{R}^d$, and for the σ-finite measure Ξ

$$\mathcal{L}_\Xi(x, y) = \int_{\mathbb{S}^{d-1}} \mathcal{L}_\theta(x, y)\mathrm{d}\Xi(\theta) \tag{25.3.7}$$

is, again, a negative definite kernel.

Consider expression (25.3.2) constructed on the basis of (25.3.7). Let us rewrite (25.3.2) in a different form. Suppose that X and Y are two random vectors in $\mathbb{R}^d$ with distributions μ and ν, respectively. We write $\mathcal{N}(X, Y)$ instead of $\mathcal{N}(\mu, \nu)$, so that

$$\mathcal{N}(X, Y) = 2E\mathcal{L}_\Xi(X, Y) - E\mathcal{L}_\Xi(X, X') - E\mathcal{L}_\Xi(Y, Y'),$$

where $X' \stackrel{d}{=} X$ and $Y' \stackrel{d}{=} Y$ are independent copies of X and Y, respectively. Note that we use the sign $\stackrel{d}{=}$ for the equality in a distribution. We have

$$\mathcal{N}(X,Y) = E\int_{\mathbb{S}^{d-1}} [4\max(\langle X,\theta\rangle, \langle Y,\theta\rangle)$$
$$-2\max(\langle X,\theta\rangle, \langle X',\theta\rangle) - 2\max(\langle Y,\theta\rangle, \langle Y',\theta\rangle)]\mathrm{d}\Xi(\theta).$$

Denote $X_\theta = \langle X,\theta\rangle$, $Y_\theta = \langle Y,\theta\rangle$. Then

$$\mathcal{N}(X,Y) = 2\int_{\mathbb{S}^{d-1}} \lim_{A\to\infty} E\int_{-A}^{A}\big(u_a(X_\theta)u_a(X'_\theta)$$
$$+u_a(Y_\theta)u_a(Y'_\theta) - 2u_a(X_\theta)u_a(Y_\theta)\big)\mathrm{d}F_A(a)\mathrm{d}\Xi(\theta).$$

But $Eu_a(X_\theta) = \Pr\{X_\theta < a\}$, and therefore

$$\mathcal{N}(X,Y) = 2\lim_{A\to\infty}\int_{\mathbb{S}^{d-1}} \mathrm{d}\Xi(\theta)\int_{-A}^{A}\Big(\Pr\{X_\theta < a\}\Pr\{X'_\theta < a\}$$
$$+\Pr\{Y_\theta < a\}\Pr\{Y'_\theta < a\} - 2\Pr\{X_\theta < a\}\Pr\{Y_\theta < a\}\Big)\mathrm{d}F_A(a)$$
$$= 2\int_{\mathbb{S}^{d-1}} \mathrm{d}\Xi(\theta)\int_{-\infty}^{\infty}\Big(F_\theta(a) - G_\theta(a)\Big)^2\mathrm{d}a,$$

where $F_\theta(a) = \Pr\{X_\theta < a\}$, $G_\theta(a) = \Pr\{Y_\theta < a\}$. So finally we have

$$\mathcal{N}(X,Y) = 2\int_{\mathbb{S}^{d-1}} \mathrm{d}\Xi(\theta)\int_{-\infty}^{\infty}\Big(F_\theta(a) - G_\theta(a)\Big)^2\mathrm{d}a. \tag{25.3.8}$$

If the support of Ξ coincides with $\mathbb{S}^{d-1}$, then $\mathfrak{N}(X,Y) = \Big(\mathcal{N}(X,Y)\Big)^{1/2}$ is a distance between the distributions of X and Y.

Let us return to the kernel

$$\mathcal{L}_\theta(x,y) = 2\max(\langle x,\theta\rangle, \langle y,\theta\rangle) - \langle x,\theta\rangle - \langle y,\theta\rangle.$$

Choose arbitrary $\theta_o \in \mathbb{S}^{d-1}$, and consider the measure

$$\Xi_o = \frac{1}{2}\big(\delta_{\theta_o} + \delta_{-\theta_o}\big),$$

where δ_{θ_o} is the measure concentrated at point θ_o. Then

$$\mathcal{L}_{\Xi_{\theta_o}}(x,y) = \int_{\mathbb{S}^{d-1}} \mathcal{L}_\theta(x,y)\mathrm{d}\Xi_o(\theta)$$
$$= \max(\langle x,\theta_o\rangle, \langle y,\theta_o\rangle) + \max(-\langle x,\theta_o\rangle, -\langle y,\theta_o\rangle)$$
$$= |\langle x-y,\theta\rangle|.$$

Now, if we have an arbitrary even measure Ξ_s on sphere $\mathbb{S}^{d-1}$, then

$$\mathcal{L}_{\Xi_s}(x,y) = \int_{\mathbb{S}^{d-1}} \mathcal{L}_\theta(x,y)\mathrm{d}\Xi_s(\theta)$$
$$= \int_{\mathbb{S}^{d-1}} |\langle x-y,\theta\rangle|\mathrm{d}\Xi_s(\theta)$$

is a negative definite kernel. Let us note that the function

$$h(z) = \int_{\mathbb{S}^{d-1}} |\langle z,\theta\rangle|\mathrm{d}\Xi_s(\theta), \quad z \in \mathbb{R}^d \tag{25.3.9}$$

is the support function of a zonoid with generating measure Ξ_s.

Summarizing all the preceding relations we may formulate the following result.

Theorem 25.3.1. Each zonoid $\mathcal{Z}$ generates a negative definite kernel on $\mathbb{R}^d$

$$\mathcal{L}_{\mathcal{Z}}(x,y) = h_{\mathcal{Z}}(x-y) = \int_{\mathbb{S}^{d-1}} |\langle x-y,\theta\rangle|\mathrm{d}\mu_{\mathcal{Z}}(\theta). \tag{25.3.10}$$

This kernel is strongly negative definite if the support of $\mu_{\mathcal{Z}}$ coincides with the whole sphere $\mathbb{S}^{d-1}$, and

$$\mathcal{N}(\mu,\nu) = 2\int_{\mathbb{R}^d}\int_{\mathbb{R}^d} \mathcal{L}_{\mathcal{Z}}(x,y)\mathrm{d}\mu(x)\mathrm{d}\nu(y) - \int_{\mathbb{R}^d}\int_{\mathbb{R}^d} \mathcal{L}_{\mathcal{Z}}(x,y)\mathrm{d}\mu(x)\mathrm{d}\mu(y)$$
$$- \int_{\mathbb{R}^d}\int_{\mathbb{R}^d} \mathcal{L}_{\mathcal{Z}}(x,y)\mathrm{d}\nu(x)\mathrm{d}\nu(y)$$

is the square of a distance between measures $\mu,\nu \in \mathcal{B}_{\mathcal{L}}$. This distance has the following representation:

$$\mathfrak{N}(\mu,\nu) = \left(\int_{\mathbb{S}^{d-1}} \mathrm{d}\mu_{\mathcal{Z}}(\theta) \int_{-\infty}^{\infty} \big(F_\theta(a) - G_\theta(a)\big)^2 \mathrm{d}a\right)^{1/2}, \tag{25.3.11}$$

where

$$\mu(\mathcal{A}) = \Pr\{X \in \mathcal{A}\}, \ \nu(\mathcal{A}) = \Pr\{Y \in \mathcal{A}\},$$
$$F_\theta(a) = \Pr\{\langle X,\theta\rangle < a\}, \ G_\theta(a) = \Pr\{\langle Y,\theta\rangle < a\}. \tag{25.3.12}$$

According to Example 25.3.2, the function $h_{\mathcal{Z}}(u)$ from (25.3.10) may be represented in the form (25.3.3). Let us investigate the connection between $\mu_{\mathcal{Z}}$ in (25.3.10) and Θ in (25.3.3). To do so, we will use the following identity:

$$|z| = \frac{2}{\pi}\int_0^{\infty}\big(1-\cos(zt)\big)\frac{dt}{t^2}. \tag{25.3.13}$$

We have

$$\begin{aligned} h_{\mathcal{Z}}(u) &= \frac{2}{\pi}\int_{\mathbb{S}^{d-1}}\int_0^{\infty}\big(1-\cos\langle u,\theta\rangle\big)\frac{dt}{t^2}d\mu_{\mathcal{Z}}(\theta) \\ &= \frac{2}{\pi}\int_{\mathbb{R}^d}\big(1-\cos\langle u,v\rangle\big)\frac{1+\|v\|^2}{\|v\|^2}d\Theta(v). \end{aligned}$$

So

$$\begin{aligned} d\Theta(v) &= \frac{2}{\pi}\frac{1}{1+t^2}dt\,d\mu(\theta), \\ v &= t\cdot\theta,\ \ \theta\in\mathbb{S}^{d-1},\ t\geq 0. \end{aligned} \tag{25.3.14}$$

If $h_{\mathcal{Z}}(u)$ is a support function of a zonoid $\mathcal{Z}$, then clearly

$$h_{\mathcal{Z}}(\tau\cdot u) = \tau h_{\mathcal{Z}}(u)$$

for all $\tau > 0$ and $u\in\mathbb{R}^d$, and, as was shown previously, $h_{\mathcal{Z}}(x-y)$ is a negative definite kernel. The inverse is also true.

Theorem 25.3.2. Suppose that f is a continuous function on $\mathbb{R}^d$ such that $f(0) = 0$, $f(-u) = f(u)$. Then the following facts are equivalent:

Fact 1. $f(\tau\cdot u) = \tau f(u)$ and $f(x-y)$ is a negative definite kernel.
Fact 2. f is a support function of a zonoid.

Proof. Previously we saw that Fact 2 implies Fact 1, and we must prove only that Fact 1 implies Fact 2. According to Example 25.3.2,

$$f(u) = \int_{\mathbb{R}^d}\big(1-\cos\langle u,v\rangle\big)d\Theta_1(v), \tag{25.3.15}$$

where

$$d\Theta_1(v) = \frac{1+\|v\|^2}{\|v\|^2}d\Theta(v),$$

and Θ is the measure from (25.3.3).

We have

$$f(\tau\cdot u) = \tau f(u) \tag{25.3.16}$$

for any $\tau > 0$, $u\in\mathbb{R}^d$. Substituting (25.3.15) into (25.3.16) and using the uniqueness of the measure Θ in (25.3.3) we obtain

$$\int_{\mathbb{R}^d}(1-\cos\langle\tau\cdot u,v\rangle)\mathrm{d}\Theta_1(v)=\tau\int_{\mathbb{R}^d}(1-\cos\langle u,v\rangle)\mathrm{d}\Theta_1(v),$$

$$(1-\cos\langle u,v\rangle)\mathrm{d}\Theta_1(v/\tau)=\tau\int_{\mathbb{R}^d}(1-\cos\langle u,v\rangle)\mathrm{d}\Theta_1(v)$$

and

$$\Theta_1(v/\tau)=\tau\Theta_1(v).$$

We write here $v=r\cdot w$ for $r>0$ and $w\in\mathbb{S}^{d-1}$. We have

$$\Theta_1(r\tau\cdot w)=\tau\Theta_1(r\cdot w)$$

and, finally, for $\tau=r$,

$$\Theta_1(r\cdot w)=\frac{1}{r}\Theta_1(w). \tag{25.3.17}$$

It is clear that representation (25.3.15) for Θ_1 of the form (25.3.17) coincides with (25.3.14).[2] □

Note that the $\mathfrak{N}$-distance can be bounded by the Hausdorf distance. Let $\mathcal{Z}_\mu$ and $\mathcal{Z}_\nu$ be two zonoids with generating measures μ and ν, respectively. The following inequality holds for their supporting functions $h(\mathcal{Z}_\mu,u)$ and $h(\mathcal{Z}_\nu,u)$:

$$|h(\mathcal{Z}_\mu,u)-h(\mathcal{Z}_\nu,u)|\le d_H(\mathcal{Z}_\mu,\mathcal{Z}_\nu).$$

Obviously, from this inequality it follows that

$$\mathcal{N}(\mu,\nu)\le 2d_H(\mathcal{Z}_\mu,\mathcal{Z}_\nu),$$

and therefore

$$\mathfrak{N}(\mu,\nu)\le(2d_H(\mathcal{Z}_\mu,\mathcal{Z}_\nu))^{1/2}. \tag{25.3.18}$$

Note that each $\mathfrak{N}$-distance generated by a zonoid is an ideal distance of degree $1/2$.

References

Beneš V, Rataj J (2004) Stochastic geometry: selected topics. Kluwer, Boston

Burger M (2000) Zonoids and conditionally positive definite functions. Portugaliae Mathematica 57:443–458

Klebanov LB (2005) $\mathfrak{N}$-distances and their applications. Karolinum Press, Prague

Schoenberg IJ (1938) Metric spaces and positive definite functions. Trans Am Math Soc 44:552–563

Ziegler GM (1995) Lectures on polytopes. Springer, New York

[2] An alternative proof of Theorem 25.3.2 is provided in Burger (2000).

Chapter 26
$\mathfrak{N}$-Distance Tests of Uniformity on the Hypersphere

The goals of this chapter are to:

- Discuss statistical tests of uniformity based on the $\mathfrak{N}$-distance theory,
- Calculate the asymptotic distribution of the test statistic.

Notation introduced in this chapter:

Notation	Description
$\widehat{X,Y}$	Smaller angle between X and Y located on unit sphere
$P_k(x)$	Legendre polynomial of order k

26.1 Introduction

Several invariant tests for uniformity of a distribution on a circle, a sphere, and a hemisphere have been proposed. In this chapter, we propose an application of $\mathfrak{N}$-distance theory for testing the hypothesis of uniformity of spherical data. The proposed procedures we discuss in this chapter have a number of advantages: consistency against all fixed alternatives, invariance of the test statistics under rotations of the sample, computational simplicity, and ease of application even in high-dimensional cases. Some new criteria of uniformity on S^{p-1} based on $\mathfrak{N}$-metrics are introduced. Particular attention is devoted to $p = 2$ (circular data) and $p = 3$ (spherical data). In these cases, the asymptotic behavior of the proposed tests under the null hypothesis is established using two approaches: the first one is based on an adaptation of methods of goodness of t-tests described in Bakshaev (2008, 2009), and the second one uses Gine theory based on Sobolev norms; see Gine (1975) and Hermans and Rasson (1985). At the end of the chapter, we present a brief comparative Monte Carlo power study for the proposed uniformity criteria. S^1 and S^2 cases are considered. Analyzed tests are compared with classical criteria

S.T. Rachev et al., *The Methods of Distances in the Theory of Probability and Statistics*,
DOI 10.1007/978-1-4614-4869-3_26,

by Gine (1975) using a variety of alternative hypotheses. Results of the simulations show that the proposed tests are powerful competitors to existing classic ones. All the results reported in this chapter were originally obtained by Bakshaev (2009). All of the proofs for propositions and theorems are provided in the last section of this chapter.

26.2 Tests of Uniformity on a Hypersphere

Consider the sample $X_1, \dots, X_n$ of observations of random variable (RV) X, where $X_i \in \mathbb{R}^p$ and $\|X_i\| = 1, i = 1, \dots, n$. Let us test the hypothesis H_0 that X has a uniform distribution on S^{p-1}.

The statistics for testing H_0 based on $\mathfrak{N}$-distance with the kernel $\mathcal{L}(x, y)$ have the form

$$T_n = n \left[\frac{2}{n} \sum_{i=1}^{n} E_Y \mathcal{L}(X_i, Y) - \frac{1}{n^2} \sum_{i,j=1}^{n} \mathcal{L}(X_i, X_j) - E\mathcal{L}(Y, Y') \right], \tag{26.2.1}$$

where X, Y, Y' are independent RVs from the uniform distribution on S^{p-1} and $E_Y \mathcal{L}(X_i, Y) = \int \mathcal{L}(X_i, y) \mathrm{d}F_Y(y)$ is a mathematical expectation calculated by Y with fixed X_i, $i = 1, \dots, n$. We should reject the null hypothesis in the case of large values of our test statistics, that is, if $T_n > c_\alpha$, where c_α can be found from the equation

$$\mathrm{Pr}_0(T_n > c_\alpha) = \alpha.$$

Here Pr_0 is the probability corresponding to the null hypothesis and α is the size of the test.

Let us consider strongly negative definite kernels of the form $\mathcal{L}(x, y) = G(\|x - y\|)$, where $\|.\|$ is the Euclidean norm. In other words, $G(.)$ depends on the length of the chord between two points on a hypersphere. As examples of such kernels we propose the following ones:

$$\mathcal{L}(x, y) = \|x - y\|^\alpha, \quad 0 < \alpha < 2,$$

$$\mathcal{L}(x, y) = \frac{\|x - y\|}{1 + \|x - y\|},$$

$$\mathcal{L}(x, y) = \log(1 + \|x - y\|^2).$$

Note that these kernels are rotation-invariant. This property implies that the mathematical expectation of the length of the chord between two independent uniformly distributed RVs Y and Y' on S^{p-1} is equal to the mean length of the chord between a fixed point and a uniformly distributed RV Y on S^{p-1}. Thus, we can rewrite (26.2.1) in the form

$$T_n = n\Big[EG(\|Y - Y'\| - \frac{1}{n^2}\sum_{i,j=1}^{n} G(\|X_i - X_j\|)\Big]. \quad (26.2.2)$$

In practice, statistics T_n with the kernel $\mathcal{L}(x, y) = \|x - y\|^\alpha$, $0 < \alpha < 2$, can be calculated using the following proposition.

Proposition 26.2.1. *In cases of $p = 2, 3$ statistic T_n will have the form*

$$T_n = \frac{(2R)^\alpha \Gamma((\alpha+1)/2)\Gamma(1/2)}{\pi\Gamma((\alpha+2)/2)} n - \frac{1}{n}\sum_{i,j=1}^{n} \|X_i - X_j\|^\alpha \quad (p = 2),$$

$$T_n = (2R)^\alpha \frac{2n}{\alpha+2} - \frac{1}{n}\sum_{i,j=1}^{n} \|X_i - X_j\|^\alpha \quad (p = 3),$$

where R is the radius of a hypersphere and $\alpha \in (0, 2)$.

In the case of $\mathcal{L}(x, y) = \|x - y\|$, test statistic (26.2.2) is very similar to Ajne's statistic A, the difference being that statistic A uses the length of the chord, whereas here we use the length of the smaller arc given by

$$A = \frac{n}{4} - \frac{1}{\pi n}\sum_{i,j=1}^{n} \psi_{ij},$$

where ψ_{ij} is the smaller of two angles between X_i and X_j, $i, j = 1, 2, \ldots, n$. One can see that Ajne's test is not consistent against all alternatives. As an example, consider the distribution on the circle concentrated in two diametrically opposite points with equal probabilities. Taking, instead of the arc, the length of the chord leads to a consistency of the $\mathfrak{N}$-distance test against all fixed alternatives:

$$\frac{T_n}{n} \xrightarrow{\Pr} \mathcal{N}(X, Y),$$

where $\mathcal{N}(X, Y)$ is the square of the $\mathfrak{N}$-distance between the probability distributions of RVs X and Y. If RVs X and Y are not identically distributed, then $\mathcal{N}(X, Y) > 0$ and $T_n \to \infty$ as $n \to \infty$.

Further, we consider the asymptotic distribution of statistics (26.2.1) under the null hypothesis. Particular attention is devoted to circular and spherical data ($p = 2, 3$). In these cases, the asymptotic behavior of the proposed tests under the null hypothesis is established using two approaches. The first is based on an adaptation of methods of goodness of t-tests described in Bakshaev (2008, 2009). The second uses Gine theory based on Sobolev norms as demonstrated in Gine (1975) and Hermans and Rasson (1985). For arbitrary dimension ($p \geq 3$) it is rather difficult from a computational point of view to establish the distribution of test statistics T_n analytically. In this case, the critical region of our criteria can be determined with the help of simulations of independent samples from the uniform distribution on S^{p-1}.

26.3 Asymptotic Distribution

26.3.1 Uniformity on a Circle

Here we consider the circle S^1 with unit length, that is, with $R = \frac{1}{2\pi}$. Let us transform the circle – and therefore our initial sample $X_1, \ldots, X_n$, $X_i = (X_{i1}, X_{i2})$, $X_{i1}^2 + X_{i2}^2 = R^2$ – to the interval $[0, 1)$ by making a cut at an arbitrary point x_0 of the circle

$$x \leftrightarrow x^*, \; x \in S^1, \; x^* \in [0, 1),$$

where x^* is the length of the smaller arc $x_0 x$.

It is easy to see that if X has a uniform distribution on S^1, then after the transformation we will get the RV X^* with uniform distribution on $[0, 1)$. Let $\mathcal{L}(x, y)$ be a strongly negative definite kernel in $\mathbb{R}^2$; then function $H(x^*, y^*)$ on $[0, 1)$ defined as

$$H(x^*, y^*) = \mathcal{L}(x, y) \tag{26.3.1}$$

is a strongly negative definite kernel on $[0, 1)$. In this case, $\mathfrak{N}$-distance statistic T_n^*, based on $H(x^*, y^*)$ for testing the uniformity on $[0, 1)$ has the form[1]

$$T_n^* = -n \int_0^1 \int_0^1 H(x^*, y^*) \mathrm{d}(F_n(x^*) - x^*) \mathrm{d}(F(y) - y),$$

where $F_n(x^*)$ is the empirical distribution function based on the sample $X_1^*, \ldots, X_n^*$, $X_i^* \in [0, 1)$, $i = 1, \ldots, n$. Due to (26.3.1), the following equality holds

$$T_n = T_n^*, \tag{26.3.2}$$

where T_n is defined by (26.2.1).

Thus, instead of testing the initial hypothesis on S^1 using T_n, we can test the uniformity on $[0, 1)$ for X^* on the basis of statistics T_n^* with the same asymptotic distribution. The limit distribution of T_n^* is established in Theorem 1 in Bakshaev (2009) and leads to the following result.

Theorem 26.3.1. *Under the null hypothesis, statistic T_n will have the same asymptotic distribution as a quadratic form:*

---

[1]See Bakshaev (2008, 2009).

$$T = \sum_{k=1}^{\infty}\sum_{j=1}^{\infty} \frac{a_{kj}}{\pi^2 kj}\zeta_k\zeta_j, \tag{26.3.3}$$

where ζ_k are independent RVs having the standard normal distribution and

$$a_{kj} = -2\int_0^1\int_0^1 H(x^*, y^*)\mathrm{d}\sin(\pi k x^*)\mathrm{d}\sin(\pi j y^*).$$

It is easy to see that in the case of rotation-invariant kernel $\mathcal{L}(x, y)$, the considered transformation of S^1 to $[0, 1)$ does not depend on the choice of the point of cut.

Proposition 26.3.1. *For the kernel $\mathcal{L}(x, y) = \|x - y\|^\alpha$, $0, \alpha < 2$, we have*

$$H(x^*, y^*) = \left(\frac{\sin \pi d}{\pi}\right)^\alpha,$$

where $d = \min(|x^* - y^*|, x^*, Y^* \in [0, 1)$.

26.3.2 Uniformity on a Sphere

In the case of a sphere, we also try to substitute the initial hypothesis of uniformity on S^2 by testing the uniformity on the unit square. Consider sphere S^2 with unit surface area, that is, $R^2 = \frac{1}{4\pi}$.

Note that if $X^* = (X_1^*, X_2^*)$ has a uniform distribution on $[0, 1)^2$, then the RV has the form $X = (X_1, X_2, X_3)$

$$X_1 = R\cos\theta_1,\ X_2 = R\sin\theta_1\cos\theta_0,\ X_3 = R\sin\theta_1\sin\theta_0, \tag{26.3.4}$$

where

$$\theta_0 = 2\pi X_1^*,\ \theta_1 = \arccos(1 - 2X_2^*)$$

has a uniform distribution on S^2.

Consider the strongly negative definite kernel $H(x^*, y^*)$ on $[0, 1)^2$ defined by

$$H(x^*, y^*) = \mathcal{L}(x, y), \tag{26.3.5}$$

where $\mathcal{L}(x, y)$ is a strongly negative definite kernel in $\mathbb{R}^3$, $x^*, y^* \in [0, 1)^2$, $x, y \in S^2$ and the correspondence between x and x^* follows from (26.3.4).

The $\mathfrak{N}$-distance statistic, based on $H(x^*, y^*)$, for testing the uniformity on $[0, 1)^2$ has the form[2]

$$T_n^* = -n\int_{[0,1)^2}\int_{[0,1)^2} H(x^*, y^*)\mathrm{d}(F_n(x^*) - x_1^* x_2^*)\mathrm{d}(F(y) - y_1 y_2),$$

---

[2]See Bakshaev (2008, 2009).

where $F_n(x^*)$ is the empirical distribution function based on the transformed sample X^*. Equations (26.3.4) and (26.3.5) imply that

$$T_n = T_n^*. \tag{26.3.6}$$

Thus, the asymptotic distribution of T_n coincides with the limit distribution of T_n^*, established in Bakshaev (2009, Theorem 2).

Theorem 26.3.2. *Under the null hypothesis, statistic T_n will have the same asymptotic distribution as the quadratic form*

$$T = \sum_{i,j,k,l=1}^{\infty} a_{ijkl}\sqrt{\alpha_{ij}\alpha_{kl}}\zeta_{ij}\zeta_{kl}, \tag{26.3.7}$$

where ζ_{ij} are independent RVs from the standard normal distribution,

$$a_{ijkl} = -\int_{[0,1)^4} H(x,y)\mathrm{d}\psi_{ij}\,\mathrm{d}\psi_{kl}, \quad x,y \in \mathbb{R}^2,$$

α_{ij}, and $\psi_{ij}(x,y)$ are eigenvalues and eigenfunctions of the integral operator A

$$Af(x) = \int_{[0,1]^2} K(x,y)f(y)\mathrm{d}y, \tag{26.3.8}$$

with the kernel

$$K(x,y) = \prod_{i=1}^{2}\min(x_i,y_i) - \prod_{i=1}^{2} x_i y_i.$$

Note that if $\mathcal{L}(x,y)$ is a rotation-invariant function on a sphere, then the values of statistics T_n and T_n^* do not depend on the choice of coordinate system on S^2. The main difficulties in applying Theorem 26.3.2 are due to the calculations of eigenfunctions of integral operator (26.3.8). One possible solution was discussed in Bakshaev (2009). Another possible solution is considered in the next subsection, where the asymptotic distribution of the proposed statistics for some strongly negative definite kernels is established with the help of Gine theory based on Sobolev tests.

26.3.3 Alternative Approach to the Limit Distribution of T_n

In this section, we propose an application of Gine theory of Sobolev invariant tests for uniformity on compact Riemannian manifolds to establish the null limit distribution of some $\mathfrak{N}$-distance statistics on the circle and sphere. We start with a brief review of Sobolev tests.[3]

Let M be a compact Riemannian manifold. The Riemannian metric determines the uniform probability measure μ on M. The intuitive idea of the Sobolev tests of uniformity is to map the manifold M into the Hilbert space $L^2(M,\mu)$ of the square-integrable functions on M by a function $t : M \to L^2(M,\mu)$ such that, if X is uniformly distributed, then the mean of $t(X)$ is 0.

The standard way of constructing such mappings t is based on the eigenfunctions of the Laplacian operator on M. For $k \geq 1$ let E_k denote the space of eigenfunctions corresponding to the kth eigenvalue, and set $d(k) = \dim E_k$. Then there is a map t_k from M into E_k given by

$$t_k(x) = \sum_{i=1}^{d(k)} f_i(x) f_i,$$

where $f_i : 1 \leq i \leq d(k)$ is any orthonormal basis of E_k. If $a1, a2, \dots$ is a sequence of real numbers such that

$$\sum_{i=1}^{\infty} a_k^2 d(k) < \infty,$$

then

$$x \mapsto t(x) = \sum_{i=1}^{\infty} a_k t_k(x)$$

defines a mapping t of M into $L^2(M,\mu)$. The resulting Sobolev statistic evaluated on observations $X_1, \dots, X_n$ on M is

$$S_n(\{a_k\}) = \sum_{i=1}^{n} \sum_{j=1}^{n} \langle t(X_i), t(X_j) \rangle,$$

where $\langle .,. \rangle$ denotes the inner product in $L^2(M,\mu)$.

The asymptotic null distribution of statistic $S_n(\{a_k\})$ is established by the following theorem.[4]

Theorem 26.3.3. *Let $X_1, \dots, X_n$ be a sequence of independent RVs with uniform distribution on M. Then*

$$S_n(\{a_k\}) \xrightarrow{d} \sum_{k=1}^{\infty} a_k^2 \xi \chi_k,$$

where $\{\chi_k\}_{k=1}^{\infty}$ is a sequence of independent RVs such that for each k, χ_k has a chi-squared distribution with $d(k)$ degrees of freedom.

---

[3]For more details see Gine (1975) and Jupp (2005).

[4]See Bakshaev (2010, Theorem 3.4).

Further, consider the $\mathfrak{N}$-distance and Sobolev tests for two special cases of a circle and a sphere.

Let M be the circle $x_1^2 + x_2^2 = 1$ in $\mathbb{R}^2$. Gine (1975) showed that in this case, Sobolev tests $S_n(\{a_k\})$ have the form

$$S_n(\{a_k\}) = \frac{2}{n} \sum_{k=1}^{\infty} \sum_{i,j=1}^{n} \cos k(X_i - X_j), \tag{26.3.9}$$

with the limit null distribution established by Theorem 26.3.3, where χ_k are independent RVs with a chi-squared distribution with $d(k) = 2$ degrees of freedom.

Consider the statistic T_n on M with strongly negative definite kernel $\mathcal{L}(x, y) = \|x - y\|$, $x, y \in \mathbb{R}^2$. From Proposition 26.2.1 we have

$$T_n = \frac{4n}{\pi} - \frac{1}{n} \sum_{i,j=1}^{n} \|X_i - X_j\| = \frac{4n}{\pi} - \frac{2}{n} \sum_{i,j=1}^{n} \sin \frac{X_i - X_j}{2}, \tag{26.3.10}$$

where $X_i - X_j$ and $\|X_i - X_j\|$ denote the length of the arc and chord between X_i and X_j, respectively.

Under the null hypothesis, the limit distribution of T_n is established by the following theorem:

Theorem 26.3.4. *If $X_1, \dots, X_n$ is a sample of independent observations of the uniform distribution on a circle with unit radius, then*

$$\frac{\pi}{4} T_n \xrightarrow{d} \sum_{k=1}^{\infty} a_k^2 \chi_k^2, \tag{26.3.11}$$

where χ_k^2 are independent RVs with a chi-squared distribution with two degrees of freedom and

$$a_k^2 = \frac{1}{2\pi} \int_0^{2\pi} \left(1 - \frac{\pi}{2} \sin \frac{x}{2}\right) \cos kx \mathrm{d}x.$$

We now consider $\mathfrak{N}$-distance and Sobolev tests on a sphere. If $M = S^2$ is the unit sphere $x_1^2 + x_2^2 + x_3^2 = 1$, then $\mathrm{d}\mu = (4\pi)^{-1} \sin\theta \mathrm{d}\theta \mathrm{d}\varphi$, where μ is the uniform distribution on S^2 and (θ, φ) are usual spherical coordinates. The general expression of Sobolev tests on a sphere has the form

$$S_n(\{a_k\}) = \frac{1}{n} \sum_{k=1}^{\infty} (2k + 1) a_k^2 \sum_{i,j=1}^{n} P_k(\cos \widehat{X_i, X_j}), \tag{26.3.12}$$

where $\widehat{X_i, X_j}$ is the smaller angle between X_i and X_j, and P_k are Legendre polynomials

$$P_k(x) = (k!2^k)^{-1}(d^k/dx^k)(x^2 - 1)^k.$$

Under the null hypothesis, the limit distribution of $S_n(\{a_k\})$ coincides with the distribution of RV

$$\sum_{k=1}^{\infty} a_k^2 \chi_{2k+1}^2, \tag{26.3.13}$$

where χ_{2k+1}^2 are independent RVs with a chi-squared distribution with $2k + 1$ degrees of freedom.

Consider the statistic T_n on S^2 with a strongly negative definite kernel $\mathcal{L}(x, y) = \|x - y\|$, $x, y \in \mathbb{R}^3$. From Proposition 26.2.1 we have

$$T_n = \frac{4n}{3} - \frac{1}{n} \sum_{i,j=1}^{n} \|X_i - X_j\| = \frac{4n}{3} - \frac{2}{n} \sum_{i,j=1}^{n} \sin \frac{\widehat{X_i, X_j}}{2}, \tag{26.3.14}$$

where $\widehat{X_i, X_j}$ and $\|X_i - X_j\|$ denote the smaller angle and chord between X_i and X_j, respectively.

The asymptotic distribution of T_n is established by the following theorem.

Theorem 26.3.5. *If $X_1, \dots, X_n$ is a sample of independent observations from the uniform distribution on S^2, then*

$$\frac{3}{4} T_n \xrightarrow{d} \sum_{k=1}^{\infty} a_k^2 \chi_{2k+1}^2, \tag{26.3.15}$$

where χ_{2k+1}^2 are independent RVs with a chi-squared distribution with $2k + 1$ degrees of freedom and

$$a_k^2 = \frac{1}{2} \int_0^{\pi} \left(1 - \frac{3}{2} \sin \frac{x}{2}\right) \sin x P_k(\cos x) \mathrm{d}x, \tag{26.3.16}$$

where $P_k(x)$ are Legendre polynomials.

26.4 Proofs

26.4.1 Proof of Proposition 26.2.1

The stated formula follows directly from (26.2.2), and the property

$$E\|Y - Y'\|^{\alpha} = E\|Y - a\|^{\alpha},$$

where Y and Y' are independent RVs uniformly distributed on S^{p-1} and a, is an arbitrary fixed point on S^{p-1}.

For the two-dimensional case calculate the expectation of the length of the chord between fixed point $a = (0, R)$ and a uniformly distributed RV Y:

$$
\begin{aligned}
E\|a-Y\|^\alpha &= \frac{1}{2\pi R}\int_0^{2\pi} R(R^2\cos^2\varphi + (R\sin^2\varphi - R)^2)^{\alpha/2}\mathrm{d}\varphi \\
&= \frac{2^{\alpha/2-1}R^\alpha}{\pi}\int_0^{2\pi}(1-\cos\varphi)^{\alpha/2}\mathrm{d}\varphi \\
&= \frac{2^{\alpha+1}R^\alpha}{\pi}\int_0^{2\pi}\sin^\alpha\varphi\mathrm{d}\varphi \\
&= \frac{(2R)^\alpha\Gamma((\alpha+1)/2)G(1/2)}{\pi\Gamma((\alpha+2)/2)}.
\end{aligned}
$$

In the case where $p = 3$, let us fix the point $a = (0, 0, R)$ and calculate the average length of the chord:

$$
\begin{aligned}
E\|a-Y\|^\alpha &= \frac{1}{4\pi R^2}\int_{-\pi}^{\pi}\int_0^{\pi} R^2\sin^2\theta\cos^2\varphi \\
&\quad + \sin^2\theta\sin^2\varphi + (\cos\theta - 1)^2)^{\alpha/2}\mathrm{d}\theta\mathrm{d}\varphi \\
&= \frac{2^{\alpha/2}R^\alpha}{4\pi}\int_{-\pi}^{\pi}\int_0^{\pi}(1-\cos\theta)^{\alpha/2}\sin\theta\mathrm{d}\theta\mathrm{d}\varphi \\
&= (2R)^\alpha\frac{2}{\alpha+2}.
\end{aligned}
$$

26.4.2 *Proof of Proposition 26.3.1*

The kernel $\mathcal{L}(x, y)$ in the case of a circle equals the length of the chord between two points $x = (x_1, x_2)$ and $y = (y_1, y_2)$ raised to the power of α. After the proposed transformation, the length of the smaller arc between x and y is equal to

$$
d = \min(|x^* - y^*|, 1 - |x^* - y^*|).
$$

The length of the chord of a circle with $R = \frac{1}{2\pi}$ based on the angle $2\pi d$ equals $\sin(\pi d)/\pi$. This completes the proof of the statement.

26.4.3 *Proof of Theorem 26.3.4*

Let us express statistic (26.3.10) in the form

$$
T_n = \frac{4}{\pi n}\sum_{i,j=1}^{n} h(X_i - X_j),
$$

where $h(x) = 1 - \frac{\pi}{2}\sin(x/2)$. The function $h(x)$ can be represented in the form of a series by a complete orthonormal sequence of functions $\{2\cos kx\}$ on $[0, 2\pi]$

$$h(x) = \sqrt{2}\sum_{k=1}^{\infty} a_k \cos kx,$$

where

$$a_k = \frac{1}{\sqrt{2}\pi}\int_0^{2\pi}\left(1 - \frac{\pi}{2}\sin\frac{x}{2}\right)\cos kx dx.$$

Note that $a_k > 0$ for all $k = 1, 2, \ldots$. After some simple calculations, we obtain

$$\int_0^{2\pi}\left(1 - \frac{\pi}{2}\sin\frac{x}{2}\right)\cos kx dx = 4\int_0^{\pi} \sin x \sin^2 kx dx - 4$$

and

$$\begin{aligned}\int_0^{\pi} \sin x \sin^2 kx dx &= -k^2\int_0^{\pi k} \sin(1/k - 2)x - \frac{k^2}{2k+1}\int_0^{\pi k} \sin\frac{x}{k} dx \\ &= \frac{4k^3}{(2k-1)(2k+1)} > 1,\ k = 1, 2, \ldots.\end{aligned}$$

Thus, statistic T_n can be rewritten in the form of Sobolev statistic (26.3.9):

$$\frac{4}{\pi}T_n = \frac{2}{n}\sum_{k=1}^{\infty}\sum_{i,j=1}^{n} \cos k(X_i - X_j),$$

where $a_k^2 = \alpha_k/\sqrt{2}$. After that, the statement of the theorem follows directly from Theorem 26.3.3.

26.4.4 Proof of Theorem 26.3.5

The proof can be done in nearly the same way as that of Theorem 26.3.4. Let us rewrite statistic T_n in the form

$$T_n = \frac{4}{3n}\sum_{i,j=1}^{n} h(\widehat{X_i, X_j}),$$

where $h(x) = 1 - (3/2)\sin(x/2)$, and then decompose $h(x)$ into a series by an orthonormal sequence of functions $\{\sqrt{2k+1}P_k(\cos x)\}$ for $x \in [0, \pi]$,

$$h(x) = \sum_{k=1}^{\infty} \sqrt{2k+1}\alpha_k P_k(\cos x),$$

where

$$\alpha_k = \frac{\sqrt{2k+1}}{4\pi} \int_0^{2\pi} \int_0^{\pi} \left(1 - \frac{3}{2}\sin\frac{\theta}{2}\right) \sin\theta P_k(\cos\theta) d\theta d\varphi.$$

As a result, statistic T_n can be expressed in the form of the Sobolev statistic (26.3.12)

$$\frac{4}{3}T_n = \frac{1}{n}\sum_{k=1}^{\infty}(2k+1)a_k^2 \sum_{i,j=1}^{n} P_k(\cos \widehat{X_i, X_j}),$$

where $\sqrt{2k+1}a_k^2 = \alpha_k$. Applying Theorem 26.3.3 we obtain the assertion of the theorem.

References

Bakshaev A (2008) Nonparametric tests based on N-distances. Lithuanian Math J 48(4):368–379

Bakshaev A (2009) Goodness of t- and homogeneity tests on the basis of N-distances. J Stat Plan Inference 139(11):3750–3758

Bakshaev A (2010) N-distance tests of uniformity on the hypersphere. Non-linear Anal Model Control 15(1):15–28

Gine EM (1975) Invariant tests for uniformity on compact Riemannian manifolds based on Sobolev norms. Ann Stat 3:1243–1266

Hermans M, Rasson JP (1985) A new Sobolev test for uniformity on the circle. Biometrika 72(3):698–702

Jupp PE (2005) Sobolev tests of goodness-of-fit of distributions on compact Riemannian manifolds. Ann Stat 33(6):2957–2966

About the Authors

Svetlozar (Zari) T. Rachev completed his Ph.D. in 1979 at Moscow State (Lomonosov) University and his Doctor of Science degree in 1986 from Steklov Mathematical Institute in Moscow. He is Professor of Finance in the College of Business at Stony Brook University in Stony Brook, NY. From January 2011 to September 2012, he held the Frey Family Chair of Quantitative Finance in the Department of Applied Mathematics & Statistics at Stony Brook University. Prior to joining the faculty of Stony Brook University in January 2011, Professor Rachev was Chair-Professor in Statistics, Econometrics, and Mathematical Finance at the University of Karlsruhe in the School of Economics and Business Engineering. He is also Professor Emeritus at the University of California, Santa Barbara, in the Department of Statistics and Applied Probability. He has published seven monographs, eight handbooks and special edited volumes, and over 300 research articles. His recently coauthored books published by Wiley in mathematical finance and financial econometrics include *Fat-Tailed and Skewed Asset Return Distributions: Implications for Risk Management, Portfolio Selection, and Option Pricing* (2005), *Operational Risk: A Guide to Basel II Capital Requirements, Models, and Analysis* (2007), *Financial Econometrics: From Basics to Advanced Modeling Techniques* (2007), and *Bayesian Methods in Finance* (2008). Professor Rachev is cofounder of Bravo Risk Management Group specializing in financial risk-management software. Bravo Group was acquired by FinAnalytica for which he currently serves as Chief Scientist.

Lev B. Klebanov is Professor of Probability and Statistics at Charles University, Prague, Czech Republic. He earned his MS degree in Applied Mathematics from St. Petersburg State University, Russia, a Ph.D. in Statistics from the St. Petersburg Branch of Steklov Mathematical Institute, Russia, under the supervision of Dr. Yuriy Vladimirovich Linnik, and a Doctor of Science degree in Statistics at St. Petersburg State University, Russia. Professor Klebanov has published eight monographs and 230 research articles.

Stoyan V. Stoyanov is Professor of Finance at EDHEC Business School and Head of Research for EDHEC-Risk Institute–Asia. Before joining EDHEC-Risk Institute

S.T. Rachev et al., *The Methods of Distances in the Theory of Probability and Statistics*,
DOI 10.1007/978-1-4614-4869-3, © Springer Science+Business Media, LLC 2013

in 2010, he was Head of Quantitative Research for FinAnalytica, a noted financial technology firm. Over his professional career, Professor Stoyanov has designed and implemented investment and risk-management models for financial institutions, codeveloped a patented system for portfolio optimization in the presence of nonnormality, and led a team of engineers designing and planning the implementation of advanced models for major financial institutions. His research focuses on probability theory, extreme risk modeling, and optimal portfolio theory. He has published over 30 articles in academic journals, contributed to many professional handbooks, and coauthored two books on financial risk assessment and portfolio optimization. He completed his Ph.D. with honors in 2005 from the School of Economics and Business Engineering (Chair of Statistics, Econometrics and Mathematical Finance) at the University of Karlsruhe.

Frank J. Fabozzi is Professor of Finance at EDHEC Business School and a member of the EDHEC-Risk Institute. He held various professorial positions in finance at Yale University's School of Management from 1994 to 2011 and from 1986 to 1992 was a visiting professor of finance and accounting at MIT's Sloan School of Management. Since fall 2011, he has been a visiting fellow in the Department of Operations Research and Financial Engineering at Princeton University, where he also served for 8 years on that department's Advisory Council. From March 2008 to June 2011, Professor Fabozzi was an Affiliated Professor at the Institute of Statistics, Econometrics and Mathematical Finance at Karlsruhe Institute of Technology in Germany. Some of his coauthored recent books include *A Probability Metrics Approach to Financial Risk Measures* (2011), *Financial Modeling with Levy Processes* (2011), *Probability and Statistics for Finance* (2010), *Simulation and Optimization Modeling in Finance* (2010), *Bayesian Methods in Finance* (2009), *Advanced Stochastic Models, Risk Assessment, and Portfolio Optimization: The Ideal Risk, Uncertainty, and Performance Measures* (2008), *Financial Econometrics* (2007), and *Robust Portfolio Optimization and Management* (2007). He earned a doctorate in economics from the City University of New York in 1972 and was awarded an honorary doctorate from Nova Southeastern University in June 1994.

Index

S.T. Rachev et al., *The Methods of Distances in the Theory of Probability and Statistics*, DOI 10.1007/978-1-4614-4869-3, © Springer Science+Business Media, LLC 2013

P

GPSR Compliance

The European Union's (EU) General Product Safety Regulation (GPSR) is a set of rules that requires consumer products to be safe and our obligations to ensure this.

If you have any concerns about our products, you can contact us on ProductSafety@springernature.com

In case Publisher is established outside the EU, the EU authorized representative is:

Springer Nature Customer Service Center GmbH
Europaplatz 3
69115 Heidelberg, Germany

Batch number: 10755248

Printed by Printforce, the Netherlands